MASS-TRANSFER OPERATIONS

Yogesh K. Dhande.

McGraw-Hill Chemical Engineering Series

BUILDING THE LITERATURE OF A PROFESSION

Fifteen prominent chemical engineers first met in New York more than 50 years ago to plan a continuing literature for their rapidly growing profession. From industry came such pioneer practitioners as Leo H. Baekeland, Arthur D. Little, Charles L. Reese, John V. N. Dorr, M. C. Whitaker, and R. S. McBride. From the universities came such eminent educators as William H. Walker, Alfred H. White, D. D. Jackson, J. H. James, Warren K. Lewis, and Harry A. Curtis. H. C. Parmelee, then editor of *Chemical and Metallurgical Engineering*, served as chairman and was joined subsequently by S. D. Kirkpatrick as consulting editor.

After several meetings, this committee submitted its report to the McGraw-Hill Book Company in September 1925. In the report were detailed specifications for a correlated series of more than a dozen texts and reference books which have since become the McGraw-Hill Series in Chemical Engineering and which became the cornerstone of the chemical engineering curriculum.

From this beginning there has evolved a series of texts surpassing by far the scope and longevity envisioned by the founding Editorial Board. The McGraw-Hill Series in Chemical Engineering stands as a unique historical record of the development of chemical engineering education and practice. In the series one finds the milestones of the subject's evolution: industrial chemistry, stoichiometry, unit operations and processes, thermodynamics, kinetics, and transfer operations.

Chemical engineering is a dynamic profession, and its literature continues to evolve. McGraw-Hill and its consulting editors remain committed to a publishing policy that will serve, and indeed lead, the needs of the chemical engineering profession during the years to come.

THE SERIES

Bailey and Ollis: *Biochemical Engineering Fundamentals*
Bennet and Myers: *Momentum, Heat, and Mass Transfer*
Beveridge and Schechter: *Optimization: Theory and Practice*
Carberry: *Chemical and Catalytic Reaction Engineering*
Churchill: *The Interpretation and Use of Rate Data—The Rate Concept*
Clarke and Davidson: *Manual for Process Engineering Calculations*
Coughanowr and Koppel: *Process Systems Analysis and Control*
Danckwerts: *Gas Liquid Reactions*
Gates, Katzer, and Schuit: *Chemistry of Catalytic Processes*
Harriott: *Process Control*
Johnson: *Automatic Process Control*
Johnstone and Thring: *Pilot Plants, Models, and Scale-up Methods in Chemical Engineering*
Katz, Cornell, Kobayashi, Poettmann, Vary, Elenbaas, and Weinaug: *Handbook of Natural Gas Engineering*
King: *Separation Processes*
Knudsen and Katz: *Fluid Dynamics and Heat Transfer*
Lapidus: *Digital Computation for Chemical Engineers*
Luyben: *Process Modeling, Simulation, and Control for Chemical Engineers*
McCabe and Smith, J. C.: *Unit Operations of Chemical Engineering*
Mickley, Sherwood, and Reed: *Applied Mathematics in Chemical Engineering*
Nelson: *Petroleum Refinery Engineering*
Perry and Chilton (Editors): *Chemical Engineers' Handbook*
Peters: *Elementary Chemical Engineering*
Peters and Timmerhaus: *Plant Design and Economics for Chemical Engineers*
Reed and Gubbins: *Applied Statistical Mechanics*
Reid, Prausnitz, and Sherwood: *The Properties of Gases and Liquids*
Sherwood, Pigford, and Wilke: *Mass Transfer*
Slattery: *Momentum, Energy, and Mass Transfer in Continua*
Smith, B. D.: *Design of Equilibrium Stage Processes*
Smith, J. M.: *Chemical Engineering Kinetics*
Smith, J. M., and Van Ness: *Introduction to Chemical Engineering Thermodynamics*
Thompson and Ceckler: *Introduction to Chemical Engineering*
Treybal: *Liquid Extraction*
Treybal: *Mass Transfer Operations*
Van Winkle: *Distillation*
Volk: *Applied Statistics for Engineers*
Walas: *Reaction Kinetics for Chemical Engineers*
Wei, Russell, and Swartzlander: *The Structure of the Chemical Processing Industries*
Whitwell and Toner: *Conservation of Mass and Energy*

MASS-TRANSFER OPERATIONS

Third Edition

Robert E. Treybal

Professor of Chemical Engineering
University of Rhode Island

McGRAW-HILL BOOK COMPANY

Auckland Bogota Guatemala Hamburg Lisbon London
Madrid Mexico New Delhi Panama Paris San Juan
São Paulo Singapore Sydney Tokyo

MASS-TRANSFER OPERATIONS

INTERNATIONAL EDITION 1981

30 *(MPM)* 20 9 8 7 6

This book was set in Times Roman by Science Typographers, Inc. The editors were Julienne V. Brown and Madelaine Eichberg; The production supervisor was Charles Hess.

Library of Congress Cataloging in Publication Data
Treybal, Robert Ewald, date
   Mass-transfer operations.

   (McGraw-Hill chemical engineering series)
   Includes bibliographical references and index.
   1. Chemical engineering.   2. Mass transfer.
I. Title.
TP156.M3T7   1980   660.2'8422   78-27876
ISBN 0-07-065176-0

When ordering this title use ISBN 0-07-066615-6

This book is dedicated
to the memory of my dear wife, Gertrude,
whose help with this edition was sorely missed.

CONTENTS

Preface   xiii

1   The Mass-Transfer Operations   1

    Classification of the Mass-Transfer Operations   2
    Choice of Separation Method   7
    Methods of Conducting the Mass-Transfer Operations   8
    Design Principles   11
    Unit Systems   12

Part 1   Diffusion and Mass Transfer

2   Molecular Diffusion in Fluids   21

    Steady-State Molecular Diffusion in Fluids at Rest
      and in Laminar Flow   26
    Momentum and Heat Transfer in Laminar Flow   38

3   Mass-Transfer Coefficients   45

    Mass-Transfer Coefficients in Laminar Flow   50
    Mass-Transfer Coefficients in Turbulent Flow   54
    Mass-, Heat-, and Momentum-Transfer Analogies   66
    Mass-Transfer Data for Simple Situations   72
    Simultaneous Mass and Heat Transfer   78

4   Diffusion in Solids   88

    Fick's-Law Diffusion   88
    Types of Solid Diffusion   93

5   Interphase Mass Transfer   104

    Equilibrium   104
    Diffusion between Phases

Material Balances 117
Stages 123

Part 2 Gas-Liquid Operations

6 Equipment for Gas-Liquid Operations 139

Gas Dispersed 139

Sparged Vessels (Bubble Columns) 140
Mechanically Agitated Vessels 146
Mechanical Agitation of Single-Phase Liquids 146
Mechanical Agitation, Gas-Liquid Contact 153
Tray Towers 158

Liquid Dispersed 186

Venturi Scrubbers 186
Wetted-Wall Towers 187
Spray Towers and Spray Chambers 187
Packed Towers 187
Mass-Transfer Coefficients for Packed Towers 202
Cocurrent Flow of Gas and Liquid 209
End Effects and Axial Mixing 209
Tray Towers vs. Packed Towers 210

7 Humidification Operations 220

Vapor-Liquid Equilibrium and Enthalpy for a Pure Substance 220
Vapor-Gas Mixtures 227
Gas-Liquid Contact Operations 241
Adiabatic Operations 242
Nonadiabatic Operation: Evaporative Cooling 263

8 Gas Absorption 275

Equilibrium Solubility of Gases in Liquids 275
One Component Transferred; Material Balances 282
Countercurrent Multistage Operation; One Component Transferred 289
Continuous-Contact Equipment 300
Multicomponent Systems 322
Absorption with Chemical Reaction 333

9 Distillation 342

Vapor-Liquid Equilibria 343
Single-Stage Operation—Flash Vaporization 363
Differential, or Simple, Distillation 367
Continuous Rectification—Binary Systems 371
Multistage Tray Towers—The Method of Ponchon and Savarit 374
Multistage Tray Towers—Method of McCabe and Thiele 402
Continuous-Contact Equipment (Packed Towers) 426
Multicomponent Systems 431
Low-Pressure Distillation 460

Part 3 Liquid-Liquid Operations

10 Liquid Extraction   477

Liquid Equilibria   479

Equipment and Flowsheets   490

Stagewise Contact   490
Stage-Type Extractors   521
Differential (Continuous-Contact) Extractors   541

Part 4 Solid-Fluid Operations

11 Adsorption and Ion Exchange   565

Adsorption Equilibria   569

Single Gases and Vapors   569
Vapor and Gas Mixtures   575
Liquids   580

Adsorption Operations   585

Stagewise Operation   585
Continuous Contact   612

12 Drying   655

Equilibrium   655

Drying Operations   661

Batch Drying   662
The Mechanisms of Batch Drying   672
Continuous Drying   686

13 Leaching   717

Unsteady-State Operation   719
Steady-State (Continuous) Operation   731
Methods of Calculation   744

Index   767

PREFACE

My purpose in presenting the third edition of this book continues to be that of the previous edition: "to provide a vehicle for teaching, either through a formal course or through self-study, the techniques of, and principles of equipment design for, the mass-transfer operations of chemical engineering." As before, these operations are largely the responsibility of the chemical engineer, but increasingly practitioners of other engineering disciplines are finding them necessary for their work. This is especially true for those engaged in pollution control and environment protection, where separation processes predominate, and in, for example, extractive metallurgy, where more sophisticated and diverse methods of separation are increasingly relied upon.

I have taken this opportunity to improve and modernize many of the explanations, to modernize the design data, and to lighten the writing as best I could. There are now included discussions of such topics as the surface-stretch theory of mass-transfer, transpiration cooling, new types of tray towers, heatless adsorbers, and the like. Complete design methods are presented for mixer-settler and sieve-tray extractors, sparged vessels, and mechanically agitated vessels for gas-liquid, liquid-liquid, and solid-liquid contact, adiabatic packed-tower absorbers, and evaporative coolers. There are new worked examples and problems for student practice. In order to keep the length of the book within reasonable limits, the brief discussion of the so-called less conventional operations in the last chapter of the previous edition has been omitted.

One change will be immediately evident to those familiar with previous editions; the new edition is written principally in the SI system of units. In order to ease the transition to this system, an important change was made: of the more than 1000 numbered equations, all but 25 can now be used with any system of consistent units, SI, English engineering, Metric engineering, cgs, or whatever. The few equations which remain dimensionally inconsistent are given in SI, and also by footnote or other means in English engineering units. All tables of engineering data, important dimensions in the text, and most student problems are treated similarly. An extensive list of conversion factors from other systems to SI is included in Chapter 1; these will cover all quantities needed for the use of this book. I hope this book will stimulate the transition to SI, the advantages of which become increasingly clear as one becomes familiar with it.

I remain as before greatly indebted to many firms and publications for permission to use their material, and most of all to the many engineers and scientists whose works provide the basis for a book of this sort. I am also indebted to Edward C. Hohmann and William R. Schowalter as well as to several anonymous reviewers who provided useful suggestions. Thanks are due to the editorial staff of the publisher, all of whom have been most helpful.

Robert E. Treybal

Robert E. Treybal passed away while this book was in production. We are grateful to Mark M. Friedman who, in handling the proofs of this book, has contributed significantly to the usability of this text.

THE MASS-TRANSFER OPERATIONS

A substantial number of the unit operations of chemical engineering are concerned with the problem of changing the compositions of solutions and mixtures through methods not necessarily involving chemical reactions. Usually these operations are directed toward separating a substance into its component parts. For mixtures, such separations may be entirely mechanical, e.g., the filtration of a solid from a suspension in a liquid, the classification of a solid into fractions of different particle size by screening, or the separation of particles of a ground solid according to their density. On the other hand, if the operations involve changes in composition of solutions, they are known as the mass-transfer operations and it is these which concern us here.

The importance of these operations is profound. There is scarcely any chemical process which does not require a preliminary purification of raw materials or final separation of products from by-products, and for these the mass-transfer operations are usually used. One can perhaps most readily develop an immediate appreciation of the part these separations play in a processing plant by observing the large number of towers which bristle from a modern petroleum refinery, in each of which a mass-transfer separation operation takes place. Frequently the major part of the cost of a process is that for the separations. These separation or purification costs depend directly upon the ratio of final to initial concentration of the separated substances, and if this ratio is large, the product costs are large. Thus, sulfuric acid is a relatively low-priced product in part because sulfur is found naturally in a relatively pure state, whereas pure uranium is expensive because of the low concentration in which it is found in nature.

The mass-transfer operations are characterized by transfer of a substance through another on a molecular scale. For example, when water evaporates from a pool into an airstream flowing over the water surface, molecules of water vapor diffuse through those of the air at the surface into the main portion of the airstream, whence they are carried away. It is not bulk movement as a result of a pressure difference, as in pumping a liquid through a pipe, with which we are primarily concerned. In the problems at hand, the mass transfer is a result of a concentration difference, or gradient, the diffusing substance moving from a place of high to one of low concentration.

CLASSIFICATION OF THE MASS-TRANSFER OPERATIONS

It is useful to classify the operations and to cite examples of each, in order to indicate the scope of this book and to provide a vehicle for some definitions of terms which are commonly used.

Direct Contact of Two Immiscible Phases

This category is by far the most important of all and includes the bulk of the mass-transfer operations. Here we take advantage of the fact that in a two-phase system of several components at equilibrium, with few exceptions the compositions of the phases are different. The various components, in other words, are differently distributed between the phases.

In some instances, the separation thus afforded leads immediately to a pure substance because one of the phases at equilibrium contains only one constituent. For example, the equilibrium vapor in contact with a liquid aqueous salt solution contains no salt regardless of the concentration of the liquid. Similarly the equilibrium solid in contact with such a liquid salt solution is either pure water or pure salt, depending upon which side of the eutectic composition the liquid happens to be. Starting with the liquid solution, one can then obtain a complete separation by boiling off the water. Alternatively, pure salt or pure water can be produced by partly freezing the solution; or, in principle at least, both can be obtained pure by complete solidification followed by mechanical separation of the eutectic mixture of crystals. In cases like these, when the two phases are first formed, they are immediately at their final equilibrium compositions and the establishment of equilibrium is not a time-dependent process. Such separations, with one exception, are not normally considered to be among the mass-transfer operations.

In the mass-transfer operations, neither equilibrium phase consists of only one component. Consequently when the two phases are initially contacted, they will not (except fortuitously) be of equilibrium compositions. The system then attempts to reach equilibrium by a relatively slow diffusive movement of the constituents, which transfer in part between the phases in the process. Separations are therefore never complete, although, as will be shown, they can be brought as near completion as desired (but not totally) by appropriate manipulations.

The three states of aggregation, gas, liquid, and solid, permit six possibilities of phase contact.

Gas-gas Since with very few exceptions all gases are completely soluble in each other, this category is not practically realized.

Gas-liquid If all components of the system distribute between the phases at equilibrium, the operation is known as *fractional distillation* (or frequently just *distillation*). In this instance the gas phase is created from the liquid by application of heat; or conversely, the liquid is created from the gas by removal of heat. For example, if a liquid solution of acetic acid and water is partially vaporized by heating, it is found that the newly created vapor phase and the residual liquid both contain acetic acid and water but in proportions at equilibrium which are different for the two phases and different from those in the original solution. If the vapor and liquid are separated mechanically from each other and the vapor condensed, two solutions, one richer in acetic acid and the other richer in water, are obtained. In this way a certain degree of separation of the original components has been accomplished.

Both phases may be solutions, each containing, however, only one common component (or group of components) which distributes between the phases. For example, if a mixture of ammonia and air is contacted with liquid water, a large portion of the ammonia, but essentially no air, will dissolve in the liquid and in this way the air-ammonia mixture can be separated. The operation is known as *gas absorption*. On the other hand, if air is brought into contact with an ammonia-water solution, some of the ammonia leaves the liquid and enters the gas phase, an operation known as *desorption* or *stripping*. The difference is purely in the direction of solute transfer.

If the liquid phase is a pure liquid containing but one component while the gas contains two or more, the operation is *humidification* or *dehumidification*, depending upon the direction of transfer (this is the exception mentioned earlier). For example, contact of dry air with liquid water results in evaporation of some water into the air (humidification of the air). Conversely, contact of very moist air with pure liquid water may result in condensation of part of the moisture in the air (dehumidification). In both cases, diffusion of water vapor through air is involved, and we include these among the mass-transfer operations.

Gas-solid Classification of the operations in this category according to the number of components which appear in the two phases is again convenient.

If a solid solution is partially vaporized without the appearance of a liquid phase, the newly formed vapor phase and the residual solid each contains all the original components, but in different proportions, and the operation is *fractional sublimation*. As in distillation, the final compositions are established by interdiffusion of the components between the phases. While such an operation is theoretically possible, practically it is not generally done because of the inconvenience of dealing with solid phases in this manner.

Not all components may be present in both phases, however. If a solid which is moistened with a volatile liquid is exposed to a relatively dry gas, the liquid leaves the solid and diffuses into the gas, an operation generally known as *drying*, sometimes as *desorption*. A homely example is drying laundry by exposure to air, and there are many industrial counterparts such as drying lumber or the removal of moisture from a wet filter cake by exposure to dry gas. In this case, the diffusion is, of course, from the solid to the gas phase. If the diffusion takes place in the opposite direction, the operation is known as *adsorption*. For example, if a mixture of water vapor and air is brought into contact with activated silica gel, the water vapor diffuses to the solid, which retains it strongly, and the air is thus dried. In other instances, a gas mixture may contain several components each of which is adsorbed on a solid but to different extents (*fractional adsorption*). For example, if a mixture of propane and propylene gases is brought into contact with activated carbon, the two hydrocarbons are both adsorbed, but to different extents, thus leading to a separation of the gas mixture.

When the gas phase is a pure vapor, as in the sublimation of a volatile solid from a mixture with one which is nonvolatile, the operation depends more on the rate of application of heat than on concentration difference and is essentially nondiffusional. The same is true of the condensation of a vapor to the condition of a pure solid, where the rate depends on the rate of heat removal.

Liquid-liquid Separations involving the contact of two insoluble liquid phases are known as *liquid-extraction operations*. A simple example is the familiar laboratory procedure: if an acetone-water solution is shaken in a separatory funnel with carbon tetrachloride and the liquids allowed to settle, a large portion of the acetone will be found in the carbon tetrachloride–rich phase and will thus have been separated from the water. A small amount of the water will also have been dissolved by the carbon tetrachloride, and a small amount of the latter will have entered the water layer, but these effects are relatively minor. As another possibility, a solution of acetic acid and acetone can be separated by adding it to the insoluble mixture of water and carbon tetrachloride. After shaking and settling, both acetone and acetic acid will be found in both liquid phases, but in different proportions. Such an operation is known as *fractional extraction*. Another form of fractional extraction can be effected by producing two liquid phases from a single-phase solution by cooling the latter below its critical solution temperature. The two phases which form will be of different composition.

Liquid-solid When all the constituents are present in both phases at equilibrium, we have the operation of *fractional crystallization*. Perhaps the most interesting examples of this are the special techniques of *zone refining*, used to obtain ultrapure metals and semiconductors, and *adductive crystallization*, where a substance, such as urea, has a crystal lattice which will selectively entrap long straight-chain molecules like the paraffin hydrocarbons but will exclude branched molecules.

Cases where the phases are solutions (or mixtures) containing but one common component occur more frequently. Selective solution of a component from a solid mixture by a liquid solvent is known as *leaching* (sometimes also as solvent extraction), and as examples we cite the leaching of gold from its ores by cyanide solutions and of cottonseed oil from the seeds by hexane. The diffusion is, of course, from the solid to the liquid phase. If the diffusion is in the opposite direction, the operation is known as *adsorption*. Thus, the colored material which contaminates impure cane sugar solutions can be removed by contacting the liquid solutions with activated carbon, whereupon the colored substances are retained on the surface of the solid carbon.

Solid-solid Because of the extraordinarily slow rates of diffusion within solid phases, there is no industrial separation operation in this category.

Phases Separated by a Membrane

These operations are used relatively infrequently, although they are rapidly increasing in importance. The membranes operate in different ways, depending upon the nature of the separation to be made. In general, however, they serve to prevent intermingling of two miscible phases. They also prevent ordinary hydrodynamic flow, and movement of substances through them is by diffusion. And they permit a component separation by selectively controlling passage of the components from one side to the other.

Gas-gas In *gaseous diffusion* or *effusion*, the membrane is microporous. If a gas mixture whose components are of different molecular weights is brought into contact with such a diaphragm, the various components of the gas pass through the pores at rates dependent upon the molecular weights. This leads to different compositions on opposite sides of the membrane and consequent separation of the mixture. In this manner large-scale separation of the isotopes of uranium, in the form of gaseous uranium hexafluoride, is carried out. In *permeation*, the membrane is not porous, and the gas transmitted through the membrane first dissolves in it and then diffuses through. Separation in this case is brought about principally by difference in solubility of the components. Thus, helium can be separated from natural gas by selective permeation through fluorocarbon-polymer membranes.

Gas-liquid These are *permeation* separations where, for example, a liquid solution of alcohol and water is brought into contact with a suitable nonporous membrane, in which the alcohol preferentially dissolves. After passage through the membrane the alcohol is vaporized on the far side.

Liquid-liquid The separation of a crystalline substance from a colloid, by contact of their solution with a liquid solvent with an intervening membrane permeable only to the solvent and the dissolved crystalline substance, is known as *dialysis*. For example, aqueous beet-sugar solutions containing undesired colloidal material are freed of the latter by contact with water with an intervening semipermeable membrane. Sugar and water diffuse through the membrane, but the larger colloidal particles cannot. *Fractional dialysis* for separating two crystalline substances in solution makes use of the difference in membrane permeability for the substances. If an electromotive force is applied across the membrane to assist in the diffusion of charged particles, the operation is *electrodialysis*. If a solution is separated from the pure solvent by a membrane which is permeable only to the solvent, the solvent diffuses into the solution, an operation known as *osmosis*. This is not a separation operation, of course, but by superimposing a pressure to oppose the osmotic pressure the flow of solvent is reversed, and the solvent and solute of a solution can be separated by *reverse osmosis*. This is one of the processes which may become important in the desalination of seawater.

Direct Contact of Miscible Phases

Because of the difficulty in maintaining concentration gradients without mixing the fluid, the operations in this category are not generally considered practical industrially except in unusual circumstances.

Thermal diffusion involves the formation of a concentration difference within a single liquid or gaseous phase by imposition of a temperature gradient upon the fluid, thus making a separation of the components of the solution possible. In this way, ^{3}He is separated from its mixture with ^{4}He.

If a condensable vapor, such as steam, is allowed to diffuse through a gas mixture, it will preferentially carry one of the components along with it, thus making a separation by the operation known as *sweep diffusion*. If the two zones within the gas phase where the concentrations are different are separated by a screen containing relatively large openings, the operation is called *atmolysis*.

If a gas mixture is subjected to a very rapid *centrifugation*, the components will be separated because of the slightly different forces acting on the various molecules owing to their different masses. The heavier molecules thus tend to accumulate at the periphery of the centrifuge. This method is also used for separation of uranium isotopes.

Use of Surface Phenomena

Substances which when dissolved in a liquid produce a solution of lowered surface tension (in contact with a gas) are known to concentrate in solution at the liquid surface. By forming a foam of large surface, as by bubbling air through the solution, and collecting the foam, the solute can be concentrated. In this manner, detergents have been separated from water, for example. The operation is known as *foam separation*. It is not to be confused with the flotation processes of the ore-dressing industries, where insoluble solid particles are removed from slurries by collection into froths.

This classification is not exhaustive but it does categorize all the major mass-transfer operations. Indeed, new operations continue to be developed, some of which can be fit into more than one category. This book includes gas-liquid, liquid-liquid, and solid-fluid operations, all of which involve direct contact of two immiscible phases and make up the great bulk of the applications of the transfer operations.

Direct and indirect operations The operations depending upon contact of two immiscible phases particularly can be further subclassified into two types. The *direct* operations produce the two phases from a single-phase solution by addition or removal of heat. Fractional distillation, fractional crystallization, and one form of fractional extraction are of this type. The *indirect* operations involve addition of a foreign substance and include gas absorption and stripping, adsorption, drying, leaching, liquid extraction, and certain types of fractional crystallization.

It is characteristic of the direct operations that the products are obtained directly, free of added substance; these operations are therefore sometimes favored over the indirect if they can be used.

If the separated products are required relatively pure, the disadvantages of the indirect operations incurred by addition of a foreign substance are several. The removed substance is obtained as a solution, which in this case must in turn be separated, either to obtain the pure substance or the added substance for reuse, and this represents an expense. The separation of added substance and product can rarely be complete, which may lead to difficulty in meeting product specifications. In any case, addition of a foreign substance may add to the problems of building corrosion-resistant equipment, and the cost of inevitable losses must be borne. Obviously the indirect methods are used only because they are, in the net, less costly than the direct methods if there is a choice. Frequently there is no choice.

When the separated substance need not be obtained pure, many of these disadvantages may disappear. For example, in ordinary drying, the water-vapor–air mixture is discarded since neither constituent need be recovered. In the production of hydrochloric acid by washing a hydrogen chloride–containing gas with water, the acid-water solution is sold directly without separation.

CHOICE OF SEPARATION METHOD

The chemical engineer faced with the problem of separating the components of a solution must ordinarily choose from several possible methods. While the choice is usually limited by the peculiar physical characteristics of the materials to be handled, the necessity for making a decision nevertheless almost always exists. Until the fundamentals of the various operations have been clearly understood, of course, no basis for such a decision is available, but it is well at least to establish the nature of the alternatives at the beginning.

One can sometimes choose between using a mass-transfer operation of the sort discussed in this book and a purely mechanical separation method. For example, in the separation of a desired mineral from its ore, it may be possible to use either the mass-transfer operation of leaching with a solvent or the purely mechanical methods of flotation. Vegetable oils can be separated from the seeds in which they occur by expression or by leaching with a solvent. A vapor can be removed from a mixture with a permanent gas by the mechanical operation of compression or by the mass-transfer operations of gas absorption or adsorption. Sometimes both mechanical and mass-transfer operations are used, especially where the former are incomplete, as in processes for recovering vegetable oils wherein expression is followed by leaching. A more commonplace example is wringing water from wet laundry followed by air drying. It is characteristic that at the end of the operation the substance removed by mechanical methods is pure, while if removed by diffusional methods it is associated with another substance.

One can also frequently choose between a purely mass-transfer operation and a chemical reaction or a combination of both. Water can be removed from an ethanol-water solution either by causing it to react with unslaked lime or by special methods of distillation, for example. Hydrogen sulfide can be separated from other gases either by absorption in a liquid solvent with or without simultaneous chemical reaction or by chemical reaction with ferric oxide. Chemical methods ordinarily destroy the substance removed, while mass-transfer methods usually permit its eventual recovery in unaltered form without great difficulty.

There are also choices to be made within the mass-transfer operations. For example, a gaseous mixture of oxygen and nitrogen may be separated by preferential adsorption of the oxygen on activated carbon, by adsorption, by distillation, or by gaseous effusion. A liquid solution of acetic acid may be separated by distillation, by liquid extraction with a suitable solvent, or by adsorption with a suitable adsorbent.

The principal basis for choice in any case is cost: that method which costs the least is usually the one to be used. Occasionally other factors also influence the decision, however. The simplest operation, while it may not be the least costly, is sometimes desired because it will be trouble-free. Sometimes a method will be discarded because of imperfect knowledge of design methods or unavailability of data for design, so that results cannot be guaranteed. Favorable previous experience with one method may be given strong consideration.

METHODS OF CONDUCTING THE MASS-TRANSFER OPERATIONS

Several characteristics of these operations influence our method of dealing with them and are described in terms which require definition at the start.

Solute Recovery and Fractionation

If the components of a solution fall into two distinct groups of quite different properties, so that one can imagine that one group of components constitutes the solvent and the other group the solute, separation according to these groups is usually relatively easy and amounts to a *solute-recovery* or *solute-removal* operation. For example, a gas consisting of methane, pentane, and hexane can be imagined to consist of methane as solvent with pentane plus hexane as solute, the solvent and solute in this case differing considerably in at least one property, vapor pressure. A simple gas-absorption operation, washing the mixture with a nonvolatile hydrocarbon oil, will easily provide a new solution of pentane plus hexane in the oil, essentially methane-free; and the residual methane will be essentially free of pentane and hexane. On the other hand, a solution consisting of pentane and hexane alone cannot be classified so readily. While the component properties differ, the differences are small, and to separate them into relatively pure components requires a different technique. Such separations are termed *fractionations*, and in this case we might use fractional distillation as a method.

Whether a solute-recovery or fractionation procedure is used may depend upon the property chosen to be exploited. For example, to separate a mixture of propanol and butanol from water by a gas-liquid contacting method, which depends on vapor pressures, requires fractionation (fractional distillation) because the vapor pressures of the components are not greatly different. But nearly complete separation of the combined alcohols from water can be obtained by liquid extraction of the solution with a hydrocarbon, using solute-recovery methods because the solubility of the alcohols as a group and water in hydrocarbons is greatly different. The separation of propanol from butanol, however, requires a fractionation technique (fractional extraction or fractional distillation, for example), because all their properties are very similar.

Unsteady-State Operation

It is characteristic of unsteady-state operation that concentrations at any point in the apparatus change with time. This may result from changes in concentrations of feed materials, flow rates, or conditions of temperature or pressure. In any case, *batch* operations are always of the unsteady-state type. In purely batch operations, all the phases are stationary from a point of view outside the apparatus, i.e., no flow in or out, even though there may be relative motion within. The familiar laboratory extraction procedure of shaking a solution with an immiscible solvent is an example. In *semibatch* operations, one phase is stationary while the other flows continuously in and out of the apparatus. As an example, we may cite the case of a drier where a quantity of wet solid is contacted continuously with fresh air, which carries away the vaporized moisture until the solid is dry.

Steady-State Operation

It is characteristic of steady-state operation that concentrations at any position in the apparatus remain constant with passage of time. This requires continuous, invariable flow of all phases into and out of the apparatus, a persistence of the flow regime within the apparatus, constant concentrations of the feed streams, and unchanging conditions of temperature and pressure.

Stagewise Operation

If two insoluble phases are first allowed to come into contact so that the various diffusing substances can distribute themselves between the phases, and if the phases are then mechanically separated, the entire operation and the equipment required to carry it out are said to constitute one *stage*, e.g., laboratory batch extraction in a separatory funnel. The operation can be carried on in continuous fashion (steady-state) or batchwise fashion, however. For separations requiring greater concentration changes, a series of stages can be arranged so that the phases flow through the assembled stages from one to the other, e.g., in countercurrent flow. Such an assemblage is called a *cascade*. In order to establish a standard for the measurement of performance, the *equilibrium, ideal*, or *theoretical*, stage is defined as one where the effluent phases are in equilibrium, so that a longer time of contact will bring about no additional change of composition. The approach to equilibrium realized in any stage is then defined as the *stage of efficiency*.

Continuous-Contact (Differential-Contact) Operation

In this case the phases flow through the equipment in continuous, intimate contact throughout, without repeated physical separation and recontacting. The nature of the method requires the operation to be either semibatch or steady-state, and the resulting change in compositions may be equivalent to that given by a fraction of an ideal stage or by many stages. Equilibrium between two phases at any position in the equipment is never established; indeed, should equilibrium occur anywhere in the system, the result would be equivalent to the effect of an infinite number of stages.

The essential difference between stagewise and continuous-contact operation can be summarized. In the case of the stagewise operation the diffusional flow of matter between the phases is allowed to reduce the concentration difference which causes the flow. If allowed to continue long enough, an equilibrium is established, after which no further diffusional flow occurs. The rate of diffusion and the time then determine the stage efficiency realized in any particular situation. On the other hand, in continuous-contact operation the departure from equilibrium is deliberately maintained, and the diffusional flow between the phases may continue without interruption. Which method will be used depends to some extent on the stage efficiency that can be practically realized. A high stage efficiency can mean a relatively inexpensive plant and one whose performance can be reliably predicted. A low stage efficiency, on the other hand, may make the continuous-contact methods more desirable for reasons of cost and certainty.

DESIGN PRINCIPLES

There are four major factors to be established in the design of any plant involving the diffusional operations: the number of equilibrium stages or their equivalent, the time of phase contact required, the permissible rate of flow, and the energy requirements.

Number of Equilibrium Stages

In order to determine the number of equilibrium stages required in a cascade to bring about a specified degree of separation, or the equivalent quantity for a continuous-contact device, the equilibrium characteristics of the system and material-balance calculations are required.

Time Requirement

In stagewise operations the time of contact is intimately connected with stage efficiency, whereas for continuous-contact equipment the time leads ultimately to the volume or length of the required device. The factors which help establish the time are several. Material balances permit calculation of the relative quantities required of the various phases. The equilibrium characteristics of the system establish the ultimate concentrations possible, and the rate of transfer of material between phases depends upon the departure from equilibrium which is maintained. The rate of transfer additionally depends upon the physical properties of the phases as well as the flow regime within the equipment.

It is important to recognize that, for a given degree of intimacy of contact of the phases, the time of contact required is independent of the total quantity of the phases to be processed.

Permissible Flow Rate

This factor enters into consideration of semibatch and steady-state operations, where it leads to the determination of the cross-sectional area of the equipment. Considerations of fluid dynamics establish the permissible flow rate, and material balances determine the absolute quantity of each of the streams required.

Energy Requirements

Heat and mechanical energies are ordinarily required to carry out the diffusional operations. Heat is necessary for the production of any temperature changes, for the creation of new phases (such as vaporization of a liquid), and for overcoming heat-of-solution effects. Mechanical energy is required for fluid and solid transport, for dispersing liquids and gases, and for operating moving parts of machinery.

The ultimate design, consequently, requires us to deal with the equilibrium characteristics of the system, material balances, diffusional rates, fluid dynamics, and energy requirements. In what follows, basic considerations of diffusion rates are discussed first (Part One) and these are later applied to specific operations. The principal operations, in turn, are subdivided into three categories, depending upon the nature of the insoluble phases contacted, gas-liquid (Part Two), liquid-liquid (Part Three), and solid-fluid (Part Four), since the equilibrium and fluid dynamics of the systems are most readily studied in such a grouping.

UNIT SYSTEMS

The principal unit system of this book is the SI (*Système International d'Unités*), but to accommodate other systems, practically all (992 of a total of 1017) numbered equations are written so that they can be used with *any* consistent set of units. In order to permit this, it is necessary to include in all expressions involving dimensions of both force and mass the conversion factor g_c, defined through Newton's second law of motion,

$$F = \frac{MA}{g_c}$$

where F = force
 M = mass
 A = acceleration

For the SI and cgs (centimeter-gram-second) system of units, g_c is unnecessary, but it can be assigned a numerical value of unity for purposes of calculation. There are four unit systems commonly used in chemical engineering, and values of g_c corresponding to these are listed in Table 1.1.

For engineering work, SI and English engineering units are probably most important. Consequently the coefficients of the 25 dimensionally inconsistent equations which cannot be used directly with any unit system are given first for SI, then by footnote or other means for English engineering units. Tables and graphs of engineering data are similarly treated.†

Table 1.2 lists the basic quantities as expressed in SI together with the unit abbreviations, Table 1.3 lists the unit prefixes needed for this book, and Table 1.4 lists some of the constants needed in several systems. Finally, Table 1.5 lists the conversion factors into SI for all quantities needed for this book. The boldface letters for each quantity represent the fundamental dimensions: **F** = force, **L** = length, **M** = mass, **mole** = mole, **T** = temperature, **Θ** = time. The list of notation at the end of each chapter gives the symbols used, their meaning, and dimensions.

---

† In practice, some departure from the standard systems is fairly common. Thus, for example, pressures are still frequently quoted in standard atmospheres, millimeters of mercury, bars, or kilograms force per square meter, depending upon the local common practice in the past.

Table 1.1 Conversion factors g_c for the common unit systems

Fundamental quantity	System			
	SI	English engineering[†]	cgs	Metric engineering[‡]
Mass M	Kilogram, kg	Pound mass, lb	Gram, g	Kilogram mass, kg
Length L	Meter, m	Foot, ft	Centimeter, cm	Meter, m
Time Θ	Second, s	Second, s, or hour, h	Second, s	Second, s
Force F	Newton, N	Pound force, lb_f	Dyne, dyn	Kilogram force, kg_f
g_c	$1 \text{ kg} \cdot \text{m/N} \cdot \text{s}^2$	$32.174 \text{ lb} \cdot \text{ft/lb}_f \cdot \text{s}^2$ or $4.1698 \times 10^8 \text{ lb} \cdot \text{ft/lb}_f \cdot \text{h}^2$	$1 \text{ g} \cdot \text{cm/dyn} \cdot \text{s}^2$	$9.80665 \text{ kg} \cdot \text{m/kg}_f \cdot \text{s}^2$

  † Note that throughout this book lb alone is used as the abbreviation pound mass and lb_f is used for pound force.
  ‡ Note that throughout this book kg alone is used as the abbreviation for kilogram mass and kg_f is used for kilogram force.

Table 1.2 Basic SI units

Force = newton, N
Length = meter, m
Mass = kilogram, kg
Mole = kilogram mole, kmol
Temperature = kelvin = K
Time = second, s
Pressure = newton/meter^2, N/m^2 = pascal, Pa
Energy = newton-meter, $N \cdot m$ = joule, J
Power = newton-meter/second, $N \cdot m/s$ = watt, W
Frequency = 1/second, s^{-1} = hertz, Hz

Table 1.3 Prefixes for SI units

Amount	Multiple	Prefix	Symbol
1 000 000	10^6	mega	M
1 000	10^3	kilo	k
100	10^2	hecto	h
10	10	deka	da
0.1	10^{-1}	deci	d
0.01	10^{-2}	centi	c
0.001	10^{-3}	milli	m
0.000 001	10^{-6}	micro	μ
0.000 000 001	10^{-9}	nano	n

Table 1.4 Constants

Acceleration of gravity†	Molar volume of ideal gases at standard conditions (STP)
9.807 m/s^2	($0°C$, 1 std atm)‡
980.7 cm/s^2	22.41 m^3/kmol
32.2 ft/s^2	22.41 l/g mol
4.17×10^8 ft/h^2	359 ft^3/lb mol
Gas constant R	Conversion factor g_c
8314 N · m/kmol · K	1 kg · m/N · s^2
1.987 cal/g mol · K	1 g · cm/dyn · s^2
82.06 atm · cm^3/g mol · K	9.80665 kg · m/kg$_f$ · s^2
0.7302 atm · ft^3/lb mol · °R	32.174 lb · ft/lb$_f$ · s^2
1545 lb$_f$ · ft/lb mol · °R	4.1698×10^8 lb · ft/lb$_f$· h^2
1.987 Btu/lb mol · °R	
847.8 kg$_f$ · m/kmol · K	

† Approximate, depends on location.
‡ Standard conditions are abbreviated STP, for standard temperature and pressure.

Table 1.5 Conversion factors to SI units

Length

Length, L
   ft(0.3048) = m
   in(0.0254) = m
   in(25.4) = mm
   cm(0.01) = m
   Å(10^{-10}) = m
   μm(10^{-6}) = m

Area, L^2
   ft^2(0.0929) = m^2
   in^2(6.452 × 10^{-4}) = m^2
   in^2(645.2) = mm^2
   cm^2(10^{-4}) = m^2

Volume, L^3
   ft^3(0.02832) = m^3
   cm^3(10^{-6}) = m^3
   l(10^{-3}) = m^3
   U.S. gal(3.285 × 10^{-3}) = m^3
   U.K. gal(4.546 × 10^{-3}) = m^3

Specific area, L^2/L^3
   (ft^2/ft^3)(3.2804) = m^2/m^3
   (cm^2/cm^3)(100) = m^2/m^3

Velocity, L/Θ
   (ft/s)(0.3048) = m/s
   (ft/min)(5.08 × 10^{-3}) = m/s
   (ft/h)(8.467 × 10^{-5}) = m/s

Table 1.5 (Continued)

Acceleration, L/Θ^2
    $(ft/s^2)(0.3048) = m/s^2$
    $(ft/h^2)(2.352 \times 10^{-8}) = m/s^2$
    $(cm/s^2)(0.01) = m/s^2$

Diffusivity, kinematic viscosity, L^2/Θ
    $(ft^2/h)(2.581 \times 10^{-5}) = m^2/s$
    $(cm^2/s)(10^{-4}) = m^2/s$
    $St(10^{-4}) = m^2/s^a$
    $cSt(10^{-6}) = m^2/s^a$

Volume rate, L^3/Θ
    $(ft^3/s)(0.02832) = m^3/s$
    $(ft^3/min)(4.72 \times 10^{-4}) = m^3/s$
    $(ft^3/h)(7.867 \times 10^{-6}) = m^3/s$
    (U.S. gal/min)$(6.308 \times 10^{-5}) = m^3/s$
    (U.K. gal/min)$(7.577 \times 10^{-5}) = m^3/s$

Mass

Mass, **M**
    $lb(0.4536) = kg$
    $ton(907.2) = kg$
    $t(1000) = kg^b$

Density, concentration, M/L^3
    $(lb/ft^3)(16.019) = kg/m^3$
    (lb/U.S. gal)$(119.8) = kg/m^3$
    (lb/U.K. gal)$(99.78) = kg/m^3$
    $(g/cm^3)(1000) = kg/m^3 = g/l$

Specific volume, L^3/M
    $(ft^3/lb)(0.0624) = m^3/kg$
    $(cm^3/g)(0.001) = m^3/kg$

Mass rate, M/Θ
    $(lb/s)(0.4536) = kg/s$
    $(lb/min)(7.56 \times 10^{-3}) = kg/s$
    $(lb/h)(1.26 \times 10^{-4}) = kg/s$

Mass rate/length, $M/L\Theta$
    $(lb/ft \cdot h)(4.134 \times 10^{-4}) = kg/m \cdot s$

Viscosity, $M/L\Theta$
    $(lb/ft \cdot s)(1.488) = kg/m \cdot s$
    $(lb/ft \cdot h)(4.134 \times 10^{-4}) = kg/m \cdot s$
    $P(0.1) = kg/m \cdot s^c$
    $cP(0.001) = kg/m \cdot s^c$
    $N \cdot s/m^2 = kg/m \cdot s$

Mass flux, mass velocity, $M/L^2\Theta$
    $(lb/ft^2 \cdot h)(1.356 \times 10^{-3}) = kg/m^2 \cdot s$
    $(g/cm^2 \cdot s)(10) = kg/m^2 \cdot s$

Molar flux, molar mass velocity, $mole/L^2\Theta$
    $(lb\ mol/ft^2 \cdot h)(1.356 \times 10^{-3}) = kmol/m^2 \cdot s$
    $(g\ mol/cm^2 \cdot s)(10) = kmol/m^2 \cdot s$

Table 1.5 (Continued)

Mass-transfer coefficient, **mole/L²Θ(F/L²)** and others
   K_g, k_g(lb mol/ft² · h · atm)(1.338×10^{-8}) = kmol/m² · s · (N/m²)
   K_L, K_c, k_L, k_c[lb mol/ft² · h · (lb mol/ft³)](8.465×10^{-5}) = kmol/m² · s · (kmol/m³)
   K_x, k_x, K_y, k_y(lb mol/ft² · h · mole fraction)(1.356×10^{-3}) = kmol/m² · s · mole fraction
   K_Y, k_Y[lb/ft² · h · (lbA/lbB)](1.356×10^{-3}) = kg/m² · s · (kg A/kg B)
   F_G, F_L(lb mol/ft² · h)(1.356×10^{-3}) = kmol/m² · s

Volumetric mass-transfer coefficient, **mole/L³Θ(F/L²)** and others
   For volumetric coefficients of the type Ka, ka, Fa, and the like, multiply the conversion factors for the coefficient by that for a.

Force

Force, **F**
   lb_f(4.448) = N
   kg_f(9.807) = N
   kp(9.807) = N[d]
   dyn(10^{-5}) = N

Interfacial tension, surface tension, **F/L**
   (lb_f/ft)(14.59) = N/m
   (dyn/cm)(10^{-3}) = N/m
   (erg/cm²)(10^{-3}) = N/m
   kg/s² = N/m

Pressure, **F/L²**
   (lb_f/ft²)(47.88) = N/m² = Pa
   (lb_f/in²)(6895) = N/m² = Pa
   std atm(1.0133×10^5) = N/m² = Pa
   inHg(3386) = N/m² = Pa
   inH₂O(249.1) = N/m² = Pa
   (dyn/cm²)(10^{-1}) = N/m² = Pa
   cmH₂O(98.07) = N/m² = Pa
   mmHg(133.3) = N/m² = Pa
   torr(133.3) = N/m² = Pa
   (kp/m²)(9.807) = N/m² = Pa
   bar(10^5) = N/m² = Pa
   (kg_f/cm²)(9.807×10^4) = N/m² = Pa

Pressure drop/length, **(F/L²)/L**
   [(lb_f/ft²)/ft](157.0) = (N/m²)/m = Pa/m
   (inH₂O/ft)(817) = (N/m²)/m = Pa/m

Energy, work, heat, **FL**
   (ft · lb_f)(1.356) = N · m = J
   Btu(1055) = N · m = J
   Chu(1900) = N · m = J[e]
   erg(10^{-7}) = N · m = J
   cal(4.187) = N · m = J
   kcal(4187) = N · m = J
   (kW · h)(3.6×10^6) = N · m = J

Enthalpy, **FL/M**
   (Btu/lb)(2326) = N · m/kg = J/kg
   (cal/g)(4187) = N · m/kg = J/kg

Table 1.5 (Continued)

Molar enthalpy, FL/mole
  (Btu/lb mol)(2326) = N · m/kmol = J/kmol
  (cal/g mol)(4187) = N · m/kmol = J/kmol

Heat capacity, specific heat, FL/MT
  (Btu/lb · °F)(4187) = N · m/kg · K = J/kg · K
  (cal/g · °C)(4187) = N · m/kg · K = J/kg · K

Molar heat capacity, FL/mole T
  (Btu/lb mol · °F)(4187) = N · m/kmol · K = J/kmol · K
  (cal/g · °C)(4187) = N · m/kmol · K = J/kmol · K

Energy flux, FL/L²Θ
  (Btu/ft² · h)(3.155) = N · m/m² · s = W/m²
  (cal/cm² · s)(4.187 × 10⁴) = N · m/m² · s = W/m²

Thermal conductivity, FL²/L²ΘT = FL/L²Θ(T/L)
  (Btu · ft/ft² · h · °F)(1.7307) = N · m/m · s · K = W/m · K
  (kcal · m/m² · h · °C)(1.163) = N · m/m · s · K = W/m · K
  (cal · cm/cm² · s · °C)(418.7) = N · m/m · s · K = W/m · K

Heat-transfer coefficient, FL/L²ΘT
  (Btu/ft² · h · °F)(5.679) = N · m/m² · s · K = W/m² · K
  (cal/cm² · s · °C)(4.187 × 10⁴) = N · m/m² · s · K = W/m² · K

Power, FL/Θ
  (ft · lb$_f$/s)(1.356) = N · m/s = W
  hp(745.7) = N · m/s = W
  (Btu/min)(4.885 × 10⁻³) = N · m/s = W
  (Btu/h)(0.2931) = N · m/s = W

Power/volume, FL/L³Θ
  (ft · lb$_f$/ft³ · s)(47.88) = N · m/m³ · s = W/m³
  (hp/1000 U.S. gal)(197) = N · m/m³ · s = W/m³

Power/mass, FL/M
  (ft · lb$_f$/lb · s)(2.988) = N · m/kg · s = W/kg

---

[a] St is the abbreviation for stokes.
[b] t is the abbreviation for metric ton (= 1000 kg).
[c] P is the abbreviation for poise.
[d] kp is the abbreviation for kilopond = kg force, kg$_f$.
[e] Chu is the abbreviation for centigrade heat unit.

In reading elsewhere, the student may come upon an empirical equation which it would be useful to convert into SI units. The procedure for this is shown in the following example.

Illustration 1.1 The minimum speed for a four-bladed paddle in an unbaffled, agitated vessel to mix two immiscible liquids is reported to be [S. Nagata et al., *Trans. Soc. Chem. Eng. Jap.*, **8**, 43 (1950)]

$$N' = \frac{30\,600}{T'^{3/2}}\left(\frac{\mu'}{\rho'}\right)^{0.111}\left(\frac{\Delta\rho'}{\rho'}\right)^{0.26}$$

where N' = impeller speed r/h = h^{-1}
T' = vessel diameter, ft
μ' = continuous-liquid viscosity, lb/ft · h
ρ' = continuous-liquid density, lb/ft^3
$\Delta\rho'$ = difference in density of liquids, lb/ft^3

Substitution of the units or dimensions of the various quantities shows that neither the dimensions nor the units on the left of the equals sign are the same, respectively, as those on the right. In other words, the equation is dimensionally inconsistent, and the coefficient 30 600 can be used only for the units listed above. Compute the coefficient required for the equation to be used with SI units: $N = s^{-1}$, $T = m$, $\mu = kg/m \cdot s$, $\rho = kg/m^3$, and $\Delta\rho = kg/m^3$.

SOLUTION The conversion factors are taken from Table 1.5. The procedure involves substitution of each SI unit together with necessary conversion factor to convert it into the unit of the equation as reported, the inverse of the usual use of Table 1.5. Thus,

$$\frac{N\,s^{-1}}{2.778 \times 10^{-4}} = N'\,h^{-1}$$

Then

$$\frac{N}{2.778 \times 10^{-4}} = \frac{30\,600}{(T/0.3048)^{3/2}} \left[\frac{\mu/(4.134 \times 10^{-4})}{16.019} \right]^{0.111} \left(\frac{\Delta\rho/16.019}{\rho/16.019} \right)^{0.26}$$

or

$$N = \frac{4.621}{T^{3/2}} \left(\frac{\mu}{\rho} \right)^{0.111} \left(\frac{\Delta\rho}{\rho} \right)^{0.26}$$

which is suitable for SI units.

ONE

DIFFUSION AND MASS TRANSFER

We have seen that most of the mass-transfer operations used for separating the components of a solution achieve this result by bringing the solution to be separated into contact with another insoluble phase. As will be developed, the rate at which a component is then transferred from one phase to the other depends upon a so-called *mass-transfer*, or rate, *coefficient* and upon the degree of departure of the system from equilibrium. The transfer stops when equilibrium is attained.

The rate coefficients for the various components in a given phase will differ from each other to the greatest extent under conditions where molecular diffusion prevails, but even then the difference is not really large. For example, gases and vapors diffusing through air will show transfer coefficients whose ratio at most may be 3 or 4 to 1. The same is true when various substances diffuse through a liquid such as water. Under conditions of turbulence, where molecular diffusion is relatively unimportant, the transfer coefficients become much more nearly alike for all components. Consequently, while in principle some separation of the components could be achieved by taking advantage of their different transfer coefficients, the degree of separation attainable in this manner is small. This is especially significant when it is considered that we frequently wish to obtain products which are nearly pure substances, where the ratio of components may be of the order of 1000 or 10 000 to 1, or even larger.

Therefore we depend almost entirely upon the differences in concentration which exist at equilibrium, and not upon the difference in transfer coefficients, for making separations. Nevertheless, the mass-transfer coefficients are of great importance, since, as they regulate the rate at which equilibrium is approached, they control the time required for separation and therefore the size and cost of the equipment used. The transfer coefficients are also important in governing the size of equipment used for entirely different purposes, such as carrying out chemical reactions. For example, the rate at which a reaction between two gases occurs on a solid catalyst is frequently governed by the rate of transfer of the gases to the catalyst surface and the rate of transfer of the product away from the catalyst.

The mass-transfer coefficients, their relationship to the phenomenon of diffusion, fluid motion, and to related rate coefficients, such as those describing heat transfer, are treated in Part One.

MOLECULAR DIFFUSION IN FLUIDS

Molecular diffusion is concerned with the movement of individual molecules through a substance by virtue of their thermal energy. The kinetic theory of gases provides a means of visualizing what occurs, and indeed it was the success of this theory in quantitatively describing the diffusional phenomena which led to its rapid acceptance. In the case of a simplified kinetic theory, a molecule is imagined to travel in a straight line at a uniform velocity until it collides with another molecule, whereupon its velocity changes both in magnitude and direction. The average distance the molecule travels between collisions is its mean free path, and the average velocity is dependent upon the temperature. The molecule thus travels a highly zigzag path, the net distance in one direction which it moves in a given time, the rate of diffusion, being only a small fraction of the length of its actual path. For this reason the diffusion rate is very slow, although we can expect it to increase with decreasing pressure, which reduces the number of collisions, and with increased temperature, which increases the molecular velocity.

The importance of the barrier molecular collision presents to diffusive movement is profound. Thus, for example, it can be computed through the kinetic theory that the rate of evaporation of water at 25°C into a complete vacuum is roughly 3.3 kg/s per square meter of water surface. But placing a layer of stagnant air at 1 std atm pressure and only 0.1 mm thick above the water surface reduces the rate by a factor of about 600. The same general mechanism prevails also for the liquid state, but because of the considerably higher molecular concentration, we find even slower diffusion rates than in gases.

The phenomenon of molecular diffusion ultimately leads to a completely uniform concentration of substances throughout a solution which may initially have been nonuniform. Thus, for example, if a drop of blue copper sulfate solution is placed in a beaker of water, the copper sulfate eventually permeates the entire liquid. The blue color in time becomes everywhere uniform, and no subsequent change occurs.

We must distinguish at the start, however, between molecular diffusion, which is a slow process, and the more rapid mixing which can be brought about by mechanical stirring and convective movement of the fluid. Visualize a tank 1.5 m in diameter into which has been placed a salt solution to a depth of 0.75 m. Imagine that a 0.75-m-deep layer of pure water has been carefully placed over the brine without disturbing the brine in any way. If the contents of the tank are left completely undisturbed, the salt will, by molecular diffusion, completely permeate the liquid, ultimately coming everywhere to one-half its concentration in the original brine. But the process is very slow, and it can be calculated that the salt concentration at the top surface will still be only 87.5 percent of its final value after 10 years and will reach 99 percent of its final value only after 28 years. On the other hand, it has been demonstrated that a simple paddle agitator rotating in the tank at 22 r/min will bring about complete uniformity in about 60 s [27]. The mechanical agitation has produced rapid movement of relatively large chunks, or eddies, of fluid characteristic of turbulent motion, which have carried the salt with them. This method of solute transfer is known as *eddy* or *turbulent diffusion*, as opposed to molecular diffusion. Of course, within each eddy, no matter how small, uniformity is achieved only by molecular diffusion, which is the ultimate process. We see then that molecular diffusion is the mechanism of mass transfer in stagnant fluids or in fluids which are moving only in laminar flow, although it is nevertheless always present even in highly developed turbulent flow.

In a two-phase system not at equilibrium, such as a layer of ammonia and air as a gas solution in contact with a layer of liquid water, spontaneous alteration through molecular diffusion also occurs, ultimately bringing the entire system to a state of equilibrium, whereupon alteration stops. At the end, we observe that the concentration of any constituent is the same throughout a phase, but it will not necessarily be the same in both phases. Thus the ammonia concentration will be uniform throughout the liquid and uniform at a different value throughout the gas. On the other hand, the chemical potential of the ammonia (or its activity if the same reference state is used), which is differently dependent upon concentration in the two phases, will be uniform everywhere throughout the system at equilibrium, and it is this uniformity which has brought the diffusive process to a halt. Evidently, then, the true driving force for diffusion is activity or chemical potential, and not concentration. In multiphase systems, however, we customarily deal with diffusional processes in each phase separately, and within one phase it is usually described in terms of that which is most readily observed, namely, concentration changes.

Molecular Diffusion

We have noted that if a solution is everywhere uniform in concentration of its constituents, no alteration occurs but that as long as it is not uniform, the solution is spontaneously brought to uniformity by diffusion, the substances moving from a place of high concentration to one of low. The rate at which a solute moves at any point in any direction must therefore depend on the concentration gradient at that point and in that direction. In describing this quantitatively, we need an appropriate measure of rate.

Rates will be most conveniently described in terms of a molar flux, or mol/(area)(time), the area being measured in a direction normal to the diffusion. In a nonuniform solution even containing only two constituents, however, both constituents must diffuse if uniformity is the ultimate result, and this leads to the use of two fluxes to describe the motion of one constituent: N, the flux relative to a fixed location in space; and J, the flux of a constituent relative to the average molar velocity of all constituents. The first of these is of importance in the applications to design of equipment, but the second is more characteristic of the nature of the constituent. For example, a fisherman is most interested in the rate at which a fish swims upstream against the flowing current to reach his baited hook (analogous to N), but the velocity of the fish relative to the stream (analogous to J) is more characteristic of the swimming ability of the fish.

The *diffusivity*, or *diffusion coefficient*, D_{AB} of a constituent A in solution in B, which is a measure of its diffusive mobility, is then defined as the ratio of its flux J_A to its concentration gradient

$$J_A = -D_{AB}\frac{\partial c_A}{\partial z} = -cD_{AB}\frac{\partial x_A}{\partial z} \qquad (2.1)$$

which is Fick's first law written for the z direction. The negative sign emphasizes that diffusion occurs in the direction of a drop in concentration. The diffusivity is a characteristic of a constituent and its environment (temperature, pressure, concentration, whether in liquid, gas, or solid solution, and the nature of the other constituents).

Consider the box of Fig. 2.1, which is separated into two parts by the partition P. Into section I, 1 kg water (A) is placed and into section II 1 kg ethanol (B) (the densities of the liquids are different, and the partition is so located that the depths of the liquids in each section are the same). Imagine the partition to be carefully removed, thus allowing diffusion of both liquids to occur. When diffusion stops, the concentration will be uniform throughout at 50 mass percent of each constituent, and the masses and moles of each constituent in the two regions will be as indicated in the figure. It is clear that while the water has diffused to the right and the ethanol to the left, there has been a net mass movement to the right, so that if the box had originally been balanced on a knife-edge, at the end of the process it would have tipped downward to the right. If the direction to the right is taken as positive, the flux N_A of A relative to the fixed position P has been positive and the flux N_B of B has been negative. For a condition of steady state, the net flux is

$$N_A + N_B = N \qquad (2.2)$$

The movement of A is made up of two parts, namely, that resulting from the bulk motion N and the fraction x_A of N which is A and that resulting from

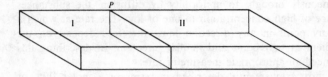

	I		II	
	kg	kmol	kg	kmol
Initially:				
H_2O	100	5.55	EtOH 100	2.17
Finally:				
H_2O:	44.08	2.45	55.92	3.10
EtOH:	44.08	0.96	55.92	1.21
Total	88.16	3.41	Total: 111.84	4.31

Figure 2.1 Diffusion in a binary solution.

diffusion J_A:

$$N_A = Nx_A + J_A \tag{2.3}$$

$$N_A = (N_A + N_B)\frac{c_A}{c} - D_{AB}\frac{\partial c_A}{\partial z} \tag{2.4}$$

The counterpart of Eq. (2.4) for B is

$$N_B = (N_A + N_B)\frac{c_B}{c} - D_{BA}\frac{\partial c_B}{\partial z} \tag{2.5}$$

Adding these gives

$$- D_{AB}\frac{\partial c_A}{\partial z} = D_{BA}\frac{\partial c_B}{\partial z} \tag{2.6}$$

or $J_A = -J_B$. If $c_A + c_B = $ const, it follows that $D_{AB} = D_{BA}$ at the prevailing concentration and temperature.

All the above has considered diffusion in only one direction, but in general concentration gradients, velocities, and diffusional fluxes exist in all directions, so that counterparts of Eqs. (2.1) to (2.6) for all three directions in the cartesian coordinate system exist. In certain solids, the diffusivity D_{AB} may also be direction-sensitive, although in fluids which are true solutions it is not.

The Equation of Continuity

Consider the volume element of fluid of Fig. 2.2, where a fluid is flowing through the element. We shall need a material balance for a component of the fluid applicable to a differential fluid volume of this type.

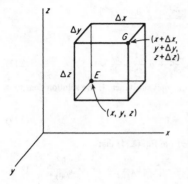

Figure 2.2 An elemental fluid volume.

The mass rate of flow of component A into the three faces with a common corner at E is

$$M_A\left[(N_{A,x})_x \, \Delta y \, \Delta z + (N_{A,y})_y \, \Delta x \, \Delta z + (N_{A,z})_z \, \Delta x \, \Delta y\right]$$

where $N_{A,x}$ signifies the x-directed flux and $(N_{A,x})_x$ its value at location x. Similarly the mass rate of flow out of the three faces with a common corner at G is

$$M_A\left[(N_{A,x})_{x+\Delta x} \, \Delta y \, \Delta z + (N_{A,y})_{y+\Delta y} \, \Delta x \, \Delta z + (N_{A,z})_{z+\Delta z} \, \Delta x \, \Delta y\right]$$

The total component A in the element is $\Delta x \, \Delta y \, \Delta z \, \rho_A$, and its rate of accumulation is therefore $\Delta x \, \Delta y \, \Delta z \, \partial \rho_A / \partial \theta$. If, in addition, A is generated by chemical reaction at the rate R_A mol/(volume)(time), its production rate is $M_A R_A \, \Delta x \, \Delta y \, \Delta z$, mass/time. Since, in general,

$$\text{Rate out} - \text{rate in} + \text{rate of accumulation} = \text{rate of generation}$$

then

$$M_A\left\{[(N_{A,x})_{x+\Delta x} - (N_{A,x})_x] \, \Delta y \, \Delta z + \left[(N_{A,y})_{y+\Delta y} - (N_{A,y})_y\right] \, \Delta x \, \Delta z\right.$$
$$\left. + [(N_{A,z})_{z+\Delta z} - (N_{A,z})_z] \, \Delta x \, \Delta y\right\} + \Delta x \, \Delta y \, \Delta z \frac{\partial \rho_A}{\partial \theta} = M_A R_A \, \Delta x \, \Delta y \, \Delta z$$

$$(2.7)$$

Dividing by $\Delta x \, \Delta y \, \Delta z$ and taking the limit as the three distances become zero gives

$$M_A\left(\frac{\partial N_{A,x}}{\partial x} + \frac{\partial N_{A,y}}{\partial y} + \frac{\partial N_{A,z}}{\partial z}\right) + \frac{\partial \rho_A}{\partial \theta} = M_A R_A \qquad (2.8)$$

Similarly, for component B

$$M_B\left(\frac{\partial N_{B,x}}{\partial x} + \frac{\partial N_{B,y}}{\partial y} + \frac{\partial N_{B,z}}{\partial z}\right) + \frac{\partial \rho_B}{\partial \theta} = M_B R_B \qquad (2.9)$$

The total material balance is obtained by adding those for A and B

$$\frac{\partial (M_A N_A + M_B N_B)_x}{\partial x} + \frac{\partial (M_A N_A + M_B N_B)_y}{\partial y} + \frac{\partial (M_A N_A + M_B N_B)_z}{\partial z} + \frac{\partial \rho}{\partial \theta} = 0 \qquad (2.10)$$

where $\rho = \rho_A + \rho_B = $ the solution density, since the mass rate of generation of A and B must equal zero.

Now the counterpart of Eq. (2.3) in terms of masses and in the x direction is

$$M_A N_{A,x} = u_x \rho_A + M_A J_{A,x} \qquad (2.11)$$

where u_x is the mass-average velocity such that

$$\rho u_x = u_{A,x} \rho_A + u_{B,x} \rho_B = M_A N_{A,x} + M_B N_{B,x} \qquad (2.12)$$

Therefore,
$$\frac{\partial(M_A N_A + M_B N_B)_x}{\partial x} = \rho \frac{\partial u_x}{\partial x} + u_x \frac{\partial \rho}{\partial x}$$

Equation (2.10) therefore becomes

$$\rho\left(\frac{\partial u_x}{\partial x} + \frac{\partial u_y}{\partial y} + \frac{\partial u_z}{\partial z}\right) + u_x \frac{\partial \rho}{\partial x} + u_y \frac{\partial \rho}{\partial y} + u_z \frac{\partial \rho}{\partial z} + \frac{\partial \rho}{\partial \theta} = 0 \qquad (2.13)$$

which is the *equation of continuity*, or a mass balance, for total substance. If the solution density is constant, it becomes

$$\frac{\partial u_x}{\partial x} + \frac{\partial u_y}{\partial y} + \frac{\partial u_z}{\partial z} = 0 \qquad (2.14)$$

Returning to the balance for component A, we see from Eq. (2.11) that

$$M_A \frac{\partial N_{A,x}}{\partial x} = u_x \frac{\partial \rho_A}{\partial x} + \rho_A \frac{\partial u_x}{\partial x} + M_A \frac{\partial J_{A,x}}{\partial x} = u_x \frac{\partial \rho_A}{\partial x} + \rho_A \frac{\partial u_x}{\partial x} - M_A D_{AB} \frac{\partial^2 c_A}{\partial x^2} \qquad (2.15)$$

Equation (2.8) then becomes

$$u_x \frac{\partial \rho_A}{\partial x} + u_y \frac{\partial \rho_A}{\partial y} + u_z \frac{\partial \rho_A}{\partial z} + \rho_A\left(\frac{\partial u_x}{\partial x} + \frac{\partial u_y}{\partial y} + \frac{\partial u_z}{\partial z}\right)$$
$$- M_A D_{AB}\left(\frac{\partial^2 c_A}{\partial x^2} + \frac{\partial^2 c_A}{\partial y^2} + \frac{\partial^2 c_A}{\partial z^2}\right) + \frac{\partial \rho_A}{\partial \theta} = M_A R_A \qquad (2.16)$$

which is the equation of continuity for substance A. For a solution of constant density, we can apply Eq. (2.14) to the terms multiplying ρ_A. Dividing by M_A, we then have

$$u_x \frac{\partial c_A}{\partial x} + u_y \frac{\partial c_A}{\partial y} + u_z \frac{\partial c_A}{\partial z} + \frac{\partial c_A}{\partial \theta} = D_{AB}\left(\frac{\partial^2 c_A}{\partial x^2} + \frac{\partial^2 c_A}{\partial y^2} + \frac{\partial^2 c_A}{\partial z^2}\right) + R_A \qquad (2.17)$$

In the special case where the velocity equals zero and there is no chemical reaction, it reduces to Fick's second law

$$\frac{\partial c_A}{\partial \theta} = D_{AB}\left(\frac{\partial^2 c_A}{\partial x^2} + \frac{\partial^2 c_A}{\partial y^2} + \frac{\partial^2 c_A}{\partial z^2}\right) \qquad (2.18)$$

This is frequently applicable to diffusion in solids and to limited situations in fluids.

In similar fashion, it is possible to derive the equations for a differential energy balance. For a fluid of constant density, the result is

$$u_x \frac{\partial t}{\partial x} + u_y \frac{\partial t}{\partial y} + u_z \frac{\partial t}{\partial z} + \frac{\partial t}{\partial \theta} = \alpha\left(\frac{\partial^2 t}{\partial x^2} + \frac{\partial^2 t}{\partial y^2} + \frac{\partial^2 t}{\partial z^2}\right) + \frac{Q}{\rho C_p} \qquad (2.19)$$

where $\alpha = k/\rho C_p$ and Q is the rate of heat generation within the fluid per unit volume from a chemical reaction. The significance of the similarities between Eqs. (2.17) and (2.19) will be developed in Chap. 3.

STEADY-STATE MOLECULAR DIFFUSION IN FLUIDS AT REST AND IN LAMINAR FLOW

Applying Eq. (2.4) to the case of diffusion only in the z direction, with N_A and N_B both constant (steady state), we can readily separate the variables, and if D_{AB} is constant, it can be integrated

$$\int_{c_{A1}}^{c_{A2}} \frac{-dc_A}{N_A c - c_A(N_A + N_B)} = \frac{1}{c D_{AB}} \int_{z_1}^{z_2} dz \qquad (2.20)$$

where 1 indicates the beginning of the diffusion path (c_A high) and 2 the end of the diffusion path (c_A low). Letting $z_2 - z_1 = z$, we get

$$\frac{1}{N_A + N_B}\ln\frac{N_A c - c_{A2}(N_A + N_B)}{N_A c - c_{A1}(N_A + N_B)} = \frac{z}{cD_{AB}} \tag{2.21}$$

or

$$N_A = \frac{N_A}{N_A + N_B}\frac{D_{AB}c}{z}\ln\frac{N_A/(N_A + N_B) - c_{A2}/c}{N_A/(N_A + N_B) - c_{A1}/c} \tag{2.22}$$

Integration under steady-state conditions where the flux N_A is not constant is also possible. Consider radial diffusion from the surface of a solid sphere into a fluid, for example. Equation (2.20) can be applied, but the flux is a function of distance owing to the geometry. Most practical problems which deal with such matters, however, are concerned with diffusion under turbulent conditions, and the transfer coefficients which are then used are based upon a flux expressed in terms of some arbitrarily chosen area, such as the surface of the sphere. These matters are considered in Chap. 3.

Molecular Diffusion in Gases

When the ideal-gas law can be applied, Eq. (2.21) can be written in a form more convenient for use with gases. Thus,

$$\frac{c_A}{c} = \frac{\bar{p}_A}{p_t} = y_A \tag{2.23}$$

where $\bar{p}_A$ = partial pressure of component A
p_t = total pressure
y_A = mole fraction concentration†

Further,

$$c = \frac{n}{V} = \frac{p_t}{RT} \tag{2.24}$$

so that Eq. (2.22) becomes

$$N_A = \frac{N_A}{N_A + N_B}\frac{D_{AB}p_t}{RTz}\ln\frac{\left[N_A/(N_A + N_B)\right]p_t - \bar{p}_{A2}}{\left[N_A/(N_A + N_B)\right]p_t - \bar{p}_{A1}} \tag{2.25}$$

or

$$N_A = \frac{N_A}{N_A + N_B}\frac{D_{AB}p_t}{RTz}\ln\frac{N_A/(N_A + N_B) - y_{A2}}{N_A/(N_A + N_B) - y_{A1}} \tag{2.26}$$

In order to use these equations the relation between N_A and N_B must be known. This is usually fixed by other considerations. For example, if methane is being cracked on a catalyst,

$$CH_4 \rightarrow C + 2H_2$$

under circumstances such that CH_4 (A) diffuses to the cracking surface and H_2 (B) diffuses back, the reaction stoichiometry fixes the relationship $N_B = -2N_A$,

---

† The component subscript A on y_A will differentiate mole fraction from the y meaning distance in the y direction.

and

$$\frac{N_A}{N_A + N_B} = \frac{N_A}{N_A - 2N_A} = -1$$

On other occasions, in the absence of chemical reaction, the ratio can be fixed by enthalpy considerations. In the case of the purely separational operations, there are two situations which frequently arise.

Steady-state diffusion of A through nondiffusing B This might occur, for example, if ammonia (A) were being absorbed from air (B) into water. In the gas phase, since air does not dissolve appreciably in water, and if we neglect the evaporation of water, only the ammonia diffuses. Thus, $N_B = 0$, $N_A = $ const,

$$\frac{N_A}{N_A + N_B} = 1$$

and Eq. (2.25) becomes

$$N_A = \frac{D_{AB}p_t}{RTz} \ln \frac{p_t - \bar{p}_{A2}}{p_t - \bar{p}_{A1}} \qquad (2.27)$$

Since $p_t - \bar{p}_{A2} = \bar{p}_{B2}$, $p_t - \bar{p}_{A1} = \bar{p}_{B1}$, $\bar{p}_{B2} - \bar{p}_{B1} = \bar{p}_{A1} - \bar{p}_{A2}$, then

$$N_A = \frac{D_{AB}p_t}{RTz} \frac{\bar{p}_{A1} - \bar{p}_{A2}}{\bar{p}_{B2} - \bar{p}_{B1}} \ln \frac{\bar{p}_{B2}}{\bar{p}_{B1}} \qquad (2.28)$$

If we let

$$\frac{\bar{p}_{B2} - \bar{p}_{B1}}{\ln(\bar{p}_{B2}/\bar{p}_{B1})} = \bar{p}_{B,M} \qquad (2.29)$$

then

$$N_A = \frac{D_{AB}p_t}{RTz\bar{p}_{B,M}}(\bar{p}_{A1} - \bar{p}_{A2}) \qquad (2.30)$$

This equation is shown graphically in Fig. 2.3. Substance A diffuses by virtue of its concentration gradient, $-d\bar{p}_A/dz$. Substance B is also diffusing relative to

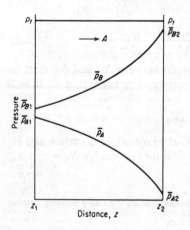

Figure 2.3 Diffusion of A through stagnant B.

the average molar velocity at a flux J_B, which depends upon $-d\bar{p}_B/dz$, but like a fish which swims upstream at the same velocity as the water flows downstream, $N_B = 0$ relative to a fixed place in space.

Steady-state equimolal counterdiffusion This is a situation which frequently pertains in distillation operations. $N_A = -N_B = $ const. Equation (2.25) becomes indeterminate, but we can go back to Eq. (2.4), which, for gases, becomes

$$N_A = (N_A + N_B)\frac{\bar{p}_A}{p_t} - \frac{D_{AB}}{RT}\frac{d\bar{p}_A}{dz} \tag{2.31}$$

or, for this case,

$$N_A = -\frac{D_{AB}}{RT}\frac{d\bar{p}_A}{dz} \tag{2.32}$$

$$\int_{z_1}^{z_2} dz = -\frac{D_{AB}}{RTN_A}\int_{\bar{p}_{A1}}^{\bar{p}_{A2}} d\bar{p}_A \tag{2.33}$$

$$N_A = \frac{D_{AB}}{RTz}(\bar{p}_{A1} - \bar{p}_{A2}) \tag{2.34}$$

This is shown graphically in Fig. 2.4.

Steady-state diffusion in multicomponent mixtures The expressions for diffusion in multicomponent systems become very complicated, but they can frequently be handled by using an *effective diffusivity* in Eq. (2.25), where the effective diffusivity of a component can be synthesized from its binary diffusivities with each of the other constituents [1].† Thus, in Eq. (2.25), $N_A + N_B$ is replaced by $\sum_{i=A}^{n} N_i$, where N_i is positive if diffusion is in the same direction as that of A and negative if in the opposite direction and D_{AB} is replaced by the effective $D_{A,m}$

$$D_{A,m} = \frac{N_A - y_A \sum_{i=A}^{n} N_i}{\sum_{i=A}^{n} \frac{1}{D_{Ai}}(y_i N_A - y_A N_i)} \tag{2.35}$$

The $D_{A,i}$ are the binary diffusivities. This indicates that $D_{A,m}$ may vary considerably from one end of the diffusion path to the other, but a linear variation with distance can usually be assumed for practical calculations [1]. A common situation is when all the N's except N_A are zero; i.e., all but one component is stagnant. Equation (2.35) then becomes [23]

$$D_{A,m} = \frac{1 - y_A}{\sum_{i=B}^{n} \frac{y_i}{D_{A,i}}} = \frac{1}{\sum_{i=B}^{n} \frac{y_i'}{D_{A,i}}} \tag{2.36}$$

† Numbered references appear at the end of the chapter.

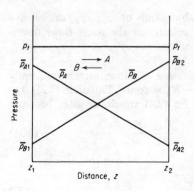

Figure 2.4 Equimolal counterdiffusion.

where y'_i is the mole fraction of component i on an A-free basis. The limitations of Eq. (2.35) and some suggestions for dealing with them have been considered [21].

Illustration 2.1 Oxygen (A) is diffusing through carbon monoxide (B) under steady-state conditions, with the carbon monoxide nondiffusing. The total pressure is 1×10^5 N/m², and the temperature 0°C. The partial pressure of oxygen at two planes 2.0 mm apart is, respectively, 13 000 and 6500 N/m². The diffusivity for the mixture is 1.87×10^{-5} m²/s. Calculate the rate of diffusion of oxygen in kmol/s through each square meter of the two planes.

SOLUTION Equation (2.30) applies. $D_{AB} = 1.87 \times 10^{-5}$ m²/s, $p_t = 10^5$ N/m², $z = 0.002$ m, $R = 8314$ N · m/kmol · K, $T = 273$ K; $\bar{p}_{A,i} = 13 \times 10^3$, $\bar{p}_{B,i} = 10^5 - 13 \times 10^3 = 87 \times 10^3$, $\bar{p}_{A2} = 6500$, $\bar{p}_{B2} = 10^5 - 6500 = 93.5 \times 10^3$, all in N/m².

$$\bar{p}_{B,M} = \frac{\bar{p}_{B1} - \bar{p}_{B2}}{\ln(\bar{p}_{B1}/\bar{p}_{B2})} = \frac{(87 - 93.5)(10^3)}{\ln(87/93.5)} = 90\,200 \text{ N/m}^2$$

$$N_A = \frac{D_{AB}p_t}{RTz\bar{p}_{B,M}}(\bar{p}_{A1} - \bar{p}_{A2}) = \frac{(1.87 \times 10^{-5})(10^5)(13 - 6.5)(10^3)}{8314(273)(0.002)(90.2 \times 10^3)}$$

$$= 2.97 \times 10^{-5} \text{ kmol/m}^2 \cdot \text{s} \quad \textbf{Ans.}$$

Illustration 2.2 Recalculate the rate of diffusion of oxygen (A) in Illustration 2.1, assuming that the nondiffusing gas is a mixture of methane (B) and hydrogen (C) in the volume ratio 2 : 1. The diffusivities are estimated to be $D_{O_2-H_2} = 6.99 \times 10^{-5}$, $D_{O_2-CH_4} = 1.86 \times 10^{-5}$ m²/s.

SOLUTION Equation (2.25) will become Eq. (2.30) for this case. $p_t = 10^5$ N/m², $T = 273$ K, $\bar{p}_{A1} = 13 \times 10^3$, $\bar{p}_{A2} = 6500$, $\bar{p}_{iM} = 90.2 \times 10^3$, all in N/m²; $z = 0.002$ m, $R = 8314$ N · m/kmol · K, as in Illustration 2.1. In Eq. (2.36), $y'_B = 2/(2 + 1) = 0.667$, $y'_C = 1 - 0.667 = 0.333$, whence

$$D_{A,m} = \frac{1}{y'_B/D_{AB} + y'_C/D_{A,c}} = \frac{1}{0.667/(1.86 \times 10^{-5}) + 0.333/(6.99 \times 10^{-5})}$$

$$= 2.46 \times 10^{-5} \text{ m}^2/\text{s}$$

Therefore Eq. (2.30) becomes

$$N_A = \frac{(2.46 \times 10^{-5})(13\,000 - 6500)}{8314(273)(0.002)(90\,200)} = 3.91 \times 10^{-5} \text{ kmol/m}^2 \cdot \text{s} \quad \textbf{Ans.}$$

Table 2.1 Diffusivities of gases at standard atmospheric pressure, 101.3 kN/m^2

System	Temp, °C	Diffusivity, m^2/s × 10^5	Ref.
H$_2$-CH$_4$	0	6.25†	3
O$_2$-N$_2$	0	1.81	3
CO-O$_2$	0	1.85	3
CO$_2$-O$_2$	0	1.39	3
Air-NH$_3$	0	1.98	26
Air-H$_2$O	25.9	2.58	7
	59.0	3.05	7
Air-ethanol	0	1.02	14
Air-n-butanol	25.9	0.87	7
	59.0	1.04	7
Air-ethyl acetate	25.9	0.87	7
	59.0	1.06	7
Air-aniline	25.9	0.74	7
	59.0	0.90	7
Air-chlorobenzene	25.9	0.74	7
	59.0	0.90	7
Air-toluene	25.9	0.86	7
	59.0	0.92	7

† For example, $D_{\text{H}_2-\text{CH}_4} = 6.25 \times 10^{-5}$ m^2/s.

Diffusivity of Gases

The diffusivity, or diffusion coefficient, D is a property of the system dependent upon temperature, pressure, and nature of the components. An advanced kinetic theory [12] predicts that in binary mixtures there will be only a small effect of composition. The dimensions of diffusivity can be established from its definition, Eq. (2.1), and are length^2/time. Most of the values for D reported in the literature are expressed as cm^2/s; the SI dimensions are m^2/s. Conversion factors are listed in Table 1.5. A few typical data are listed in Table 2.1; a longer list is available in "The Chemical Engineers' Handbook" [18]. For a complete review, see Ref. 17.

Expressions for estimating D in the absence of experimental data are based on considerations of the kinetic theory of gases. The Wilke-Lee modification [25] of the Hirschfelder-Bird-Spotz method [11] is recommended for mixtures of nonpolar gases or of a polar with a nonpolar gas†

$$D_{AB} = \frac{10^{-4}\left(1.084 - 0.249\sqrt{1/M_A + 1/M_B}\,\right)T^{3/2}\sqrt{1/M_A + 1/M_B}}{p_t(r_{AB})^2 f(kT/\varepsilon_{AB})}$$

(2.37)

† The listed units must be used in Eq. (2.37). For D_{AB}, p_t, and T in units of feet, pounds force, hours, and degrees Rankine and all other quantities as listed above, multiply the right-hand side of Eq. (2.37) by 334.7.

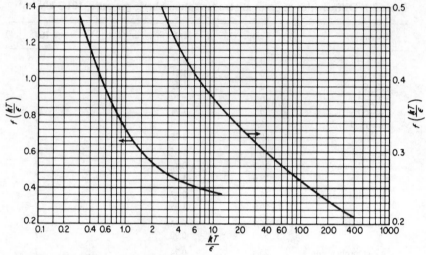

Figure 2.5 Collision function for diffusion.

where D_{AB} = diffusivity, m^2/s
 T = absolute temperature, K
M_A, M_B = molecular weight of A and B, respectively, kg/kmol
 p_t = abs pressure, N/m^2
 r_{AB} = molecular separation at collision, nm = $(r_A + r_B)/2$
 ε_{AB} = energy of molecular attraction = $\sqrt{\varepsilon_A \varepsilon_B}$
 k = Boltzmann's constant
$f(kT/\varepsilon_{AB})$ = collision function given by Fig. 2.5

The values of r and ε, such as those listed in Table 2.2, can be calculated from other properties of gases, such as viscosity. If necessary, they can be estimated for each component empirically [25]

$$r = 1.18v^{1/3} \tag{2.38}$$

$$\frac{\varepsilon}{k} = 1.21 T_b \tag{2.39}$$

where v is the molal volume of liquid at normal boiling point, $m^3/kmol$ (estimate from Table 2.3), and T_b is the normal boiling point in Kelvins. In using Table 2.3, the contributions for the constituent atoms are added together. Thus, for toluene, C_7H_8, $v = 7(0.0148) + 8(0.0037) - 0.015 = 0.1182$. Diffusion through air, when the constituents of the air remain in fixed proportions, is handled as if the air were a single substance.

Illustration 2.3 Estimate the diffusivity of ethanol vapor, C_2H_5OH, (A), through air (B) at 1 std atm pressure, 0°C.

SOLUTION $T = 273$ K, $p_t = 101.3$ kN/m^2, $M_A = 46.07$, $M_B = 29$. From Table 2.2 for air, $\varepsilon_B/k = 78.6$, $r_B = 0.3711$ nm. Values for ethanol will be estimated through Eqs. (2.38) and

Table 2.2 Force constants of gases as determined from viscosity data

Gas	ε/k, K	r, nm	Gas	ε/k, K	r, nm
Air	78.6	0.3711	HCl	344.7	0.3339
CCl_4	322.7	0.5947	He	10.22	0.2551
CH_3OH	481.8	0.3626	H_2	59.7	0.2827
CH_4	148.6	0.3758	H_2O	809.1	0.2641
CO	91.7	0.3690	H_2S	301.1	0.3623
CO_2	195.2	0.3941	NH_3	558.3	0.2900
CS_2	467	0.4483	NO	116.7	0.3492
C_2H_6	215.7	0.4443	N_2	71.4	0.3798
C_3H_8	237.1	0.5118	N_2O	232.4	0.3828
C_6H_6	412.3	0.5349	O_2	106.7	0.3467
Cl_2	316	0.4217	SO_2	335.4	0.4112

† From R. A. Svehla, *NASA Tech. Rept.* R-132, Lewis Research Center, Cleveland, Ohio, 1962.

(2.39). From Table 2.3, v_A = 2(0.0148) + 6(0.0037) + 0.0074 = 0.0592, whence r_A = 1.18(0.0592)$^{1/3}$ = 0.46 nm. The normal boiling point is $T_{b,A}$ = 351.4 K and ε_A/k = 1.21(351.4) = 425.

$$r_{AB} = \frac{0.46 + 0.3711}{2} = 0.416 \qquad \frac{\varepsilon_{AB}}{k} = \sqrt{425(78.6)} = 170.7$$

$$\frac{kT}{\varepsilon_{AB}} = \frac{273}{170.7} = 1.599$$

Fig. 2.5:

$$f\left(\frac{kT}{\varepsilon_{AB}}\right) = 0.595 \qquad \sqrt{\frac{1}{M_A} + \frac{1}{M_B}} = 0.237$$

Eq. (2.37):

$$D_{AB} = \frac{10^{-4}[1.084 - 0.249(0.237)](273^{3/2})(0.237)}{(101.3 \times 10^3)(0.416)^2(0.595)}$$

$$= 1.05 \times 10^{-5} \text{ m}^2/\text{s}$$

The observed value (Table 2.1) is 1.02×10^{-5} m²/s.

Table 2.3 Atomic and molecular volumes

Atomic volume, $m^3/1000$ atoms $\times 10^3$		Molecular volume, $m^3/kmol \times 10^3$		Atomic volume, $m^3/1000$ atoms $\times 10^3$		Molecular volume, $m^3/kmol \times 10^3$	
Carbon	14.8	H_2	14.3	Oxygen	7.4	NH_3	25.8
Hydrogen	3.7	O_2	25.6	In methyl esters	9.1	H_2O	18.9
Chlorine	24.6	N_2	31.2	In higher esters	11.0	H_2S	32.9
Bromine	27.0	Air	29.9	In acids	12.0	COS	51.5
Iodine	37.0	CO	30.7	In methyl ethers	9.9	Cl_2	48.4
Sulfur	25.6	CO_2	34.0	In higher ethers	11.0	Br_2	53.2
Nitrogen	15.6	SO_2	44.8	Benzene ring: subtract	15	I_2	71.5
In primary amines	10.5	NO	23.6	Naphthalene ring: subtract	30		
In secondary amines	12.0	N_2O	36.4				

Equation (2.37) shows D varying almost as $T^{3/2}$ (although a more correct temperature variation is given by considering also the collision function of Fig. 2.5) and inversely as the pressure, which will serve for pressures up to about 1500 kN/m² (15 atm) [19].

The coefficient of self-diffusion, or D for a gas diffusing through itself, can be determined experimentally only by very special techniques involving, for example, the use of radioactive tracers. It can be estimated from Eq. (2.37) by setting A = B.

Molecular Diffusion in Liquids

The integration of Eq. (2.4) to put it in the form of Eq. (2.22) requires the assumption that D_{AB} and c are constant. This is satisfactory for binary gas mixtures but not in the case of liquids, where both may vary considerably with concentration. Nevertheless, in view of our very meager knowledge of the D's, it is customary to use Eq. (2.22), together with an average c and the best average D_{AB} available. Equation (2.22) is also conveniently written†

$$N_A = \frac{N_A}{N_A + N_B} \frac{D_{AB}}{z} \left(\frac{\rho}{M}\right)_{av} \ln \frac{N_A/(N_A + N_B) - x_{A2}}{N_A/(N_A + N_B) - x_{A1}} \qquad (2.40)$$

where ρ and M are the solution density and molecular weight, respectively. As for gases, the value of $N_A/(N_A + N_B)$ must be established by the circumstances prevailing. For the most commonly occurring cases, we have, as for gases:

1. *Steady-state diffusion of* A *through nondiffusing* B. N_A = const, $N_B = 0$, whence

$$N_A = \frac{D_{AB}}{z x_{BM}} \left(\frac{\rho}{M}\right)_{av} (x_{A1} - x_{A2}) \qquad (2.41)$$

where

$$x_{BM} = \frac{x_{B2} - x_{B1}}{\ln (x_{B2}/x_{B1})} \qquad (2.42)$$

2. *Steady-state equimolal counterdiffusion.* $N_A = -N_B$ = const.

$$N_A = \frac{D_{AB}}{z} (c_{A1} - c_{A2}) = \frac{D_{AB}}{z} \left(\frac{\rho}{M}\right)_{av} (x_{A1} - x_{A2}) \qquad (2.43)$$

Illustration 2.4 Calculate the rate of diffusion of acetic acid (A) across a film of nondiffusing water (B) solution 1 mm thick at 17°C when the concentrations on opposite sides of the film are, respectively, 9 and 3 wt % acid. The diffusivity of acetic acid in the solution is 0.95×10^{-9} m²/s.

SOLUTION Equation (2.41) applies. $z = 0.001$ m, $M_A = 60.03$, $M_B = 18.02$. At 17°C, the

---

† The component subscript on x_A indicates mole fraction A, to distinguish it from x meaning distance in the x direction.

density of the 9% solution is 1012 kg/m^3. Therefore,

$$x_{A1} = \frac{0.09/60.03}{0.09/60.03 + 0.91/18.02} = \frac{0.0015}{0.0520} = 0.0288 \text{ mole fraction acetic acid}$$

$$x_{B1} = 1 - 0.0288 = 0.9712 \text{ mole fraction water}$$

$$M = \frac{1}{0.0520} = 19.21 \text{ kg/kmol} \qquad \frac{\rho}{M} = \frac{1012}{19.21} = 52.7 \text{ kmol/m}^3$$

Similarly the density of the 3% solution is 1003.2 kg/m^3, $x_{A2} = 0.0092$, $x_{B2} = 0.9908$, $M = 18.40$, and $\rho/M = 54.5$.

$$\left(\frac{\rho}{M}\right)_{av} = \frac{52.7 + 54.5}{2} = 53.6 \text{ kmol/m}^3 \qquad x_{BM} = \frac{0.9908 - 0.9712}{\ln(0.9908/0.9712)} = 0.980$$

Eq. (2.41):

$$N_A = \frac{0.95 \times 10^{-9}}{0.001(0.980)} 53.6(0.0288 - 0.0092) = 1.018 \times 10^{-6} \text{ kmol/m}^2 \cdot \text{s} \quad \textbf{Ans.}$$

Diffusivity of Liquids

The dimensions for diffusivity in liquids are the same as those for gases, length^2/time. Unlike the case for gases, however, the diffusivity varies appreciably with concentration. A few typical data are listed in Table 2.4, and larger lists are available [6, 8, 10, 15, 17].

Estimates of the diffusivity in the absence of data cannot be made with anything like the accuracy with which they can be made for gases because no sound theory of the structure of liquids has been developed. For dilute solutions of nonelectrolytes, the empirical correlation of Wilke and Chang [23, 24] is recommended.†

$$D_{AB}^0 = \frac{(117.3 \times 10^{-18})(\varphi M_B)^{0.5} T}{\mu v_A^{0.6}} \tag{2.44}$$

where D_{AB}^0 = diffusivity of A in very dilute solution in solvent B, m^2/s

M_B = molecular weight of solvent, kg/kmol

T = temperature, K

μ = solution viscosity, kg/m · s

v_A = solute molal volume at normal boiling point, m^3/kmol

= 0.0756 for water as solute

φ = association factor for solvent

= 2.26 for water as solvent [9]

= 1.9 for methanol as solvent

= 1.5 for ethanol as solvent

= 1.0 for unassociated solvents, e.g., benzene and ethyl ether

The value of v_A may be the true value [9] or, if necessary, estimated from the data of Table 2.3, except when water is the diffusing solute, as noted above. The

---

† The listed units must be used for Eq. (2.44). For D, μ, and T in units of feet, hours, pounds mass, and degrees Rankine, with v_A as listed above, multiply the right-hand side of Eq. (2.44) by 5.20×10^7.

Table 2.4 Liquid diffusivities [8]

Solute	Solvent	Temp, °C	Solute concentration, kmol/m^3	Diffusivity,† m^2/s × 10^9
Cl$_2$	Water	16	0.12	1.26
HCl	Water	0	9	2.7
			2	1.8
		10	9	3.3
			2.5	2.5
		16	0.5	2.44
NH$_3$	Water	5	3.5	1.24
		15	1.0	1.77
CO$_2$	Water	10	0	1.46
		20	0	1.77
NaCl	Water	18	0.05	1.26
			0.2	1.21
			1.0	1.24
			3.0	1.36
			5.4	1.54
Methanol	Water	15	0	1.28
Acetic acid	Water	12.5	1.0	0.82
			0.01	0.91
		18.0	1.0	0.96
Ethanol	Water	10	3.75	0.50
			0.05	0.83
		16	2.0	0.90
n-Butanol	Water	15	0	0.77
CO$_2$	Ethanol	17	0	3.2
Chloroform	Ethanol	20	2.0	1.25

† For example, D for Cl$_2$ in water is 1.26×10^{-9} m^2/s.

association factor for a solvent can be estimated only when diffusivities in that solvent have been experimentally measured. If a value of φ is in doubt, the empirical correlation of Scheibel [20] can be used to estimate D. There is also some doubt about the ability of Eq. (2.44) to handle solvents of very high viscosity, say 0.1 kg/m · s (100 cP) or more. Excellent reviews of all these matters are available [4, 19].

The diffusivity in concentrated solutions differs from that in dilute solutions because of changes in viscosity with concentration and also because of changes in the degree of nonideality of the solution [16]

$$D_A \mu = \left(D_{BA}^0 \mu_A\right)^{x_A} \left(D_{AB}^0 \mu_B\right)^{x_B} \left(1 + \frac{d \log \gamma_A}{d \log x_A}\right) \qquad (2.45)$$

where D_{AB}^0 is the diffusivity of A at infinite dilution in B and D_{BA}^0 the diffusivity of B at infinite dilution in A. The activity coefficient γ_A can typically be obtained from vapor-liquid equilibrium data as the ratio (at ordinary pressures) of the real to ideal partial pressures of A in the vapor in equilibrium with a liquid of concentration x_A:

$$\gamma_A = \frac{\bar{p}_A}{x_A p_A} = \frac{y_A p_t}{x_A p_A} \tag{2.46}$$

and the derivative $(d \log \gamma_A)/(d \log x_A)$ can be obtained graphically as the slope of a graph of $\log \gamma_A$ vs. $\log x_A$.

For strong electrolytes dissolved in water, the diffusion rates are those of the individual ions, which move more rapidly than the large, undissociated molecules, although the positively and negatively charged ions must move at the same rate in order to maintain electrical neutrality of the solution. Estimates of these effects have been thoroughly reviewed [8, 19] but are beyond the scope of this book.

Illustration 2.5 Estimate the diffusivity of mannitol, $CH_2OH(CHOH)_4CH_2OH$, $C_6H_{14}O_6$, in dilute solution in water at 20°C. Compare with the observed value, 0.56×10^{-9} m²/s.

SOLUTION From the data of Table 2.3

$$v_A = 0.0148(6) + 0.0037(14) + 0.0074(6) = 0.185$$

For water as solvent, $\varphi = 2.26$, $M_B = 18.02$, $T = 293$ K. For dilute solutions, the viscosity μ may be taken as that for water, $0.001\ 005$ kg/m · s. Eq. (2.44):

$$D_{AB} = \frac{(117.3 \times 10^{-18})[2.26(18.02)]^{0.5}(293)}{0.001\ 005(0.185)^{0.6}} = 0.601 \times 10^{-9} \text{ m}^2/\text{s} \quad \textbf{Ans.}$$

Illustration 2.6 Estimate the diffusivity of mannitol in dilute water solution at 70°C and compare with the observed value, 1.56×10^{-9} m²/s.

SOLUTION At 20°C, the observed $D_{AB} = 0.56 \times 10^{-9}$ m²/s, and $\mu = 1.005 \times 10^{-3}$ kg/m · s (Illustration 2.5). At 70°C, the viscosity of water is 0.4061×10^{-3} kg/m · s. Equation (2.44) indicates that $D_{AB}\mu/T$ should be constant:

$$\frac{D_{AB}(0.4061 \times 10^{-3})}{70 + 273} = \frac{(0.56 \times 10^{-9})(1.005 \times 10^{-3})}{20 + 273}$$

$$D_{AB} = 1.62 \times 10^{-9} \text{ m}^2/\text{s at 70°C} \quad \textbf{Ans.}$$

Applications of Molecular Diffusion

While the flux relative to the average molar velocity J always means transfer *down* a concentration gradient, the flux N need not. For example, consider the dissolution of a hydrated salt crystal such as $Na_2CO_3 \cdot 10H_2O$ into pure water at 20°C. The solution in contact with the crystal surface contains Na_2CO_3 and H_2O at a concentration corresponding to the solubility of Na_2CO_3 in H_2O, or 0.0353 mole fraction Na_2CO_3 and 0.9647 mole fraction H_2O. For the Na_2CO_3, transfer is from the crystal surface at a concentration 0.0353 outward to 0 mole fraction Na_2CO_3 in the bulk liquid. But the water of crystallization which dissolves must transfer outward in the ratio 10 mol H_2O to 1 mol Na_2CO_3 from a concentration at the crystal surface of 0.9647 to 1.0 mole fraction in the bulk liquid, or transfer *up* a concentration gradient. Application of Eq. (2.40) confirms this. The expressions developed for the rate of mass transfer under conditions where molecular diffusion defines the mechanism of mass transfer (fluids which are stagnant or in laminar flow) are of course directly applicable to the experimental measurement of the diffusivities, and they are extensively used for this.

In the practical applications of the mass-transfer operations, the fluids are always in motion, even in batch processes, so that we do not have stagnant fluids. While occasionally the moving fluids are entirely in laminar flow, more frequently the motion is turbulent. If the fluid is in contact with a solid surface, where the fluid velocity is zero, there will be a region in predominantly laminar flow adjacent to the surface. Mass transfer must then usually take place through the laminar region, and molecular diffusion predominates there. When two immiscible fluids in motion are in contact and mass transfer occurs between them, there may be no laminar region, even at the interface between the fluids.

In practical situations like these, it has become customary to describe the mass-transfer flux in terms of mass-transfer coefficients. The relationships of this chapter are then rarely used directly to determine mass-transfer rates, but they are particularly useful in establishing the form of the mass-transfer coefficient-rate equations and in computing the mass-transfer coefficients for laminar flow.

MOMENTUM AND HEAT TRANSFER IN LAMINAR FLOW

In the flow of a fluid past a phase boundary, such as that through which mass transfer occurs, there will be a velocity gradient within the fluid, which results in a transfer of momentum through the fluid. In some cases there is also a transfer of heat by virtue of a temperature gradient. The processes of mass, momentum, and heat transfer under these conditions are intimately related, and it is useful to consider this point briefly.

Momentum Transfer

Consider the velocity profile for the case of a gas flowing past a flat plate, as in Fig. 2.6. Since the velocity at the solid surface is zero, there must necessarily be a layer (the laminar sublayer) adjacent to the surface where the flow is predominantly laminar. Within this region, the fluid can be imagined as being made up of thin layers sliding over each other at increasing velocities at increasing distances from the plate. The force per unit area parallel to the surface, or shearing stress τ, required to maintain their velocities is proportional to the velocity gradient,

$$\tau g_c = -\mu \frac{du}{dz} \qquad (2.47)$$

where μ is the viscosity and z is measured as increasing in the direction toward

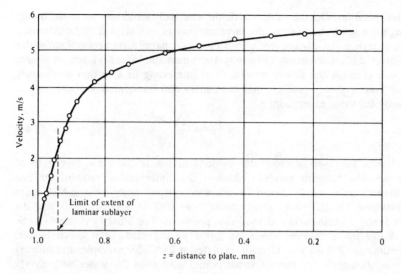

Figure 2.6 Velocity profile, flow of air along a flat plate. [*Page, et al., Ind. Eng. Chem.*, **44**, 424 (1952).]

the surface. This can be written

$$\tau g_c = -\frac{\mu}{\rho}\frac{d(u\rho)}{dz} = -\nu\frac{d(u\rho)}{dz} \tag{2.48}$$

where ν is the kinematic viscosity, μ/ρ.

The kinematic viscosity has the same dimensions as diffusivity, length²/time, while the quantity $u\rho$ can be looked upon as a volumetric momentum concentration. The quantity τg_c is the rate of momentum transfer per unit area, or flux of momentum. Equation (2.48) is therefore a rate equation analogous to Eq. (2.1) for mass flux. In the transfer of momentum in this manner there is of course no bulk flow of fluid from one layer to the other in the z direction. Instead, molecules in one layer, in the course of traveling in random directions, will move from a fast-moving layer to an adjacent, more slowly moving layer, thereby transmitting momentum corresponding to the difference in velocities of the layers. Diffusion in the z direction occurs by the same mechanism. At high molecular concentrations, such as in gases at high pressures or even more so in liquids, the molecular diameter becomes appreciable in comparison with the molecular movement between collisions, and momentum can be imagined as being transmitted directly through the molecules themselves [2]. Visualize, for example, a number of billiard balls arranged in a group in close contact with each other on a table. A moving cue ball colliding with one of the outermost balls of the packed group will transmit its momentum very rapidly to one of the balls on the opposite side of the group, which will then be propelled from its original position. On the other hand, the cue ball is unlikely to move bodily through the group because of the large number of collisions it will experience. Thus, at high molecular concentrations the direct parallelism between molecular diffusivity and momentum diffusivity (or kinematic viscosity) breaks down: diffusion is much the slower process. It is interesting to note that a relatively simple kinetic theory predicts that both mass and momentum diffusivities are given by the same expression,

$$D_{AA} = \frac{\mu_A}{\rho_A} = \frac{w\lambda}{3} \qquad (2.49)$$

where w is the average molecular velocity and λ is the mean free path of molecule. The Schmidt number, which is the dimensionless ratio of the two diffusivities, $Sc = \mu/\rho D$, should by this theory equal unity for a gas. A more sophisticated kinetic theory gives values from 0.67 to 0.83, which is just the range found experimentally at moderate pressures. For binary gas mixtures, Sc may range up to 5. For liquids, as might be expected, Sc is much higher: approximately 297 for self-diffusion in water at 25°C, for example, and ranging into the thousands for more viscous liquids and even for water with slowly diffusing solutes.

Heat Transfer

When a temperature gradient exists between the fluid and the plate, the rate of heat transfer in the laminar region of Fig. 2.6 is

$$q = -k\frac{dt}{dz} \qquad (2.50)$$

where k is the thermal conductivity of the fluid. This can also be written

$$q = -\frac{k}{C_p\rho}\frac{d(tC_p\rho)}{dz} = -\alpha\frac{d(tC_p\rho)}{dz} \qquad (2.51)$$

where C_p is the specific heat at constant pressure. The quantity $tC_p\rho$ may be looked upon as a volumetric thermal concentration, and $\alpha = k/C_p\rho$ is the thermal diffusivity, which, like momentum and mass diffusivities, has dimensions length^2/time. Equation (2.51) is therefore a rate equation analogous to the corresponding equations for momentum and mass transfer.

In a gas at relatively low pressure the heat energy is transferred from one position to another by the molecules traveling from one layer to another at a lower temperature. A simplified kinetic theory leads to the expression

$$\alpha = \frac{k}{C_p\rho} = \tfrac{1}{3}w\lambda\frac{C_v}{C_p} \qquad (2.52)$$

Equation (2.49) and (2.52) would give the dimensionless ratio $\nu/\alpha = C_p\mu/k$ equal to C_p/C_v. A more advanced kinetic theory modifies the size of the ratio, known as the Prandtl number Pr, and experimentally it has the range 0.65 to 0.9

for gases at low pressure, depending upon the molecular complexity of the gas. At high molecular concentrations, the process is modified. Thus for most liquids, Pr is larger (Pr = 7.02 for water at 20°C, for example).

The third dimensionless group, formed by dividing the thermal by the mass diffusivity, is the Lewis number, Le = α/D = Sc/Pr, and it plays an important part in problems of simultaneous heat and mass transfer, as will be developed later.

We can summarize this brief discussion of the similarity between momentum, heat, and mass transfer as follows. An elementary consideration of the three processes leads to the conclusion that in certain simplified situations there is a direct analogy between them. In general, however, when three- rather than one-dimensional transfer is considered, the momentum-transfer process is of a sufficiently different nature for the analogy to break down. Modification of the simple analogy is also necessary when, for example, mass and momentum transfer occur simultaneously. Thus, if there were a net mass transfer toward the surface of Fig. 2.6, the momentum transfer of Eq. (2.48) would have to include the effect of the net diffusion. Similarly, mass transfer must inevitably have an influence on the velocity profile. Nevertheless, even the limited analogies which exist are put to important practical use.

NOTATION FOR CHAPTER 2

Consistent units in any system may be used, except in Eqs. (2.37) to (2.39) and (2.44).

c	concentration, mol/volume, mole L^3
C_p.	heat capacity at constant pressure, FL/MT
C_v	heat capacity at constant volume, FL/MT
d	differential operator
D	diffusivity, L^2/Θ
D^0	diffusivity for a solute at infinite dilution, L^2/Θ
f	function
g_c	conversion factor, $ML/F\Theta^2$
J	flux of diffusion relative to the molar average velocity, mole/$L^2\Theta$
k	thermal conductivity, FL/LTΘ
$\mathbf{k}$	Boltzmann's constant, 1.38×10^{-16} erg/K
$\ln$	natural logarithm
Le	Lewis number = $k/\rho D C_p$, dimensionless
M	molecular weight, M/mole
n	number of moles, dimensionless
N	molar flux relative to a fixed surface, mole/$L^2\Theta$
p	vapor pressure, F/L^2
$\bar{p}$	partial pressure, F/L^2
p_t	total pressure, F/L^2
Pr	Prandtl number = $C_p\mu/k$, dimensionless
q	flux of heat, FL/$L^2\Theta$
r	molecular separation at collision, nm
R	universal gas constant, FL/mole T
R_i	rate of production of component i, mole/$L^3\theta$
Sc	Schmidt number = $\mu/\rho D$, dimensionless

T	absolute temperature, T
T_b	normal boiling point, K
u	linear velocity, L/θ
v	liquid molar volume, $m^3/kmol$
V	volume, L^3
w	average molar velocity, L/θ
x	(with no subscript) distance in the x direction, L
x_i	mole-fraction concentration of component i in a liquid
y	(with no subscript) distance in the y direction, L
y_i	mole-fraction concentration of component i in a gas
y_i'	mole-fraction concentration of component i, diffusing-solute-free basis
z	distance in the z direction, L
α	thermal diffusivity, L^2/θ
γ	activity coefficient, dimensionless
∂	partial differential operator
Δ	difference
ε	energy of molecular attraction, ergs
θ	time
λ	mean free path of a molecule, L
μ	viscosity, $M/L\theta$
ν	kinematic viscosity or momentum diffusivity $= \mu/\rho$, L^2/θ
ρ	density, M/L^3
τ	shearing stress, F/L^2
φ	dissociation factor for a solvent, dimensionless

Subscripts

A	component A
B	component B
i	component i
n	the last of n components
m	effective
M	mean
x	in the x direction
y	in the y direction
z	in the z direction
1	beginning of diffusion path
2	end of diffusion path

REFERENCES

1. Bird, R. B., W. E. Stewart, and E. N. Lightfoot: "Transport Phenomena," Wiley, New York, 1960.
2. Bosworth, R. C. L.: "Physics in the Chemical Industry," Macmillan, London, 1950.
3. Chapman, S., and T. G. Cowling: "Mathematical Theory of Non-uniform Gases," Cambridge University Press, Cambridge, 1939.
4. Ertl, H., R. K. Ghai, and F. A. L. Dullien: *AIChE J.*, **19**, 881 (1973); **20**, 1 (1974).
5. Gainer, J. L., and A. B. Metzner: Transport Phenomena, *AIChE–IChE Symp. Ser.*, **6**, 74, 1965.
6. Ghai, R. K., H. Ertl, and F. A. Dullien: *Nat. Aux. Publ. Serv.*, *Doc.* 02172, Microfiche Publications, New York, 1973.
7. Gilliland, E. R.: *Ind. Eng. Chem.*, **26**, 681 (1934).
8. Harned, H. S., and B. B. Owen: "The Physical Chemistry of Electrolytic Solutions," 3d ed., Reinhold, New York, 1958.

9. Hayduck, W., and H. Laudie: *AIChE J.*, **20**, 611 (1974).
10. Himmelblau, P. M.: *Chem. Rev.*, **64**, 527 (1964).
11. Hirschfelder, J. O., R. B. Bird, and E. L. Spotz: *Trans. ASME*, **71**, 921 (1949); *Chem. Rev.*, **44**, 205 (1949).
12. Hirschfleder, J. O., C. F. Curtis, and R. B. Bird: "Molecular Theory of Gases and Liquids," Wiley, New York, 1954.
13. Hiss, T. G., and E. L. Cussler: *AIChE J.*, **19**, 698 (1973).
14. "International Critical Tables," vol. V, McGraw-Hill Book Company, New York, 1929.
15. Johnson, P. A., and A. L. Babb: *Chem. Rev.*, **56**, 387 (1956).
16. Leffler, J., and H. T. Cullinan: *Ind. Eng. Chem. Fundam.*, **9**, 84 (1970).
17. Marrero, T. R., and E. A. Mason: *J. Phys. Chem. Ref. Data*, **1**, 3 (1972).
18. Perry, R. H., and C. H. Chilton (eds).: "The Chemical Engineers' Handbook," 5th ed., pp. 3-222 to 3-225, McGraw-Hill Book Company, New York, 1973.
19. Reid, R. C., J. M. Prausnitz, and T. K. Sherwood: "The Properties of Gases and Liquids," 3d ed., McGraw-Hill Book Company, New York, 1977.
20. Scheibel, E. G.: *Ind. Eng. Chem.*, **46**, 2007 (1954).
21. Shain, S. S.: *AIChE J.*, **7**, 17 (1961).
22. Stuel, L. I., and G. Thodos: *AIChE J.*, **10**, 266 (1964).
23. Wilke, C. R.: *Chem. Eng. Progr.*, **45**, 218 (1949).
24. Wilke, C. R., and P. Chang: *AIChE J.*, **1**, 264 (1955).
25. Wilke, C. R., and C. Y. Lee: *Ind. Eng. Chem.*, **47**, 1253 (1955).
26. Wintergeist, E.: *Ann. Phys.*, **4**, 323 (1930).
27. Wood, J. C., E. R. Whittemore, and W. L. Badger: *Chem. Met. Eng.*, **27**, 1176 (1922).

PROBLEMS

2.1 In an oxygen-nitrogen gas mixture at 1 std atm, 25°C, the concentrations of oxygen at two planes 2 mm apart are 10 and 20 vol %, respectively. Calculate the flux of diffusion of the oxygen for the case where:

(a) The nitrogen is nondiffusing. **Ans.:** 4.97×10^{-5} kmol/m$^2 \cdot$ s.

(b) There is equimolar counterdiffusion of the two gases.

2.2 Repeat the calculations of Prob. 2.1 for a total pressure of 1000 kN/m^2.

2.3 Estimate the diffusivities of the following gas mixtures:

(a) Acetone-air, STP.† **Ans.:** 9.25×10^{-6} m^2/s.

(b) Nitrogen–carbon dioxide, 1 std atm, 25°C.

(c) Hydrogen chloride-air, 200 kN/m^2, 25°C. **Ans.:** 9.57×10^{-6} m^2/s.

(d) Toluene-air, 1 std atm, 30°C. Reported value [Gilliland, *Ind. Eng. Chem.*, **26**, 681 (1934)] = 0.088 cm^2/s.

(e) Aniline-air, STP. Observed value = 0.0610 cm^2/s (Gilliland, *loc. cit.*).

2.4 The diffusivity of carbon dioxide in helium is reported to be 5.31×10^{-5} m^2/s at 1 std atm, 3.2°C. Estimate the diffusivity at 1 std atm, 225°C. Reported value = 14.14×10^{-5} m^2/s [Seager, Geertson, and Giddings, *J. Chem. Eng. Data*, **8**, 168 (1963)].

2.5 Ammonia is diffusing through a stagnant gas mixture consisting of one-third nitrogen and two-thirds hydrogen by volume. The total pressure is 30 lb$_f$/in^2 abs (206.8 kN/m^2) and the temperature 130°F (54°C). Calculate the rate of diffusion of the ammonia through a film of gas 0.5 mm thick when the concentration change across the film is 10 to 5% ammonia by volume. **Ans.:** 2.05×10^{-4} kmol/m$^2 \cdot$ s.

2.6 Estimate the following liquid diffusivities:

(a) Ethanol in dilute water solution, 10°C.

† STP (standard temperature and pressure is 0°C and 1 std atm.

(b) Carbon tetrachloride in dilute solution in methanol, 15°C (observed value = 1.69×10^{-5} cm^2/s). Ans.: 1.49×10^{-9} m^2/s.

2.7 The diffusivity of bromoform in dilute solution in acetone at 25°C is listed in Ref. 14, p. 63, as 2.90×10^{-5} cm^2/s. Estimate the diffusivity of benzoic acid in dilute solution in acetone at 25°C. Reported value [Chang and Wilke, *J. Phys. Chem.*, **59**, 592 (1955)] = 2.62×10^{-5} cm^2/s.

Ans.: 2.269×10^{-9} m^2/s.

2.8 Calculate the rate of diffusion of NaCl at 18°C through a stagnant film of water 1 mm thick when the concentrations are 20 and 10%, respectively, on either side of the film. Ans.: 3.059×10^{-6} kmol/m$^2 \cdot$ s.

2.9 At 1 std atm, 100°C, the density of air = 0.9482 kg/m^3, the viscosity = 2.18×10^{-5} kg/m $\cdot$ s, thermal conductivity = 0.0317 W/m $\cdot$ K, and specific heat at constant pressure = 1.047 kJ/kg $\cdot$ K. At 25°C, the viscosity = 1.79×10^{-5} kg/m $\cdot$ s.

(a) Calculate the kinematic viscosity at 100°C, m^2/s.

(b) Calculate the thermal diffusivity at 100°C, m^2/s.

(c) Calculate the Prandtl number at 100°C.

(d) Assuming that for air at 1 std atm, Pr = Sc and that Sc = const with changing temperature, calculate D for air at 25°C. Compare with the value of D for the system O_2-N_2 at 1 std atm, 25°C, Table 2.1.

2.10 Ammonia is being cracked on a solid catalyst according to the reaction

$$2NH_3 \rightarrow N_2 + 3H_2$$

At one place in the apparatus, where the pressure is 1 std atm abs and the temperature 200°C, the analysis of the bulk gas is 33.33% NH$_3$ (A), 16.67% N$_2$ (B), and 50.00% H$_2$ (C) by volume. The circumstances are such that NH$_3$ diffuses from the bulk-gas stream to the catalyst surface, and the products of the reaction diffuse back, as if by molecular diffusion through a gas film in laminar flow 1 mm thick. Estimate the local rate of cracking, kg NH$_3$/(m^2 catalyst surface) $\cdot$ s, which might be considered to occur if the reaction is diffusion-controlled (chemical reaction rate very rapid) with the concentration of NH$_3$ at the catalyst surface equal to zero.

Ans.: 0.0138 kg/m$^2 \cdot$ s.

2.11 A crystal of copper sulfate $CuSO_4 \cdot 5H_2O$ falls through a large tank of pure water at 20°C. Estimate the rate at which the crystal dissolves by calculating the flux of $CuSO_4$ from the crystal surface to the bulk solution. Repeat by calculating the flux of water. *Data and assumptions:* Molecular diffusion occurs through a film of water uniformly 0.0305 mm thick, surrounding the crystal. At the inner side of the film, adjacent to the crystal surface, the concentration of copper sulfate is its solubility value, 0.0229 mole fraction $CuSO_4$ (solution density = 1193 kg/m^3). The outer surface of the film is pure water. The diffusivity of $CuSO_4$ is 7.29×10^{-10} m^2/s.

THREE
MASS-TRANSFER COEFFICIENTS

We have seen that when a fluid flows past a solid surface under conditions such that turbulence generally prevails, there is a region immediately adjacent to the surface where the flow is predominantly laminar. With increasing distance from the surface, the character of the flow gradually changes, becoming increasingly turbulent, until in the outermost regions of the fluid fully turbulent conditions prevail. We have noted also that the rate of transfer of dissolved substance through the fluid will necessarily depend upon the nature of the fluid motion prevailing in the various regions.

In the turbulent region, particles of fluid no longer flow in the orderly manner found in the laminar sublayer. Instead, relatively large portions of the fluid, called *eddies*, move rapidly from one position to the other with an appreciable component of their velocity in the direction perpendicular to the surface past which the fluid is flowing. These eddies bring with them dissolved material, and the eddy motion thus contributes considerably to the mass-transfer process. Since the eddy motion is rapid, mass transfer in the turbulent region is also rapid, much more so than that resulting from molecular diffusion in the laminar sublayer. Because of the rapid eddy motion, the concentration gradients existing in the turbulent region will bc smaller than those in the film, and Fig. 3.1 shows concentration gradients of this sort. In the experiment for which these are the data, air in turbulent motion flowed past a water surface, and water evaporated into the air. Samples of the air were taken at various distances from the surface, and the water-vapor concentration was determined by analysis. At the water surface, the water concentration in the gas was the same as the vapor pressure of pure water at the prevailing temperature. It was not possible to sample the gas very close to the water surface, but the rapid change in concentration in the region close to the surface and the slower change in the

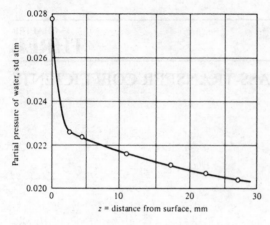

Figure 3.1 Evaporation of water into air [61].

outer turbulent region are nevertheless unmistakable. It is important also to note the general similarity of data of this sort to the velocity distribution shown in Fig. 2.6.

It is also useful to compare the data for mass transfer with similar data for heat transfer. Thus, in Fig. 3.2 are plotted the temperatures at various distances from the surface when air flowed past a heated plate. The large temperature gradient close to the surface and the lesser gradient in the turbulent region are again evident. It will generally be convenient to keep the corresponding heat-transfer process in mind when the mass-transfer process is discussed, since in many instances the methods of reasoning used to describe the latter are borrowed directly from those found successful with the former.

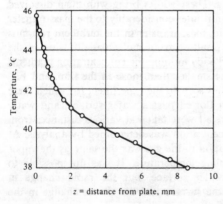

Figure 3.2 Heat transfer, flow of air past a heated plate [48].

Mass-Transfer Coefficients

The mechanism of the flow process involving the movements of the eddies in the turbulent region is not thoroughly understood. On the other hand, the mechanism of molecular diffusion, at least for gases, is fairly well known, since it can be described in terms of a kinetic theory to give results which agree well with experience. It is therefore natural to attempt to describe the rate of mass transfer through the various regions from the surface to the turbulent zone in the same manner found useful for molecular diffusion. Thus, the $D_{AB}c/z$ of Eq. (2.22), which is characteristic of molecular diffusion, is replaced by F, a mass-transfer coefficient [5, 12]. For binary solutions,

$$N_A = \frac{N_A}{N_A + N_B} F \ln \frac{N_A/(N_A + N_B) - c_{A2}/c}{N_A/(N_A + N_B) - c_{A1}/c} \tag{3.1}$$

where c_A/c is the mole-fraction concentration, x_A for liquids, y_A for gases.† As in molecular diffusion, the ratio $N_A/(N_A + N_B)$ is ordinarily established by nondiffusional considerations.

Since the surface through which the transfer takes place may not be plane, so that the diffusion path in the fluid may be of variable cross section, N is defined as the *flux at the phase interface*, or *boundary*, where substance leaves or enters the phase for which F is the mass-transfer coefficient. N_A is positive when c_{A1} is at the beginning of the transfer path and c_{A2} at the end. In any case, one of these concentrations will be at the phase boundary. The manner of defining the concentration of A in the fluid will influence the value of F, and this is usually established arbitrarily. If mass transfer takes place between a phase boundary and a large quantity of unconfined fluid, as, for example, when a drop of water evaporates while falling through a great volume of air, the concentration of diffusing substance in the fluid is usually taken as the constant value found at large distances from the phase boundary. If the fluid is in a confining duct, so that the concentration is not constant along any position of the transfer path, the bulk-average concentration $\bar{c}_A$, as found by mixing all the fluid passing a given point, is usually used. In Fig. 3.3, where a liquid evaporates into the flowing gas, the concentration c_A of the vapor in the gas varies continuously from c_{A1} at the liquid surface to the value at $z = Z$. In this case c_{A2} in Eq. (3.1) would usually be taken as $\bar{c}_A$, defined by

$$\bar{c}_A = \frac{1}{\bar{u}_x S} \int_0^S u_x c_A \, dS \tag{3.2}$$

where $u_x(z) =$ velocity distribution in gas across the duct ($=$ time average of u_x for turbulence)
$\bar{u}_x =$ bulk-average velocity (volumetric rate/duct cross section)
$S =$ duct cross-sectional area

---

† Equation (3.1) is identical with the result obtained by combining Eqs. (21.4-11), (21.5-27), and (21.5-47) of Ref. 7 or by combining Eqs. (3.4) and (5.37) of Ref. 54, and is applicable to both low mass-transfer fluxes and to high fluxes as corrected through the film theory.

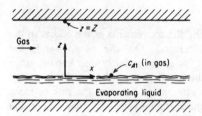

Figure 3.3 Mass transfer to a confined fluid.

In any case, one must know how the mass-transfer coefficient is defined in order to use it properly.

The F of Eq. (3.1) is a *local* mass-transfer coefficient, defined for a particular location on the phase-boundary surface. Since the value of F depends on the local nature of the fluid motion, which may vary along the surface, an average value F_{av} is sometimes used in Eq. (3.1) with constant c_{A1} and c_{A2}, which takes into account these variations in F. The effect of variation in c_{A1} and c_{A2} on the flux must be accounted for separately.

For multicomponent systems, there may be sufficiently important interaction between the components for the form of the binary system equation not to be exact [63]. Nevertheless, Eq. (3.1) can serve as a suitable approximation if $N_A + N_B$ is replaced by $\sum_{i=1}^{n} N_i$, where n is the number of components.

The two situations noted in Chap. 2, equimolar counterdiffusion and transfer of one substance through another which is not transferred, occur so frequently that special mass-transfer coefficients are usually used for them. These are defined by equations of the form

$$\text{Flux} = (\text{coefficient})(\text{concentration difference})$$

Since concentration may be defined in a number of ways and standards have not been established, we have a variety of coefficients for each situation:

Transfer of A through nontransferring $B[N_B = 0, N_A/(N_A + N_B) = 1]$:

$$N_A = \begin{cases} k_G(\bar{p}_{A1} - \bar{p}_{A2}) = k_y(y_{A1} - y_{A2}) = k_c(c_{A1} - c_{A2}) & \text{gases} \quad (3.3) \\ k_x(x_{A1} - x_{A2}) = k_L(c_{A1} - c_{A2}) & \text{liquids} \quad (3.4) \end{cases}$$

Equimolar countertransfer $[N_A = -N_B, N_A/(N_A + N_B) = \infty]$:

$$N_A = \begin{cases} k_G'(\bar{p}_{A1} - \bar{p}_{A2}) = k_y'(y_{A1} - y_{A2}) = k_c'(c_{A1} - c_{A2}) & \text{gases} \quad (3.5) \\ k_x'(x_{A1} - x_{A2}) = k_L'(c_{A1} - c_{A2}) & \text{liquids} \quad (3.6) \end{cases}$$

An expression of this sort was suggested as early as 1897 [45] for the dissolution of solids in liquids. These expressions are, of course, analogous to the definition of a heat-transfer coefficient h: $q = h(t_1 - t_2)$. Whereas the concept of the heat-transfer coefficient is generally applicable, at least in the absence of mass transfer, the coefficients of Eqs. (3.3) and (3.4) are more restricted. Thus, k_c

of Eq. (3.3) can be considered as a replacement of D_{AB}/z in an integration of Eq. (2.1), and the bulk-flow term of Eq. (2.4) has been ignored in equating this to N_A. The coefficients of Eqs. (3.3) and (3.4) are therefore generally useful only for low mass-transfer rates. Values measured under one level of transfer rate should be converted into F for use with Eq. (3.1) before being applied to another. To obtain the relation between F and the k's, note that for gases, for example, F replaces $D_{AB}p_t/RTz$ in Eq. (2.25) and that k_G replaces $D_{AB}p_t/RTz\bar{p}_{B,M}$ in Eq. (2.30). From this it follows that $F = k_G\bar{p}_{B,M}$. In this manner the conversions of Table 3.1 have been obtained. Since the bulk-flow term $N_A + N_B$ of Eq. (2.4) is zero for equimolal countertransfer, $F = k_y'$ (gases) and $F = k_x'$ (liquids), and Eqs. (3.5) and (3.6) are identical with Eq. (3.1) for this case.

Table 3.1 Relations between mass-transfer coefficients

Rate equation		Units of coefficient
Equimolal counterdiffusion	Diffusion of A through nondiffusing B	
Gases		
$N_A = k_G' \Delta\bar{p}_A$	$N_A = k_G \Delta\bar{p}_A$	$\dfrac{\text{Moles transferred}}{(\text{Area})(\text{time})(\text{pressure})}$
$N_A = k_y' \Delta y_A$	$N_A = k_y \Delta y_A$	$\dfrac{\text{Moles transferred}}{(\text{Area})(\text{time})(\text{mole fraction})}$
$N_A = k_c' \Delta c_A$	$N_A = k_c \Delta c_A$	$\dfrac{\text{Moles transferred}}{(\text{Area})(\text{time})(\text{mol/vol})}$
	$W_A = k_Y \Delta Y_A$	$\dfrac{\text{Mass transferred}}{(\text{Area})(\text{time})(\text{mass A/mass B})}$

Conversions:
$$F = k_G\bar{p}_{B,M} = k_y\frac{\bar{p}_{B,M}}{p_t} = k_c\frac{\bar{p}_{B,M}}{RT} = \frac{k_Y}{M_B} = k_G'p_t = k_y' = k_c'\frac{p_t}{RT} = k_c'c$$

Liquids		
$N_A = k_L' \Delta c_A$	$N_A = k_L \Delta c_A$	$\dfrac{\text{Moles transferred}}{(\text{Area})(\text{time})(\text{mol/vol})}$
$N_A = k_x' \Delta x_A$	$N_A = k_x \Delta x_A$	$\dfrac{\text{Moles transferred}}{(\text{Area})(\text{time})(\text{mole fraction})}$

Conversions:
$$F = k_x x_{B,M} = k_L x_{B,M}c = k_L'c = k_L'\frac{\rho}{M} = k_x'$$

Many data on mass transfer where $N_A/(N_A + N_B)$ is neither unity nor infinity have nevertheless been described in terms of the k-type coefficients. Before these can be used for other situations, they must be converted into F's.

In a few limited situations mass-transfer coefficients can be deduced from theoretical principles. In the great majority of cases, however, we depend upon direct measurement under known conditions, for use later in design.

MASS-TRANSFER COEFFICIENTS IN LAMINAR FLOW

In principle, at least, we do not need mass-transfer coefficients for laminar flow, since molecular diffusion prevails, and the relationships of Chap. 2 can be used to compute mass-transfer rates. A uniform method of dealing with both laminar and turbulent flow is nevertheless desirable.

Mass-transfer coefficients for laminar flow should be capable of computation. To the extent that the flow conditions are capable of description and the mathematics remains tractable, this is so. These are, however, severe requirements, and frequently the simplification required to permit mathematical manipulation is such that the results fall somewhat short of reality. It is not our purpose to develop these methods in detail, since they are dealt with extensively elsewhere [6, 7]. We shall choose one relatively simple situation to illustrate the general technique and to provide some basis for considering turbulent flow.

Mass Transfer from a Gas into a Falling Liquid Film

Figure 3.4 shows a liquid falling in a thin film in laminar flow down a vertical flat surface while being exposed to a gas A, which dissolves in the liquid. The liquid contains a uniform concentration c_{A0} of A at the top. At the liquid

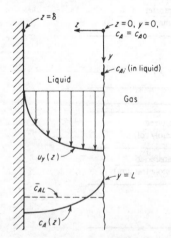

Figure 3.4 Falling liquid film.

surface, the concentration of the dissolved gas is $c_{A, i}$, in equilibrium with the pressure of A in the gas phase. Since $c_{A, i} > c_{A0}$, gas dissolves in the liquid. The problem is to obtain the mass-transfer coefficient k_L, with which the amount of gas dissolved after the liquid falls the distance L can be computed.

The problem is solved by simultaneous solution of the equation of continuity (2.17) for component A with the equations describing the liquid motion, the Navier-Stokes equations. The simultaneous solution of this formidable set of partial differential equations becomes possible only when several simplifying assumptions are made. For present purposes, assume the following:

1. There is no chemical reaction. R_A of Eq. (2.17) = 0.
2. Conditions do not change in the x direction (perpendicular to the plane of the paper, Fig. 3.4). All derivatives with respect to x of Eq. (2.17) = 0.
3. Steady-state conditions prevail. $\partial c_A / \partial \theta = 0$.
4. The rate of absorption of gas is very small. This means that u_z in Eq. (2.17) due to diffusion of A is essentially zero.
5. Diffusion of A in the y direction is negligible in comparison with the movement of A downward due to bulk flow. Therefore, $D_{AB} \, \partial^2 c_A / \partial y^2 = 0$.
6. Physical properties (D_{AB}, ρ, μ) are constant.

Equation (2.17) then reduces to

$$u_y \frac{\partial c_A}{\partial y} = D_{AB} \frac{\partial^2 c_A}{\partial z^2} \tag{3.7}$$

which states that any A added to the liquid running down at any location z, over an increment in y, got there by diffusion in the z direction. The equations of motion under these conditions reduce to

$$\mu \frac{d^2 u_y}{dz^2} + \rho g = 0 \tag{3.8}$$

The solution to Eq. (3.8), with the conditions that $u_y = 0$ at $z = \delta$ and that $du_y / dz = 0$ at $z = 0$, is well known

$$u_y = \frac{\rho g \delta^2}{2\mu} \left[1 - \left(\frac{z}{\delta} \right)^2 \right] = \frac{3}{2} \bar{u}_y \left[1 - \left(\frac{z}{\delta} \right)^2 \right] \tag{3.9}$$

where $\bar{u}_y$ is the bulk-average velocity. The film thickness is then

$$\delta = \left(\frac{3 \bar{u}_y \mu}{\rho g} \right)^{1/2} = \left(\frac{3 \mu \Gamma}{\rho^2 g} \right)^{1/3} \tag{3.10}$$

where Γ is the mass rate of liquid flow per unit of film width in the x direction. Substituting Eq. (3.9) into (3.7) gives

$$\frac{3}{2} \bar{u}_y \left[1 - \left(\frac{z}{\delta} \right)^2 \right] \frac{\partial c_A}{\partial y} = D_{AB} \frac{\partial^2 c_A}{\partial z^2} \tag{3.11}$$

which is to be solved under the following conditions:

1. At $z = 0$, $c_A = c_{A, i}$ at all values of y.
2. At $z = \delta$, $\partial c_A / \partial z = 0$ at all values of y, since no diffusion takes place into the solid wall.
3. At $y = 0$, $c_A = c_{A0}$ at all values of z.

The solution results in a general expression (an infinite series) giving c_A for any z and y, thus providing a concentration distribution $c_A(z)$ at $y = L$, as shown in Fig. 3.4. The bulk-average $\bar{c}_{A, L}$ at $y = L$ can then be found in the manner of Eq. (3.2). The result is [31]

$$\frac{c_{A, i} - \bar{c}_{A, L}}{c_{A, i} - c_{A0}} = 0.7857 e^{-5.1213\eta} + 0.1001 e^{-39.318\eta} + 0.03599 e^{-105.64\eta} + \cdots \tag{3.12}$$

where $\eta = 2D_{AB}L/3\delta^2 \bar{u}_y$. The total rate of absorption is then $\bar{u}_y \delta(\bar{c}_{A, L} - c_{A0})$ per unit width of liquid film.

Alternatively, to obtain a local mass-transfer coefficient, we can combine Eq. (2.4) for the case of negligible bulk flow in the z direction ($N_A + N_B = 0$) with Eq. (3.4), keeping in mind that mass-transfer coefficients use fluxes at the phase interface ($z = 0$)

$$N_A = -D_{AB}\left(\frac{\partial c_A}{\partial z}\right)_{z=0} = k_L(c_{A, i} - \bar{c}_{A, L}) \tag{3.13}$$

In this case, however, because of the nature of the series which describes c_A, the derivative is undefined at $z = 0$. It is better therefore to proceed with an average coefficient for the entire liquid-gas surface. The rate at which A is carried by the liquid at any y, per unit width in the x direction, is $\bar{u}_y \delta \bar{c}_A$ mol/time. Over a distance dy, per unit width, therefore, the rate of solute absorption is, in mol/time,

$$\bar{u}_y \delta \, d\bar{c}_A = k_L(c_{A, i} - \bar{c}_A) \, dy \tag{3.14}$$

$$\bar{u}_y \delta \int_{\bar{c}_A = c_{A0}}^{\bar{c}_A = \bar{c}_{A, L}} \frac{d\bar{c}_A}{c_{A, i} - \bar{c}_A} = \int_0^L k_L \, dy = k_{L, \text{av}} \int_0^L dy \tag{3.15}$$

$$k_{L, \text{av}} = \frac{\bar{u}_y \delta}{L} \ln \frac{c_{A, i} - c_{A0}}{c_{A, i} - \bar{c}_{A, L}} \tag{3.16}$$

which defines the average coefficient. Now for small rates of flow or long times of contact of the liquid with the gas (usually for film Reynolds numbers $\text{Re} = 4\Gamma/\mu$ less than 100), only the first term of the series of Eq. (3.12) need be used. Substituting in Eq. (3.16) gives

$$k_{L, \text{av}} = \frac{\bar{u}_y \delta}{L} \ln \frac{e^{5.1213\eta}}{0.7857} = \frac{\bar{u}_y \delta}{L}(0.241 + 5.1213\eta) \approx 3.41 \frac{D_{AB}}{\delta} \tag{3.17}$$

$$\frac{k_{L, \text{av}} \delta}{D_{AB}} = \text{Sh}_{\text{av}} \approx 3.41 \tag{3.18}$$

where Sh represents the Sherwood number, the mass-transfer analog to the Nusselt number of heat transfer. A similar development for large Reynolds numbers or short contact time [53] leads to

$$k_{L,\,av} = \left(\frac{6D_{AB}\Gamma}{\pi\rho\delta L}\right)^{1/2} \tag{3.19}$$

$$Sh_{av} = \left(\frac{3}{2\pi}\frac{\delta}{L}\, Re\, Sc\right)^{1/2} \tag{3.20}$$

The product Re Sc is the Péclet number Pe.

These average k_L's can be used to compute the total absorption rate. Thus the average flux $N_{A,\,av}$ for the entire gas-liquid surface, per unit width, is the difference in rate of flow of A in the liquid at $y = L$ and at $y = 0$, divided by the liquid surface. This can be used with some mean concentration difference

$$N_{A,\,av} = \frac{\bar{u}_y\delta}{L}(\bar{c}_{A,\,L} - c_{A0}) = k_{L,\,av}(c_{A,\,i} - \bar{c}_A)_M \tag{3.21}$$

Substitution for $k_{L,\,av}$ from Eq. (3.16) shows that the logarithmic average of the difference at the top and bottom of the film is required

$$(c_{A,\,i} - \bar{c}_A)_M = \frac{(c_{A,\,i} - c_{A0}) - (c_{A,\,i} - \bar{c}_{A,\,L})}{\ln[(c_{A,\,i} - c_{A0})/(c_{A,\,i} - \bar{c}_{A,\,L})]} \tag{3.22}$$

The experimental data show that the $k_{L,\,av}$ realized may be substantially larger than the theoretical values, even for low mass-transfer rates, owing to ripples and waves not considered in the analysis which form at values of Re beginning at about 25 [20]. The equations do apply for Re up to 1200 if ripples are suppressed by addition of wetting agents [37]. Rapid absorption, as for very soluble gases, produces important values of u_z, and this will cause further discrepancies due to alteration of the velocity profile in the film. The velocity profile may also be altered by flow of the gas, so that even in the simplest case, when both fluids move, k_L should depend on both flow rates.

Illustration 3.1 Estimate the rate of absorption of CO_2 into a water film flowing down a vertical wall 1 m long at the rate of 0.05 kg/s per meter of width at 25°C. The gas is pure CO_2 at 1 std atm. The water is essentially CO_2-free initially.

SOLUTION The solubility of CO_2 in water at 25°C, 1 std atm, is $c_{A,\,i} = 0.0336$ kmol/m³ soln; $D_{AB} = 1.96 \times 10^{-9}$ m²/s; soln density $\rho = 998$ kg/m³; and viscosity $\mu = 8.94 \times 10^{-4}$ kg/m · s. $\Gamma = 0.05$ kg/m · s, $L = 1$ m.

$$\delta = \left(\frac{3\mu\Gamma}{\rho^2 g}\right)^{1/3} = \left[\frac{3(8.94 \times 10^{-4})(0.05)}{(998)^2(9.807)}\right]^{1/3} = 2.396 \times 10^{-4}\text{ m}$$

$$Re = \frac{4\Gamma}{\mu} = \frac{4(0.05)}{8.94 \times 10^{-4}} = 203$$

Consequently Eq. (3.19) should apply:

$$k_{L, av} = \left(\frac{6D_{AB}\Gamma}{\pi\rho\delta L} \right)^{1/2} = \left[\frac{6(1.96 \times 10^{-9})(0.05)}{\pi(998)(2.396 \times 10^{-4})(1)} \right]^{1/2}$$

$$= 2.798 \times 10^{-5} \text{ kmol/m}^2 \cdot \text{s} \cdot (\text{kmol/m}^3)$$

$$\bar{u}_y = \frac{\Gamma}{\rho\delta} = \frac{0.05}{998(2.396 \times 10^{-4})} = 0.209 \text{ m/s}$$

At the top, $c_{A, i} - \bar{c}_A = c_{A, i} - c_{A0} = c_{A, i} = 0.0336$ kmol/m^3. At the bottom, $c_{A, i} - \bar{c}_{A, L} = 0.0336 - \bar{c}_{A, L}$ kmol/m^3. The flux of absorption is given by Eqs. (3.21) and (3.22):

$$\frac{0.209(2.396 \times 10^{-4})\bar{c}_{A, L}}{1} = \frac{(2.798 \times 10^{-5})[0.0336 - (0.0336 - \bar{c}_{A, L})]}{\ln[0.0336/ (0.0336 - \bar{c}_{A, L})]}$$

Therefore

$$\bar{c}_{A, L} = 0.01438 \text{ kmol/m}^3$$

and the rate of absorption is estimated to be

$$\bar{u}_y (\bar{c}_{A, L} - c_{A0}) = 0.209(2.396 \times 10^{-4})(0.01438 - 0)$$

$$= 7.2 \times 10^{-7} \text{ kmol/s} \cdot (\text{m of width})$$

The actual value may be substantially larger.

MASS-TRANSFER COEFFICIENTS IN TURBULENT FLOW

Most practically useful situations involve turbulent flow, and for these it is generally not possible to compute mass-transfer coefficients because we cannot describe the flow conditions mathematically. Instead, we rely principally on experimental data. The data are limited in scope, however, with respect to circumstances and situations as well as to range of fluid properties. Therefore it is important to be able to extend their applicability to situations not covered experimentally and to draw upon knowledge of other transfer processes (of heat, particularly) for help.

To this end, there are many theories which attempt to interpret or explain the behavior of mass-transfer coefficients, such as the film, penetration, surface-renewal, and other theories. They are all speculations, and are continually being revised. It is helpful to keep in mind that transfer coefficients, for both heat and mass transfer, are expedients used to deal with situations which are not fully understood. They include in one quantity effects which are the result of both molecular and turbulent diffusion. The relative contribution of these effects, and indeed the detailed character of the turbulent diffusion itself, differs from one situation to another. The ultimate interpretation or explanation of the transfer coefficients will come only when the problems of the fluid mechanics are solved, at which time it will be possible to abandon the concept of the transfer coefficient.

Eddy Diffusion

It will be useful first to describe briefly an elementary view of fluid turbulence, as a means of introducing the definitions of terms used in describing transfer

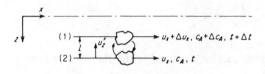

Figure 3.5 Eddy diffusion.

under turbulent conditions. Turbulence is characterized by motion of the fluid particles which is irregular with respect both to direction and time. Thus, for a fluid flowing turbulently in a duct (Fig. 3.5), the flow is, in the net, in the axial (or x) direction. At any location 2 within the central region of the cross section, the time average of the velocity may be u_x, but at any instant the velocity will actually be $u_x + u_{ix}'$, where u_{ix}' is the *deviating* or *fluctuating* velocity. Values of u_{ix}' will vary with time through a range of positive and negative values, the time average being zero, although u_x', the square root of the time average of $(u_{ix}')^2$, will be finite. Although the time-average value of $u_z = 0$, since the net flow is axially directed, the deviating velocity in the z direction will be u_{iz}' at any instant.

The turbulent fluid is imagined to consist of lumps of fluid, or eddies, of a great size range [15, 28, 59]. The largest eddies, whose size in the case of flow in a pipe is of the order of the pipe radius, contain only perhaps 20 percent of the turbulent kinetic energy. These eddies, buffeted by smaller eddies, produce smaller and smaller eddies to which they transfer their energy. The medium, so-called *energy-containing eddies* make the largest contribution to the turbulent kinetic energy. The smallest eddies, which ultimately dissipate their energy through viscosity, are substantially reenergized by larger eddies, and a state of equilibrium is established. This range of the energy spectrum has lost all relationship to the means by which the turbulence was originally produced and depends only upon the rate at which energy is supplied and dissipated. For this reason it is called the *universal range*. Turbulence characteristic of this range is isotropic, *ie.*,

$$u_x' = u_y' = u_z' \tag{3.23}$$

Kolmogoroff [34] defined the velocity u_d' and length scale l_d of these small eddies in terms of the power input per unit mass of the fluid, P/m:

$$u_d' = \left(\frac{\nu P g_c}{m} \right)^{1/4} \tag{3.24}$$

$$l_d = \left(\frac{\nu^3 m}{P g_c} \right)^{1/4} \tag{3.25}$$

or, after elimination of ν,

$$\frac{P g_c}{m} = \frac{u_d'^3}{l_d} \tag{3.26}$$

At equilibrium, when the energy lost from the medium-sized eddies must be at the same rate as that used to produce the small eddies, a similar relation should apply for the medium-sized eddies:

$$\frac{Pg_c}{m} = A \frac{u'^3}{l_e} \tag{3.27}$$

where A is a constant of the order of unity.

The size of the smallest eddies can be estimated in the case of pipe flow in the following manner.

Illustration 3.2 Consider the flow of water at 25°C in a 25-mm-ID tube at an average velocity $\bar{u} = 3$ m/s. $\mu = 8.937 \times 10^{-4}$ kg/m · s, $\rho = 997$ kg/m^3.

$$d = 0.025 \text{ m} \qquad \text{Re} = \frac{d\bar{u}\rho}{\mu} = 83\,670$$

At this Reynolds number, the Fanning friction factor for smooth tubes ("The Chemical Engineers' Handbook," 5th ed., p. 5-22, McGraw-Hill, New York, 1973) is $f = 0.0047$.

$$\Delta p_t = \text{pressure drop} = \frac{2\rho f L u^2}{dg_c}$$

where L = length of pipe. Take $L = 1$ m, whence

$$\Delta p_t = \frac{2(997)(0.0047)(1)(3)^2}{0.025(1)} = 3374 \text{ N/m}^2$$

The power expended equals the product of volumetric flow rate and pressure drop:

$$P = \frac{\pi}{4} d^2 \bar{u} \, \Delta p_t = \frac{\pi}{4}(0.025)^2(3)(3374) = 4.97 \text{ N} \cdot \text{m/s for 1-m pipe}$$

$$\text{Associated mass} = \frac{\pi}{4} d^2 L \rho = \frac{\pi}{4}(0.025)^2(1)(997) = 0.489 \text{ kg}$$

$$\frac{P}{m} = \frac{4.97}{0.489} = 10.16 \text{ N} \cdot \text{m/kg} \cdot \text{s}$$

Eq. (3.25):

$$l_d = \left[\left(\frac{8.937 \times 10^{-4}}{997} \right)^3 \frac{1}{10.16(1)} \right]^{1/4} = 1.63 \times 10^{-5} \text{ m}$$

Eq. (3.24):

$$u'_d = \left[\frac{(8.937 \times 10^{-4})(10.16)(1)}{997} \right]^{1/4} = 0.0549 \text{ m/s}$$

A simpler view of turbulence, however, is helpful in understanding some of the concepts used in mass transfer. In Fig. 3.5, consider a second location 1, where the x velocity is larger than at 2 by an amount $\Delta u_x = -l \, du_x/dz$. The distance l, the *Prandtl mixing length*, is defined such that $u'_x = \Delta u_x = -l \, du_x/dz$. Owing to the z-directed fluctuating velocity, a particle of fluid, an eddy, may move from 2 to 1 at a velocity u'_{iz}, but it will be replaced by an eddy of equal volume moving from 1 to 2. It is imagined, in the Prandtl theory, that the eddies retain their identity during the interchange but blend into the fluid at their new location. This is clearly a great oversimplification in view of what was said before.

The mass velocity of eddy interchange is then $\rho u'_{iz}$, and thanks to the different velocities at 1 and 2 there will be a momentum transfer flux $\rho u'_{iz} u'_{ix}$. If u'_{iz} and u'_{ix} are essentially equal, there results an average shear stress due to the turbulent eddy interchange,

$$\tau_{\text{turb}} g_c = \rho |u'_{iz} u'_{ix}| = \rho u'_z u'_x = \rho (u'_x)^2 = \rho l^2 \left(\frac{du_x}{dz} \right)^2 \tag{3.28}$$

Molecular motion, of course, still also contributes to the shear stress, as given by Eq. (2.47), so that the *total* shear stress becomes

$$\tau g_c = -\mu \frac{du_x}{dz} + \rho l^2 \left(\frac{du_x}{dz} \right)^2 = -\left[\mu + \rho l^2 \left(-\frac{du_x}{dz} \right) \right] \frac{du_x}{dz} = -\left(\frac{\mu}{\rho} + E_v \right) \frac{d(u_x \rho)}{dz}$$

$$\tag{3.29}$$

E_v is the eddy momentum diffusivity, length^2/time. While μ or $\nu = \mu/\rho$ is a constant for a given fluid at fixed temperature and pressure, E_v will depend on the local degree of turbulence. In the various regions of the duct, ν predominates near the wall, while in the turbulent core E_v predominates, to an extent depending upon the degree of turbulence.

The eddies bring about a transfer of dissolved solute, as we have mentioned before. The average concentration gradient between 1 and 2 in Fig. 3.5 is $\Delta c_A / l$, proportional to a local gradient, $- dc_A / dz$. The flux of A due to the interchange, $u'_z \Delta c_A$, and the concentration gradient can be used to define an eddy diffusivity of mass E_D, length^2/time

$$E_D = \frac{b_1 u'_z \Delta c_A}{\Delta c_A / l} = \frac{J_{A,\,\text{turb}}}{- dc_A / dz} \tag{3.30}$$

where b_1 is a proportionality constant. The *total* flux of A, due both to molecular and eddy diffusion, then will be

$$J_A = -(D_{AB} + E_D) \frac{dc_A}{dz} \tag{3.31}$$

As in the case of momentum transfer, D is a constant for a particular solution at fixed conditions of temperature and pressure, while E_D depends on the local turbulence intensity. D predominates in the region near the wall, E_D in the turbulent core.

In similar fashion, an eddy thermal diffusivity E_H, length^2/time, can be used to describe the flux of heat as a result of a temperature gradient,

$$E_H = \frac{b_2 u'_z \Delta(\rho C_p t)}{\Delta(\rho C_p t)/l} = \frac{q_{\text{turb}}}{- d(\rho C_p t)/dz} \tag{3.32}$$

where b_2 is a proportionality constant. The *total* heat flux due to conduction and eddy motion is

$$q = -(k + E_H \rho C_p) \frac{dt}{dz} = -(\alpha + E_H) \frac{d(t \rho C_p)}{dz} \tag{3.33}$$

As before, α for a given fluid is fixed for fixed conditions of temperature and pressure, but E_H varies with the degree of turbulence and hence with location.

The three eddy diffusivities of momentum, heat, and mass can be computed from measured velocity, temperature, and concentration gradients, respectively, in that order of increasing difficulty.

Momentum eddy diffusivity For turbulent flow in circular pipes, Re = 50 000 to 350 000, the data of the turbulent core ($z_+ > 30$) yield [9]

$$\frac{2E_v}{d\,\bar{u}\,(f/2)^{0.5}} = 0.063\left[1 + \left(\frac{2z}{d}\right)^2 - 2\left(\frac{2z}{d}\right)^4\right] \tag{3.34}$$

where z is the distance from the center and f is the Fanning friction factor. This shows that the maximum E_v occurs at a distance $z = d/4$. Thus, for water at 25°C in a smooth 5-cm-diameter tube at Re = 150 000, the average velocity $\bar{u} = 2.69$ m/s, $f = 0.004$, and the maximum E_v calculates to be 2.1 × 10^{-4} m²/s, which is 238 times the kinematic viscosity ν. For the region closer to the wall ($z_+ < 5$), it is generally accepted that

$$\frac{E_v}{\nu} = \left(\frac{z_+}{K}\right)^3 \tag{3.35}$$

where

$$z_+ = \left(\frac{d}{2} - z\right)\frac{\bar{u}}{\nu}\left(\frac{f}{2}\right)^{0.5}$$

a dimensionless distance from the wall. The value of K has not been firmly established and may be in the range 8.9 [15] to 14.5 [40]. It is noteworthy that Eq. (3.35) indicates that eddy motion persists right up to the wall. It is also noteworthy that Murphree [43] deduced the proportionality between E_v and z_+^3 as early as 1932. For a *transition region* ($z_+ = 5$ to 30) [15],

$$\frac{E_v}{\nu} = \left(\frac{z_+}{11}\right)^2 \tag{3.36}$$

We can now estimate where the eddy and molecular momentum diffusivities are equal, a location which Levich suggested as the outer limit of the viscous sublayer [38]. For $E_v = \nu$, $z_+ = 11$, and for the same conditions as before (water at 25°C, Re = 150 000, 5-cm tube), this occurs at $d/2 - z = 0.082$ mm from the wall. The small distance emphasizes the difficulty in obtaining reliable data.

Heat and mass eddy diffusivity The evidence is that with Prandtl and Schmidt numbers close to unity, as for most gases, the eddy diffusivities of heat and mass are equal to the momentum eddy diffusivity for all regions of turbulence [15]. For turbulent fluids where Prandtl and Schmidt numbers exceed unity, the ratios E_H/E_v and E_D/E_v will vary with location relative to the wall and in the turbulent core will lie generally in the range 1.2 to 1.3, with E_D and E_H essentially equal [44, 62]. For $z_+ = 0$ to 45, with Pr and Sc > 1, a critical analysis of the theoretical and experimental evidence [44] led to

$$\frac{E_D}{\nu} = \frac{E_H}{\nu} = \frac{0.00090z_+^3}{\left(1 + 0.0067z_+^2\right)^{0.5}} \tag{3.37}$$

which, as z_+ approaches zero, becomes

$$\frac{E_D}{\nu} = \frac{E_H}{\nu} = \left(\frac{z_+}{10.36}\right)^3 \tag{3.38}$$

The region close to the wall where molecular diffusivity exceeds eddy diffusivity of mass has been termed the *diffusional sublayer* [38], and we can estimate its extent in particular instances. For example, for flow of water at 25°C, in the pipe considered above, $\nu = 8.964 \times 10^{-7}$ m²/s, and typically for solutes in water, D is of the order of 10^{-9} m²/s. When E_D is set equal to D, Eq. (3.37) yields $z_+ = 1.08$; so the diffusional sublayer lies well within the viscous sublayer, the more so the smaller the D and the larger the Sc. At this z_+ for the pipe considered above and Re = 150 000, the distance from the wall is 0.008 mm.

Mass-transfer coefficients covering the transfer of material from an interface to the turbulent zone will depend upon the total diffusivity $D + E_D$. Success in estimating this theoretically in the manner used for laminar flow will depend on knowledge of how the ratio E_D/D varies throughout the diffusion path. Knowledge of how the ratio E_v/ν varies will be of direct help only if $E_D = E_v$, which is evidently not true generally. But knowledge of the heat-transfer coefficients, which depend upon E_H/ν, should be useful in predicting the mass-transfer coefficients, since E_H and E_D are evidently the same except for the liquid metals.

Film Theory

This is the oldest and most obvious picture of the meaning of the mass-transfer coefficient, borrowed from a similar concept used for convective heat transfer. When a fluid flows turbulently past a solid surface, with mass transfer occurring from the surface to the fluid, the concentration-distance relation is as shown by the full curve of Fig. 3.6, the shape of which is controlled by the continuously varying ratio of E_D to D. The film theory postulates that the concentration will follow the broken curve of the figure, such that the entire concentration difference $c_{A1} - c_{A2}$ is attributed to molecular diffusion within an "effective" film of thickness z_F. It was early recognized [39] that if this theory were to have useful application, the film would have to be very thin, so that the

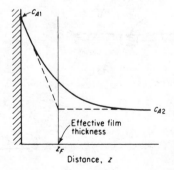

Distance, z **Figure 3.6** Film theory.

quantity of solute within the film would be small relative to the amount passing through it, or ne concentration gradient would have to be set up quickly. The concentration gradient in the film is that characteristic of steady state.

In Eq. (3.1), it is clear that F has merely replaced the group $D_{AB}c/z$ of Eq. (2.22). The film theory states that z of Eq. (2.21) is z_F, the effective film thickness; this thickness depends upon the nature of the flow conditions. Similarly, the z's of Eqs. (2.26), (2.30), (2.34), (2.40), (2.41), and (2.43) are interpreted to be z_F, incorporated into the k's of Eqs. (3.3) to (3.6).

The film theory therefore predicts that F and the k-type mass-transfer coefficients for different solutes being transferred under the same fluid-flow conditions are directly proportional to the D's for the solutes. On the other hand, we observe for turbulent flow a much smaller dependency, proportional to D^n, where n may be anything from nearly zero to 0.8 or 0.9, depending upon the circumstances. The simple film theory, in conflict with our knowledge of turbulent flow (see above), has therefore been largely discredited. Nevertheless it does well in handling the effect of high mass-transfer flux [Eq. (3.1)] and the effect of mass transfer on heat transfer and in predicting the effect of chemical reaction rate on mass transfer [14]. Mass- and heat-transfer coefficients are still frequently called *film coefficients*.

Mass Transfer at Fluid Surfaces

Because the velocity of a fluid surface, as in the contact of a gas and liquid, is not zero, a number of theories have been developed to replace the film theory.

Penetration theory Higbie [27] emphasized that in many situations the time of exposure of a fluid to mass transfer is short, so that the concentration gradient of the film theory, characteristic of steady state, would not have time to develop. His theory was actually conceived to describe the contact of two fluids, as in Fig. 3.7. Here, as Higbie depicted it in Fig. 3.7a, a bubble of gas rises through a liquid which absorbs the gas. A particle of the liquid b, initially at the top of the

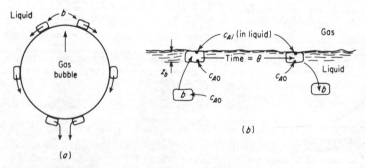

Figure 3.7 Penetration theory.

bubble, is in contact with the gas for the time θ required for the bubble to rise a distance equal to its diameter while the liquid particle slips along the surface of the bubble. An extension to cases where the liquid may be in turbulent motion, as in Fig. 3.7b shows an eddy b rising from the turbulent depths of the liquid and remaining exposed for a time θ to the action of the gas. In this theory the time of exposure is taken as constant for all such eddies or particles of liquid.

Initially, the concentration of dissolved gas in the eddy is uniformly c_{A0}, and internally the eddy is considered to be stagnant. When the eddy is exposed to the gas at the surface, the concentration in the liquid at the gas-liquid surface is $c_{A,i}$, which may be taken as the equilibrium solubility of the gas in the liquid. During the time θ, the liquid particle is subject to unsteady-state diffusion or penetration of solute in the z direction, and, as an approximation, Eq. (2.18) may be applied

$$\frac{\partial c_A}{\partial \theta} = D_{AB}\frac{\partial^2 c_A}{\partial z^2} \tag{3.39}$$

For short exposure times, and with slow diffusion in the liquid, the molecules of dissolving solute are never able to reach the depth z_b corresponding to the thickness of the eddy, so that from the solute point of view, z_b is essentially infinite. The conditions on Eq. (3.39) then are

$$c_A = \begin{cases} c_{A0} & \text{at } \theta = 0 & \text{for all } z \\ c_{A,i} & \text{at } z = 0 & \text{for } \theta > 0 \\ c_{A0} & \text{at } z = \infty & \text{for all } \theta \end{cases}$$

Solving Eq. (3.39) and proceeding in the manner described earlier for a falling liquid film provides the average flux over the time of exposure

$$N_{A,\,av} = 2(c_{A,\,i} - c_{A0})\sqrt{\frac{D_{AB}}{\pi\theta}} \tag{3.40}$$

and comparison with Eq. (3.4) shows

$$k_{L,\,av} = \sqrt{\frac{2 D_{AB}}{\pi\theta}} \tag{3.41}$$

with $k_{L,\,av}$ proportional to $D_{AB}^{0.5}$ for different solutes under the same circumstances. This indicated dependence on D is typical of short exposure times, where the depth of solute penetration is small relative to the depth of the absorbing pool [compare Eq. (3.19)]. As pointed out above, experience shows a range of exponents on D from nearly zero to 0.8 or 0.9.

Surface-renewal theories Danckwerts [13] pointed out that the Higbie theory with its constant time of exposure of the eddies of fluid at the surface is a special case of what may be a more realistic picture, where the eddies are exposed for varying lengths of time. The liquid-gas interface is then a. mosaic of surface elements of different exposure-time histories, and since the rate of solute penetration depends upon exposure time, the average rate for a unit surface area must be determined by summing up the individual values. On the assumption that the chance of a surface element's being replaced by another is quite independent of how long it has been in the surface, and if s is the fractional rate of replacement of elements, Danckwerts found

$$N_{A, av} = (c_{A, i} - c_{A0})\sqrt{D_{AB}s} \tag{3.42}$$

and therefore $\qquad k_{L, av} = \sqrt{D_{AB}s} \tag{3.43}$

Danckwerts pointed out that all theories of this sort, derived with the original boundary conditions on Eq. (3.39), will lead to $k_{L, av}$ proportional to $D_{AB}^{0.5}$, regardless of the nature of the surface-renewal rate s which may apply. Subsequently there have been many modifications of this approach [26, 35].

Combination film–surface-renewal theory Dobbins [17, 18], concerned with the rate of absorption of oxygen into flowing streams and rivers, pointed out that the film theory ($k_L \propto D_{AB}$) assumes a time of exposure of the surface elements sufficiently long for the concentration profile within the film to be characteristic of steady state, whereas the penetration and surface-renewal theories ($k_L \propto D_{AB}^{0.5}$) assume the surface elements to be essentially infinitely deep, the diffusing solute never reaching the region of constant concentration below. The observed dependence, $k_L \propto D_{AB}^n$, with n dependent upon circumstances, might be explained by allowing for a finite depth of the surface elements or eddies.

Accordingly, Dobbins replaced the third boundary condition on Eq. (3.39) by $c_A = c_{A0}$ for $z = z_b$, where z_b is finite. With Danckwerts' rate of renewal of the surface elements, he obtained

$$k_{L, av} = \sqrt{D_{AB}s} \, \coth\sqrt{\frac{sz_b^2}{D_{AB}}} \tag{3.44}$$

This is shown in Fig. 3.8. For rapid penetration (D_{AB} large), the rate of surface small (s small), or for thin surface elements, the mass-transfer coefficient takes on the character described by the film theory; whereas for slow penetration or

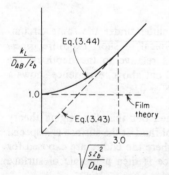

Figure 3.8 Combination film–surface-renewal theory. *(After Dobbins [17].)*

rapid renewal it follows Eq. (3.43). Consequently $k_L \propto D_{AB}^n$, where n may have any value between the limits 0.5 and 1.0, which could account for many observations. Toor and Marchello [59] have made similar suggestions.

King [33, 36] has proposed modifications which embody the notion of surface renewal with a dampening of the eddy diffusivities near the surface owing to surface tension, and these lead to values of n less than 0.5.

Surface-stretch theory Lightfoot and his coworkers [2, 3, 29, 50, 56] in a development which shows great promise, have extended the penetration-surface-renewal concepts to situations where the interfacial surface through which mass transfer occurs changes periodically with time. An example is a liquid drop such as ethyl acetate rising through a denser liquid such as water, with mass transfer of a solute such as acetic acid from the water to the drop. Such a drop, if relatively large, wobbles and oscillates, changing its shape as shown schematically in Fig. 3.9. If the central portion of the drop is thoroughly turbulent, the mass-transfer resistance of the drop resides in a surface layer of varying thickness. Similar situations occur while drops and bubbles form at nozzles and when liquid surfaces are wavy or rippled. For these cases the theory leads to

$$k_{L,\,av} = \frac{(A/A_r)\sqrt{D_{AB}/\pi\theta_r}}{\sqrt{\int_0^{\theta/\theta_r}\left(\dfrac{A}{A_r}\right)^2 d\theta}} \tag{3.45}$$

where A = time-dependent interface surface
$\quad\;\; A_r$ = reference value of A, defined for each situation
$\quad\;\; \theta_r$ = const, with dimensions of time, defined for each situation; e.g., for drop formation θ_r might be drop-formation time

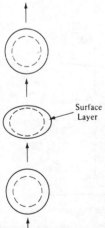

Figure 3.9 Liquid drop rising through another liquid. *(After Rose and Kintner [50].)*

Surface
Layer

The integral of Eq. (3.45) can be evaluated once the periodic nature of the surface variation is established, and this has been done for a number of situations.

Flow Past Solids; Boundary Layers

For the theories discussed above, where the interfacial surface is formed between two fluids, the velocity at that surface will not normally be zero. But when one of the phases is a solid, the fluid velocity parallel to the surface at the interface must of necessity be zero, and consequently the two circumstances are inherently different.

In Fig. 3.10 a fluid with uniform velocity u_0 and uniform solute concentration c_{A0} meets up with a flat solid surface AK. Since the velocity u_x is zero at the surface and rises to u_0 at some distance above the plate, the curve $ABCD$ separates the region of velocity u_0 from the region of lower velocity, called the *boundary layer*. The boundary layer may be characterized by laminar flow, as below curve AB, but if the velocity u_0 is sufficiently large, for values of $\mathrm{Re}_x = xu_0\rho/\mu$ greater than approximately 5×10^5, depending upon the turbulence intensity of the incident stream, the flow in the bulk of the boundary layer will be turbulent, as below the curve CD. Below the turbulent boundary layer there will be a thinner region, the viscous sublayer, extending from the plate to the curve FG. Boundary layers also occur for flow at the entrance to circular pipes, for flow along the outside of cylinders, and the like. When the surface over which the fluid passes is curved convexly in the direction of flow, as in flow at right angles to the outside of a cylinder and past a sphere, well-developed boundary layers form at very low rates of flow. At higher rates of flow, however, the boundary layer separates from the surface, and eddies form in the wake behind the object.

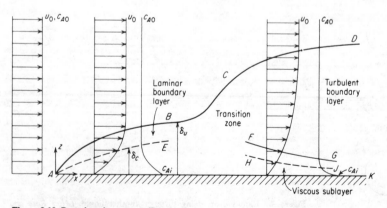

Figure 3.10 Boundary layer on a flat plate.

If mass transfer from the surface into the fluid occurs, as with a solid subliming into a gas or a solid dissolving into a liquid, the solute concentration in the fluid at the solid surface is everywhere $c_{A, i}$, greater than c_{A0}. There will be a curve AE and also one HJ, which separate regions of uniform concentration c_{A0} from regions of higher values of c_A, corresponding to a concentration boundary layer. In the region where only a laminar velocity boundary layer exists, the equations of motion and of mass transfer can be solved simultaneously to yield the concentration profile and, from its slope at the surface, the laminar mass-transfer coefficient [7]. This is a fairly complex problem particularly if account is taken of the influence of the flux of mass of A in the z direction on the velocity profile. If this influence is negligible, and if mass transfer begins at the leading edge A, it develops that the thickness of the velocity boundary layer δ_u and that of the concentration boundary layer δ_c are in the ratio $\delta_u / \delta_c = Sc^{1/3}$. Similar considerations apply if there is heat transfer between the plate and the fluid.

In laminar flow with low mass-transfer rates and constant physical properties past a solid surface, as for the two-dimensional laminar boundary layer of Fig. 3.10, the momentum balance or equation of motion (Navier-Stokes equation) for the x direction becomes [7]

$$u_x \frac{\partial u_x}{\partial x} + u_z \frac{\partial u_x}{\partial z} = \nu \left(\frac{\partial^2 u_x}{\partial x^2} + \frac{\partial^2 u_x}{\partial z^2} \right) \tag{3.46}$$

If there is mass transfer without chemical reaction, Eq. (2.17) gives the equation of continuity for substance A

$$u_x \frac{\partial c_A}{\partial x} + u_z \frac{\partial c_A}{\partial z} = D_{AB} \left(\frac{\partial^2 c_A}{\partial x^2} + \frac{\partial^2 c_A}{\partial z^2} \right) \tag{3.47}$$

and if there is heat transfer between the fluid and the plate, Eq. (2.19) provides

$$u_x \frac{\partial t}{\partial x} + u_z \frac{\partial t}{\partial z} = \alpha \left(\frac{\partial^2 t}{\partial x^2} + \frac{\partial^2 t}{\partial z^2} \right) \tag{3.48}$$

which are to be solved simultaneously, together with the equation of continuity (2.14). It is clear that these equations are all of the same form, with u_x, c_A, and t and the three diffusivities of momentum ν, mass D_{AB}, and heat α replacing each other in appropriate places.

In solving these, dimensionless forms of the variables are usually substituted

$$\frac{u_x - (u_{x, z=0} = 0)}{u_0 - (u_{x, z=0} = 0)} \qquad \frac{c_A - c_{A, i}}{c_{A0} - c_{A, i}} \qquad \text{and} \qquad \frac{t - t_i}{t_0 - t_i}$$

Then the boundary conditions become identical. Thus, for Fig. 3.10, at $z = 0$ all three dimensionless variables are zero, and at $z = \infty$, all three equal unity. Consequently the form of the solutions, which provide dimensionless velocity, concentration, and temperature profiles, are the same. Indeed, if all three diffusivities are equal, so that $Sc = Pr = 1$, the profiles in dimensionless form are identical. The initial slopes of the concentration, temperature, and velocity profiles provide the means of computing the corresponding local transfer coefficients

$$N_A = -D_{AB}\left(\frac{\partial c_A}{\partial z}\right)_{z=0} = k_L(c_{A,i} - c_{A0}) \tag{3.49}$$

$$q = -\alpha\left[\frac{\partial(tC_p\rho)}{\partial z}\right]_{z=0} = h(t_i - t_0) \tag{3.50}$$

$$\tau_i g_c = \nu\left[\frac{\partial(u_x\rho)}{\partial z}\right]_{z=0} = \frac{f}{2}u_0(\rho u_0 - 0) \tag{3.51}$$

where f is the dimensionless friction factor and $fu_0/2$ might be considered a momentum-transfer coefficient. When the coefficients are computed and arranged as dimensionless groups, the results are all of the same form, as might be expected. In particular, for the flat plate of Fig. 3.10 at low mass-transfer rates,

$$\frac{Nu}{Re_x\,Pr^{1/3}} = \frac{Sh}{Re_x\,Sc^{1/3}} = \frac{f}{2} = 0.332\,Re_x^{-1/2} \tag{3.52}$$

and the average coefficients provide Nu_{av} and Sh_{av} given by the same expressions with 0.332 replaced by 0.664. These show the mass-transfer coefficients to vary as $D_{AB}^{2/3}$, which is typical of the results of boundary-layer calculations. Equation (3.52) is applicable for cases of low mass-transfer rate, with heat or mass transfer beginning at the leading edge of the plate, and for Re_x up to approximately 80 000; for heat transfer, $Pr > 0.6$; and for constant t_0 and c_{A0}.

In the regions where a turbulent boundary layer and a viscous sublayer exist, the calculations for the transfer coefficients depend upon the expression chosen for the variation of the eddy diffusivities with distance from the wall. Alternatively, a great reliance is put upon experimental measurements. Results will be summarized later.

MASS-, HEAT-, AND MOMENTUM-TRANSFER ANALOGIES

There are many more data available for pressure-drop friction and heat transfer than for mass transfer. The similarity of Eqs. (3.46) to (3.48) for the transfer processes of momentum, mass, and heat, and the identical solutions, Eq. (3.52), lead to the possibility of deducing the mass-transfer characteristics for other situations from knowledge of the other two.

For the case of the flat plate of Fig. 3.10 and other similar situations, the friction factor corresponds to a skin-friction drag along the surface. For flow where separation occurs, such as for flow past a sphere or at right angles to a cylinder or any bluff object, one might expect the friction factor based on total drag, which includes not only skin friction but form drag due to flow separation as well, to follow a function different from that of the mass- and heat-transfer groups.

In the case of turbulent flow, the differential equations will contain time-averaged velocities and in addition the eddy diffusivities of momentum, mass, and heat transfer. The resulting equations cannot be solved for lack of information about the eddy diffusivities, but one might expect results of the form

$$\frac{\text{Sh}}{\text{Re Sc}} = \psi_1\left(\frac{f}{2}, \text{Sc}, \frac{E_v}{E_D}\right) = \psi_2\left(\frac{f}{2}, \text{Sc}, \frac{E_D}{v}\right) \qquad (3.53)$$

$$\frac{\text{Nu}}{\text{Re Pr}} = \psi_1\left(\frac{f}{2}, \text{Pr}, \frac{E_v}{E_H}\right) = \psi_2\left(\frac{f}{2}, \text{Pr}, \frac{E_H}{v}\right) \qquad (3.54)$$

Successful completion of an analogy requires, therefore, knowledge of how the ratios E_v/v, E_D/E_v, and E_H/E_v vary with distance from the fluid interface. It is customary arbitrarily to set $E_D/E_v = E_H/E_v = 1$, despite experimental evidence to the contrary and to make some arbitrary decision about E_v/v. With these assumptions, however, experimental velocity profiles permit prediction of the profiles and coefficients for mass and heat transfer. Further, assuming only $E_D = E_H$ (which is much more reasonable) permits information on heat transfer to be converted directly for use in mass-transfer calculations and vice versa.

Let us sum up and further generalize what can be done with the analogies:

1. For analogous circumstances, temperature and concentration profiles in dimensionless form and heat- and mass-transfer coefficients in the form of dimensionless groups, respectively, are given by the same functions. To convert equations or correlations of data on heat transfer and temperatures to corresponding mass transfer and concentrations, the dimensionless groups of the former are replaced by the corresponding groups of the latter. Table 3.2 lists the commonly appearing dimensionless groups. The limitations are:

   a. The flow conditions and geometry must be the same.

   b. Most heat-transfer data are based on situations involving no mass transfer. Use of the analogy would then produce mass-transfer coefficients corresponding to no net mass transfer, in turn corresponding most closely to k'_G, k'_c, or k'_y ($= F$). Sherwood numbers are commonly written in terms of any of the coefficients, but when derived by replacement of Nusselt numbers for use where the net mass transfer is not zero, they should be taken as $\text{Sh} = Fl/cD_{AB}$, and the F used with Eq. (3.1). Further, the result will be useful only in the absence of chemical reaction.

   c. The boundary conditions which would be used to solve the corresponding differential equations must be analogous. For example, in the case of the mass-transfer problem of the falling film of Fig. 3.4, the analogous circumstances for heat transfer would require heat transfer from the gas (not from the solid wall) to the liquid, the wall to be impervious to the transfer of heat, heat transfer beginning at the same value of y as mass transfer (in this case at $y = 0$), a constant temperature for the liquid-gas interface, and constant fluid properties.

   d. For turbulent flow, $E_H = E_D$ at any location.

Table 3.2 Corresponding dimensionless groups of mass and heat transfer

No.	Mass transfer	Heat transfer
1	$\dfrac{c_A - c_{A1}}{c_{A2} - c_{A1}}$	$\dfrac{t - t_1}{t_2 - t_1}$
2	Reynolds number $$\mathrm{Re} = \frac{lu\rho}{\mu}$$	Reynolds number $$\mathrm{Re} = \frac{lu\rho}{\mu}$$
3	Schmidt number $$\mathrm{Sc} = \frac{\mu}{\rho D_{AB}} = \frac{\nu}{D_{AB}}$$	Prandtl number $$\mathrm{Pr} = \frac{C_p \mu}{k} = \frac{\nu}{\alpha}$$
4	Sherwood number $$\mathrm{Sh} = \frac{Fl}{c D_{AB}}, \frac{k_G \bar{p}_{B,M} RTl}{p_t D_{AB}},$$ $$\frac{k_c \bar{p}_{B,M} l}{p_t D_{AB}}, \frac{k_c' l}{D_{AB}}, \frac{k_y' RTl}{p_t D_{AB}}, \text{ etc.}$$	Nusselt number $$\mathrm{Nu} = \frac{hl}{k}$$
5	Grashof number† $$\mathrm{Gr}_D = \frac{gl^3 \, \Delta\rho}{\rho}\left(\frac{\rho}{\mu}\right)^2$$	Grashof number† $$\mathrm{Gr}_H = gl^3 \beta \, \Delta t \left(\frac{\rho}{\mu}\right)^2$$
6	Peclet number $$\mathrm{Pe}_D = \mathrm{Re}\,\mathrm{Sc} = \frac{lu}{D_{AB}}$$	Peclet number $$\mathrm{Pe}_H = \mathrm{Re}\,\mathrm{Pr} = \frac{C_p lu\rho}{k} = \frac{lu}{\alpha}$$
7	Stanton number $$\mathrm{St}_D = \frac{\mathrm{Sh}}{\mathrm{Re}\,\mathrm{Sc}} = \frac{\mathrm{Sh}}{\mathrm{Pe}_H} = \frac{F}{cu},$$ $$\frac{F}{G}, \frac{k_G \bar{p}_{B,M} M_{av}}{\rho u}, \text{ etc.}$$	Stanton number $$\mathrm{St}_D = \frac{\mathrm{Nu}}{\mathrm{Re}\,\mathrm{Pr}} = \frac{\mathrm{Nu}}{\mathrm{Pe}_H} = \frac{h}{C_p u\rho}$$
8	$j_D = \mathrm{St}_D \, \mathrm{Sc}^{2/3}$	$j_H = \mathrm{St}_H \, \mathrm{Pr}^{2/3}$

† The Grashof number appears in cases involving natural convection; $\Delta\rho = |\rho_1 - \rho_2|$, $\Delta t = |t_1 - t_2|$, in the same phase.

2. Friction factors and velocity profiles can be expected to correlate with the corresponding heat- and mass-transfer quantities only if $E_v = E_H = E_D$ in turbulent flow and in extension to viscous sublayers only if $E_v/\nu = E_H/\alpha = E_D/D_{AB}$. For either laminar or turbulent flow, the friction factors must indicate skin friction and not form drag as well. In general, it is safest to avoid the friction analogy to mass and heat transfer.

These analogies are particularly useful in extending the very considerable amount of information on heat transfer to yield corresponding information on mass transfer, which is more likely to be lacking. Alternatively, local mass-transfer coefficients can be measured with relative ease through sublimation or dissolution of solids, and these can be converted into the analogous local heat-transfer coefficients, which are difficult to obtain [57].

Illustration 3.3 What is the heat-transfer analog to Eq. (3.12)?

SOLUTION The equation remains the same with the dimensionless concentration ratio replaced by $(t_i - t_L)/(t_i - t_0)$; the dimensionless group

$$\eta = \frac{2D_{AB}L}{3\delta^2 \bar{u}_y} = \frac{2}{3} \frac{D_{AB}}{\delta \bar{u}_y} \frac{L}{\delta} = \frac{2}{3 \, Pe_D} \frac{L}{\delta}$$

for mass transfer is replaced by

$$\eta = \frac{2}{3 \, Pe_H} \frac{L}{\delta} = \frac{2}{3} \frac{\alpha}{\delta \bar{u}_y} \frac{L}{\delta} = \frac{2\alpha L}{3\delta^2 \bar{u}_y} \quad \text{Ans.}$$

Illustration 3.4 For flow of a fluid at right angles to a circular cylinder, the average heat-transfer coefficient (averaged around the periphery) for fluid Reynolds numbers in the range 1 to 4000 is given by [19]

$$Nu_{av} = 0.43 + 0.532 \, Re^{0.5} \, Pr^{0.31}$$

where Nu and Re are computed with the cylinder diameter and fluid properties are taken at the mean of the cylinder and bulk-fluid temperatures.

Estimate the rate of sublimation of a cylinder of uranium hexafluoride, UF_6, 6 mm diameter, exposed to an airstream flowing at a velocity of 3 m/s. The surface temperature of the solid is 43°C, at which temperature the vapor pressure of UF_6 is 400 mmHg (53.32 kN/m^2). The bulk air is at 1 std atm pressure, 60°C.

SOLUTION The analogous expression for the mass-transfer coefficient is

$$Sh_{av} = 0.43 + 0.532 \, Re^{0.5} \, Sc^{0.31}$$

Along the mass-transfer path (cylinder surface to bulk air) the average temperature is 51.5°C, and the average partial pressure of UF_6 is 200 mmHg (26.66 kN/m^2), corresponding to $26.66/101.33 = 0.263$ mole fraction UF_6, 0.737 mole fraction air. The corresponding physical properties of the gas, at the mean temperature and composition, are estimated to be: density = 4.10 kg/m^3, viscosity = 2.7×10^{-5} kg/m · s, diffusivity = 9.04×10^{-6} m^2/s.

$$Re = \frac{d \, u\rho}{\mu} = \frac{(6 \times 10^{-3})(3)(4.10)}{2.7 \times 10^{-5}} = 2733 \qquad Sc = \frac{\mu}{\rho D} = \frac{2.7 \times 10^{-5}}{4.10(9.04 \times 10^{-6})} = 0.728$$

$$Sh_{av} = 0.43 + 0.532(2733)^{0.5}(0.728)^{0.31} = 25.6 = \frac{F_{av}d}{cD_{AB}}$$

$$c = \frac{1}{22.41} \frac{273.2}{273.2 + 51.5} = 0.0375 \, kmol/m^3$$

$$F_{av} = \frac{Sh_{av}cD_{AB}}{d} = \frac{25.6(0.0375)(9.04 \times 10^{-6})}{6 \times 10^{-3}} = 1.446 \times 10^{-3} \, kmol/m^2 \cdot s$$

In this case, N_B(air) = 0, so that $N_A/(N_A + N_B) = 1.0$. Since $c_A/c = 53.32/101.33 = 0.526$ mole fraction, Eq. (3.1) provides

$$N_A = 1.44 \times 10^{-3} \ln \frac{1 - 0}{1 - 0.526} = 1.08 \times 10^{-3} \, kmol \, UF_6/m^2 \cdot s$$

which is the mass-transfer flux based on the entire cylinder area.

Note 1 The calculated N_A is an instantaneous flux. Mass transfer will rapidly reduce the cylinder diameter, so that the Reynolds number and hence F will change with time. Furthermore, the surface will not remain in the form of a circular cylinder owing to the variation of the local F about the perimeter, so that the empirical correlation for Nu_{av} and Sh_{av} will then no longer apply.

Note 2 Use of k_c or k_G computed from a simple definition of Sh produces incorrect results because of the relatively high mass-transfer rate, which is not allowed for in the heat-transfer correlation. Thus, $k_c = k_G RT = Sh_{av} D_{AB}/d = 25.6(9.04 \times 10^{-6})/(6 \times 10^{-3}) = 0.0386$ kmol/m² · s · (kmol/m³) or $k_G = 0.0386/8.314(273.2 + 51.5) = 1.43 \times 10^{-5}$ kmol/m² · s · (kN/m²). This leads [Eq. (3.3)] to $N_A = (1.43 \times 10^{-5})(53.32 - 0) = 7.62 \times 10^{-4}$ kmol/m² · s, which is too low. However, if k_G is taken to be $1.43 \times 10^{-5} p_t/\bar{p}_{B, M}$, the correct answer will be given by Eq. (3.3).

Turbulent Flow in Circular Pipes

A prodigious amount of effort has been spent in developing the analogies between the three transport phenomena for this case, beginning with Reynolds' original concept of the analogy between momentum and heat transfer in 1874 [49]. The number of well-known relationships proposed is in the dozens, and new ones of increasing complexity are continually being suggested. Space limitations will not permit discussion of this development, and we shall confine ourselves to a few of the results.

Before any extensive measurements of the eddy diffusivities had been made, Prandtl had assumed that the fluid in turbulent flow consisted only of a viscous sublayer where E_v and E_H were both zero and a turbulent core where molecular diffusivities were negligibly important. The resulting analogy between heat and momentum transfer, as we might expect, is useful only for cases where Pr is essentially unity. A remarkably simple empirical modification of Prandtl's analogy by Colburn [11], on the other hand, represents the data even for large Pr very well (Colburn analogy)

$$St_{H, av} = \frac{h_{av}}{C_p \bar{u}_x \rho} = \frac{\frac{1}{2}f}{Pr^{2/3}} \tag{3.55}$$

The group $St_{H, av} Pr^{2/3}$ is called j_H

$$St_{H, av} Pr^{2/3} = \frac{h_{av}}{C_p \bar{u}_x \rho} Pr^{2/3} = j_H = \frac{1}{2}f = \psi(Re) \tag{3.56}$$

The mass-transfer (Chilton-Colburn) analogy [10] is

$$St_{D, av} Sc^{2/3} = \frac{F_{av}}{c\bar{u}_x} Sc^{2/3} = j_D = \frac{1}{2}f = \psi(Re) \tag{3.57}$$

which, as will be shown, agrees well with experimental data when Sc is not too large.

Analogies between the transport phenomena can be developed from the velocity, concentration, and temperature distributions from which the eddy diffusivities of Eqs. (3.34) to (3.38) were obtained. These data are not extremely accurate near the wall, however, and the resulting analogies then depend upon the empirical expressions chosen to describe the data, as in the equations mentioned. One of the relatively simple, yet quite successful, of these is the pair [22]

$$St_{H,\,av} = \frac{f/2}{1.20 + 11.8(f/2)^{0.5}(Pr - 1)Pr^{-1/3}}$$ (3.58)

$$St_{D,\,av} = \frac{f/2}{1.20 + 11.8(f/2)^{5}(Sc - 1)Sc^{-1/3}}$$ (3.59)

which are useful for Pr = 0.5 to 600 and Sc up to 3000.

We have seen, however, that the ratios E_D/D and E_H/α are not equal to E_v/ν, and particularly for large Sc one would not anticipate an analogy between heat and mass transfer on the one hand and momentum transfer on the other. The experimental data bear this out.

Experimental data have come from so-called wetted-wall towers (Fig. 3.11) and flow of liquids through soluble pipes. In Fig. 3.11, a volatile pure liquid is permitted to flow down the inside surface of a circular pipe while a gas is blown upward or downward through the central core. Measurement of the rate of evaporation of the liquid into the gas stream over the known surface permits calculation of the mass-transfer coefficients for the gas phase. Use of different gases and liquids provides variation of Sc. In this way, Sherwood and Gilliland

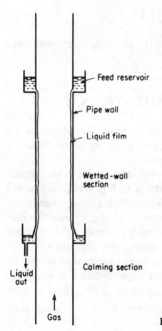

Figure 3.11 Wetted-wall tower.

[52] covered values of Re from 2000 to 35 000, Sc from 0.6 to 2.5, and gas pressures from 0.1 to 3 atm. Linton and Sherwood [41] caused water to flow through a series of pipes made by casting molten benzoic acid and other sparingly soluble solids. In this way the range of Sc was extended to 3000. Those data are empirically correlated by

$$Sh_{av} = 0.023 \ Re^{0.83} \ Sc^{1/3} \quad (3.60)$$

although the data for gases only are better correlated with 0.44 replacing $\frac{1}{3}$ as the exponent on Sc, perhaps because of the influence of ripples in the liquid film affecting the gas mass transfer [24]. Over the range Re = 5000 to 200 000, the friction factor in smooth pipes can be expressed as

$$\tfrac{1}{2}f = 0.023 \ Re^{-0.2} \quad (3.61)$$

Substitution in Eq. (3.60) then gives

$$\frac{Sh_{av}}{Re^{1.03} \ Sc} Sc^{2/3} = \tfrac{1}{2}f \quad (3.62)$$

which can be compared with Eq. (3.57). The heat-transfer analog of Eq. (3.60) is satisfactory for Pr = 0.7 to 120 [15].

As turbulence intensity increases, theory predicts that the exponent on Re should approach unity [38]. Thus [44], for liquids with Sc > 100,

$$Sh_{av} = 0.0149 \ Re^{0.88} \ Sc^{1/3} \quad (3.63)$$

and the heat- mass-transfer analogy applies [Eq. (3.63) yields values of Sh_{av} slightly low for Sc = 20 000 to 100 000].

MASS-TRANSFER DATA FOR SIMPLE SITUATIONS

Table 3.3 provides a few of the correlations taken from the literature for relatively simple situations. Others more appropriate to particular types of mass-transfer equipment will be introduced as needed.

Experimental data are usually obtained by blowing gases over various shapes wet with evaporating liquids or causing liquids to flow past solids which dissolve. Average, rather than local, mass-transfer coefficients are usually obtained. In most cases, the data are reported in terms of k_G, k_c, k_L, and the like, coefficients applicable to the binary systems used with $N_B = 0$, without details concerning the actual concentrations of solute during the experiments. Fortunately, the solute concentrations during the experiments are usually fairly low, so that if necessary, conversion of the data to the corresponding F is usually at least approximately possible by simply taking $\bar{p}_{B,M}/p_t$, $x_{B,M}$, etc., equal to unity (see Table 3.1).

Particularly when fluids flow past immersed objects, the local mass-transfer coefficient varies with position on the object, due especially to the separation of the boundary layer from the downstream surfaces to form a wake. This phenomenon has been studied in great detail for some shapes, e.g., cylinders [8]. The average mass-transfer coefficient in these cases can sometimes best be correlated by adding the contributions of the laminar boundary layer and the wake. This is true for the second entry of item 5, Table 3.3, for example, where these contributions correspond respectively to the two Reynolds-number terms.

For immersed objects the turbulence level of the incident fluid also has an important effect [8]. Thus, for example, a sphere falling through a quiet fluid will result in a mass-transfer coefficient different from that where the fluid flows past a stationary sphere. In most cases the distinction has not yet been fully established. Surface roughness, which usually increases the coefficient, has also been incompletely studied, and the data of Table 3.3 are for smooth surfaces, for the most part.

When mass or heat transfer occurs, it necessarily follows that the physical properties of the fluid vary along the transfer path. This problem has been dealt with theoretically (see, for example, Refs. 25 and 47), but in most cases empirical correlations use properties which are the average of those at the ends of the transfer path. The problem is most serious for natural convection, less so for forced.

In many cases, particularly when only gases have been studied, the range of Schmidt numbers covered is relatively small; sometimes only one Sc (0.6 for air-water vapor) is studied. The j_D of Eq. (3.57) has been so successful in dealing with turbulent flow in pipes that many data of small Sc range have been put in this form without actually establishing the validity of the $\frac{2}{3}$ exponent on Sc for the situation at hand. Extension to values of Sc far outside the original range of the experiments, particularly for liquids, must therefore be made with the understanding that a considerable approximation may be involved. Some data, as for flow past spheres, for example, cannot be put in the j_D form, and it is likely that many of the correlations using j_D may not be very general.

For situations not covered by the available data, it seems reasonable to assume that the mass-transfer coefficients can be estimated as functions of Re from the corresponding heat-transfer data, if available. If necessary, in the absence of a known effect of Pr, the heat-transfer data can be put in the j_H form and the analog completed by equating j_D to j_H at the same Re. This should serve reasonably well if the range of extrapolation of Sc (or Pr) is not too large.

Illustration 3.5 We wish to estimate the rate at which water will evaporate from a wetted surface of unusual shape when hydrogen at 1 std atm, 38°C, is blown over the surface at a superficial velocity of 15 m/s. No mass-transfer measurements have been made, but heat-transfer measurements indicate that for air at 38°C, 1 std atm, the heat-transfer coefficient h between air and the surface is given empirically by

$$h = 2.3 G'^{0.6}$$

where G' is the superficial air mass velocity, 21.3 kg/m^2 · s. Estimate the required mass-transfer

Table 3.3 Mass transfer† for simple situations

Fluid motion	Range of conditions	Equation	Ref.
1. Inside circular pipes	Re = 4000–60 000 Sc = 0.6–3000	$j_D = 0.023\,Re^{-0.17}$ $Sh = 0.023\,Re^{0.83}\,Sc^{1/3}$	41, 52
	Re = 10 000 – 400 000 Sc > 100	$j_D = 0.0149\,Re^{-0.12}$ $Sh = 0.0149\,Re^{0.88}\,Sc^{1/3}$	44
2. Unconfined flow parallel to flat plates‡	Transfer begins at leading edge $Re_x < 50\,000$	$j_D = 0.664\,Re_x^{-0.5}$	32
	$Re_x = 5 \times 10^5 – 3 \times 10^7$ Pr = 0.7–380	$Nu = 0.037\,Re_x^{0.8}\,Pr_0^{0.43}\left(\dfrac{Pr_0}{Pr_i}\right)^{0.25}$	65
	$Re_x = 2 \times 10^4 – 5 \times 10^5$ Pr = 0.7–380	Between above and $Nu = 0.0027\,Re_x\,Pr_0^{0.43}\left(\dfrac{Pr_0}{Pr_i}\right)^{0.25}$	
3. Confined gas flow parallel to a flat plate in a duct	$Re_e = 2600–22\,000$	$j_D = 0.11\,Re_e^{-0.29}$	46
4. Liquid film in wetted-wall tower, transfer between liquid and gas	$\dfrac{4\Gamma}{\mu} = 0$–1200, ripples suppressed	Eqs. (3.18)–(3.22)	20, 37
	$\dfrac{4\Gamma}{\mu} = 1300$–8300	$Sh = (1.76 \times 10^{-5})\left(\dfrac{4\Gamma}{\mu}\right)^{1.506} Sc^{0.5}$	

5.	Perpendicular to single cylinders	$Re = 400\text{–}25\,000$ $Sc = 0.6\text{–}2.6$	$\dfrac{k_G P_t}{G_M} Sc^{0.56} = 0.281\, Re^{-0.4}$	5
		$Re' = 0.1\text{–}10^5$ $Pr = 0.7\text{–}1500$	$Nu = (0.35 + 0.34\, Re'^{0.5} + 0.15\, Re'^{0.58})\, Pr^{0.3}$	16, 21, 42
6.	Past single spheres	$Sc = 0.6\text{–}3200$	$Sh = Sh_0 + 0.347(Re''\, Sc^{0.5})^{0.62}$	
		$Re''\, Sc^{0.5} = 1.8\text{–}600\,000$	$Sh_0 = \begin{cases} 2.0 + 0.569(Gr_D\, Sc)^{0.250} & Gr_D\, Sc < 10^8 \\ 2.0 + 0.0254(Gr_D\, Sc)^{0.333}\, Sc^{0.244} & Gr_D\, Sc > 10^8 \end{cases}$	55
7.	Through fixed beds of pellets§	$Re'' = 90\text{–}4000$ $Sc = 0.6$	$j_D = j_H = \dfrac{2.06}{\epsilon}\, Re''^{-0.575}$	
		$Re'' = 5000\text{–}10\,300$ $Sc = 0.6$	$j_D = 0.95 j_H = \dfrac{20.4}{\epsilon}\, Re''^{-0.815}$	4, 23, 64
		$Re'' = 0.0016\text{–}55$ $Sc = 168\text{–}70\,600$	$j_D = \dfrac{1.09}{\epsilon}\, Re''^{-2/3}$	
		$Re'' = 5\text{–}1500$ $Sc = 168\text{–}70\,600$	$j_D = \dfrac{0.250}{\epsilon}\, Re''^{-0.31}$	

† Average mass-transfer coefficients throughout, for constant solute concentrations at the phase surface. Generally, fluid properties are evaluated at the average conditions between the phase surface and the bulk fluid. The heat-mass-transfer analogy is valid throughout.

‡ Mass-transfer data for this case scatter badly but are reasonably well represented by setting $j_D = j_H$.

§ For fixed beds, the relation between ϵ and d_p is $a = 6(1 - \epsilon)/d_p$, where a is the specific solid surface, surface per volume of bed. For mixed sizes [58]

$$d_p = \frac{\displaystyle\sum_{i=1}^{n} n_i d_{pi}^3}{\displaystyle\sum_{i=1}^{n} n_i d_{pi}^2}$$

coefficient. Physical-property data are:

	38°C, 1 std atm	
	Air	H_2
Density ρ, kg/m^3	0.114	0.0794
Viscosity μ, kg/m · s	1.85×10^{-5}	9×10^{-6}
Thermal conductivity k		
W/m · K	0.0273	0.1850
Heat capacity C_p, kJ/kg · K	1.002	14.4
Diffusivity with water		
vapor, m^2/s		7.75×10^{-5}

SOLUTION The experimental heat-transfer data do not include effects of changing Prandtl number. It will therefore be necessary to assume that the j_H group will satisfactorily describe the effect.

$$j_H = \frac{h}{C_p \rho u} \text{Pr}^{2/3} = \frac{h}{C_p G'} \text{Pr}^{2/3} = \psi(\text{Re})$$

$$h = \frac{C_p G'}{\text{Pr}^{2/3}} \psi(\text{Re}) = 21.3 G'^{0.6} \quad \text{for air}$$

The $\psi(\text{Re})$ must be compatible with $21.3 G'^{0.6}$. Therefore, let $\psi(\text{Re}) = b \, \text{Re}^n$ and define Re as lG'/μ, where l is a characteristic linear dimension of the solid body.

$$h = \frac{C_p G'}{\text{Pr}^{2/3}} b \, \text{Re}^n = \frac{C_p G'}{\text{Pr}^{2/3}} b \left(\frac{lG'}{\mu} \right)^n = \frac{bC_p}{\text{Pr}^{2/3}} \left(\frac{l}{\mu} \right)^{G'^{1+n}} = 21.3 G'^{0.6}$$

$$1 + n = 0.6 \qquad n = -0.4$$

$$\frac{bC_p}{\text{Pr}^{2/3}} \left(\frac{l}{\mu} \right)^{-0.4} = 21.3 \quad \text{and}$$

$$b = 21.3 \frac{\text{Pr}^{2/3}}{C_p} \left(\frac{l}{\mu} \right)^{0.4}$$

Using the data for air at 38°C, 1 std atm, gives

$$\text{Pr} = \frac{C_p \mu}{k} = \frac{1002(1.85 \times 10^{-5})}{0.0273} = 0.68$$

$$b = \frac{21.3(0.68)^{2/3}}{1002} \left(\frac{1}{1.85 \times 10^{-5}} \right)^{0.4} = 1.285 l^{0.4}$$

$$j_H = \frac{h}{C_p G'} \text{Pr}^{2/3} = \frac{1.285 l^{0.4}}{\text{Re}^{0.4}} = \psi(\text{Re})$$

The heat- mass-transfer analogy will be used to estimate the mass-transfer coefficient ($j_D = j_H$).

$$j_D = \frac{k_G \bar{p}_{\text{B}, M} M_{\text{av}}}{\rho \mu} \text{Sc}^{2/3} = \psi(\text{Re}) = \frac{1.285 l^{0.4}}{\text{Re}^{0.4}}$$

$$k_G \bar{p}_{\text{B}, M} = F = \frac{1.285 l^{0.4} \rho u}{\text{Re}^{0.4} M_{\text{av}} \text{Sc}^{2/3}} = \frac{1.285 (\rho u)^{0.6} \mu^{0.4}}{M_{\text{av}} \text{Sc}^{2/3}}$$

For H_2-H_2O, 38°C, 1 std atm, Sc $= \mu/\rho D = (9 \times 10^{-6})/0.0794(7.75 \times 10^{-5}) = 1.465$. At $u = 15$ m/s, and assuming the density, molecular weight, and viscosity of the gas are essentially those of H_2,

$$k_G \bar{p}_{B, M} = F = \frac{1.285[0.0794(15)]^{0.6}(9 \times 10^{-6})^{0.4}}{2.02(1.465)^{2/3}} = 0.00527 \text{ kmol/m}^2 \cdot \text{s} \quad \textbf{Ans.}$$

Flux Variation with Concentration

In many situations the concentrations of solute in the bulk fluid, and even at the fluid interface, may vary in the direction of flow. Further, the mass-transfer coefficients depend upon fluid properties and rate of flow, and if these vary in the direction of flow, the coefficients will also. The flux N_A of Eqs. (3.1) and (3.3) to (3.6) is therefore a local flux and will generally vary with distance in the direction of flow. This problem was dealt with, in part, in the development leading to Illustration 3.1. In solving problems where something other than the local flux is required, allowance must be made for these variations. This normally requires some considerations of material balances, but there is no standard procedure. An example is offered below, but it must be emphasized that generally some sort of improvisation for the circumstances at hand will be required.

Illustration 3.6 Nickel carbonyl is to be produced by passing carbon monoxide gas downward through a bed of nickel spheres, 12.5 mm diam. The bed is 0.1 m^2 in cross section and is packed so that there are 30% voids. Pure CO enters at 50°C, 1 std atm, at the rate of 2×10^{-3} kmol/s. The reaction is

$$Ni + 4CO \rightarrow Ni(CO)_4$$

For present purposes, the following simplifying assumptions will be made:

1. The reaction is very rapid, so that the partial pressure of CO at the metal surface is essentially zero. The carbonyl forms as a gas, which diffuses as fast as it forms from the metal surface to the bulk-gas stream. The rate of reaction is controlled entirely by the rate of mass transfer of CO from the bulk gas to the metal surface and that of the Ni(CO)$_4$ to the bulk gas.
2. The temperature remains at 50°C and the pressure at 1 std atm throughout.
3. The viscosity of the gas = 2.4×10^{-5} kg/m · s, and the Schmidt number = 2.0 throughout.
4. The size of the nickel spheres remains constant.

Estimate the bed depth required to reduce the CO content of the gas to 0.5%.

SOLUTION Let A = CO, B = Ni(CO)$_4$. $N_B = -N_A/4$ and $N_A/(N_A + N_B) = \frac{4}{3}$. In Eq. (3.1), $c_{A2}/c = y_{A,i}$ at the metal interface = 0, and $c_{A1}/c = y_A$ = mole fraction CO in the bulk gas. Equation (3.1) therefore becomes

$$N_A = \frac{4}{3} F \ln \frac{4/3}{4/3 - y_A} = \frac{4}{3} F \ln \frac{4}{4 - 3y_A} \quad (3.64)$$

where N_A is the mass-transfer flux based on the metal surface. Since y_A varies with position in the bed, Eq. (3.64) gives a local flux, and to determine the total mass transfer, allowance must be made for the variation of y_A and F with bed depth.

Let G = kmol gas/(m^2 bed cross section) · s flowing at any depth z from the top. The CO content of this gas is $y_A G$ kmol CO/(m^2 bed) · s. The change in CO content in passing through a depth dz of bed is $-d(y_A G)$ kmol/m$^2 \cdot$ s. If a is the specific metal surface, m^2/m^3 bed, a depth dz and a cross section of 1 m^2 (volume $= dz$ m^3) has a metal surface $a\, dz$ m^2. Therefore, $N_A = -d(y_A G)/a\, dz$ kmol/(m^2 metal surface) $\cdot$ s. For each kmol of CO consumed, $\frac{1}{4}$ kmol Ni(CO)$_4$ forms, representing a net loss of $\frac{3}{4}$ kmol per kilomole of CO consumed. The CO consumed through bed depth dz is therefore $(G_0 - G)(\frac{4}{3})$ kmol, where G_0 is the molar superficial mass velocity at the top, and the CO content at depth z is $G_0 - (G_0 - G)(\frac{4}{3})$. Therefore

$$y_A = \frac{G_0 - (G_0 - G)(4/3)}{G} \qquad G = \frac{G_0}{4 - 3y_A} \qquad d(y_A G) = \frac{4 G_0\, dy_A}{(4 - 3y_A)^2}$$

Substituting in Eq. (3.64) gives

$$-\frac{4 G_0\, dy_A}{(4 - 3y_A)^2 a\, dz} = \tfrac{4}{3} F \ln \frac{4}{4 - 3y_A} \tag{3.65}$$

F is given by item 7 of Table 3.3 and is dependent upon y_A. At depth z, the CO mass velocity $= [G_0 - (G_0 - G)(\frac{4}{3})]28.0$, and the Ni(CO)$_4$ mass velocity $= [(G_0 - G)(\frac{1}{3})]170.7$, making a total mass velocity $G' = 47.6 G_0 - 19.6 G$ kg/m$^2 \cdot$ s. Substituting for G leads to

$$G' = 47.6 G_0 - \frac{19.6 G_0}{4 - 3y_A} = G_0 \left(47.6 - \frac{19.6}{4 - 3y_A}\right) \text{ kg/m}^2 \cdot \text{s}$$

With 12.5-mm spheres, $d_p = 0.0125$ m; $\mu = 2.4 \times 10^{-5}$ kg/m $\cdot$ s.

$$\text{Re}'' = \frac{d_p G'}{\mu} = \frac{0.0125 G_0 [47.6 - 19.6/(4 - 3y_A)]}{2.4 \times 10^{-5}} = G_0 \left(2.48 - \frac{1.02}{4 - 3y_A}\right)(10^4)$$

With $G_0 = (2 \times 10^{-3})/0.1 = 0.02$ kmol/m$^2 \cdot$ s and y_A in the range 1 to 0.005, the range of Re is 292 to 444. Therefore, Table 3.3,

$$j_D = \frac{F}{G} \text{Sc}^{2/3} = \frac{2.06}{\varepsilon} \text{Re}''^{-0.575}$$

For Sc $= 2$ and $\varepsilon = 0.3$ void fraction, this becomes

$$F = \frac{2.06}{0.3(2)^{2/3}} \frac{G_0}{4 - 3y_A} \left[G_0 \left(2.48 - \frac{1.02}{4 - 3y_A}\right)(10^4) \right]^{-0.575} \tag{3.66}$$

Equation (3.66) can be substituted in Eq. (3.65). Since $a = 6(1 - \varepsilon)/d_p = 6(1 - 0.3)/0.0125 = 336$ m^2/m^3 and $G_0 = 0.02$ kmol/m$^2 \cdot$ s, the result, after consolidation and rearrangement for integration, is

$$Z = \int_0^Z dz = -0.0434 \int_{1.0}^{0.005} \frac{[2.48 - 1.02/(4 - 3y_A)]^{0.575}}{(4 - 3y_A)\ln[4/(4 - 3y_A)]} dy_A$$

The integration is most readily done numerically, or graphically by plotting the integrand as ordinate vs. y_A as abscissa and determining the area under the resulting curve between the indicated limits. As a result, $Z = 0.154$ m. **Ans.**

SIMULTANEOUS MASS AND HEAT TRANSFER

Mass transfer may occur simultaneously with the transfer of heat, either as a result of an externally imposed temperature difference or because of the absorption or evolution of heat which generally occurs when a substance is transferred from one phase to another. In such cases, within one phase, the heat transferred is a result not only of the conduction (convection) by virtue of the temperature

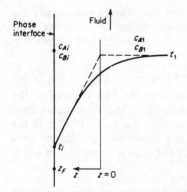

Figure 3.12 Effect of mass transfer on heat transfer [12].

difference which would happen in the absence of mass transfer but also includes the sensible heat carried by the diffusing matter.

Consider the situation shown in Fig. 3.12. Here a fluid consisting of substances A and B flows past a second phase under conditions causing mass transfer. The total mass transferred is given by the following variation of Eq. (3.1):

$$N_A + N_B = F \ln \frac{N_A/(N_A + N_B) - c_{A,i}/c}{N_A/(N_A + N_B) - c_{A1}/c} \qquad (3.67)$$

As usual, the relationship between N_A and N_B is fixed by other considerations. As a result of the temperature difference, there is a heat flux described by the ordinary heat-transfer coefficient h in the absence of mass transfer. If we think in terms of the film theory, this heat flux is $h(-dt/dz)z_F$. The total sensible heat flux q_s to the interface must include in addition the sensible heat brought there by the movement of matter through the temperature difference. Thus,

$$q_s = h\left(-\frac{dt}{dz}\right)z_F + (N_A M_A C_{p,A} + N_B M_B C_{p,B})(t - t_i) \qquad (3.68)$$

Rearranging and integrating gives

$$\int_{t_i}^{t_1} \frac{dt}{q_s - (N_A M_A C_{p,A} + N_B M_B C_{p,B})(t - t_i)} = \frac{1}{hz_F} \int_{z_F}^{0} dz \qquad (3.69)$$

$$q_s = \frac{N_A M_A C_{p,A} + N_B M_B C_{p,B}}{1 - e^{-(N_A M_A C_{p,A} + N_B M_B C_{p,B})/h}}(t_1 - t_i) \qquad (3.70)$$

The term multiplying the temperature difference of Eq. (3.70) may be considered as a heat-transfer coefficient corrected for mass transfer [1, 12]; it will be larger if the mass transfer is in the same direction as the heat transfer, smaller if the two are in opposite directions. Equation (3.70) is corrected for high mass-transfer flux. It can be applied for condensation of component A in the presence of noncondensing B ($N_B = 0$) or for multicomponent mixtures [51], for which $\sum N_i M_i C_{pi}$ is then used in the corrected coefficient.

The *total* heat release at the interface q_t will then include additionally the effect produced when the transferred mass passes through the interface. This may be a latent heat of vaporization, a heat of solution, or both, depending upon the circumstances. Thus,

$$q_t = q_s + \lambda_A N_A + \lambda_B N_B \tag{3.71}$$

where λ is the molar heat evolution. In some cases the heat released at the interface continues to flow to the left in Fig. 3.12, owing to a temperature drop in the adjacent phase. In others, where the mass transfer in the fluid is in the direction opposite the sensible-heat transfer, it is possible that the diffusing mass may carry heat from the interface into the fluid as fast as it is released, in which case no heat enters the adjacent phase.

Illustration 3.7 An air–water-vapor mixture flows upward through a vertical copper tube, 25.4 mm OD, 1.65 mm wall thickness, which is surrounded by flowing cold water. As a result, the water vapor condenses and flows as a liquid down the inside of the tube. At one level in the apparatus, the average velocity of the gas is 4.6 m/s, its bulk-average temperature 66°C, the pressure 1 std atm, and the bulk-average partial pressure of water vapor = 0.24 std atm. The film of condensed liquid is such that its heat-transfer coefficient = 11 400 W/m² · K. The cooling water has a bulk-average temperature of 24°C and a heat-transfer coefficient = 570 W/m² · K (see Fig. 3.13). Compute the local rate of condensation of water from the airstream.

SOLUTION For the metal tube, ID = 25.4 − 1.65(2) = 22.1 mm = 0.0221 m. Av diam = (0.0254 + 0.0221)/2 = 0.0238 m.

For the gas mixture, A = water, B = air. $N_B = 0$, $N_A/(N_A + N_B) = 1$; $y_{A1} = 0.24$, $y_{A, i} = p_{A, i}$ = vapor pressure of water at the interface temperature t_i. $M_{av} = 0.24(18.02) + 0.76(29) = 26.4$; $\rho = (26.4/22.41)[273/(273 + 66)] = 0.950$ kg/m³; $\mu = 1.75 \times 10^{-5}$ kg/m · s. $C_{p, A} = 1880$, C_p of the mixture = 1145 J/kg · K. Sc = 0.6, Pr = 0.75.

$$G' = \text{mass velocity} = u\rho = 4.6(0.950) = 4.37 \text{ kg/m}^2 \cdot \text{s}$$

$$G = \text{molar mass velocity} = \frac{G'}{M_{av}} = \frac{4.37}{26.4} = 0.1652$$

$$\text{Re} = \frac{dG'}{\mu} = \frac{0.0221(4.37)}{1.75 \times 10^{-5}} = 5510$$

The mass-transfer coefficient is given by item 1, Table 3.3:

$$j_D = \text{St}_D \text{ Sc}^{2/3} = \frac{F}{G} \text{Sc}^{2/3} = 0.023 \text{ Re}^{-0.17} = 0.023(5510)^{-0.17} = 0.00532$$

$$F = \frac{0.00532 G}{\text{Sc}^{2/3}} = \frac{0.00532(0.1652)}{(0.60)^{2/3}} = 1.24 \times 10^{-3} \text{ kmol/m}^2 \cdot \text{s}$$

The heat-transfer coefficient in the absence of mass transfer will be estimated through $j_D = j_H$.

$$j_H = \text{St}_H \text{ Pr}^{2/3} = \frac{h}{C_p G'} \text{Pr}^{2/3} = 0.00532$$

$$h = \frac{0.00532(1145)(4.37)}{(0.75)^{2/3}} = 32.3 \text{ W/m}^2 \cdot \text{K}$$

The sensible-heat transfer flux to the interface is given by Eq. (3.70), with $N_B = 0$. This is combined with Eq. (3.71) to produce

$$q_t = \frac{N_A(18.02)(1880)}{1 - e^{-N_A(18.02)(1880)/32.3}} (66 - t_i) + \lambda_{A, i} N_A \tag{3.72}$$

where $\lambda_{A, i}$ is the molal latent heat of vaporization of water at t_i. All the heat arriving at the interface is carried to the cold water. U is the overall heat-transfer coefficient, interface to cold

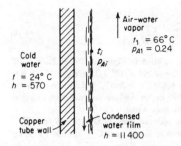

Figure 3.13 Illustration 3.7.

water, based on the inside tube surface

$$\frac{1}{U} = \frac{1}{11\,400} + \frac{0.00165}{381}\frac{22.1}{23.8} + \frac{1}{570}\frac{22.1}{25.4}$$

where 381 is the thermal conductivity of the copper.

$$U = 618 \text{ W/m}^2 \cdot \text{K} \qquad q_t = 618(t_i - 24) \tag{3.73}$$

The rate of mass transfer is given by Eq. (3.67) with $N_B = 0$, $c_{A,i}/c = p_{A,i}$, $c_{A1}/c = y_{A1} = 0.24$.

$$N_A = 1.24 \times 10^{-3} \ln\frac{1 - p_{A,i}}{1 - 0.24} \tag{3.74}$$

Equations (3.72) to (3.74) are solved simultaneously by trial with the vapor-pressure curve for water, which relates $p_{A,i}$ and t_i, and the latent-heat data. It is easiest to assume t_i, which is checked when q_t from Eqs. (3.72) and (3.73) agree.

As a final trial, assume $t_i = 42.2°C$, whence $p_{A,i} = 0.0806$ std atm, $\lambda_{A,i} = 43.4 \times 10^6$ J/kmol, and [Eq. (3.74)] $N_A = 2.45 \times 10^{-4}$ kmol/m$^2 \cdot$ s. By Eq. (3.73) $q_t = 11\,240$; by Eq. (3.72) $q_t = 11\,470$ W/m^2, a sufficiently close check. The local rate of condensation of water is therefore 2.45×10^{-4} kmol/m$^2 \cdot$ s. **Ans.**

Note In this case, the true q_s [Eq. (3.68)] $= 875$ W/m^2, whereas the uncorrected h gives $32.3(66 - 42.2) = 770$ W/m^2.

Equation (3.70) can also be used for calculations of *transpiration cooling,* a method of cooling porous surfaces which are exposed to very hot gases by forcing a cold gas or evaporating liquid through the surface into the gas stream.

Illustration 3.8 Air at 600°C, 1 std atm, flows past a flat, porous surface. Saturated steam at 1.2 std atm flows through the surface into the airstream in order to keep the temperature at the surface at 260°C. The air velocity is such that h would be 1100 W/m$^2 \cdot$ K if no steam were used. (a) What rate of steam flow is required? (b) Repeat if instead of steam liquid water at 25°C is forced through the surface.

SOLUTION (a) A = water, B = air, $N_B = 0$. The positive direction is taken to be from the bulk gas to the surface. Let $H_{A,i}$ = enthalpy of steam at the interface and $H_{A,S}$ = initial enthalpy of steam. Equation (3.70) becomes†

$$q_s = N_A M_A (H_{A,S} - H_{A,i}) = \frac{N_A M_A C_{p,A}(t_1 - t_i)}{1 - e^{-N_A M_A C_{p,A}/h}}$$

† Radiation contributions to the heat transfer from the gas to the surface are negligible; it is assumed that the emissivity of the air and its small water-vapor content is negligible.

which reduces to

$$N_A = \frac{-h}{M_A C_{p,A}} \ln\left[1 - \frac{C_{p,A}(t_1 - t_i)}{H_{A,S} - H_{A,i}}\right] \qquad (3.75)$$

$H_{A,S}$ = enthalpy of saturated steam at 1.2 std atm, relative to liquid at 0°C = 2.684×10^6 J/kg. $H_{A,i}$ = enthalpy of steam at 1 std atm, 260°C = 2.994×10^6 J/kg. M_A = 18.02, h = 1100, $C_{p,A}$ from 260 to 600°C = 2090 J/kg · K, t_1 = 600°C, t_i = 260°C. Substitution into Eq. (3.75) provides N_A = – 0.0348 kmol/m² · s or –0.627 kg/m² · s (the negative sign indicates that the mass flux is in the negative direction, into the gas). **Ans.**

The convective heat-transfer coefficient has been reduced from 1100 to 572 W/m² · K.

(b) The enthalpy of liquid water at 25°C relative to liquid at 0°C = 1.047×10^5 J/kg. Substitution of this instead of 2.684×10^6 into Eq. (3.75) yields N_A = – 6.42×10^{-3} kmol/m² · s or –0.1157 kg/m² · s. **Ans.**

NOTATION FOR CHAPTER 3

Any consistent set of units may be used.

a	specific surface of a fixed bed of pellets, pellet surface/volume of bed, L^2/L^3
A	mass-transfer surface, L^2
b_1, b_2	const
c	solute concentration (if subscripted), molar density of a solution (if not subscripted), **mole/L^3**
$\bar{c}$	bulk-average concentration, **mole/L^3**
C_p	heat capacity at constant pressure, **FL/MT**
d	differential operator
	diameter, L
d_c	diameter of a cylinder, L
d_e	equivalent diameter of a noncircular duct = 4(cross-sectional area)/perimeter, L
d_p	diameter of a sphere; for a nonspherical particle, diameter of a sphere of the same surface as the particle, L
D	molecular diffusivity, L^2/Θ
E_D	eddy mass diffusivity, L^2/Θ
E_H	eddy thermal diffusivity, L^2/Θ
E_c	eddy momentum diffusivity, L^2/Θ
f	friction factor; Fanning friction factor for flow through pipes; dimensionless
F	mass-transfer coefficient, **mole/$L^2\Theta$**
g	acceleration due to gravity, L^2/Θ
g_c	conversion factor, **ML/FΘ^2**
G	molar mass velocity, **mole/$L^2\Theta$**
G'	mass velocity, **M/$L^2\Theta$**
Gr	Grashof number, dimensionless
h	heat-transfer coefficient, **FL/$L^2\Theta$T**
H	enthalpy, **FL/M**
j_D	mass-transfer dimensionless group, $St_D Sc^{2/3}$
j_H	heat-transfer dimensionless group, $St_H Pr^{2/3}$
J	mass-transfer flux relative to molar average velocity, **mole/$L^2\Theta$**
k	thermal conductivity, **FL^2/$L^2\Theta$T**
k_c, k_G, k_x, k_y, etc.	mass-transfer coefficients **mole/$L^2\Theta$(concentration difference)** (see Table 3.1)
K	const
l	length; Prandtl mixing length; L
l_d	length scale of eddies in the universal range, L

l_e	length scale of medium-sized eddies, L
L	length of a wetted-wall tower, L
m	mass, M
M	molecular weight, M/mole
n	a number, dimensionless
N	mass-transfer flux at, and relative to, a phase boundary, mole/$L^2\Theta$
Nu	Nusselt number, dimensionless
p	vapor pressure, F/L^2
$\bar{p}$	partial pressure, F/L^2
p_t	total pressure, F/L^2
P	power, FL/Θ
Pe	Peclet number, dimensionless
Pr	Prandtl number, dimensionless
q	heat-transfer flux, $FL/L^2\Theta$
R	universal gas constant, FL/mole T
Re	Reynolds number, dimensionless
Re'	Reynolds number for flow outside a cylinder, $d_c G'/\mu$, dimensionless
Re''	Reynolds number for flow past a sphere, $d_p G'/\mu$, dimensionless
Re_e	Reynolds number for flow in a noncircular duct, $d_e G'/\mu$, dimensionless
Re_x	Reynolds number computed with x as the length dimension, dimensionless
s	fractional rate of surface-element replacement, Θ^{-1}
S	cross-sectional area of a duct, L^2
Sc	Schmidt number, dimensionless
Sh	Sherwood number, dimensionless
St	Stanton number, dimensionless
t	temperature, T
u	local velocity, L/Θ
$\bar{u}$	bulk-average velocity, L/Θ
u'	root-mean-square deviating velocity, L/Θ
u_i'	instantaneous deviating velocity, L/Θ
u_d'	velocity of eddies in the universal range, L/Θ
x, y, z (no subscript)	distance in the $x, y,$ and z direction, respectively, L
x_A	concentration of component A in a liquid, mole fraction
y_A	concentration of component A in a gas, mole fraction
z	distance from the center of a pipe, L
z_b	depth of penetration (film and surface-renewal theories), L
z_F	effective film thickness (film theory), L
z_+	dimensionless distance from the wall $\left(\dfrac{d}{2} - z\right)\dfrac{u}{\nu}\left(\dfrac{f}{2}\right)^{0.5}$ for a pipe
α	thermal diffusivity $= k/C_p\rho$, L^2/θ
β	volumetric coefficient of expansion, T^{-1}
Γ	mass rate of flow/unit width, $M/L\Theta$
δ	thickness of a layer, L
θ	time, Θ
λ	molar heat evolution on passing through a surface, FL/mole
μ	viscosity, $M/L\theta$
ρ	density, M/L^3
τ	shear stress, F/L^2
φ	area of surface elements (surface-renewal theory), $L^2/L^2\Theta$
ψ_1, ψ_2	functions

Subscripts

A	component A
B	component B

c	concentration
D	for mass transfer
H	for heat transfer
i	interface; instantaneous when used with velocity
M	logarithmic mean
r	reference condition
s	sensible heat
t	total
turb	turbulent
x, y, z	in the x, y, z directions, respectively
0	approach, or initial, value
1	at beginning of mass-transfer path
2	at end of mass-transfer path

REFERENCES

1. Ackerman, G.: *Ver. Dtsch. Ing. Forschungs.*, **382**, 1 (1937).
2. Angelo, J. B., and E. N. Lightfoot: *AIChE J.*, **14**, 531 (1968).
3. Angelo, J. B., E. N. Lightfoot, and D. W. Howard: *AIChE J.*, **12**, 751 (1966).
4. Barker, J. J.: *Ind. Eng. Chem.*, **57**(4), 43; (5), 33 (1965).
5. Bedingfield, C. H., and I. B. Drew: *Ind. Eng. Chem.*, **42**, 1164 (1950).
6. Bennett, C. O., and J. E. Myers: "Momentum, Heat, and Mass Transfer," 2d ed., McGraw-Hill Book Company, New York, 1974.
7. Bird, R. B., W. E. Stewart, and E. N. Lightfoot: "Transport Phenomena," Wiley, New York, 1960.
8. Boulos, M. I., and D. C. T. Pei: *Can. J. Chem. Eng.*, **51**, 673 (1973).
9. Brinkworth, B. J., and P. C. Smith: *Chem. Eng. Sci.*, **24**, 787 (1969).
10. Chilton, T. H., and A. P. Colburn: *Ind. Eng. Chem.*, **26**, 1183 (1934).
11. Colburn, A. P.: *Trans. AIChE*, **29**, 174 (1933).
12. Colburn, A. P., and T. B. Drew: *Trans. AIChE*, **33**, 197 (1937).
13. Danckwerts, P. V.: *AIChE J*, **1**, 456 (1955); *Ind. Eng. Chem.*, **43**, 1460 (1951).
14. Danckwerts, P. V.: "Gas-Liquid Reactions," McGraw-Hill Book Company, New York, 1970.
15. Davies, J. T.: "Turbulence Phenomena," Academic, New York, 1972.
16. Davis, A. H.: *Phil. Mag.*, **47**, 1057 (1924).
17. Dobbins, W. E.: in J. McCabe and W. W. Eckenfelder (eds.), "Biological Treatment of Sewage and Industrial Wastes," pt. 2-1, Reinhold, New York, 1956.
18. Dobbins, W. E.: *Int. Conf. Water Pollution Res., London, 1962*, p. 61, Pergamon, New York, 1964.
19. Eckert, E. R. G., and R. M. Drake: "Heat and Mass Transfer," 2d ed., McGraw-Hill Book Company, New York, 1959.
20. Emmet, R. E., and R. L. Pigford: *Chem. Eng. Prog.*, **50**, 87 (1954).
21. Fand, R. M.: *Int. J. Heat Mass Transfer*, **8**, 995 (1965).
22. Friend, W. L., and A. B. Metzner: *AIChE J.*, **4**, 393 (1958).
23. Gupta, A. S., and G. Thodos: *AIChE J.*, **9**, 751 (1963); *Ind. Eng. Chem. Fundam.*, **3**, 218 (1964).
24. Hakki, T. A., and J. A. Lamb: *Trans. Inst. Chem. Eng. Lond.*, **50**, 115 (1972).
25. Hanna, O. T.: *AIChE J.*, **8**, 278 (1962).
26. Harriott, P.: *Chem. Eng. Sci.*, **17**, 149 (1962).
27. Higbie, R.: *Trans. AIChE*, **31**, 365 (1935).
28. Hinze, J. O.: "Turbulence," McGraw-Hill Book Company, New York, 1959.
29. Howard, D. W., and E. N. Lightfoot: *AIChE J.*, **14**, 458 (1968).
30. Hughmark, G. A.: *AIChE J.*, **17**, 902 (1971).
31. Johnstone, H. F., and R. L. Pigford: *Trans. AIChE*, **38**, 25 (1952).

32. Kays, W. M.: "Convective Heat and Mass Transfer," McGraw-Hill Book Company, New York, 1966.
33. King, C. J.: *Ind. Eng. Chem. Fundam.*, 5, 1 (1966).
34. Kolmogoroff, A. N.: *C. R. Acad. Sci. U.R.R.S.*, 30, 301 (1941); 32, 16 (1941).
35. Koppel, L. B., R. D. Patel, and J. T. Holmes: *AIChE J.*, 12, 941 (1966).
36. Kozinski, A. A., and C. J. King: *AIChE J.*, 12, 109 (1966).
37. Lamourelle, A. P., and O. C. Sandall: *Chem. Eng. Sci.*, 27, 1035 (1972).
38. Levich, V. G.: "Physicochemical Hydrodynamics," Prentice-Hall, Englewood Cliffs, N.J., 1962.
39. Lewis, W. K., and W. Whitman: *Ind. Eng. Chem.*, 16, 1215 (1924).
40. Lin, C. S., R. W. Moulton, and G. L. Putnam: *Ind. Eng. Chem.*, 45, 636 (1953).
41. Linton, W. H., and T. K. Sherwood: *Chem. Eng. Prog.*, 46, 258 (1950).
42. McAdams, W. H.: "Heat Transmission," 3d ed., McGraw-Hill Book Company, New York, 1954.
43. Murphree, E. V.: *Ind. Eng. Chem.*, 24, 726 (1932).
44. Notter, R. H., and C. A. Sleicher: *Chem. Eng. Sci.*, 26, 161 (1971).
45. Noyes, A. A., and W. R. Whitney: *J. Am. Chem. Soc.*, 19, 930 (1897).
46. O'Brien, L. J., and L. F. Stutzman: *Ind. Eng. Chem.*, 42, 1181 (1950).
47. Olander, D. R.: *Int. J. Heat Mass Transfer*, 5, 765 (1962).
48. Page, F., W. G. Schlinger, D. K. Breaux, and B. H. Sage: *Ind. Eng. Chem.*, 44, 424 (1952).
49. Reynolds, O.: "Scientific Papers of Osborne Reynolds," vol. II, Cambridge University Press, New York, 1901.
50. Rose, P. M., and R. C. Kintner: *AIChE J.*, 12, 530 (1966).
51. Schrodt, J. T.: *AIChE J.*, 19, 753 (1973).
52. Sherwood, T. K., and E. R. Gilliland: *Ind. Eng. Chem.*, 26, 516 (1934).
53. Sherwood, T. K., and R. L. Pigford: "Absorption and Extraction," 2d ed., McGraw-Hill Book Company, New York, 1952.
54. Sherwood, T. K., R. L. Pigford, and C. R. Wilke: "Mass Transfer," McGraw-Hill Book Company, New York, 1975.
55. Steinberger, R. L., and R. E. Treybal: *AIChE J.*, 6, 227 (1960).
56. Stewart, W. E., J. B. Angelo, and E. N. Lightfoot: *AIChE J.*, 16, 771 (1970).
57. Stynes, S. K., and J. E. Myers: *AIChE J.*, 10, 437 (1964).
58. Tan, A. Y., B. D. Prushen, and J. A. Guin: *AIChE J.*, 21, 396 (1975).
59. Taylor, G. I.: *Rep. Mem. Br. Advis. Comm. Aeronaut.* 272, 423 (1916).
60. Toor, R. L., and J. M. Marchello: *AIChE J.*, 4, 97 (1958).
61. Towle, W. L., and T. K. Sherwood: *Ind. Eng. Chem.*, 31, 457 (1939).
62. Trinite, M., and P. Valentin: *Int. J. Heat Mass Transfer*, 15, 1337 (1972).
63. Von Behren, G. L., W. O. Jones, and D. T. Wasan: *AIChE J.*, 18, 25 (1972).
64. Wilson, E. J., and C. J. Geankoplis: *Ind. Eng. Chem. Fundam.*, 5, 9 (1966).
65. Zhukauskas, A. A., and A. B. Ambrazyavichyas: *Int. J. Heat Mass Transfer*, 3, 305 (1961).

PROBLEMS

3.1 Estimate the mass-transfer coefficient and effective film thickness to be expected in the absorption of ammonia from air by a 2 N sulfuric acid solution in a wetted-wall tower under the following circumstances:

Airflow = 41.4 g/min (air only)
Av partial pressure ammonia in air = 30.8 mmHg
Total pressure = 760 mmHg
Av gas temp = 25°C
Av liquid temp = 25°C
Tower diam = 1.46 cm

For absorption of ammonia in sulfuric acid of this concentration, the entire mass-transfer resistance lies within the gas, and the partial pressure of ammonia at the interface is negligible. *Note*: The circumstances correspond to run 47 of Chambers and Sherwood [*Trans. AIChE*, **33**, 579 (1937)], who observed $d/z_F = 16.6$.

3.2 Powell [*Trans. Inst. Chem. Eng. Lond.*, **13**, 175 (1935); **18**, 36 (1940)] evaporated water from the outside of cylinders into an airstream flowing parallel to the axes of the cylinders. The air temperature was 25°C, and the total pressure standard atmospheric. The results are given by

$$\frac{wl}{p_w - \bar{p}_A} = 3.17 \times 10^{-8}(ul)^{0.8}$$

where w = water evaporated, g/cm$^2 \cdot$ s

$\bar{p}_A$ = partial pressure of water in airstream, mmHg

p_w = water-vapor pressure at surface temperature, mmHg

u = velocity of airstream, cm/s

l = length of cylinder, cm

(a) Transform the equation into the form $j_D = \psi(\text{Re}_l)$, where Re_l is a Reynolds number based on the cylinder length.

(b) Calculate the rate of sublimation from a cylinder of napthalene 0.075 m diam by 0.60 m long (3 in diam by 24 in long) into a stream of pure carbon dioxide at a velocity of 6 m/s (20 ft/s) at 1 std atm, 100°C. The vapor pressure of naphthalene at the surface temperature may be taken as 1330 N/m^2 (10 mmHg) and the diffusivity in carbon dioxide as 5.15×10^{-6} m^2/s (0.0515 cm^2/s) at STP. Express the results as kilograms naphthalene evaporated per hour. **Ans.**: 0.386 kg/h.

3.3 Water flows down the inside wall of a wetted-wall tower of the design of Fig. 3.11, while air flows upward through the core. In a particular case, the ID is 25 mm (or 1 in), and dry air enters at the rate 7.0 kg/m$^2 \cdot$ s of inside cross section (or 5000 lb/ft$^2 \cdot$ h). Assume the air is everywhere at its average temperature, 36°C, the water at 21°C, and the mass-transfer coefficient constant. Pressure = 1 std atm. Compute the average partial pressure of water in the air leaving if the tower is 1 m (3 ft) long. **Ans.**: 1620 N/m^2.

3.4 Winding and Cheney [*Ind. Eng. Chem.*, **40**, 1087 (1948)] passed air through a bank of rods of naphthalene. The rods were of "streamline" cross section, arranged in staggered array, with the air flowing at right angles to the axes of the rods. The mass-transfer coefficient was determined by measuring the rate of sublimation of the naphthalene. For a particular shape, size, and spacing of the rods, with air at 37.8°C (100°F), 1 std atm, the data could be correlated by†

$$k_G = 3.58 \times 10^{-9}G'^{0.56}$$

where G' is the superficial mass velocity, kg/m$^2 \cdot$ s, and k_G the mass-transfer coefficient, kmol/m$^2 \cdot$ s $\cdot$ (N/m^2). Estimate the mass-transfer coefficient to be expected for evaporation of water into hydrogen gas for the same geometrical arrangement when the hydrogen flows at a superficial velocity of 15.25 m/s (50 ft/s), 37.8°C (100°F), 2 std atm pressure. D for naphthalene-air, 1 std atm, 37.8°C = 7.03×10^{-6} m^2/s; for water-hydrogen, 0°C, 1 std atm, 7.5×10^{-5} m^2/s.

3.5 A mixing tank 3 ft (0.915 m) in diameter contains water at 25°C to a depth of 3 ft (0.915 m). The liquid is agitated with a rotating impeller at an intensity of 15 hp/1000 gal (2940 W/m^3). Estimate the scale of the smallest eddies in the universal range.

3.6 The free-fall terminal velocity of water drops in air at 1 std atm pressure is given by [53]:

Diam, mm	0.05	0.2	0.5	1.0	2.0	3.0
Velocity, m/s	0.055	0.702	2.14	3.87	5.86	7.26
ft/s	0.18	2.3	7.0	12.7	19.2	23.8

† For G in lb/ft$^2 \cdot$ h and k_G in lb mol/ft$^2 \cdot$ h $\cdot$ atm the coefficient is 0.00663.

A water drop of initial diameter 1.0 mm falls in quiet dry air at 1 std atm, 38°C (100°F). The liquid temperature may be taken as 14.4°C (58°F). Assume that the drop remains spherical and that the atmospheric pressure remains constant at 1 std atm.

(a) Calculate the initial rate of evaporation.

(b) Calculate the time and distance of free fall for the drop to evaporate to a diameter of 0.2 mm. Ans.: 103.9 s, 264 m.

(c) Calculate the time for the above evaporation, assuming that the drop is suspended without motion (as from a fine thread) in still air. Ans.: 372 s.

3.7 The temperature variation of rate of a chemical reaction which involves mass transfer is sometimes used to determine whether the rate of mass transfer or that of the chemical reaction controls, or is the dominating mechanism. Consider a fluid flowing through a 25-mm-ID (1.0-in) circular tube, where the transferred solute is ammonia in dilute solution. Compute the mass-transfer coefficient for each of the following cases:

(a) The fluid is a dilute solution of ammonia in air, 25°C, 1 std atm, flowing at a Reynolds number = 10 000. $D_{AB} = 2.26 \times 10^{-5}$ m^2/s (0.226 cm^2/s).

(b) Same as part (a) (same mass velocity), but temperature = 35°C.

(c) The fluid is a dilute solution of ammonia in liquid water, 25°C, flowing at a Reynolds number = 10 000. $D_{AB} = 2.65 \times 10^{-9}$ m^2/s (2.65 × 10^{-5} cm^2/s).

(d) Same as part (c) (same mass velocity) but temperature = 35°C.

In the case of both gas and liquid, assuming that the mass-transfer coefficient follows an Arrhenius-type equation, compute the "energy of activation" of mass transfer. Are these high or low in comparison with the energy of activation of typical chemical reactions? Note that, for dilute solutions, the identity of the diffusing solute need not have been specified in order to obtain the "energy of activation" of mass transfer. What other method might be used to determine whether reaction rate or mass-transfer rate controls?

3.8 A gas consisting of 50% air and 50% steam, by volume, flows at 93.0°C (200°F), 1 std atm pressure, at 7.5 m/s (25 ft/s) average velocity through a horizontal square duct 0.3 m (1 ft) wide. A horizontal copper tube, 25.4 mm (1 in) OD, 1.65 mm (0.065 in) wall thickness, passes through the center of the duct from one side to the other at right angles to the duct axis, piercing the duct walls. Cold water flows inside the tube at an average velocity 3.0 m/s (10 ft/s), av temp = 15.6°C (60°F). Estimate the rate of condensation on the outside of the copper tube. Ans.: 6.92 × 10^{-4} kg/s.

Data: McAdams [42, p. 338] gives the heat-transfer coefficient for the condensate film on the tube as

$$h_{av} = 1.51 \left(\frac{k^3 \rho^2 g}{\mu^2} \right)^{1/3} \left(\frac{\mu L}{4W} \right)^{1/3}$$

where, in any consistent unit system, W = mass rate of condensate and L = tube length. The fluid properties are to be evaluated at the mean temperature of the condensate film. Data for the group $(k^3 \rho^2 g/\mu^2)^{1/3}$ for water, in English engineering units, are listed by McAdams [42, table A-27, p. 484] for convenience. They must be multiplied by 5.678 for use with SI units.

3.9 A hollow, porous cylinder, OD 25 mm (1 in), ID 15 mm, is fed internally with liquid diethyl ether at 20°C. The ether flows radially outward and evaporates on the outer surface. Nitrogen, initially ether-free, at 100°C, 1 std atm, flows at right angles to the cylinder at 3 m/s (10 ft/s), carrying away the evaporated ether. The ether flow is to be that which will just keep the outer surface of the cylinder wet with liquid (since mass-transfer rates vary about the periphery, the cylinder will be rotated slowly to keep the surface uniformly wet). Compute the surface temperature and the rate of ether flow, kg/s per meter of cylinder length.

Ans.: Flow rate = 9.64 × 10^{-4} kg/s · m.

Data: The heat capacity C_p for liquid ether = 2282, vapor = 1863 J/kg · K. Thermal conductivity of the vapor at 0°C = 0.0165, at 85°C = 0.0232 W/m · K. Latent heat of vaporization at −17.8°C = 397.8, at 4.4°C = 391.7, at 32.2°C = 379.2 kJ/kg. Other data can be obtained from "The Chemical Engineers' Handbook."

FOUR
DIFFUSION IN SOLIDS

It was indicated in Chap. 1 that certain of the diffusional operations such as leaching, drying, and adsorption and membrane operations such as dialysis, reverse osmosis, and the like involve contact of fluids with solids. In these cases, some of the diffusion occurs in the solid phase and may proceed according to any of several mechanisms. While in none of these is the mechanism as simple as in the diffusion through solutions of gases and liquids, nevertheless with some exceptions it is usually possible to describe the transfer of diffusing substance by means of the same basic law used for fluids, Fick's law.

FICK'S-LAW DIFFUSION

When the concentration gradient remains unchanged with passage of time, so that the rate of diffusion is constant, Fick's law can be applied in the form used in Chap. 2 when the diffusivity is independent of concentration and there is no bulk flow. Thus N_A, the rate of diffusion of substance A per unit cross section of solid, is proportional to the concentration gradient in the direction of diffusion, $- dc_A/dz$,

$$N_A = - D_A \frac{dc_A}{dz} \tag{4.1}$$

where D_A is the diffusivity of A through the solid. If D_A is constant, integration of Eq. (4.1) for *diffusion through a flat slab* of thickness z results in

$$N_A = \frac{D_A(c_{A1} - c_{A2})}{z} \tag{4.2}$$

which parallels the expressions obtained for fluids in a similar situation. Here

c_{A1} and c_{A2} are the concentrations at opposite sides of the slab. For other solid shapes, the rate is given by

$$w = N_A S_{av} = \frac{D_A S_{av}(c_{A1} - c_{A2})}{z} \tag{4.3}$$

with appropriate values of the average cross section for diffusion S_{av} to be applied. Thus, for *radial diffusion through a solid cylinder* of inner and outer radii a_1 and a_2, respectively, and of length l,

$$S_{av} = \frac{2\pi l(a_2 - a_1)}{\ln (a_2/a_1)} \tag{4.4}$$

and

$$z = a_2 - a_1 \tag{4.5}$$

For *radial diffusion through a spherical shell* of inner and outer radii a_1 and a_2,

$$S_{av} = 4\pi a_1 a_2 \tag{4.6}$$

$$z = a_2 - a_1 \tag{4.7}$$

Illustration 4.1 Hydrogen gas at 2 std atm, 25°C, flows through a pipe made of unvulcanized neoprene rubber, with ID and OD 25 and 50 mm, respectively. The solubility of the hydrogen is reported to be 0.053 cm^3 (STP)/cm^3 · atm, and the diffusivity of hydrogen through the rubber to be 1.8×10^{-6} cm^2/s. Estimate the rate of loss of hydrogen by diffusion per meter of pipe length.

SOLUTION At 2 std atm hydrogen pressure the solubility is 0.053(2) = 0.106 m^3 H$_2$ (STP)/(m^3 rubber). Therefore concentration c_{A1} at the inner surface of the pipe = 0.106/22.41 = 4.73 $\times$ 10^{-3} kmol H$_2$/m^3. At the outer surface, $c_{A2} = 0$, assuming the resistance to diffusion of H$_2$ away from the surface is negligible.

$$D_A = 1.8 \times 10^{-10} \text{ m}^2/\text{s} \qquad l = 1 \text{ m}$$

$$z = a_2 - a_1 = \frac{50 - 25}{2(1000)} = 0.0125 \text{ m}$$

Eq. (4.4):

$$S_{av} = \frac{2\pi(1)(50 - 25)}{2(1000) \ln (50/25)} = 0.1133 \text{ m}^2$$

Eq. (4.3):

$$w = \frac{(1.8 \times 10^{-10})(0.1133)[(4.73 \times 10^{-3}) - 0]}{0.0125}$$

$$= 7.72 \times 10^{-12} \text{ kmol H}_2/\text{s for 1 m length}$$

which corresponds to 5.6 $\times$ 10^{-5} g H$_2$/m · h. **Ans.**

Unsteady-State Diffusion

Since solids are not so readily transported through equipment as fluids, the application of batch and semibatch processes and consequently unsteady-state diffusional conditions arise much more frequently than with fluids. Even in continuous operation, e.g., a continuous drier, the history of each solid piece as it passes through equipment is representative of the unsteady state. These cases are therefore of considerable importance.

Where there is no bulk flow, and in the absence of chemical reaction, Fick's second law, Eq. (2.18), can be used to solve problems of unsteady-state diffusion by integration with appropriate boundary conditions. For some simple cases, Newman [10] has summarized the results most conveniently.

1. *Diffusion from a slab with sealed edges.* Consider a slab of thickness $2a$, with sealed edges on four sides, so that diffusion can take place only toward and from the flat parallel faces, a cross section of which is shown in Fig. 4.1. Suppose initially the concentration of solute throughout the slab is uniform, c_{A0}, and that the slab is immersed in a medium so that the solute will diffuse out of the slab. Let the concentration at the surfaces be $c_{A, \infty}$, invariant with passage of time. If the diffusion were allowed to continue indefinitely, the concentration would fall to the uniform value $c_{A, \infty}$, and $c_{A0} - c_{A, \infty}$ is a measure of the amount of solute removed. On the other hand, if diffusion from the slab were stopped at time θ, the distribution of solute would be given by the curve marked c, which by internal diffusion would level off to the uniform concentration $c_{A, \theta}$, where $c_{A, \theta}$ is the average concentration at time θ. The quantity $c_{A, \theta} - c_{A, \infty}$ is a measure of the amount of solute still unremoved. The fraction unremoved E is given by integration of Eq. (2.18),

$$E = \frac{c_{A, \theta} - c_{A, \infty}}{c_{A, \theta} - c_{A, \infty}} = f\left(\frac{D\theta}{a^2}\right)$$

$$= \frac{8}{\pi^2}\left(e^{-D\theta\pi^2/4a^2} + \frac{1}{9}e^{-9D\theta\pi^2/4a^2} + \frac{1}{25}e^{-25D\theta\pi^2/4a^2} + \cdots\right) = E_a \quad (4.8)$$

The function is shown graphically in Fig. 4.2.

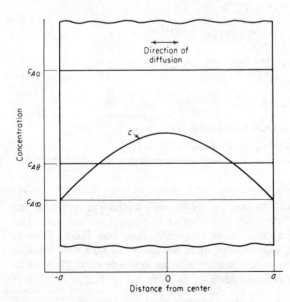

Figure 4.1 Unsteady-state diffusion in a slab.

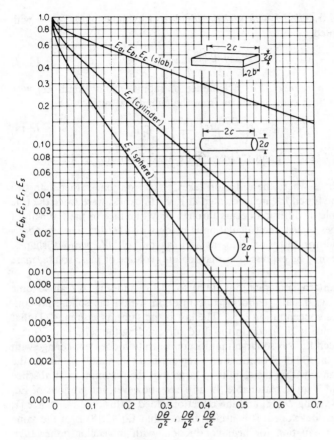

Figure 4.2 Unsteady-state diffusion.

2. *Diffusion from a rectangular bar with sealed ends.* For a rectangular bar of thickness $2a$ and width $2b$, with sealed ends,

$$E = f\left(\frac{D\theta}{a^2}\right) f\left(\frac{D\theta}{b^2}\right) = E_a E_b \qquad (4.9)$$

3. *Diffusion for a rectangular parallelepiped.* For a brick-shaped bar, of dimensions $2a$, $2b$, and $2c$, with diffusion from all six faces,

$$E = f\left(\frac{D\theta}{a^2}\right) f\left(\frac{d\theta}{b^2}\right) f\left(\frac{d\theta}{c^2}\right) = E_a E_b E_c \qquad (4.10)$$

4. *Diffusion from a sphere.* For a sphere of radius a,

$$E = f'\left(\frac{D\theta}{a^2}\right) = E_s \qquad (4.11)$$

5. *Diffusion from a cylinder with sealed ends.* For a cylinder of radius a, with plane ends sealed,

$$E = f''\left(\frac{D\theta}{a^2}\right) = E_r \qquad (4.12)$$

6. *Diffusion from a cylinder.* For a cylinder of radius a and length $2c$, with diffusion from both ends as well as from the cylindrical surface,

$$E = f\left(\frac{D\theta}{c^2}\right) f''\left(\frac{D\theta}{a^2}\right) = E_c E_r \qquad (4.13)$$

The functions $f'(D\theta/a^2)$ and $f''(D\theta/a^2)$ are also shown in Fig. 4.2.

For solid shapes where the diffusion takes place from one rather than two opposite faces, the functions are calculated as if the thickness were twice the true value. For example, if diffusion occurs through only one face of the flat slab of thickness $2a$, edges sealed, the calculation is made with $D\theta/4a^2$. The equations can also be used for diffusion into, as well as out of, the various shapes. Concentrations c may be expressed as mass solute/volume or mass solute/mass solid.

It is important to note that Eqs. (4.8) to (4.13) and Fig. 4.2 all assume constant diffusivity, initially uniform concentration within the solid, and constancy of the edge concentration $c_{A,\infty}$. The last is the same as assuming (1) that there is no resistance to diffusion in the fluid surrounding the solid and (2) that the quantity of such fluid is so large that its concentration does not change with time or (3) that the fluid is continuously replenished. In many instances, the diffusional resistance within the solid is so large that the assumption of absence of any in the fluid is quite reasonable. In any case, however, integrations of Eq. (2.18) involving varying D [18] and the effect of added diffusional resistance [1, 3, 8, 10] have been developed. It is useful to note that Eq. (2.18) is of the same form as Fourier's equation for heat conduction, with molecular rather than thermal diffusivity and concentration rather than temperature. Consequently the lengthy catalog of solutions to the problems of heat transfer of Carslaw and Jaeger [2] can be made applicable to diffusion by appropriate substitutions. Crank's book [3] deals particularly with problems of this sort for diffusion.

Illustration 4.2 A 5% agar gel containing a uniform urea concentration of 5 g/100 cm^3 is molded in the form of a 3-cm cube. One face of the cube is exposed to a running supply of fresh water, into which the urea diffuses. The other faces are protected by the mold. The temperature is 5°C. At the end of 68 h, the average urea concentration in the gel has fallen to 3 g/100 cm^3. The resistance to diffusion may be considered as residing wholly within the gel. (*a*) Calculate the diffusivity of the urea in the gel. (*b*) How long would it have taken for the average concentration to fall to 1 g/100 cm^3? (*c*) Repeat part (*b*) for the case where two opposite faces of the cube are exposed.

SOLUTION (*a*) c_A can be calculated in terms of g/100 cm^3. $c_{A0} = 5$ g/100 cm^3; $c_{A,\theta} = 3$, $c_{A,\infty} = 0$ since pure water was the leaching agent. $a = \frac{3}{2} = 1.5$ cm; $\theta = 68(3600) = 245\,000$ s.

$$\frac{c_{A,\theta} - c_{A,\infty}}{c_{A0} - c_{A,\infty}} = \frac{3}{5} = 0.6 = E$$

The abscissa read from Fig. 4.2, which is $D\theta/4a^2$ for diffusion from only one exposed face, is 0.128

$$D = \frac{0.128(4)(a^2)}{\theta} = \frac{0.128(4)(1.5)^2}{245\,000} = 4.70 \times 10^{-6}\,cm^2/s = 4.70 \times 10^{-10}\,m^2/s$$

(b) For $c_{A,\theta} = 1$ g/100 cm^3

$$\frac{c_{A,\theta} - c_{A,\infty}}{c_{A0} - c_{A,\infty}} = \frac{1}{5} = 0.20 \qquad \frac{D\theta}{4a^2} = 0.568 \qquad \text{from Fig. 4.2}$$

$$\theta = \frac{0.568(4)(a^2)}{D} = \frac{0.568(4)(1.5)^2}{4.70 \times 10^{-6}} = 1\,087\,000\,s = 302\,h$$

(c) For two opposite faces exposed, $a = 1.5$ cm, $c_{A,\theta}/c_{A0} = 0.2$, and $D\theta/a^2 = 0.568$.

$$\theta = \frac{0.568a^2}{D} = \frac{0.568(1.5)^2}{4.70 \times 10^{-6}} = 222\,000\,s = 61.5\,h$$

TYPES OF SOLID DIFFUSION

The structure of the solid and its interaction with the diffusing substance have a profound influence on how diffusion occurs and on the rate of transport.

Diffusion through Polymers

In many respects, diffusion of solutes through certain types of polymeric solids is more like diffusion through liquid solutions than any of the other solid-diffusion phenomena, at least for the permanent gases as solutes. Imagine two bodies of a gas (e.g., H_2) at different pressures separated by a polymeric membrane (e.g., polyethylene). The gas dissolves in the solid at the faces exposed to the gas to an extent usually describable by Henry's law, concentration directly proportional to pressure. The gas then diffuses from the high- to the low-pressure side in a manner usually described as *activated*: the polymeric chains are in a state of constant thermal motion, and the diffusing molecules jump from one position to another over a potential barrier [4, 11]. A successful jump requires that a hole or passage of sufficient size be available, and this in turn depends on the thermal motion of the polymer chains. The term "activated" refers to the temperature dependence of the diffusivity, which follows an Arrhenius-type expression,

$$D_A = D_0 e^{-H_D/RT} \tag{4.14}$$

where H_D is the energy of activation and D_0 is a constant. For simple gases, D_A is usually reasonably independent of concentration. It may, however, be a strong function of pressure of molding the polymer [5]. For the permanent gases diffusivities may be of the order of 10^{-10} m^2/s.

Particularly for large molecules, the size and shape of the diffusing molecules (as measured by molecular volume, branched as opposed to straight-chain structure, etc.) determine the hole size required. Solvents sometimes diffuse by *plasticizing* the polymers†, and consequently the rate may be very much higher

---

† Plasticizers are organic compounds added to high polymers which solvate the polymer molecule. This results in easier processing and a product of increased flexibility and toughness.

for good solvents, e.g., benzene or methyl ethyl ketone in polymeric rubbers, than for the permanent gases. Such diffusivities are frequently strongly dependent upon solute concentration in the solid. With certain oxygenated polymers, e.g., cellulose acetate, such solutes as water, ammonia, and alcohols form hydrogen bonds with the polymer and move from one set of bonding sites to another. Solutes which cannot form hydrogen bonds are then excluded.

Unfortunately (because it necessarily adds complicated dimensions and awkward units) it has become the practice to describe the diffusional characteristics in terms of a quantity P, the permeability. Since at the two faces of a membrane the equilibrium solubility of the gas in the polymer is directly proportional to the pressure, Eq. (4.2) can be converted into†

$$V_A = \frac{D_A s_A (\bar{p}_{A1} - \bar{p}_{A2})}{z}$$ (4.15)

where V_A = diffusional flux, cm³ gas (STP)/cm² · s
D_A = diffusivity of A, cm²/s
$\bar{p}_A$ = partial pressure of diffusing gas, cmHg
s_A = solubility coefficient or Henry's law constant, cm³ gas (STP)/ (cm³ solid) · cmHg
z = thickness of polymeric membrane, cm

Permeability is then defined as†

$$P = D_A s_A$$ (4.16)

where P = permeability, cm³ gas (STP)/cm² · s · (cmHg/cm).

Commercial application of these principles has been made for separating hydrogen from waste refinery gases in shell-and-tube devices which resemble in part the common heat exchanger. However, in this use the polymeric-fiber tubes are only 30 μm OD, and there are 50 million of them in a shell roughly 0.4 m in diameter [6].

Illustration 4.3 Calculate the rate of diffusion of carbon dioxide, CO_2, through a membrane of vulcanized rubber 1 mm thick at 25°C if the partial pressure of the CO_2 is 1 cmHg on one side and zero on the other. Calculate also the permeability of the membrane for CO_2. At 25°C, the solubility coefficient is 0.90 cm³ gas (STP)/cm³ atm. The diffusivity is 1.1×10^{-10} m²/s.

SOLUTION Since 1 std atm = 76.0 cmHg pressure, the solubility coefficient in terms of cmHg is $0.90/76.0 = 0.01184$ cm³ gas (STP)/cm³ · cmHg = s_A. $z = 0.1$ cm, $\bar{p}_{A1} = 1$ cmHg, $\bar{p}_{A2} = 0$, $D_A = (1.1 \times 10^{-10})(10^4) = 1.1 \times 10^{-6}$ cm²/s. Eq. (4.15):

$$V_A = \frac{(1.1 \times 10^{-6})(0.01184)(1.0 - 0)}{0.1}$$

$$= 0.13 \times 10^{-6} \text{ cm}^3 \text{ (STP)/cm}^2 \cdot \text{s} \quad \textbf{Ans.}$$

† Equations (4.15) and (4.16) are actually dimensionally consistent and may be used with any set of consistent units. Thus if SI units are used [$D = $ m²/s, $s = $ m³/m³ · (N/m²), $\bar{p} = $ N/m², $z = $ m, $V = $ m³/m² · s], the units of P will be m³/m² · s · (N/m²)/m. Similarly in English engineering units [$D = $ ft²/h, $s = $ ft³/ft³ · (lb_f/ft²), $\bar{p} = $ lb_f/ft², $z = $ ft, $V = $ ft³/ft² · s] the units of P will be ft³/ft² · s · [(lb_f/ft²)/ft]. But nowhere do consistent units of any well-known system ever seem to be used, unfortunately.

Eq. (4.16):

$$P = (1.1 \times 10^{-6})(0.01184) = 0.13 \times 10^{-7} \text{ cm}^3 \text{ (STP)}/\text{cm}^2 \cdot \text{s} \cdot \text{(cmHg/cm)}$$

Diffusion through Crystalline Solids

The mechanisms of diffusion vary greatly depending upon the crystalline structure and the nature of the solute [16, 17]. For crystals with lattices of cubic symmetry, the diffusivity is isotropic, but not so for noncubic crystals. Particularly, but not necessarily, in metals the principal mechanisms are the following:

1. *Interstitial mechanism.* Interstitial sites are places between the atoms of a crystal lattice. Small diffusing solute atoms may pass from an interstitial site to the next when the matrix atoms of the crystal lattice move apart temporarily to provide the necessary space. Carbon diffuses through α- and γ-iron in this manner.
2. *Vacancy mechanism.* If lattice sites are unoccupied (vacancies), an atom in an adjacent site may jump into such a vacancy.
3. *Interstitialcy mechanism.* In this case a large atom occupying an interstitial site pushes one of its lattice neighbors into an interstitial position and moves into the vacancy thus produced.
4. *Crowd-ion mechanism.* An extra atom in a chain of close-packed atoms can displace several atoms in the line from their equilibrium position, thus producing a diffusion flux.
5. *Diffusion along grain boundaries.* The diffusivity in a single-crystal metal is always substantially smaller than that for a multicrystalline sample because the latter has diffusion along the grain boundaries (crystal interfaces) and dislocations.

Diffusion in Porous Solids

The solid may be in the form of a porous barrier or membrane separating two bodies of fluid, as in the case of gaseous diffusion. Here the solute movement may be by *diffusion* from one fluid body to the other by virtue of a concentration gradient, or it may be *hydrodynamic* as a result of a pressure difference. Alternatively, in the case of adsorbents, catalyst pellets [13], solids to be dried, ore particles to be leached, and the like, the solid is normally completely surrounded by a single body of fluid, and inward and outward movement of solute through the pores of the solid is solely by diffusion. Diffusive movement may be within the fluid filling the pores or may also involve *surface* diffusion of adsorbed solute.

The pores of the solid may be *interconnected* (accessible to fluid from both ends of the pores), *dead-end* (connected to the outside of the solid only from one end), or *isolated* (inaccessible to external fluid). The pores in most solids are neither straight nor of constant diameter. Catalyst particles manufactured by pressing powders containing *micropores* into pellets, with *macropores* surrounding the powder particles of a different order of magnitude in size, are said to be *bidisperse* [19].

Diffusion, constant total pressure Imagine the pores of a solid to be all of the same length and diameter, filled with a binary solution at constant pressure, with a concentration gradient of the components over the length of the pores. Then within the solution filling the pores, Eq. (2.26) for gases or Eq. (2.40) for liquids would apply. The fluxes N_A and N_B would be based upon the cross-sectional area of the pores, and the distance would be the length of the pores l. The ratio of fluxes $N_A/(N_A + N_B)$ would, as usual, depend upon nondiffusional matters, fixed perhaps because in leaching of a solute A the solvent B would be nondiffusing, or fixed by the stoichiometry of a catalyzed reaction on the surface of the pores. Since the length of the various pores and their cross-sectional area are not constant, it is more practical to base the flux upon the gross external surface of the membrane, pellet, or whatever the nature of the solid, and the length upon some arbitrary but readily measured distance z such as membrane thickness, pellet radius, and the like. Since the fluxes will be smaller than the true values based on true fluid cross section and length, an *effective* diffusivity $D_{AB, \text{eff}}$, smaller than the true D_{AB}, must be used, ordinarily determined by experiment [13]. For the same type of processes in a given solid, presumably the ratio $D_{AB}/D_{AB, \text{eff}}$ will be constant, and once measured it can be applied to all solutes. When $N_B = 0$, Eqs. (4.1) to (4.13) would be expected to apply with $D_{AB, \text{eff}}$ replacing D_A.

Illustration 4.4 Porous alumina spheres, 10 mm diameter, 25% voids, were thoroughly impregnated with an aqueous potassium chloride, KCl, solution, concentration 0.25 g/cm³. When immersed in pure running water, the spheres lost 90% of their salt content in 4.75 h. The temperature was 25°C. At this temperature the average diffusivity of KCl in water over the indicated concentration range is 1.84×10^{-9} m²/s.

Estimate the time for removal of 90% of the dissolved solute if the spheres had been impregnated with potassium chromate, K_2CrO_4, solution at a concentration 0.28 g/cm³, when immersed in a running stream of water containing 0.02 g K_2CrO_4/cm³. The average diffusivity of K_2CrO_4 in water at 25°C is 1.14×10^{-9} m²/s.

SOLUTION For these spheres, $a = 0.005$ m, and for the KCl diffusion, $\theta = 4.75(3600) = 17\,000$ s. When the spheres are surrounded by pure water, the ultimate concentration in the spheres $c_{A, \infty} = 0$.

$$\frac{c_{A, \theta} - c_{A, \infty}}{c_{A0} - c_{A, \infty}} = 0.1 \quad \text{for 90\% removal of KCl}$$

From Fig. 4.2, $D_{\text{eff}}\, \theta/a^2 = 0.18$, where D_{eff} is the effective diffusivity.

$$D_{\text{eff}} = \frac{0.18a^2}{\theta} = \frac{0.18(0.005)^2}{17\,000} = 2.65 \times 10^{-10} \text{ m}^2/\text{s}$$

$$D_{AB}/D_{\text{eff}} = \frac{1.84 \times 10^{-9}}{2.65 \times 10^{-10}} = 6.943$$

For the K_2CrO_4 diffusion, $c_{A0} = 0.28$, $c_{A, \infty} = 0.02$, and $c_{A, \theta} = 0.1(0.28) = 0.028$ g/cm³.

$$E = \frac{c_{A, \theta} - c_{A, \infty}}{c_{A0} - c_{A, \infty}} = \frac{0.028 - 0.02}{0.28 - 0.02} = 0.0308 = E_s$$

Fig. 4.2:

$$D_{eff}\frac{\theta}{a^2} = 0.30$$

$$D_{eff} = \frac{D_{AB}}{6.943} = \frac{1.14 \times 10^{-9}}{6.943} = 1.642 \times 10^{-10} m^2/s$$

$$\theta = \frac{0.30a^2}{D_{eff}} = \frac{0.30(0.005)^2}{1.642 \times 10^{-10}} = 45\,700\,s = 12.7\,h \quad \textbf{Ans.}$$

In steady-state diffusion of gases, there are two types of diffusive movement, depending on the ratio of pore diameter d to the mean free path of the gas molecules λ. If the ratio d/λ is greater than approximately 20, ordinary molecular diffusion predominates and

$$N_A = \frac{N_A}{N_A + N_B} \frac{D_{AB, eff} p_t}{RTz} \ln \frac{N_A/(N_A + N_B) - y_{A2}}{N_A/(N_A + N_B) - y_{A1}} \quad (2.26)$$

$D_{AB, eff}$, like D_{AB}, varies inversely as p_t and approximately directly as $T^{3/2}$ (see Chap. 2). If, however, the pore diameter and the gas pressure are such that the molecular mean free path is relatively large, d/λ less than about 0.2, the rate of diffusion is governed by the collisions of the gas molecules with the pore walls and follows Knudsen's law. Since molecular collisions are unimportant under these conditions, each gas diffuses independently. In a straight circular pore of diameter d and length l

$$N_A = \frac{d\bar{u}_A}{3RTl}(\bar{p}_{A1} - \bar{p}_{A2}) \quad (4.17)$$

where u_A is the mean molecular velocity of A. Since the kinetic theory of gases provides

$$u_A = \left(\frac{8g_c RT}{\pi M_A}\right)^{1/2} \quad (4.18)$$

we have $N_A = \left(\frac{8g_c RT}{\pi M_A}\right)^{1/2}\frac{d}{3RTl}(\bar{p}_{A1} - \bar{p}_{A2}) = \frac{D_{K,A}(\bar{p}_{A1} - \bar{p}_{A2})}{RTl} \quad (4.19)$

where $D_{K,A}$ is the Knudsen diffusion coefficient

$$D_{K,A} = \frac{d}{3}\left(\frac{8g_c RT}{\pi M_A}\right)^{1/2} \quad (4.20)$$

Since normally d is not constant and the true l is unknown, l in Eq. (4.19) is ordinarily replaced by z, the membrane thickness, and $D_{K,A}$ by $D_{K,A,eff}$, the effective Knudsen diffusivity, which is determined by experiment. $D_{K,A,eff}$ is independent of pressure and varies as $(T/M)^{1/2}$. For binary gas mixtures,

$$\frac{N_B}{N_A} = -\left(\frac{M_A}{M_B}\right)^{1/2} \quad (4.21)$$

In addition, for a given solid [15], $D_{AB, eff}/D_{K, A, eff} = D_{AB}/D_{K, A}$. The mean free path λ can be estimated from the relation

$$\lambda = \frac{3.2\mu}{p_t}\left(\frac{RT}{2\pi g_c M}\right)^{0.5} \tag{4.22}$$

In the range d/λ from roughly 0.2 to 20, a transition range, both molecular and Knudsen diffusion have influence, and the flux is given by [7, 12, 15]†

$$N_A = \frac{N_A}{N_A + N_B}\frac{D_{AB, eff}\, p_t}{RTz}\ln\frac{\dfrac{N_A}{N_A + N_B}\left(1 + \dfrac{D_{AB, eff}}{D_{K, A, eff}}\right) - y_{A2}}{\dfrac{N_A}{N_A + N_B}\left(1 + \dfrac{D_{AB, eff}}{D_{K, A, eff}}\right) - y_{A1}} \tag{4.23}$$

It has been shown [15] that Eq. (4.23) reverts to Eq. (2.26) for conditions under which molecular diffusion prevails ($D_{K, A, eff} \gg D_{AB, eff}$) and to Eq. (4.19) when Knudsen diffusion prevails ($D_{AB, eff} \gg D_{K, A, eff}$). Further, for open-ended pores, Eq. (4.21) applies throughout the transition range for solids whose pore diameters are of the order of 10 μm or less.

Illustration 4.5 The effective diffusivities for passage of hydrogen and nitrogen at 20°C through a 2-mm-thick piece of unglazed porcelain were measured by determining the countercurrent diffusion fluxes at 1.0 and 0.01 std atm pressure [15]. Equation (4.23), solved simultaneously for the two measurements, provided the diffusivities, $D_{H_2-N_2, eff} = 5.3 \times 10^{-6}$ m²/s at 1.0 std atm and $D_{K, H_2, eff} = 1.17 \times 10^{-5}$ m²/s. Estimate (a) the equivalent pore diameter of the solid and (b) the diffusion fluxes for O_2-N_2 mixtures at a total pressure of 0.1 std atm, 20°C, with mole fractions of $O_2 = 0.8$ and 0.2 on either side of the porcelain.

SOLUTION (a) $D_{H_2-N_2}$ at 20°C, 1 std atm $= 7.63 \times 10^{-5}$ m²/s. Therefore $D_{true}/D_{eff} = (7.63 \times 10^{-5})/(5.3 \times 10^{-6}) = 14.4$. Since this ratio is strictly a matter of the geometry of the solid, it should apply to all gas mixtures at any condition. Therefore $D_{K, H_2} = (1.17 \times 10^{-5})(14.4) = 1.684 \times 10^{-4}$ m²/s. From Eq. (4.20)

$$d = 3D_{K, H_2}\left(\frac{\pi M_{H_2}}{8g_c RT}\right)^{0.5} = 3(1.684 \times 10^{-4})\left[\frac{\pi(2.02)}{8(1)(8314)(293)}\right]^{0.5}$$

$$= 2.88 \times 10^{-7}\,m = 0.288\,\mu m \quad \textbf{Ans.}$$

(b) Let A = O_2, B = N_2. Table 2.1: D_{AB}, STP $= 1.81 \times 10^{-5}$ m²/s. Therefore at 0.1 std atm, 20°C,

$$D_{AB} = 1.81 \times 10^{-5}\frac{1}{0.1}\left(\frac{293}{273}\right)^{1.5} = 2.01 \times 10^{-4}\,m^2/s$$

$$D_{AB, eff} = \frac{2.01 \times 10^{-4}}{14.4} = 1.396 \times 10^{-5}\,m^2/s$$

Eq. (4.20):

$$D_{K, A} = D_{K, H_2}\left(\frac{M_{H_2}}{M_A}\right)^{0.5} = 1.684 \times 10^{-4}\left(\frac{2.02}{32}\right)^{0.5} = 4.23 \times 10^{-5}\,m^2/s$$

$$D_{K, A, eff} = \frac{4.23 \times 10^{-5}}{14.4} = 2.94 \times 10^{-6}\,m^2/s$$

† Equation (4.23) is known as the "dusty gas" equation because a porous medium consisting of a random grouping of spheres in space was used as a model for its derivation.

Since the equivalent pore diameter is less than 10 μm, Eq. (4.21) will apply.

$$\frac{N_B}{N_A} = - \left(\frac{M_A}{M_B} \right)^{0.5} = - \left(\frac{32}{28.02} \right)^{0.5} = -1.069$$

$$\frac{N_A}{N_A + N_B} = \frac{1}{1 + N_B/N_A} = \frac{1}{1 - 1.069} = -14.49$$

$y_{A1} = 0.8$, $y_{A2} = 0.2$, $p_t = 10\,133$ N/m^2, $z = 0.002$ m, $D_{AB,\,eff}/D_{K,\,A,\,eff} = (1.396 \times 10^{-5})/(2.94 \times 10^{-6}) = 4.75$. Eq. (4.23):

$$N_{O_2} = \frac{-14.49(1.396 \times 10^{-5})(10\,133)}{8314(293)(0.002)} \ln \frac{-14.49(1 + 4.75) - 0.2}{-14.49(1 + 4.75) - 0.8}$$

$$= 3.01 \times 10^{-6} \text{ kmol/m}^2 \cdot \text{s}$$

$$N_{N_2} = (3.01 \times 10^{-6})(-1.069) = -3.22 \times 10^{-6} \text{ kmol/m}^2 \cdot \text{s} \quad \textbf{Ans.}$$

Knudsen diffusion is not known for liquids, but important reductions in diffusion rates occur when the molecular size of the diffusing solute becomes significant relative to the pore size of the solid [14].

Surface diffusion is a phenomenon accompanying adsorption of solutes onto the surface of the pores of the solid. It is an activated diffusion [see Eq. (4.14)], involving the jumping of adsorbed molecules from one adsorption site to another. It can be described by a two-dimensional analog of Fick's law, with surface concentration expressed, for example, as mol/area instead of mol/volume. Surface diffusivities are typically of the order of 10^{-7} to 10^{-9} m^2/s at ordinary temperatures for physically adsorbed gases [13] (see Chap. 11). For liquid solutions in adsorbent resin particles, surface diffusivities may be of the order of 10^{-12} m^2/s [9].

Hydrodynamic flow of gases If there is a difference in absolute pressure across a porous solid, a hydrodynamic flow of gas through the solid will occur. Consider a solid consisting of uniform straight capillary tubes of diameter d and length l reaching from the high- to low-pressure side. At ordinary pressures, the flow of gas in the capillaries may be either laminar or turbulent, depending upon whether the Reynolds number $d\,u\rho/\mu$ is below or above 2100. For present purposes, where velocities are small, flow will be laminar. For a single gas, this can be described by Poiseuille's law for a compressible fluid obeying the ideal gas law

$$N_A = \frac{d^2 g_c}{32 \mu l R T} p_{t,\,av}(p_{t1} - p_{t2}) \tag{4.24}$$

where

$$p_{t,\,av} = \frac{p_{t1} - p_{t2}}{2} \tag{4.25}$$

This assumes that the entire pressure difference is the result of friction in the pores and ignores entrance and exit losses and kinetic-energy effects, which is satisfactory for present purposes. Since the pores are neither straight nor of constant diameter, just as with diffusive flow it is best to base N_A on the gross external cross section of the solid and write Eq. (4.24) as†

$$N_A = \frac{k}{RTz} p_{t,\,av}(p_{t1} - p_{t2}) \qquad (4.26)$$

If conditions of pore diameter and pressure occur for which Knudsen flow prevails ($d/\lambda < 0.2$), the flow will be described by Knudsen's law, Eqs. (4.17) to (4.20). There will be of course a range of conditions for a transition from hydrodynamic to Knudsen flow. If the gas is a mixture with different compositions and different total pressure on either side of the porous solid, the flow may be a combination of hydrodynamic, Knudsen, and diffusive. Younquist [20] reviews these problems.

Illustration 4.6 A porous carbon diaphragm 1 in (25.4 mm) thick of average pore diameter 0.01 cm permitted the flow of nitrogen at the rate of 9.0 ft³ (measured at 1 std atm, 80°F)/ft² · min (0.0457 m³/m² · s, 26.7°C) with a pressure difference across the diaphragm of 2 inH₂O (50.8 mmH₂O). The temperature was 80°F and the downstream pressure 0.1 std atm. Calculate the flow to be expected at 250°F (121°C) with the same pressure difference.

SOLUTION The viscosity of nitrogen = 1.8×10^{-5} kg/m · s at 26.7°C (300 K). At 0.1 std atm, $p_t = 10\,133$ N/m². $M_{N_2} = 28.02$. Eq. (4.22):

$$\lambda = \frac{3.2\mu}{p_t} \left(\frac{RT}{2\pi g_c M} \right)^{0.5} = \frac{3.2(1.8 \times 10^{-5})}{10\,133} \left[\frac{8314(300)}{2\pi(1)(28.02)} \right]^{0.5}$$

$$= 6.77 \times 10^{-7} \text{ m}$$

With $d = 10^{-4}$ m, $d/\lambda = 148$, Knudsen flow will not occur, and Poiseuille's law, Eq. (4.26), applies and will be used to calculate k. N₂ flow corresponding to 9.0 ft³/ft² · min at 300 K, 1 std atm = 0.0457 m³/m² · s or

$$0.0457 \frac{273}{300} \frac{1}{22.41} = 1.856 \times 10^{-3} \text{ kmol/m}^2 \cdot \text{s} = N_A$$

$$p_{t1} - p_{t2} = (2 \text{ inH}_2\text{O}) \frac{3386}{13.6} = 498 \text{ N/m}^2$$

$$p_{t,\,av} = 10\,133 + \frac{498}{2} = 10\,382 \text{ N/m}^2$$

Eq. (4.26):

$$k = \frac{N_A RTz}{p_{t,\,av}(p_{t1} - p_{t2})} = \frac{(1.856 \times 10^{-3})(8314)(300)(0.0254)}{10\,382(498)}$$

$$= 2.27 \times 10^{-5} \text{ m}^4/\text{N} \cdot \text{s}$$

At 250°F (121°C = 393 K), the viscosity of nitrogen = 2.2×10^{-5} kg/m · s. This results in a new k [Eq. (4.24)]:

$$k = \frac{(2.27 \times 10^{-5})(1.8 \times 10^{-5})}{2.2 \times 10^{-5}} = 1.857 \times 10^{-5} \text{ m}^4/\text{N} \cdot \text{s}$$

Eq. (4.26):

$$N_A = \frac{(1.857 \times 10^{-5})(10\,382)(498)}{8314(393)(0.0254)} = 1.157 \times 10^{-3} \text{ kmol/m}^2 \cdot \text{s}$$

$$= 7.35 \text{ ft}^3/\text{ft}^2 \cdot \text{min, measured at 121°C, 1 std atm} \quad \textbf{Ans.}$$

---

† Equation (4.26) is sometimes written in terms of volume rate V of gas flow at the average pressure per unit cross section of the solid and a permeability P

$$p_{t,\,av}V = \frac{Pp_{t,\,av}(p_{t1} - p_{t2})}{z}$$

NOTATION FOR CHAPTER 4

Any consistent set of units may be used, except as noted.

a	one-half thickness; radius, L
b	one-half width, L
c	concentration, mole/L^3
	one-half length, L
d	pore diameter, L
	differential operator
D	molecular diffusivity, L^2/Θ [in Eqs. (4.15) to (4.16), cm^2/s]
D_{eff}	effective diffusivity, L^2/Θ
D_K	Knudsen diffusivity, L^2/Θ
E	fraction of solute removal, dimensionless
f, f', f''	functions
g_c	conversion factor, ML/FΘ^2
H_D	energy of activation, FL/mole
k	const, L^4/FΘ
l	length, pore length, L
M	molecular weight, M/mole
N	molar flux, mole/L$^2\Theta$
p	partial pressure, F/L^2 [in Eq. (4.15), cmHg]
p_t	total pressure, F/L^2
P	permeability, L^3/L$^2\Theta$[(F/L^2)/L] [in Eq. (4.16), cm^3 gas (STP)/cm^2 · s · (cmHg/cm)]
R	universal gas constant, FL/mole T
s	solubility coefficient or Henry's-law constant, L^3/L^3(F/L^2) [in Eqs. (4.15) to (4.16), cm^3 gas (STP) cm^3 · (cmHg)]
S	cross-sectional area, L^2
T	absolute temperature, T
u	average velocity, L/Θ
$\bar{u}$	mean molecular velocity, L/Θ
V	flux of diffusion, L^3/L$^2\Theta$ [in Eq. (4.15), cm^3 gas (STP)/cm^2 · s]
w	rate of diffusion, mole/Θ
y	concentration, mole fraction
z	thickness of membrane or pellet, L
	distance in direction of diffusion, L
θ	time, Θ
λ	molecular mean free path, L
μ	viscosity, M/LΘ
π	3.1416
ρ	density, M/L^3

Subscripts

av	average
A, B	components A, B
θ	at time Θ
0	initial (at time zero)
1, 2	positions 1, 2
∞	at time ∞; at equilibrium

REFERENCES

1. Barrer, R. M.: "Diffusion in and through Solids," Cambridge University Press, London, 1941.
2. Carslaw, H. S., and J. C. Jaeger: "Conduction of Heat in Solids," 2d ed., Oxford University Press, Fair Lawn, N.J., 1959.

3. Crank, J.: "The Mathematics of Diffusion," Oxford University Press, Fair Lawn, N.J., 1956.
4. Crank, J., and G. S. Park (eds.): "Diffusion in Polymers," Academic, New York, 1968.
5. Dale, W. C., and C. E. Rogers: *AIChE J.*, **19**, 445 (1973).
6. E. I. du Pont de Nemours and Co., Inc.: *du Pont Petrol. Chem. News*, no. 172, 1972; no. 174, 1973.
7. Evans, R. B., B. M. Watson, and E. A. Mason: *J. Chem. Phys.*, **33**, 2076 (1961).
8. Jost, W.: "Diffusion in Solids, Liquids, and Gases," Academic, New York, 1960.
9. Komiyama, H., and J. M. Smith: *AIChE J.*, **20**, 1110, (1974).
10. Newman, A. B.: *Trans. AIChE*, **27**, 203, 310 (1931).
11. Rogers, C. E., M. Fels, and N. N. Li, *Recent Dev. Sep. Sci.*, **2**, 107 (1972).
12. Rothfield, L. B.: *AIChE J.*, **9**, 19 (1963).
13. Satterfield, C. N.: "Mass Transfer in Heterogeneous Catalysis," M.I.T. Press, Cambridge, Mass., 1970.
14. Satterfield, C. N., C. K. Colton, and W. H. Pitcher, Jr.: *AIChE J.*, **19**, 628 (1973).
15. Scott, D. S., and F. A. L. Dullien: *AIChE J.*, **8**, 713 (1962).
16. Shewman, P. G.: "Diffusion in Solids," McGraw-Hill Book Company, New York, 1963.
17. Stark, J. P.: "Solid State Diffusion," Wiley-Interscience, New York, 1976.
18. Van Arsdel, W. B.: *Chem. Eng. Prog.*, **43**, 13 (1947).
19. Wakao, N., and J. M. Smith: *Chem. Eng. Sci.*, **17**, 825 (1962).
20. Youngquist, G. R.: *Ind. Eng. Chem.*, **62**(8), 53 (1970).

PROBLEMS

4.1 Removal of soybean oil impregnating a porous clay by contact with a solvent for the oil has been shown to be a matter of internal diffusion of the oil through the solid [Boucher, Brier, and Osburn, *Trans. AIChE*, **38**, 967 (1942)]. Such a clay plate, $\frac{1}{16}$ in thick, 1.80 in long, and 1.08 in wide (1.588 by 45.7 by 27.4 mm), thin edges sealed, was impregnated with soybean oil to a uniform concentration 0.229 kg oil/kg dry clay. It was immersed in a flowing stream of pure tetrachloroethylene at 120°F (49°C), whereupon the oil content of the plate was reduced to 0.048 kg oil/kg dry clay in 1 h. The resistance to diffusion may be taken as residing wholly within the plate, and the final oil content of the clay may be taken as zero when contacted with pure solvent for an infinite time.

(*a*) Calculate the effective diffusivity.

(*b*) A cylinder of the same clay 0.5 in (12.7 mm) diameter, 1 in (25.4 mm) long, contains an initial uniform concentration of 0.17 kg oil/kg clay. When immersed in a flowing stream of pure tetrachloroethylene at 49°C, to what concentration will the oil content fall in 10 h? **Ans.:** 0.0748.

(*c*) Recalculate part (*b*) for the cases where only one end of the cylinder is sealed and where neither end is sealed.

(*d*) How long will it take for the concentration to fall to 0.01 kg oil/kg clay for the cylinder of part (*b*) with neither end sealed? **Ans.:** 41 h.

4.2 A slab of clay, like that used to make brick, 50 mm thick, was dried from both flat surfaces with the four thin edges sealed, by exposure to dry air. The initial uniform moisture content was 15%. The drying took place by internal diffusion of the liquid water to the surface, followed by evaporation at the surface. The diffusivity may be assumed to be constant with varying water concentration and uniform in all directions. The surface moisture content was 3%. In 5 h the average moisture content had fallen to 10.2%.

(*a*) Calculate the effective diffusivity.

(*b*) Under the same drying conditions, how much longer would it have taken to reduce the average water content to 6%?

(*c*) How long would be required to dry a sphere of 150 mm radius from 15 to 6% under the same drying conditions?

(*d*) How long would be required to dry a cylinder 0.333 m long by 150 mm diameter, drying from all surfaces, to a moisture content of 6%? **Ans.:** 47.5 h.

4.3 A spherical vessel of steel walls 2 mm thick is to contain 1 l of pure hydrogen at an absolute pressure of 1.3×10^6 N/m^2 (189 lb$_f$/in^2) and temperature 300°C. The internal surface will be at the saturation concentration of hydrogen; the outer surface will be maintained at zero hydrogen content. The solubility is proportional to $p_t^{1/2}$, where p_t is the hydrogen pressure; and at 1 std atm, 300°C, the solubility is 1 ppm (parts per million) by weight. At 300°C, the diffusivity of hydrogen in the steel = 5×10^{-10} m^2/s. The steel density is 7500 kg/m^3 (468 lb/ft^3).

(a) Calculate the rate of loss of hydrogen when the internal pressure is maintained at 1.3×10^6 N/m^2, expressed as kg/h.

(b) If no hydrogen is admitted to the vessel, how long will be required for the internal pressure to fall to one-half its original value? Assume the linear concentration gradient in the steel is always maintained and the hydrogen follows the ideal-gas law at prevailing pressures. **Ans.:** 7.8 h.

4.4 An unglazed porcelain plate 5 mm thick has an average pore diameter of 0.2 μm. Pure oxygen gas at an absolute pressure of 20 mmHg, 100°C, on one side of the plate passed through at a rate 0.093 cm^3 (at 20 mmHg, 100°C)/cm$^2 \cdot$ s when the pressure on the downstream side was so low as to be considered negligible. Estimate the rate of passage of hydrogen gas at 25°C and a pressure of 10 mmHg abs, with a negligible downstream pressure. **Ans.:** 1.78×10^{-6} kmol/m$^2 \cdot$ s.

FIVE

INTERPHASE MASS TRANSFER

Thus far we have considered only the diffusion of substances within a single phase. In most of the mass-transfer operations, however, two insoluble phases are brought into contact in order to permit transfer of constituent substances between them. Therefore we are now concerned with the simultaneous application of the diffusional mechanism for each phase to the combined system. We have seen that the rate of diffusion within each phase is dependent upon the concentration gradient existing within it. At the same time the concentration gradients of the two-phase system are indicative of the departure from equilibrium which exists between the phases. Should equilibrium be established, the concentration gradients and hence the rate of diffusion will fall to zero. It is necessary, therefore, to consider both the diffusional phenomena and the equilibria in order to describe the various situations fully.

EQUILIBRIUM

It is convenient first to consider the equilibrium characteristics of a particular operation and then to generalize the result for others. As an example, consider the gas-absorption operation which occurs when ammonia is dissolved from an ammonia-air mixture by liquid water. Suppose a fixed amount of liquid water is placed in a closed container together with a gaseous mixture of ammonia and air, the whole arranged so that the system can be maintained at constant temperature and pressure. Since ammonia is very soluble in water, some ammonia molecules will instantly transfer from the gas into the liquid, crossing the interfacial surface separating the two phases. A portion of the ammonia molecules escapes back into the gas, at a rate proportional to their concentration in the liquid. As more ammonia enters the liquid, with consequent increase in concentration within the liquid, the rate at which ammonia returns to the gas increases, until eventually the rate at which it enters the liquid exactly equals that at which it leaves. At the same time, through the mechanism of diffusion, the concentrations throughout each phase become uniform. A dynamic equilibrium now exists, and while ammonia molecules continue to transfer back and forth from one phase to the other, the net transfer falls to zero. The concentrations within each phase no longer change. To the observer who cannot see the individual molecules the diffusion has apparently stopped.

If we now inject additional ammonia into the container, a new set of equilibrium concentrations will eventually be established, with higher concentrations in each phase than were at first obtained. In this manner we can eventually obtain the complete relationship between the equilibrium concentrations in both phases. If the ammonia is designated as substance A, the equilibrium concentrations in the gas and liquid, y_A and x_A mole fractions, respectively, give rise to an *equilibrium-distribution curve* of the type shown in Fig. 5.1. This curve results irrespective of the amounts of water and air that we start with and is influenced only by the conditions, such as temperature and pressure, imposed upon the three-component system. It is important to note that at equilibrium the concentrations in the two phases are not equal; instead the chemical potential of the ammonia is the same in both phases, and it will be recalled (Chap. 2) that it is equality of chemical potentials, not concentrations, which causes the net transfer of solute to stop.

The curve of Fig. 5.1 does not of course show all the equilibrium concentrations existing within the system. For example, the water will partially vaporize into the gas phase, the components of the air will also dissolve to a small extent in the liquid, and equilibrium concentrations for these substances will also be established. For the moment we need not consider these equilibria, since they

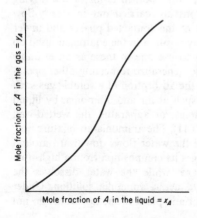

Figure 5.1 Equilibrium distribution of a solute between a gas and a liquid phase at constant temperature.

Mole fraction of A in the liquid = x_A

Mole fraction of A in the gas = y_A

are of minor importance to the discussion at hand. Obviously also, concentration units other than mole fractions may be used to describe the equilibria.

Generally speaking, whenever a substance is distributed between two insoluble phases, a dynamic equilibrium of this type can be established. The various equilibria are peculiar to the particular system considered. For example, replacement of the water in the example considered above with another liquid such as benzene or with a solid adsorbent such as activated carbon or replacement of the ammonia with another solute such as sulfur dioxide will each result in new curves not at all related to the first. The equilibrium resulting for a two-liquid-phase system bears no relation to that for a liquid-solid system. A discussion of the characteristic shapes of the equilibrium curves for the various situations and the influence of conditions such as temperature and pressure must be left for the studies of the individual unit operations. Nevertheless the following principles are common to all systems involving the distribution of substances between two insoluble phases:

1. At a fixed set of conditions, referring to temperature and pressure, there exists a set of equilibrium relationships which can be shown graphically in the form of an equilibrium-distribution curve for each distributed substance by plotting the equilibrium concentrations of the substance in the two phases against each other.
2. For a system in equilibrium, there is no net diffusion of the components between the phases.
3. For a system not in equilibrium, diffusion of the components between the phases will occur so as to bring the system to a condition of equilibrium. If sufficient time is available, equilibrium concentrations will eventually prevail.

DIFFUSION BETWEEN PHASES

Having established that departure from equilibrium provides the driving force for diffusion, we can now study the rates of diffusion in terms of the driving forces. Many of the mass-transfer operations are carried out in steady-flow fashion, with continuous and invariant flow of the contacted phases and under circumstances such that concentrations at any position in the equipment used do not change with time. It will be convenient to use one of these as an example with which to establish the principles and to generalize respecting other operations later. For this purpose, let us consider the absorption of a soluble gas such as ammonia (substance A) from a mixture such as air and ammonia, by liquid water as the absorbent, in one of the simplest of apparatus, the wetted-wall tower previously described in Chap. 3 (Fig. 3.11). The ammonia-air mixture may enter at the bottom and flow upward while the water flows downward around the inside of the pipe. The gas mixture changes its composition from a high- to a low-solute concentration as it flows upward, while the water dissolves the ammonia and leaves at the bottom as an aqueous ammonia solution. Under steady-state conditions, the concentrations at any point in the apparatus do not change with passage of time.

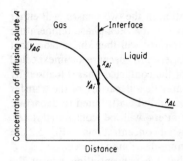

Figure 5.2 The two-resistance concept.

Local Two-Phase Mass Transfer

Let us investigate the situation at a particular level along the tower, e.g., at a point midway between top and bottom. Since the solute is diffusing from the gas phase into the liquid, there must be a concentration gradient in the direction of mass transfer within each phase. This can be shown graphically in terms of the distance through the phases, as in Fig. 5.2, where a section through the two phases in contact is shown. It will be assumed that no chemical reaction occurs. The concentration of A in the main body of the gas is $y_{A, G}$ mole fraction, and it falls to $y_{A, i}$ at the interface. In the liquid, the concentration falls from $x_{A, i}$ at the interface to $x_{A, L}$ in the bulk liquid. The bulk concentrations $y_{A, G}$ and $x_{A, L}$ are clearly not equilibrium values, since otherwise diffusion of the solute would not occur. At the same time, these bulk concentrations cannot be used directly with a mass-transfer coefficient to describe the rate of interphase mass transfer, since the two concentrations are differently related to the chemical potential, which is the real "driving force" of mass transfer.

To get around this problem, Lewis and Whitman [8, 17] assumed that the only diffusional resistances are those residing in the fluids themselves. There is then no resistance to solute transfer across the interface separating the phases, and as a result the concentrations $y_{A, i}$ and $x_{A, i}$ are equilibrium values, given by the system's equilibrium-distribution curve.†

The reliability of this theory has been the subject of a great amount of study. A careful review of the results indicates that departure from concentration equilibrium at the interface must be a rarity. It has been shown theoretically [11] that departure from equilibrium can be expected if mass-transfer rates are very high, much higher than are likely to be encountered in any practical situation. Careful measurements of real situations when interfaces are clean and conditions carefully controlled [4, 12] verify the validity of the assumption.

Unexpectedly large and small transfer rates between two phases nevertheless occur, and these have frequently been incorrectly attributed to departure from the equilibrium assumption. For example, the heat of transfer of a solute,

---

† This has been called the "two-film theory," improperly so because it is not related to, or dependent upon, the film theory of mass-transfer coefficients described in Chap. 3. A more appropriate name would be the "two-resistance theory."

resulting from a difference in the heats of solution in the two phases, will either raise or lower the interface temperature relative to the bulk-phase temperature [13], and the equilibrium distribution at the interface will then differ from that which can be assumed for the bulk-phase temperature. Because of unexpected phenomena near the interface in one or both of the contacted phases (caused by the presence of surfactants and the like) there may be a departure of the transfer rates from the expected values which may improperly be attributed to departure from equilibrium at the interface. These matters will be considered later. Consequently, in ordinary situations the interfacial concentrations of Fig. 5.2 are those corresponding to a point on the equilibrium-distribution curve.

Referring again to Fig. 5.2, it is clear that the concentration rise at the interface, from $y_{A,i}$ to $x_{A,i}$, is not a barrier to diffusion in the direction gas to liquid. They are equilibrium concentrations and hence correspond to equal chemical potentials of substance A in both phases at the interface.

The various concentrations can also be shown graphically, as in Fig. 5.3, whose coordinates are those of the equilibrium-distribution curve. Point P represents the two bulk-phase concentrations and point M those at the interface. For steady-state mass transfer, the rate at which A reaches the interface from the gas must equal that at which it diffuses to the bulk liquid, so that no accumulation or depletion of A at the interface occurs. We can therefore write the flux of A in terms of the mass-transfer coefficients for each phase and the concentration changes appropriate to each (the development will be done in terms of the k-type coefficients, since this is simpler, and the results for F-type coefficients will be indicated later). Thus, when k_y and k_x are the locally applicable coefficients,

$$N_A = k_y(y_{A,G} - y_{A,i}) = k_x(x_{A,i} - x_{A,L}) \tag{5.1}$$

and the differences in y's or x's are considered the driving forces for the mass transfer. Rearrangement as

$$\frac{y_{A,G} - y_{A,i}}{x_{A,L} - x_{A,i}} = -\frac{k_x}{k_y} \tag{5.2}$$

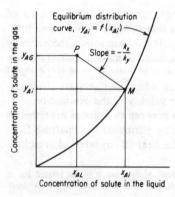

Figure 5.3 Departure of bulk-phase concentrations from equilibrium.

provides the slope of the line PM. If the mass-transfer coefficients are known, the interfacial concentrations and hence the flux N_A can be determined, either graphically by plotting the line PM or analytically by solving Eq. (5.2) with an algebraic expression for the equilibrium-distribution curve,

$$y_{A, i} = f(x_{A, i}) \qquad (5.3)$$

Local Overall Mass-Transfer Coefficients

In experimental determinations of the rate of mass transfer, it is usually possible to determine the solute concentrations in the bulk of the fluids by sampling and analyzing. Successful sampling of the fluids at the interface, however, is ordinarily impossible, since the greatest part of the concentration differences, such as $y_{A, G} - y_{A, i}$, takes place over extremely small distances. Any ordinary sampling device will be so large in comparison with this distance that it is impossible to get close enough to the interface. Sampling and analyzing, then, will provide $y_{A, G}$ and $x_{A, L}$ but not $y_{A, i}$ and $x_{A, i}$. Under these circumstances, only an overall effect, in terms of the bulk concentrations, can be determined. The bulk concentrations are, however, not by themselves on the same basis in terms of chemical potential.

Consider the situation shown in Fig. 5.4. Since the equilibrium-distribution curve for the system is unique at fixed temperature and pressure, then y_A^*, in equilibrium with $x_{A, L}$, is as good a measure of $x_{A, L}$ as $x_{A, L}$ itself, and moreover it is on the same basis as $y_{A, G}$. The entire two-phase mass-transfer effect can then be measured in terms of an overall mass-transfer coefficient K_y,

$$N_A = K_y(y_{A, G} - y_A^*) \qquad (5.4)$$

From the geometry of the figure,

$$y_{A, G} - y_A^* = (y_{A, G} - y_{A, i}) + (y_{A, i} - y_A^*) = (y_{A, G} - y_{A, i}) + m'(x_{A, i} - x_{A, L}) \qquad (5.5)$$

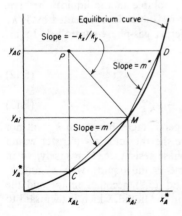

Figure 5.4 Overall concentration differences.

where m' is the slope of the chord CM. Substituting for the concentration differences the equivalents (flux/coefficient) as given by Eqs. (5.1) and (5.4) gives

$$\frac{N_A}{K_y} = \frac{N_A}{k_y} + \frac{m'N_A}{k_x} \tag{5.6}$$

or

$$\frac{1}{K_y} = \frac{1}{k_y} + \frac{m'}{k_x} \tag{5.7}$$

This shows that the relationship between the individual-phase transfer coefficients and the overall coefficient takes the form of addition of resistances (hence "two-resistance" theory). In similar fashion, x_A^* is a measure of $y_{A, G}$ and can be used to define another overall coefficient K_x

$$N_A = K_x(x_A^* - x_{A, L}) \tag{5.8}$$

and it is readily shown that

$$\frac{1}{K_x} = \frac{1}{m''k_y} + \frac{1}{k_x} \tag{5.9}$$

where m'' is the slope of the chord MD in Fig. 5.4. Equations (5.7) and (5.9) lead to the following relationships between the mass-transfer resistances:

$$\frac{\text{Resistance in gas phase}}{\text{Total resistance, both phases}} = \frac{1/k_y}{1/K_y} \tag{5.10}$$

$$\frac{\text{Resistance in liquid phase}}{\text{Total resistance, both phases}} = \frac{1/k_x}{1/K_x} \tag{5.11}$$

Assuming that the numerical values of k_x and k_y are roughly the same, the importance of the slope of the equilibrium-curve chords can readily be demonstrated. If m' is small (equilibrium-distribution curve very flat), so that at equilibrium only a small concentration of A in the gas will provide a very large concentration in the liquid (solute A is very soluble in the liquid), the term m'/k_x of Eq. (5.7) becomes minor, the major resistance is represented by $1/k_y$, and it is said that the rate of mass transfer is gas-phase-controlled. In the extreme, this becomes

$$\frac{1}{K_y} \approx \frac{1}{k_y} \tag{5.12}$$

or

$$y_{A, G} - y_A^* \approx y_{A, G} - y_{A, i} \tag{5.13}$$

Under such circumstances, even fairly large percentage changes in k_x will not significantly affect K_y, and efforts to increase the rate of mass transfer would best be directed toward decreasing the gas-phase resistance. Conversely, when m'' is very large (solute A relatively insoluble in the liquid), with k_x and k_y nearly equal, the first term on the right of Eq. (5.9) becomes minor and the major resistance to mass transfer resides within the liquid, which is then said to control the rate. Ultimately, this becomes

$$\frac{1}{K_x} \approx \frac{1}{k_x} \tag{5.14}$$

$$x_A^* - x_{A,L} \approx x_{A,i} - x_{A,L} \tag{5.15}$$

In such cases efforts to effect large changes in the rate of mass transfer are best directed to conditions influencing the liquid coefficient k_x. For cases where k_x and k_y are not nearly equal, Fig. 5.4 shows that it will be the relative size of the ratio k_x/k_y and of m' (or m'') which will determine the location of the controlling mass-transfer resistance.

It is sometimes useful to note that the effect of temperature is much larger for liquid mass-transfer coefficients than for those for gases (see Prob. 3.7, for example). Consequently a large effect of temperature on the overall coefficient, when it is determined experimentally, is usually a fairly clear indication that the controlling mass-transfer resistance is in the liquid phase.

For purposes of establishing the nature of the two-resistance theory and the overall coefficient concept, gas absorption was chosen as an example. The principles are applicable to any of the mass-transfer operations, however, and can be applied in terms of k-type coefficients using any concentration units, as listed in Table 3.1 (the F-type coefficients are considered separately below). For each case, the values of m' and m'' must be appropriately defined and used consistently with the coefficients. Thus, if the generalized phases are termed E (with concentrations expressed as i) and R (with concentrations expressed as j),

$$\frac{1}{K_E} = \frac{1}{k_E} + \frac{m'}{k_R} \tag{5.16}$$

$$\frac{1}{K_R} = \frac{1}{m''k_E} + \frac{1}{k_R} \tag{5.17}$$

$$m' = \frac{i_{A,i} - i_A^*}{j_{A,i} - j_{A,R}} \qquad m'' = \frac{i_{A,E} - i_{A,i}}{j_A^* - j_{A,i}} \tag{5.18}$$

$$i_A^* = f(j_{A,R}) \qquad i_{A,E} = f(j_A^*) \tag{5.19}$$

where f is the equilibrium-distribution function.

Local Coefficients—General Case

When we deal with situations which do not involve either diffusion of only one substance or equimolar counterdiffusion, or if mass-transfer rates are large, the F-type coefficients should be used. The general approach is the same, although the resulting expressions are more cumbersome than those developed above. Thus, in a situation like that shown in Figs. 5.2 to 5.4, the mass-transfer flux is

$$N_A = \frac{N_A}{\Sigma N} F_G \ln \frac{N_A/\Sigma N - y_{A,i}}{N_A/\Sigma N - y_{A,G}} = \frac{N_A}{\Sigma N} F_L \ln \frac{N_A/\Sigma N - x_{A,L}}{N_A/\Sigma N - x_{A,i}} \tag{5.20}$$

where F_G and F_L are the gas- and liquid-phase coefficients for substance A and

$\Sigma N = N_A + N_B + N_C + \cdots$. Equation (5.20) then becomes [5]

$$\frac{N_A/\Sigma N - y_{A,i}}{N_A/\Sigma N - y_{A,G}} = \left(\frac{N_A/\Sigma N - x_{A,L}}{N_A/\Sigma N - x_{A,i}} \right)^{F_L/F_G} \tag{5.21}$$

The interfacial compositions $y_{A,i}$ and $x_{A,i}$ can be found by plotting Eq. (5.21) (with $y_{A,i}$ replaced by y_A and $x_{A,i}$ by x_A) on the distribution diagram (Fig. 5.3) and determining the intersection of the resulting curve with the distribution curve. This is, in general, a trial-and-error procedure, since $N_A/\Sigma N$ may not be known, and must be done in conjunction with Eq. (5.20). In the special cases for diffusion only of A and of equimolar counterdiffusion in two-components phases, no trial and error is required.

We can also define the overall coefficients F_{OG} and F_{OL} as

$$N_A = \frac{N_A}{\Sigma N} F_{OG} \ln \frac{N_A/\Sigma N - y_A^*}{N_A/\Sigma N - y_{A,G}} = \frac{N_A}{\Sigma N} F_{OL} \ln \frac{N_A/\Sigma N - x_{A,L}}{N_A/\Sigma N - x_A^*} \tag{5.22}$$

By a procedure similar to that used for the K's, it can be shown that the overall and individual phase F's are related:

$$\exp\left[\frac{N_A}{(N_A/\Sigma N)F_{OG}} \right] = \exp\left[\frac{N_A}{(N_A/\Sigma N)F_G} \right]$$
$$+ m' \frac{N_A/\Sigma N - x_{A,L}}{N_A/\Sigma N - y_{A,G}} \left\{ 1 - \exp\left[-\frac{N_A}{(N_A/\Sigma N)F_L} \right] \right\} \tag{5.23}$$

$$\exp\left[-\frac{N_A}{(N_A/\Sigma N)F_{OL}} \right] = \frac{1}{m''} \left(\frac{N_A/\Sigma N - y_{A,G}}{N_A/\Sigma N - x_{A,L}} \right) \left\{ 1 - \exp\left[\frac{N_A}{(N_A/\Sigma N)FdG} \right] \right\}$$
$$+ \exp\left[-\frac{N_A}{(N_A/\Sigma N)F_L} \right] \tag{5.24}$$

where $\exp Z$ means e^Z. These fortunately simplify for the two important special cases:

1. *Diffusion of one component* ($\Sigma N = N_A$, $N_A/\Sigma N = 1.0$)†

$$e^{N_A/F_{OG}} = e^{N_A/F_G} + m' \frac{1 - x_{A,L}}{1 - y_{A,G}} (1 - e^{-N_A/F_L}) \tag{5.25}$$

$$e^{-N_A/F_{OL}} = \frac{1}{m''} \frac{1 - y_{A,G}}{1 - x_{A,L}} (1 - e^{N_A/F_G}) + e^{-N_A/F_L} \tag{5.26}$$

† Equations (5.25) and (5.26) can also be written in the following forms:

$$\frac{1}{F_{OG}} = \frac{1}{F_G} \frac{(1 - y_A)_{iM}}{(1 - y_A)_{*M}} + \frac{m'(1 - x_A)_{iM}}{F_L(1 - y_A)_{*M}} \tag{5.25a}$$

$$\frac{1}{F_{OL}} = \frac{1}{m'' F_G} \frac{(1 - y_A)_{iM}}{(1 - x_A)_{*M}} + \frac{1}{F_L} \frac{(1 - x_A)_{iM}}{(1 - x_A)_{*M}} \tag{5.26a}$$

where
$\quad (1 - y_A)_{iM}$ = logarithmic mean of $1 - y_{A,G}$ and $1 - y_{A,i}$
$\quad (1 - y_A)_{*M}$ = logarithmic mean of $1 - y_{A,G}$ and $1 - y_A^*$
$\quad (1 - x_A)_{iM}$ = logarithmic mean of $1 - x_{A,L}$ and $1 - x_{A,i}$
$\quad (1 - x_A)_{*M}$ = logarithmic mean of $1 - x_{A,L}$ and $1 - x_A^*$

INTERPHASE MASS TRANSFER 113

2. *Equimolar counterdiffusion* $[\Sigma N = 0 \ (F_G = k'_y, F_L = k'_x)]$

$$\frac{1}{F_{OG}} = \frac{1}{F_G} + \frac{m'}{F_L} \tag{5.27}$$

$$\frac{1}{F_{OL}} = \frac{1}{m'' F_G} + \frac{1}{F_L} \tag{5.28}$$

Use of Local Overall Coefficients

The concept of overall mass-transfer coefficients is in many ways similar to that of overall heat-transfer coefficients in heat-exchanger design. And, as is the practice in heat transfer, the overall coefficients of mass transfer are frequently synthesized through the relationships developed above from the individual coefficients for the separate phases. These can be taken, for example, from the correlations of Chap. 3 or from those developed in later chapters for specific types of mass-transfer equipment. It is important to recognize the limitations inherent in this procedure [6].

The hydrodynamic circumstances must be the same as those for which the correlations were developed. Especially in the case of two fluids, where motion in one may influence motion in the other, the individual correlations may make inadequate allowances for the effect on the transfer coefficients. There are, however, other important effects. Sometimes the transfer of a solute, e.g., the transfer of ethyl ether from water into air, results in large interfacial-tension gradients at the interface, and these in turn may result in an astonishingly rapid motion of the liquid at the interface (*interfacial turbulence* or *Marangoni effect*) [10]. In such a case the bulk-phase Reynolds numbers used in Chap. 3 no longer represent the turbulence level at the interface, and the mass-transfer rates are likely to be substantially larger than expected. In other cases, surface-active agents in a liquid, even when at extraordinarily low average concentrations, are concentrated at the interface, where they may (1) partly block the interface to solute transfer [15], (2) make the interfacial liquid layers more rigid [1], or (3) interact with a transferring solute [3]. In any case, the mass-transfer rates are reduced.

The mere presence of a mass-transfer resistance in one phase must have no influence on that in the other: the resistances must not interact, in other words. This may be important, since the individual-phase coefficient correlations are frequently developed under conditions where mass-transfer resistance in the second phase is nil, as when the second phase is a pure substance. The analysis of such situations is dependent upon the mechanism assumed for mass transfer (film, surface-renewal, etc., theories), and has been incompletely developed.

Average Overall Coefficients

As will be shown, in practical mass-transfer apparatus the bulk concentrations in the contacted phases normally vary considerably from place to place, so that a point such as P in Figs. 5.3 and 5.4 is just one of an infinite number which forms a curve on these figures. Thus, in the countercurrent wetted-wall gas absorber considered earlier, the solute concentrations in both fluids are small at the top and large at the bottom. In such situations, we can speak of an average overall coefficient applicable to the entire device. Such an average coefficient can be synthesized from constituent individual-phase coefficients by the same equations as developed above for local overall coefficients only, provided the quantity $m'k_E/k_R$ (or $m''k_R/k_E$) remains everywhere constant [6]. Of course, variations in values of the coefficients and slope m' may occur in such a way that they are self-compensating. But in practice, since average phase coefficients, assumed constant for the entire apparatus, are the only ones usually available, average overall coefficients generally will have meaning only when $m' = m'' = $ const, in other words, for cases of straight-line equilibrium-distribution curves. Obviously, also, the same hydrodynamic regime must exist throughout the apparatus for an average overall coefficient to have meaning.

Illustration 5.1 A wetted-wall absorption tower, 1 in (2.54 cm) ID, is fed with water as the wall liquid and an ammonia-air mixture as the central-core gas. At a particular level in the tower, the ammonia concentration in the bulk gas is 0.80 mole fraction, that in the bulk liquid 0.05 mole fraction. The temperature = 80°F (26.7°C), the pressure 1 std atm. The rates of flow are such that the local mass-transfer coefficient in the liquid, from a correlation obtained with dilute solutions, is $k_L = 0.34$ mol/ft² · h · (lb mol/ft³) = 2.87×10^{-5} kmol/m² · s · (kmol/m³), and the local Sherwood number for the gas is 40. The diffusivity of ammonia in air = 0.890 ft²/h = 2.297×10^{-5} m²/s. Compute the local mass-transfer flux for the absorption of ammonia, ignoring the vaporization of water.

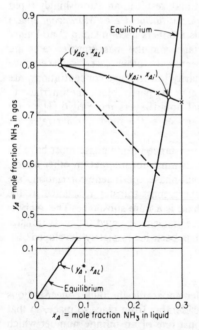

Figure 5.5 Construction for Illustration 5.1.

SOLUTION $y_{A, G} = 0.80$, $x_{A, L} = 0.05$ mole fraction ammonia. Because of the large concentration of ammonia in the gas, F's rather than k's will be used. Notation is that of Chap. 3.

Liquid From Table 3.1, $F_L = k_L x_{B, M_c}$. Since the molecular weight of ammonia and water are so nearly the same and the density of a dilute solution is practically that of water, the molar density $c = 1000/18 = 55.5$ kmol/m^3. Since the k_L was determined for dilute solutions, where $x_{B, M}$ is practically 1.0,

$$F_L = (2.87 \times 10^{-5})(1.0)(55.5) = 1.590 \times 10^{-3} \text{ kmol/m}^2 \cdot \text{s}$$

Gas

$$\text{Sh} = \frac{F_G d}{c D_A} = 40 \qquad d = 0.254 \text{ m} \qquad c = \frac{1}{22.41} \frac{273}{273 + 26.7} = 0.04065 \text{ kmol/m}^3$$

$$D_A = 2.297 \times 10^{-5} \text{ m}^2/\text{s}$$

$$F_G = \frac{40 c D_A}{d} = \frac{40(0.04065)(2.297 \times 10^{-5})}{0.0254} = 1.471 \times 10^{-3} \text{ kmol/m}^2 \cdot \text{s}$$

Mass-transfer flux The equilibrium-distribution data for ammonia are taken from those at 26.7°C in "The Chemical Engineers' Handbook," 5th ed., p. 3-68:

NH$_3$ mole fraction x_A	NH$_3$ partial pressure		$y_A = \dfrac{\bar{p}_A}{1.0133 \times 10^5}$
	lb$_f$/in^2	N/m$^2 = \bar{p}_A$	
0	0	0	0
0.05	1.04	7 171	0.0707
0.10	1.98	13 652	0.1347
0.25	8.69	59 917	0.591
0.30	13.52	93 220	0.920

These are plotted as the equilibrium curve in Fig. 5.5.

For transfer of only one component, $N_A/\Sigma N = 1.0$. Equation (5.21), with $y_{A, i}$ and $x_{A, i}$ replaced by y_A and x_A, becomes

$$y_A = 1 - (1 - y_{A, G})\left(\frac{1 - x_{A, L}}{1 - x_A}\right)^{F_L/F_G} = 1 - (1 - 0.8)\left(\frac{1 - 0.05}{1 - x_A}\right)^{1.078}$$

from which the following are computed:

x_A	0.05	0.15	0.25	0.30
y_A	0.80	0.780	0.742	0.722

These are plotted as x's on Fig. 5.5, and the resulting curve intersects the equilibrium curve to give the interface compositions, $x_{A, i} = 0.274$, $y_{A, i} = 0.732$. Therefore Eq. (5.20) provides

$$N_A = 1(1.590 \times 10^{-3}) \ln \frac{1 - 0.05}{1 - 0.274} = 1(1.471 \times 10^{-3}) \ln \frac{1 - 0.732}{1 - 0.80}$$

$$= 4.30 \times 10^{-4} \text{ kmol NH}_3 \text{ absorbed/m}^2 \cdot \text{s, local flux} \quad \textbf{Ans.}$$

Overall coefficient Although it is not particularly useful in this case, the overall coefficient will be computed to demonstrate the method. The gas concentration in equilibrium with the bulk liquid ($x_{A, L}$ = 0.05) is, from Fig. 5.5, y_A^* = 0.0707. The chord slope m' is therefore

$$m' = \frac{y_{A, i} - y_A^*}{x_{A, i} - x_{A, L}} = \frac{0.732 - 0.0707}{0.274 - 0.05} = 2.95$$

Note that unless the equilibrium curve is straight, this cannot be obtained without first obtaining ($y_{A, i}, x_{A, i}$), in which case there is no need for the overall coefficient. In Eq. (5.25), normally we would obtain F_{OG} by trial, by assuming N_A. Thus (with an eye on the answer obtained above), assume N_A = 4.30 × 10^{-4}. Then Eq. (5.25) becomes

$$\frac{e^{4.30 \times 10^{-4}}}{F_{OG}} = e^{(4.30 \times 10^{-4})/(1.471 \times 10^{-3})}$$

$$+ 2.95 \frac{1 - 0.05}{1 - 0.80} [1 - e^{-(4.30 \times 10^{-4})/(1.590 \times 10^{-3})}]$$

$$= 4.66$$

$$F_{OG} = 2.78 \times 10^{-4} \text{ kmol/m}^2 \cdot \text{s}$$

As a check of the trial value of N_A, Eq. (5.22) is

$$N_A = 1(2.78 \times 10^{-4}) \ln \frac{1 - 0.0707}{1 - 0.80} = 4.30 \times 10^{-4} \quad \text{(check)}$$

Use of k-type coefficients There are, of course, k's which are consistent with the F's and which will produce the correct result. Thus, $k_y = F_{GP_t}/\bar{p}_{B, M}$ = 6.29 × 10^{-3} and $k_x = F_L/x_{B, M}$ = 1.885 × 10^{-3} will produce the same result as above. But these k's are specific for the concentration levels at hand, and the $\bar{p}_{B, M}$ and $x_{B, M}$ terms, which correct for the bulk-flow flux [the $N_A + N_B$ term of Eq. (2.4)], cannot be obtained until $x_{A, i}$ and $y_{A, i}$ are first obtained as above.

However, if it had been assumed that the concentrations were dilute and that the bulk-flow terms were negligible, the Sherwood number *might* have been (incorrectly) interpreted as

$$\text{Sh} = \frac{k_y RTd}{p_t D_A} = 40 = \frac{k_y(8314)(273.2 + 26.7)(0.0254)}{(1.0133 \times 10^5)(2.297 \times 10^{-5})}$$

$$k_y = 1.47 \times 10^{-3} \text{ kmol/m}^2 \cdot \text{s} \cdot \text{(mole fraction)}$$

and, in the case of the liquid,

$$k_x = k_L c = (2.87 \times 10^{-5})(55.5) = 1.59 \times 10^{-3} \text{ kmol/m}^2 \cdot \text{s} \cdot \text{(mole fraction)}$$

These k's are suitable for small driving forces but are unsuitable here. Thus

$$-\frac{k_x}{k_y} = -\frac{1.59 \times 10^{-3}}{1.47 \times 10^{-3}} = -1.08$$

and a line of this slope (the dashed line of Fig. 5.5) drawn from ($x_{A, L}, y_{A, G}$) intersects the equilibrium curve at (x_A = 0.250, y_A = 0.585). If this is interpreted to be ($x_{A, i}, y_{A, i}$), the calculated flux would be

$$N_A = k_x(x_{A, i} - x_{A, L}) = (1.59 \times 10^{-3})(0.250 - 0.05) = 3.17 \times 10^{-4}$$

or

$$N_A = k_y(y_{A, G} - y_{A, i}) = (1.47 \times 10^{-3})(0.8 - 0.585) = 3.17 \times 10^{-4}$$

The corresponding overall coefficient, with m' = (0.585 − 0.0707)/(0.250 − 0.05) = 2.57, would be

$$\frac{1}{K_y} = \frac{1}{k_y} + \frac{m'}{k_x} = \frac{1}{1.47 \times 10^{-3}} + \frac{2.57}{1.59 \times 10^{-3}}$$

$$K_y = 4.35 \times 10^{-4} \text{ kmol/m}^2 \cdot \text{s} \cdot \text{(mole fraction)}$$

$$N_A = K_y(y_{A, G} - y_A^*) = (4.35 \times 10^{-4})(0.8 - 0.0707) = 3.17 \times 10^{-4}$$

This value of N_A is of course incorrect. However, had the gas concentration been low, say 0.1 mole fraction NH$_3$, these k's would have been satisfactory and would have given the same flux as the F's.

MATERIAL BALANCES

The concentration-difference driving forces discussed above, as we have said, are those existing at one position in the equipment used to contact the immiscible phases. In the case of a steady-state process, because of the transfer of solute from one phase to the other, the concentration within each phase changes as it moves through the equipment. Similarly, in the case of a batch process, the concentration in each phase changes with time. These changes produce corresponding variations in the driving forces, and these can be followed with the help of material balances. In what follows, all concentrations are the bulk-average values for the indicated streams.

Steady-State Cocurrent Processes

Consider any mass-transfer operation whatsoever conducted in a steady-state cocurrent fashion, as in Fig. 5.6, in which the apparatus used is represented simply as a rectangular box. Let the two insoluble phases be identified as phase E and phase R, and for the present consider only the case where a single substance A diffuses from phase R to phase E during their contact. The other constituents of the phases, solvents for the diffusing solutes, are then considered not to diffuse.

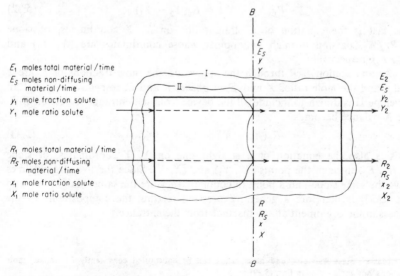

E_1 moles total material /time
E_S moles non-diffusing material /time
y_1 mole fraction solute
Y_1 mole ratio solute

R_1 moles total material / time
R_S moles non-diffusing material / time
x_1 mole fraction solute
X_1 mole ratio solute

Figure 5.6 Steady-state cocurrent processes.

At the entrance to the device in which the phases are contacted, phase R contains R_1 mol per unit time of total substances, consisting of nondiffusing solvent R_S mol per unit time and diffusing solute A, whose concentration is x_1 mole fraction. As phase R moves through the equipment, A diffuses to phase E and consequently the total quantity of R falls to R_2 mol per unit time at the exit, although the rate of flow of nondiffusing solvent R_S is the same as at the entrance. The concentration of A has fallen to x_2 mole fraction. Similarly, phase E at the entrance contains E_1 mol per unit time total substances, of which E_S mol is nondiffusing solvent, and an A concentration of y_1 mole fraction. Owing to the accumulation of A to a concentration y_2 mole fraction, phase E increases in amount to E_2 mol per unit time at the exit, although the solvent content E_S has remained constant.

Envelope I, the closed, irregular line drawn about the equipment, will help to establish a material balance for substance A, since an accounting for substance A must be made wherever the envelope is crossed by an arrow representing a flowing stream. The A content of the entering R phase is R_1x_1; that of the E phase is E_1y_1. Similarly the A content of the leaving streams is R_2x_2 and E_2y_2, respectively. Thus

$$R_1x_1 + E_1y_1 = R_2x_2 + E_2y_2 \qquad (5.29)$$

or

$$R_1x_1 - R_2x_2 = E_2y_2 - E_1y_1 \qquad (5.30)$$

but

$$R_1x_1 = R_S \frac{x_1}{1-x_1} = R_SX_1 \qquad (5.31)$$

where X_1 is the mole-ratio concentration of A at the entrance, mol A/mol non-A. The other terms can be similarly described, and Eq. (5.30) becomes

$$R_S(X_1 - X_2) = E_S(Y_2 - Y_1) \qquad (5.32)$$

The last is the equation of a straight line on X, Y coordinates, of slope $- R_S/E_S$, passing through two points whose coordinates are (X_1, Y_1) and (X_2, Y_2), respectively.

At any section B-B through the apparatus, the mole fractions of A are x and y and the mole ratios X and Y, in phases R and E, respectively, and if an envelope II is drawn to include all the device from the entrance to section B-B, the A balance becomes

$$R_S(X_1 - X) = E_S(Y - Y_1) \qquad (5.33)$$

This is also the equation of a straight line on X, Y coordinates, of slope $- R_S/E_S$, through the points (X_1, Y_1) and (X, Y). Since the two straight lines have the same slope and a point in common, they are the same straight line and Eq. (5.33) is therefore a general expression relating the compositions of the phases in the equipment at any distance from the entrance.†

---

† Masses, mass fractions, and mass ratios can be substituted consistently for moles, mole fractions, and mole ratios in Eqs. (5.29) to (5.33).

Since X and Y represent concentrations in the two phases, the equilibrium relationship may also be expressed in terms of these coordinates. Figure 5.7 shows a representation of the equilibrium relationship as well as the straight line QP of Eqs. (5.32) and (5.33). The line QP, called an *operating line*, should not be confused with the driving-force lines of the earlier discussion. At the entrance to the apparatus, for example, the mass-transfer coefficients in the two phases may give rise to the driving-force line KP, where K represents the interface compositions at the entrance and the distances KM and MP represent, respectively, the driving forces in phase E and phase R. Similarly, at the exit, point L may represent the interface composition and LQ the line representative of the driving forces. If the apparatus were longer than that indicated in Fig. 5.6, so that eventually an equilibrium between the two phases is established, the corresponding equilibrium compositions X_e and Y_e would be given by an extension of the operating line to the intersection with the equilibrium curve at T. The driving forces, and therefore the diffusion rate, at this point will have fallen to zero. Should the diffusion be in the opposite direction, i.e., from phase E to phase R, the operating line will fall on the opposite side of the equilibrium curve, as in Fig. 5.8.

It must be emphasized that the graphical representation of the operating line as a *straight* line is greatly dependent upon the units in which the concentrations of the material balance are expressed. The representations of Figs. 5.7 and 5.8 are straight lines because the mole-ratio concentrations are based on the unchanging quantities E_S and R_S. If Eq. (5.30) were plotted on mole-fraction coordinates, or if any concentration unit proportional to mole fractions such as partial pressure were used, the nature of the operating *curve* obtained would be

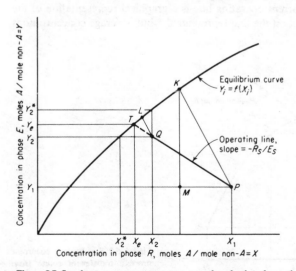

Figure 5.7 Steady-state cocurrent process, transfer of solute from phase R to phase E.

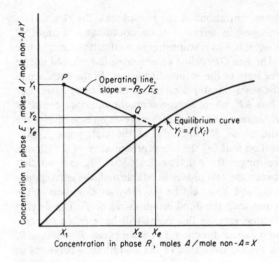

Figure 5.8 Steady-state cocurrent process, transfer of solute from phase E to phase R.

indicated as in Fig. 5.9. Extrapolation to locate the ultimate equilibrium conditions at T is of course much more difficult than for mole-ratio coordinates. On the other hand, in any operation where the total quantities of each of the phases E and R remain constant while the compositions change owing to diffusion of several components, a diagram in terms of mole fraction will provide a straight-line operating line, as Eq. (5.30) indicates (let $E_2 = E_1 = E$; $R_2 = R_1 = R$). If all the components diffuse so that the total quantities of each phase do not remain constant, the operating line will generally be curved.

To sum up, the cocurrent operating line is a graphical representation of the material balance. A point on the line represents the bulk-average concentrations

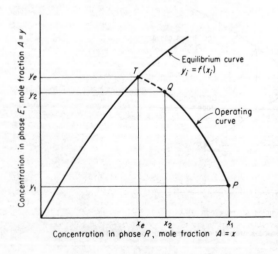

Figure 5.9 Steady-state cocurrent process, transfer of solute from phase R to phase E.

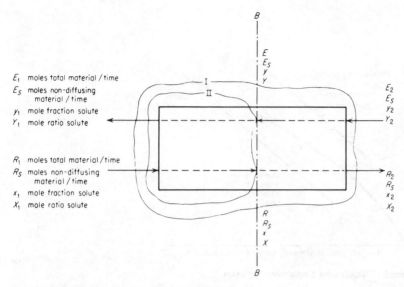

Figure 5.10 Steady-state countercurrent process.

of the streams in contact with each other at any section in the apparatus. Consequently the line passes from the point representing the streams entering the apparatus to that representing the effluent streams.

Steady-State Countercurrent Processes

If the same process as previously considered is carried out in countercurrent fashion, as in Fig. 5.10, where the subscripts 1 indicate that end of the apparatus where phase R enters and 2 that end where phase R leaves, the material balances become, by envelope I,†

$$E_2 y_2 + R_1 x_1 = E_1 y_1 + R_2 x_2 \qquad (5.34)$$

and

$$R_S(X_1 - X_2) = E_S(Y_1 - Y_2) \qquad (5.35)$$

and, for envelope II,

$$Ey + R_1 x_1 = E_1 y_1 + Rx \qquad (5.36)$$

and

$$R_S(X_1 - X) = E_S(Y_1 - Y) \qquad (5.37)$$

Equations (5.36) and (5.37) give the general relationship between concentrations in the phases at any section, while Eqs. (5.34) and (5.35) establish the entire material balance. Equation (5.35) is that of a straight line on X, Y coordinates,

† Masses, mass ratios, and mass fractions can be substituted consistently for moles, mole ratios, and mole fractions in Eqs. (5.34) to (5.37).

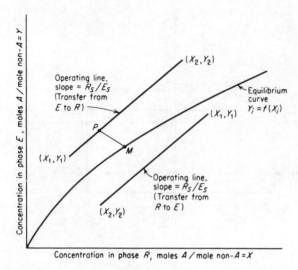

Concentration in phase R, moles A / mole non-A = X

Figure 5.11 Steady-state countercurrent process.

of slope R_S/E_S, through points of coordinates (X_1, Y_1), (X_2, Y_2), as shown in Fig. 5.11. The line will lie above the equilibrium-distribution curve if diffusion proceeds from phase E to phase R, below for diffusion in the opposite direction. For the former case, at a point where the concentrations in the phases are given by the point P, the driving-force line may be indicated by line PM, whose slope depends upon the relative diffusional resistances of the phases. The driving forces obviously change in magnitude from one end of the equipment to the other. If the operating line should touch the equilibrium curve anywhere, so that the contacted phases are in equilibrium, the driving force and hence the rate of mass transfer would become zero and the time required for a finite material transfer would be infinite. This can be interpreted in terms of a limiting ratio of flow rates of the phases for the concentration changes specified.

As in the cocurrent case, linearity of the operating line depends upon the method of expressing the concentrations. The operating lines of Fig. 5.11 are straight because the mole-ratio concentrations X and Y are based on the quantities R_S and E_S, which are stipulated to be constant. If for this situation mole fractions (or quantities such as partial pressures, which are proportional to mole fractions) are used, the operating lines are curved, as indicated in Fig. 5.12. However, for some operations should the total quantity of each of the phases E and R be constant while the compositions change, the mole-fraction diagram would provide the straight-line operating lines, as Eq. (5.34) would indicate (let $E = E_1 = E_2$; $R = R_1 = R_2$). As in the cocurrent case, however, the operating line will generally be curved if all the components diffuse so that the total quantities of each phase do not remain constant.

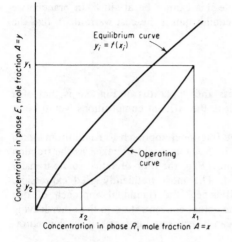

Figure 5.12 Steady-state countercurrent process, transfer of solute from phase R to phase E.

The countercurrent operating line is a graphical representation of the material balance, passing from the point representing the streams at one end of the apparatus to the point representing the streams at the other. A point on the line represents the bulk-average concentrations of the streams passing each other at any section in the apparatus. It is useful to rearrange Eq. (5.36) to read

$$Ey - Rx = E_1 y_1 - R_1 x_1 \qquad (5.38)$$

The left-hand side represents the *net* flow rate of solute A to the left at section B-B in Fig. 5.10. The right-hand side is the net flow rate of solute out at the left end of the apparatus, or the difference in solute flow, out $-$ in, and for steady state this is constant. Since the section B-B was taken to represent any section of the apparatus, at every point on the operating line the *net* flow rate of solute is constant, equal to the net flow out at the end.

The advantage of the countercurrent method over the cocurrent is that, for the former, the average driving force for a given situation will be larger, resulting in either smaller equipment for a given set of flow conditions or smaller flows for a given equipment size.

STAGES

A *stage* is defined as any device or combination of devices in which two insoluble phases are brought into intimate contact, where mass transfer occurs between the phases tending to bring them to equilibrium, and where the phases are then mechanically separated. A process carried out in this manner is a *single-stage* process. An *equilibrium*, *ideal*, or *theoretical*, *stage* is one where the time of contact between phases is sufficient for the effluents indeed to be in equilibrium, and although in principle this cannot be attained, in practice we can frequently approach so close to equilibrium if the cost warrants it, that the difference is unimportant.

Continuous Cocurrent Processes

Quite obviously the cocurrent process and apparatus of Fig. 5.6 is that of a single stage, and if the stage were ideal, the effluent compositions would be at point T on Fig. 5.7 or 5.8.

A *stage efficiency* is defined as the fractional approach to equilibrium which a real stage produces. Referring to Fig. 5.7, this might be taken as the fraction which the line QP represents of the line TP, or the ratio of actual solute transfer to that if equilibrium were attained. The most frequently used expression, however, is the Murphree stage efficiency, the fractional approach of one leaving stream to equilibrium with the actual concentration in the other leaving stream [9]. Referring to Fig. 5.7, this can be expressed in terms of the concentrations in phase E or in phase R,

$$E_{ME} = \frac{Y_2 - Y_1}{Y_2^* - Y_1} \qquad E_{MR} = \frac{X_1 - X_2}{X_1 - X_2^*} \qquad (5.39)$$

These definitions are somewhat arbitrary since, in the case of a truly cocurrent operation such as that of Fig. 5.7, it would be impossible to obtain a leaving concentration in phase E higher than Y_e or one in phase R lower than X_e. They are nevertheless useful, as will be developed in later chapters. The two Murphree efficiencies are not normally equal for a given stage, and they can be simply related only when the equilibrium relation is a straight line. Thus, for a straight equilibrium line of slope $m = (Y_2^* - Y_2)/(X_2 - X_2^*)$, it can be shown that

$$E_{ME} = \frac{E_{MR}}{E_{MR}(1 - S) + S} = \frac{E_{MR}}{E_{MR}(1 - 1/A) + 1/A} \qquad (5.40)$$

The derivation is left to the student (Prob. 5.6). Here $A = R_S/mE_S$ is called the *absorption factor*, and its reciprocal, $S = mE_S/R_S$, is the *stripping factor*. As will be shown, both have great economic importance.

Batch Processes

It is characteristic of batch processes that while there is no flow of the phases into and out of the equipment used, the concentrations within each phase change with time. When initially brought into contact, the phases will not be of equilibrium compositions, but they will approach equilibrium with passage of time. The material-balance equation (5.33) for the cocurrent steady-state operation then shows the relation between the concentrations X and Y in the phases which coexist at any time after the start of the operation, and Figs. 5.7 and 5.8 give the graphical representation of these compositions. Point T on these figures represents the ultimate compositions which are obtained at equilibrium. The batch operation is a single stage.

Cascades

A group of stages interconnected so that the various streams flow from one to the other is called a cascade. Its purpose is to increase the extent of mass transfer over and above that which is possible with a single stage. The fractional overall stage efficiency of a cascade is then defined as the number of equilibrium stages to which the cascade is equivalent divided by the number of real stages.

Two or more stages connected so that the flow is cocurrent between stages will, of course, never be equivalent to more than one equilibrium stage, although the overall stage efficiency can thereby be increased. For effects greater than one equilibrium stage, they can be connected for cross flow or countercurrent flow.

Cross-Flow Cascades

In Fig. 5.13 each stage is represented simply by a circle, and within each the flow is cocurrent. The R phase flows from one stage to the next, being contacted in each stage by a fresh E phase. There may be different flow rates of the E phase to each stage, and each stage may have a different Murphree stage efficiency. The material balances are obviously merely a repetition of that for a single stage and the construction on the distribution diagram is obvious. Cross flow is used sometimes in adsorption, leaching, drying, and extraction operations but rarely in the others.

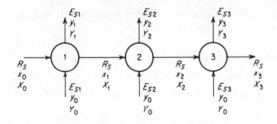

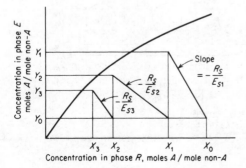

Concentration in phase R, moles A / mole non-A

Figure 5.13 A cross-flow cascade of three real stages.

Countercurrent Cascades

These are the most efficient arrangements, requiring the fewest stages for a given change of composition and ratio of flow rates, and they are therefore most frequently used. Refer to Fig. 5.14, where a cascade of N_p equilibrium stages is shown. The flow rates and compositions are numbered corresponding to the effluent from a stage, so that Y_2 is the concentration in the E phase leaving stage 2, etc. Each stage is identical in its action to the cocurrent process of Fig. 5.6; yet the cascade as a whole has the characteristics of the countercurrent process of Fig. 5.10. The cocurrent operating lines for the first two stages are written beneath them on the figure, and since the stages are ideal, the effluents are in equilibrium (Y_2 in equilibrium with X_2, etc.). The graphical relations are shown in Fig. 5.15. Line PQ is the operating line for stage 1, MN for stage 2, etc., and the coordinates (X_1, Y_1) fall on the equilibrium curve because the stage is ideal. Line ST is the operating line for the entire cascade, and points such as B, C, etc., represent compositions of streams passing each other between stages. We can therefore determine the number of equilibrium stages required for a countercurrent process by drawing the stairlike construction $TQBNC \cdots S$ (Fig. 5.15). If anywhere the equilibrium curve and cascade operating line should touch, the stages become *pinched* and an infinite number are required to bring about the desired composition change. If the solute transfer is from E to R, the entire construction falls above the equilibrium curve of Fig. 5.15.

For most cases, because of either a curved operating line or equilibrium curve, the relation between number of stages, compositions, and flow ratio must be determined graphically, as shown. For the *special* case where both are straight, however, with the equilibrium curve continuing straight to the origin of the distribution graph, an analytical solution can be developed which will be most useful.

In Fig. 5.14, a solute balance for stages $n + 1$ through N_p is

$$E_S(Y_{n+1} - Y_{N_p+1}) = R_S(X_n - X_{N_p}) \qquad (5.41)$$

If the equilibrium-curve slope is $m = Y_{n+1}/X_{n+1}$, and if the absorption factor $A = R_S/mE_S$, then

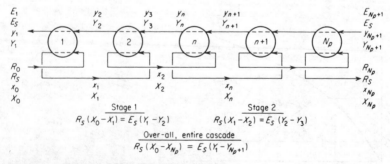

Figure 5.14 Countercurrent multistage cascade.

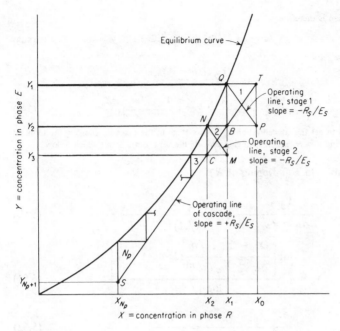

Figure 5.15 Countercurrent multistage cascade, solute transfer from phase R to phase E.

by substitution and rearrangement, Eq. (5.41) becomes

$$X_{n+1} - AX_n = \frac{Y_{N_p+1}}{m} - AX_{N_p} \tag{5.42}$$

This is a linear first-order finite-difference equation, whose solution is handled much like that of ordinary differential equations [18]. Thus, putting it in operator form gives

$$(D - A)X_n = \frac{Y_{N_p+1}}{m} - AX_{N_p} \tag{5.43}$$

where the operator D indicates the finite difference. The characteristic equation is then

$$M - A = 0 \tag{5.44}$$

from which $M = A$. Hence the general solution (here a little different from ordinary differential equations) is

$$X_n = C_1 A^n \tag{5.45}$$

with C_1 a constant. Since the right-hand side of Eq. (5.43) is a constant, the particular solution is $X = C_2$, where C_2 is a constant. Substituting this into the original finite-difference equation (5.42) provides

$$C_2 - AC_2 = \frac{Y_{N_p+1}}{m} - AX_{N_p} \tag{5.46}$$

from which

$$C_2 = \frac{Y_{N_p+1}/m - AX_{N_p}}{1 - A} \tag{5.47}$$

The complete solution is therefore

$$X_n = C_1 A^n + \frac{Y_{N_p+1}/m - AX_{N_p}}{1 - A}$$ (5.48)

To determine C_1, we set $n = 0$:

$$C_1 = X_0 - \frac{Y_{N_p+1}/m - AX_{N_p}}{1 - A}$$

and therefore

$$X_n = \left(X_0 - \frac{Y_{N_p+1}/m - AX_{N_p}}{1 - A} \right) A^n + \frac{Y_{N_p+1}/m - AX_{N_p}}{1 - A}$$ (5.49)

This result is useful to get the concentration X_n at any stage in the cascade, knowing the terminal concentrations. Putting $n = N_p$ and rearranging provide the very useful forms which follow.

For transfer from R to E (stripping of R)

$A \neq 1$:

$$\frac{X_0 - X_{N_p}}{X_0 - Y_{N_p+1}/m} = \frac{(1/A)^{N_p+1} - 1/A}{(1/A)^{N_p+1} - 1}$$ (5.50)

$$N_p = \frac{\log\left[\frac{X_0 - Y_{N_p+1}/m}{X_{N_p} - Y_{N_p+1}/m}(1 - A) + A \right]}{\log 1/A}$$ (5.51)

$A = 1$:

$$\frac{X_0 - X_{N_p}}{X_0 - Y_{N_p+1}/m} = \frac{N_p}{N_p + 1}$$ (5.52)

$$N_p = \frac{X_0 - X_{N_p}}{X_{N_p} - Y_{N_p+1}/m}$$ (5.53)

For transfer from E to R (absorption into R). A similar treatment yields:

$A \neq 1$:

$$\frac{Y_{N_p+1} - Y_1}{Y_{N_p+1} - mX_0} = \frac{A^{N_p+1} - A}{A^{N_p+1} - 1}$$ (5.54)

$$N_p = \frac{\log\left[\frac{Y_{N_p+1} - mX_0}{Y_1 - mX_0}\left(1 - \frac{1}{A}\right) + \frac{1}{A} \right]}{\log A}$$ (5.55)

$A = 1$:

$$\frac{Y_{N_p+1} - Y_1}{Y_{N_p+1} - mX_0} = \frac{N_p}{N_p + 1}$$ (5.56)

$$N_p = \frac{Y_{N_p+1} - Y_1}{Y_1 - mX_0}$$ (5.57)

These are called the Kremser-Brown-Souders (or simply Kremser) equations, after those who derived them for gas absorption [7, 14] although apparently Turner [16] had used them earlier for leaching and solids washing. They are plotted in Fig. 5.16, which then becomes very convenient for quick solutions. We

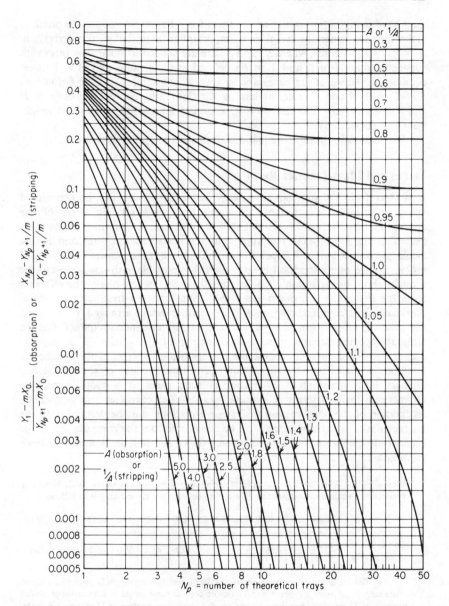

Figure 5.16 Number of theoretical stages for countercurrent cascades, with Henry's law equilibrium and constant absorption or stripping factors. [*After Hachmuth and Vance, Chem. Eng. Prog.*, **48**, *523, 570, 617 (1952).*]

shall have many opportunities to use them for different operations. In order to make them so generally useful, it is important to note that in the derivation, $A = R_S/mE_S$ = const, where R_S and E_S are the nonsolute molar flow rates with concentrations (X in R and Y in E) defined as mole ratios and $m = Y/X$ at equilibrium. However, as necessary to keep A constant, A can be defined as R/mE, with R and E either as total molar flow rates, concentrations (x in R and y in E) as mole fractions, $m = y/x$, or as total mass flow rates and weight fractions, respectively.

We shall delay using the equations until specific opportunity arises.

Stages and Mass-Transfer Rates

It is clear from the previous discussion that each process can be considered either in terms of the number of stages it represents or in terms of the appropriate mass-transfer rates. A batch or continuous cocurrent operation, for example, is a single-stage operation, but the stage efficiency realized in the available contact time will depend upon the average mass-transfer rates prevailing. A change of composition greater than that possible with one stage can be brought about by repetition of the cocurrent process, where one of the effluents from the first stage is brought again into contact with a fresh treating phase. Alternatively, a countercurrent multistage cascade can be arranged. If, however, the countercurrent operation is carried out in continuous-contact fashion without repeated separation and recontacting of the phases in a stepwise manner, it is still possible to describe the operation in terms of the number of equilibrium stages to which it is equivalent. But in view of the differential changes in composition which occur in such cases, it is more correct to characterize them in terms of average mass-transfer coefficients or equivalent. Integration of the point mass-transfer rate equations must be delayed until the characteristics of each operation can be considered, but the computation of the number of equilibrium stages requires only the equilibrium and material-balance relationships.

Illustration 5.2 When a certain sample of moist soap is exposed to air at 75°C, 1 std atm pressure, the equilibrium distribution of moisture between the air and soap is as follows:

Wt % moisture in soap	0	2.40	3.76	4.76	6.10	7.83	9.90	12.63	15.40	19.02
Partial pressure water in air, mmHg	0	9.66	19.20	28.4	37.2	46.4	55.0	63.2	71.9	79.5

(a) Ten kilograms of wet soap containing 16.7% moisture by weight is placed in a vessel containing 10 m³ moist air whose initial moisture content corresponds to a water-vapor partial pressure of 12 mmHg. After the soap has reached a moisture content of 13.0%, the air in the vessel is entirely replaced by fresh air of the original moisture content and the system is then allowed to reach equilibrium. The total pressure and temperature are maintained at 1 std atm and 75°C, respectively. What will be the ultimate moisture content of the soap?

(b) It is desired to dry the soap from 16.7 to 4% moisture continuously in a countercurrent stream of air whose initial water-vapor partial pressure is 12 mmHg. The pressure and temperature will be maintained throughout at 1 std atm and 75°C. For 1 kg initial wet soap per hour, what is the minimum amount of air required per hour?

(c) If 30% more air than that determined in part (b) is used, what will be the moisture content of the air leaving the drier? To how many equilibrium stages will the process be equivalent?

SOLUTION (a) The process is a batch operation. Since air and soap are mutually insoluble, with water distributing between them, Eq. (5.32) applies. Let the E phase be air-water and the R phase be soap-water. Define Y as kg water/kg dry air and X as kg water/kg dry soap. It will be necessary to convert the equilibrium data to these units.

For a partial pressure of moisture $= \bar{p}$ mmHg and a total pressure of 760 mmHg, $\bar{p}/(760 - \bar{p})$ is the mole ratio of water to dry air. Multiplying this by the ratio of molecular weights of water to air then gives $Y = [\bar{p}/(760 - \bar{p})](18.02/29)$. Thus, at $\bar{p} = 71.9$ mmHg,

$$Y = \frac{71.9}{760 - 71.9} \frac{18.02}{29} = 0.0650 \text{ kg water/kg dry air}$$

In similar fashion the percent moisture content of the soap is converted into units of X. Thus, at 15.40% moisture,

$$X = \frac{15.40}{100 - 15.40} = 0.182 \text{ kg water/kg dry soap}$$

In this manner all the equilibrium data are converted into these units and the data plotted as the equilibrium curve of Fig. 5.17.

First operation Initial air, $\bar{p} = 12$ mmHg:

$$Y_1 = \frac{12}{760 - 12} \frac{18.02}{29} = 0.00996 \text{ kg water/kg dry air}$$

Initial soap, 16.7% water:

$$S_1 = \frac{16.7}{100 - 16.7} = 0.2 \text{ kg water/kg dry soap}$$

Final soap, 13% water:

$$X_2 = \frac{13.0}{100 - 13.0} = 0.1493 \text{ kg water/kg dry soap}$$

$$R_S = 10(1 - 0.167) = 8.33 \text{ kg dry soap}$$

E_S, the mass of dry air, is found by use of the ideal-gas law,

$$E_S = 150 \frac{760 - 12}{760} \frac{273}{273 + 75} \frac{29}{22.41} = 10.1 \text{ kg dry air}$$

$$\text{Slope of operating line} = \frac{-R_S}{E_S} = \frac{-8.33}{10.1} = -0.825$$

From point P, coordinates (X_1, Y_1), in Fig. 5.17, the operating line of slope -0.825 is drawn, to reach the abscissa $X_2 = 0.1493$ at point Q. The conditions at Q correspond to the end of the first operation.

Second operation $X_1 = 0.1493$, $Y_1 = 0.00996$, which locate point S in the figure. The operating line is drawn parallel to the line PQ since the ratio $-R_S/E_S$ is the same as in the first operation. Extension of the line to the equilibrium curve at T provides the final value of $X_2 = 0.103$ kg water/kg dry soap.

$$\text{Final moisture content of soap} = \frac{0.103}{1.103} 100 = 9.33\%$$

(b) Equation (5.35) and Fig. 5.10 apply

$$R_S = 1(1 - 0.167) = 0.833 \text{ kg dry soap/h}$$

Entering soap: $X_1 = 0.20$ kg water/kg dry soap

Leaving soap: $X_2 = \dfrac{0.04}{1 - 0.04} = 0.0417$ kg water/kg dry soap

Entering air: $Y_2 = 0.00996$ kg water/kg dry air

$$\text{Slope of operating line} = \frac{R_S}{E_S} = \frac{\text{kg dry soap/h}}{\text{kg dry air/h}}$$

Point D, coordinates (X_2, Y_2), is plotted in Fig. 5.17. The operating line of least slope giving rise to equilibrium conditions will indicate the least amount of air usable. Such a line is DG, which provides equilibrium conditions at $X_1 = 0.20$. The corresponding value of $Y_1 = 0.068$ kg water/kg dry air is read from the figure at G. Eq. (5.35):

$$0.833(0.20 - 0.0417) = E_S(0.068 - 0.00996)$$

$$E_S = 2.27 \text{ kg dry air/h}$$

This corresponds to

$$\frac{2.27}{29} 22.41 \frac{760}{760 - 12} \frac{273 + 75}{273} = 2.27 \text{ m}^3/\text{kg dry soap}$$

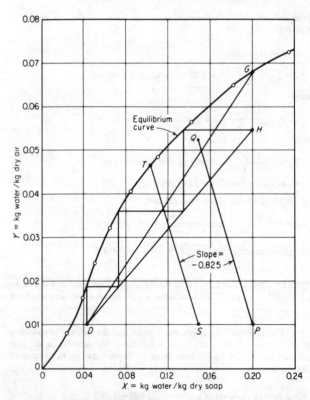

Figure 5.17 Solution to Illustration 5.2.

(c) $E_S = 1.30(2.27) = 2.95$ kg dry air/h. Eq. (5.35):

$$0.833(0.20 - 0.0417) = 2.95(Y_1 - 0.00996)$$

$$Y_1 = 0.0547 \text{ kg water/kg dry air}$$

This corresponds to operating line DH, where H has the coordinates $X_1 = 0.20$, $Y_1 = 0.0547$. The equilibrium stages are marked on the figure and total three.

NOTATION FOR CHAPTER 5

Any consistent set of units may be used.

A	substance A, the diffusing solute
A	absorption factor, dimensionless
e	2.7183
E	phase E; total rate of flow of phase E, mole/Θ
E_S	rate of flow of nondiffusing solvent in phase E, mole/Θ
E_{ME}	fractional Murphree stage efficiency for phase E
E_{MR}	fractional Murphree stage efficiency for phase R
f	equilibrium-distribution function
F_G	gas-phase mass-transfer coefficient, mole/$L^2\Theta$
F_L	liquid-phase mass-transfer coefficient, mole/$L^2\Theta$
F_{OG}	overall gas mass-transfer coefficient, mole/$L^2\Theta$
F_{OL}	overall liquid mass-transfer coefficient, mole/$L^2\Theta$
i	generalized concentration in phase R
j	generalized concentration in phase E
k_x	liquid mass-transfer coefficient, mole/$L^2\Theta$(mole fraction)
k_y	gas mass-transfer coefficient, mole/$L^2\Theta$(mole fraction)
K_x	overall liquid mass-transfer coefficient, mole/$L^2\Theta$(mole fraction)
K_y	overall gas mass-transfer coefficient, mole/$L^2\Theta$(mole fraction)
m	equilibrium-curve slope, dimensionless
m', m''	slopes of chords of the equilibrium curve, dimensionless
n	stage n
N	mass-transfer flux, mole/$L^2\Theta$
N_p	total number of ideal stages in a cascade
R	phase R; total rate of flow of phase R, mole/Θ
R_S	rate of flow of nondiffusing solvent in phase R, mole/Θ
S	stripping factor, dimensionless
x	concentration of A in a liquid or in phase R, mole fraction
X	concentration of A in phase R, mole A/mole non-A
y	concentration of A in a gas or in phase E, mole fraction
Y	concentration of A in phase E, mole A/mole non-A

Subscripts

A	substance A
e	equilibrium
E	phase E
G	gas
i	interface
L	liquid
n	stage n
O	overall
R	phase R

0   entering stage 1
1, 2   positions 1 and 2; stages 1 and 2

Superscript

*   in equilibrium with bulk concentration in other phase

REFERENCES

1. Davies, T. J.: *Adv. Chem. Eng.*, **4**, 1 (1963).
2. Garner, F. H., and A. H. P. Skelland: *Ind. Eng. Chem.*, **48**, 51 (1956).
3. Goodridge, F., and I. D. Robb: *Ind. Eng. Chem. Fundam.*, **4**, 49 (1965).
4. Gordon, K. F., and T. K. Sherwood: *Chem. Eng. Prog. Symp. Ser.*, **50**(10), 15 (1954); *AIChE J.*, **1**, 129 (1955).
5. Kent, E. R., and R. L. Pigford: *AIChE J.*, **2**, 363 (1956).
6. King, C. J.: *AIChE J.*, **10**, 671 (1964).
7. Kremser, A.: *Natl. Petrol. News*, **22**(21), 42 (1930).
8. Lewis, W. K., and W. G. Whitman: *Ind. Eng. Chem.*, **16**, 1215 (1924).
9. Murphree, E. V.: *Ind. Eng. Chem.*, **24**, 519 (1932).
10. Sawistowski, H.: in C. Hanson (ed.), "Recent Advances in Liquid-Liquid Extraction," p. 293, Pergamon, New York, 1971.
11. Schrage, R. W.: "A Theoretical Study of Interphase Mass Transfer," Columbia University Press, New York, 1953.
12. Scriven, L. E., and R. L. Pigford: *AIChE J.*, **4**, 439 (1958); **5**, 397 (1959).
13. Searle, R., and K. F. Gordon: *AIChE J.*, **3**, 490 (1957).
14. Souders, M., and G. G. Brown: *Ind. Eng. Chem.*, **24**, 519 (1932).
15. Thompson, D. W.: *Ind. Eng. Chem. Fundam.*, **9**, 243 (1970).
16. Turner, S. D.: *Ind. Eng. Chem.*, **21**, 190 (1929).
17. Whitman, W. G.: *Chem. Met. Eng.*, **29**, 147 (1923).
18. Wylie, C. R., Jr.: "Advanced Engineering Mathematics," 3d ed., McGraw-Hill Book Company, New York, 1966.

PROBLEMS

5.1 Repeat the calculations for the local mass-transfer flux of ammonia in Illustration 5.1 on the assumption that the gas pressure is 2 std atm. The gas mass velocity, bulk-average gas and liquid concentration, liquid flow rate, and temperature are unchanged.

5.2 In a certain apparatus used for the absorption of sulfur dioxide, SO_2, from air by means of water, at one point in the equipment the gas contained 10% SO_2 by volume and was in contact with liquid containing 0.4% SO_2 (density = 990 kg/m³ = 61.8 lb/ft³). The temperature was 50°C and the total pressure 1 std atm. The overall mass-transfer coefficient based on gas concentrations was $K_G = 7.36 \times 10^{-10}$ kmol/m² · s · (N/m²) = 0.055 lb mol SO_2/ft² · h · atm. Of the total diffusional resistance, 47% lay in the gas phase, 53% in the liquid. Equilibrium data at 50°C are:

kg SO_2/100 kg water	0.2	0.3	0.5	0.7
Partial pressure SO_2, mmHg	29	46	83	119

(*a*) Calculate the overall coefficient based on liquid concentrations in terms of mol/vol.

(*b*) Calculate the individual mass-transfer coefficient for the gas, expressed as k_G mol/(area)(time)(pressure), k_y mol/(area)(time)(mole fraction), and k_c mol/(area)(time)(mole/vol), and for the liquid expressed as k_L mole/(area)(time)(mole/vol) and k_x mol/(area)(time)(mole fraction).

(c) Determine the interfacial compositions in both phases.

5.3 The equilibrium partial pressure of water vapor in contact with a certain silica gel on which water is adsorbed is, at 25°C, as follows:

Partial pressure of water, mmHg	0	2.14	4.74	7.13	9.05	10.9	12.6	14.3	16.7
lb water/100 lb dry gel = kg/100 kg	0	5	10	15	20	25	30	35	40

(a) Plot the equilibrium data as $\bar{p}$ = partial pressure of water vapor against x = wt fraction water in the gel.

(b) Plot the equilibrium data as X = mol water/mass dry gel, Y = mol water vapor/mol dry air, for a total pressure of 1 std atm.

(c) When 10 lb (4.54 kg) of silica gel containing 5 wt% adsorbed water is placed in a flowing airstream containing a partial pressure of water vapor = 12 mmHg, the total pressure is 1 std atm and the temperature 25°C. When equilibrium is reached, what mass of additional water will the gel have adsorbed? Air is not adsorbed.

(d) One pound mass (0.454 kg) of silica gel containing 5 wt% adsorbed water is placed in a vessel in which there are 400 ft³ (11.33 m³) moist air whose partial pressure of water is 15 mmHg. The total pressure and temperature are kept at 1 std atm and 25°C, respectively. At equilibrium, what will be the moisture content of the air and gel and the mass of water adsorbed by the gel? Ans.: 0.0605 kg water.

(e) Write the equation of the operating line for part (d) in terms of X and Y. Convert this into an equation in terms of $\bar{p}$ and x, and plot the operating curve on $\bar{p}$, x coordinates.

(f) If 1 lb (0.454 kg) of silica gel containing 18% adsorbed water is placed in a vessel containing 500 ft³ (14.16 m³) dry air and the temperature and pressure are maintained at 25°C and 1 std atm, respectively, what will be the final equilibrium moisture content of the air and the gel.

(g) Repeat part (f) for a total pressure of 2 std atm. Note that the equilibrium curve in terms of X and Y previously used is not applicable.

5.4 The equilibrium adsorption of benzene vapor on a certain activated charcoal at 33.3°C is reported as follows:

Benzene vapor adsorbed, cm³ (STP)/g charcoal	15	25	40	50	65	80	90	100
Partial pressure benzene, mmHg	0.0010	0.0045	0.0251	0.115	0.251	1.00	2.81	7.82

(a) A nitrogen–benzene vapor mixture containing 1.0% benzene by volume is to be passed countercurrently at the rate 100 ft³/min (4.72 × 10⁻² m³/s) to a moving stream of the activated charcoal so as to remove 95% of the benzene from the gas in a continuous process. The entering charcoal contains 15 cm³ benzene vapor (at STP) adsorbed per gram charcoal. The temperature and total pressure are to be maintained at 33.3°C and 1 std atm, respectively, throughout. Nitrogen is not adsorbed. What is the least amount of charcoal which can be used per unit time? If twice as much is used, what will be the concentration of benzene adsorbed upon the charcoal leaving?

(b) Repeat part (a) for a cocurrent flow of gas and charcoal.

(c) Charcoal which has adsorbed upon it 100 cm³ (at STP) benzene vapor per gram is to be stripped at the rate 100 lb/h (45.4 kg/h) of its benzene content to a concentration of 55 cm³ adsorbed benzene per gram charcoal by continuous countercurrent contact with a stream of pure nitrogen gas at 1 std atm. The temperature will be maintained at 33.3°C. What is the minimum rate of nitrogen flow? What will be the benzene content of the exit gas if twice as much nitrogen is fed? What will be the number of stages? Ans.: 3 theoretical stages.

5.5 A mixture of hydrogen and air, 4 std atm pressure, bulk-average concentration 50% H_2 by volume, flows through a circular reinforced tube of vulcanized rubber with respective ID and OD of 25.4 and 28.6 mm (1.0 and 1.125 in) at an average velocity 3.05 m/s (10 ft/s). Outside the pipe, hydrogen-free air at 1 std atm flows at right angles to the pipe at a velocity 3.05 m/s (10 ft/s). The temperature is everywhere 25°C. The solubility of hydrogen in the rubber is reported as 0.053 cm^3 H_2 (STP)/(cm^3 rubber) · (atm partial pressure), and its diffusivity in the rubber as 0.18×10^{-5} cm^2/s. The diffusivity of H_2–air $= 0.611$ cm^2/s at 0°C, 1 std atm. Estimate the rate of loss of hydrogen from the pipe per unit length of pipe. Ans.: 4.5×10^{-11} kmol/(m of pipe length) · s.

5.6 Assuming that the equilibrium curve of Fig. 5.7 is straight, and of slope m, derive the relation between the Murphree stage efficiencies E_{ME} and E_{MR}.

5.7 If the equilibrium-distribution curve of the cross-flow cascade of Fig. 5.13 is everywhere straight, and of slope m, make a solute material balance about stage $n + 1$ and by following the procedure used to derive Eq. (5.51) show that

$$N_p = \frac{\log\left[(X_0 - Y_0/m)/(X_{N_p} - Y_0/m)\right]}{\log(S + 1)}$$

where S is the stripping factor, mE_S/R_S, constant for all stages, and N_p is the total number of stages.

5.8 A single-stage liquid-extraction operation with a linear equilibrium-distribution curve ($Y = mX$ at equilibrium) operates with a stripping factor $S = mE_S/R_S = 1.0$ and has a Murphree stage efficiency $E_{ME} = 0.6$. In order to improve the extraction with a minimum of plant alteration, the effluents from the stage will both be led to another stage identical in construction, also of efficiency $E_{ME} = 0.6$. The combination, two stages in series, is in effect a single stage. It is desired to compute the overall Murphree stage efficiency E_{MEO} of the combination. A direct derivation of the relation between E_{ME} and E_{MEO} is difficult. The result is most readily obtained as follows:

(a) Refer to Fig. 5.7. Define a stage efficiency for one stage as $E = (Y_2 - Y_1)/(Y_e - Y_1)$. For a linear equilibrium-distribution curve, derive the relation between E_{ME} and E.

(b) Derive the relation between E_O, the value for the combined stages in series, and E.

(c) Compute E_{MEO}. Ans. = 0.8824.

TWO

GAS-LIQUID OPERATIONS

The operations which include humidification and dehumidification, gas absorption and desorption, and distillation in its various forms all have in common the requirement that a gas and a liquid phase be brought into contact for the purpose of a diffusional interchange between them.

The order listed above is in many respects that of increasing complexity of the operations, and this is therefore the order in which they will be considered. Since in humidification the liquid is a pure substance, concentration gradients exist and diffusion of matter occurs only in the gas phase. In absorption, concentration gradients exist in both the liquid and the gas, and diffusion of at least one component occurs in both. In distillation, all the substances comprising the phases diffuse. These operations are also characterized by an especially intimate relationship between heat and mass transfer. The evaporation or condensation of a substance introduces consideration of latent heats of vaporization, sometimes heats of solution as well. In distillation, the new phase necessary for mass-transfer separation is created from the original by addition or withdrawal of heat. Our discussion must necessarily include consideration of these important heat quantities and their effects.

In all these operations, the equipment used has as its principal function the contact of the gas and liquid in as efficient a fashion as possible, commensurate with the cost. In principle, at least, any type of equipment satisfactory for one of these operations is also suitable for the others, and the major types are indeed used for all. For this reason, our discussion begins with equipment.

EQUIPMENT FOR
GAS-LIQUID OPERATIONS

The purpose of the equipment used for the gas-liquid operations is to provide intimate contact of the two fluids in order to permit interphase diffusion of the constituents. The rate of mass transfer is directly dependent upon the interfacial surface exposed between the phases, and the nature and degree of dispersion of one fluid in the other are therefore of prime importance. The equipment can be broadly classified according to whether its principal action is to disperse the gas or the liquid, although in many devices both phases become dispersed.

GAS DISPERSED

In this group are included those devices, such as sparged and agitated vessels and the various types of tray towers, in which the gas phase is dispersed into bubbles or foams. Tray towers are the most important of the group, since they produce countercurrent, multistage contact, but the simpler vessel contactors have many applications.

Gas and liquid can conveniently be contacted, with gas dispersed as bubbles, in agitated vessels whenever multistage, countercurrent effects are not required. This is particularly the case when a chemical reaction between the dissolved gas and a constituent of the liquid is required. The carbonation of a lime slurry, the chlorination of paper stock, the hydrogenation of vegetable oils, the aeration of fermentation broths, as in the production of penicillin, the production of citric acid from beet sugar by action of microorganisms, and the aeration of activated sludge for biological oxidation are all examples. It is perhaps significant that in most of them solids are suspended in the liquids. Because the more complicated countercurrent towers have a tendency to clog with such solids and because solids can be suspended in the liquids easily in agitated vessels, the latter are usually more successful in such service.†

The gas-liquid mixture can be mechanically agitated, as with an impeller, or, as in the simplest design, agitation can be accomplished by the gas itself in sparged vessels. The operation may be batch, semibatch with continuous flow of gas and a fixed quantity of liquid, or continuous with flow of both phases.

SPARGED VESSELS (BUBBLE COLUMNS)

A *sparger* is a device for introducing a stream of gas in the form of small bubbles into a liquid. If the vessel diameter is small, the sparger, located at the bottom of the vessel, may simply be an open tube through which the gas issues into the liquid. For vessels of diameter greater than roughly 0.3 m, it is better to use several orifices for introducing the gas to ensure better gas distribution. In that case, the orifices may be holes, from 1.5 to 3 mm ($\frac{1}{16}$ to $\frac{1}{4}$ in) in diameter, drilled in a pipe distributor placed horizontally at the bottom of the vessel. Porous plates made of ceramics, plastics, or sintered metals are also used, but their fine pores are more readily plugged with solids which may be present in the gas or the liquid.

The purpose of the sparging may be contacting the sparged gas with the liquid. On the other hand, it may be simply a device for agitation. It can provide the gentlest of agitation, used for example in washing nitroglycerin with water; it can provide vigorous agitation as in the *pachuca tank* (which see). Air agitation in the extraction of radioactive liquids offers the advantage of freedom from moving parts, but it may require decontamination of the effluent air. There is no standardization of the depth of liquid; indeed, very deep tanks, 15 m (50 ft) or more, may be advantageous [117] despite the large work of compression required for the gas.

Gas-Bubble Diameter

The size of gas bubbles depends upon the rate of flow through the orifices, the orifice diameter, the fluid properties, and the extent of turbulence prevailing in the liquid. What follows is for cases where turbulence in the liquid is solely that generated by the rising bubbles and when orifices are horizontal and sufficiently separated to prevent bubbles from adjacent orifices from interfering with each other (at least approximately $3d_p$ apart).

---

† For cases where very low gas-pressure drop is required, as in absorption of sulfur dioxide from flue gas by limestone slurries, venturi scrubbers (which see) are frequently used.

Very slow gas flow rate [118], $Q_{Go} < [20(\sigma d_o g_c)^5/(g\,\Delta\rho)^2\rho_L^3]^{1/6}$ For waterlike liquids, the diameter can be computed by equating the bouyant force on the immersed bubble, $(\pi/6)d_p^3\,\Delta\rho g/g_c$, which tends to lift the bubble away from the orifice, to the force $\pi d_o\sigma$ due to surface tension, which tends to retain it at the orifice. This provides

$$d_p = \left(\frac{6d_o\sigma g_c}{g\,\Delta\rho}\right)^{1/3} \tag{6.1}$$

which has been tested for orifice diameters up to 10 mm. For large liquid viscosities, up to 1 kg/m · s (1000 cP) [32],

$$d_p = 2.312\left(\frac{\mu_L Q_{Go}}{\rho_L g}\right)^{1/4} \tag{6.2}$$

Intermediate flow rates, $Q_{Go} > [20(\sigma d_o g_c)^5/(g\,\Delta\rho)^2\rho_L^3]^{1/6}$ but $Re_o < 2100$ These bubbles are larger than those described above, although still fairly uniform, and they form in chains rather than separately. For air-water [71]

$$d_p = 0.0287 d_o^{1/2}\,Re_o^{1/3} \tag{6.3}$$

where d_p and d_o are in meters† and $Re_o = d_o V_o \rho_G/\mu_G = 4w_o/\pi d_o \mu_G$. For other gases and liquids [118]

$$d_p = \left(\frac{72\rho_L}{\pi^2 g\,\Delta\rho}\right)^{1/5} Q_{Go}^{0.4} \tag{6.4}$$

Large gas rates [71], $Re_o = 10\,000$ to $50\,000$ Jets of gas which rise from the orifice break into bubbles at some distance from the orifice. The bubbles are smaller than those described above and nonuniform in size. For air-water and orifice diameters 0.4 to 1.6 mm

$$d_p = 0.0071\,Re_o^{-0.05} \tag{6.5}$$

with d_p in meters.‡ For the *transition range* ($Re_o = 2100$ to $10\,000$) there is no correlation of data. It is suggested that d_p for air-water can be approximated by the straight line on log-log coordinates between the points given by d_p at $Re_o = 2100$ and at $Re_o = 10\,000$.

Rising Velocity (Terminal Velocity) of Single Bubbles [83]

Typically, the steady-state rising velocity of single gas bubbles, which occurs when the bouyant force equals the drag force on the bubbles, varies with the bubble diameter as shown in Fig. 6.1.

---

† For d_p and d_o in feet, the coefficient of Eq. (6.3) is 0.052.
‡ For d_p in feet, the coefficient of Eq. (6.5) is 0.0233.

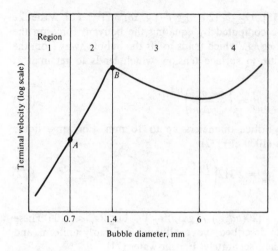

Figure 6.1 Rising (terminal) velocity of single gas bubbles.

Region 1, $d_p < 0.7$ mm The bubbles are spherical, and they behave like rigid spheres, for which the terminal velocity is given by Stokes' law

$$V_t = \frac{g d_p^2 \, \Delta\rho}{18\mu_L} \tag{6.6}$$

Region 2, 0.7 mm $< d_p < 1.4$ mm The gas within the bubble circulates, so that the surface velocity is not zero. Consequently the bubble rises faster than rigid spheres of the same diameter. There is no correlation of data; it is suggested that V_t may be estimated as following the straight line on Fig. 6.1 drawn between points A and B. Coordinates of these points are given respectively by Eqs. (6.6) and (6.7).

Regions 3 (1.4 mm $< d_p < 6$ mm) and 4 ($d_p > 6$ mm) The bubbles are no longer spherical and in rising follow a zigzag or helical path. For region 4 the bubbles have a spherically shaped cap. In both these regions for liquids of low viscosity [80]

$$V_t = \sqrt{\frac{2\sigma g_c}{d_p \rho_L} + \frac{g d_p}{2}} \tag{6.7}$$

Swarms of Gas Bubbles

The behavior of large numbers of bubbles crowded together is different from that of isolated bubbles. Rising velocities are smaller because of crowding, and bubble diameter may be altered by liquid turbulence, which causes bubble breakup, and by coalescence of colliding bubbles.

Gas Holdup

By *gas holdup* φ_G is meant the volume fraction of the gas-liquid mixture in the vessel which is occupied by the gas. If the superficial gas velocity, defined as the volume rate of gas flow divided by the cross-sectional area of the vessel, is V_G, then V_G/φ_G can be taken as the true gas velocity relative to the vessel walls. If the liquid flows upward, cocurrently with the gas, at a velocity relative to the vessel walls $V_L/(1 - \varphi_G)$, the relative velocity of gas and liquid, or *slip velocity*, is

$$V_S = \frac{V_G}{\varphi_G} - \frac{V_L}{1 - \varphi_G} \tag{6.8}$$

Equation (6.8) will also give the slip velocity for countercurrent flow of liquid if V_L for the downward liquid flow is assigned a negative sign.

The holdup for sparged vessels, correlated through the slip velocity, is shown in Fig. 6.2. This is satisfactory for no liquid flow ($V_L = 0$), for cocurrent

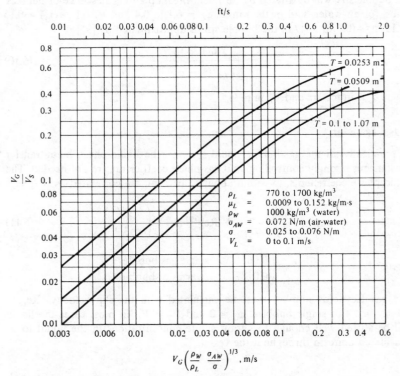

Figure 6.2 Slip velocity, sparged vessels. [*G. A. Hughmark, Ind. Eng. Chem. Process Des. Dev.*, **6**, 218 (1967), *with permission of the American Chemical Society.*]

liquid flow up to $V_L = 0.1$ m/s (0.3 ft/s), and very probably also for small countercurrent liquid flow [1]. For large countercurrent velocities, holdup is best treated differently [17]. The holdup increases with increasing liquid viscosity, passes through a maximum at $\mu_L = 0.003$ kg/m $\cdot$ s (3 cP), and then decreases [40].

Specific Interfacial Area

If unit volume of a gas-liquid mixture contains a gas volume φ_G made up of n bubbles of diameter d_p, then $n = \varphi_G/(\pi d_p^3/6)$. If the interfacial area in the unit volume is a, then $n = a/\pi d_p^2$. Equating the two expressions for n provides the specific area

$$a = \frac{6\varphi_G}{d_p} \tag{6.9}$$

For low velocities, the bubble size may be taken as that produced at the orifices of the sparger, corrected as necessary for pressure. If the liquid velocity is large, the bubble size will be altered by turbulent breakup and coalescence of bubbles. Thus, for air-water, and in the ranges $\varphi_G = 0.1$ to 0.4 and $V_L/(1 - \varphi_G) = 0.15$ to 15 m/s (0.5 to 5 ft/s), the bubble size is approximated by [95]

$$d_p = \frac{2.344 \times 10^{-3}}{\left[V_L/(1 - \varphi_G)\right]^{0.67}} \tag{6.10}$$

where d_p is in meters and V_L in m/s.†

Mass Transfer

In practically all the gas-bubble liquid systems, the liquid-phase mass-transfer resistance is strongly controlling, and gas-phase coefficients are not needed. The liquid-phase coefficients, to within about 15 percent, are correlated by [61]

$$\mathrm{Sh}_L = \frac{F_L d_p}{c D_L} = 2 + b' \, \mathrm{Re}_G^{0.779} \, \mathrm{Sc}_L^{0.546} \left(\frac{d_p g^{1/3}}{D_L^{2/3}} \right)^{0.116} \tag{6.11}$$

where
$$b' = \begin{cases} 0.061 & \text{single gas bubbles} \\ 0.0187 & \text{swarms of bubbles} \end{cases}$$

The gas Reynolds number must be calculated with the slip velocity: $\mathrm{Re}_G = d_p V_S \rho_L/\mu_L$. For single bubbles, $\varphi_G = 0$ and $V_S = V_t$. In most cases the liquid will be stirred well enough to permit solute concentrations in the liquid to be considered uniform throughout the vessel.

---

† For d_p in feet and V_L in ft/s, the coefficient of Eq. (6.10) is 0.01705.

Power

The power supplied to the vessel contents, which is responsible for the agitation and creation of large interfacial area, is derived from the gas flow. Bernoulli's equation, a mechanical-energy balance written for the gas between location o (just above the sparger orifices) and location s (at the liquid surface), is

$$\frac{V_s^2 - V_o^2}{2g_c} + (Z_s - Z_o)\frac{g}{g_c} + \int_o^s \frac{dp}{\rho_G} + W + H_f = 0 \qquad (6.12)$$

The friction loss H_f and V_s can be neglected, and the gas density can be described by the ideal-gas law, whereupon

$$W = \frac{V_o^2}{2g_c} + \frac{p_o}{\rho_{Go}} \ln \frac{p_o}{p_s} + (Z_o - Z_s)\frac{g}{g_c} \qquad (6.13)$$

W is the work done by the gas on the vessel contents, per unit mass of gas. Work for the gas compressor will be larger, to account for friction in the piping and orifices, any dust filter for the gas that may be used, and compressor inefficiency.

Illustration 6.1 A vessel 1.0 m in diameter and 3.0 m deep (measured from gas sparger at the bottom to liquid overflow at the top) is to be used for stripping chlorine from water by sparging with air. The water will flow continuously upward at the rate 8×10^{-4} m^3/s (1.7 ft^3/min). Airflow will be 0.055 kg/s (7.28 lb/min) at 25°C. The sparger is in the form of a ring, 25 cm in diameter, containing 50 orifices, each 3.0 mm in diameter. Estimate the specific interfacial area and the mass-transfer coefficient.

SOLUTION The gas flow rate per orifice = 0.055/50 = 0.0011 kg/s = $w_o \cdot d_o$ = 0.003 m, and $\mu_G = 1.8 \times 10^{-5}$ kg/m · s. $Re_o = 4w_o/\pi d_o\mu G$ = 26 000; Eq. (6.5): $d_o = 4.27 \times 10^{-3}$ m at the orifice. The pressure at the orifice, ignoring the gas holdup, is that corresponding to 3.0 m of water (density 1000 kg/m^3) or 101.3 kN/m^2 std atm + 3.0(1000)(9.81)/1000 = 130.8 kN/m^2. The average pressure in the column (for 1.5 m of water) is 116.0 kN/m^2. Therefore

$$\rho_G = \frac{29}{22.41} \frac{273}{298} \frac{116.0}{101.3} = 1.358 \text{ kg/m}^3$$

$T = 1.0$ m, and the vessel cross-sectional area = $\pi T^2/4 = 0.785$ m^2.

$$V_G = \frac{0.055}{0.785(1.358)} = 0.0516 \text{ m/s} \quad \text{and} \quad V_L = \frac{8 \times 10^{-4}}{0.785} = 0.00102 \text{ m/s}$$

Since $\rho_L = 1000$ kg/m^3 and $\sigma = 0.072$ N/m for the liquid, the abscissa of Fig. 6.2 is 0.0516 m/s. Hence (Fig. 6.2), $V_G/V_S = 0.11$ and $V_S = V_G/0.11 = 0.469$ m/s. Eq. (6.8):

$$0.469 = \frac{0.0516}{\varphi_G} - \frac{0.00102}{1 - \varphi_G}$$

$$\varphi_G = 0.1097$$

Equation (6.10) is not to be used in view of the small V_L. Therefore the bubble diameter at the average column pressure, calculated from d_p at the orifice, is

$$d_p = \left[(0.00427)^3 \frac{130.8}{116.0} \right]^{1/3} = 0.00447 \text{ m}$$

Eq. (6.9): $a = 147.2$ m^2/m^3. **Ans.**

At 25°C, D_L for Cl_2 in $H_2O = 1.44 \times 10^{-9}$ m^2/s; $g = 9.81$ m/s^2. In Eq. (6.11), $d_p g^{1/3}/D_L^{2/3} = 7500$. The viscosity of water $\mu_1 = 8.937 \times 10^{-4}$ kg/m · s, and $Re_G = d_p V_S \rho_L/\mu_L = 2346$, $Sc_L = \mu_L/\rho_L D_L = 621$, and hence $Sh_L = 746$. For dilute solutions of Cl_2 in H_2O the molar density $c = 1000/18.02 = 55.5$ kmol/m^3. Therefore $F_L = (cD_L Sh_L)/d_p = 0.0127$ kmol/m^2 · s. **Ans.**

MECHANICALLY AGITATED VESSELS

Mechanical agitation of a liquid, usually by a rotating device, is especially suitable for dispersing solids, liquids, or gases into liquids, and it is used for many of the mass-transfer operations. Agitators can produce very high turbulence intensities (u'/u averages as high as 0.35 [104] ranging up to 0.75 in the vicinity of an agitating impeller [31, 74], contrasted with about 0.04 for turbulent flow in pipes), which not only produce good mass-transfer coefficients but are necessary for effective dispersion of liquids and gases. High liquid velocities, particularly desirable for suspending solids, are readily obtained.

It will be useful first to establish the characteristics of these vessels when used simply for stirring single-phase liquids, as in blending, for example. Applications to particular mass-transfer operations will be considered separately later.

MECHANICAL AGITATION OF SINGLE-PHASE LIQUIDS

Typical agitated vessels are vertical circular cylinders; rectangular tanks are unusual, although not uncommon in certain liquid-extraction applications. The liquids are usually maintained at a depth of one to two tank diameters.

Impellers

There are literally scores of designs. The discussion here is limited to the most popular, as shown in Fig. 6.3. These are usually mounted on an axially arranged, motor-driven shaft, as in Fig. 6.4. In the smaller sizes, particularly, the impeller and shaft may enter the vessel at an angle to the vessel axis, with the motor drive clamped to the rim of the vessel.

The *marine-type propeller*, Fig. 6.3a, is characteristically operated at relatively high speed, particularly in low-viscosity liquids, and is especially useful for its high liquid-circulating capacity. The ratio of impeller diameter to vessel diameter, d_i/T, is usually set at 1 : 5 or less. The liquid flow is axial, and the propeller is turned so that it produces downward flow toward the bottom of the vessel. In describing the propeller, *pitch* refers to the ratio of the distance advanced per revolution by a free propeller operating without slip to the propeller diameter; *square pitch*, which is most common in agitator designs, means a pitch equal to unity. Propellers are more frequently used for liquid-blending operations than for mass-transfer purposes.

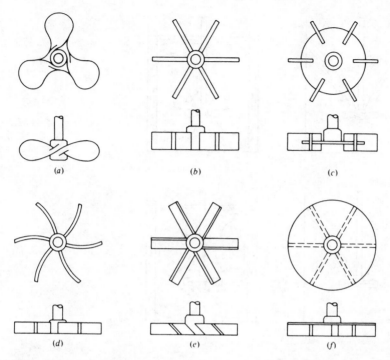

Figure 6.3 Impellers, with typical proportions: (*a*) marine-type propellers; (*b*) flat-blade turbine, $w = d_i/5$, (*c*) disk flat-blade turbine, $w = d_i/5$, $d_d = 2d_i/3$, $B = d_i/4$; (*d*) curved-blade turbine, $w = d_i/8$; (*e*) pitched-blade turbine, $w = d_i/8$; and (*f*) shrouded turbine, $w = d_i/8$.

Turbines, particularly the flat-blade designs of Fig. 6.3*b* and *c*, are frequently used for mass-transfer operations.† The curved blade design (Fig. 6.3*d*) is useful for suspension of fragile pulps, crystals, and the like and the pitched-blade turbine (Fig. 6.3*e*) more frequently for blending liquids. The shrouded impeller of Fig. 6.3*f* has limited use in gas-liquid contracting. Flow of liquid from the impeller is radial except for the pitched-blade design, where it is axial. Although the best ratio d_i/T depends upon the application, a value of 1 : 3 is common. Turbines customarily operate with peripheral speeds of the order of 2.5 to 4.6 m/s (450 to 850 ft/min), depending upon the service.

Vortex Formation and Prevention

Typical flow patterns for single-phase Newtonian liquids of moderate viscosity are shown in Fig. 6.4. For an axially located impeller operating at low speeds in

---

† These designs have been in use for many years. G. Agricola's "De re metallica" (H. C. and L. H. Hoover, translators, from the Latin of 1556, book 8, p. 299, Dover, New York, 1950) shows a woodcut of six-bladed flat-blade turbine impellers being used to suspend a gold ore in water.

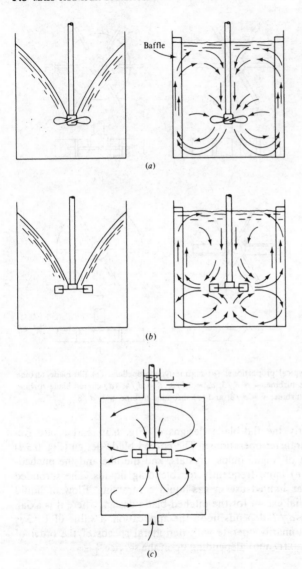

Figure 6.4 Liquid agitation in presence of a gas-liquid interface, with and without wall baffles: (*a*) marine impeller and (*b*) disk flat-blade turbines; (*c*) in full vessels without a gas-liquid interface (continuous flow) and without baffles.

an open vessel with a gas-liquid surface, the liquid surface is level and the liquid circulates about the axis. As the impeller speed is increased to produce turbulent conditions, the power required to turn the impeller increases and a vortex begins to form around the shaft. At higher speeds the vortex eventually reaches the impeller, as in the left-hand sketches of Fig. 6.4a and b. Air is drawn into the liquid, the impeller operates partly in air, and the power required drops. The drawing of gas into the liquid is frequently undesirable, and, in addition, vortex formation leads to difficulties in scaling up of model experiments and pilot-plant studies, so that steps are usually taken to prevent vortices. A possible exception is when a solid that is difficult to wet is to be dissolved in a liquid, e.g., powdered boric acid in water, when it is useful to pour the solid into the vortex.

Vortex formation can be prevented by the following means:

1. *Open tanks with gas-liquid surfaces.*†
   a. Operation only in the laminar range for the impeller (Re 10 to 20). This is usually impractically slow for mass-transfer purposes.
   b. Off-center location of the impeller on a shaft entering the vessel at an angle to the vessel axis. This is used relatively infrequently in permanent installations, and is used mostly in small-scale work with impellers where the agitator drive is clamped to the rim of the vessel.
   c. Installation of baffles. This is by far the most common method. Standard baffling [78] consists of four flat, vertical strips arranged radially at 90° intervals around the tank wall, extending for the full liquid depth, as in the right-hand sketches of Fig. 6.4a and b. The standard baffle width is usually $T/12$, less frequently $T/10$ ("10 percent baffles"). Baffles are sometimes set at a clearance from the vessel wall of about one-sixth the baffle width to eliminate stagnant pockets in which the solids can accumulate. They may be used as supports for helical heating or cooling coils immersed in the liquid. The condition of *fully baffled turbulence* is considered to exist for these vessels at impeller Reynolds numbers above 10 000.‡ The presence of baffles reduces swirl and increases the vertical liquid currents in the vessel, as shown on the right of Fig. 6.4. The power required to turn the impeller is also increased.
2. *Closed tanks, operated full, with no gas-liquid surface* [70]. This is especially convenient for cases where the liquid flows continuously through the vessel, as in Fig. 6.4c. A circular flow pattern is superimposed upon the axial flow directed toward the center of the impeller. The arrangement is not practical, of course, for gas-liquid contact, where baffles should be used.

---

† Addition of minute amounts of certain polymers to the liquid can inhibit vortex formation during bottom drainage of unagitated tanks [R. J. Gordon, et al., *Nature Phys. Sci.*, **231**, 25 (1971); *J. Appl. Polymer Sci.*, **16**, 1629 (1972); *AIChE J.*, **22**, 947 (1976)].

‡ At very high speeds, gas can still be drawn into the impeller despite the presence of baffles. For flat-blade turbines with four blades in tanks with standard baffles, this will occur when [27]

$$\frac{d_i N^2}{g} \frac{d_i^2 w}{T^2 Z} \left(\frac{C}{Z} \right)^{2/3} > 0.005$$

Fluid Mechanics

Discussions of the characteristics of agitated vessels frequently involve considerations of similarity, and this is particularly important when it is desired to obtain simple laws describing vessels of different sizes. For example, with proper regard to similarity and by describing results in terms of dimensionless groups, it is possible to experiment with a vessel filled with air and expect the results for power, degree of turbulence, and the like, to be applicable to vessels filled with liquid [52]. Three types of similarity are significant when dealing with liquid motion:

1. *Geometric similarity* refers to linear dimensions. Two vessels of different sizes are geometrically similar if the ratios of corresponding dimensions on the two scales are the same, and this refers to tank, baffles, impellers, and liquid depths. If photographs of two vessels are completely superimposable, they are geometrically similar.
2. *Kinematic similarity* refers to motion and requires geometric similarity and the same ratio of velocities for corresponding positions in the vessels. This is especially important for mass-transfer studies.
3. *Dynamic similarity* concerns forces and requires all force ratios for corresponding positions to be equal in kinematically similar vessels.

The complex motion in an agitated vessel cannot be put into complete analytical form, but the equations of motion can be described in terms of dimensionless groups as follows [5]:

$$f\left(\frac{\bar{L} u \rho}{\mu}, \frac{u^2}{\bar{L} g}, \frac{\Delta p \, g_c}{\rho u^2} \right) = 0 \qquad (6.14)$$

where $\bar{L}$ is some characteristic length and Δp is a pressure difference. The first of these groups is the ratio of inertial to viscous forces, the familiar Reynolds number. If the characteristic length is taken as d_i and the velocity proportional to the impeller-tip speed $\pi N d_i$, when the π is omitted the group becomes the impeller Reynolds number $\text{Re} = d_i^2 N \rho_L / \mu_L$.

The second group, the ratio of inertial to gravity forces, important when the liquid surface is wavy or curved, as with a vortex, is the *Froude number*. With u and $\bar{L}$ defined as before, it becomes $\text{Fr} = d_i N^2 / g$.

The third is the ratio of pressure differences resulting in flow to inertial forces. If Δp is taken as force/area, the group can be recognized as a *drag coefficient*, which is usually written $F_D / A(\rho u^2 / 2 g_c)$. The group can be developed into a dimensionless *power number* as follows. Power required by the impeller is the product of Δp and the volumetric liquid-flow rate Q_L, so that

$$\Delta p = \frac{P}{Q_L} \qquad (6.15)$$

The fluid velocity is proportional [104] to $N d_i^2 / T$ or, for geometrically similar vessels, to $N d_i$, so that Q_L is proportional to $N d_i^3$ [52]. Substitution of these transforms the group into $\text{Po} = P g_c / \rho N^3 d_i^5$.

For geometrically similar vessels, therefore, Eq. (6.14) can be written

$$f(\text{Re, Fr, Po}) = 0 \qquad (6.16)$$

For nongeometrical similarity, a variety of dimension ratios is also required [4, 102]

$$\frac{T}{d_i}, \frac{Z}{d_i}, \frac{C}{d_i}, \frac{l}{d_i}, \frac{w}{d_i}, \frac{nb}{T}$$

For two-phase dispersions, other groups such as the *Weber number* We = $\rho u^2 d_p / \sigma g_c$, the ratio of inertial to surface forces, may be significant. *Sherwood* and *Schmidt numbers* are of course important in mass transfer. For dynamic similarity in two sizes of vessels operated with a vortex, the Reynolds and Froude numbers must be the same for both vessels. Since the impellers would be geometrically similar but unequally sized, it becomes impossible to specify the impeller speeds in the two vessels containing the same liquid to accomplish this. Thus, equal Reynolds numbers require

$$\frac{N_1}{N_2} = \left(\frac{d_{i2}}{d_{i1}} \right)^2$$

while equal Froude numbers require

$$\frac{N_1}{N_2} = \left(\frac{d_{i2}}{d_{i1}} \right)^{0.5}$$

where 1 and 2 designate the two sizes, and the two conditions cannot be met simultaneously. For this reason most work with open vessels has been done with baffles to prevent vortex formation. Closed vessels operated full cannot form a vortex, and baffles for these are unnecessary. For these nonvortexing cases, the Froude number is no longer important; Eq. (6.16) becomes

$$f(\text{Re, Po}) = 0 \qquad (6.17)$$

Power in nonvortexing systems For geometrically similar vessels operated with single-phase newtonian liquids, the power characteristics of the impellers of Fig. 6.3 are shown in Fig. 6.5. These curves represent Eq. (6.17) and were obtained experimentally. At Reynolds numbers less than about 10, the curves have a slope of -1, and at very high Reynolds numbers the power number becomes constant. The curves are thus quite analogous to those for the drag coefficient of an immersed body or the friction factor for pipe flow.

All the energy in the liquid stream from an impeller is dissipated through turbulence and viscosity to heat [93]. The velocity gradient in the liquid, which sets the shear rate and hence the turbulence intensity, is greatest at the impeller tip, less at a distance relatively removed. At high Reynolds numbers for all the curves of Fig. 6.5, where the power number becomes constant, the impeller power becomes independent of liquid viscosity and varies only as $\rho_L N^3 d_i^5$. Since it has already been shown that for geometrically similar vessels Q varies as Nd_i^3, at constant power number P is proportional to $QN^2 d_i^2$ for a given liquid. Consequently a small impeller operating at high speed produces less flow and

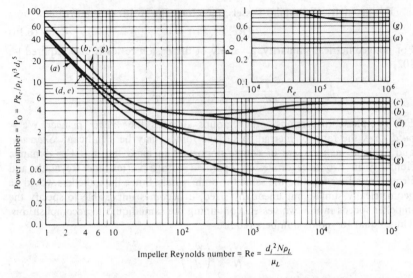

Impeller Reynolds number = Re = $\dfrac{d_i^2 N \rho_L}{\mu_L}$

Figure 6.5 Power for agitation impellers immersed in single-phase liquids, baffled vessels with a gas-liquid surface [except curves (c) and (g)]. Curves correspond to impellers of Fig. 6.3: (a) marine impellers, (b) flat-blade turbines, $w = d_i/5$, (c) disk flat-blade turbines with and without a gas-liquid surface, (d) curved-blade turbines, (e) pitched-blade turbines, (g) flat-blade turbines, no baffles, no gas-liquid interface, no vortex.

Notes on Fig. 6.5

1. The power P is only that imparted to the liquid by the impeller. It is not that delivered to the motor drive, which additionally includes losses in the motor and speed-reducing gear. These may total 30 to 40 percent of P. A stuffing box where the shaft enters a covered vessel causes additional losses.
2. All the curves are for axial impeller shafts, with liquid depth Z equal to the tank diameter T.
3. Curves a to e are for open vessels, with a gas-liquid surface, fitted with four baffles, $b = T/10$ to $T/12$.
4. Curve a is for marine propellers, $d_i/T \approx \frac{1}{3}$, set a distance $C = d_i$ or greater from the bottom of the vessel. The effect of changing d_i/T is apparently felt only at very high Reynolds numbers and is not well established.
5. Curves b to e are for turbines located at a distance $C = d_i$ or greater from the bottom of the vessel. For disk flat-blade turbines, curve c, there is essentially no effect of d_i/T in the range 0.15 to 0.50. For open types, curve b, the effect of d_i/T may be strong, depending upon the group nb/T [8].
6. Curve g is for disk flat-blade turbines operated in unbaffled vessels filled with liquid, covered, so that no vortex forms [70]. If baffles are present, the power characteristics at high Reynolds numbers are essentially the same as curve b for baffled open vessels, with only a slight increase in power [88].
7. For very deep tanks, two impellers are sometimes mounted on the same shaft, one above the other. For the flat-blade disk turbines, at a spacing equal to $1.5 d_i$ or greater, the combined power for both will approximate twice that for a single turbine [4].

more turbulence than a large impeller operating at slow speed at the same power. Smaller higher-speed impellers are thus especially suited for dispersing liquids and gases, while the larger impellers are particularly effective for suspending solids.

MECHANICAL AGITATION, GAS-LIQUID CONTACT

A typical arrangement is shown in Fig. 6.6*a*. The baffled vessel, with liquid depth approximating the tank diameter, must be provided with adequate freeboard to allow for the gas holdup during gas flow. Gas is best introduced below the impeller through a ring-shaped sparger of diameter equal to, or somewhat smaller than, that of the impeller [92], arranged with holes on top. Holes may be 3 to 6.5 mm ($\frac{1}{8}$ to $\frac{1}{4}$ in) in diameter, in number to provide a hole Reynolds number Re_o of the order of 10 000 or more, although this is not especially important. The distance between holes should not be less than d_p, the gas-bubble diameter as given for example by Eq. (6.5). When time of contact must be relatively large, deep vessels can be used, as in Fig. 6.6*b*, whereupon multiple impellers (preferably not more than two [93]) are used to redisperse gas bubbles which coalesce, thereby maintaining a large interfacial area. In very large aeration tanks for activated sludge and other fermentations, where volumes may

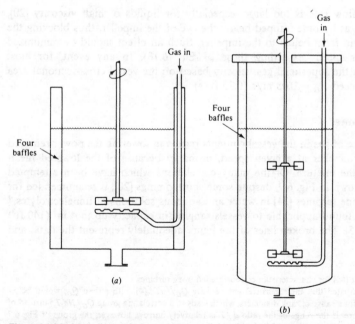

Figure 6.6 Mechanically agitated vessels for gas-liquid contact: (*a*) standard, (*b*) multiple impellers for deep tanks.

be as large as 90 m³ (24 000 gal), liquid depths must usually be kept not larger than about 3 to 5 m to save on air-compression costs. The large tank-diameter–liquid-depth ratios which then develop require multiple mixers and points for introducing the gas.

Impellers

Both open and disk flat-blade turbines are used extensively, particularly because of the high discharge velocities normal to the flow of gas which they maintain. Especially in the larger sizes, the disk type is preferred.† They are best specified with $d_i/T = 0.25$ to 0.4 and set off the bottom of the vessel a distance equal to the impeller diameter. In some cases, especially designed impellers can be used to induce the gas flow from the space above the liquid down into the agitated mass [81, 115].

For production of effective gas dispersions with disk flat-blade turbines, the impeller speed should exceed that given by [122]

$$\frac{Nd_i}{(\sigma g g_c/\rho_L)^{0.25}} = 1.22 + 1.25\frac{T}{d_i} \tag{6.18}$$

Gas Flow

If the gas flow rate is too large, especially for liquids of high viscosity [20], bubbles of gas become trapped below the eye of the impeller, thus blocking the flow of liquid from below to the impeller. Such an effect should be minimized for impeller speeds exceeding those of Eq. (6.18). In any event, for most installations the superficial gas velocity based on the vessel cross-sectional area does not exceed $V_G = 0.08$ m/s (0.25 ft/s).

Impeller Power

The presence of gas in the vessel contents results in lowering the power required to turn an impeller at a given speed, probably because of the lowered mean density of the mixture. Of the many correlations which have been attempted [86], that shown in Fig. 6.7, despite some shortcomings [20], is recommended for disk flat-blade turbines [24] in water and aqueous solutions of nonelectrolytes.‡ It has been found applicable to vessels ranging in capacity up to 4 m³ (140 ft³) and larger [5]. The broken lines of the figure adequately represent the data, and

---

† See Miller [84] for characteristics of four-bladed open turbines.

‡ It is anticipated that the superficial gas velocity $Q_G/(\pi T^2/4)$, rather than Q_G, would better correlate data for a range of tank diameters, which leads to a correlating group Q_G/Nd_iT^2 instead of that of the figure. If the range of the ratio d_i/T is relatively narrow, however, the group of Fig. 6.7 will serve.

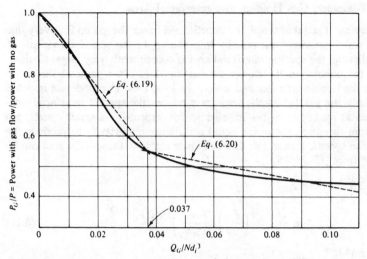

Figure 6.7 Impeller power for gassed vessels, disk flat-blade turbines, $d_i/T = 0.25$ to 0.33. [*Adapted from Calderbank, Trans. Inst. Chem. Eng., Lond.,* **36**, 443 (1958).]

their equations are

$$
\frac{P_G}{P} = \begin{cases} 1 - 12.2\dfrac{Q_G}{Nd_i^3} & \dfrac{Q_G}{Nd_i^3} < 0.037 \qquad (6.19) \\[3mm] 0.62 - 1.85\dfrac{Q_G}{Nd_i^3} & \dfrac{Q_G}{Nd_i^3} > 0.037 \qquad (6.20) \end{cases}
$$

where P_G is the power with gas and P that with no gas flowing. Gassed power levels of 600 to 1000 W/m^3 (12.5 to 20 ft $\cdot$ lb$_f$/ft$^3 \cdot$ s or 3 to 5 hp per 1000 gal) are common, although values as large as 4 times as much have been used. Furthermore, the ratio of power required to volume of vessel decreases with vessel size. Power characteristics for two impellers, as in Fig. 6.6*b*, have not been studied; a conservative estimate would be that for one gassed plus that for one ungassed impeller [5]. In many instances an appreciable amount of power will be supplied to the liquid by way of the gas [see Eq. 6.13)]. Aqueous electrolytes, because of the smaller bubble-coalescence rates which they produce, exhibit different power requirements [55].

To safeguard against overloading the agitator motor if the flow of gas should suddenly be stopped, the motor and drive should be sized as if there were no gas. An excellent review provides many details [57].

Bubble Diameter, Gas Holdup, and Interfacial Area

If it is turning at suitable speed, the impeller will break the gas up into very fine bubbles, which are swept out radially from the tip. In the immediate vicinity of the impeller tip, the specific interfacial area is consequently very large. Coalescnece of the bubbles in the slower-moving liquid in other parts of the vessel increases the bubble diameter and lowers the local area. For moderate impeller speeds, only the gas fed to the impeller produces the specific interfacial area designated as a_o, but at higher impeller speeds, especially in small vessels, gas drawn from the space above the liquid contributes importantly to produce the larger mean specific area a. For disk flat-blade turbines, the specific area can be estimated from [23, 24, 84]:

$Re^{0.7}\left(\dfrac{Nd_i}{V_G}\right)^{0.3} < 30\,000$:

$$a_o = 1.44\left[\left(\frac{P_G}{v_L}\right)^{0.4}\left(\frac{\rho_L}{\sigma^3 g_c}\right)^{0.2}\right]\left(\frac{V_G}{V_t}\right)^{1/2} \tag{6.21}$$

$Re^{0.7}\left(\dfrac{Nd_i}{V_G}\right)^{0.3} > 30\,000$ (surface aeration significant):

$$\frac{a}{a_o} = 8.33 \times 10^{-5}\, Re^{0.7}\left(\frac{Nd_i}{V_G}\right)^{0.3} - 1.5 \tag{6.22}$$

The quantity in the square brackets of Eq. (6.21) has the net dimension $\mathbf{L}^{-1}$. The impeller Reynolds number is computed with liquid density and viscosity.

The mean bubble diameter for disk flat-blade turbines is given by

$$d_p = K\left[\left(\frac{v_L}{P_G}\right)^{0.4}\left(\frac{\sigma^3 g_c}{\rho_L}\right)^{0.2}\right]\varphi_G{}^m\left(\frac{\mu_G}{\mu_L}\right)^{0.25} \tag{6.23}$$

where for electrolyte solutions $K = 2.25$, $m = 0.4$ and for aqueous solutions of alcohols and tentatively for other organic solutes in water $K = 1.90$, $m = 0.65$.

The gas holdup can then be estimated by substituting Eqs. (6.21) to (6.23) into Eq. (6.9). Thus, for $Re^{0.7}\,(Nd_i/V_G)^{0.3} < 30\,000$, we obtain

$$\varphi_G = \left[0.24K\left(\frac{\mu_G}{\mu_L}\right)^{0.25}\left(\frac{V_G}{V_t}\right)^{1/2}\right]^{1/(1-m)} \tag{6.24}$$

The use of Eqs. (6.21) to (6.24) requires a trial-and-error procedure. To initiate this, it is frequently satisfactory to try $V_t = 0.21$ m/s, or the equivalent in other units, to be corrected through Eqs. (6.6) to (6.7).

Mass-Transfer Coefficients

The evidence is that the mass-transfer resistance lies entirely within the liquid phase [23, 24, 101], and for gas bubbles of diameter likely to be encountered in agitated vessels

$$Sh_L = 2.0 + 0.31\, Ra^{1/3} \tag{6.25}$$

where Ra is the *Rayleigh number* $d_p^3 \, \Delta\rho \, g / D_L \mu_L$. Equation (6.25) indicates that k_L is independent of agitator power. This is because the $\Delta\rho$ is so large for gas-liquid mixtures that bubble motion due to g rather than to P_G determines the rate of mass transfer. There is some evidence to the contrary [69].

Illustration 6.2 Water containing dissolved air is to be deoxygenated by contact with nitrogen in an agitated vessel 1 m (3.28 ft) in diameter. The water will enter continuously at the bottom of the vessel and leave through an overflow pipe set in the side of the vessel at an appropriate height. Assume that the ungassed water would have a depth of 1 m (3.28 ft).

$$\text{Water rate} = 9.5 \times 10^{-4} \, \text{m}^3/\text{s} \, (1.99 \, \text{ft}^3/\text{min})$$

$$\text{Gas rate} = 0.061 \, \text{kg/s} \, (8.07 \, \text{lb/min}) \qquad \text{temp} = 25°\text{C}$$

Specify the arrangement to be used and estimate the average gas-bubble diameter, gas holdup, specific interfacial area, and mass-transfer coefficient to be expected.

SOLUTION Use four vertical wall baffles, 100 mm wide, set at 90° intervals. Use a 305-mm-diameter (1-ft), six-bladed disk flat-blade turbine impeller, arranged axially, 300 mm from the bottom of the vessel.

The sparger underneath the impeller will be in the form of a 240-mm-diameter ring made of 12.7-mm ($\frac{1}{2}$-in) tubing drilled in the top with 3.18-mm-diameter ($\frac{1}{8}$-in) holes (d_o = 0.00316 m). The viscosity of N_2 is 1.8×10^{-5} kg/m · s at 25°C. For an orifice Reynolds number of 35 000 = $4w_o/\pi d_o \mu_G$ = $4w_o/\pi(0.00316)(1.8 \times 10^{-5})$, w_o = 1.564×10^{-3} kg/s gas flow through each orifice. Therefore, use $0.061/(1.564 \times 10^{-3})$ = 39 holes at approx $\pi(240)/39$ = 16-mm intervals around the sparger ring.

At 25°C, the viscosity of water is 8.9×10^{-4} kg/m · s, the surface tension is 0.072 N/m, and the density is 1000 kg/m^3. In Eq. (6.18) d_i = 0.305 m, T = 1 m, g = 9.81 m/s^2, and g_c = 1, whence N = 2.84 r/s (170 r/min) minimum impeller speed for effective dispersion. Use N = 5 r/s (300 r/min).

Re = $d_i^2 N \rho_L / \mu_L$ = 523 000. Fig. 6.5: Po = 5.0, and the no-gas power = P = $5.0 \rho_L N^3 d_i^5 / g_c$ = 1650 W:

The gas at the sparger, and hence at the impeller, is at a head of 0.7 m of water, or 101 330 + 0.7(1000)(9.81) = 108.2 × 10^{-3} N/m^2 abs. Mol wt N_2 = 28.02.

$$Q_G = \frac{0.061}{28.02}(22.41)\frac{298}{273}\frac{101.3}{108.2} = 0.050 \, \text{m}^3/\text{s}$$

With N = 5 r/s, Q_G/Nd_i^3 = 0.35. In Fig. 6.7 the abscissa is off scale. Take P_G/P = 0.43, whence P_G = 0.43(1650) = 710 W (0.95 hp) delivered by the impeller to the gassed mixture. *Note:* Power for the ungassed mixture is 1650 W (2.2 hp), and allowance for inefficiencies of motor and speed reducer may bring the power input to the motor to 2250 W (3 hp), which should be specified.

V_G = $0.050/[\pi(1)^2/4]$ = 0.0673 m/s superficial gas velocity. Re$^{0.7}(Nd_i/V_G)^{0.3}$ = 26 100. The interfacial area is given by Eq. (6.21).

P_G = 710 W, v_L = $\pi(1)^3/4$ = 0.785 m^3, g_c = 1. Equation (6.21) provides

$$a = \frac{106.8}{V_t^{0.5}} \tag{6.26}$$

with a in m^2/m^3 and V_t in m/s. For d_p [Eq. (6.23)], use K = 2.25, m = 0.4, and μ_G = 1.8 × 10^{-5} kg/m · s. Equation (6.23) becomes

$$d_p = 2.88 \times 10^{-3} \, \varphi_G^{0.4} \tag{6.27}$$

Substitution of Eqs. (6.26) and (6.27) into (6.9) provides

$$\varphi_G = \frac{7.067 \times 10^{-3}}{V_t^{0.833}} \tag{6.28}$$

Equations (6.27) and (6.28) are to be solved simultaneously with the relationship for V_t. After several trials, choose $V_t = 0.240$ m/s, whence $\varphi_G = 0.0232$, $d_p = 6.39 \times 10^{-4}$ m = 0.639 mm, whence Eq. (6.6) provides $V_t = 0.250$ m/s, a satisfactory check. From Eq. (6.26), $a = 214$ m^2/m^3.

For N_2 in water at 25°C, $D_L = 1.9 \times 10^{-9}$ m^2/s, and Ra = 1.514×10^{-6}. Equation (6.25) provides Sh$_L = 37.6 = F_L d_p / c D_L$. Since $c = 1000/18.02 = 55.5$ kmol/m^3, $F_L = 6.20 \times 10^{-3}$ kmol/m$^2 \cdot$ s.

Since a well-designed turbine impeller provides very thorough mixing of the liquid, it can reasonably be assumed that the concentration throughout the liquid phase in such devices is quite uniform. Therefore the benefits of counter-current flow for continuous operation cannot be had with a single tank and agitator.

Multistage Absorption

Multistage arrangements, with counterflow of gas and liquid, can be made with multiple vessels and agitators piped to lead the gas from the top of one tank to the bottom of the next and the liquid from vessel to vessel in the opposite direction. A more acceptable alternative is to place the agitated vessels one atop the other, with the agitators on a single shaft. Towers of this sort are regularly used for liquid extraction, as shown in Fig. 10.52. However a laboratory tower of that design containing 12 agitated compartments was studied for the absorption of ammonia from air into water [113]. The absorption rates were comparable to those obtained with towers packed with 25-mm (1-in) Raschig rings. The design should be particularly advantageous where liquid rates are low, where packing is inadequately wet, or where the liquid contains suspended solids tending to clog conventional packed and tray towers.

TRAY TOWERS

Tray towers are vertical cylinders in which the liquid and gas are contacted in stepwise fashion on trays or plates, as shown schematically for one type (bubble-cap trays) in Fig. 6.8. The liquid enters at the top and flows downward by gravity. On the way, it flows across each tray and through a downspout to the tray below. The gas passes upward through openings of one sort or another in the tray, then bubbles through the liquid to form a froth, disengages from the froth, and passes on to the next tray above. The overall effect is a multiple countercurrent contact of gas and liquid, although each tray is characterized by a cross flow of the two. Each tray of the tower is a stage, since on the tray the fluids are brought into intimate contact, interphase diffusion occurs, and the fluids are separated.

The number of equilibrium stages (theoretical trays) in a column or tower is dependent only upon the difficulty of the separation to be carried out and is determined solely from material balances and equilibrium considerations. The stage or tray efficiency, and therefore the number of real trays, is determined by

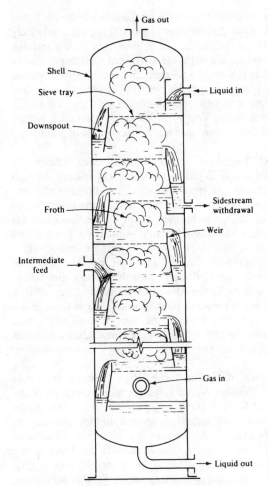

Figure 6.8 Schematic section through sieve-tray tower.

the mechanical design used and the conditions of operation. The diameter of the tower, on the other hand, depends upon the quantities of liquid and gas flowing through the tower per unit time. Once the number of equilibrium stages, or theoretical trays, required has been determined, the principal problem in the design of the tower is to choose dimensions and arrangements which will represent the best compromise between several opposing tendencies, since it is generally found that conditions leading to high tray efficiencies will ultimately lead to operational difficulties.

For stage or tray efficiencies to be high the time of contact should be long to permit the diffusion to occur, the interfacial surface between phases must be made large, and a relatively high intensity of turbulence is required to obtain high mass-transfer coefficients. In order to provide long contact time, the liquid pool on each tray should be deep, so that bubbles of gas will require a relatively long time to rise through the liquid. When the gas bubbles only slowly through the openings on the tray, the bubbles are large, the interfacial surface per unit of gas volume is small, the liquid is relatively quiescent, and much of it may even pass over the tray without having contacted the gas. On the other hand, when the gas velocity is relatively high, it is dispersed very thoroughly into the liquid, which in turn is agitated into a froth. This provides large interfacial surface areas. For high tray efficiencies, therefore, we require deep pools of liquid and relatively high gas velocities.

These conditions, however, lead to a number of difficulties. One is the mechanical entrainment of droplets of liquid in the rising gas stream. At high gas velocities, when the gas is disengaged from the froth, small droplets of liquid will be carried by the gas to the tray above. Liquid carried up the tower in this manner reduces the concentration change brought about by the mass transfer and consequently adversely affects the tray efficiency. And so the gas velocity may be limited by the reduction in tray efficiency due to liquid entrainment.

Furthermore, great liquid depths on the tray and high gas velocities both result in high pressure drop for the gas in flowing through the tray, and this in turn leads to a number of difficulties. In the case of absorbers and humidifiers, high pressure drop results in high fan power to blow or draw the gas through the tower, and consequently high operating cost. In the case of distillation, high pressure at the bottom of the tower results in high boiling temperatures, which in turn may lead to heating difficulties and possibly damage to heat-sensitive compounds.

Ultimately, purely mechanical difficulties arise. High pressure drop may lead directly to a condition of *flooding*. With a large pressure difference in the space between trays, the level of liquid leaving a tray at relatively low pressure and entering one of high pressure must necessarily assume an elevated position in the downspouts, as shown in Fig. 6.8. As the pressure difference is increased due to the increased rate of flow of either gas or liquid, the level in the downspout will rise further to permit the liquid to enter the lower tray. Ultimately the liquid level may reach that on the tray above. Further increase in either flow rate then aggravates the condition rapidly, and the liquid will fill the entire space between the trays. The tower is then flooded, the tray efficiency falls to a low value, the flow of gas is erratic, and liquid may be forced out of the exit pipe at the top of the tower.

For liquid-gas combinations which tend to foam excessively, high gas velocities may lead to a condition of *priming*, which is also an inoperative situation. Here the foam persists throughout the space between trays, and a great deal of liquid is carried by the gas from one tray to the tray above. This is an exaggerated condition of entrainment. The liquid so carried recirculates between trays, and the added liquid-handling load increases the gas pressure drop sufficiently to lead to flooding.

We can summarize these opposing tendencies as follows. Great depths of liquid on the trays lead to high tray efficiencies through long contact time but also to high pressure drop per tray. High gas velocities, within limits, provide good vapor-liquid contact through excellence of dispersion but lead to excessive entrainment and high pressure drop. Several other undesirable conditions may occur. If liquid rates are too low, the gas rising through the openings of the tray may push the liquid away (*coning*), and contact of the gas and liquid is poor. If the gas rate is too low, much of the liquid may rain down through the openings of the tray (*weeping*), thus failing to obtain the benefit of complete flow over the trays; and at very low gas rates, none of the liquid reaches the downspouts (*dumping*). The relations between these conditions are shown schematically in Fig. 6.9, and all types of trays are subject to these difficulties in some form. The various arrangements, dimensions, and operating conditions chosen for design are those which experience has proved to be reasonably good compromises. The general design procedure involves a somewhat empirical application of them, followed by computational check to ensure that pressure drop and flexibility, i.e., ability of the tower to handle more or less than the immediately expected flow quantities, are satisfactory.

A great variety of tray designs have been and are being used. During the first half of this century, practically all towers were fitted with bubble-cap trays, but new installations now use either sieve trays or one of the proprietary designs which have proliferated since 1950.

General Characteristics

Certain design features common to the most frequently used tray designs will be dealt with first.

Shell and trays The tower may be made of any number of materials, depending upon the corrosion conditions expected. Glass, glass-lined metal, impervious carbon, plastics, even wood but most frequently metals are used. For metal towers, the shells are usually cylindrical for reasons of cost. In order to facilitate cleaning, small-diameter towers are fitted with hand holes, large towers with manways about every tenth tray [66].

The trays are usually made of sheet metals, of special alloys if necessary, the thickness governed by the anticipated corrosion rate. The trays must be stiffened and supported (see, for example, Fig. 6.14) and must be fastened to the shell to prevent movement owing to surges of gas, with allowance for thermal expansion. This can be arranged by use of tray-support rings with slotted bolt holes to which the trays are bolted. Large trays must be fitted with manways (see Fig. 6.14) so that a person can climb from one tray to another during repair and cleaning [126]. Trays should be installed level to within 6 mm ($\frac{1}{4}$ in) to promote good liquid distribution.

Tray spacing Tray spacing is usually chosen on the basis of expediency in construction, maintenance, and cost and later checked to be certain that adequate insurance against flooding and excessive entrainment is present. For special cases where tower height is an important consideration, spacings of

Table 6.1 Recommended general conditions and dimensions for tray towers

1. Tray spacing

Tower diameter T		Tray spacing t	
m	ft	m	in
		0.15	6 minimum
1 or less	4 or less	0.50	20
1–3	4–10	0.60	24
3–4	10–12	0.75	30
4–8	12–24	0.90	36

2. Liquid flow

    a. Not over 0.015 m³/(m diam) · s (0.165 ft³/ft · s) for single-pass cross-flow trays
    b. Not over 0.032 m³/(m weir length) · s (0.35 ft³/ft · s) for others

3. Downspout seal

    a. Vacuum, 5 mm minimum, 10 mm preferred ($\frac{1}{4}-\frac{1}{2}$ in)
    b. Atmospheric pressure and higher, 25 mm minimum, 40 mm preferred (1–1.5 in)

4. Weir length for straight, rectangular weirs, cross-flow trays, 0.6T to 0.8T, 0.7T typical

Weir length W	Distance from center of tower	Tower area used by one downspout, %
0.55T	0.4181T	3.877
0.60T	0.3993T	5.257
0.65T	0.2516T	6.899
0.70T	0.3562T	8.808
0.75T	0.3296T	11.255
0.80T	0.1991T	14.145

5. Typical pressure drop per tray

Total pressure	Pressure drop
35 mmHg abs	3 mmHg or less
1 std atm	500–800 N/m² (0.07–0.12 lb$_f$/in²)
2×10^6 N/m²	1000 N/m²
300 lb$_f$/in²	0.15 lb$_f$/in²

15 cm (6 in) have been used. For all except the smallest tower diameters, 50 cm (20 in) would seem to be a more workable minimum from the point of view of cleaning the trays. Table 6.1 summarizes recommended values.

Tower diameter The tower diameter and consequently its cross-sectional area must be sufficiently large to handle the gas and liquid rates within the region of satisfactory operation of Fig. 6.9. For a given type of tray at flooding, the

Figure 6.9 Operating characteristics of sieve trays.

superficial velocity of the gas V_F (volumetric rate of gas flow Q per net cross section for flow A_n) is related to fluid densities by[†]

$$V_F = C_F \left(\frac{\rho_L - \rho_G}{\rho_G} \right)^{1/2} \qquad (6.29)$$

The net cross section A_n is the tower cross section A_t minus the area taken up by the downspouts (A_d in the case of a cross-flow tray as in Fig. 6.8). C_F is an empirical constant, the value of which depends on the tray design. Some appropriately smaller value of V is used for actual design; for nonfoaming liquids this is typically 80 to 85 percent of V_F (75 percent or less for foaming liquids), subject to check for entrainment and pressure-drop characteristics. Ordinarily the diameter so chosen will be adequate, although occasionally the liquid flow may be limiting. A well-designed single-pass cross-flow tray can ordinarily be expected to handle up to 0.015 m³/s of liquid per meter of tower diameter ($q/T = 0.015$ m³/m · s = 0.165 ft³/ft · s). For most installations, considerations of cost make it impractical to vary the tower diameter from one end of the tower to the other to accommodate variations in gas or liquid flow, and maximum flow quantities are used to set a uniform diameter. When variation in flows are such that a 20 percent difference in diameter is indicated for the upper and lower sections, two diameters will probably be economical [108].

The tower diameter required may be decreased by use of increased tray spacing, so that tower cost, which depends on height as well as diameter, passes through a minimum at some optimum tray spacing.

Downspouts The liquid is led from one tray to the next by means of downspouts, or downcomers. These may be circular pipes or preferably portions of the tower cross section set aside for liquid flow by vertical plates, as in Fig. 6.8. Since the liquid is agitated into a froth on the tray, adequate residence time must

---

[†] Equation (6.29) is empirical; the values of C_F depend on the units used as well as the tray design.

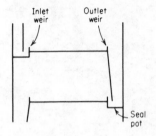

Figure 6.10 Seal-pot arrangement.

be allowed in the downspout to permit disengaging the gas from the liquid, so that only clear liquid enters the tray below. The downspout must be brought close enough to the tray below to seal into the liquid on that tray (item 3, Table 6.1), thus preventing gas from rising up the downspout to short-circuit the tray above. Seal pots and seal-pot dams (inlet weirs) may be used [126], as in Fig. 6.10, but they are best avoided (see below), especially if there is a tendency to accumulate sediment. If they are used, weep holes (small holes through the tray) in the seal pot should be used to facilitate draining the tower on shutdown.

Weirs The depth of liquid on the tray required for gas contacting is maintained by an overflow (outlet) weir, which may or may not be a continuation of the downspout plate. Straight weirs are most common; multiple V-notch weirs maintain a liquid depth which is less sensitive to variations in liquid flow rate and consequently also from departure of the tray from levelness; circular weirs, which are extensions of circular pipes used as downspouts, are not recommended. Inlet weirs (Fig. 6.10) may result in a hydraulic jump of the liquid [67] and are not generally recommended. In order to ensure reasonably uniform distribution of liquid flow on a single-pass tray, a weir length of from 60 to 80 percent of the tower diameter is used. Table 6.1 lists the percentage of the tower cross section taken up by downspouts formed from such weir plates.

Liquid flow Several of the schemes commonly used for directing the liquid flow on trays are shown in Fig. 6.11. Reverse flow can be used for relatively small towers, but by far the most common arrangement is the single-pass cross-flow tray of Fig. 6.11*b*. For large-diameter towers, radial or split flow (Fig. 6.11*c* and *d*) can be used, although every attempt is usually made to use the cross-flow tray because of its lower cost. A recent variant of Fig. 6.11*c* uses ring-shaped downspouts concentric with the tower shell instead of the pipes of that figure [110]. For trays fitted with bubble caps (see below) requiring the liquid to flow great distances leads to undesirable liquid-depth gradients. For towers of very large diameter containing such trays, cascade designs of several levels each with its own weir (Fig. 6.11*e*) have been used, but their cost is considerable. Commercial columns up to 15 m (50 ft) in diameter have been built. Two-pass

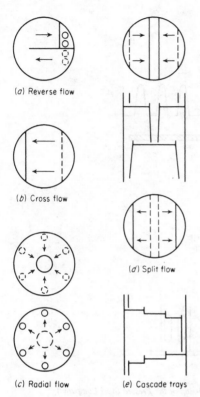

(a) Reverse flow

(b) Cross flow

(c) Radial flow

(d) Split flow

(e) Cascade trays

Figure 6.11 Tray arrangements. Arrows show direction of liquid flow.

trays are common for diameters of 3 to 6 m, and more passes for larger diameters.

Bubble-Cap Trays

On these trays (Figs. 6.12 and 6.13) chimneys or risers lead the gas through the tray and underneath caps surmounting the risers. A series of slots is cut into the rim or skirt of each cap, and the gas passes through them to contact the liquid which flows past the caps. The liquid depth is such that the caps are covered or nearly so. Detailed design characteristics are available [13, 46, 119].

Bubble caps have been used for over 150 years, and during 1920–1950 practically all new tray towers used them. They offer the distinct advantage of being able to handle very wide ranges of liquid and gas flow rates satisfactorily (the ratio of design rate to minimum rate is the *turndown ratio*). They have now been abandoned for new installations because of their cost, which is roughly double that for sieve, counterflow, and valve trays.

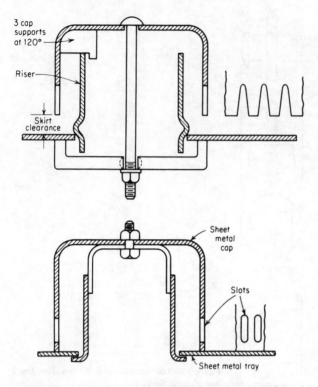

3 cap
supports
at 120°

Riser

Skirt
clearance

Sheet
metal
cap

Slots

Sheet metal tray

Figure 6.12 Typical bubble-cap designs. *(With permission of the Pressed Steel Company.)*

Sieve (perforated) trays

These trays have been known almost as long as bubble-cap trays, but they fell out of favor during the first half of this century. Their low cost, however, has now made them the most important of tray devices.

A vertical section and a plan view are shown schematically in Fig. 6.14. The principal part of the tray is a horizontal sheet of perforated metal, across which the liquid flows, with the gas passing upward through the perforations. The gas, dispersed by the perforations, expands the liquid into a turbulent froth, characterized by a very large interfacial surface for mass transfer. The trays are subject to flooding because of backup of liquid in the downspouts or excessive entrainment (priming), as described earlier.

Design of Sieve Trays

The diameter of the tower must be chosen to accommodate the flow rates, the details of the tray layout must be selected, estimates must be made of the gas-pressure drop and approach to flooding, and assurance against excessive weeping and entrainment must be established.

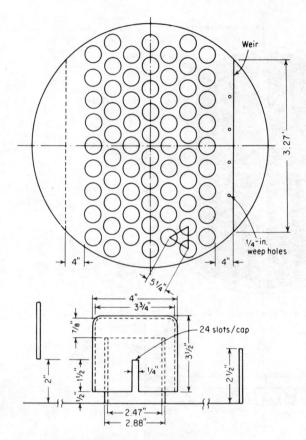

Figure 6.13 Typical bubble-cap tray arrangement.

Tower diameter The flooding constant C_F of Eq. (6.29) has been correlated for the data available on flooding [44, 47]. The original curves can be represented by†

$$C_F = \left[\alpha \log \frac{1}{(L'/G')(\rho_G/\rho_L)^{0.5}} + \beta \right] \left(\frac{\sigma}{0.020} \right)^{0.2} \tag{6.30}$$

Typically, for perforations of 6 mm diameter, the best tray efficiencies result from values of $V\sqrt{\rho_G}$ in the range 0.7 to 2.2 (with V and ρ_G in SI units; for V in ft/s and ρ_G in lb/ft^3, the range is 0.6 to 1.8) [2].

Perforations and active area Hole diameters from 3 to 12 mm ($\frac{1}{8}$ to $\frac{1}{2}$ in) are commonly used, 4.5 mm ($\frac{3}{16}$ in) most frequently [45], although holes as large as 25 mm have been successful [72]. For most installations, stainless steel or other alloy perforated sheet is used, rather than carbon steel, even though not necessarily required for corrosion resistance. Sheet thickness [25] is usually less than

---

† Equation (6.30) is empirical. See Table 6.2 for values of α and β.

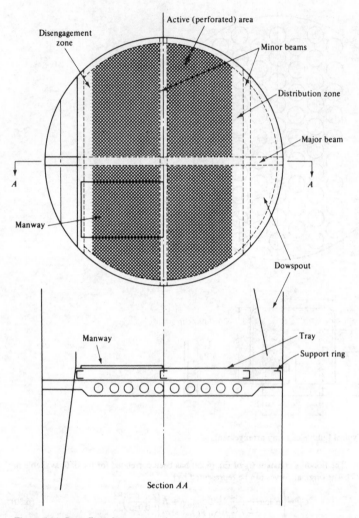

Figure 6.14 Cross-flow sieve tray.

one-half the hole diameter for stainless steel, less than one diameter for carbon steel or copper alloys. Table 6.2 lists typical values.

The holes are placed in the corners of equilateral triangles at distances between centers (pitch) of from 2.5 to 5 hole diameters. For such an arrangement

$$\frac{A_o}{A_a} = \frac{\text{hole area}}{\text{active area}} = 0.907 \left(\frac{d_0}{p'} \right)^2 \qquad (6.31)$$

Typically, as in Fig. 6.14, the peripheral tray support, 25 to 50 mm (1 to 2 in) wide, and the beam supports will occupy up to 15 percent of the cross-sectional area of the tower; the distribution

Table 6.2 Recommended dimensions for sieve-tray towers

1. Flooding constant C_F [Eqs. (6.29) and (6.30)], $d_o < 6$ mm ($\frac{1}{4}$ in)

Range of $\dfrac{A_o}{A_a}$	Range of $\dfrac{L'}{G'}\left(\dfrac{\rho_G}{\rho_L}\right)^{0.5}$	Units of t	Units of σ	Units of V_F	α, β
> 0.1	0.01–0.1, use values at 0.1				
	0.1–1.0	m	N/m	m/s	$\alpha = 0.0744t + 0.01173$ $\beta = 0.0304t + 0.015$
		in	dyn/cm $\times 10^{-3}$	ft/s	$\alpha = 0.0062t + 0.0385$ $\beta = 0.00253t + 0.050$
< 0.1	Multiply α and β by $5A_o/A_a + 0.5$				

2. Hole diameter and plate thickness

Hole diameter		Plate thickness/hole diameter	
mm	in	Stainless steel	Carbon steel
3.0	$\frac{1}{8}$	0.65	
4.5	$\frac{3}{16}$	0.43	
6.0	$\frac{1}{4}$	0.32	
9.0	$\frac{3}{8}$	0.22	0.5
12.0	$\frac{1}{2}$	0.16	0.38
15.0	$\frac{5}{8}$	0.17	0.3
18.0	$\frac{3}{4}$	0.11	0.25

3. Liquid depth

50 mm (2 in) minimum, 100 mm (4 in) maximum

4. Typical active area

Tower diameter		$\dfrac{A_a}{A_t}$
m	ft	
1	3	0.65
1.25	4	0.70
2	6	0.74
2.5	8	0.76
3	10	0.78

zone for liquid entering the tray and the disengagement zone for disengaging foam (which are sometimes omitted) use 5 percent or more [47, 66]; and downspouts require additional area (Table 6.1). The remainder is available for active perforations (active area A_a). Typical values of A_a are listed in Table 6.2.

Liquid depth Liquid depths should not ordinarily be less than 50 mm (2 in), to ensure good froth formation; depths of 150 mm (6 in) have been used [126], but 100 mm is a more common maximum. These limits refer to the sum of the weir height h_W plus the crest over the weir h_1, calculated as clear liquid although in the perforated area the equivalent clear-liquid depth will be smaller than this. Liquid depths as great as 1.5 m have been studied [56] but are not used in ordinary operations.

Weirs Refer to Fig. 6.15, where a schematic representation of a cross-flow tray is shown. The crest of liquid over a straight rectangular weir can be estimated by the well-known Francis formula†

$$\frac{q}{W_{\text{eff}}} = 1.839h_1^{3/2} \tag{6.32}$$

where q = rate of liquid flow, m^3/s
W_{eff} = effective length of the weir, m
h_1 = liquid crest over the weir, m

Because the weir action is hampered by the curved sides of the circular tower, it is recommended [39] that W_{eff} be represented as a chord of the circle of diameter T, a distance h_1 farther from the center than the actual weir, as in Fig. 6.16. Equation (6.32) can then be rearranged to‡

$$h_1 = 0.666\left(\frac{q}{W}\right)^{2/3}\left(\frac{W}{W_{\text{eff}}}\right)^{2/3} \tag{6.33}$$

The geometry of Fig. 6.16 leads to

$$\left(\frac{W_{\text{eff}}}{W}\right)^2 = \left(\frac{T}{W}\right)^2 - \left\{\left[\left(\frac{T}{W}\right)^2 - 1\right]^{0.5} + \frac{2h_1}{T}\frac{T}{W}\right\}^2 \tag{6.34}$$

For $W/T = 0.7$, which is typical, Eq. (6.33) can be used with $W_{\text{eff}} = W$ for $h_1/W = 0.055$ or less with a maximum error of only 2 percent in h_1, which is negligible.

Pressure drop for the gas For convenience, all gas-pressure drops will be expressed as heads of clear liquid of density ρ_L on the tray. The pressure drop for the gas h_G is the sum of the effects for flow of gas through the dry plate and those caused by the presence of liquid:

$$h_G = h_D + h_L + h_R \tag{6.35}$$

where h_D = dry-plate pressure drop
h_L = pressure drop resulting from depth of liquid on tray
h_R = residual pressure drop

Although Fig. 6.15 shows a change in liquid depth Δ, half of which should be used in Eq. (6.35), in practice this is so small, except for the very largest trays, that it can reasonably be neglected. It can be roughly estimated, if necessary [62]. Estimates of the gas-pressure drop summarized here are based on a critical study [79] of all available data and the generally accepted methods. Most of the data are for the air-water system, and extension to others is somewhat uncertain.

Dry pressure drop h_D This is calculated on the basis that it is the result to a loss in pressure on entrance to the perforations, friction within the short tube formed by the perforation owing to plate

---

† The coefficient 1.839 applies only for meter-second units. For q in ft^3/s, W_{eff} in feet, and h_1 in inches the coefficient is 0.0801.

‡ The coefficient 0.666 applies only for meter-second units. For q in ft^3/s, W and W_{eff} in feet, and h_1 in inches the coefficient is 5.38.

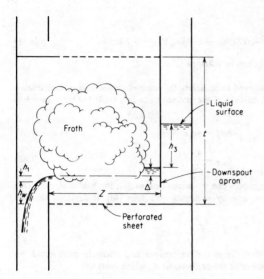

Figure 6.15 Schematic diagram of a cross-flow sieve tray.

thickness, and an exit loss [64]

$$\frac{2h_D g \rho_L}{V_o^2 \rho_G} = C_o \left[0.40 \left(1.25 - \frac{A_o}{A_n} \right) + \frac{4lf}{d_o} + \left(1 - \frac{A_o}{A_n} \right)^2 \right]$$ (6.36)

The Fanning friction factor f is taken from a standard chart [15]. C_o is an orifice coefficient which depends upon the ratio of plate thickness to hole diameter [82]. Over the range $l/d_o = 0.2$ to 2.0

$$C_o = 1.09 \left(\frac{d_o}{l} \right)^{0.25}$$ (6.37)

Hydraulic head h_L In the perforated region of the tray, the liquid is in the form of a froth. The equivalent depth of clear liquid h_L is an estimate of that which would obtain if the froth collapsed. That is usually less than the height of the outlet weir, decreasing with increased gas rate. Some methods of estimating h_L use a specific *aeration factor* to describe this [87]. In Eq. (6.38), which is the recommended relationship [49], the effect of the factor is included as a function of the variables

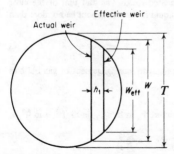

Figure 6.16 Effective weir length.

which influence it†

$$h_L = 6.10 \times 10^{-3} + 0.725 h_W - 0.238 h_W V_a \rho_G^{0.5} + 1.225 \frac{q}{z} \tag{6.38}$$

where z is the average flow width, which can be taken as $(T + W)/2$.

Residual gas-pressure drop h_R This is believed to be largely the result of overcoming surface tension as the gas issues from a perforation. A balance of the internal force in a static bubble required to overcome surface tension is

$$\frac{\pi d_p^2}{4} \Delta p_B = \pi d_p \sigma \tag{6.39}$$

or

$$\Delta p_B = \frac{4\sigma}{d_p} \tag{6.40}$$

where Δp_B is the excess pressure in the bubble due to surface tension. But the bubble of gas grows over a finite time when the gas flows, and by averaging over time [43], it develops that the appropriate value is Δp_R

$$\Delta p_R = \frac{6\sigma}{d_p} \tag{6.41}$$

Since the bubbles do not really issue singly from the perforations into relatively quiet liquid, we substitute as an approximation the diameter of the perforations d_o, which leads to

$$h_R = \frac{\Delta p_R g_c}{\rho_L g} = \frac{6\sigma g_c}{\rho_L d_o g} \tag{6.42}$$

Other methods of dealing with $h_L + h_R$ have been proposed (see, for example, Ref. 33).

Comparison of observed data with values of h_G calculated by these methods shows a standard deviation of 14.7 percent [79]

Pressure loss at liquid entrance h_2 The flow of liquid under the downspout apron as it enters the tray results in a pressure loss which can be estimated as equivalent to three velocity heads [26, 47]

$$h_2 = \frac{3}{2g} \left(\frac{q}{A_{da}} \right)^2 \tag{6.43}$$

where A_{da} is the smaller of two areas, the downspout cross section or the free area between the downspout apron and the tray. Friction in the downspout is negligible.

Backup in the downspout Refer to Fig. 6.15. The distance h_3, the difference in liquid level inside and immediately outside the downspout, will be the sum of the pressure losses resulting from liquid and gas flow for the tray above

$$h_3 = h_G + h_2 \tag{6.44}$$

Since the mass in the downspout will be partly froth carried over the weir from the tray above, not yet disengaged, whose average density can usually be estimated roughly as half that of the clear liquid, safe design requires that the level of equivalent clear liquid in the downspout be no more than half the tray spacing. Neglecting Δ, the requirement is

$$h_W + h_1 + h_3 < \frac{t}{2} \tag{6.45}$$

For readily foaming systems or where high liquid viscosity hampers disengagement of gas bubbles the backup should be less.

† SI units must be used in Eq. (6.38). For h_L and h_W in inches, V_a in ft/s, ρ_G in lb/ft³, q in ft³/s, and z in feet

$$h_L = 0.24 + 0.725 h_W - 0.29 h_W V_a \rho_G^{0.5} + 4.48 \frac{q}{z}$$

Weeping If the gas velocity through the holes is too small, liquid will drain through them and contact on the tray for that liquid will be lost. In addition, for cross-flow trays, such liquid does not flow the full length of the tray below. The data on incipient weeping are meager, particularly for large liquid depths, and in all likelihood there will always be *some* weeping. A study of the available data [79] led to the following as the best representation of V_{ow}, the minimum gas velocity through the holes below which excessive weeping is likely:

$$\frac{V_{ow}\mu_G}{\sigma g_c} = 0.0229\left(\frac{\mu_G^2}{\sigma g_c \rho_G d_o}\frac{\rho_L}{\rho_G}\right)^{0.379}\left(\frac{l}{d_o}\right)^{0.293}\left(\frac{2A_a d_o}{\sqrt{3}\,p'^3}\right)^{2.8/(Z/d_o)^{0.724}} \tag{6.46}$$

Available data for h_L in the range 23 to 48 mm (0.9 to 1.9 in) do not indicate h_L to be of important influence. It may be for larger depths.

Liquid entrainment When liquid is carried by the gas up to the tray above, the entrained liquid is caught in the liquid on the upper tray. The effect is cumulative, and liquid loads on the upper trays of a tower can become excessive. A convenient definition of the degree of entrainment is the fraction of the liquid entering a tray which is carried to the tray above [44]

$$\text{Fractional entrainment} = E = \frac{\text{moles liquid entrained/ (area)(time)}}{L + \text{moles liquid entrained/ (area)(time)}}$$

The important influence of entrainment on tray efficiency will be considered later. Figure 6.17 represents a summary of sieve-tray entrainment data [44, 47] with an accuracy of ± 20 percent.

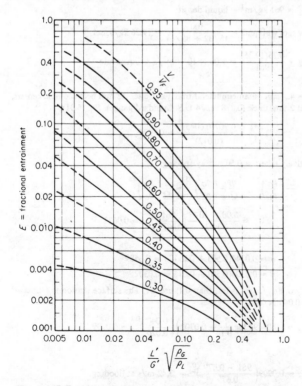

Figure 6.17 Entrainment, sieve trays [44]. *(From Petro/Chemical Engineering, with permission.)*

Under certain conditions, sieve trays are subject to lateral oscillations of the liquid, which may slosh from side to side or from the center to the sides and back. The latter type especially may seriously increase entrainment. The phenomenon is related to froth height relative to tower diameter [8, 9] and a change from frothing to bubbling [96] but is incompletely characterized.

Illustration 6.3 A dilute aqueous solution of methanol is to be stripped with steam in a sieve-tray tower. The conditions chosen for design are:

Vapor. 0.100 kmol/s (794 lb mol/h), 18 mol % methanol
Liquid. 0.25 kmol/s (1984 lb mol/h), 15 mass % methanol
Temperature. 95°C
Pressure. 1 std atm

Design a suitable cross-flow tray.

SOLUTION Mol wt methanol = 32, mol wt water = 18. Av mol wt gas = 0.18(32) + 0.82(18) = 20.5 kg/kmol.

$$\rho_G = \frac{20.5}{22.41} \frac{273}{273 + 95} = 0.679 \text{ kg/m}^3 = \text{gas density}$$

$$Q = 0.10(22.41)\frac{273 + 95}{273} = 3.02 \text{ m}^3/\text{s} = \text{vapor rate}$$

$$\rho_L = 961 \text{ kg/m}^3 = \text{liquid density}$$

$$\text{Av mol wt liquid} = \frac{100}{15/32 + 85/18} = 19.26 \text{ kg/kmol}$$

$$q = \frac{0.25(19.26)}{961} = 5.00 \times 10^{-3} \text{ m}^3/\text{s} = \text{liquid rate}$$

Perforations Take $d_o = 4.5$ mm on an equilateral-triangular pitch 12 mm between hole centers, punched in sheet metal 2 mm thick (0.078 in, 14 U.S. Standard gauge). Eq. (6.31):

$$\frac{A_o}{A_a} = \frac{0.907(0.0045)^2}{(0.012)^2} = 0.1275$$

Tower diameter Tentatively take $t = 0.50$ m tray spacing.

$$\frac{L'}{G'}\left(\frac{\rho_G}{\rho_L}\right)^{0.5} = \frac{q\rho_L}{Q\rho_G}\left(\frac{\rho_G}{\rho_L}\right)^{0.5} = \frac{q}{Q}\left(\frac{\rho_L}{\rho_G}\right)^{0.5}$$

$$= \frac{5.00 \times 10^{-3}}{3.02}\left(\frac{961}{0.679}\right)^{0.5} = 0.0622$$

Table 6.2:

$$\alpha = 0.0744(0.50) + 0.01173 = 0.0489$$

$$\beta = 0.0304(0.50) + 0.015 = 0.0302$$

In Eq. (6.30), since $(L'/G')(\rho_G/\rho_L)^{0.5}$ is less than 0.1, use 0.1. The surface tension is estimated as 40 dyn/cm, or $\sigma = 0.040$ N/m.

$$C_F = \left(0.04893 \log\frac{1}{0.1} + 0.0302\right)\left(\frac{0.040}{0.020}\right)^{0.2} = 0.0909$$

Eq. (6.29):

$$V_F = 0.0909\left(\frac{961 - 0.679}{0.679}\right)^{0.5} = 3.42 \text{ m/s at flooding}$$

Use 80% of flooding velocity. $V = 0.8(3.42) = 2.73$ m/s, based on A_n.

$$A_n = \frac{Q}{V} = \frac{3.02}{2.73} = 1.106 \text{ m}^2$$

Tentatively choose a weir length $W = 0.7T$. Table 6.1: the tray area used by one downspout = 8.8%,

$$A_t = \frac{1.106}{1 - 0.088} = 1.213 \text{ m}^2$$

$T = [4(1.213)/\pi]^{0.5} = 1.243$ m, say 1.25 m. Corrected $A_t = \pi(1.25)^2/4 = 1.227$ m^2

$W = 0.7(1.25) = 0.875$ m (final)

$A_d = 0.088(1.227) = 0.1080$ m^2 downspout cross section

$A_a = A_t - 2A_d$ − area taken by (tray support + disengaging and distributing zones)

For a design similar to that shown in Fig. 6.14, with a 40-mm-wide support ring, beams between downspouts, and 50-mm-wide disengaging and distributing zones, these areas total 0.222 m^2.

$$A_a = 1.227 - 2(0.1080) - 0.222 = 0.789 \text{ m}^2 \text{ for perforated sheet}$$

$$\frac{q}{W} = \frac{5.00 \times 10^{-3}}{0.875} = 5.71 \times 10^{-3} \text{ m}^3/\text{m} \cdot \text{s} \quad \text{OK}$$

Weir crest h_1 and weir height h_W Try $h_1 = 25$ mm $= 0.025$ m. $h_1/T = 0.025/1.25 = 0.02$. $T/W = 1/0.7 = 1.429$. Eq. (6.34): $W_{\text{eff}}/W = 0.938$.

Eq. (6.33): $h_1 = 0.666[(5.00 \times 10^{-3})/0.875]^{2/3}(0.938)^{2/3} = 0.0203$ m. Repeat with $h_1 = 0.0203$: $W_{\text{eff}}/W = 0.9503$; h_1[Eq. (6.33)] = 0.0205 m. OK. Set weir height $h_W = 50$ mm = 0.05 m.

Dry pressure drop h_D Eq. (6.37): $C_o = 1.09(0.0045/0.002)^{0.25} = 1.335$.

$$A_o = 0.1275A_a = 0.1275(0.789) = 0.1006 \text{ m}^2$$

$$V_o = \frac{Q}{A_o} = \frac{3.02}{0.1006} = 30.0 \text{ m/s} \qquad \mu_G = 0.0125 \text{ cP} = 1.25 \times 10^{-5} \text{ kg/m} \cdot \text{s}$$

$$\text{Hole Reynolds number} = \frac{d_o V_o \rho_G}{\mu_G} = \frac{0.0045(30.0)(0.679)}{1.25 \times 10^{-5}} = 7330$$

$f = 0.008$ ("The Chemical Engineers' Handbook," 5th ed., fig. 5.26). $g = 9.807$ m^2/s; $l = 0.002$ m; Eq. (6.36): $h_D = 0.0564$ m.

Hydraulic head h_L

$$V_a = \frac{Q}{A_a} = \frac{3.01}{0.789} = 3.827 \text{ m/s}$$

$$z = \frac{T + W}{2} = \frac{1.25 + 0.875}{2} = 1.063 \text{ m}$$

Eq. (6.38):

$$h_L = 0.0106 \text{ m}$$

Residual pressure drop h_R Eq. (6.42):

$$h_R = \frac{6(0.04)(1)}{961(0.0045)(9.807)} = 5.66 \times 10^{-3} \text{ m}$$

Total gas-pressure drop h_G

Eq. (6.35):

$$h_G = 0.0564 + 0.0106 + 5.66 \times 10^{-3} = 0.0727 \text{ m}$$

Pressure loss at liquid entrance h_2 The downspout apron will be set at $h_W - 0.025 = 0.025$ m above the tray. The area for liquid flow under the apron $= 0.025 W = 0.0219$ m^2. Since this is smaller than A_d, $A_{da} = 0.0219$ m^2. Eq. (6.43):

$$h_2 = \frac{3}{2(9.807)} \left(\frac{5.0 \times 10^{-3}}{0.0219} \right)^2 = 7.97 \times 10^{-3} \text{ m}$$

Backup in downspout Eq. (6.44): $h_3 = 0.0727 + 7.97 \times 10^{-3} = 0.0807$ m.

Check on flooding $h_W + h_1 + h_3 = 0.1512$, which is well below $t/2 = 0.25$ m. Therefore the chosen t is satisfactory.

Weeping velocity For $W/T = 0.7$, the weir is set $0.3296T = 0.412$ m from the center of the tower. Therefore $Z = 2(0.412) = 0.824$ m. All other quantities in Eq. (6.46) have been evaluated, and the equation then yields $V_{ow} = 8.71$ m/s. The tray will not weep excessively until the gas velocity through the holes V_o is reduced to close to this value.

Entrainment

$$\frac{V}{V_F} = 0.8 \qquad \frac{L'}{G'} \left(\frac{\rho_G}{\rho_L} \right)^{0.5} = 0.0622 \text{ (see tower diameter)}$$

Fig. 6.17: $E = 0.05$. The recycling of liquid resulting from such entrainment is too small to influence the tray hydraulics appreciably.

The mass-transfer efficiency is estimated in Illustration 6.4.

Proprietary Trays

Although for many decades the basic design of tray-tower internals remained relatively static, recent years have seen a great many innovations, only a few of which can be mentioned here.

Linde trays These designs have involved improvements both in the perforation design and the tray arrangements [108, 120]. Figure 6.18a shows the *slotted tray*, an alteration in the perforation pattern to influence the flow of liquid. The slots, distributed throughout the tray, not only reduce the hydraulic gradient in large trays (over 10 m diameter [34]) but are also so deployed that they influence the direction of liquid flow to eliminate stagnant areas and achieve, as nearly as possible, desirable plug flow of liquid across the tray. Figure 6.18b shows a *bubbling promoter*, or inclined perforated area at the liquid entrance to the tray. This reduces excessive weeping and produces more uniform froth throughout the tray. The multiple downspouts of Fig. 6.19a are not sealed in the liquid on the tray below; instead the liquid is delivered through slots in the bottom closure to spaces between the downspouts on the tray below. The parallel-flow tray of Fig. 6.19b is so designed that the liquid on all trays in one-half the tower flows from right to left, and on the trays in the other half from left to right. Such an arrangement approximates the so-called Lewis case II [75] and results in an improved Murphree tray efficiency.

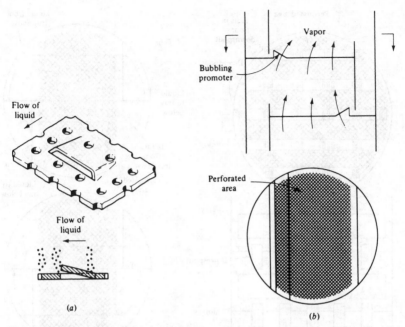

Figure 6.18 Linde sieve trays: (*a*) slotted sieve tray; (*b*) bubbling promoter, cross-flow tray. (*From Chemical Engineering Progress, with permission.*)

Valve trays These are sieve trays with large (roughly 35- to 40-mm-diameter) variable openings for gas flow. The perforations are covered with movable caps which rise as the flow rate of gas increases. At low gas rates and correspondingly small openings, the tendency to weep is reduced. At high gas rates the gas-pressure drop remains low but not as low as that for sieve trays. The Glitsch Ballast Tray valve design is shown in Fig. 6.20; other well-known designs are the Koch Flexitray, the Nutter Float-Valve [10], and the Wyatt valve [114]. Tray efficiencies seem to be roughly the same as those for sieve trays with 6-mm ($\frac{1}{4}$-in) perforations [2, 13].

Counterflow trays These tray-resembling devices differ from conventional trays in that there are no ordinary downspouts: liquid and vapor flow countercurrently through the same openings. *Turbo-Grids* [109] are sheet metal stamped with slotted openings to form the tray, which is installed so alternate trays have the openings at right angles. *Kittel trays*, used extensively in Europe [60, 98, 110], are slotted double trays made of expanded metal arranged so that liquid movement on the tray is influenced by the passage of the gas. *Ripple trays* [65] are perforated trays bent into a sinusoidal wave, with alternate trays installed with the waves at right angles. *Leva trays* [73] are basically very short wetted-wall towers between close-spaced trays, with no weirs and hence no controlled

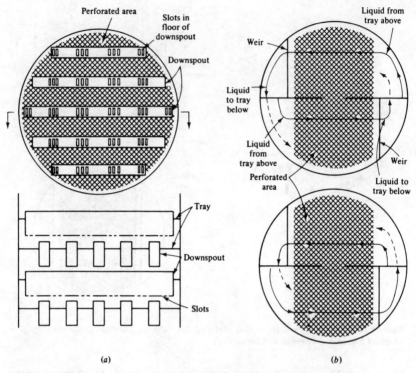

Figure 6.19 Linde tray arrangements: (*a*) multiple downspouts; (*b*) parallel-flow tray.

liquid depth on the trays. Since the gas-pressure drop is low, the trays are especially suited to vacuum distillation.

In most cases the design of proprietary trays is best left to the supplier.

Tray Efficiency

Tray efficiency is the fractional approach to an equilibrium stage (see Chap. 5) which is attained by a real tray. Ultimately, we require a measure of approach to equilibrium of all the vapor and liquid from the tray, but since the conditions at various locations on the tray may differ, we begin by considering the local, or *point*, efficiency of mass transfer at a particular place on the tray surface.

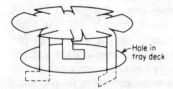

Figure 6.20 Glitsch Ballast Tray valve design (schematic). *(Fritz W. Glitsch and Sons, Inc.)*

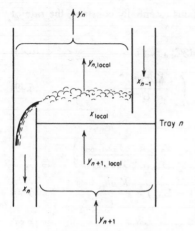

Figure 6.21 Tray efficiency.

Point efficiency Figure 6.21 is a schematic representation of one tray of a multitray tower. The tray n is fed from tray $n-1$ above by liquid of average composition x_{n-1} mole fraction of transferred component, and it delivers liquid of average composition x_n to the tray below. At the place under consideration, a pencil of gas of composition $y_{n+1, \text{local}}$ rises from below, and as a result of mass transfer, leaves with a concentration $y_{n, \text{local}}$. At the place in question, it is assumed that the local liquid concentration x_{local} is constant in the vertical direction. The point efficiency is then defined by

$$\mathbf{E}_{OG} = \frac{y_{n, \text{local}} - y_{n+1, \text{local}}}{y^*_{\text{local}} - y_{n+1, \text{local}}} \tag{6.47}$$

Here y^*_{local} is the concentration in equilibrium with x_{local}, and Eq. (6.47) then represents the change in gas concentration which actually occurs as a fraction of that which would occur if equilibrium were established. The subscript G signifies that gas concentrations are used, and the O emphasizes that $\mathbf{E}_{OG}$ is a measure of the overall resistance to mass transfer for both phases. As the gas passes through the openings of the tray and through the liquid and foam, it encounters several hydrodynamic regimes, each with different rates of mass transfer. The danger in attempting to describe the entire effect in terms of a single quantity, in cases of this sort, was pointed out in Chap. 5, but present information permits nothing better.

Consider that the gas rises at a rate G mol/(area)(time). Let the interfacial surface between gas and liquid be a area/volume of liquid-gas foam. As the gas rises a differential height dh_L, the area of contact is $a \, dh_L$ per unit area of tray. If, while of concentration y, it undergoes a concentration change dy in this height, and if the total quantity of gas remains essentially constant, the rate of solute transfer is $G \, dy$:

$$G \, dy = K_y(a \, dh_L)(y_{\text{local}}^* - y) \tag{6.48}$$

Then

$$\int_{y_{n+1,\,\text{local}}}^{y_{n,\,\text{local}}} \frac{dy}{y_{\text{local}}^* - y} = \int_0^{h_L} \frac{K_y a \, dh_L}{G} \tag{6.49}$$

Since y_{local}^* is constant for constant x_{local},

$$-\ln\frac{y_{\text{local}}^* - y_{n,\,\text{local}}}{y_{\text{local}}^* - y_{n+1,\,\text{local}}} = -\ln\left(1 - \frac{y_{n,\,\text{local}} - y_{n+1,\,\text{local}}}{y_{\text{local}}^* - y_{n+1,\,\text{local}}}\right)$$

$$= -\ln(1 - \mathbf{E}_{OG}) = \frac{K_y a h_L}{G} \tag{6.50}$$

Therefore

$$\mathbf{E}_{OG} = 1 - e^{-K_y a h_L/G} = 1 - e^{N_{tOG}} \tag{6.51}$$

The exponent on e is simplified to N_{tOG}, the number of overall gas-transfer units. Just as K_y contains both gas and liquid resistance to mass transfer, so also is N_{tOG} made up of the transfer units for the gas NT_G and those for the liquid NT_L. As will be shown in Chap. 8, these can be combined in the manner of Eq. (5.7):

$$\frac{1}{N_{tOG}} = \frac{1}{N_{tG}} + \frac{mG}{L}\frac{1}{N_{tL}} \tag{6.52}$$

The terms on the right represent, respectively, the gas and liquid mass-transfer resistances, which must be obtained experimentally. For example, by contacting a gas with a pure liquid on a tray, so that vaporization of the liquid occurs without mass-transfer resistance in the liquid, the vaporization efficiency provides N_{tG}, which can then be correlated in terms of fluid properties, tray design, and operating conditions. Values of N_{tL} obtained through absorption of relatively insoluble gases into liquids (see Chap. 5) can be similarly correlated.

The m of Eq. (6.52) is a local average value for cases where the equilibrium distribution curve is not straight. Refer to Fig. 6.22. It was shown in Chap. 5 that for a given x_{local} and y_{local} the correct slope for adding the resistances is that of a chord m'. For the situation shown in Fig. 6.21, the gas composition changes, leading to a change in m' from an initial value at $y_{n+1,\,\text{local}}$ to a final value at $y_{n,\,\text{local}}$. It has been shown [97] that the correct average m for Eq. (6.52) can be adequately approximated by

$$m = \frac{m'_{y_{n+1}} + m'_{y_n}}{2} \tag{6.53}$$

The location of a point of coordinates $(x_{\text{local}}, y_{\text{local}}^*)$ must be obtained by trial, to fit the definition of E_{OG}, Eq. (6.47). Equations (6.51) and (6.52) show that the lower the solubility of the vapor (the larger the m), the lower the tray efficiency will be.

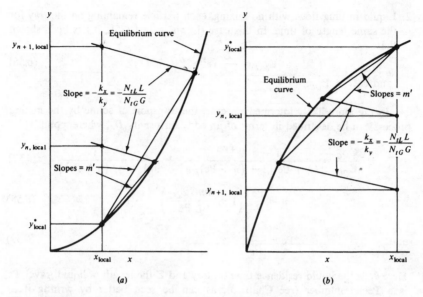

Figure 6.22 Values of m'_{local}: (a) for gas absorption and (b) for distillation and stripping.

Murphree tray efficiency The bulk-average concentrations of all the local pencils of gas of Fig. 6.21 are y_{n+1} and y_n. The Murphree efficiency of the entire tray is then [see Eq. (5.39)]

$$\mathbf{E}_{MG} = \frac{y_n - y_{n+1}}{y_n^* - y_{n+1}} \qquad (6.54)$$

where y_n^* is the value in equilibrium with the leaving liquid of concentration x_n.

The relationship between $\mathbf{E}_{MG}$ and $\mathbf{E}_{OG}$ can then be derived by integrating the local $\mathbf{E}_{OG}$'s over the surface of the tray. Clearly, if all the gas entering were uniformly mixed and fed uniformly to the entire tray cross section, and if the mechanical contacting of gas and liquid were everywhere uniform, the uniformity of concentration of exit gas $y_{n+1, local}$ would then depend on the uniformity of liquid concentration on the tray. Liquid on the tray is splashed about by the action of the gas, some of it even being thrown backward in the direction from which it enters the tray (back mixing). The two extreme cases which might be visualized are:

1. Liquid completely back-mixed, everywhere of uniform concentration x_n. In this case, Eq. (6.54) reverts to Eq. (6.47), and

$$\mathbf{E}_{MG} = \mathbf{E}_{OG} \qquad (6.55)$$

2. Liquid in plug flow, with no mixing, each particle remaining on the tray for the same length of time. In this case (Lewis' case I [75]), it has been shown that

$$\mathbf{E}_{MG} = \frac{L}{mG} \left(e^{\mathbf{E}_{OG}mG/L} - 1 \right) \tag{6.56}$$

and $\mathbf{E}_{MG} > \mathbf{E}_{OG}$.

In the more likely intermediate case, the transport of solute by the mixing process can be described in terms of an eddy diffusivity D_E, whereupon [21]

$$\frac{\mathbf{E}_{MG}}{\mathbf{E}_{OG}} = \frac{1 - e^{-(\eta + \text{Pe})}}{(\eta + \text{Pe})[1 + (\eta + \text{Pe})/\eta]} + \frac{e^{\eta} - 1}{\eta[1 + \eta/(\eta + \text{Pe})]} \tag{6.57}$$

where

$$\eta = \frac{\text{Pe}}{2} \left[\left(1 + \frac{4mG\mathbf{E}_{OG}}{L \, \text{Pe}} \right)^{0.5} - 1 \right] \tag{6.58}$$

and

$$\text{Pe} = \frac{Z^2}{D_E \theta_L} \tag{6.59}$$

Here θ_L is the liquid residence time on tray and Z the length of liquid travel. Pe is a Péclet number (see Chap. 3), as can be seen better by writing it as $(Z/D_E)(Z/\theta_L)$, whence Z/θ_L becomes the average liquid velocity. Pe = 0 corresponds to complete back mixing ($D_E = \infty$), while Pe = ∞ corresponds to plug flow ($D_E = 0$). Large values of Pe result when mixing is not extensive and for large values of Z (large tower diameters). Although point efficiencies cannot exceed unity, it is possible for Murphree efficiencies.

Entrainment A further correction is required for the damage done by entrainment. Entrainment represents a form of back mixing, which acts to destroy the concentration changes produced by the trays. It can be shown [29] that the Murphree efficiency corrected for entrainment is

$$\mathbf{E}_{MGE} = \frac{\mathbf{E}_{MG}}{1 + \mathbf{E}_{MG}[E/(1 - E)]} \tag{6.60}$$

Data Experimental information on N_{tG}, N_{tL}, and D_E is needed to use the relations developed above. Many data have been published, particularly for bubble-cap trays, and an excellent review is available [47]. Perhaps the best-organized information results from a research program sponsored by the American Institute of Chemical Engineers [21]. Most of the work was done with bubble-cap trays, but the relatively fewer data together with subsequent information from sieve trays [50] indicate that the empirical expressions below represent the performance of sieve trays reasonably well and even valve trays as a first approximation [47]. No attempt will be made here to outline in detail the range of conditions covered by these expressions, which should be used with caution, especially if the original report is not consulted.

Sieve trays†

$$N_{tG} = \frac{0.776 + 4.57h_W - 0.238V_a\rho_G^{0.5} + 104.6q/Z}{Sc_G^{0.5}} \tag{6.61}$$

$$N_{tL} = 40\,000D_L^{0.5}(0.213V_a\rho_G^{0.5} + 0.15)\theta_L \tag{6.62}$$

$$D_E = \left(3.93 \times 10^{-3} + 0.0171V_a + \frac{3.67q}{Z} + 0.1800h_W\right)^2 \tag{6.63}$$

$$\theta_L = \frac{\text{vol liquid on tray}}{\text{vol liquid rate}} = \frac{h_L zZ}{q} \tag{6.64}$$

There is good evidence [6, 7] that flow patterns on typical trays are not so well defined as Eq. (6.57) implies and that areas where the liquid is channeled and others where it is relatively stagnant may exist. These effects adversely alter the Murphree efficiency. There is further evidence that surface-tension changes with changes in liquid composition can strongly influence efficiencies [22, 127]. Particularly in large-diameter columns the vapor entering a tray may not be thoroughly mixed, and this requires separate treatment for converting E_{OG} to E_{MG} [35]. Multicomponent mixtures introduce special problems in computing tray efficiencies [58, 111, 125].

Illustration 6.4 Estimate the efficiency of the sieve tray of Illustration 6.3.

SOLUTION From the results of Illustration 6.3 we have

Vapor rate = 0.100 kmol/s	liquid rate = 0.25 kmol/s
$\rho_G = 0.679$ kg/m^3	$h_L = 0.0106$ m
$q = 5.00 \times 10^{-3}$ m^3/s	$h_W = 0.05$ m
$V_a = 3.827$ m/s	$Z = 0.824$ m
$z = 1.063$ m	$E = 0.05$

$$y_{n+1,\,\text{local}} = 0.18 \text{ mole fraction methanol}$$

† Equations (6.61) to (6.63) are empirical and can be used only with SI units. For D_E and D_L in ft^2/h, h_L and h_W in inches, q in ft^3/s, V_a in ft/s, z and Z in feet, θ_L in hours, and ρ_G in lb/ft^3, Eqs. (6.61) to (6.64) become

$$N_{tG} = \frac{0.776 + 0.116h_W - 0.290V_a\rho_G^{0.5} + 9.72q/Z}{Sc_G^{0.5}} \tag{6.61a}$$

$$N_{tL} = (7.31 \times 10^5)D_L^{0.5}(0.26V_a\rho_G^{0.5} + 0.15)\theta_L \tag{6.62a}$$

$$D_E = \left(0.774 + 1.026V_a + \frac{67.2q}{Z} + 0.900h_W\right)^2 \tag{6.63a}$$

$$\theta_L = \frac{2.31 \times 10^{-3}h_L zZ}{q} \tag{6.64a}$$

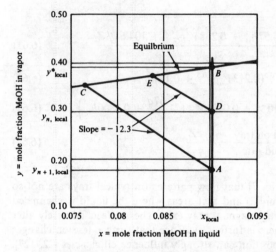

Figure 6.23 Illustration 6.4. Determination of m.

In addition, by the methods of Chap. 2, the following can be estimated: $Sc_G = 0.865$, $D_L = 5.94 \times 10^{-9}$ m^2/s. Substitution of these data in Eqs. (6.61) to (6.64) yields

$$\theta_L = 1.857 \text{ s} \qquad N_{tG} = 0.956$$
$$N_{tL} = 4.70 \qquad D_E = 0.0101 \text{ m}^2/\text{s}$$

For 15 mass % methanol, $x_{local} = (15/32)/(15/32 + 85/18) = 0.0903$ mole fraction methanol. The relevant portion of the equilibrium curve is plotted in Fig. 6.23 with data from "The Chemical Engineers' Handbook," 4th ed., p. 13-5. Point A is plotted with coordinates $(x_{local}, y_{n+1, local})$. At B, the ordinate is y^*_{local}. Line AC is drawn with slope

$$-\frac{N_{tL}L}{N_{tG}G} = -\frac{4.70}{0.956}\frac{0.25}{0.100} = -12.3$$

In this case it is evident that the equilibrium curve is so nearly straight that chord BC (not shown) will essentially coincide with it, as will all other relevant chords from B (as from B to E), and m equals the slope of the equilibrium curve, 2.50.

Equation (6.52) then yields $N_{tOG} = 0.794$, and Eq. (6.51) $E_{OG} = 0.548$. [This provides through Eq. (6.74), although unnecessary for what follows, $y_{n, local} = 0.297$, plotted at D, Fig. 6.23.] Equations (6.57) to (6.59) then give

$$Pe = 36.2 \qquad \eta = 0.540 \qquad E_{MG} = 0.716$$

and with $E = 0.05$, Eq. (6.60) yields $E_{MGE} = 0.70$. **Ans.**

Overall tray efficiency Another method of describing the performance of a tray tower is through the overall tray efficiency,

$$E_O = \frac{\text{number of ideal trays required}}{\text{number of real trays required}} \qquad (6.65)$$

While reliable information on such an efficiency is most desirable and convenient to use, it must be obvious that so many variables enter into such a measure that really reliable values of E_O for design purposes would be most difficult to come by. As will be shown in Chap. 8, E_O can be derived from the individual tray efficiencies in certain simple situations.

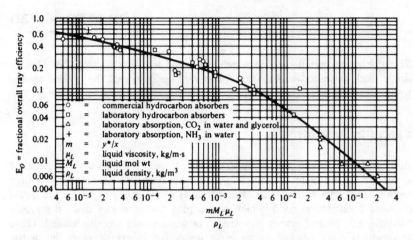

Figure 6.24 Overall tray efficiencies of bubble-cap tray absorbers. For μ_L in centipoises and ρ_L in lb/ft³, use as abscissa $6.243 \times 10^{-5}\, mM_L\mu_L/\rho_L$. (*After O'Connell* [90].)

Provided that only standard tray designs are used and operation is within the standard ranges of liquid and gas rates, some success in correlating E_O with conditions might be expected. O'Connell [90] was successful in doing this for absorption and distillation in bubble-cap-tray towers, and his correlations are shown in Figs. 6.24 and 6.25. They must be used with great caution, but for rough estimates they are most useful, even for sieve and valve trays.

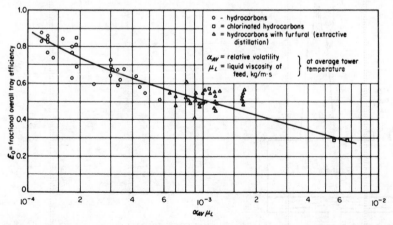

Figure 6.25 Overall tray efficiencies of bubble-cap tray distillation towers separating hydrocarbons and similar mixtures. For μ_L in centipoises, use as abscissa $\alpha_{av}\mu_L \times 10^{-3}$. (*After O'Connell* [90].)

LIQUID DISPERSED

This group includes devices in which the liquid is dispersed into thin films or drops, such as wetted-wall towers, sprays and spray towers, the various packed towers, and the like. The packed towers are the most important of the group.

VENTURI SCRUBBERS

In these devices, which are similar to ejectors, the gas is drawn into the throat of a venturi by a stream of absorbing liquid sprayed into the convergent duct section, as shown in Fig. 6.26. The device is used especially where the liquid contains a suspended solid, which would plug the otherwise more commonly used tray and packed towers, and where low gas-pressure drop is required. These applications have become increasingly important in recent years, as in the absorption of sulfur dioxide from furnace gases with slurries of limestone, lime, or magnesia [51, 59, 68, 116]. Some very large installations (10 m diameter) are in service for electric utilities. The cocurrent flow produces only a single stage

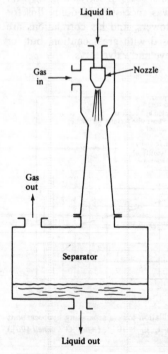

Figure 6.26 Venturi scrubber.

[54], but this becomes less important when a chemical reaction occurs, as in the case of the sulfur dioxide absorbers. Multistage countercurrent effects can be obtained by using several venturis [3]. The device is also used for removing dust particles from gases [12].

WETTED-WALL TOWERS

A thin film of liquid running down the inside of a vertical pipe, with gas flowing either cocurrently or countercurrently, constitutes a wetted-wall tower. Such devices have been used for theoretical studies of mass transfer, as described in Chap. 3, because the interfacial surface between the phases is readily kept under control and is measurable. Industrially, they have been used as absorbers for hydrochloric acid, where absorption is accompanied by a very large evolution of heat [63]. In this case the wetted-wall tower is surrounded with rapidly flowing cooling water. Multitube devices have also been used for distillation, where the liquid film is generated at the top by partial condensation of the rising vapor. Gas-pressure drop in these towers is probably lower than in any other gas-liquid contacting device, for a given set of operating conditions.

SPRAY TOWERS AND SPRAY CHAMBERS

The liquid can be sprayed into a gas stream by means of a nozzle which disperses the liquid into a fine spray of drops. The flow may be countercurrent, as in vertical towers with the liquid sprayed downward, or parallel, as in horizontal spray chambers (see Chap. 7). These devices have the advantage of low pressure drop for the gas but also have a number of disadvantages. There is a relatively high pumping cost for the liquid, owing to the pressure drop through the spray nozzle. The tendency for entrainment of liquid by the gas leaving is considerable, and mist eliminators will almost always be necessary. Unless the diameter/length ratio is very small, the gas will be fairly thoroughly mixed by the spray and full advantage of countercurrent flow cannot be taken. Ordinarily, however, the diameter/length ratio cannot be made very small since then the spray would quickly reach the walls of the tower and become ineffective as a spray.

PACKED TOWERS

Packed towers, used for continuous contact of liquid and gas in both counter-current and cocurrent flow, are vertical columns which have been filled with packing or devices of large surface, as in Fig. 6.27. The liquid is distributed over, and trickles down through, the packed bed, exposing a large surface to contact the gas.

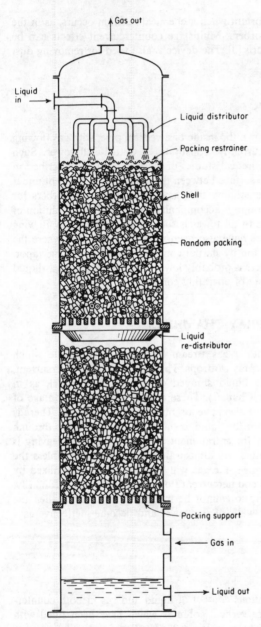

Figure 6.27 Packed tower.

Packing

The tower packing, or *fill*, should offer the following characteristics:

1. Provide for large interfacial surface between liquid and gas. The surface of the packing per unit volume of packed space a_p should be large but not in a microscopic sense. Lumps of coke, for example, have a large surface thanks to their porous structure, but most of this would be covered by the trickling film of liquid. The specific packing surface a_p in any event is almost always larger than the interfacial liquid-gas surface.
2. Possess desirable fluid-flow characteristics. This ordinarily means that the fractional void volume ϵ, or fraction of empty space, in the packed bed should be large. The packing must permit passage of large volumes of fluid through small tower cross sections without loading or flooding (see below) and with low pressure drop for the gas. Furthermore, gas-pressure drop should be largely the result of skin friction if possible, since this is more effective than form drag in promoting high values of the mass-transfer coefficients (see Wetted-wall towers).
3. Be chemically inert to fluids being processed.
4. Have structural strength to permit easy handling and installation.
5. Represent low cost.

Packings are of two major types, random and regular.

Random Packings

Random packings are simply dumped into the tower during installation and allowed to fall at random. In the past such readily available materials as broken stone, gravel, or lumps of coke were used, but although inexpensive, they are not desirable for reasons of small surface and poor fluid-flow characteristics. Random packings most frequently used at present are manufactured, and the common types are shown in Fig. 6.28. Raschig rings are hollow cylinders, as shown, of diameters ranging from 6 to 100 mm ($\frac{1}{4}$ to 4 in) or more. They may be made of chemical stoneware or porcelain, which are useful in contact with most liquids except alkalies and hydrofluoric acid; of carbon, which is useful except in strongly oxidizing atmospheres; of metals; or of plastics. Plastics must be especially carefully chosen, since they may deteriorate rapidly with certain organic solvents and with oxygen-bearing gases at only slightly elevated temperatures. Thin-walled metal and plastic packings offer the advantage of lightness in weight, but in setting floor-loading limits it should be anticipated that the tower may inadvertently fill with liquid. Lessing rings and others with internal partitions are less frequently used. The saddle-shaped packings, Berl and Intalox saddles, and variants of them, are available in sizes from 6 to 75 mm ($\frac{1}{4}$ to 3 in), made of chemical stoneware or plastic. Pall rings, also known as Flexirings, Cascade rings, and as a variant, Hy-Pak, are available in metal and plastic. Tellerettes are available in the shape shown and in several modifications, made

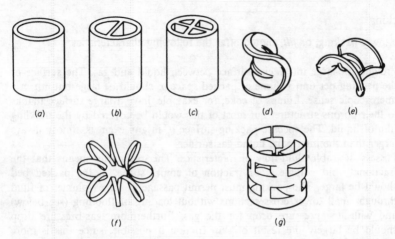

Figure 6.28 Some random tower packings: (a) Raschig rings, (b) Lessing ring, (c) partition ring, (d) Berl saddle *(courtesy of Maurice A. Knight)*, (e) Intalox saddle *(Chemical Processing Products Division, Norton Co.)*, (f) Tellerette *(Ceilcote Company, Inc.)*, and (g) pall ring *(Chemical Processing Products Division, Norton Co.)*.

of plastic. Generally the random packings offer larger specific surface (and larger gas-pressure drop) in the smaller sizes, but they cost less per unit volume in the larger sizes. As a rough guide, packing sizes of 25 mm or larger are ordinarily used for gas rates of 0.25 m³/s (≈ 500 ft³/min), 50 mm or larger for gas rates of 1 m³/s (2000 ft³/min). During installation the packings are poured into the tower to fall at random, and in order to prevent breakage of ceramic or carbon packings, the tower may first be filled with water to reduce the velocity of fall.

Regular Packings

These are of great variety. The counterflow trays already considered are a form of regular packing, as are the arrangements of Fig. 6.29. The regular packings offer the advantages of low pressure drop for the gas and greater possible fluid flow rates, usually at the expense of more costly installation than random packing. Stacked Raschig rings are economically practical only in very large sizes. There are several modifications of the expanded metal packings [105]. Wood grids, or *hurdles*, are inexpensive and frequently used where large void volumes are required, as with tar-bearing gases from coke ovens or liquids carrying suspended solid particles. Knitted or otherwise woven wire screen, rolled as a fabric into cylinders (Neo-Kloss) or other metal gauzelike arrangements (Koch-Sulzer, Hyperfil, and Goodloe packings) provide a large interfacial surface of contacted liquid and gas and a very low gas-pressure drop, especially useful for vacuum distillation [10, 77]. *Static mixers* were originally designed as *line mixers*, for mixing two fluids in cocurrent flow. There are several designs,

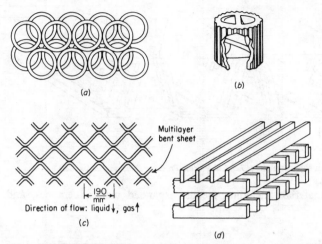

Figure 6.29 Regular, or stacked, packings: (*a*) Raschig rings, stacked staggered (top view), (*b*) double spiral ring *(Chemical Processing Products Division, Norton Co.)*, (*c*) section through expanded-metal-lath packing, (*d*) wood grids.

but in general they consist of metal eggcratelike devices installed in a pipe to cause a multitude of splits of cocurrently flowing fluids into left- and right-hand streams, breaking each stream down into increasingly smaller streams [85]. These devices have been shown to be useful for countercurrent gas-liquid contact [121], with good mass-transfer characteristics at low gas-pressure drop.

Tower Shells

These may be of wood, metal, chemical stoneware, acidproof brick, glass, plastic, plastic- or glass-lined metal, or other material depending upon the corrosion conditions. For ease of construction and strength they are usually circular in cross section.

Packing Supports

An open space at the bottom of the tower is necessary for ensuring good distribution of the gas into the packing. Consequently the packing must be supported above the open space. The support must, of course, be sufficiently strong to carry the weight of a reasonable height of packing, and it must have ample free area to allow for flow of liquid and gas with a minimum of restriction. A bar grid, of the sort shown in Fig. 6.27, can be used [112], but specially designed supports which provide separate passageways for gas and liquid are preferred. Figure 6.30 shows one variety, whose free area for flow is of the order of 85 percent and which can be made in various modifications and of

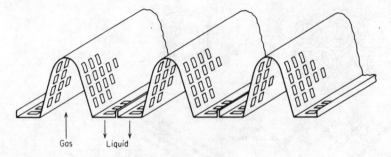

Figure 6.30 Multibeam support plate. *(Chemical Process Products Division, Norton Co.)*

many different materials including metals, expanded metals, ceramics, and plastics.

Liquid Distribution

The importance of adequate initial distribution of the liquid at the top of the packing is indicated in Fig. 6.31. Dry packing is of course completely ineffective for mass transfer, and various devices are used for liquid distribution. Spray nozzles generally result in too much entrainment of liquid in the gas to be useful. The arrangement shown in Fig. 6.27 or a ring of perforated pipe can be used in small towers. For large diameters, a distributor of the type shown in Fig. 6.32 can be used, and many other arrangements are available. It is generally considered necessary to provide at least five points of introduction of liquid for each 0.1 m² (1 ft²) of tower cross section for large towers ($d \geqslant 1.2$ m = 4 ft) and a greater number for smaller diameters.

Random Packing Size and Liquid Redistribution

In the case of random packings, the packing density, i.e., the number of packing pieces per unit volume, is ordinarily less in the immediate vicinity of the tower

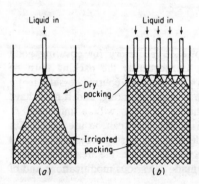

Figure 6.31 Liquid distribution and packing irrigation: (*a*) inadequate, (*b*) adequate.

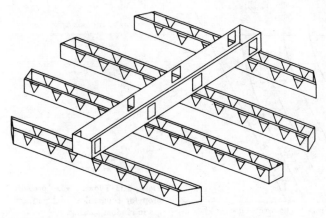

Figure 6.32 Weir-trough liquid distributor. *(Chemicals Process Products Division, Norton Co.)*

walls, and this leads to a tendency of the liquid to segregate toward the walls and the gas to flow in the center of the tower (channeling). This tendency is much less pronounced when the diameter of the individual packing pieces is smaller than at least one-eighth the tower diameter, but it is recommended that, if possible, the ratio $d_p/T = 1 : 15$. Even so it is customary to provide for redistribution of the liquid at intervals varying from 3 to 10 times the tower diameter, but at least every 6 or 7 m. Knitted mesh packings placed under a packing support (Fig. 6.30) make good redistributors [41]. With proper attention to liquid distribution, packed towers are successfully built to diameters of 6 or 7 m or more.

Packing Restrainers

These are necessary when gas velocities are high, and they are generally desirable to guard against lifting of packing during a sudden gas surge. Heavy screens or bars may be used. For heavy ceramic packing, heavy bar plates resting freely on the top of the packing may be used. For plastics and other lightweight packings, the restrainer is attached to the tower shell.

Entrainment Eliminators

Especially at high gas velocities, the gas leaving the top of the packing may carry off droplets of liquid as a mist. This can be removed by mist eliminators, through which the gas must pass, installed above the liquid inlet. A layer of mesh (of wire, Teflon, polyethylene, or other material) especially knitted with 98 to 99 percent voids, roughly 100 mm thick, will collect virtually all mist particles [18, 124]. Other types of eliminators include cyclones and venetian-blind arrangements [53]. A meter of dry random packing is very effective.

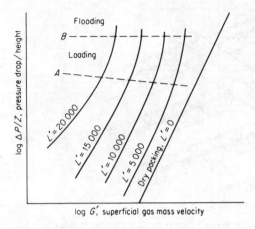

Figure 6.33 Typical gas pressure drop for counterflow of liquid and gas in random packings.

Countercurrent Flow of Liquid and Gas through Packing

For most random packings, the pressure drop suffered by the gas is influenced by the gas and liquid flow rates in a manner similar to that shown in Fig. 6.33. The slope of the line for dry packing is usually in the range 1.8 to 2.0, indicating turbulent flow for most practical gas velocities.

At a fixed gas velocity, the gas-pressure drop increases with increased liquid rate, principally because of the reduced free cross section available for flow of gas resulting from the presence of the liquid. In the region of Fig. 6.33 below A, the liquid holdup, i.e., the quantity of liquid contained in the packed bed, is reasonably constant with changing gas velocity, although it increases with liquid rate. In the region between A and B, the liquid holdup increases rapidly with gas rate, the free area for gas flow becomes smaller, and the pressure drop rises more rapidly. This is known as *loading*. As the gas rate is increased to B at fixed liquid rate, one of a number of changes occurs: (1) a layer of liquid, through which the gas bubbles, may appear at the top of the packing; (2) liquid may fill the tower, starting at the bottom or at any intermediate restriction such as a packing support, so that there is a change from gas–continuous liquid–dispersed to liquid–continuous gas–dispersed (inversion); or (3) slugs of foam may rise rapidly upward through the packing. At the same time, entrainment of liquid by the effluent gas increases rapidly, and the tower is *flooded*. The gas pressure drop then increases very rapidly. The change in conditions in the region A to B of Fig. 6.33 is gradual, and initial loading and flooding are frequently determined by the change in slope of the pressure-drop curves rather than through any visible effect. It is not practical to operate a tower in a flooded condition; most towers operate just below, or in the lower part of, the loading region.

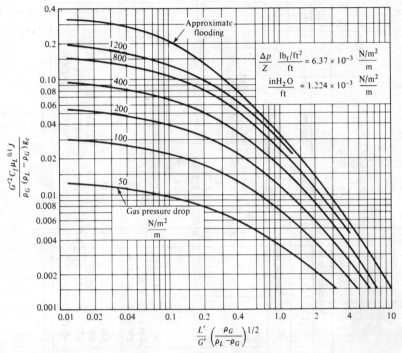

Figure 6.34 Flooding and pressure drop in random-packed towers. For SI units $g_c = 1$, C_f from Table 6.3, and use $J = 1$. For $G' = lb/ft^2 \cdot h$, $\rho = lb/ft^3$, $\mu_L = cP$, $g_c = 4.18 \times 10^8$, C_f from Table 6.3, and use $J = 1.502$. [*Coordinates of Eckert* [38], *Chemical Process Products Division, Norton Co.*)

Flooding and Loading

Flooding conditions in random packings depend upon the method of packing (wet or dry) and settling of the packing [76]. The upper curve of Fig. 6.34 correlates the flooding data for most random packings reasonably well. More specific data are in the handbooks [47] or are available from the manufacturers. The limit of loading cannot readily be correlated.

Typically, absorbers and strippers are designed for gas-pressure drops of 200 to 400 N/m^2 per meter of packed depth (0.25 to 0.5 inH$_2$O/ft), atmospheric-pressure fractionators from 400 to 600 (N/m^2)/m, and vacuum stills for 8 to 40 (N/m^2)/m (0.01 to 0.05 inH$_2$O/ft) [38]. Values of C_f which characterize the packings are given in Table 6.3 (these and other values in Table 6.3 change with changes in manufacturing procedures, so that the manufacturers should be consulted before completing final designs). Flooding velocities for regular or stacked packings will generally be considerably greater than for random packing.

Table 6.3 Characteristics of random packings†

Packing	Nominal size, mm (in)										
	6 ($\frac{1}{4}$)	9.5 ($\frac{3}{8}$)	13 ($\frac{1}{2}$)	16 ($\frac{5}{8}$)	19 ($\frac{3}{4}$)	25 (1)	32 ($1\frac{1}{4}$)	38 ($1\frac{1}{2}$)	50 (2)	76 (3)	89 ($3\frac{1}{2}$)
				Raschig ring							
Ceramic:											
Wall thickness, mm	0.8	1.6	2.4	2.4	2.4	3	4.8	4.8	6	9.5	
C_f	1600	1000	580	380	255	155	125	95	65	37	
C_D			909	749	457	301		181.8	135.6		
ε	0.73	0.68	0.63	0.68	0.73	0.73	0.74	0.71	0.74	0.78	
a_p, m²/m³ (ft²/ft³)	787 (240)	508 (155)	364 (111)	328 (100)	262 (80)	190 (58)	148 (45)	125 (38)	92 (28)	62 (19)	
Metal:											
0.8-mm wall:											
C_f	700	390	300	170	155	115					
ε	0.69		0.84		0.88	0.92					
a_p, m²/m³ (ft²/ft³)	774 (236)		420 (128)		274 (83.5)	206 (62.7)					
1.6-mm wall:											
C_f			410	290	220	137	110	83	57	32	
C_D			688	431	485	304		172.9	133.5		
ε			0.73		0.78	0.85	0.87	0.90	0.92	0.95	
a_p, m²/m³ (ft²/ft³)			387 (118)		236 (71.8)	186 (56.7)	162 (49.3)	135 (41.2)	103 (31.4)	68 (20.6)	

196

Pall rings

Plastic:							
C_f				97	52	40	25	16
C_D				207	105.2	61.8	47.5	23.9
e				0.87	0.90	0.91	0.92	0.92
a, m²/m³ (ft²/ft³)				341 (104)	206 (63)	128 (39)	102 (31)	85 (26)
Metal:							
C_f				70	48	28	20	16
C_D				133.4	95.5	56.6	36.5	
e				0.93	0.94	0.95	0.96	
a, m²/m³ (ft²/ft³)				341 (104)	206 (63)	128 (39)	102 (31)	
Flexirings:							
C_f				78	45	28	22	18
e				0.92	0.94	0.96	0.96	0.97
a, m²/m³ (ft²/ft³)				345 (105)	213 (65)	131 (40)	115 (35)	92 (28)
Hy-pak:‡							
C_f					45		18	15
C_D					88.1		28.7	26.6
e					0.96		0.97	0.97

Table 6.3 Continued

Packing	Nominal size, mm (in)										
	6 ($\frac{1}{4}$)	9.5 ($\frac{3}{8}$)	13 ($\frac{1}{2}$)	16 ($\frac{5}{8}$)	19 ($\frac{3}{4}$)	25 (1)	32 ($1\frac{1}{4}$)	38 ($1\frac{1}{2}$)	50 (2)	76 (3)	89 ($3\frac{1}{2}$)
Berl saddles											
Ceramic:											
C_f	900		240		170	110		65	45		
C_D			508		295	184					
ε	0.60		0.63		0.66	0.69		0.75	0.72		
a, m²/m³ (ft²/ft³)	899 (274)		466 (142)		269 (82)	249 (76)		144 (44)	105 (32)		
Intalox saddles											
Ceramic:											
C_f	725	330	200		145	98		52	40	22	
C_D			399		256	241.5		96.2	71.3	40.6	
ε	0.75		0.78		0.77	0.775		0.81	0.79		
a, m²/m³ (ft²/ft³)	984 (300)		623 (190)		335 (102)	256 (78)		195 (59.5)	118 (36)		
Plastic:											
C_f						33			21	16	
C_D						96.7			56.5	30.1	
ε						0.91			0.93	0.94	
a, m²/m³ (ft²/ft³)						207 (63)			108 (33)	89 (27)	

Super Intalox

Ceramic:‡		
C_f	60	30
C_D	123	63.3
e	0.79	0.81
q_p, m²/m³ (ft²/ft³)	253 (77)	105 (32)

Plastic:‡			
C_f	33	21	16
C_D	79.5	53.5	30.1
e	0.90	0.93	0.94
q_p, m²/m³ (ft²/ft³)	207 (63)	108 (33)	89 (27)

Tellerettes

		67-mm	95-mm (R)
Plastic:			
C_f	40	20	
e	0.87	0.93	0.92
q_p, m³/m³ (ft²/ft³)	180 (55)	112 (34)	112 (34)

† Data are for wet-dumped packing from Chemical Process Products Division, Norton Co.; Koch Engineering Co.; Ceilcote Co.; *Chemical Engineering Progress*, and *Chemical Engineering*.
‡ Sizes, 1, 2, and 3.

Pressure Drop for Single-Phase Flow

The pressure drop suffered by a single fluid in flowing through a bed of packed solids such as spheres, cylinders, gravel, sand, etc., when it alone fills the voids in the bed, is reasonably well correlated by the Ergun equation [42]

$$\frac{\Delta p}{Z} \frac{g_c \varepsilon^3 d_p \rho_g}{(1 - \varepsilon) G'^2} = \frac{150(1 - \varepsilon)}{\text{Re}} + 1.75 \tag{6.66}$$

This is applicable equally well to flow of gases and liquids. The term on the left is a friction factor. Those on the right represent contributions to the friction factor, the first for purely laminar flow, the second for completely turbulent flow. There is a gradual transition from one type of flow to the other as a result of the diverse character of the void spaces, the two terms of the equation changing their relative importance as the flow rate changes. $\text{Re} = d_p G'/\mu$, and d_p is the effective diameter of the particles, the diameter of a sphere of the same surface/volume ratio as the packing in place. If the specific surface is a_p, the surface per unit volume of the particles is $a_p/(1 - \varepsilon)$, and from the properties of a sphere

$$d_p = \frac{6(1 - \varepsilon)}{a_p} \tag{6.67}$$

This will not normally be the same as the nominal size of the particles.

For flow of gases at G' greater than about 0.7 kg/m² · s (500 lb/ft² · h), the first term on the right of Eq. (6.66) is negligible. For a specific type and size of manufactured tower packing, Eq. (6.66) can then be simplified to the empirical expression

$$\frac{\Delta p}{Z} = C_D \frac{G'^2}{\rho_G} \tag{6.68}$$

Values of C_D for SI units are listed in Table 6.3.†

Pressure Drop for Two-Phase Flow

For simultaneous countercurrent flow of liquid and gas, the pressure-drop data of various investigators show wide discrepancies due to differences in packing density and manufacture, such as changes in wall thickness. Estimates therefore cannot be expected to be very accurate. For most purposes the generalized correlation of Fig. 6.34 will serve. The values of C_f are found in Table 6.3. More elaborate data for particular packings are available from the manufacturers.

Illustration 6.5 Sulfur dioxide is to be removed from gas with the characteristics of air by scrubbing with an aqueous ammonium salt solution in a tower packed with 25-mm ceramic Intalox saddles. The gas, entering at the rate 0.80 m³/s (1690 ft³/min) at 30°C, 1 bar, contains 7.0% SO_2, which is to be nearly completely removed. The washing solution will enter at the rate

† For G' in lb/ft² · h, ρ_G in lb/ft³, and $\Delta p/Z$ in inH₂O/ft, multiply the values of C_D from Table 6.3 by 1.405×10^{-10}.

3.8 kg/s (502 lb/min) and has a density 1235 kg/m³ (77 lb/ft³), viscosity 2.5×10^{-3} kg/m · s (2.5 cP). (a) Choose a suitable tower diameter. (b) If the irrigated packed height is 8.0 m (26.25 ft), and if 1 m of 25-mm ceramic Intalox saddles is used above the liquid inlet as an entrainment separator, estimate the power requirement to overcome the gas-pressure drop. The overall efficiency of fan and motor is 60%.

SOLUTION (a) Since the larger flow quantities are at the bottom for an absorber, the diameter will be chosen to accommodate the bottom conditions.

$$\text{Av mol wt gas in} = 0.07(64) + 0.93(29) = 31.45 \text{ kg/kmol}$$

$$\text{Gas in} = 0.80 \frac{273}{303} \frac{0.987}{1.0} \frac{1}{22.41} = 0.0317 \text{ kmol/s}$$

or

$$0.0317(31.45) = 0.998 \text{ kg/s}$$

$$\rho_G = \frac{0.998 \text{ kg/s}}{0.80 \text{ m}^3/\text{s}} = 1.248 \text{ kg/m}^3$$

Assuming essentially complete absorption, SO_2 removed $= 0.0317(0.07)(64) = 0.1420$ kg/s.

$$\text{Liquid leaving} = 3.8 + 0.1420 = 3.94 \text{ kg/s}$$

$$\frac{L'}{G'}\left(\frac{\rho_G}{\rho_L}\right)^{0.5} = \frac{3.94}{0.998}\left(\frac{1.248}{1235}\right)^{0.5} = 0.125$$

In Fig. 6.34 use a gas-pressure drop of 400 (N/m²)/m. The ordinate is therefore 0.061. For 25-mm ceramic Intalox saddles, $C_f = 98$ (Table 6.3). Therefore,

$$G' = \left[\frac{0.061\rho_G(\rho_L - \rho_G)g_c}{C_f\mu_L^{0.1}J}\right]^{0.5} = \left[\frac{0.061(1.248)(1235 - 1.248)}{98(2.5 \times 10^{-3})^{0.1}(1.0)}\right]^{0.5}$$

$$= 1.321 \text{ kg/m}^2 \cdot \text{s}$$

The tower cross-sectional area is therefore (0.998 kg/s)/(1.321 kg/m² · s) = 0.756 m².

$$T = [4(0.756)/\pi]^{0.5} = 0.98 \text{ m, say } 1.0 \text{ m diameter } \textbf{Ans.}$$

The corresponding cross-sectional area is 0.785 m².

(b) The pressure drop for the 8.0 m of irrigated packing is approximately 400(8) = 3200 N/m².

For the dry packing, gas flow rate = $G' = (0.998 - 0.1420)/0.785 = 1.09$ kg/m² · s at a pressure of $100\,000 - 3200 = 96\,800$ N/m². The gas density is

$$\rho_G = \frac{29}{22.41} \frac{273}{303} \frac{96\,800}{101\,330} = 1.114 \text{ kg/m}^3$$

Table 6.3: C_D for this packing = 241.5 for SI units. Eq. (6.68):

$$\frac{\Delta p}{Z} = \frac{241.5(1.09)^2}{1.114} = 258 \text{ N/m}^2 \text{ for 1 m of packing}$$

The total pressure drop for packing is therefore 3200 + 258 = 3458 N/m². To estimate fan power, to this must be added the pressure drop for the packing supports and liquid distributor (negligible for a well-designed apparatus) and inlet expansion and outlet contraction losses for the gas. For a gas velocity in the inlet and outlet pipe of 7.5 m/s, the expansion and outlet-contraction losses would amount at most to 1.5 velocity heads, or $1.5V^2/2g_c = 1.5(7.5)^2/2(1) = 42$ N · m/kg or (42 N · m/kg)(1.24 kg/m³) = 52.1 N/m². The fan power output for the tower is therefore estimated as

$$\frac{[(3458 + 52) \text{ N/m}^2][(0.998 - 0.147)\text{kg/s}]}{1.114 \text{ kg/m}^3} = 2681 \text{ N} \cdot \text{m/s} = 2.681 \text{ kW}$$

The power for the fan motor is then 2.681/0.6 = 4.469 kW = 6 hp.

MASS-TRANSFER COEFFICIENTS FOR PACKED TOWERS

When a packed tower is operated in the usual manner as a countercurrent absorber or stripper for transfer of solute between the gas and liquid, the rate of solute transfer can be computed from measured values of the rate of gas and liquid flow and the bulk concentrations of solute in the entering and leaving streams. As explained in Chap. 5, because of the impossibility of measuring solute concentrations at the gas-liquid interface, the resulting rates of mass transfer can be expressed only as overall coefficients, rather than as coefficients for the individual fluids. Further, since the interfacial area between gas and liquid is not directly measured in such experiments, the *flux* of mass transfer cannot be determined, but instead only the rate as the product of the flux and the total interfacial area. By dividing these rates by the volume of the packing, the results appear as *volumetric overall coefficients*, $K_x a$, $K_y a$, $K_G a$, $F_{OG} a$, $F_{OL} a$, etc., where a is the interfacial surface per unit packed volume.

The individual fluid mass-transfer coefficients (k_x, k_y, F_L, F_G) and the interfacial area a which make up these overall volumetric coefficients are differently dependent upon fluid properties, flow rates, and type of packing. The overall volumetric coefficients are therefore useful only in the design of towers filled with the same packing and handling the same chemical system at the same flow rates and concentrations as existed during the measurements. For general design purposes, the individual coefficients and the interfacial area are necessary.

To obtain individual coefficients, the general approach has been to choose experimental conditions such that the resistance to mass transfer in the gas phase is negligible in comparison with that in the liquid. This is the case for the absorption or desorption of very insoluble gases, e.g., oxygen or hydrogen in water [see Eq. (5.14)] [106]. Measurements in such systems then lead to values of $k_x a$, $k_L a$, and $F_L a$, which can be correlated in terms of system variables. There are evidently no systems involving absorption or desorption where the solute is so soluble in the liquid that the liquid-phase resistance is entirely negligible. But by subtracting the known liquid resistance from the overall resistances (see Chap. 5), it is possible to arrive at the gas-phase coefficients $k_y a$, $k_G a$, and $F_G a$ and to correlate them in terms of system variables.

Another approach to obtaining pure gas-phase coefficients is to make measurements when a pure liquid evaporates into a gas. Here there is no liquid resistance since there is no concentration gradient within the liquid. The resulting volumetric coefficients $k_y a$ and $F_G a$, however, do not agree with those obtained in the manner first described above. The reason is the different effective interfacial areas, as explained below.

Liquid Holdup

Holdup refers to the liquid retained in the tower as films wetting the packing and as pools caught in the crevices between packing particles. It is found that the total holdup ϕ_{Lt} is made up of two parts,

$$\phi_{Lt} = \phi_{L0} + \phi_{Ls} \tag{6.69}$$

where ϕ_{Ls} is the static and ϕ_{Lo} the operating, or moving, holdup, each expressed as volume liquid/packed volume. The moving holdup consists of liquid which continuously moves through the packing, continuously replaced regularly and rapidly by new liquid flowing from above. On stopping the gas and liquid flow, the moving holdup drains from the packing. The static holdup is liquid retained as pools in protected interstices in the packing, largely stagnant, and only slowly replaced by fresh liquid. On stopping the flows, the static holdup does not drain.

When absorption or desorption of a solute occurs, involving transfer of a solute between the bulk liquid and gas, the liquid of the static holdup rapidly comes to equilibrium with the adjacent gas, and thereafter its interfacial surface contributes nothing to mass transfer except as it is slowly replaced. For absorption and desorption, therefore, the smaller area offered by the moving holdup is effective. When evaporation or condensation occurs with the liquid phase a single, pure component, however, the area offered by the total holdup is effective since the liquid then offers no resistance to mass transfer.

Mass Transfer

For most of the standard packings, data for $K_G a$ or the equivalent H_{tOG} are available in handbooks [47] or from manufacturers' bulletins for specific systems. For some of the packings, data for $k_G a$ (or H_{tG}) and $k_L a$ (or H_{tL}) are also available, and attempts have been made to correlate these generally in terms of operating conditions [30, 46, 47, 94].

For *Raschig rings* and *Berl saddles*, Shulman and coworkers [107] have established the nature of the area-free mass-transfer coefficients k_G by passing gases through beds filled with packings made with naphthalene, which sublimes into the gas. By comparing these with $k_G a$'s from aqueous absorption [48] and other systems the interfacial areas for absorption and vaporization were obtained. The data on $k_L a$'s [106] then provided the correlation for k_L, the liquid-phase coefficient. Their work is summarized as follows.

For Raschig rings and Berl saddles, the gas-phase coefficient is given by†

$$\frac{F_G \, \text{Sc}_G^{2/3}}{G} = \frac{k_G p_{\text{B}, M} \, \text{Sc}_G^{2/3}}{G} = 1.195 \left[\frac{d_s G'}{\mu_G (1 - \varepsilon_{Lo})} \right]^{-0.36} \tag{6.70}$$

where ε_{Lo}, the operating void space in the packing, is given by

$$\varepsilon_{Lo} = \varepsilon - \phi_{Lt} \tag{6.71}$$

and d_s is the diameter of a sphere of the same surface as a single packing particle

---

† The j form of Eq. (6.70), with the $\frac{2}{3}$ exponent on Sc_G, correlates the data well, although it is inconsistent with the surface-renewal-theory approach, which is generally accepted for fluid interfaces, and with the correlation of Eq. (6.72).

(not the same as d_p). The fluid properties should be evaluated at average conditions between interface and bulk gas. The liquid coefficient is given by

$$\frac{k_L d_s}{D_L} = 25.1 \left(\frac{d_s L'}{\mu_L} \right)^{0.45} Sc_L^{0.5} \qquad (6.72)$$

Since the liquid data were obtained at very low solute concentrations, k_L can be converted to F_L through $F_L = k_L c$, where c is the molar density of the solvent liquid. There may be an additional effect not included in Eq. (6.72). If concentration changes result in an increase in surface tension as the liquid flows down the column, the liquid film on the packing becomes *stabilized* and the mass-transfer rate may become larger. If the surface tension decreases, the film may break up into rivulets and the mass-transfer rate decreases [127]. Definite conclusions cannot yet be drawn [16].

The interfacial areas for absorption and desorption with water or very dilute aqueous solutions a_{AW} are given for conditions below loading by Shulman in an extensive series of graphs, well represented by the empirical expressions of Table 6.4. For absorption or desorption with nonaqueous liquids, the area is a_A, given by

$$a_A = a_{AW} \frac{\varphi_{Lo}}{\varphi_{LoW}} \qquad (6.73)$$

and for contact of a gas with a pure liquid, as in vaporization, the areas are

$$a_V = 0.85 a_A \frac{\varphi_{Lt}}{\varphi_{Lo}} \qquad (6.74)$$

$$a_{VW} = 0.85 a_{AW} \frac{\varphi_{LtW}}{\varphi_{LoW}} \qquad (6.75)$$

The subscript W indicates water as the liquid. The holdup data are summarized in Table 6.5. Although these data are based on work with absorption, desorption, and vaporization, the evidence is that the same coefficients apply to distillation in packed towers [89]. In distillation, a_A's should be used.

Illustration 6.6 A tower packed with 38-mm (1.5-in) Berl saddles is to be used for absorbing benzene vapor from a dilute mixture with inert gas. The circumstances are:

Gas Av mol wt = 11, viscosity = 10^{-5} kg/m · s (0.01 cP); 27°C, p_t = 107 kN/m² (803 mmHg), D_G = 1.30 × 10^{-5} m²/s (0.504 ft²/h), flow rate G' = 0.716 kg/m² · s (528 lb/ft² · h).

Liquid Av mol wt = 260, viscosity = 2 × 10^{-3} kg/m · s (2.0 cP), density = 840 kg/m³ (52.4 lb/ft³), surface tension = 3.0 × 10^{-2} N/m (30 dyn/cm), D_L = 4.71 × 10^{-10} m²/s (1.85 × 10^{-5} ft²/h), flow rate L' = 2.71 kg/m² · s (2000 lb/ft² · h).
Compute the volumetric mass-transfer coefficients.

Table 6.4 Interfacial area for absorption and desorption, aqueous liquids†

For conditions below loading, $a_{AW} = m(808 G'/\rho_G^{0.5})^n L'^p$; use only for SI units: L' and G' kg/m² · s [or (lb/h ft²)(0.001356)] and ρ_G kg/m³ [or (lb/ft³)(16.019)]; $a_{AW} = $ m²/m³; divide by 3.281 for ft²/ft³; the original data cover L' up to 6.1 kg/m² · s (4500 lb/ft² · h); extrapolation to $L = $ 10.2 kg/m² · s (7500 lb/ft² · h) has been suggested [107]

Packing	Nominal size		Range of L'		m	n	p
	mm	in	kg/m² · s	lb/ft² · h			
Raschig	13	0.5	0.68–2.0	500–1500	28.01	$0.2323L' - 0.30$	−1.04
rings			2.0–6.1	1500–4500	14.69	$0.01114L' + 0.148$	−0.111
	25	1	0.68–2.0	500–1500	34.42	0	0.552
			2.0–6.1	1500–4500	68.2	$0.0389L' - 0.0793$	−0.47
	38	1.5	0.68–2.0	500–1500	36.5	$0.0498L' - 0.1013$	0.274
			2.0–6.1	1500–4500	40.11	$0.01091L' - 0.022$	0.140
	50	2	0.68–2.0	500–1500	31.52	0	0.481
			2.0–6.1	1500–4500	34.03	0	0.362
Berl							
saddles	13‡	0.5‡	0.68–2.0	500–1500	16.28	0.0529	0.761
			2.0–6.1	1500–4500	25.61	0.0529	0.170
	25§	1§	0.68–2.0	500–1500	52.14	$0.0506L' - 0.1029$	0
			2.0–6.1	1500–4500	73.0	$0.0310L' - 0.0630$	−0.359
	38§	1.5§	0.68–2.0	500–1500	40.6	−0.0508	0.455
			2.0–6.1	1500–4500	62.4	$0.0240L' - 0.0996$	−0.1355

† Data of Shulman, et al. [107].

‡ For $G' < 1.08$ kg/m² · s (800 lb/ft² · h) only. For higher values, see Shulman, *AIChE J.*, **1**, 253 (1955), figs. 16 and 17.

§ It has been shown that constants for $L' = 0.68$ to 2.0 may apply over the range $L' = 0.68$ to 6.1 [99]. This may reflect some change in the shape of the packing over the years.

SOLUTION For the gas,

$$\rho_G = \frac{11}{22.41} \frac{1.07 \times 10^5}{1.0133 \times 10^5} \frac{273}{273 + 27} = 0.472 \text{ kg/m}^3$$

$$Sc_G = \frac{\mu_G}{\rho_G D_G} = \frac{1 \times 10^{-5}}{0.472(1.30 \times 10^{-5})} = 1.630$$

$$G' = 0.716 \text{ kg/m}^2 \cdot \text{s} \qquad G = \frac{0.716}{11} = 0.0651 \text{ kmol/m}^2 \cdot \text{s}$$

For the liquid,

$$Sc_L = \frac{\mu_L}{\rho_L D_L} = \frac{2 \times 10^{-3}}{840(4.77 \times 10^{-10})} = 4990$$

Table 6.5 Liquid holdup in packed towers

$\varphi_{Lt} = \varphi_{Lo} + \varphi_{Ls}$ $\varphi_{LtW} = \varphi_{LoW} + \varphi_{LsW}$ $\varphi_{Lo} = \varphi_{LoW}H$ Use only SI units: L' in kg/m² · s = (lb/ft² · s)(0.001356); ρ_L in kg/m³ = (lb/ft³)(16.019); μ_L in kg/m · s = cP(0.001); σ in N/m = (dyn/cm)(0.001); φ is dimensionless

Packing	Nominal size mm	Nominal size in	d_s, m	φ_{Ls}	Water, ordinary temperatures	μ_L, kg/m · s	H
Ceramic Raschig rings	13	0.5	0.01774	$\dfrac{0.0486\mu_L^{0.02}\sigma^{0.99}}{d_s^{1.21}\rho_L^{0.37}}$	$\beta = 1.508d_s^{0.376}$ $\varphi_{LtW} = \dfrac{2.47\times10^{-4}}{d_s^{1.21}}$	<0.012	$\dfrac{975.7L'^{0.57}\mu_L^{0.13}}{\rho_L^{0.84}}(2.024L'^{-0.430}-1)\left(\dfrac{\sigma}{0.073}\right)^{0.1737-0.262\log L'}$
	25	1	0.0356				
	38	1.5	0.0530		$\varphi_{LtW} = \dfrac{(2.09\times10^{-6})(737.5L')^{\beta}}{d_s^2}$	>0.012	$\dfrac{2168L'^{0.57}\mu_L^{0.31}}{\rho_L^{0.84}}(2.024L'^{-0.430}-1)\left(\dfrac{\sigma}{0.073}\right)^{0.1737-0.262\log L'}$
	50	2	0.0725				
Carbon Raschig rings	25	1	0.01301	$\dfrac{0.0237\mu_L^{0.02}\sigma^{0.23}}{d_s^{1.21}\rho_L^{0.37}}$	$\beta = 1.104d_s^{0.376}$ $\varphi_{LtW} = \dfrac{5.94\times10^{-4}}{d_s^{1.21}}$	<0.012	$\dfrac{407.9L'^{0.57}\mu_L^{0.13}}{\rho_L^{0.84}}(1.393L'^{-0.315}-1)\left(\dfrac{\sigma}{0.073}\right)^{0.1737-0.262\log L'}$
	38	1.5	0.0543				
	50	2	0.0716		$\varphi_{LtW} = \dfrac{(7.34\times10^{-6})(737.5L')^{\beta}}{d_s^2}$	>0.012	$\dfrac{901L'^{0.57}\mu_L^{0.31}}{\rho_L^{0.84}}(1.393L'^{-0.315}-1)\left(\dfrac{\sigma}{0.073}\right)^{0.1737-0.262\log L'}$
Ceramic Berl saddles	13	0.5	0.31622	$\dfrac{4.23\times10^{-3}\mu_L^{0.04}\sigma^{0.55}}{d_s^{1.56}\rho_L^{0.37}}$	$\beta = 1.508d_s^{0.376}$ $\varphi_{LtW} = \dfrac{5.014\times10^{-5}}{d_s^{1.56}}$	<0.020	$\dfrac{1404L'^{0.57}\mu_L^{0.13}}{\rho_L^{0.84}}(3.24L'^{-0.413}-1)\left(\dfrac{\sigma}{0.073}\right)^{0.2817-0.262\log L'}$
	25	1	0.0320				
	38	1.5	0.0472		$\varphi_{LtW} = \dfrac{(2.32\times10^{-6})(737.5L')^{\beta}}{d_s^2}$	>0.020	$\dfrac{2830L'^{0.57}\mu_L^{0.31}}{\rho_L^{0.84}}(3.24L'^{-0.413}-1)\left(\dfrac{\sigma}{0.073}\right)^{0.2817-0.262\log L'}$

† Data of Shulman, et al. [107].

Holdup Table 6.5:

$$L' = 2.71 \text{ kg/m}^2 \cdot \text{s} \qquad d_s = 0.0472 \text{ m} \qquad \beta = 1.508 d_s^{0.376} = 0.478$$

$$\varphi_{LsW} = \frac{5.014 \times 10^{-5}}{d_s^{1.56}} = 5.86 \times 10^{-3} \text{ m}^3/\text{m}^3$$

$$\varphi_{LtW} = \frac{(2.32 \times 10^{-6})(737.5L')^\beta}{d_s^2} = 0.0394 \text{ m}^3/\text{m}^3$$

$$\varphi_{LoW} = \varphi_{LtW} - \varphi_{LsW} = 0.0335 \text{ m}^3/\text{m}^3$$

$$H = \frac{1404 L'^{0.57} \mu_L^{0.13}}{\rho_L^{0.84}(3.24 L'^{0.413} - 1)}\left(\frac{\sigma}{0.073}\right)^{0.2817 - 0.262 \log L'}$$

$$= 0.855$$

$$\varphi_{Lo} = \varphi_{LoW} H = 0.0335(0.855) = 0.0286$$

$$\varphi_{Ls} = \frac{4.23 \times 10^{-3} \mu_L^{0.04} \sigma^{0.55}}{d_s^{1.56} \rho_L^{0.37}} = 4.65 \times 10^{-3} \text{ m}^3/\text{m}^3$$

$$\varphi_{Lt} = \varphi_{Lo} + \varphi_{Ls} = 0.0286 + 4.65 \times 10^{-3} = 0.0333 \text{ m}^3/\text{m}^3$$

Interfacial area Table 6.4:

$$m = 62.4 \qquad n = 0.0240 L' - 0.0996 = -0.0346$$

$$p = -0.1355 \qquad a_{AW} = m\left(\frac{808 G'}{\rho_G^{0.5}}\right)^n L'^p = 43.8 \text{ m}^2/\text{m}^3$$

Eq. (6.73):

$$a_A = a_{AW}\frac{\varphi_{Lo}}{\varphi_{LoW}} = \frac{43.8(0.0286)}{0.0335} = 37.4 \text{ m}^2/\text{m}^3$$

Table 6.3:

$$\varepsilon = 0.75$$

Eq. (6.71):

$$\varepsilon_{Lo} = \varepsilon - \varphi_{Lt} = 0.75 - 0.0333 = 0.717$$

Eq. (6.70):

$$\frac{F_G(1.630)^{2/3}}{0.0651} = 1.195\left[\frac{0.0472(0.716)}{(1 \times 10^{-5})(1 - 0.717)}\right]^{-0.36}$$

$$F_G = 1.914 \times 10^{-3} \text{ kmol/m}^2 \cdot \text{s} = k_G p_{B, M}$$

Eq. (6.72):

$$\frac{k_L(0.0472)}{4.77 \times 10^{-10}} = 25.1\left[\frac{0.0472(2.71)}{0.002}\right]^{0.45}(4990)^{0.5}$$

$$k_L = 1.156 \times 10^{-4} \text{ kmol/m}^2 \cdot \text{s} \cdot (\text{kmol/m}^3)$$

Since the data for k_L were taken at very low concentrations, this can be converted into $F_L = k_L c$, where $c = 840/260 = 3.23 \text{ kmol/m}^3$ molar density. Therefore $F_L = (1.156 \times 10^{-4})(3.23) = 3.73 \times 10^{-4} \text{ kmol/m}^2 \cdot \text{s} = k_L x_{B, M} c$. The volumetric coefficients are

$$F_G a_A = (1.914 \times 10^{-3})(37.4) = 0.0716 \text{ kmol/m}^3 \cdot \text{s} = k_g a_A p_{B, M}$$

$$F_L a_A = (3.73 \times 10^{-4})(37.4) = 0.0140 \text{ kmol/m}^3 \cdot \text{s} = k_L a_A x_{B, M} c$$

Illustration 6.7 A tower packed with 50-mm (2-in) ceramic Raschig rings is to be used for dehumidification of air by countercurrent contact with water. At the top of the tower, the conditions are to be:

Water: flow rate = 5.5 kg/m² · s (4060 lb/ft² · h), temp = 15°C.
Air: flow rate = 1.10 kg/m² · s (811 lb/ft² · h), temp = 20°C, 1 std atm, essentially dry.
Compute the transfer coefficients for the top of the tower.

SOLUTION For the gas, $G' = 1.10$ kg/m² · s, $G = 1.10/29 = 0.0379$ kmol/m² · s, $\mu_G = 1.8 \times 10^{-5}$ kg/m · s, $Sc_G = 0.6$ for air-water vapor, and

$$\rho_G = \frac{29}{22.41}\frac{273}{293} = 1.206 \text{ kg/m}^3$$

For the liquid, $L' = 5.5$ kg/m² · s. Table 6.5:

$$d_s = 0.0725 \text{ m} \qquad \beta = 1.508 d_s^{0.376} = 0.562$$

$$\varphi_{LtW} = \frac{(2.09 \times 10^{-6})(737.5L')^\beta}{d_s^2} = 0.0424 \text{ m}^3/\text{m}^3$$

$$\varphi_{LsW} = \frac{2.47 \times 10^{-8}}{d_s^{1.21}} = 5.91 \times 10^{-3} \text{ m}^3/\text{m}^3$$

$$\varphi_{LoW} = \varphi_{LtW} - \varphi_{LsW} = 0.0424 - 5.91 \times 10^{-3} = 0.0365 \text{ m}^3/\text{m}^3$$

Table 6.4:

$$m = 34.03, n = 0, p = 0.362.$$

$$a_{AW} = m\left(\frac{808G'}{\rho_G^{0.5}}\right)^n L'^p = 63.1 \text{ m}^2/\text{m}^3$$

Eq. (6.75):

$$a_{VW} = 0.85a_{AW}\frac{\phi_{LtW}}{\varphi_{LoW}} = 0.85(63.1)\frac{0.0424}{0.0365} = 62.3 \text{ m}^2/\text{m}^3$$

Table 6.3:

$$\varepsilon = 0.74, \varepsilon_{Lo} = \varepsilon - \varphi_{LtW} = 0.74 - 0.0424 = 0.698.$$

Eq. (6.70):

$$\frac{F_G(0.6)^{2/3}}{0.0379} = 1.195\left[\frac{0.0725(1.10)}{(1.8 \times 10^{-5})(1 - 0.698)}\right]^{-0.36} = 0.0378$$

$$F_G = 2.01 \times 10^{-3} \text{ kmol/m}^2 \cdot \text{s}$$

$$F_G a_{VW} = (2.01 \times 10^{-3})(62.3) = 0.125 \text{ kmol/m}^3 \cdot \text{s} = k_G a_{VW} p_{B,M}$$

Since the liquid is pure water, it has no mass-transfer coefficient. However, for such processes (see Chap. 7), we need convection heat-transfer coefficients for both gas and liquid. These can be estimated, in the absence of directly applicable data, through the heat- mass-transfer analogy. Thus, from Eq. (6.70), and assuming $j_D = j_H$,

$$j_H = \frac{h_G}{C_p G'}\text{Pr}_G^{2/3} = 0.0378$$

where h_G is the gas-phase heat-transfer coefficient. For air, $C_p = 1005$ N · m/kg · K, $\text{Pr}_G = 0.74$.

$$h_G = \frac{0.0378 C_p G'}{\text{Pr}_G^{2/3}} = \frac{0.0378(1005)(1.10)}{(0.74)^{2/3}} = 51.1 \text{ W/m}^2 \cdot \text{K}$$

Similarly, the heat-transfer analog of Eq. (6.72) is

$$\text{Nu} = \frac{h_L d_s}{k_{\text{th}}} = 25.1\left(\frac{d_s L'}{\mu_L}\right)^{0.45}\text{Pr}_L^{0.5}$$

where h_L is the liquid-phase heat-transfer coefficient. $k_{th} = 0.587$ W/m · K and $Pr_L = C_p \mu_L / k_{th} = 4187(1.14 \times 10^{-3})/0.587 = 8.1$. Therefore

$$h_L = 25.1 \frac{k_{th}}{d_s} \left(\frac{d_s L'}{\mu_L} \right)^{0.45} Pr_L{}^{0.5} = \frac{25.1(0.587)}{0.0725} \left[\frac{0.0725(5.5)}{1.14 \times 10^{-4}} \right]^{0.45} (8.1)^{0.5}$$

$$= 8071 \text{ W/m}^2 \cdot \text{K}$$

The corresponding volumetric coefficients are

$$h_G a_{VW} = 51.1(62.3) = 3183 \text{ W/m}^3 \cdot \text{K}$$

$$h_L a_{VW} = 8071(62.3) = 503\,000 \text{ W/m}^3 \cdot \text{K}$$

COCURRENT FLOW OF GAS AND LIQUID

Cocurrent flow of gas and liquid (usually downward) through packed beds is used for catalytic chemical reaction between components of the fluids, the catalyst usually being some active substance supported upon granular ceramic material. Such arrangements are known as *trickle-bed reactors*. For the strictly diffusional separation processes, where the packing is inactive and merely provides a means of producing interfacial surface, it should be recalled that cocurrent flow can produce no better than one theoretical stage of separation. This is sufficient for substantially complete separation where chemical reaction accompanies the diffusional separation, e.g., in the absorption of hydrogen sulfide into a strongly basic solution such as aqueous sodium hydroxide. For these situations, there is the advantage that flooding is impossible: there is no upper limit for the permissible flow rates. In the absence of reaction, however, cocurrent flow rarely is useful.

Practically all the available data concern conditions suitable for catalysis, with small packing particles (small spheres, cylinders, and the like) which are the catalyst supports. Very complete reviews of pressure drop and mass-transfer for such situations are available [100, 103].

END EFFECTS AND AXIAL MIXING

In addition to that occurring within the packed volume, there will be mass transfer where the liquid is introduced by sprays or other distributors and where it drips off the lower packing support. These constitute the *end effects*. For proper evaluation of the packing itself, correction of experimental data must be made for these effects, usually by operating at several packing heights followed by extrapolation of the results to zero height. New equipment designed with uncorrected data will be conservatively designed if shorter than the experimental equipment but underdesigned if taller. Alternatively, equipment designed with corrected data will include a small factor of safety represented by the end effects. The data represented by Eqs. (6.69) to (6.75) and Tables 6.4 and 6.5 are believed to be reasonably free of end effects.

As will be clear in later chapters, the equations generally used for design, which are the same as those used for computing mass-transfer coefficients from laboratory measurements, are based on the assumption that the gas and liquid flow in piston, or plug, flow, with a flat velocity profile across the tower and with each portion of the gas or liquid having the same residence time in the tower. Such is not actually the case. Nonuniformity of packing and maldistribution of liquid may lead to channeling, or regions where the liquid flow is abnormally great; the liquid in the static holdup moves forward much more slowly than that in the moving holdup; drops of liquid falling from a packing piece may be blown upward by the gas. Similarly the resistance to gas flow at the tower walls and in relatively dry regions of the packing is different from that elsewhere; the downward movement of liquid induces downward movement of the gas. As a result, the purely piston-type flow does not occur, and there is a relative movement within each fluid parallel to the axis of the tower, variously described as axial mixing, axial dispersion, and back mixing. *Axial dispersion* is the spreading of the residence time in unidirectional flow owing to the departure from purely piston-type, or plug, flow; the fluid particles move forward but at different speeds. *Back mixing* is the backward flow in a direction opposite to that of the net flow, caused by frictional drag of one fluid upon the other, spray of liquid into the gas (entrainment), and the like. The transport of solute by these mechanisms is usually described in terms of an eddy diffusivity. For gas-liquid packed towers [19, 28, 123] many of the data scatter badly. For 25-mm Berl saddles and Raschig rings, and 50-mm Raschig rings, axial mixing in the liquid is more important than that in the gas [36]. Such transport of solute reduces the concentration differences for interphase mass transfer. The true mass-transfer coefficients are undoubtedly larger than those we customarily use, since the latter have not been corrected for axial mixing. The effect is strong for spray towers, which are relatively poor devices for separation operations as a result. Fortunately the effects appear not to be large for packed towers under ordinary circumstances.

In tray towers, entrainment of liquid in the gas is a form of back mixing, and there are back mixing and axial mixing on the trays, which we have already considered. In sparged vessels, the liquid is essentially completely back-mixed to a uniform solute concentration. Both phases are largely completely mixed in mechanically agitated vessels.

TRAY TOWERS VS. PACKED TOWERS

The following may be useful in considering a choice between the two major types of towers.

1. *Gas-pressure drop.* Packed towers will ordinarily require a smaller pressure drop. This is especially important for vacuum distillation.
2. *Liquid holdup.* Packed towers will provide a substantially smaller liquid

holdup. This is important where liquid deterioration occurs with high temperatures and short holding times are essential. It is also important in obtaining sharp separations in batch distillation.

3. *Liquid/gas ratio.* Very low values of this ratio are best handled in tray towers. High values are best handled in packed towers.

4. *Liquid cooling.* Cooling coils are more readily built into tray towers; and liquid can more readily be removed from trays, to be passed through coolers and returned, than from packed towers.

5. *Side streams.* These are more readily removed from tray towers.

6. *Foaming systems.* Packed towers operate with less bubbling of gas through the liquid and are the more suitable.

7. *Corrosion.* Packed towers for difficult corrosion problems are likely to be less costly.

8. *Solids present.* Neither type of tower is very satisfactory. Agitated vessels and venturi scrubbers are best but provide only a single stage. If multistage countercurrent action is required, it is best to remove the solids first. Dust in the gas can be removed by a venturi scrubber at the bottom of a tower. Liquids can be filtered or otherwise clarified before entering a tower.

9. *Cleaning.* Frequent cleaning is easier with tray towers.

10. *Large temperature fluctuations.* Fragile packings (ceramic, graphite) tend to be crushed. Trays or metal packings are satisfactory.

11. *Floor loading.* Plastic packed towers are lighter in weight than tray towers, which in turn are lighter than ceramic or metal packed towers. In any event, floor loadings should be designed for accidental complete filling of the tower with liquid.

12. *Cost.* If there is no overriding consideration, cost is the major factor to be taken into account.

NOTATION FOR CHAPTER 6

Any consistent set of units may be used, except as noted.

a	average specific interfacial surface for mass transfer, area/volume, L^2/L^3
a_A	specific interfacial surface for absorption, desorption, and distillation, area/volume, L^2/L^3
a_p	specific surface of packing, area/volume, L^2/L^3
a_V	specific interfacial surface for contact of a gas with a pure liquid, area/volume, L^2/L^3
A	projected area, L^2
A_a	active area, area of perforated sheet, L^2
A_d	downspout cross-sectional area, L^2
A_{da}	smaller of the two areas, A_d or free area between downspout apron and tray, L^2
A_n	net tower cross-sectional area for gas flow $= A_t - A_d$ for cross-flow trays, L^2
A_o	area of perforations, L^2
A_t	tower cross-sectional area, L^2
b	baffle width, L
B	blade length, L
c	molar density of the liquid, mole/L^3
	solute concentration, mole/L^3

C	distance from impeller to tank bottom, L
C_D	empirical constant, Eq. (6.68) and Table 6.3
C_f	characterization factor of packing, two-phase flow, empirical constant, Table 6.3
C_F	flooding constant for trays, Eqs. (6.29) and (6.30), empirical
C_o	orifice coefficient, dimensionless
C_p	specific heat at constant pressure, FL/MT
d_d	disk diameter, L
d_i	impeller diameter, L
d_o	orifice or perforation diameter, L
d_p	average bubble diameter, L
	nominal diameter of tower-packing particle, L
d_s	diameter of sphere of same surface as a single packing particle, L
D	diffusivity, L^2/Θ
D_E	eddy diffusivity of back mixing, L^2/Θ
E	fractional entrainment, entrained liquid/(entrained liquid + net liquid flow), mole/mole or M/M
E_{MG}	Murphree gas-phase stage efficiency, fractional
E_{MGE}	Murphree gas-phase stage efficiency corrected for entrainment, fractional
E_O	overall tray efficiency of a tower, fractional
E_{OG}	point gas-phase tray efficiency, fractional
f	function
	Fanning friction factor, dimensionless
F	mass-transfer coefficient, mole/$L^2\Theta$
F_D	drag force, F
Fr	impeller Froude number $= d_i N^2/g$, dimensionless
g	acceleration of gravity, L/Θ^2
g_c	conversion factor, $ML/F\Theta^2$
G	superficial molar gas mass velocity, mole/$L^2\Theta$
G'	superficial gas mass velocity, $M/L^2\Theta$
h	heat-transfer coefficient, $FL/L^2T\Theta$
h_D	dry-plate gas-pressure drop as head of clear liquid, L
h_G	gas-pressure drop as head of clear liquid, L
h_L	gas-pressure drop due to liquid holdup on tray, as head of clear liquid, L
h_R	residual gas-pressure drop as head of clear liquid, L
h_W	weir height, L
h_1	weir crest, L
h_2	head loss owing to liquid flow under downspout apron, L
h_3	backup of liquid in downspout, L
H	correction factor (holdup, packed towers), Table 6.5
H_f	mechanical energy lost to friction, FL/M
H_{tG}	height of gas-phase transfer unit, L
H_{tL}	height of liquid-phase transfer unit, L
H_{tOG}	overall height of a transfer unit, L
j_D	mass-transfer group $= F_G \, Sc_G^{2/3}/G$, dimensionless
j_H	heat-transfer group $= h \, Pr_G^{2/3}/C_p G'$, dimensionless
J	conversion factor, Fig. 6.31
k_G	gas-phase mass-transfer coefficient, mole/$\Theta L^2(F/L^2)$
k_L	liquid-phase mass-transfer coefficient, mole/$\Theta L^2(mole/L^3)$
k_{th}	thermal conductivity, $FL^2/\Theta L^2T$
k_x	liquid-phase mass-transfer coefficient, mole/$\Theta L^2(mole \ fraction)$
k_y	gas-phase mass-transfer coefficient, mole/$\Theta L^2(mole \ fraction)$
K_G	overall gas-phase mass-transfer coefficient, mole/$\Theta L^2(F/L^2)$
K_y	overall gas-phase mass-transfer coefficient, mole/$\Theta L^2(mole \ fraction)$
l	plate thickness, L

ln   natural logarithm

log   common logarithm

L superficial liquid molar mass velocity, $\text{mole}/L^2\Theta$

L' superficial liquid mass velocity, $M/L^2\Theta$

$\bar{L}$ characteristic length, L

m average slope of a chord of equilibrium curve, Eqs. (6.52) to (6.59), mole fraction in gas/mole fraction in liquid
empirical constant, Table 6.4

m' slope of a chord of equilibrium curve, Eq. (6.53), mole fraction gas/mole fraction in liquid

n number of baffles, dimensionless
tray number, dimensionless
empirical constant, Table 6.4

N impeller speed, Θ^{-1}

N_{tG} number of gas-phase transfer units, dimensionless

N_{tL} number of liquid-phase transfer units, dimensionless

N_{tOG} number of overall gas-phase transfer units, dimensionless

Nu   Nusselt number $= hd_s/k_{th}$, dimensionless

p pressure, F/L^2
empirical constant, Table 6.4

p' pitch of perforations, L

Δp pressure difference, F/L^2

Δp_R residual gas-pressure drop, F/L^2

P power delivered by an impeller, no gas flow, FL/Θ

P_G power delivered by an impeller with gas flow, FL/Θ

Pe   Péclet number for liquid mixing, Eq. (6.59), dimensionless

Po   power number $= Pg_c/\rho LN^3d_i^5$, dimensionless

Pr   Prandtl number $= C_p\mu/k_{th}$, dimensionless

q volumetric liquid flow rate, L^3/Θ

Q volumetric flow rate, L^3/Θ

Q_{Go} volumetric gas flow rate per orifice, L^3/Θ

Ra   Rayleigh number $= d_p^3\,\Delta p\,g/D_L\mu_L$, dimensionless

Re   impeller Reynolds number $= d_i^2N\rho_L/\mu_L$, dimensionless
Reynolds number for flow through packing $= d_pG'/\mu$, Eq. (6.66), dimensionless

Re_G gas Reynolds number $= d_pV_S\rho_L/\mu_L$, dimensionless

Re_o orifice Reynolds number $= d_oV_o\rho_G/\mu_G = 4w_o/\pi d_o\mu_G$, dimensionless

Sc   Schmidt number $= \mu/\rho D$, dimensionless

Sh   Sherwood number $= Fd_p/cD$, dimensionless

T tank diameter, tower diameter, L

t tray spacing, L

u average velocity, L/Θ

u' root-mean-square fluctuating velocity, L/Θ

v_L liquid volume, L^3

V superficial velocity based on tower cross section; for tray towers, based on A_n, L/Θ

V_a gas velocity based on A_a, L/Θ

V_F flooding velocity based on A_n, L/Θ

V_o velocity through an orifice, L/Θ

V_{ow} minimum gas velocity through perforations below which excessive weeping occurs, L/Θ

V_S slip velocity, L/Θ

V_t terminal settling velocity of a single bubble, L/Θ

w blade width, L

w_o mass rate of flow per orifice, M/Θ

W weir length, L
work done by gas/unit mass, FL/M

We   Weber number $= \rho u^2d_p/\sigma g_c$, dimensionless

x	concentration in liquid, mole fraction
y	concentration in gas, mole fraction
z	average flow width for liquid on a tray, L
Z	liquid depth in an agitated vessel; depth of packing; length of travel on a tray, L
α	empirical constant, Table 6.2
β	empirical constant: for flooding velocity, Table 6.2; for holdup in packing, Table 6.5
ε	fractional void volume in a dry packed bed, volume voids/volume bed, dimensionless
η	defined by Eq. (6.58), dimensionless
θ_L	time of residence of a liquid on a tray, Θ
μ	viscosity, $M/L\Theta$
ρ	density, M/L^3
$\Delta\rho$	difference in density, M/L^3
σ	surface tension F/L
φ_G	gas holdup, volume fraction, dimensionless
φ_L	liquid holdup, fraction of packed volume, dimensionless

Subscripts

av	average
AW	air-water
F	at flooding
G	gas
L	liquid
n	tray number
o	orifice; operating or moving (holdup and packing void space)
s	surface; static (holdup in packing)

Superscript

*	in equilibrium with bulk liquid

REFERENCES

1. Akita, K., and F. Yoshida: *Ind. Eng. Chem. Process Des. Dev.*, **12**, 76 (1973).
2. Anderson, R. H., G. Garrett, and M. Van Winkle: *Ind. Eng. Chem. Process Des. Dev.*, **15**, 96 (1976).
3. Anon.: *Chem. Eng.*, **72**, 86 (Aug. 16, 1965).
4. Bates, R. L., P. L. Fondy, and R. R. Corpstein: *Ind. Eng. Chem. Process Des. Dev.*, **2**, 310 (1963).
5. Bates, R. L., P. L. Fondy, and J. G. Fenic: in V. W. Uhl and J. B. Gray (eds.), "Mixing," vol. 1, chap. 3, Academic, New York, 1966.
6. Bell, R. L.: *AIChE J.*, **18**, 491, 498 (1972).
7. Bell, R. L., and R. B. Solari: *AIChE J.*, **20**, 688 (1974).
8. Biddulph, M. W.: *AIChE J.*, **21**, 41 (1975).
9. Biddulph, M. W., and D. J. Stephens: *AIChE J.*, **20**, 60 (1974).
10. Billet, R.: *Br. Chem. Eng.*, **14**, 489 (1969).
11. Billet, R.: *Chem. Eng.*, **79**, 68 (Feb. 21, 1972); *Chem. Eng. Prog.*, **63**(9), 55 (1967).
12. Boll, R. H.: *Ind. Eng. Chem. Fundam.*, **12**, 40 (1973).
13. Bolles, W. L.: in B. D. Smith, "Design of Equilibrium Stage Processes," chap. 14, McGraw-Hill Book Company, New York, 1963.
14. Bolles, W. L.: *Chem. Eng. Prog.*, **72**(9), 43 (1976).
15. Boucher, D. F., and G. E. Alves: in R. H. Perry, and C. Chilton (eds.), "The Chemical Engineers' Handbook," 5th ed., p. 5-22, McGraw-Hill Book Company, New York, 1973.

16. Boyes, A. P., and A. B. Ponter: *Ind. Eng. Chem. Process Des. Dev.*, **10**, 140 (1971).
17. Bridge, A. G., L. Lapidus, and J. C. Elgin: *AIChE J.*, **10**, 819 (1964).
18. Brinck, J. A., W. F. Burggrabe, and C. E. Greenwell: *Chem. Eng. Prog.*, **64**(11), 82 (1968).
19. Brittan, M. I., and E. T. Woodburn: *AIChE J.*, **12**, 541 (1966).
20. Bruijn, W., K. van't Riet, and J. M. Smith: *Trans. Inst. Chem. Eng. Lond.*, **52**, 88 (1974).
21. "Bubble-Tray Design Manual," American Institute of Chemical Engineers, New York, 1958.
22. Butcher, K. L., and M. S. Medani: *Trans. Inst. Chem. Eng. Lond.*, **49**, 225 (1971).
23. Calderbank, P. H.: *Trans. Inst. Chem. Eng. Lond.*, **36**, 443 (1958); **37**, 173 (1959).
24. Calderbank, P. H.: in V. W. Uhl and J. B. Gray (eds.), "Mixing," vol. 2, chapt. 6, Academic, New York, 1967.
25. Chase, J. D.: *Chem. Eng.* **74**(10), 105 (July 31, 1967); (18), 139 (Aug. 28, 1967).
26. Cicalese, J. J., J. A. Davies, P. J. Harrington, G. S. Houghland, A. J. L. Huchinson, and T. G. Walsh: *Proc. Am. Petrol. Inst.*, **26**(III), 180 (1946).
27. Clark, M. W., and T. Vermeulen: *AIChE J.*, **10**, 420 (1964).
28. Co, P., and R. Bibaud: *Can. J. Chem. Eng.*, **49**, 727 (1971).
29. Colburn, A. P.: *Ind. Eng. Chem.*, **28**, 526 (1936).
30. Cornell, D., W. G. Knapp, and J. R. Fair: *Chem. Eng. Prog.*, **56**(7), 68 (1960).
31. Cutter, L. A.: *AIChE J.*, **12**, 35 (1966).
32. Davidson, J. F., and B. G. Schuler: *Trans. Inst. Chem. Eng. Lond.*, **38**, 144 (1960).
33. Davy, C. A. E., and G. G. Haselden: *AIChE J.*, **21**, 1218 (1975).
34. Delnicki, W. V.: private communication, May 23, 1975.
35. Diener, D. A.: *Ind. Eng. Chem. Process Des. Dev.*, **6**, 499 (1967).
36. Dunn, W. E., T. Vermeulen, C. R. Wilke, and T. T. Word: *Ind. Eng. Chem. Fundam.*, **16**, 116 (1977).
37. Eckert, J. S.: *Chem. Eng. Progr.*, **66**(3), 39 (1970).
38. Eckert, J. S.: *Chem. Eng.*, **82**, 70 (Apr. 14, 1975).
39. Edmister, W. C.: *Petr. Eng.*, **1948**(12), 193.
40. Eissa, S. H., and K. Schügerl: *Chem. Eng. Sci.*, **30**, 1251 (1975).
41. Ellis, S. R. M., and D. P. Bayley: *Chem. Eng.*, **1974**(291), 714.
42. Ergun, S.: *Chem. Eng. Prog.*, **48**, 89 (1952).
43. Eversole, W. G., G. H. Wagner, and E. Stackhouse: *Ind. Eng. Chem.*, **33**, 1459 (1941).
44. Fair, J. R.: *Petro/Chem. Eng.*, **33**, 210 (September 1961).
45. Fair, J. R.: *Chem. Eng.*, **72**, 116 (July 5, 1965).
46. Fair, J. R.: *Chem. Eng.*, **76**, 90 (July 14, 1970).
47. Fair, J. R.: in R. H. Perry and C. H. Chilton (eds.), "The Chemical Engineers' Handbook," 5th ed., sec. 18, McGraw-Hill Book Company, New York, 1973.
48. Fellinger, L.: reported by T. K. Sherwood and R. L. Pigford, "Absorption and Extraction," 2d ed., McGraw-Hill Book Company, New York, 1952.
49. Foss, A. S., and J. A. Gerster: *Chem. Eng. Prog.*, **59**, 35 (1963).
50. Gerster, J. A.: *Chem. Eng. Prog.*, **59**, 28 (1956).
51. Gleason, R. J., and J. D. McKenna: *Chem. Eng. Prog. Symp. Ser.*, **68**(126), 119 (1972).
52. Günkel, A., and M. E. Weber: *AIChE J.*, **21**, 931 (1975).
53. Hanf, E. B.: *Chem. Eng. Prog.*, **67**(11), 54 (1971).
54. Harris, L. S., and G. S. Haun: *Chem. Eng. Prog.*, **60**(5), 100 (1964).
55. Hasson, I. T. M., and C. W. Robinson: *AIChE J.*, **23**, 48 (1977).
56. Haug, H. F.: *Chem. Eng. Sci.*,, **31**, 295 (1976).
57. Hicks, R. W., and L. E. Gates: *Chem. Eng.*, **83**, 141 (July 19, 1976).
58. Holland, C. D., et al.: *Chem. Eng. Sci.*, **25**, 431 (1970); **26**, 1723 (1971).
59. Hollands, K. G. T., and K. C. Goel: *Ind. Eng. Chem. Fundam.*, **14**, 16 (1975).
60. Hoppe, K., G. Krüger, and H. Ikier: *Br. Chem. Eng.*, **12**, 715, 1381 (1967).
61. Hughmark, G. A.: *Ind. Eng. Chem. Process Des. Dev.*, **6**, 218 (1967).
62. Hughmark, G. A., and H. E. O'Connell: *Chem. Eng. Prog.*, **53**, 127 (1957).
63. Hulswitt, C., and J. A. Mraz: *Chem. Eng.*, **79**, 80 (May 15, 1972).
64. Hunt, C. d'A., D. N. Hanson, and C. R. Wilke: *AIChE J.*, **1**, 441 (1955).

65. Hutchinson, M. H., and R. F. Baddour: *Chem. Eng. Prog.*, **52**, 503 (1956).
66. Interess, E.: *Chem. Eng.*, **78**, 167 (Nov. 15, 1971).
67. Jamison, R. H.: *Chem. Eng. Prog.*, **65** (3), 46 (1969).
68. Kerr, C. P.: *Ind. Eng. Chem. Process Des. Dev.*, **13**, 222 (1974).
69. Koetsier, W. T., D. Thoenes, and J. F. Frankena: *Chem. Eng. J.*, **5**, 61, 71 (1973).
70. Laity, D. S., and R. E. Treybal: *AIChE J.*, **3**, 176 (1957).
71. Leibson, I., E. G. Holcomb, A. G. Cocoso, and J. J. Jacmie: *AIChE J.*, **2**, 296 (1956).
72. Lemieux, E. J., and L. J. Scotti: *Chem. Eng. Prog.*, **65**(3), 52 (1969).
73. Leva, M.: *Chem. Eng. Prog.*, **67**(3), 65 (1971).
74. Levins, D. M., and J. R. Glastonbury: *Trans. Inst. Chem. Eng. Lond.*, **50**, 32 (1972).
75. Lewis, W. K., Jr.: *Ind. Eng. Chem.*, **28**, 399 (1936).
76. Lobo, W. E., L. Friend, F. Hashmall, and F. Zenz: *Trans. AIChE*, **41**, 693 (1945).
77. Lubowicz, R. E., and P. Reich: *Chem. Eng. Prog.*, **67**(3), 59 (1971).
78. Mack, D. E., and A. E. Kroll: *Chem. Eng. Prog.*, **44**, 189 (1948).
79. Madigan, C. M.: M. Ch. Eng. dissertation, New York University, 1964.
80. Maneri, C. C., and H. D. Mendelson: *AIChE J.*, **14**, 295 (1968).
81. Martin, G. Q.: *Ind. Eng. Chem. Process Des. Dev.*, **11**, 397 (1972).
82. McAllister, R. A., P. H. McGinnis, and C. A. Planck: *Chem. Eng. Sci.*, **9**, 25 (1958).
83. Mendelson, H. D.: *AIChE J.*, **13**, 250 (1967).
84. Miller, D. N.: *AIChE J.*, **20**, 445 (1974).
85. Morris, W. D., and J. Benyan: *Ind. Eng. Chem. Process Des. Dev.*, **15**, 338 (1976).
86. Nagata, S.: "Mixing, Principles and Applications," Halsted, New York, 1975.
87. Nemunaitis, R. R.: *Hydrocarbon Process.*, **55**(11), 235 (1971).
88. Nienow, A. W., and D. Miles: *Ind. Eng. Chem. Process Des. Dev.*, **10**, 41 (1971).
89. Norman, W. S., T. Cakaloz, A. Z. Fiesco, and D. H. Sutcliffe: *Trans. Inst. Chem. Eng. Lond.*, **41**, 61 (1963).
90. O'Connell, H. E.: *Trans. AIChE*, **42**, 741 (1946).
91. Oldshue, J. Y.: *Biotech. Bioeng.*, **8**, 3 (1966).
92. Oldshue, J. Y., and F. L. Connelly: *Chem. Eng. Prog.*, **73**(3), 85 (1977).
93. Oldshue, J. Y., and J. H. Rushton: *Chem. Eng. Prog.*, **49**, 161, 267 (1953).
94. Onda, K., H. Takeuchi, and Y. Okumoto: *J. Chem. Eng. Jap.*, **1**, 56 (1968).
95. Petrick, M.: *U.S. At. Energy Comm.* ANL-658, 1962.
96. Pinczewski, W. V., N. D. Benke, and C. J. D. Fell: *AIChE J.*, **21**, 1019, 1210 (1975).
97. Plank, C. A., and E. R. Gerhard: *Ind. Eng. Chem. Process Des. Dev.*, **2**, 34 (1963).
98. Pollard, B.: *Chem. Ind. Lond.*, **1958**, 1414.
99. Raal, J. D., and M. K. Khurana: *Can. J. Chem. Eng.*, **51**, 162 (1973).
100. Reiss, L. P.: *Ind. Eng. Chem. Process Des. Dev.*, **6**, 486 (1967).
101. Resnick, W., and B. Gal-Or: *Adv. Chem. Eng.*, **7**, 295 (1968).
102. Rushton, J. H., E. W. Costich, and H. J. Everett: *Chem. Eng. Prog.*, **46**, 395, 467 (1950).
103. Satterfield, C. N.: *AIChE J.*, **21**, 209 (1975).
104. Schwartzberg, H. G., and R. E. Treybal: *Ind. Eng. Chem. Fundam.*, **7**, 1, 6 (1968).
105. Scofield, R. C.: *Chem. Eng. Prog.*, **46**, 405 (1950).
106. Sherwood, T. K., and F. A. L. Holloway: *Trans. AIChE*, **36**, 39 (1940).
107. Shulman, H. L., et al.: *AIChE J.*, **1**, 247, 253, 259 (1955); **3**, 157 (1957); **5**, 280 (1959); **6**, 175, 469 (1960); **9**, 479 (1963); **13**, 1137 (1967); **17**, 631 (1971).
108. Smith, V. C., and W. V. Delnicki: *Chem. Eng. Prog.*, **71**(8), 68 (1975).
109. Staff, Shell Development Co.: *Chem. Eng. Prog.*, **50**, 57 (1954).
110. Stahl, Apparate- und Gerätebau GmbH, Viernheim, Germany, *Bull.*, 1975.
111. Standart, G. L.: *Chem. Eng. Sci.*, **26**, 985 (1971).
112. Strek, F., J. Werner, and A. Paniuticz: *Int. Chem. Eng.*, **9**, 464 (1969).
113. Sullivan, G. A., and R. E. Treybal: *Chem. Eng. J.*, **1**, 302 (1970).
114. Thorngren, J. T.: *Ind. Eng. Chem. Process Des. Dev.*, **11**, 428 (1972).
115. Tupuwala, H. H., and G. Hamer: *Trans. Inst. Chem. Eng. Lond.*, **52**, 113 (1974).
116. Uchida, S., and C. Y. Wen: *Ind. Eng. Chem. Process Des. Dev.*, **12**, 437 (1973).

117. Urza, I. J., and M. J. Jackson: *Ind. Eng. Chem. Process Des. Dev.*, **14**, 106 (1975).
118. Van Krevelen, W., and P. J. Hoftijzer: *Chem. Eng. Prog.*, **46**(1), 29 (1950).
119. Van Winkle, M.: "Distillation," McGraw-Hill Book Company, New York, 1967.
120. Weiler, D. W., B. L. England, and W. V. Delnicki: *Chem. Eng. Prog.*, **69**(10), 67 (1973).
121. Weinstein, B., and A. H. Williams: *79th Natl. Meet. AIChE, Houston, 1975.*
122. Westerterp, K. R., L. L. van Dierendonck, and J. A. de Kraa: *Chem. Eng. Sci.*, **18**, 157, 495 (1963).
123. Woodburn, E. T.: *AIChE J.*, **20**, 1003 (1974).
124. York, O. H., and E. W. Poppele: *Chem. Eng. Prog.*, **59**(6), 45 (1963).
125. Young, G. C., and J. H. Weber: *Ind. Eng. Chem. Process Des. Dev.*, **11**, 440 (1972).
126. Zenz, F. A.: *Chem. Eng.*, **79**, 120 (Nov. 13, 1972).
127. Zuiderweg, F. J., and A. Harmens: *Chem. Eng. Sci.*, **9**, 89 (1958).

PROBLEMS

6.1 A gas bubble released below the surface of a deep pool of liquid will expand as it rises to the surface because of the reduction of pressure to which it is subject.

(a) Derive an expression for the time of rise of a bubble of initial diameter d_{p0} released at a depth Z below the surface of a liquid of density ρ_L. Atmospheric pressure is p_A. Assume that the bubble assumes the terminal velocity corresponding to its diameter instantly.

(b) If an air bubble is initially 0.50 mm in diameter and released at a depth of 10 m in water at 25°C, compute the time of rise to the surface. Atmospheric pressure is standard.

Ans.: 54 s.

(c) Repeat for a 2.0-mm-diameter bubble.

6.2 Calculate the power imparted to the vessel contents by the gas in the case of the sparged vessel of Illustration 6.1.

Ans.: 1.485 kW.

6.3 Calculate the power imparted to the vessel by the gas and the total power for the agitated vessel of Illustration 6.2.

6.4 Petroleum oils which contain small amounts of suspended water droplets, resulting from washing the oils with water after chemical treatment, are cloudy in appearance. Sparging with air evaporates the moisture and the oil is "brightened," i.e., made clear. Presumably the air bubbles pick up water after diffusion of dissolved water through the oil.

A batch of a petroleum-oil product is to be brightened at 80°C with air. The oil viscosity at that temperature is 10 cP, its specific gravity 0.822, and its surface tension 20 dyn/cm; av mol wt = 320. The oil will be contained in a vessel 1.25 m in diameter, liquid depth 2.5 m. Air will be introduced at the rate of 0.06 kg/s at the bottom through a sparger ring 400 mm in diameter, containing 50 holes each 6.5 mm in diameter.

Estimate the gas-bubble size, the gas holdup, the specific interfacial area, and the mass-transfer coefficient for water in the liquid at the gas-liquid surface.

Ans.: $a = 216.6$ m^2/m^3, $k_L = 5.71 \times 10^{-5}$ kmol/m$^2 \cdot$ s $\cdot$ (kmol/m^3).

6.5 A baffled fermenter tank, 5.0 ft (1.525 m) in diameter, which will contain 1600 gal (6.056 m^3) of a beet-sugar solution, will be used to produce citric acid by action of the microorganism *Aspergillus niger* at 25°C. The vessel will be agitated with two 28-in-diameter (0.71-m) flat-blade disk turbines, one located 2 ft (0.61 m) above the bottom of the tank, the other 6 ft (1.83 m) from the bottom, on the same shaft, turning at 92 r/min. Sterilized air will be introduced below the lower impeller at 2.1 ft/min (0.01067 m/s) superficial velocity based on the tank cross-sectional area. The solution sp gr = 1.038, the viscosity 1.4 cP (1.4×10^{-3} kg/m $\cdot$ s).

(a) Estimate the agitator power required.

Ans.: 5.4 kW.

(b) If the air is introduced through five ½-in, schedule 40, open pipes (ID = 0.622 in = 15.8 mm), estimate the power imparted to the tank contents by the air.

6.6 A baffled fermenter [10 ft (3.05 m) diameter, liquid depth 10 ft (3.05 m)] contains nutrient liquid of sp gr 1.1, viscosity 5 cP, surface tension 50 dyn/cm. The liquid contains dissolved electrolytes. It will be aerated at the rate 360 ft/h (0.0305 m/s) superficial air velocity based on the tank cross section and agitated with a flat-blade disk turbine 4.5 ft (1.37 m) in diameter turning at 60 r/min and located 3 ft (0.91 m) from the bottom of the tank. The temperature will be 80°F (26.7°C). The diffusivity of oxygen in water at 25°C = 2.5×10^{-5} cm²/s. Estimate the diffusivity of oxygen in the solution by assuming that the quantity $D_L \mu_L / t$ = const, where t = absolute temperature.

Compute the agitator power required, the interfacial area, and the mass-transfer coefficient for oxygen transfer.

Ans.: $P = 16.2$ hp = 12.1 kW, $k_L = 0.31$ lb mol/ft² · h · (lb mol/ft³) = 2.63×10^{-5} kmol/ m² · s · (kmol/m³).

6.7 A baffled fermentation tank for a pilot plant is to be 1.5 m in diameter and will contain liquid to a depth 2.0 m. The flat-blade disk turbine impeller, 0.5 m diameter, will be located $\frac{1}{3}$ m from the bottom of the vessel. Air at a superficial velocity of 0.06 m/s will be introduced beneath the impeller. The temperature will be 27°C, and the liquid properties may be taken as those of water. Small-scale tests indicate that a suitable agitator power will be 0.5 kW/m³ of liquid. At what speed should the impeller be turned?

6.8 A sieve-tray tower is to be designed for stripping an aniline-water solution with steam. The circumstances at the top of the tower, which are to be used to establish the design, are

$$\text{Temperature} = 98.5°C \qquad \text{pressure} = 745 \text{ mmHg abs}$$

Liquid:

$$\text{Composition} = 7.00 \text{ mass \% aniline}$$

$$\text{Rate} = 6.3 \text{ kg/s} = 50\,000 \text{ lb/h} \qquad \text{density} = 961 \text{ kg/m}^3 = 60 \text{ lb/ft}^3$$

$$\text{Viscosity} = 3 \times 10^{-4} \text{ kg/m} \cdot \text{s} = 0.3 \text{ cP}$$

$$\text{Surface tension} = 0.058 \text{ N/m} = 58 \text{ dyn/cm}$$

$$\text{Aniline diffusivity} = 52 \times 10^{-10} \text{ m}^2/\text{s} = 52 \times 10^{-6} \text{ cm}^2/\text{s (est)}$$

Vapor:

$$\text{Composition} = 3.6 \text{ mole \% aniline} \qquad \text{rate} = 3.15 \text{ kg/s} = 25\,000 \text{ lb/h}$$

$$\text{Aniline diffusivity} = 1.261 \times 10^{-5} \text{ m}^2/\text{s} = 0.1261 \text{ cm}^2/\text{s (est)}$$

The equilibrium data [Griswold, et al.: *Ind. Eng. Chem.*, **32**, 878 (1940)] indicate that $m = 0.0636$ at this concentration.

(a) Design a suitable cross-flow tray for such a tower. Report details respecting perforation size and arrangement, tower diameter, tray spacing, weir length and height, downspout seal, pressure drop for the gas, height of liquid in the downspout, and entrainment in the gas. Check for excessive weeping.

(b) Estimate the tray efficiency for the design reported in part (a).

6.9 A gas containing methane, propane, and butane is to be scrubbed countercurrently in a sieve-tray tower with a hydrocarbon oil to absorb principally the butane. It is agreed to design a tray for the circumstances existing at the bottom of the tower, where the conditions are

$$\text{Pressure} = 350 \text{ kN/m}^2 \text{ (51 lb}_f/\text{in}^2 \text{ abs)} \qquad \text{temperature} = 38°C$$

Gas: 0.25 kmol/s (1984 lb mol/h), containing 85% methane, 10% propane, and 5% butane by volume.

Liquid: 0.15 kmol/s (1190 lb mol/h), av mol wt = 150, density = 849 kg/m³ (53 lb/ft³), surface tension = 0.025 N/m (25 dyn/cm), viscosity = 0.00160 kg/m · s (1.60 cP).

(a) Design a suitable sieve tray, and check for weeping and flooding by downspout backup.

(b) Estimate the tray efficiency for butane absorption, corrected for entrainment. The average molecular diffusivities may be taken as 3.49×10^{-6} m²/s for the gas and 1.138×10^{-9} m²/s for the liquid.

6.10 A packed tower is to be designed for the countercurrent contact of a benzene-nitrogen gas mixture with kerosene to wash out the benzene from the gas. The circumstances are:

Gas in: 1.50 m^3/s (53 ft^3/s), containing 5 mol % benzene, at 25°C, 1.1 × 10^5 N/m^2 (16 lb_f/in^2).

Gas out: substantially pure nitrogen.

Liquid in: 4.0 kg/s (8.82 lb/s), density = 800 kg/m^3 (50 lb/ft^3), viscosity = 0.0023 kg/m · s.

The packing will be 50-mm (2-in) metal Pall rings, and the tower diameter will be set to produce 400 N/m^2 per meter of gas-pressure drop (0.5 in H_2O/ft) for irrigated packing.

(a) Calculate the tower diameter to be used.

(b) Assume that, for the diameter chosen, the irrigated packed depth will be 6 m (19.7 ft) and that 1 m of unirrigated packing will be placed over the liquid inlet to act as entrainment separator. The blower-motor combination to be used at the gas inlet will have an overall efficiency of 60%. Calculate the power required to blow the gas through the packing.

6.11 A small water-cooling tower, 1 m diameter, packed with 76-mm (3-in) ceramic Raschig rings, is fed with water at the rate 28 kg/m^2 · s (20 600 lb/ft^2 · h), in at 40°C and out at 25°C. The water is contacted with air (30°C, 1 std atm, essentially dry) drawn upward countercurrently to the water flow. Neglecting evaporation of the water, estimate the rate of airflow which would flood the tower.

SEVEN

HUMIDIFICATION OPERATIONS

The operations considered in this chapter are concerned with the interphase transfer of mass and of energy which result when a gas is brought into contact with a pure liquid in which it is essentially insoluble. While the term *humidification operations* is used to characterize these in a general fashion, the purpose of such operations may include not only humidification of the gas but dehumidification and cooling of the gas, measurement of its vapor content, and cooling of the liquid as well. The matter transferred between phases in such cases is the substance constituting the liquid phase, which either vaporizes or condenses. As in all mass-transfer problems, it is necessary for a complete understanding of the operation to be familiar with the equilibrium characteristics of the systems. But since the mass transfer in these cases will invariably be accompanied by a simultaneous transfer of heat energy as well, some consideration must also be given to the enthalpy characteristics of the systems.

VAPOR-LIQUID EQUILIBRIUM AND ENTHALPY FOR A PURE SUBSTANCE

As indicated above, the substance undergoing interphase transfer in these operations is the material constituting the liquid phase, which diffuses in the form of a vapor. The equilibrium vapor-pressure characteristics of the liquid are therefore of importance.

Vapor-Pressure Curve

Every liquid exerts an equilibrium pressure, the vapor pressure, to an extent depending upon the temperature. When the vapor pressures of a liquid are plotted against the corresponding temperatures, a curve like *TBDC* (Fig. 7.1) results. The vapor-pressure curve for each substance is unique, but each exhibits characteristics generally similar to that in the figure. The curve separates two areas of the plot, representing respectively, conditions where the substance exists wholly in the liquid state and wholly in the vapor state. If the conditions imposed upon the substance are in the liquid-state area, such as at point *A*, the substance will be entirely liquid. Under all conditions in the lower area, such as those at point *E*, the substance is entirely a vapor. At all conditions corresponding to points on the curve *TBDC*, however, liquid and vapor may coexist in any proportions indefinitely. Liquid and vapor represented by points on the vapor-pressure curve are called *saturated liquid* and *saturated vapor*, respectively. Vapor or gas at a temperature above that corresponding to saturation is termed *superheated*. The vapor-pressure curve has two abrupt endpoints, at *T* and *C*. Point *T*, from which originate curves *LT* and *ST* separating the conditions for the solid state from those for the liquid and vapor, is the *triple point*, at which all three states of aggregation may coexist. Point *C* is the *critical point*, or *state*, whose coordinates are the *critical pressure* and *critical temperature*. At the critical point, distinction between the liquid and vapor phases disappears, and all the properties of the liquid, such as density, viscosity, refractive index, etc., are identical with those of the vapor. The substance at a temperature above the critical is called a *gas*, and it will then not be liquefied regardless of how high a pressure may be imposed. This distinction between a gas and a vapor, however,

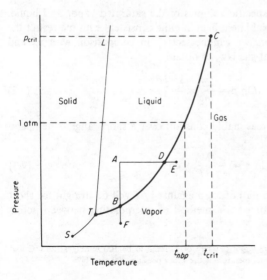

Figure 7.1 Vapor pressure of a pure liquid.

is not always strictly adhered to, and the term "gas" is frequently used to designate merely a condition relatively far removed from the vapor-pressure curve. The temperature corresponding to each pressure on the curve is termed the *boiling point* of the liquid at the pressure in question, and that corresponding to 1 std atm in particular is known as the *normal boiling point*, as at t_{nbp} in Fig. 7.1.

Whenever a process involves bringing a sample of fluid across the vapor-pressure curve, such as the isobaric process ADE or the isothermal process ABF, there will be a change of phase. This will be accompanied by the evolution (for condensation) or absorption (for vaporization) of the *latent heat of vaporization* at constant temperature, for example, at points B or D in the processes mentioned above. Heat added or given up with changing temperatures is called *sensible heat*.

Interpolation between data For such common liquids as water, many refrigerants, and others, the vapor-pressure-temperature curve has been established at many points. For most liquids, however, only relatively few data are available, so that it is necessary frequently to interpolate between, or extrapolate beyond, the measurements. The curve on arithmetic coordinates (Fig. 7.1) is very inconvenient for this because of the curvature, and some method of linearizing the curve is needed. Most of the common methods stem from the Clausius-Clapeyron equation, which relates the slope of the vapor-pressure curve to the latent heat of vaporization

$$\frac{dp}{dT} = \frac{\lambda'}{T(v_G - v_L)} \tag{7.1}$$

where v_G and v_L are molal specific volumes of the saturated vapor and liquid, respectively, and λ' is the molal latent heat in units consistent with the rest of the equation. As a simplification, we can neglect v_L in comparison with v_G and express the latter by the ideal-gas law, to obtain

$$d \ln p = \frac{dp}{p} = \frac{\lambda' \, dT}{RT^2} \tag{7.2}$$

and if λ' can be considered reasonably constant over a short range of temperature,

$$\ln p = -\frac{\lambda'}{RT} + \text{const} \tag{7.3}$$

Equation (7.3) suggests that a plot of $\log p$ against $1/T$ will be straight for short temperature ranges. It also suggests a method of interpolating between points listed in a table of data.

Illustration 7.1 A table lists the vapor pressure of benzene to be 100 mmHg at 26.1°C and 400 mmHg at 60.6°C. At what temperature is the vapor pressure 200 mmHg?

SOLUTION At 26.1°C, $1/T = 1/299.1$ K^{-1}; at 60.6°C, $1/T = 1/333.6$ K^{-1}.

$$\frac{1/299.1 - 1/T}{1/299.1 - 1/333.6} = \frac{\log 100 - \log 200}{\log 100 - \log 400}$$

$$T = 315.4 \text{ K} = 42.4°\text{C}$$

The correct value is 42.2°C. Linear interpolation would have given 37.6°C.

Reference-Substance Plots [21]

Equation (7.2) can be rewritten for a second substance, a *reference substance*, at the same temperature,

$$d \ln p_r = \frac{\lambda_r' \, dT}{RT^2} \tag{7.4}$$

where the subscript r denotes the reference substance. Dividing Eq. (7.2) by (7.4) provides

$$\frac{d \ln p}{d \ln p_r} = \frac{\lambda'}{\lambda_r'} = \frac{M\lambda}{M_r\lambda_r} \tag{7.5}$$

which, upon integration, becomes

$$\log p = \frac{M\lambda}{M_r\lambda_r} \log p_r + \text{const} \tag{7.6}$$

Equation (7.6) suggests that a linear graph will result if $\log p$ as ordinate is plotted against $\log p_r$ for the reference substance as abscissa, where for each plotted point the vapor pressures are taken at the same temperature. Such a plot is straight over larger temperature ranges (but not near the critical temperature) than that based on Eq. (7.3), and, moreover, the slope of the curve gives the ratio of the latent heats at the same temperature. The reference substance chosen is one whose vapor-pressure data are well known.

Illustration 7.2 (a) Plot the vapor pressure of benzene over the range 15 to 180°C using water as the reference substance according to Eq. (7.6). (b) Determine the vapor pressure of benzene at 100°C. (c) Determine the latent heat of vaporization of benzene at 25°C.

SOLUTION (a) Logarithmic graph paper is marked with scales for the vapor pressure of benzene and for water, as in Fig. 7.2. The vapor pressure of benzene at 15.4°C is 60 mmHg, and that for water at this temperature is 13.1 mmHg. These pressures provide the coordinates of the lowest point on the plot. In similar fashion, additional data for benzene are plotted with the help of a steam table, thus providing the curve shown. The line is very nearly straight over the temperature range used.

(b) At 100°C, the vapor pressure of water is 760 mmHg. Entering the plot at this value for the abscissa, the vapor pressure of benzene is read as 1400 mmHg. Alternatively the abscissa can be marked with the temperatures corresponding to the vapor pressures of water, as shown, thus eliminating the necessity of referring to the steam table.

(c) The slope of the curve at 25°C is 0.775. (*Note:* This is most conveniently determined with a millimeter rule. If the coordinates are used, the slope will be $\Delta \log p / \Delta \log p_r$.) At 25°C,

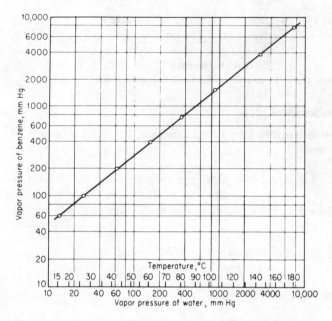

Figure 7.2 Reference-substance plot for the vapor pressure of benzene.

the latent heat of vaporization of water is 2443 kN · m/kg. From Eq. (7.6)

$$\frac{\lambda M}{\lambda_r M_r} = \frac{78.05}{2\ 443\ 000(18.02)} = 0.775$$

$$\lambda = 437\ \text{kN} \cdot \text{m/kg for benzene at } 25°\text{C}$$

(The accepted value is 434 kN · m/kg.)

Enthalpy

The internal energy U of a substance is the total energy residing in the substance owing to the motion and relative position of the constituent atoms and molecules. Absolute values of internal energy are not known, but numerical values relative to some arbitrarily defined standard state for the substance can be computed. The sum of the internal energy and the product of pressure and volume of the substance, when both quantities are expressed in the same units, is defined as the *enthalpy* of the substance.

$$H = U + pv$$

In a batch process at constant pressure, where work is done only in expansion against the pressure, the heat absorbed by the system is the gain in enthalpy,

$$Q = \Delta H = \Delta(U + pv) \tag{7.7}$$

In a steady-state continuous-flow process, the net transfer of energy to the system as heat and work will be the sum of its gains in enthalpy and potential and kinetic energies. It frequently happens that the changes in potential and kinetic energies are insignificant in comparison with the enthalpy change and that there is no mechanical work done. In such cases, Eq. (7.7) can be used to compute the heat added to the system, and such a calculation is termed a *heat balance*. In *adiabatic* operations, where no exchange of heat between the system and its surroundings occurs, the heat balance becomes simply an equality of enthalpies in the initial and final condition.

Absolute values of the enthalpy of a substance, like the internal energy, are not known. However, by arbitrarily setting the enthalpy of a substance at zero when it is in a convenient reference state, relative values of enthalpy at other conditions can be calculated. To define the reference state, the temperature, pressure, and state of aggregation must be established. For the substance water, the ordinary steam tables list the relative enthalpy at various conditions referred to the enthalpy of the substance at 0°C, the equilibrium vapor pressure at this temperature, and in the liquid state. For other substances, other reference conditions may be more convenient.

Figure 7.3 is a graphical representation of the relative enthalpy of a typical substance where the liquid, vapor, and gaseous states are shown. The data are

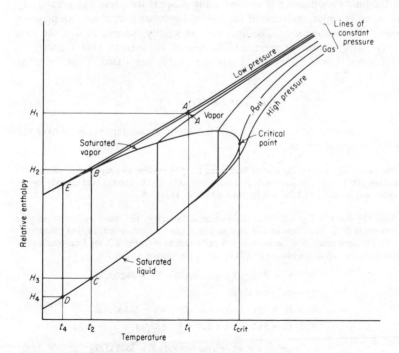

Figure 7.3 Typical enthalpy-temperature diagram for a pure substance.

most conveniently shown on lines of constant pressure. The curves marked "saturated liquid" and "saturated vapor," however, cut across the constant-pressure lines and show the enthalpies for these conditions at temperatures and pressures corresponding to the equilibrium vapor-pressure relationship for the substance. The vertical distance between the saturated-vapor and -liquid curves, such as the distance BC, represents the latent heat of vaporization at the corresponding temperature. The latent heat thus decreases with increased temperature, becoming zero at the critical point. In the vapor state at low pressures, the enthalpy is essentially a function of temperature; at all pressures where the ideal-gas law can be used to describe the pvt relation, the lines of constant pressure are superimposed and the enthalpy is independent of pressure. Except near the critical temperature, the enthalpy of the liquid is also substantially independent of pressure until exceedingly high pressures are reached.

The change in enthalpy between two conditions, such as those at A and D, may be taken simply as the difference in ordinates corresponding to the points. Thus, to calculate the enthalpy of the substance in the superheated condition at point A relative to the saturated liquid at D, or $H_1 - H_4$, we can add the enthalpy change $H_1 - H_2$, the sensible heat of the vapor from the saturation temperature t_2 at the same pressure to the superheated condition at A; $H_2 - H_3$, the latent heat of vaporization at t_2; and $H_3 - H_4$, the sensible heat of the liquid from the final condition at D to the boiling point at the prevailing pressure t_2. For a liquid or vapor, the slope of the constant-pressure lines at any temperature is termed the *heat capacity*. The lines are not strictly straight, so that the heat capacity changes with temperature. By use of an average heat capacity or average slope, however, sensible heats are readily calculated. Thus, referring again to Fig. 7.3,

$$H_1 - H_2 = C(t_1 - t_2)$$

where C is the average heat capacity of the vapor at constant pressure over the indicated temperature range.

Illustration 7.3 Compute the heat evolved when 10 kg of benzene as a superheated vapor at 94 mmHg, 100°C, is cooled and condensed to a liquid at 10°C. The average heat capacity for the vapor may be taken as 1.256 and for the liquid 1.507 kJ/kg · K.

SOLUTION Refer to Fig. 7.2. When the pressure is 94 mmHg, the saturation temperature for benzene is 25°C. The latent heat of vaporization at this temperature is 434 kJ/kg (Illustration 7.2). The initial condition corresponds to a point such as A on Fig. 6.3, the final condition to point D, the path of the process to $ABCD$. Using the notation of Fig. 7.3,

$$H_1 - H_2 = C(t_1 - t_2) = 1.256(100 - 25) = 94.2 \text{ kJ/kg}$$

$$H_2 - H_3 = 434 \text{ kJ/kg}$$

$$H_3 - H_4 = C(t_2 - t_4) = 1.507(25 - 10) = 22.6 \text{ kJ/kg}$$

$$H_1 - H_4 = 94.2 + 434 + 22.6 = 550.8 \text{ kJ/kg}$$

Heat evolved for 10 kg benzene = 10(550.8) = 5508 kJ/kg

VAPOR-GAS MIXTURES

In what follows, the term *vapor* will be applied to that substance, designated as substance A, in the vaporous state which is relatively near its condensation temperature at the prevailing pressure. The term *gas* will be applied to substance B, which is a relatively highly superheated gas.

Absolute Humidity

While the common concentration units (partial pressure, mole fraction, etc.) which are based on total quantity are useful, when operations involve changes in vapor content of a vapor-gas mixture without changes in the gas content, it is more convenient to use a unit based on the unchanging amount of gas. The ratio mass of vapor/mass of gas is the *absolute humidity* Y'. If the quantities are expressed in moles, the ratio is the *molal* absolute humidity Y. Under conditions where the ideal-gas law applies,

$$Y = \frac{y_A}{y_B} = \frac{\bar{p}_A}{\bar{p}_B} = \frac{\bar{p}_A}{p_t - \bar{p}_A} \frac{\text{moles A}}{\text{moles B}}$$

$$Y' = Y\frac{M_A}{M_B} = \frac{\bar{p}_A}{p_t - \bar{p}_A} \frac{M_A}{M_B} \frac{\text{mass A}}{\text{mass B}} \tag{7.8}$$

In many respects the molal ratio is the more convenient, thanks to the ease with which moles and volumes can be interrelated through the gas law, but the mass ratio has nevertheless become firmly established in the humidification literature. The mass absolute humidity was first introduced by Grosvenor [7] and is sometimes called the *Grosvenor humidity*.

Illustration 7.4 In a mixture of benzene vapor (A) and nitrogen gas (B) at a total pressure of 800 mmHg and a temperature of 60°C, the partial pressure of benzene is 100 mmHg. Express the benzene concentration in other terms.

SOLUTION $\bar{p}_A = 100$, $\bar{p}_B = 800 - 100 = 700$ mmHg.

(a) Mole fraction. Since the pressure fraction and mole fraction are identical for gas mixtures, $y_A = \bar{p}_A/p_t = 100/800 = 0.125$ mole fraction benzene. The mole fraction nitrogen = $y_B = 1 - 0.125 = 700/800 = 0.875$.

(b) Volume fraction of benzene equals the mole fraction, 0.125.

(c) Absolute humidity.

$$Y = \frac{y_A}{y_B} = \frac{\bar{p}_A}{\bar{p}_B} = \frac{0.125}{0.875} = \frac{100}{700} = 0.143 \text{ mol benzene/mol nitrogen}$$

$$Y' = Y\frac{M_A}{M_B} = 0.143\frac{78.05}{28.08} = 0.398 \text{ kg benzene/kg nitrogen}$$

Saturated Vapor-Gas Mixtures

If an insoluble dry gas B is brought into contact with sufficient liquid A, the liquid will evaporate into the gas until ultimately, at equilibrium, the partial pressure of A in the vapor-gas mixture reaches its saturation value, the vapor pressure p_A at the prevailing temperature. So long as the gas can be considered insoluble in the liquid, the partial pressure of vapor in the saturated mixture is independent of the nature of the gas and total pressure (except at very high pressures) and is dependent only upon the temperature and identity of the liquid. However, the saturated molal absolute humidity $Y_s = p_A/(p_t - p_A)$ will depend upon the total pressure, and the saturated absolute humidity $Y'_s = Y_s M_A/M_B$ upon the identity of the gas as well. Both saturated humidities become infinite at the boiling point of the liquid at the prevailing total pressure.

Illustration 7.5 A gas (B)–benzene (A) mixture is saturated at 1 std atm, 50°C. Calculate the absolute humidity if B is (a) nitrogen and (b) carbon dioxide.

SOLUTION Since the mixture is saturated, the partial pressure of benzene, $\bar{p}_A$, equals the equilibrium vapor pressure p_A of benzene at 50°C. From Fig. 7.2, $p_A = 275$ mmHg, or 0.362 std atm.

(a)

$$Y_s = \frac{p_A}{p_t - p_A} = \frac{0.362}{1 - 0.362} = 0.567 \text{ kmol C}_6\text{H}_6/\text{kmol N}_2$$

$$Y'_s = \frac{Y_s M_A}{M_B} = \frac{0.567(78.05)}{28.02} = 1.579 \text{ kg C}_6\text{H}_6/\text{kg N}_2$$

(b)

$$Y_s = \frac{p_A}{p_t - p_A} = \frac{0.362}{1 - 0.362} = 0.567 \text{ kmol C}_6\text{H}_6/\text{kmol CO}_2$$

$$Y'_s = \frac{Y_s M_A}{M_B} = \frac{0.567(78.05)}{44.01} = 1.006 \text{ kg C}_6\text{H}_6/\text{kg CO}_2$$

Unsaturated Vapor-Gas Mixtures

If the partial pressure of the vapor in a vapor-gas mixture is for any reason less than the equilibrium vapor pressure of the liquid at the same temperature, the mixture is unsaturated.

Dry-bulb temperature This is the temperature of a vapor-gas mixture as ordinarily determined by immersion of a thermometer in the mixture.

Relative saturation *Relative saturation*, also called *relative humidity*, expressed as a percentage is defined as $100\bar{p}_A/p_A$, where p_A is the vapor pressure at the dry-bulb temperature of the mixture. For any vapor, the graphical representation of conditions of constant relative saturation can easily be constructed on a vapor-pressure–temperature chart, as in Fig. 7.4a, by dividing the ordinates of the vapor-pressure curve into appropriate intervals. Thus the curve for 50 percent relative saturation shows a vapor partial pressure equal to one-half the equilibrium vapor pressure at any temperature. A reference-substance plot, such as Fig. 7.2, could also be used for this.

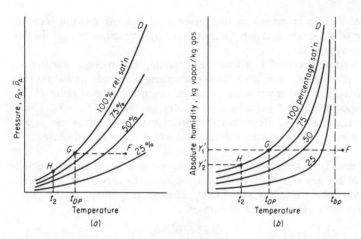

Figure 7.4 Forms of psychrometric charts.

Percentage saturation *Percentage saturation*, or *percentage absolute humidity*, is defined as $100 Y / Y_s$ and $100 Y' / Y_s'$, where the saturated values are computed at the dry-bulb temperature of the mixture. Graphical representation of the quantity for any vapor can be made on a chart of Y vs. t (in which case the chart must be limited to a single total pressure) or one of Y' vs. t (for a single total pressure and a specific gas), as in Fig. 7.4b. On this chart, the saturation humidities are plotted from vapor-pressure data with the help of Eq. (7.8), to give curve GD. The curve for humidities at 50 percent saturation is plotted at half the ordinate of curve GD, etc. All the curves of constant percentage saturation reach infinity at the boiling point of the liquid at the prevailing pressure.†

Dew point This is the temperature at which a vapor-gas mixture becomes saturated when cooled at constant total pressure out of contact with a liquid. For example, if an unsaturated mixture such as that at F (Fig. 7.4) is cooled at constant pressure out of contact with liquid, the path of the cooling process follows the line FG, the mixture becoming more nearly saturated as the temperature is lowered, and fully saturated at t_{DP}, the dew-point temperature. All mixtures of absolute humidity Y_1' on this figure have the same dew point. If the temperature is reduced only an infinitesimal amount below t_{DP}, vapor will condense as a liquid dew. This is used as a method of humidity determination: a

---

† For this reason curves of constant relative saturation are sometimes drawn on absolute-humidity-temperature charts. Since relative saturation and percentage saturation are not numerically equal for an unsaturated mixture, the position of such curves must be computed by the methods of Illustration 7.6.

shiny metal surface is cooled in the presence of the gas mixture, and the appearance of a fog which clouds the mirrorlike surface indicates that the dew point has been reached.

If the mixture is cooled to a lower temperature, the vapor-gas mixture will continue to precipitate liquid, itself always remaining saturated, until at the final temperature t_2 (Fig. 7.4) the residual vapor-gas mixture will be at point H. The mass of vapor condensed per unit mass of dry gas will be $Y_1' - Y_2'$. Except under specially controlled circumstances supersaturation will not occur, and no vapor-gas mixture whose coordinates lie to the left of curve GD will result.

Humid volume The humid volume v_H of a vapor-gas mixture is the volume of unit mass of dry gas and its accompanying vapor at the prevailing temperature and pressure. For a mixture of absolute humidity Y' at t_G and p_t, total pressure, the ideal-gas law gives the humid volume as

$$v_H = \left(\frac{1}{M_B} + \frac{Y'}{M_A} \right) 22.41 \frac{t_G + 273}{273} \frac{1.013 \times 10^5}{p_t} = 8315 \left(\frac{1}{M_B} + \frac{Y'}{M_A} \right) \frac{t_G + 273}{p_t}$$

(7.9)

where v_H is in m^3/kg, t_G in degrees Celsius, and $p_t = N/m^2$.† The humid volume of a saturated mixture is computed with $Y' = Y_s'$ and that for a dry gas with $Y' = 0$. These values can then be plotted against temperature on a psychrometric chart. For partially saturated mixtures, v_H can be interpolated between values for 0 and 100 percentage saturation at the same temperature according to percentage saturation. When the mass of *dry* gas in a mixture is multiplied by the humid volume, the volume of *mixture* results.

Humid heat The humid heat C_S is the heat required to raise the temperature of unit mass of gas and its accompanying vapor one degree at constant pressure. For a mixture of absolute humidity Y',

$$C_S = C_B + Y' C_A$$

(7.10)

Provided neither vaporation nor condensation occurs, the heat in Btu required to raise the temperature of a mass of W_B dry gas *and* its accompanying vapor an amount Δt will be

$$Q = W_B C_S \Delta t$$

(7.11)

Enthalpy The (relative) enthalpy of a vapor-gas mixture is the sum of the (relative) enthalpies of the gas and of the vapor content. Imagine unit mass of a gas containing a mass Y' of vapor at dry-bulb temperature t_G. If the mixture is unsaturated, the vapor is in a superheated state and we can calculate the

---

† For v_H in ft^3/lb, t_G in degrees Fahrenheit, and p_t in atmospheres Eq. (7.9) becomes

$$v_H = 0.730 \left(\frac{1}{M_B} + \frac{Y'}{M_A} \right) \frac{t_G + 460}{p_t}$$

enthalpy relative to the reference states gas and saturated liquid at t_0. The enthalpy of the gas alone is $C_B(t_G - t_0)$. The vapor at t_G is at a condition corresponding to point A on Fig. 7.3, and its reference state corresponds to point D. If t_{DP} is the dew point of the mixture (t_2 in Fig. 7.3) and λ_{DP} the latent heat of vaporization of the vapor at that temperature, the enthalpy per unit mass of vapor will be $C_A(t_G - t_{DP}) + \lambda_{DP} + C_{A, L}(t_{DP} - t_0)$. Then the total enthalpy for the mixture, per unit mass of dry gas, is

$$H' = C_B(t_G - t_0) + Y'\left[C_A(t_G - t_{DP}) + \lambda_{DP} + C_{A, L}(t_{DP} - t_0)\right] \quad (7.12)$$

Refer again to Fig. 7.3. For the low pressures ordinarily encountered in humidification work, the point A which actually lies on a line of constant pressure corresponding to the partial pressure of the vapor in the mixture can, for all practical purposes, be considered as lying on the line whose pressure is the saturation pressure of the vapor at the reference temperature, or at A'. The vapor enthalpy can then be computed by following the path $A'ED$ and becomes, per unit mass of vapor, $C_A(t_G - t_0) + \lambda_0$, where λ_0 is the latent heat of vaporization at the reference temperature. The enthalpy of the mixture, per unit mass of dry gas, is then

$$H' = C_B(t_G - t_0) + Y'\left[C_A(t_G - t_0) + \lambda_0\right] = C_S(t - t_0) + Y'\lambda_0 \quad (7.13)$$

Occasionally different reference temperatures are chosen for the dry gas and for the vapor. Note that the enthalpy H' for a mixture can be increased by increasing the temperature at constant humidity, by increasing the humidity at constant temperature, or by increasing both. Alternatively, under certain conditions H' may remain constant as t and Y' vary in opposite directions.

By substitution of Y'_s and the appropriate humid heat in Eq. (7.13), the enthalpy of saturated mixtures H'_s can be computed and plotted against temperature on the psychrometric chart. Similarly H for the dry gas can be plotted. Enthalpies for unsaturated mixtures can then be interpolated between the saturated and dry values at the same temperature according to the percentage humidity.

The System Air-Water

While psychrometric charts for any vapor-gas mixture can be prepared when circumstances warrant, the system air-water occurs so frequently that unusually complete charts for this mixture are available. Figure 7.5a and 7.5b shows two versions of such a chart, for SI and English engineering units, respectively, prepared for a total pressure of 1 std atm. For convenient reference, the various equations representing the curves are listed in Table 7.1. It should be noted that all the quantities (absolute humidity, enthalpies, humid volumes) are plotted against temperature. For the enthalpies, gaseous air and saturated liquid water at 0°C (32°F) were the reference conditions, so that the chart can be used in conjunction with the steam tables. The data for enthalpy of saturated air were then plotted with two enthalpy scales to provide for the large range of values

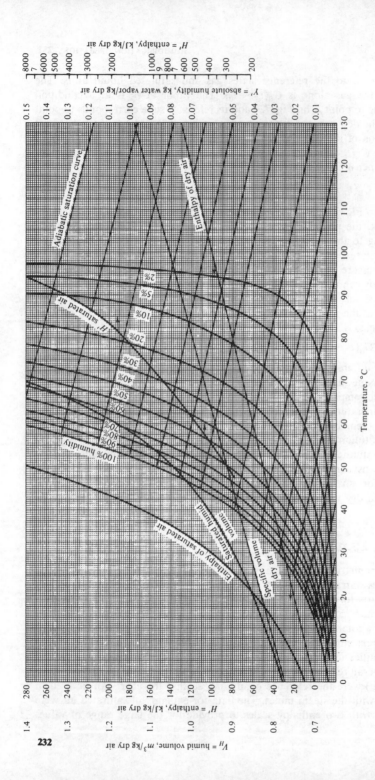

Figure 7.5 (a) Psychrometric chart for air-water vapor, 1 std atm abs, in SI units.

232

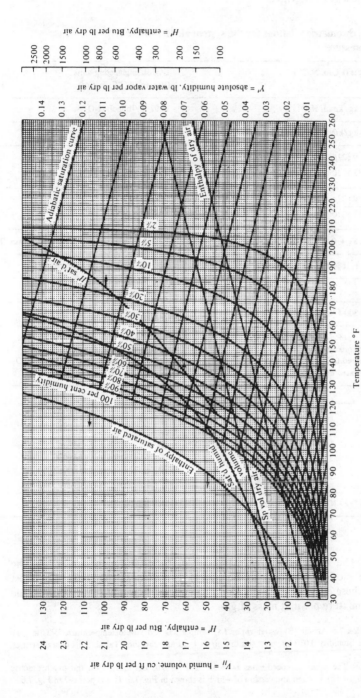

Figure 7.5 (b) Psychrometric chart for air-water vapor, 1 std atm abs, in English engineering units.

Table 7.1 Psychrometric relations for the system air (B)–water (A) at 1 std atm pressure

	SI units (kg, m, N, °C)		English engineering units (Btu, ft³, lb, °F, lb_f/in²)	
M_A	18.02 kg/kmol, H_2O		18.02 lb/lb mol, H_2O	
M_B	28.97 kg/kmol, air		28.97 lb/lb mol, air	
Y'	$\dfrac{0.622\bar{p}_{H_2O}}{1.0133 \times 10^5 - \bar{p}_{H_2O}}$ kg H_2O/kg air		$\dfrac{0.622\bar{p}_{H_2O}}{14.696 - \bar{p}_{H_2O}}$ lb H_2O/lb air	
Y'_s	$\dfrac{0.622 p_{H_2O}}{1.0133 \times 10^5 - p_{H_2O}}$ kg H_2O/kg air		$\dfrac{0.622 p_{H_2O}}{14.696 - p_{H_2O}}$ lb H_2O/lb air	
v_H	$(0.00283 + 0.00456 Y')(t_G + 273)$ m³ mixture/kg air		$(0.0252 + 0.0405 Y')(t_G + 460)$ ft³ mixture/lb air	
C_S	$1005 + 1884 Y'$ J for mixture/(kg air) · °C		$0.24 + 0.45 Y'$ Btu for mixture/(lb air) · °F	
t_0	0°C		32°F	
λ_0	2 502 300 J/kg		1075.8 Btu/lb	
H	$(1005 + 1884 Y')t_G + 2\,502\,300\,Y'$ J for mixture/kg air, referred to gaseous air and saturated liquid H_2O, 0°C		$(0.24 + 0.45 Y')(t_G - 32) + 1075.8\,Y'$ Btu for mixture/lb air, referred to gaseous air and liquid H_2O, 32°F	
H_s	t, °C	H'_s, J/kg	t, °F	H'_s, Btu/lb
	0	9 479	32	4.074
	10	29 360	40	7.545
	20	57 570	60	18.780
	30	100 030	80	36.020
	40	166 790	100	64.090
	50	275 580	120	112.00
	60	461 500	140	198.40
h_G/k'_Y	950 J/kg · K		0.227 Btu/lb · °F	

necessary. The series of curves marked "adiabatic-saturation curves" on the chart were plotted according to Eq. (7.21), to be considered later. For most purposes these can be considered as curves of constant enthalpy for the vapor-gas mixture per unit mass of gas.

Illustration 7.6 An air (B)–water-vapor (A) sample has a dry-bulb temperature 55°C and an absolute humidity 0.030 kg water/kg dry air at 1 std atm pressure. Tabulate its characteristics.

SOLUTION The point of coordinates $t_G = 55°C$, $Y' = 0.030$ is located on the psychrometric chart (Fig. 7.5a), a schematic version of which is shown in Fig. 7.6. This is point D in Fig. 7.6.

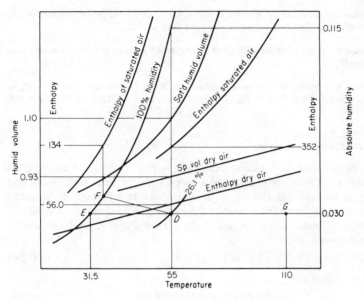

Figure 7.6 Solutions to Illustrations 7.6 and 7.7.

(a) By vertical interpolation between the adjacent curves of constant percent humidity, the sample has a percentage humidity = 26.1%. Alternatively, the saturation humidity at 55°C is $Y_s' = 0.115$, and the percentage humidity at D is therefore $(0.030/0.115)100 = 26.1\%$.

(b) The molal absolute humidity = $Y = Y'(M_B/M_A) = 0.030(28.97/18.02) = 0.0482$ kmol water/kmol dry air.

(c) The partial pressure of water vapor in the sample, by Eq. (7.8), is

$$\bar{p}_A = \frac{Yp_t}{1+Y} = \frac{0.0482(1.0133 \times 10^5)}{1.0482} = 4660 \text{ N/m}^2$$

(d) The vapor pressure of water at 55°C = 118 mmHg or $118(133.3) = 15\,730$ N/m² = p_A. The relative humidity = $\bar{p}_A(100)/p_A = 4660(100)/15\,730 = 29.6\%$.

(e) Dew point. From point D proceed at constant humidity to the saturation curve at point E, at which the dew point temperature is 31.5°C.

(f) Humid volume. At 55°C, the specific volume of dry air is 0.93 m³/kg. The humid volume of saturated air = 1.10 m³/kg dry air. Interpolating for 26.1% humidity,

$$v_H = 0.93 + (1.10 - 0.93)(0.261) = 0.974 \text{ m}^3/\text{kg dry air}$$

(g) Humid heat, Eq. (7.10):

$$C_S = C_B + Y'C_A = 1005 + 0.030(1884) = 1061.5 \text{ J (for wet air)}/(\text{kg dry air}) \cdot \text{K}$$

(h) Enthalpy. At 55°C, the enthalpy of dry air is 56 000 J/kg dry air; that for saturated air is 352 000 N · m/kg dry air. Interpolating for 26.1% humidity gives

$$H' = 56\,000 + (352\,000 - 56\,000)(0.261) = 133\,300 \text{ J/kg dry air}$$

Alternatively, Eq. (7.13) or Table 7.1:

$$H' = C_S(t_G - t_0) + Y'\lambda_0 = (1005 + 1884Y')t_G + 2\,502\,300Y'$$
$$= [1005 + 1884(0.030)]55 + 2\,502\,300(0.030) = 133.4 \text{ kJ/kg dry air}$$

As another alternative line DF is drawn parallel to the adjacent adiabatic-saturation curves. At F, the enthalpy is 134 kJ/kg dry air, or nearly the same as at D.

Illustration 7.7 If 100 m³ of the moist air of Illustration 7.6 is heated to 110°C, how much heat is required?

SOLUTION After heating the mixture will be at point G of Fig. 7.6. The mass of dry air $= W_B = 100/v_H = 100/0.974 = 102.7$ kg. Eq. (7.11): $Q = W_B C_S \Delta t = 102.7(1061.5)(110 - 55) = 6.00 \times 10^6$ J.

Adiabatic-Saturation Curves

Consider the operation indicated schematically in Fig. 7.7. Here the entering gas is contacted with liquid, for example, in a spray, and as a result of diffusion and heat transfer between gas and liquid the gas leaves at conditions of humidity and temperature different from those at the entrance. The operation is adiabatic inasmuch as no heat is gained or lost to the surroundings. A mass balance for substance A gives

$$L' = G'_S(Y'_2 - Y'_1) \tag{7.14}$$

An enthalpy balance is

$$G'_S H'_1 + L' H_L = G'_S H'_2 \tag{7.15}$$

therefore

$$H'_1 + (Y'_2 - Y'_1)H_L = H'_2 \tag{7.16}$$

This can be expanded by the definition of H' given in Eq. (7.13),

$$C_{S1}(t_{G1} - t_0) + Y'_1\lambda_0 + (Y'_2 - Y'_1)C_{A,L}(t_L - t_0) = C_{S2}(t_{G2} - t_0) + Y'_2\lambda_0 \tag{7.17}$$

In the special case where the leaving gas-vapor mixture is saturated, and therefore at conditions t_{as}, Y'_{as}, H'_{as}, and the liquid enters at t_{as}, the gas is humidified by evaporation of liquid and cooled. Equation (7.17) becomes, on expansion of the humid-heat terms,

$$C_B(t_{G1} - t_0) + Y'_1 C_A(t_{G1} - t_0) + Y'_1\lambda_0 + (Y'_{as} - Y'_1)C_{A,L}(t_{as} - t_0)$$
$$= C_B(t_{as} - t_0) + Y'_{as}C_A(t_{as} - t_0) + Y'_{as}\lambda_0 \tag{7.18}$$

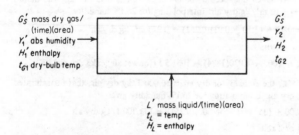

G'_S mass dry gas/(time)(area)
Y'_1 abs humidity
H'_1 enthalpy
t_{G1} dry-bulb temp

G'_S
Y'_2
H'_2
t_{G2}

L' mass liquid/(time)(area)
t_L = temp
H_L = enthalpy

Figure 7.7 Adiabatic gas-liquid contact.

By subtracting $Y_1' C_A t_{as}$ from both sides and simplifying this becomes

$$(C_B + Y_1' C_A)(t_{G1} - t_{as}) = C_{S1}(t_{G1} - t_{as})$$
$$= (Y_{as}' - Y_1')[C_A(t_{as} - t_0) + \lambda_0 - C_{A,L}(t_{as} - t_0)]$$

$$(7.19)$$

Reference to Fig. 7.3 shows the quantity in brackets to be equal to λ_{as}. Consequently,

$$C_{S1}(t_{G1} - t_{as}) = (Y_{as}' - Y_1')\lambda_{as} \tag{7.20}$$

or

$$t_{G1} - t_{as} = (Y_{as}' - Y_1')\frac{\lambda_{as}}{C_{S1}} \tag{7.21}$$

This is the equation of a curve on the psychrometric chart, the "adiabatic-saturation curve"† which passes through the points (Y_{as}', t_{as}) on the 100 percent saturation curve and (Y_1', t_{G1}). Since the humid heat C_{S1} contains the term Y_1', the curve is not straight but slightly concave upward. For any vapor-gas mixture there is an *adiabatic-saturation temperature* t_{as} such that if contacted with liquid at t_{as}, the gas will become humidified and cooled. If sufficient contact time is available, the gas will become saturated at (Y_{as}', t_{as}) but otherwise will leave unsaturated at (Y_2', t_{G2}), a point on the adiabatic-saturation curve for the initial mixture. Eventually, as Eq. (7.20) indicates, the sensible heat given up by the gas in cooling equals the latent heat required to evaporate the added vapor.

The psychrometric chart (Fig. 7.5) for air-water contains a family of adiabatic-saturation curves, as previously noted. Each point on the curve represents a mixture whose adiabatic-saturation temperature is at the intersection of the curve with the 100 percent humidity curve.

Illustration 7.8 Air at 83°C, $Y' = 0.030$ kg water/kg dry air, 1 std atm is contacted with water at the adiabatic-saturation temperature and is thereby humidified and cooled to 90% saturation. What are the final temperature and humidity of the air?

SOLUTION The point representing the original air is located on the psychrometric chart (Fig. 7.5a). The adiabatic-saturation curve through the point reaches the 100% saturation curve at 40°C, the adiabatic-saturation temperature. This is the water temperature. On this curve, 90% saturation occurs at 41.5°C, $Y' = 0.0485$ kg water/kg air, the outlet-air conditions.

Wet-Bulb Temperature

The wet-bulb temperature is the steady-state temperature reached by a small amount of liquid evaporating into a large amount of unsaturated vapor-gas mixture. Under properly controlled conditions it can be used to measure the humidity of the mixture. For this purpose a thermometer whose bulb has been

---

† The adiabatic-saturation curve is nearly one of constant enthalpy per unit mass of dry gas. As Eq. (7.16) indicates, H_{as}' differs from H_1' by the enthalpy of the evaporated liquid at its entering temperature t_{as}, but this difference is usually unimportant.

covered with a wick kept wet with the liquid is immersed in a rapidly moving stream of the gas mixture. The temperature indicated by this thermometer will ultimately reach a value lower than the dry-bulb temperature of the gas if the latter is unsaturated, and from a knowledge of this value the humidity is computed.

Consider a drop of liquid immersed in a rapidly moving stream of unsaturated vapor-gas mixture. If the liquid is initially at a temperature higher than the gas dew point, the vapor pressure of the liquid will be higher at the drop surface than the partial pressure of vapor in the gas, and the liquid will evaporate and diffuse into the gas. The latent heat required for the evaporation will at first be supplied at the expense of the sensible heat of the liquid drop, which will then cool down. As soon as the liquid temperature is reduced below the dry-bulb temperature of the gas, heat will flow from the gas to the liquid, at an increasing rate as the temperature difference becomes larger. Eventually the rate of heat transfer from the gas to the liquid will equal the rate of heat requirement for the evaporation, and the temperature of the liquid will remain constant at some low value, the wet-bulb temperature t_w. The mechanism of the wet-bulb process is essentially the same as that governing the adiabatic saturation, except that in the case of the former the humidity of the gas is assumed not to change during the process.

Refer to Fig. 7.8, sketched in the manner of the film theory, where a drop of liquid is shown already at the steady-state conditions and the mass of gas is so large as it passes the drop that its humidity is not measurably affected by the evaporation. Since both heat and mass transfer occur simultaneously, Eq. (3.71)

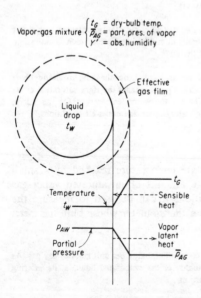

Vapor-gas mixture $\begin{cases} t_G & = \text{dry-bulb temp.} \\ \bar{p}_{AG} & = \text{part. pres. of vapor} \\ Y' & = \text{abs. humidity} \end{cases}$

Effective gas film

Liquid drop t_w

Temperature t_w

t_G

Sensible heat

p_{AW}

Partial pressure

Vapor latent heat

$\bar{p}_{AG}$

Figure 7.8 The wet-bulb temperature.

applies with $q_t = 0$ since no heat passes through the gas-liquid interface, and $N_B = 0$. Therefore†

$$q_s = \frac{N_A M_A C_A}{1 - e^{-N_A M_A C_A / h_G}}(t_G - t_w) \approx h_G(t_G - t_w) \qquad (7.22)$$

and the approximation of the right-hand side is usually satisfactory since ordinarily the rate of mass transfer is small. Further,

$$N_A = F \ln \frac{1 - p_{A,w}/p_t}{1 - \bar{p}_{A,G}/p_t} \approx k_G(\bar{p}_{A,G} - p_{A,w}) \qquad (7.23)$$

where the approximation on the right is usually satisfactory since N_A is small [the form of Eq. (7.23) reflects the fact that N_A is negative if q_s is taken to be positive]. $p_{A,w}$ is the vapor pressure of A at t_w. Substituting Eqs. (7.22) and (7.23) into Eq. (3.71) with N_B and q_t equal to zero, we get

$$h_G(t_G - t_w) + \lambda_w M_A k_G(\bar{p}_{A,G} - p_{A,w}) = 0 \qquad (7.24)$$

where λ_w is the latent heat at the wet-bulb temperature per unit of mass. From this,

$$t_G - t_w = \frac{\lambda_w M_A k_G(p_{A,w} - p_{A,G})}{h_G} = \frac{\lambda_w M_B \bar{p}_{B,M} k_G(Y'_w - Y')}{h_G} \qquad (7.25)$$

where $\bar{p}_{B,M}$ is the average partial pressure of the gas. Since (Table 3.1) $M_B \bar{p}_{B,M} k_G = k_Y$, Eq. (7.25) becomes

$$t_G - t_w = \frac{\lambda_w(Y'_w - Y')}{h_G/k_Y} \qquad (7.26)$$

which is the form of the relationship commonly used. The quantity $t_G - t_w$ is the *wet-bulb depression*.

In order to use Eq. (7.26) for determination of Y, it is necessary to have at hand appropriate values of h_G/k_Y, the *psychrometric ratio*.‡ Values of h_G and k_Y can be estimated independently for the particular shape of the wetted surface by correlations like those of Table 3.3, using the heat- mass-transfer analogy if necessary. Alternatively, experimental values of the ratio can be employed. Henry and Epstein [10] have critically examined the data and methods of

---

† For very careful measurements, the possibility of the liquid surface's receiving heat by radiation from either the gas itself or from the surroundings must also be considered. Assuming that the source of radiation is at temperature t_G, we have

$$q_s = (h_G + h_R)(t_G - t_w)$$

where the radiative heat transfer is described by an equivalent convection-type coefficient h_R. In wet-bulb thermometry, the effect of radiation can be minimized by using radiation shields and maintaining a high velocity of gas to keep h_G relatively high (at least 5 to 6 m/s in the case of air–water-vapor mixtures at ordinary temperatures). The relative size of h_G and h_R in any case can be estimated by standard methods [13]. It is necessary to observe the additional precaution of feeding the wick surrounding the thermometer bulb with an adequate supply of liquid preadjusted as nearly as practicable to the wet-bulb temperature.

‡ The quantity $h_G/k_Y C_S$ is also sometimes termed the *psychrometric ratio*.

measurement and have produced some measurements of their own. For flow of gases past cylinders, such as wet-bulb thermometers, and past single spheres, the results for 18 vapor-gas systems are well correlated by

$$\frac{h_G}{k_Y C_S} = \left(\frac{\text{Sc}}{\text{Pr}} \right)^{0.567} = \text{Le}^{0.567} \qquad (7.27)$$

for rates of flow which are turbulent, independent of the Reynolds number. A large range of Lewis numbers Le, 0.335 to 7.2, is made possible by using not only surfaces wetted with evaporating liquids but also cylinders and spheres cast from volatile solids, which provide large Schmidt numbers.

Vapor-air systems are the most important. For dilute mixtures, where $C_S = C_B$, and with Pr for air taken as 0.707, Eq. (7.27) becomes (SI units)†

$$\frac{h_G}{k_Y} = 1223 \, \text{Sc}^{0.567} \qquad (7.28)$$

For the system *air–water vapor*, for which Dropkin's [3] measurements are generally conceded to be the most authoritative, a thorough analysis [23, 24] led to the value $h_G/k_Y = 950 \, \text{N} \cdot \text{m/kg} \cdot \text{K}$,† which is recommended for this system. It agrees closely with Eq. (7.28).

It will be noted that Eq. (7.26) is identical with Eq. (7.21) for the adiabatic-saturation temperature, but with replacement of C_{S1} by h_G/k_Y. These are nearly equal for air–water vapor at moderate humidities, and for many practical purposes the adiabatic-saturation curves of Fig. 7.5 can be used instead of Eq. (7.26). This is *not* the case for most other systems.

The Lewis relation We have seen that for the system air–water vapor, h_G/k_Y is approximately equal to C_S, or, approximately, $h_G/k_Y C_S = 1$. This is the so-called *Lewis relation* (after W. K. Lewis). Not only does it lead to near equality of the wet-bulb and adiabatic-saturation temperatures (as in the case of air–water vapor) but also to other simplifications to be developed later. It can be shown, through consideration of Eqs. (3.31) and (3.33), with $J_A = N_A$ and equality of the eddy diffusivities E_D and E_H, that the Lewis relation will be followed only if the thermal and molecular diffusivities are identical, or if Sc = Pr, or Le = 1. This is, of course, the conclusion also reached from the empirical equation (7.27). Le is essentially unity for air–water vapor but not for most other systems.

Illustration 7.9 For an air–water-vapor mixture of dry-bulb temperature 65°C, a wet-bulb temperature 35°C was determined under conditions such that the radiation coefficient can be considered negligible. The total pressure was 1 std atm. Compute the humidity of the air.

SOLUTION At $t_w = 35$°C, $\lambda_w = 2\,419\,300$ J/kg, and $Y'_w = 0.0365$ kg H_2O/kg dry air

---

† With h_G/k_Y expressed as Btu/lb · °F, the coefficient of Eq. (7.28) for air mixtures becomes 0.292 and for air–water vapor $h_G/k_Y = 0.227$.

(Fig. 7.5a); $h_G/k_Y = 950$ J/kg · K, $t_G = 65°$C. Eq. (7.26):

$$65 - 35 = \frac{2\,419\,300(0.0365 - Y')}{950}$$

$$Y' = 0.0247 \text{ kg H}_2\text{O/kg air}$$

Alternatively as an approximation, the adiabatic-saturation curve for $t_{as} = 35°$C in Fig. 7.5a is followed to a dry-bulb temperature 65°C, where Y' is read as 0.0238 kg H$_2$O/kg air.

Illustration 7.10 Estimate the wet-bulb and adiabatic-saturation temperatures for a toluene-air mixture of 60°C dry-bulb temperature, $Y' = 0.050$ kg vapor/kg air, 1 std atm.

SOLUTION **Wet-bulb temperature** $t_G = 60°$C, $Y' = 0.050$ kg toluene/kg air. $D_{AB} = 0.92 \times 10^{-5}$ m^2/s at 59°C, 1 std atm. At 60°C, ρ for air = 1.060 kg/m^3 and $\mu = 1.95 \times 10^{-5}$ kg/m · s.

Sc *should* be calculated for mean conditions between those of the gas-vapor mixture and the wet-bulb saturation conditions. However, for the dilute mixture considered here, the bulk-gas value of Sc is satisfactory and is essentially independent of temperature

$$\text{Sc} = \frac{\mu}{\rho D_{AB}} = \frac{1.95 \times 10^{-5}}{1.060(0.92 \times 10^{-5})} = 2.00$$

Eq. (7.28): $h_G/k_Y = 1223(2.00)^{0.567} = 1812$ J/kg · K (observed value = 1842). Eq. (7.26):

$$60 - t_w = \frac{\lambda_w}{1812}(Y'_w - 0.050)$$

Solution for t_w is by trial and error. Try $t_w = 35°$C. $p_{A,w} = 46.2$ mmHg, $Y'_w = [46.2/(760 - 46.2)]$ 92/29 = 0.2056, $\lambda_w = (96.6$ cal/gm)(4187) = 404 460 J/kg. The equation provides $t_w = 25.3°$C instead of the 35°C assumed. Upon repeated trials, t_w is computed to be 31.8°C. **Ans.**

Adiabatic-saturation temperature $t_{G1} = 60°$C, $Y'_1 = 0.05$, C for toluene vapor = 1256 J/kg · K. $C_{S1} = 1005 + 1256(0.05) = 1067.8$ J/kg · K. Eq. (7.21):

$$60 - t_{as} = (Y'_{as} - 0.05)\frac{\lambda_{as}}{1067.8}$$

In the same fashion as the wet-bulb temperature, t_{as} is calculated by trial and found to be 25.7°C. **Ans.**

GAS-LIQUID CONTACT OPERATIONS

Direct contact of a gas with a pure liquid may have any of several purposes:

1. Adiabatic operations.

   a. *Cooling a liquid.* The cooling occurs by transfer of sensible heat and also by evaporation. The principal application is cooling of water by contact with atmospheric air (water cooling).
   b. *Cooling a hot gas.* Direct contact provides a nonfouling heat exchanger which is very effective, providing the presence of some of the vapor of the liquid is not objectionable.
   c. *Humidifying a gas.* This can be used for controlling the moisture content of air for drying, for example.

    d. Dehumidifying a gas. Contact of a warm vapor-gas mixture with a cold liquid results in condensation of the vapor. There are applications in air conditioning, recovery of solvent vapors from gases used in drying, and the like.

2. Nonadiabatic operations.

    a. Evaporative cooling. A liquid or gas inside a pipe is cooled by water flowing in a film about the outside, the latter in turn being cooled by direct contact with air.

    b. Dehumidifying a gas. A gas-vapor mixture is brought into contact with refrigerated pipes, and the vapor condenses upon the pipes.

Although operations of this sort are simple in the sense that mass transfer is confined to the gas phase (there can be no mass transfer within the pure liquid), they are nevertheless complex owing to the large heat effects which accompany evaporation or condensation.

ADIABATIC OPERATIONS

These are usually carried out in some sort of packed tower, frequently with countercurrent flow of gas and liquid. General relationships will be developed first, to be particularized for specific operations.

Fundamental Relationships

Refer to Fig. 7.9, which shows a tower of unit cross-sectional area. A mass balance for substance A over the lower part of the tower (envelope I) is

$$L' - L'_1 = G'_S(Y' - Y'_1) \tag{7.29}$$

or

$$dL' = G'_S \, dY' \tag{7.30}$$

Similarly, an enthalpy balance is

$$L'H_L + G'_S H'_1 = L'_1 H_{L1} + G'_S H' \tag{7.31}$$

These can be applied to the entire tower by putting subscript 2 on the unnumbered terms.

    The rate relationships are fairly complex and will be developed in the manner of Olander [20]. Refer to Fig. 7.10, which represents a section of the tower of differential height dZ and shows the liquid and gas flowing side by side, separated by the gas-liquid interface. The changes in temperature, humidity, etc., are all differential over this section.

    The interfacial surface of the section is dS. If the specific interfacial surface per packed volume is a (not the same as the packing surface a_p), since the volume of packing per unit cross section is dZ, then $dS = a \, dZ$. If the packing is incompletely wetted by the liquid, the surface for mass transfer a_M, which is the liquid-gas interface, will be smaller than that for heat transfer a_H, since heat

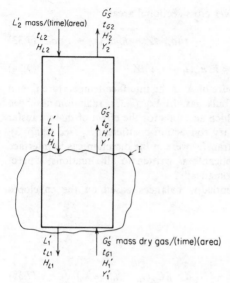

Figure 7.9 Continuous countercurrent adiabatic gas-liquid contact.

transfer may also occur between the packing and the fluids. Note that a_M corresponds to a_V of Chap. 6. The transfer rates are then:

Mass, as mass rate per tower cross-sectional area:

$$N_A M_A a_M \, dZ = -G_S' \, dY' = M_A F_G\left(\ln\frac{1 - \bar{p}_{A,\,i}/p_t}{1 - p_{A,\,G}/p_t}\right) a_M \, dZ \qquad (7.32)$$

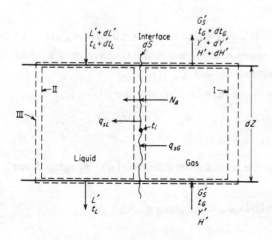

Figure 7.10 Differential section of a packed tower.

Sensible heat, as energy rate per tower cross-sectional area:

$$Gas: \quad q_{sG}a_H \, dZ = \frac{N_A M_A C_A}{1 - e^{-N_A M_A C_A/h_G}}(t_G - t_i)a_H \, dZ = h'_G a_H(t_G - t_i) \, dZ \quad (7.33)$$

$$Liquid: \qquad\qquad q_{sL}a_H \, dZ = h_L a_H(t_i - t_L) \, dZ \qquad\qquad (7.34)$$

In Eq. (7.32), $\bar{p}_{A,\,i}$ is the vapor pressure of A at the interface temperature t_i, and $\bar{p}_{A,\,G}$ is the partial pressure in the bulk gas. In Eq. (7.33), radiation has been neglected, and the coefficient h'_G, which accounts for the effect of mass transfer on heat transfer, replaces the ordinary convection coefficient h_G (see Chap. 3). The rate equations are written as if transfer were in the direction gas to interface to liquid, but they are directly applicable as written to all situations; correct signs for the fluxes will develop automatically.

We now require a series of enthalpy balances based on the envelopes sketched in Fig. 7.10.

Envelope I:

Rate enthalpy in $= G'_S H'$

Rate enthalpy out $= G'_S(H' + dH') - (G'_S \, dY')[C_A(t_G - t_0) + \lambda_0] \qquad (7.35)$

The second term is the enthalpy of the transferred vapor [recall that N_A and $G_S \, dY$ have opposite signs in Eq. (7.32)].

Rate in $-$ rate out $=$ heat-transfer rate

$$G'_S H' - G'_S(H' + dH') + (G'_S \, dY')[C_A(t_G - t_0) + \lambda_0] = h'_G a_H(t_G - t_i) \, dZ$$
$$(7.36)$$

If dH', obtained by differentiation of Eq. (7.13), is substituted, this reduces to

$$- G'_S C_S \, dt_G = h'_G a_H(t_G - t_i) \, dZ \qquad (7.37)$$

Envelope II:

Rate enthalpy in $= (L' + dL')C_{A,\,L}(t_L + dt_L - t_0) + (- G'_S \, dY')C_{A,\,L}(t_i - t_0)$

Here the second term is the enthalpy of the material transferred, now a liquid.

Rate enthalpy out $= L'C_{A,\,L}(t_L - t_0)$

Rate out $=$ rate in $+$ heat-transfer rate

$$L'C_{A,\,L}(t_L - t_0) = (L' + dL')C_{A,\,L}(t_L + dt_L - t_0) - (G'_S \, dY')C_{A,\,L}(t_i - t_0)$$
$$+ h_L a_H(t_i - t_L) \, dZ \qquad (7.38)$$

If Eq. (7.30) is substituted and the second-order differential $dY' \, dt_L$ ignored, this becomes

$$L'C_{A,\,L} \, dt_L = (G'_S C_{A,\,L} \, dY' - h_L a_H \, dZ)(t_i - t_L) \qquad (7.39)$$

Envelope III:

$$\text{Rate enthalpy in} = G_S' H' + (L' + dL') C_{A,L}(t_L + dt_L - t_0)$$
$$\text{Rate enthalpy out} = L' C_{A,L}(t_L - t_0) + G_S'(H' + dH')$$
$$\text{Rate in} = \text{rate out (adiabatic operation)}$$
$$G_S' H' + (L' + dL') C_{A,L}(t_L + dt_L - t_0) = L' C_{A,L}(t_L - t_0) + G_S'(H' + dH')$$
$$(7.40)$$

Substitutions of Eq. (7.30) and the differential of Eq. (7.13) for dH' are made, and the term $dH' \, dt_L$ is ignored, whereupon this becomes

$$L' C_{A,L} \, dt_L = G_S' \{ C_S \, dt_G + [C_A(t_G - t_0) - C_{A,L}(t_L - t_0) + \lambda_0] \, dY' \}$$
$$(7.41)$$

These will now be applied to the adiabatic operations.

Water Cooling with Air

This is without question the most important of the operations. Water, warmed by passage through heat exchangers, condensers, and the like, is cooled by contact with atmospheric air for reuse. The latent heat of water is so large that only a small amount of evaporation produces large cooling effects. Since the rate of mass transfer is usually small, the temperature level is generally fairly low, and the Lewis relation applies reasonably well for the air-water system, the relationships of the previous section can be greatly simplified by making reasonable approximations.

Thus, if the sensible-heat terms of Eq. (7.41) are ignored in comparison with the latent heat, we have

$$L' C_{A,L} \, dt_L = G_S' C_S \, dt_G + G_S' \lambda_0 \, dY' \approx G_S' \, dH$$
$$(7.42)$$

Here the last term on the right ignores the Y' which appears in the definition of C_S. Integrating, on the further assumption that L' is essentially constant (little evaporation), gives

$$L' C_{A,L}(t_{L2} - t_{L1}) = G_S'(H_2' - H_1')$$
$$(7.43)$$

This enthalpy balance can be represented graphically by plotting the gas enthalpy H' against t_L, as in Fig. 7.11. The line ON on the chart represents Eq. (7.43), and it passes through the points representing the terminal conditions for the two fluids. Insofar as $L_2' - L_1'$ is small in comparison with L', the line is straight and of slope $L' C_{A,L} / G_S'$. The equilibrium curve in the figure is plotted for conditions of the gas at the gas-liquid interface, i.e., the enthalpy of saturated gas at each temperature.

If the mass-transfer rate is small, as it usually is, Eq. (7.32) can be written as

$$G_S' \, dY' = k_Y a_M (Y_i' - Y') \, dZ$$
$$(7.44)$$

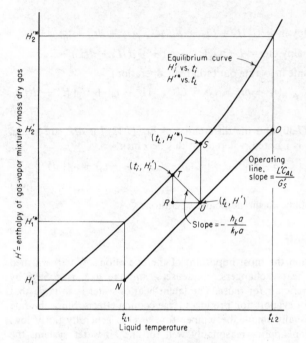

Figure 7.11 Operating diagram for a water cooler.

and Eq. (7.37) as

$$G_S' C_S \, dt_G = h_G a_H (t_i - t_G) \, dZ \qquad (7.45)$$

When the sensible heat of the transferred vapor is ignored, Eq. (7.39) becomes

$$L' C_{A, L} \, dt_L = h_L a_H (t_L - t_i) \, dZ \qquad (7.46)$$

Substituting Eqs. (7.44) and (7.45) into (7.42) gives

$$G_S' \, dH' = h_G a_H (t_i - t_G) \, dZ + \lambda_0 k_Y a_M (Y_i' - Y') \, dZ \qquad (7.47)$$

If $h_G a_H / C_S k_Y a_M = r$, this becomes

$$G_S' \, dH' = k_Y a_M [(C_S r t_i + \lambda_0 Y_i') - (C_S r t_G + \lambda_0 Y')] \, dZ \qquad (7.48)$$

For the special case where $r = 1$ [16, 17] the terms in parentheses are gas enthalpies. The restriction that $r = 1$ requires Le = 1 (air-water), and $a_M = a_H$ $= a$ (the latter will be true only for thoroughly irrigated tower filling; even for air-water contacting, values of r as high as 2 have been observed with low liquid rates [9]). With these understood, Eq. (7.48) is

$$G_S' \, dH' = k_Y a (H_i' - H') \, dZ \qquad (7.49)$$

which is remarkable in that the mass-transfer coefficient is used with an enthalpy driving force. Combining Eqs. (7.42), (7.46), and (7.49) then provides

$$G'_S \, dH' = k_Y a(H'_i - H') \, dZ = h_L a(t_L - t_i) \, dZ \tag{7.50}$$

At a position in the apparatus corresponding to point U on the operating line (Fig. 7.11), point T represents the interface conditions and the distance TR the enthalpy driving force $H'_i - H'$ within the gas phase. By making constructions like the triangle RTU at several places along the operating line, corresponding H'_i and H' values can be obtained. Equation (7.50) then provides, assuming $k_Y a$ is constant,

$$\int_{H'_1}^{H'_2} \frac{dH'}{H'_i - H'} = \frac{k_Y a}{G'_S} \int_0^Z dZ = \frac{k_Y a Z}{G'_S} \tag{7.51}$$

The integral can be evaluated graphically and the packed height Z computed. The enthalpy integral of Eq. (7.51) is sometimes given another interpretation. Thus,

$$\int_{H'_1}^{H'_2} \frac{dH'}{H'_i - H'} = \frac{H'_2 - H'_1}{(H'_i - H')_{av}} = N_{tG} \tag{7.52}$$

where the middle part of the equation is the number of times the average driving force divides into the enthalpy change. This is a measure of the difficulty of enthalpy transfer, called the *number of gas-enthalpy transfer units* N_{tG}. Consequently,

$$Z = H_{tG} N_{tG} \tag{7.53}$$

where the height of a gas-enthalpy transfer unit $= H_{tG} = G'_S / k_Y a$. H_{tG} is frequently preferred over $k_Y a$ as a measure of packing performance since it is less dependent upon rates of flow and has the simple dimension of length.

As discussed in Chap. 5, an overall driving force representing the enthalpy difference for the bulk phases but expressed in terms of H' can be used, such as the vertical distance SU (Fig. 7.11). This requires a corresponding overall coefficient and leads to overall numbers and heights of transfer units:†

$$N_{tOG} = \int_{H'_1}^{H'_2} \frac{dH'}{H'^* - H'} = \frac{K_Y a Z}{G'_S} = \frac{Z}{H_{tOG}} \tag{7.54}$$

The use of Eq. (7.53) is satisfactory (see Chap. 5) only if the equilibrium enthalpy curve of Fig. 7.11 is straight, which is not strictly so, or if $h_L a$ is infinite, so that the interface temperature equals the bulk-liquid temperature. Although the few data available indicate that $h_L a$ is usually quite large (see, for

† The water-cooling-tower industry frequently uses Eq. (7.54) in another form:

$$\frac{K_Y a Z}{L'} = \int_{t_{L1}}^{t_{L2}} \frac{dt_L}{H'^* - H'}$$

which results from combining Eqs. (7.42) and (7.54) and setting $C_{A,L}$ for water $= 1$.

example, Illustration 6.7), there are uncertainties owing to the fact that many have been taken under conditions such that $h_G a_H / C_S k_Y a_M = r$ was *not* unity even though assumed to be so. In any case, it frequently happens that, for cooling-tower packings, only $K_Y a$ or H_{tOG}, and not the individual phase coefficients, are available.

Just as with concentrations (Chap. 5), an operating line on the enthalpy coordinates of Fig. 7.11 which anywhere touches the equilibrium curve results in a zero driving force and consequently an infinite interfacial surface, or infinite height Z, to accomplish a given temperature change in the liquid. This condition would then represent the limiting ratio of L'/G_S' permissible. It is also clear that point N, for example, will be below the equilibrium curve so long as the entering-air enthalpy H_1' is less than the saturation enthalpy $H_1'^*$ for air at t_{L1}. Since the enthalpy H' is for most practical purposes only a function of the adiabatic-saturation temperature (or, for air-water, the wet-bulb temperature), the entering-air wet-bulb temperature must be below t_{L1} but its dry-bulb tempera- ture need not be. For this reason, it is perfectly possible to cool water to a value of t_{L1} less than the entering-air dry-bulb temperature t_{G1}. It is also possible to operate a cooler with entering air saturated, so long as its temperature is less than t_{L1}. The difference between the exit-liquid temperature and the entering-air wet-bulb temperature, $t_{L1} - t_{w1}$, called the *wet-bulb temperature approach*, is then a measure of the driving force available for diffusion at the lower end of the equipment. In the design of cooling towers, this is ordinarily specified to be from 2.5 to 5°C, with t_{w1} set at the "5 percent wet-bulb temperature" (the wet-bulb temperature which is exceeded only 5 percent of the time on the average during the summer months).

Makeup fresh water in recirculating water systems must be added to replace losses from entrainment (drift, or windage), evaporation losses, and blowdown. Windage losses can be estimated as 0.1 to 0.3 percent of the recirculation rate for induced-draft towers. If makeup water introduces dissolved salts (hardness) which will otherwise accumulate, a small amount of water is deliberately discarded (blowdown) to keep the salt concentration at some predetermined level. A calculation is demonstrated in Illustration 7.11. Chlorine treatment of the water to control algae and slime and addition of chromate-phosphate mixtures to inhibit corrosion have been common in the past, but restrictions on discharges to the environment by the blowdown have led to the use of nonchro- mate inhibitors. Many other practical details are available [14, 15].

The use of overall mass-transfer coefficients does not distinguish between convective and evaporative cooling of the liquid and will not permit computa- tion of the humidity or dry-bulb temperature of the exit air. The air will ordinarily be very nearly saturated, and for purposes of estimating makeup requirements it may be so assumed. The temperature-humidity history of the air as it passes through the tower can be estimated by a graphical method on the $H' t_L$ diagram (Fig. 7.11) if $h_L a$ and $k_Y a$ are known [17] but approach of the gas to saturation is very critical to the computations, and it is recommended instead that these be done by the methods outlined later (see page 255) which make no assumptions. Cooling towers for systems other than air-water (Le $\neq$ 1) or when $a_M \neq a_H$ must also be treated by the general methods discussed later. Some cooling towers use a cross flow of air and water, for which methods of computation are also available [11, 22, 27].

Illustration 7.11 A plant requires 15 kg/s (1984 lb/min) of cooling water to flow through its distillation-equipment condensers, thereby removing 270 W (55 270 Btu/min) from the condensers. The water will leave the condensers at 45°C. It is planned to cool the water for reuse by contact with air in an induced-draft cooling tower. The design conditions are as follows: entering air at 30°C dry-bulb, 24°C wet-bulb temperature, water to be cooled to within 5°C of the inlet-air wet-bulb temperature, i.e., to 29°C; an air/water-vapor ratio of 1.5 times the minimum. Makeup water will come from a well at 10°C, hardness 500 ppm dissolved solids. The circulating water is not to contain more than 2000 ppm hardness. For the packing to be used, $K_Y a$ is expected to be 0.90 kg/m^3 · s · ΔY (202 lb/ft^2 · h · ΔY), for a liquid rate at least 2.7 kg/m^2 · s and a gas rate 2.0 kg/m^2 · s (1991 and 1474 lb/ft^2 · h, respectively). Compute the dimensions of the packed section and the makeup-water requirement.

SOLUTION Refer to Fig. 7.12, which represents the flowsheet of the operation. The entering air humidity and enthalpy are taken from Fig. 7.5a. The operating diagram, Fig. 7.13, contains the saturated-air–enthalpy curve, and on this plot is point N representing the condition at the bottom of the tower ($t_{L1} = 29$°C, $H_1' = 72\,000$ N · m/kg dry air). The operating line will pass through N and end at $t_{L2} = 45$°C. For the minimum value of G_S', the operating line will have the least slope which causes it to touch the equilibrium curve and will consequently pass through point O, where $H_2' = 209\,500$ N · m/kg dry air. The slope of the line $O'N$ is therefore

$$\frac{L'C_{A,L}}{G_{S,\,min}'} = \frac{15(4187)}{G_{S,\,min}'} = \frac{209\,500 - 72\,000}{45 - 29}$$

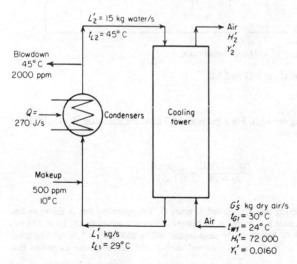

Figure 7.12 Flowsheet for Illustration 7.11.

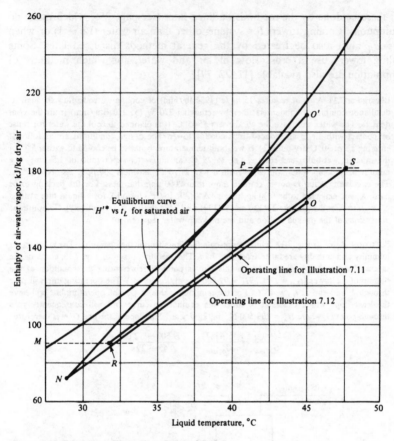

Figure 7.13 Solutions to Illustrations 7.11 and 7.12.

whence $G'_{S,\,min} = 7.31$ kg dry air/s. For a gas rate of 1.5 times the minimum, $G'_S = 1.5(7.31) = 10.97$ kg dry air/s. Therefore

$$\frac{H'_2 - 72\,000}{45 - 29} = \frac{15(4187)}{10.97}$$

and $H'_2 = 163\,600$ N · m/kg dry air, plotted at point O. The operating line is therefore line ON. For a liquid rate at least 2.7 kg/m² · s, the tower cross section would be $15/2.7 = 5.56$ m². For a gas rate of at least 2.0 kg/m² · s, the cross section will be $10.97/2.0 = 5.50$ m². The latter will therefore be used, since the liquid rate will then exceed the minimum to ensure that $K_Y a = 0.90$.

Basis: 1 m² cross section, $G'_S = 2.0$ kg/m² · s. The driving force $H'^* - H'$ is obtained at frequent intervals of t_L from Fig. 7.13 as follows

t_L, °C	$\chi H'^*$(equilibrium curve), J/kg	H' (operating line), J/kg	$\dfrac{10^5}{H'^* - H'}$
29	100 000	72 000	3.571
32.5	114 000	92 000	4.545
35	129 800	106 500	4.292
37.5	147 000	121 000	3.846
40	166 800	135 500	3.195
42.5	191 000	149 500	2.410
45	216 000	163 500	1.905

The data of the last two columns are plotted against each other, H' as abscissa, and the area under the curve is 3.25. From Eq. (7.54)

$$3.25 = \frac{K_Y a Z}{G'_S} = \frac{0.90 Z}{2.0}$$

$$z = 7.22 \text{ m (23.7 ft) packed height } \textbf{Ans.}$$

Note: In this case $N_{tOG} = 3.25$, and $H_{tOG} = G'_S / K_Y a = 2.0/0.90 = 2.22$ m.

Makeup-water requirement For present purposes, define

E = evaporation rate, kg/h
W = windage loss, kg/h
B = blowdown rate, kg/h
M = makeup rate, kg/h
x_C = weight fraction hardness in circulated water
x_M = wt fraction hardness in makeup water

For continuous makeup and continuous blowdown, a total material balance is

$$M = B + E + W$$

and a hardness balance is

$$M x_M = (B + W) x_C$$

Elimination of M results in

$$B = E \frac{x_M}{x_C - x_M} - W$$

Assuming the outlet air ($H' = 163\ 500$ N · m/kg) is essentially saturated, $Y'_2 = 0.0475$. The approximate rate of evaporation is then

$$E = 2.0(5.50)(0.0475 - 0.0160) = 0.3465 \text{ kg/s}$$

The windage loss is estimated as 0.2 percent of the circulation rate,

$$W = 0.002(15) = 0.03 \text{ kg/s}$$

Since the weight fractions x_C and x_M are proportional to the corresponding ppm values, the blowdown rate is

$$B = 0.3465 \frac{500}{2000 - 500} - 0.03 = 0.0855 \text{ kg/s}$$

The makeup rate is then estimated to be

$$M = B + E + W = 0.0855 + 0.3465 + 0.03 = 0.462 \text{ kg/s (3670 lb/h)} \quad \textbf{Ans.}$$

Illustration 7.12 In the cooler of Illustration 7.11, to what temperature would the water be cooled if, after the tower was built and operated at the design L' and G_S' values, the entering air entered at dry-bulb temperature $t_{G1} = 32°C$, wet-bulb temperature $t_{W1} = 28°C$?

SOLUTION For the new conditions, $H_1' = 90\,000$ J/kg, and the new operating line must originate on the broken line at M in Fig. 7.13. Since in all likelihood the heat load on the plant condensers, which is ultimately transferred to the air, will remain the same, the change in air enthalpy will be the same as in Illustration 7.11:

$$H_2' - 90\,000 = 163\,600 - 72\,000$$

$$H_2 = 181\,600 \text{ J/kg}$$

The new operating line must therefore end on the broken line at P in Fig. 7.13, and because of the same ratio L'/G_S' as in Illustration 7.11 it must be parallel to the original line NO. Since at the same flow rates as in Illustration 7.11 the value of H_{tOG} remains the same, for the same packing depth N_{tOG} remains at 3.25. The new operating line RS is therefore located by trial so that $N_{tOG} = 3.25$. The temperature at $R = t_{L1}$ is 31.7°C, which is the temperature to which the water will be cooled. **Ans.**

Dehumidification of Air–Water Vapor

If a warm vapor-gas mixture is contacted with cold liquid so that the humidity of the gas is greater than that at the gas-liquid interface, vapor will diffuse toward the liquid and the gas will be dehumidified. In addition, sensible heat can be transferred as a result of temperature differences within the system. For air–water-vapor mixtures (Le = 1) contacted with cold water, the methods of water cooling apply with only obvious modification. The operating line on the gas-enthalpy–liquid-temperature graph will be above the equilibrium curve, the driving force is $H' - H'^*$, and Eq. (7.54) can be used with this driving force. For all other systems, for which Le $\neq$ 1, the general methods below must be used.

Recirculating Liquid–Gas Humidification-Cooling

This is a special case where the liquid enters the equipment at the adiabatic-saturation temperature of the entering gas. This can be achieved by continuously reintroducing the exit liquid to the contactor immediately, without addition or removal of heat on the way, as in Fig. 7.14. The development which follows applies to any liquid-gas system, regardless of the Lewis number. In such a system, the temperature of the entire liquid will fall to, and remain at, the adiabatic-saturation temperature. The gas will be cooled and humidified, following along the path of the adiabatic-saturation curve on the psychrometric chart which passes through the entering-gas conditions. Depending upon the degree of contact, the gas will approach more or less closely equilibrium with the liquid, or its adiabatic-saturation conditions. This supposes that the makeup liquid enters

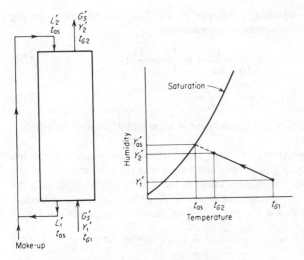

Figure 7.14 Recirculating liquid, gas humidification-cooling.

at the adiabatic-saturation temperature also, but for most purposes the quantity of evaporation is so small relative to the total liquid circulation that minor deviations from this temperature for the makeup liquid can be ignored.

As has been shown previously, the enthalpy of the gas is practically a function only of its adiabatic-saturation temperature, which remains constant throughout the operation. The enthalpy of the liquid at constant temperature is also constant, so that an operating "line" on a plot such as Fig. 7.11 would be merely a single point on the equilibrium curve. This diagram therefore cannot be used for design purposes. The temperature and humidity changes, which lie entirely within the gas phase, can be used, however, and these are shown schematically in Fig. 7.14. If mass transfer is used as a basis for design, Eq. (7.44) becomes

$$G_S' \, dY = k_Y a (Y_{as}' - Y') \, dZ \tag{7.55}$$

$$\int_{Y_1'}^{Y_2'} \frac{dY'}{Y_{as}' - Y'} = \frac{k_Y a}{G_S'} \int_0^Z dZ \tag{7.56}$$

and since Y_{as}' is constant,

$$\ln \frac{Y_{as}' - Y_1'}{Y_{as}' - Y_2'} = \frac{k_Y a Z}{G_S'} \tag{7.57}$$

Equation (7.57) can be used directly, or it can be rearranged by solving for G_S' and multiplying each side by $Y_2' - Y_1'$ or its equivalent,

$$G_S'(Y_2' - Y_1') = \frac{k_Y a Z [(Y_{as}' - Y_1') - (Y_{as}' - Y_2')]}{\ln [(Y_{as}' - Y_1')/(Y_{as}' - Y_2')]} = k_Y a Z (\Delta Y')_{av} \tag{7.58}$$

where $(\Delta Y')_{av}$ is the logarithmic average of the humidity-difference driving forces at the ends of the equipment. Alternatively,

$$N_{tG} = \frac{Y'_2 - Y'_1}{(\Delta Y')_{av}} = \ln \frac{Y'_{as} - Y'_1}{Y'_{as} - Y'_2} \tag{7.59}$$

and

$$H_{tG} = \frac{G'_S}{k_Y a} = \frac{Z}{N_{tG}} \tag{7.60}$$

where N_{tG} is the number of gas-phase transfer units and H_{tG} the corresponding height of a transfer unit.†

In contacting operations of this sort, where one phase approaches equilibrium with the other under conditions such that the characteristics of the latter do not change, the maximum change in the first phase corresponds to the operation of one theoretical stage (see Chap. 5). Since the humidity in adiabatic equilibrium with the liquid is Y'_{as}, the Murphree gas-phase stage efficiency is

$$E_{MG} = \frac{Y'_2 - Y'_1}{Y'_{as} - Y'_1} = 1 - \frac{Y'_{as} - Y'_2}{Y'_{as} - Y'_1} = 1 - e^{-k_Y aZ/G'_S} = 1 - e^{-N_{tG}} \tag{7.61}$$

If heat transfer is used as the basis for design, similar treatment of Eq. (7.45) leads to

$$G'_S C_{S1}(t_{G1} - t_{G2}) = \frac{h_G aZ[(t_{G1} - t_{as}) - (t_{G2} - t_{as})]}{\ln[(t_{G1} - t_{as})/(t_{G2} - t_{as})]} = h_G aZ(\Delta t)_{av} \tag{7.62}$$

where $h_G a$ is the volumetric-heat-transfer coefficient of sensible-heat transfer between the bulk of the gas and the liquid surface.

Illustration 7.13 A horizontal spray chamber (Fig. 7.19) with recirculated water is used for adiabatic humidification and cooling of air. The active part of the chamber is 2 m long and has cross section of 2 m². With an air rate 3.5 m³/s at dry-bulb temperature 65.0°C, $Y' = 0.0170$ kg water/kg dry air, the air is cooled and humidified to a dry-bulb temperature 42.0°C. If a duplicate spray chamber operated in the same manner were to be added in series with the existing chamber, what outlet conditions could be expected for the air?

SOLUTION For the existing chamber, the adiabatic-saturation line on Fig. 7.5 for $t_{G1} = 65.0°C$, $Y'_1 = 0.0170$ shows $t_{as} = 32.0°C$, $Y'_{as} = 0.0309$, and at $t_{G2} = 42.0°C$, $Y'_2 = 0.0265$. From Eq. (7.57) with $Z = 2$ m:

$$\ln \frac{0.0309 - 0.0170}{0.0309 - 0.0265} = \frac{k_Y a(2)}{G'_S}$$

$$\frac{k_Y a}{G'_S} = 0.575 \text{ m}^{-1}$$

---

† To be entirely consistent with the definition of Chap. 8, Eq. (7.60) should read $H_{tG} = G'_S/k_Y a(1 - y_A)$. The value of $1 - y_A$ in the present application, however, is ordinarily very close to unity.

For the extended chamber, with $Z = 4$ m, $k_Y a/G'_S$ and t_{as} will remain the same. Therefore, Eq. (7.57):

$$\ln \frac{0.0309 - 0.0170}{0.0309 - Y'_2} = 0.575(4)$$

$$Y'_2 = 0.0295$$

With the same adiabatic-saturation curve as before, $t_{G2} = 34.0°C$.

General Methods

For all other countercurrent operations, and even for those discussed above when the approximations are not appropriate or when Le $\neq 1$, we must return to the equations developed earlier. Equating the right-hand sides of Eqs. (7.39) and (7.41) provides

$$t_i = t_L$$
$$+ \frac{G'_S\{C_S(dt_G/dZ) + [C_A t_G - C_{A,L} t_L + (C_{A,L} - C_A)t_0 + \lambda_0](dY'/dZ)\}}{G'_S C_{A,L}(dY'/dZ) - h_L a_H}$$
(7.63)

The humidity gradient in this expression is obtained from Eq. (7.32):

$$\frac{dY'}{dZ} = -\frac{M_A F_G a_M}{G'_S} \ln \frac{1 - \bar{p}_{A,i}/p_t}{1 - \bar{p}_{A,G}/p_t} = -\frac{M_A F_G a_M}{G'_S} \ln \frac{Y' + M_A/M_B}{Y'_i + M_A/M_B}$$
(7.64)

$$\approx -\frac{M_A k_G a_M}{G'_S}(\bar{p}_{A,G} - \bar{p}_{A,i}) \approx -\frac{k_Y a_M}{G'_S}(Y' - Y'_i)$$
(7.64a)

where the approximations of Eq. (7.64a) are suitable for low vapor concentrations. The temperature gradient is taken from Eq. (7.37):

$$\frac{dt_G}{dZ} = -\frac{h'_G a_H(t_G - t_i)}{G'_S C_S} \approx -\frac{h_G a_H(t_G - t_i)}{G'_S C_S}$$
(7.65)

where $h_G a_H$ rather than $h'_G a_H$ may be used at low transfer rates. Unless a_H and a_M are separately known (which is not usual), there will be difficulty in evaluating h'_G exactly. Here we must assume $a_H = a_M$:

$$h'_G a_H = \frac{N_A M_A C_A a_H}{1 - e^{-N_A M_A C_A a_H/h_G a_H}} \approx \frac{N_A M_A C_A a_M}{1 - e^{-N_A M_A C_A a_M/h_G a_H}}$$
$$\approx -\frac{G'_S C_A(dY'/dZ)}{1 - e^{G'_S C_A(dY'/dZ)/h_G a_H}}$$
(7.66)

The effect of the approximation is not normally important.

Equations (7.64) and (7.65) are integrated numerically, using a procedure outlined in Illustration 7.14. Extensive trial and error is required, since t_i from Eq. (7.63) is necessary before $\bar{p}_{A,i}$ (or Y'_i) can be computed. The t_i-$\bar{p}_{A,i}$ relationship is that of the vapor-pressure curve. If at any point in the course of the calculations $\bar{p}_{A, G}$ (or Y') at a given t_G calculates to be larger than the corresponding saturation vapor concentration, a fog may form in the gas phase, in which case the entire analysis is invalid and new conditions at the terminals of the tower must be chosen.

Illustration 7.14 A producer gas, 65% N_2, 35% CO, initially dry at 1 std atm and 315°C, flowing at a rate of 5 m³/s (10 600 ft³/min), is to be cooled to 27°C by countercurrent contact with water entering at 18°C. A tower packed with 50-mm ceramic Raschig rings will be used, with $L_2'/G_2' = 2.0$. Specify the diameter of a suitable tower and the packed height.

SOLUTION For the entering gas, $Y_1' = 0$, $M_B = 28.0$, $p_t = 1$ std atm, $t_{G1} = 315°C$, whence

$$\rho_{G1} = \frac{28.0}{22.41} \frac{273}{273 + 315} = 0.580 \text{ kg/m}^3$$

therefore   Gas in $= 5(0.580) = 2.90$ kg/s

At the top of the tower, $t_{L2} = 18°C$, $\rho_{L2} = 1000$ kg/m³, $\mu_{L2} = 1.056 \times 10^{-3}$ kg/m · s, and $t_{G2} = 27°C$. Since the outlet gas is likely to be nearly saturated ($Y' = 0.024$), estimate $Y_2' = 0.022$, to be checked later. The outlet gas rate $= 2.90(1.022) = 2.964$ kg/s.

$$M_{av} = \frac{1.022}{1/28.0 + 0.022/18.02} = 27.7 \text{ kg/kmol}$$

$$\rho_{G2} = \frac{27.7}{22.41} \frac{273}{273 + 27} = 1.125 \text{ kg/m}^3$$

$$\frac{L'}{G'} \left(\frac{\rho_G}{\rho_L - \rho_G} \right)^{1/2} = 2 \left(\frac{1.125}{1000 - 1.125} \right)^{1/2} = 0.0671$$

Fig. 6.34: at a gas-pressure drop of 400 (N/m²)/m, the ordinate is $0.073 = G'^2 C_f \mu_L^{0.1} J / \rho_G (\rho_L - \rho_G) g_c$. Table 6.3: $C_f = 65$; $J = 1$, $g_c = 1$, whence $G_2' = 1.583$ kg/m² · s (tentative).

The tower cross section is $2.964/1.583 = 1.87$ m² (tentative). The corresponding diameter is 1.54, say 1.50 m, for which the cross section is $\pi(1.50)^2/4 = 1.767$ m² (final). $G_s' = 2.90/1.767 = 1.641$ kg/m² · s (final). $G_2' = 2.964/1.767 = 1.677$ kg/m² · s (final); $L_2' = 2(1.677) = 3.354$ kg/m² · s.

Calculations will be started at the bottom, for which L_1' and t_{L1} must be known. An overall water balance [Eq. (7.29)] is

$$3.354 - L_1' = 1.641(0.022 - 0)$$

$$L_1' = 3.319 \text{ kg/m}^2 \cdot \text{s}$$

The heat capacities of CO and N_2 are $C_B = 1089$, and that of water vapor is $C_A = 1884$ J/kg · K. $C_{S1} = 1089$, $C_{S2} = 1089 + 1884(0.022) = 1130$ N · m/(kg dry gas) · K. For convenience use as base temperature $t_0 = 18°C$, for which $\lambda_0 = 2.46 \times 10^6$ J/kg. $C_{AL} = 4187$ J/kg · K. An overall enthalpy balance [Eq. (7.32)]:

$$3.354(4187)(18 - 18) + 1.641(1089)(315 - 18) = 3.319(4187)$$
$$\times (t_{L1} - 18) + 1.641[1130(27 - 18) + 0.022(2.46 \times 10^6)]$$

$t_{L1} = 49.2°C$, for which $\mu_L = 0.557 \times 10^{-3}$ kg/m · s, $\rho_L = 989$ kg/m³, thermal conductivity = 0.64 W/m · K, Prandtl number $\text{Pr}_L = 3.77$. For the entering gas, the viscosity is 0.0288×10^{-3} kg/m · s, the diffusivity of water vapor 0.8089×10^{-4} m²/s, $\text{Sc}_G = 0.614$, and $\text{Pr}_G = 0.74$.

With the data of Chap. 6 and the methods of Illustration 6.7, we obtain $a = a_{VW} = 53.1$ m²/m³, $F_G a = 0.0736$ kmol/m³ · s, $h_G a = 4440$, and $h_L a = 350\ 500$ W/m³ · K.

At the bottom, t_i will be estimated by trial and error. After several trials, assume $t_i = 50.3°C$. $\bar{p}_{A, i} = 93.9$ mmHg/760 $= 0.1235$ std atm. Eq. (7.64):

$$\frac{dY'}{dZ} = -\frac{18.02(0.0736)}{1.641} \ln \frac{1 - 0.1235}{1 - 0} = 0.1066 \text{ (kg H}_2\text{O/kg dry gas)/m}$$

Eq. (7.66)

$$h'_G a = \frac{-1.641(1884)(0.1066)}{1 - \exp[1.641(1844)(0.1066)/4440]} = 4280 \text{ W/m}^3 \cdot \text{K}$$

Eq. (7.65)

$$\frac{dt_G}{dZ} = \frac{-4280(315 - 50.3)}{1.641(1089)} = -630°\text{C/m}$$

For use in Eq. (7.63):

$$C_A t_G - C_{A, L} t_L + (C_{A, L} - C_A)t_0 + \lambda_0 = 1884(315) - 4187(49.2)$$

$$+ (4187 - 1884)(18) + 2.46 \times 10^6 = 2.889 \times 10^6 \text{ J/kg}$$

Eq. (7.63):

$$t_i = 49.2 + \frac{1.641[1089(-630) + (2.889 \times 10^6)(0.1066)]}{1.641(4187)(0.1066) - 350\,500} = 50.3°\text{C} \quad \text{(check)}$$

A suitably small increment in gas temperature is now chosen, and with the computed gradients assumed constant over a small range, the conditions at the end of the increment are computed. For example, take $\Delta t_G = -30°\text{C}$.

$$\Delta Z = \frac{\Delta t_G}{dt_G/dZ} = \frac{-30}{-630} = 0.0476 \text{ m}$$

$$t_G = t_G(\text{at } Z = 0) + \Delta t_G = 315 - 30 = 285°\text{C}$$

$$Y' = Y'(\text{at } Z = 0) + \frac{dY'}{dZ} \Delta Z = 0 + 0.1066(0.0476)$$

$$= 5.074 \times 10^{-3} \text{ kg H}_2\text{O/kg dry gas}$$

$$\bar{p}_{A, G} = \frac{Y'}{Y' + M_A/M_B} = \frac{5.074 \times 10^{-3}}{5.074 \times 10^{-3} + 18/28} = 7.83 \times 10^{-3} \text{ std atm}$$

$$C_S = 1089 + 1884(5.074 \times 10^{-3}) = 1099 \text{ J/kg} \cdot \text{K}$$

L' is calculated by a water balance [Eq. (7.29)] over the increment:

$$L' - 3.319 = 1.641(5.074 \times 10^{-3} - 0)$$

$$L' = 3.327 \text{ kg/m}^2 \cdot \text{s}$$

t_L is calculated by an enthalpy balance [Eq. (7.31)]:

$$3.327(4187)(t_L - 18) + 1.641(1089)(315 - 18) = 3.319(4187)(49.2 - 18)$$

$$+ 1.641[1099(285 - 18) + (5.074 \times 10^{-3})(2.46 \times 10^6)]$$

therefore $\qquad\qquad\qquad t_L = 47.1°\text{C}$

New gradients dY'/dZ and dt_G/dZ at this level in the tower and another interval are then computed in the same manner. The process is repeated until the gas temperature falls to 27°C. The intervals of Δt_G chosen must be small as the gas approaches saturation. The entire computation is readily adapted to a digital computer. The computed $t_G - Y'$ results are shown in Fig. 7.15, along with the saturation humidity curve. At $t_G = 27°\text{C}$, the value of Y'_2 was calculated to be 0.0222, which is considered sufficiently close to the value assumed earlier, 0.022. The sum of the ΔZ's $= Z = 1.54$ m. This relatively small packed depth emphasizes the effectiveness of this type of operation.

The direct contact of a vapor-laden gas with cold liquid of the same composition as the vapor can be useful as a means of vapor recovery. For this, a flowsheet of the sort shown in Fig. 7.16 can be employed and the tower designed

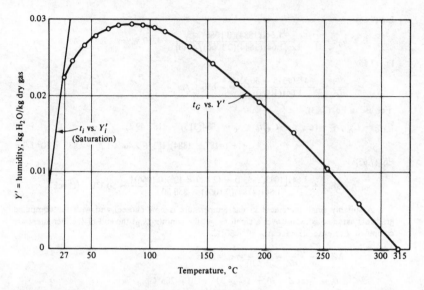

Figure 7.15 Temperatures and humidities for Illustration 7.14.

through Eqs. (7.63) to (7.69). The method is direct, whereas alternative methods of vapor recovery require additional operations for their completion. For example, absorption of the vapor into a solvent must be followed by distillation or stripping to recover the solvent and obtain the recovered solute; adsorption onto a solid similarly requires additional steps. With gas-vapor mixtures other than air–water vapor, however, the molecular diffusivity is likely to be less than the thermal diffusivity, so that $Le = Sc/Pr = \alpha/D_{AB}$ may exceed 1.0. Heat transfer is then faster than mass transfer, the path of the gas on a psychrometric chart tends to enter the supersaturation region above the saturation curve, and fog may result if suitable nucleation conditions are present [25, 26]. The difficulty is

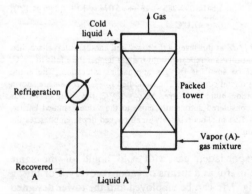

Figure 7.16 Recovery of vapor from a gas mixture by direct contact with a cold liquid.

alleviated if the entering vapor-gas mixture is sufficiently superheated. An example is given in Prob. 7.11.

Equipment

Any of the gas-liquid contact devices described in Chap. 6 are applicable to the operations described here, and conventional packed and tray towers are very effective in these services.

Water-cooling towers Air and water are low-cost substances, and where large volumes must be handled, as in many water-cooling operations, equipment of low initial cost and low operating cost is essential. The framework and internal packing are frequently of redwood, a material which is very durable in continuous contact with water, or Douglas fir. Impregnation of the wood under pressure with fungicides such as coal-tar creosote, pentachlorophenols, acid copper chromate, and the like, is commonplace. Siding for the towers is commonly redwood, asbestos cement, glass-reinforced polyester plastic, and the like. Towers have been built entirely of plastic. The internal packing ("fill") is usually a modified form of hurdle (see Chap. 6), horizontal slats arranged staggered with alternate tiers at right angles. Plastic packing may be polypropylene, molded in a grid or other form [5]. A great many arrangements are used [12, 15]. The void space is very large, usually greater than 90 percent, so that the gas-pressure drop will be as low as possible. The air-water interfacial surface consequently includes not only that of the liquid films which wet the slats (or other packing) but also the surface of the droplets which fall as rain from each tier of packing to the next.

The common arrangements are shown schematically in Fig. 7.17. Of the *natural-circulation towers* (Fig. 7.17a and b) the atmospheric towers depend on prevailing winds for air movement. The natural-draft design [6, 8] ensures more positive air movement even in calm weather by depending upon the displacement of the warm air inside the tower by the cooler outside air. Fairly tall chimneys are then required. Both these tower types must be relatively tall in order to operate at a small wet-bulb-temperature approach. Natural-draft equipment is used commonly in the southwestern United States and in the Middle East, where the humidity is usually low, in parts of Europe where air temperatures are generally low, and with increasing frequency everywhere as energy for fan power becomes more costly.

Mechanical-draft towers may be of the forced-draft type (Fig. 7.17c), where the air is blown into the tower by a fan at the bottom. These are particularly subject to recirculation of the hot, humid discharged air into the fan intake owing to the low discharge velocity, which materially reduces the tower effectiveness. Induced draft, with the fan at the top, avoids this and also permits more uniform internal distribution of air. The arrangements of Fig. 7.17d and e are most commonly used, and a more detailed drawing is shown in Fig. 7.18. Liquid rates are ordinarily in the range $L' = 0.7$ to 3.5 kg/m$^2 \cdot$ s (500 to 2500

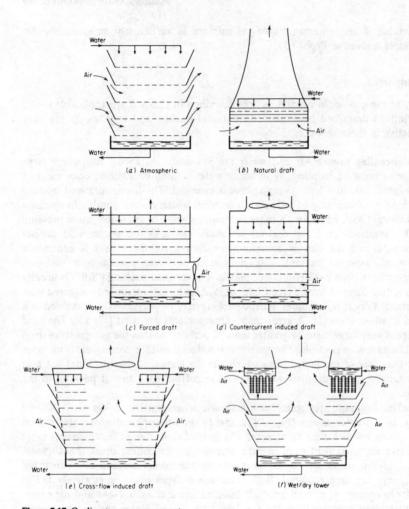

Figure 7.17 Cooling-tower arrangements.

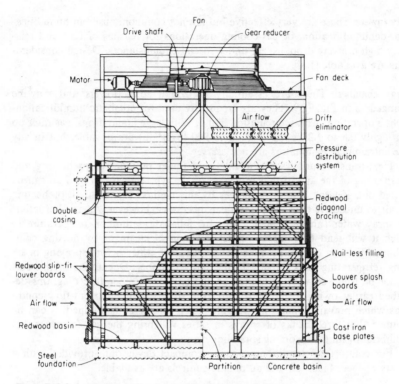

Figure 7.18 Induced-draft cooling tower. *(The Marley Co., Inc.)*

lb/ft² · h), and superficial air rates are of the order of $G_S' = 1.6$ to 2.8 kg/m² · s (1200 to 2100 lb/ft² · h), whereupon the air pressure drop is ordinarily less than 250 N/m² (25 mmH$_2$O). If fogging is excessive, finned-type heat exchangers can be used to evaporate the fog by heat from the hot water to be cooled, as in Fig. 7.17*f*.

Mass-transfer rates The correlations of mass-transfer coefficients for the standard packings discussed in Chap. 6 are suitable for the operations discussed here (see particularly Illustration 6.7). Additional data for humidification with Berl saddles [9]; Intalox saddles, and Pall rings [18†] are available. Data for some of the special tower fillings generally used for water-cooling towers are available in texts specializing in this type of equipment [15, 19].

---

† The data for H$_2$O–H$_2$ and H$_2$O–CO$_2$ should be used with caution since H-t_L diagrams were used in interpreting the data despite the fact that Le is substantially different from unity for these systems.

Tray towers These are very effective but are not commonly used in humidification, dehumidification, or gas-cooling operations for reasons of cost and relatively high pressure drop, except under special circumstances. Design considerations are available [1, 4].

Spray chambers These are essentially horizontal spray towers and may be arranged as in Fig. 7.19. They are frequently used for adiabatic humidification-cooling operations with recirculating liquid. With large liquid drops, gas rates up to roughly 0.8 to 1.2 kg/m$^2 \cdot$ s (600 to 900 lb/ft$^2 \cdot$ h) are possible, but in any case entrainment eliminators are necessary.

Heat-transfer surfaces at the inlet and outlet provide for preheating and afterheating of the air, so that processes of the type shown in Fig. 7.20 can be carried out. If large humidity changes by this method are required, preheating the air to unusually high temperatures is necessary, however. As an alternative, the spray water can be heated above the adiabatic-saturation temperature to which it will tend to come by direct injection of steam or by heating coils. Dehumidification can be practiced by cooling the water before spraying or by using refrigerating coils directly in the spray chamber. Operations of this sort cannot be followed with assurance on the enthalpy-temperature diagrams described earlier owing to the departure from strictly countercurrent-flow conditions which prevail. When an adequate spray density is maintained, it can be assumed that three banks of sprays in series will bring the gas to substantial equilibrium with the incoming spray liquid.

For comfort air conditioning, many compact devices are provided with a variety of these facilities, and automatic controls are available.

Spray ponds These are sometimes used for water cooling where close approach to the air wet-bulb temperature is not required. Spray ponds are essentially

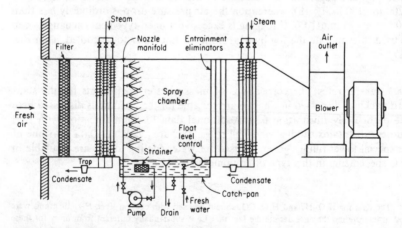

Figure 7.19 Schematic arrangement of a spray chamber.

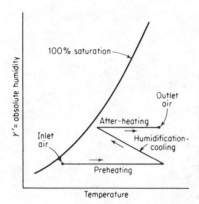

Figure 7.20 A simple conditioning process.

fountains, where the water is sprayed upward into the air and allowed to fall back into a collection basin. They are subject to high windage losses of water.

Dehumidification by other methods Adsorption, e.g., methods using activated silica gel, alumina, or molecular sieves as adsorbents (see Chap. 11), and washing gases with water solutions containing dissolved substances which appreciably lower the partial pressure of the water (see Chap. 8) are other commonly used dehumidification processes, particularly when very dry gases are required.

NONADIABATIC OPERATION: EVAPORATIVE COOLING

In evaporative cooling, a fluid is cooled while it flows through a tube. Water flows in a film or spray about the outside of the tube, and air is blown past the water to carry away the heat removed from the tube-side fluid. Advantage is taken of the large heat-transfer rate resulting when the spray water is evaporated into the airstream. A schematic arrangement is shown in Fig. 7.21a. The tube-side fluid usually flows through a bank of tubes in parallel, as shown in Fig. 7.21b and c. Since the water is recirculated from top to bottom of the heat exchanger, the temperature t_{L2} at which it enters is the same as that at which it leaves, t_{L1}. While the water temperature does not remain constant as it passes through the device, it does not vary greatly from the terminal value.

Figure 7.22 shows the temperature and gas-enthalpy profiles through a typical section of the exchanger. In the present analysis [24] the heat-transfer system will be divided at the bulk-water temperature t_L, corresponding to a saturation gas enthalpy $H_1'^*$. The overall heat-transfer coefficient U_o based on the outside tube surface, from tube fluid to bulk water, is then given by

$$\frac{1}{U_o} = \frac{d_o}{d_i h_T} + \frac{d_o}{d_{av}} \frac{z_m}{k_m} + \frac{1}{h_L'}$$ (7.67)

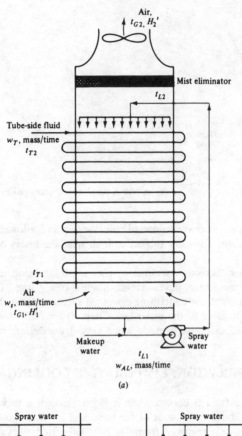

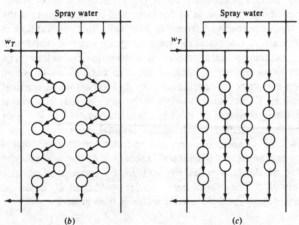

Figure 7.21 Evaporative cooler: (a) schematic arrangement, (b) and (c) arrangements of tube-side fluid flow.

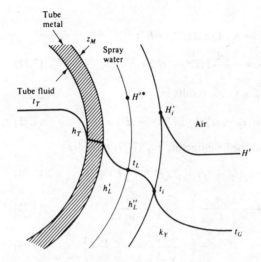

Figure 7.22 Temperature and enthalpy gradients, evaporative cooler.

The overall coefficient K_Y for use with gas enthalpies, from $H_1'^*$ to H', is

$$\frac{1}{K_Y} = \frac{1}{k_Y} + \frac{m}{h_L''} \tag{7.68}$$

where m is the slope of a chord such as chord TS on Fig. 7.11:

$$m = \frac{H'^* - H_i'}{t_L - t_i}$$

Since the variation in t_L is small, for all practical purposes m can be taken as constant,

$$m = \frac{dH'^*}{dt_L} \tag{7.69}$$

If A_o is the outside surface of all the tubes, $A_o x$ is the area from the bottom to the level where the bulk-water temperature is t_L. Here x is the fraction of the heat-transfer surface to that level. For a differential portion of the exchanger, the heat loss by the tube-side fluid is

$$w_T C_T \, dt_L = U_o A_o \, dx (t_T - t_L)$$

whence

$$\frac{dt_T}{dx} = \frac{U_o A_o}{w_T C_T} (t_T - t_L) \tag{7.70}$$

The heat lost by the cooling water is

$$w_{A,L} C_{A,L} \, dt_L = K_Y A_o \, dx (H'^* - H') - U_o A_o \, dx (t_T - t_L)$$

or

$$\frac{dt_L}{dx} = \frac{K_Y A_o}{w_{A,L} C_{A,L}} (H'^* - H') - \frac{U_o A_o}{w_{A,L} C_{A,L}} (t_T - t_L) \tag{7.71}$$

The heat gained by the air is

$$w_S \, dH' = K_Y A_o \, dx (H'^* - H')$$

or

$$\frac{dH'}{dx} = \frac{K_Y A_o}{w_S} (H'^* - H') \tag{7.72}$$

Subtraction of Eq. (7.71) from Eq. (7.70) results in

$$\frac{d(t_T - t_L)}{dx} + \alpha_1 (t_T - t_L) + \beta_1 (H'^* - H') = 0 \tag{7.73}$$

Multiplying Eq. (7.71) by m and then subtracting Eq. (7.72) provides

$$\frac{d(H'^* - H')}{dx} + \alpha_2 (t_T - t_L) + \beta_2 (H'^* - H') = 0 \tag{7.74}$$

where

$$\alpha_1 = -\left(\frac{U_o A_o}{w_T C_T} + \frac{U_o A_o}{w_{A,L} C_{A,L}} \right) \qquad \alpha_2 = \frac{m U_o A_o}{w_T C_T} \tag{7.75}$$

and

$$\beta_1 = \frac{K_Y A_o}{W_{A,L} C_{A,L}} \qquad \beta_2 = -\left(\frac{m K_Y A_o}{w_{A,L} C_{A,L}} - \frac{K_Y A_o}{w_S} \right) \tag{7.76}$$

Integration of Eqs. (7.73) and (7.74), with U_o and K_Y assumed constant, yields

$$t_T - t_L = M_1 e^{r_1 x} + M_2 e^{r_2 x} \tag{7.77}$$

$$H'^* - H' = N_1 e^{r_1 x} + N_2 e^{r_2 x} \tag{7.78}$$

For use with average driving forces, these yield

$$Q = w_T C_T (t_{T2} - t_{T1}) = w_S (H_2' - H_1') = U_o A_o (t_T - t_L)_{av} = K_Y A_o (H'^* - H')_{av} \tag{7.79}$$

where

$$(t_T - t_L)_{av} = \frac{M_1}{r_1} (e^{r_1} - 1) + \frac{M_2}{r_2} (e^{r_2} - 1) \tag{7.80}$$

$$(H'^* - H')_{av} = \frac{N_1}{r_1} (e^{r_1} - 1) + \frac{N_2}{r_2} (e^{r_2} - 1) \tag{7.81}$$

In these expressions, r_1 and r_2 are the roots of the quadratic equation

$$r^2 + (\alpha_1 + \beta_2) r + (\alpha_1 \beta_2 - \alpha_2 \beta_1) = 0 \tag{7.82}$$

and

$$N_j = \frac{M_j (r_j + \alpha_1)}{\beta_1} \qquad j = 1, 2 \tag{7.83}$$

These expressions permit the design of the evaporative cooler provided values of the various transfer coefficients are available. The available data [24] are meager despite the widespread use of these devices.

For h_T use standard correlation for sensible-heat changes or vapor condensation [13] for flow inside tubes.

For tubes 19.05 mm OD ($\frac{3}{4}$ in) on 38-mm (1.5-in) center, triangular pitch (Fig. 7.21b and c), $\Gamma/d_o = 1.36$ to 3.0 kg $H_2O/m^2 \cdot s$ (1000 to 2200 lb/ft$^2 \cdot$ h), $G'_{S, \, min} = 0.68$ to 5 kg/m$^2 \cdot$ s (500 to 3700 lb/ft$^2 \cdot$ h), water 15 to 70°C, in kilograms, meters, newtons, seconds, and kelvins and t_L in degrees Celsius†

$$h'_L = (982 + 15.58t_L)\left(\frac{\Gamma}{d_o}\right)^{1/3} \qquad W/m^2 \cdot K \qquad (7.84)$$

$$h''_L = 11\ 360\ W/m^2 \cdot K \qquad (7.85)$$

$$k_Y = 0.0493[G'_{S,\ min}(1 + Y'_{av})]^{0.905} \qquad kg/m^2 \cdot s \cdot \Delta Y' \qquad (7.86)$$

For other tube diameters, since the general relationship is expected to be

$$Sh = const\ Re^n\ Sc^{n'}$$

k_Y can be expected to vary as $G^n d_o{}^{n-1}$, or in this case as $d_o{}^{-0.095}$

Illustration 7.15 An evaporative cooler of the sort shown in Fig. 7.21a and b is to be used to cool oil. The oil, flowing at a rate 4.0 kg/s in the tubes, will enter at 95°C. Its average properties are density = 800 kg/m³, viscosity = 0.005 kg/m · s, thermal conductivity = 0.1436 W/m · K, and heat capacity = 2010 J/kg · K.

The cooler consists of a rectangular vertical shell, 0.75 m wide, fitted with 400 Admiralty Metal tubes, 19.05 mm OD, 1.65 mm wall thickness, 3.75 m long, on equilateral-triangular centers 38 mm apart. The tubes are arranged in horizontal rows of 20 tubes through which the oil flows in parallel, in stacks 10 tubes tall, as in Fig. 7.21b. Cooling water will be recycled at the rate 10 kg/s. The design air rate is to be 2.3 kg dry air/s, entering at 30°C dry-bulb temperature, standard atmospheric pressure, humidity 0.01 kg H₂O/kg dry air.

Estimate the temperature to which the oil will be cooled.

SOLUTION The minimum area for airflow is on a plane through the diameters of tubes in a horizontal row. The free area is

$$[0.75 - 20(0.01905)](3.75) = 1.384\ m^2$$

$$w_S = 2.3\ kg/s \qquad G'_{S,\ min} = \frac{2.3}{1.384} = 1.662\ kg/m^2 \cdot s$$

$Y'_1 = 0.01$ kg H₂O/kg dry air. Estimate $Y'_2 = 0.06$. $Y'_{av} = (0.01 + 0.06)/2 = 0.035$. Eq. (7.86):

$$k_Y = 0.0493[1.162(1 + 0.035)]^{0.905} = 0.05826\ kg/m^2 \cdot s \cdot \Delta Y'$$

Fig. 7.5:

$$H'_i(\text{satd at } t_w = 19.5°C) = 56\ 000\ J/kg$$

Outside surface of tubes = $A_o = 400\pi(0.01905)(3.75) = 89.8\ m^2$. The cooling water is distributed over 40 tubes. Since the tubes are staggered,

$$\Gamma = \frac{10}{40(2)(3.75)} = 0.0333\ kg/m \cdot s \qquad \Gamma/d_o = \frac{0.0333}{0.01905} = 1.750\ kg/m^2 \cdot s$$

Tentatively take av $t_L \approx t_{L1} = t_{L2} = 28°C$. Eq. (7.84):

$$h'_L = [982 + 15.58(28)](1.750)^{1/3} = 1709\ W/m^2 \cdot K$$

† For units of Btu, hours, feet, pounds mass, and degrees Fahrenheit Eqs. (7.84) to (7.86) become

$$h'_L = 13.72(1 + 0.0123t_L)\left(\frac{\Gamma}{d_o}\right)^{1/3} \qquad Btu/ft^2 \cdot h \cdot °F$$

$$h''_L = 2000\ Btu/ft^2 \cdot h \cdot °F$$

$$k_Y = 0.0924[G'_{S,\ min}(1 + Y'_{av})]^{0.905} \qquad lb/ft^2 \cdot h \cdot \Delta Y'$$

Eq. (7.85):

$$h_L'' = 11\ 360\ \text{W/m}^2 \cdot \text{K}$$

Fig. 7.5:

$$m = \frac{dH'^*}{dt_L} \text{ at } 28°C = 5000\ \text{J/kg} \cdot \text{K}$$

Eq. (7.68):

$$\frac{1}{K_Y} = \frac{1}{0.05826} + \frac{5000}{11\ 360}$$

Tube ID = d_i = $[19.05 - 2(1.65)]/1000 = 0.01575$ m. Inside cross section, each tube = $\pi(0.01575)^2/4 = 1.948 \times 10^{-4}$ m^2. G_T' = tube-fluid mass velocity = $4.0/20\ (1.948 \times 10^{-4})$ = 1026.7 kg/m$^2 \cdot$ s;

$$\text{Re}_T = \frac{d_i G_T'}{\mu_T} = \frac{0.01575(1026.7)}{0.005} = 3234 \qquad \text{Pr}_T = \frac{C_T \mu_T}{k_T} = \frac{2010(0.005)}{0.1436} = 70.0$$

From a standard correlation (fig. 10-8, p. 10-14, "The Chemical Engineers' Handbook," 5th ed.), $h_T = 364$ W/m$^2 \cdot$ K.

$$d_{\text{av}} \text{ for tubes} = \frac{0.01905 + 0.01575}{2} = 0.01740\ \text{m} \qquad k_m = 112.5\ \text{W/m} \cdot \text{K}$$

Eq. (7.67):

$$\frac{1}{U_o} = \frac{0.01905}{0.01575(364)} + \frac{0.01905}{0.01740}\frac{0.00165}{112.5} + \frac{1}{1709}$$

$$U_o = 255\ \text{W/m}^2 \cdot \text{K}$$

$w_T = 4.0$ kg/s, $w_{A,\,L} = 10$ kg/s, $w_S = 2.3$ kg/s; $C_T = 2010$, $C_{A,\,L} = 4187$ N $\cdot$ m/kg $\cdot$ K. Equations (7.75) and (7.76) then yield

$$\alpha_1 = -3.393 \qquad \alpha_2 = 18\ 794 \qquad \beta_1 = 12.08 \times 10^{-5} \qquad \beta_2 = 1.402$$

With these values, the roots of Eq. (7.82) are

$$r_1 = 3.8272 \qquad r_2 = -1.8362$$

Since t_L is unknown, proceed as follows. Eq. (7.83):

$$N_1 = \frac{M_1(3.8272 - 3.393)}{12.08 \times 10^{-5}} = 3594 M_1 \tag{7.87}$$

$$N_2 = \frac{M_2(-1.8362 - 3.393)}{12.08 \times 10^{-5}} = -43\ 288 M_2 \tag{7.88}$$

From Eq. (7.77) at $x = 1$ (top)

$$95 - t_{L2} = M_1 e^{3.8272} + M_2 e^{-1.8362} = 45.934 M_1 + 0.15942 M_2 \tag{7.89}$$

From Eq. (7.78) at $x = 0$ (bottom)

$$H_1'^* - 56\ 000 = N_1 + N_2 \tag{7.90}$$

Eq. (7.80):

$$(t_T - t_L)_{\text{av}} = \frac{M_1}{3.8272}(e^{3.872} - 1) + \frac{M_2}{-1.8362}(e^{-1.8362} - 1)$$

$$= 11.741 M_1 + 0.4578 M_2 \tag{7.91}$$

Eq. (7.81):

$$(H'^* - H')_{\text{av}} = 11.741 N_1 + 0.4578 N_2 \tag{7.92}$$

Equation (7.79), with Eqs. (7.91) and (7.92) for the average temperature and enthalpy differences, gives

$$U_o A_o (t_T - t_L)_{av} = K_Y A_o (H'^* - H')_{av}$$
$$255(89.8)(11.741 M_1 + 0.4578 M_2) = 0.0568(89.8)(11.741 N_1 + 0.4578 N_2) \qquad (7.93)$$

Elimination of the M's and the N's by solving Eqs. (7.87) to (7.90) and (7.93) simultaneously (with $t_{L2} = t_{L1}$) results in

$$H_1'^* = 106\,560 - 532.2 t_{L1}$$

This is solved simultaneously with the saturated-enthalpy curve of Fig. 7.5, whence $t_{L1} = 28°C$, thus checking the value assumed at the start. $H_1'^* = 91\,660$ J/kg. Equations (7.87) to (7.90) then provide $M_1 = 1.461$, $M_2 = -0.7204$, and from Eq. (7.91) $(t_T - t_L)_{av} = 16.83°C$. Eq. (7.79):

$$Q = 4.0(2010)(95 - t_{T1}) = 255(89.8)(16.83) = 2.3(H_2' - 56\,000)$$

$$t_{T1} = 47.0°C = \text{the outlet temperature} \quad \textbf{Ans.}$$

$$H_2' = 223\,500 \text{ J/kg}$$

If the outlet air is essentially saturated, this corresponds to $Y' = 0.0689$ and the average $Y' = (0.01 + 0.0689)/2 = 0.0395$, which is considered sufficiently close to the 0.035 assumed earlier.

Note: It is estimated that in the absence of the spray water, all other quantities as above, the oil would be cooled to 81°C.

NOTATION FOR CHAPTER 7

Any consistent set of units may be used, except as noted.

a	specific interfacial surface, based on volume of packed section, L^2/L^3
a_H	specific interfacial surface for heat transfer, L^2/L^3
a_M	specific interfacial surface for mass transfer, L^2/L^3
A_o	outside surface of tubes, L^2
c	molar density, mole/L^3
C	heat capacity of a gas or vapor, unless otherwise indicated, at constant pressure, FL/MT
C_S	humid heat, heat capacity of a vapor-gas mixture per unit mass of dry gas content, FL/MT
C_T	heat capacity of tube-side fluid, FL/MT
d_{av}	mean diameter of a tube, L
d_i	inside diameter of a tube, L
d_o	outside diameter of a tube, L
D_{AB}	molecular diffusivity of vapor A in mixture with gas B, L^2/Θ
E_D	eddy diffusivity of mass, L^2/Θ
E_H	eddy diffusivity of heat, L^2/Θ
E_{MG}	Murphree gas-phase stage efficiency, fractional
F	mass-transfer coefficient, mole/$L^2\Theta$
G	molar mass velocity, mole/$L^2\Theta$
G_S'	superficial mass velocity of dry gas, $M/L^2\Theta$
h	convective heat-transfer coefficient, $FL/L^2T\Theta$
h'	convective heat-transfer coefficient corrected for mass transfer, $FL/L^2T\Theta$
h_L', h_L''	water heat-transfer coefficients, evaporative cooler, $FL/L^2T\Theta$
h_R	radiation heat-transfer coefficient in convection form, $FL/L^2T\Theta$
H	enthalpy, FL/M
H'	enthalpy of a vapor-gas mixture per unit mass of dry gas, FL/M

H_{tG}	height of a gas transfer unit, L
H_{tOG}	overall height of a gas-enthalpy transfer unit, L
J_A	flux of mass for no bulk flow, $\mathbf{mole/L^2\Theta}$
k	thermal conductivity, $\mathbf{FL^2/L^2T\Theta}$
k_G	gas-phase mass-transfer coefficient, $\mathbf{mole/L^2\Theta(F/L^2)}$
k_Y	gas-phase mass-transfer coefficient, $\mathbf{M/L^2\Theta(M/M)}$
K_Y	overall gas-phase mass-transfer coefficient, $\mathbf{M/L^2\Theta(M/M)}$
ln	natural logarithm
log	common logarithm
L'	superficial mass velocity of liquid, $\mathbf{M/L^2\Theta}$
Le	Lewis number, Sc/Pr, dimensionless
m	slope of a chord of the saturated-enthalpy curve on a H'^*-t_L diagram; dH'^*/dt_L, $\mathbf{FL/MT}$
M	mol wt, $\mathbf{M/mole}$
M_1, M_2	quantities defined by Eq. (7.83)
N	mass-transfer flux, $\mathbf{mole/L^2\Theta}$
N_{tG}	number of gas transfer units, dimensionless
N_{tOG}	number of overall gas transfer units, dimensionless
N_1, N_2	quantities defined by Eq. (7.83)
p	vapor pressure, F/L^2
$\bar{p}$	partial pressure, F/L^2
$\bar{p}_{B,M}$	mean partial pressure of nondiffusing gas, F/L^2
Pr	Prandtl number, $C\mu/k$, dimensionless
q_s	sensible-heat-transfer flux, $\mathbf{FL/L^2\Theta}$
q_T	total-heat-transfer flux, $\mathbf{FL/L^2\Theta}$
Q	evaporative-cooler heat load, $\mathbf{FL/\Theta}$; gain in enthalpy, $\mathbf{FL/M}$
r	$h_G a_H/C_S k_Y a_M$ [not in Eq. (7.82)], dimensionless
r_1, r_2	roots of Eq. (7.82)
R	universal gas constant, $\mathbf{FL/mole\ K}$
S	interfacial surface, L^2
Sc	Schmidt number, $\mu/\rho D_{AB}$, dimensionless
Sh	Sherwood number, $k_Y d_o/M_B c D_{AB}$, dimensionless
t	temperature, T
t_{as}	adiabatic saturation temperature, T
t_{DP}	dew-point temperature, T
t_W	wet-bulb temperature, T
t_0	reference temperature, T
T	absolute temperature, T
U	internal energy, FL/M
U_o	overall heat-transfer coefficient based on outside tube surface, $\mathbf{FL/L^2T\Theta}$
v	molal specific volume, $L^3/\mathbf{mole}$
v_H	humid volume, volume vapor-gas mixture/mass dry gas, L^3/M
w	mass rate, $\mathbf{M/\Theta}$
w_S	mass rate of dry gas, M/Θ
W_B	mass of dry gas, M
x	fraction of heat-transfer surface traversed in the direction of airflow, dimensionless
Y'	absolute humidity, mass vapor/mass dry gas, M/M
z_m	metal thickness, L
Z	length or height of active part of equipment, L
α	thermal diffusivity, L^2/Θ
α_1, α_2	quantities defined by Eq. (7.75)
β_1, β_2	quantities defined by Eq. (7.76)
Γ	mass rate of spray water per tube per tube length (Fig. 7.21c); use twice the tube length for Fig. 7.21b, M/LΘ

Δ	difference
λ	latent heat of vaporization, FL/M
λ′	molal latent heat of vaporization, $FL/mole$
μ	viscosity, $M/L\Theta$
ρ	density, M/L^3

Subscripts

0	reference condition
1, 2	positions 1, 2
as	adiabatic saturation
av	average
A	substance A, the vapor
bp	boiling point
B	substance B, the gas
G	pertaining to the gas
i	interface
L	pertaining to the liquid
m	metal
min	minimum
nbp	normal boiling point
o	overall; outside
r	reference substance
s	saturation
T	tube-side fluid
W	wet-bulb temperature

Superscript

*	at saturation

REFERENCES

1. Barrett, E. C., and S. G. Dunn: *Ind. Eng. Chem. Process Des. Dev.*, 13, 353 (1974).
2. "Cooling Towers," American Institute of Chemical Engineers, New York, 1972.
3. Dropkin, D.: *Cornell Univ. Eng. Expt. Stn. Bull.*, 23, 1936; 26, 1939.
4. Fair, J. R.: *Chem. Eng.*, 81, 91 (June 12, 1972).
5. Fuller, A. L., A. L. Kohl, and E. Butcher: *Chem. Eng. Prog.*, 53, 501 (1957).
6. Furzer, I. A.: *Ind. Eng. Chem. Process Des. Dev.*, 7, 558, 561 (1968).
7. Grosvenor, W. M.: *Trans. AIChE*, 1, 184 (1908).
8. Hayashi, Y., E. Hirai, and M. Okubo: *Heat Transfer Jap. Res.*, 2(2), 1 (1973).
9. Hensel, S. L., and R. E. Treybal: *Chem. Eng. Prog.*, 48, 362 (1952).
10. Henry, H. C., and N. Epstein: *Can. J. Chem. Eng.*, 48, 595, 602, 609 (1970).
11. Inazume, H., and S. Kageyama: *Chem. Eng. Sci.*, 30, 717 (1975).
12. Kelly, N. W., and L. K. Swenson: *Chem. Eng. Prog.*, 52, 263 (1956).
13. Kreith, F.: "Principles of Heat Transfer," International Textbook, Scranton, Pa., 1958.
14. Kuehmister, A. M.: *Chem. Eng.*, 78, 112 (May 31, 1971).
15. McKelvey, K. K., and M. Brooke: "The Industrial Cooling Tower," Van Nostrand, Princeton, N.J., 1959.
16. Merkel, F.: *Ver. Dtsch. Forschungsarb.*, 275 (1925).
17. Mickley, H. S.: *Chem. Eng. Prog.*, 45, 739 (1949).
18. Nemunaitis, R. R., and J. S. Eckert: *Chem. Eng. Prog.*, 71 (8), 60 (1975).
19. Norman, W. S.: "Absorption, Distillation, and Cooling Towers," Wiley, New York, 1961.

20. Olander, D. R.: *AIChE J.*, **6**, 346 (1960); *Ind. Eng. Chem.*, **53**, 121 (1961).
21. Othmer, D. F., et al.: *Ind. Eng. Chem.*, **32**, 841 (1940); **34**, 952 (1942).
22. Park, J. E., and J. M. Vance: *Chem. Eng. Prog.*, **67** (7), 55 (1971).
23. Parker, R. O.: D. Eng. Sci. dissertation, New York University, 1959; The Psychrometric Ratio for the System Air-Water Vapor, *1st Interam. Congr. Chem. Eng., San Juan, 1961.*
24. Parker, R. O., and R. E. Treybal: *Chem. Eng. Prog. Symp. Ser.*, **57**(32), 138 (1960).
25. Schrodt, J. T.: *Ind. Eng. Chem. Process Des. Dev.*, **11**, 20 (1972).
26. Steinmeyer, D. E.: *Chem. Eng. Prog.*, **68**(7), 64 (1972).
27. Wnek, W. J., and R. H. Snow: *Ind. Eng. Chem. Process Des. Dev.*, **11**, 343 (1972).

PROBLEMS

7.1 Prepare a logarithmic reference-substance plot of the vapor pressure of acetone over a temperature range of 10°C to its critical temperature, 235°C, with water as reference substance. With the help of the plot, determine (a) the vapor pressure of acetone at 65°C, (b) the temperature at which acetone has a vapor pressure of 500 torr, and (c) the latent heat of vaporization of acetone at 40°C (accepted value = 536.1 kN · m/kg).

7.2 A mixture of nitrogen and acetone vapor at 800 mmHg total pressure, 25°C, has a percentage saturation of 80%. Calculate (a) the absolute molal humidity, (b) the absolute humidity, kg acetone/kg nitrogen, (c) the partial pressure of acetone, (d) the relative humidity, (e) the volume percent acetone, and (f) the dew point.

7.3 In a plant for the recovery of acetone which has been used as a solvent, the acetone is evaporated into a stream of nitrogen gas. A mixture of acetone vapor and nitrogen flows through a duct, 0.3 by 0.3 m cross section. The pressure and temperature at one point in the duct are 800 mmHg, 40°C, and at this point the average velocity is 3.0 m/s. A wet-bulb thermometer (wick wet with acetone) indicates a temperature at this point of 27°C. Calculate the kilograms of acetone per second carried by the duct.

Ans.: 0.194 kg/s.

7.4 Would you expect the wet-bulb temperature of hydrogen-water vapor mixtures to be equal to, greater than, or less than the adiabatic-saturation temperature? Take the pressure as 1 std atm. Diffusivity for $H_2O-H_2 = 7.5 \times 10^{-5}$ m²/s at 0°C, 1 std atm. Heat capacity of $H_2 = 14\,650$ N · m/kg · K, thermal conductivity = 0.173 W/m · K.

7.5 An air–water-vapor mixture, 1 std atm, 180°C, flows in a duct (wall temperature = 180°C) at 3 m/s average velocity. A wet-bulb temperature, measured with an ordinary, unshielded thermometer covered with a wetted wick (9.5 mm outside diameter) and inserted in the duct at right angles to the duct axis, is 52°C. Under these conditions, the adiabatic-saturation curves of the psychrometric chart do not approximate wet-bulb lines, radiation to the wet bulb and the effect of mass transfer on heat transfer are not negligible, and the k-type (rather than F) mass-transfer coefficients should not be used.

(a) Make the best estimate you can of the humidity of the air, taking these matters into consideration.

(b) Compute the humidity using the ordinary wet-bulb equation (7.26) with the usual h_G/k_Y, and compare.

7.6 Prepare a psychrometric chart for the mixture acetone-nitrogen at a pressure of 800 mmHg over the ranges −15 to 60°C, $Y' = 0$ to 3 kg vapor/kg dry gas. Include the following curves, all plotted against temperature: (a) 100, 75, 50, and 25% humidity; (c) dry and saturated humid volumes; (c) enthalpy of dry and saturated mixtures expressed as N · m/kg dry gas, referred to liquid acetone and nitrogen gas at −15°C; (d) wet-bulb curves for $t_W = 25$°C; (e) adiabatic-saturation curves $t_{as} = 25$ and 40°C.

7.7 A drier requires 1.50 m³/s (3178 ft³/min) of air at 65°C, 20% humidity. This is to be prepared from air at 27°C dry-bulb, 18°C wet-bulb temperatures by direct injection of steam into the airstream followed by passage of the air over steam-heated finned tubes. The available steam is saturated at 35 000 N/m² (5.08 lb$_f$/in²) gauge. Compute the kilograms of steam per second required (a) for direct injection and (b) for the heat exchanger.

    Ans.: (a) 0.0463 kg/s; (b) 0.0281 kg/s.

7.8 For the purpose of carrying out a catalytic reaction of acetone vapor with another reagent, an acetone-nitrogen mixture will be produced by passing nitrogen upward through a tower packed with 38-mm (1.5-in) Berl saddles, irrigated with liquid acetone recirculated from the bottom to the top of the tower, with addition of makeup acetone as needed. The process is to be one of adiabatic gas humidification, as in Fig. 7.14. The N_2 will be preheated to the necessary temperature and introduced at the rate 1.5 kg/s (1106 lb/h). The desired ultimate humidity is 0.6 kg acetone/kg N_2, and the circulated liquid acetone is to be at 25°C, the adiabatic-saturation temperature. The surface tension of acetone at 25°C is 23.2 dyn/cm.

    (a) To what temperature should the N_2 be preheated?

    (b) Specify the liquid-circulation rate and the dimensions of a suitable packed section of the tower.

    Ans: 7.1 kg/s; 1.5 m diam, 1.1 m depth.

    (c) Another possible process would involve heating the recirculating liquid, admitting the gas at ordinary temperatures. Are there advantages to this alternative? What portion of this chapter could be used for the design of the tower?

7.9 A recently installed induced-draft cooling tower was guaranteed by the manufacturer to cool 2000 U.S. gal/min (0.1262 m³/s) of 43°C water to 30°C when the available air has a wet-bulb temperature 24°C. A test on the tower, when operated at full fan capacity, provided the following data:

Inlet water, 0.1262 m³/s, 46.0°C
Outlet water, 25.6°C
Inlet air, 24.0°C dry-bulb, 15.6°C wet-bulb temperature
Outlet air, 37.6°C, essentially saturated

    (a) What is the fan capacity, m³/s?

    (b) Can the tower be expected to meet the guarantee conditions? Note that to do so, N_{tOG} in the test must at least equal the guarantee value if H_{tOG} is unchanged.

7.10 It is desired to dehumidify 1.2 m³/s (2540 ft³/min) of air, available at 38.0°C dry-bulb, 30.0°C wet-bulb temperatures, to a wet-bulb temperature of 15.0°C in a countercurrent tower using water chilled to 10.0°C. The packing will be 50-mm (2-in) Raschig rings. To keep entrainment at a minimum, G_S' will be 1.25 kg air/m² · s (922 lb/ft² · h), and a liquid rate of 1.5 times the minimum will be used.

    (a) Specify the cross section and height of the packed portion of the tower.

    Ans.: $Z = 1.013$ m.

    (b) What will be the temperature of the outlet water?

7.11 A new process involves a product wet with methanol, and it is planned to dry the product by a continuous evaporation of the methanol into a stream of hot carbon dioxide (CO_2 rather than air in order to avoid an explosive mixture). One of the schemes being considered is that of Fig. 7.16. The tower will be packed with 25-mm ceramic Raschig rings. The CO_2-methanol mixture from the drier, at the rate 0.70 kg/(m² tower cross section) · s (516 lb/ft² · h) total flow, 1 std atm, 80°C, will contain 10% methanol vapor by volume. The refrigerated liquid methanol will enter the tower at −15°C at a rate of 4.75 kg/m² · s (3503 lb/ft² · h). The methanol of the effluent gas is to be reduced to 2.0% by volume.

    (a) Calculate the value of the coefficients $F_G a$, $h_G a$, and $h_L a$ at the top of the tower. For this purpose assume that the outlet gas is at −9°C. Take the surface tension of methanol as 25 dyn/cm.

    Ans.: $F_G a_o = 0.0617$ kmol/m³ · s, $h_G a_o = 2126$, $h_L a_o = 32\,900$ W/m³ · K.

(*b*) Compute the required depth of packing. Take the base temperature for enthalpies as −15°C, at which temperature the latent heat of vaporization of methanol is 1197.5 kJ/kg. Take $C_{A,L} = 2470$, $C_A = 1341$, $C_B = 846$ J/kg · K. In order to simplify the computations, assume that the interfacial area, void space, gas viscosity, Sc_G, and Pr_G all remain constant. Then F_G varies as $G/G'^{0.36}$. Since $h_L a$ is very large, assume also that $t_i = t_L$. For present purposes, ignore any solubility of CO_2 in methanol.

Ans.: 0.64 m.

GAS ABSORPTION

Gas absorption is an operation in which a gas mixture is contacted with a liquid for the purposes of preferentially dissolving one or more components of the gas and to provide a solution of them in the liquid. For example, the gas from by-product coke ovens is washed with water to remove ammonia and again with an oil to remove benzene and toluene vapors. Objectionable hydrogen sulfide is removed from such a gas or from naturally occurring hydrocarbon gases by washing with various alkaline solutions in which it is absorbed. Valuable solvent vapors carried by a gas stream can be recovered for reuse by washing the gas with an appropriate solvent for the vapors. Such operations require mass transfer of a substance from the gas stream to the liquid. When mass transfer occurs in the opposite direction, i.e., from the liquid to the gas, the operation is called desorption, or stripping. For example, the benzene and toluene are removed from the absorption oil mentioned above by contacting the liquid solution with steam, whereupon the vapors enter the gas stream and are carried away, and the absorption oil can be used again. Since the principles of both absorption and desorption are basically the same, we can study both operations at the same time.

Ordinarily, these operations are used only for solute recovery or solute removal. Separation of solutes from each other to any important extent requires the fractionation techniques of distillation.

EQUILIBRIUM SOLUBILITY OF GASES IN LIQUIDS

The rate at which a gaseous constituent of a mixture will dissolve in an absorbent liquid depends upon the departure from equilibrium which exists, and therefore it is necessary to consider the equilibrium characteristics of gas-liquid systems. A very brief discussion of such matters was presented in Chap. 5, but some elaboration will be required here.

Two-Component Systems

If a quantity of a single gas and a relatively nonvolatile liquid are brought to equilibrium in the manner described in Chap. 5, the resulting concentration of dissolved gas in the liquid is said to be the *gas solubility* at the prevailing temperature and pressure. At fixed temperature, the solubility concentration will increase with pressure in the manner, for example, of curve *A*, Fig. 8.1, which shows the solubility of ammonia in water at 30°C.

Different gases and liquids yield separate solubility curves, which must ordinarily be determined experimentally for each system. If the equilibrium pressure of a gas at a given liquid concentration is high, as in the case of curve *B* (Fig. 8.1), the gas is said to be relatively insoluble in the liquid, while if it is low, as for curve *C*, the solubility is said to be high. But these are relative matters

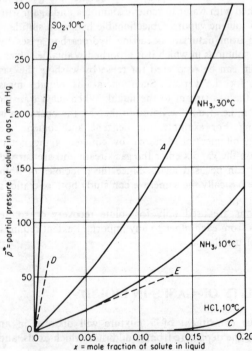

Figure 8.1 Solubilities of gases in water.

only, for it is possible to produce any ultimate gas concentration in the liquid if sufficient pressure is applied, so long as the liquefied form of the gas is completely soluble in the liquid.

The solubility of any gas is influenced by the temperature, in a manner described by van't Hoff's law of mobile equilibrium: if the temperature of a system at equilibrium is raised, that change will occur which will absorb heat. Usually, but not always, the solution of a gas results in an evolution of heat, and it follows that in most cases the solubility of a gas decreases with increasing temperature. As an example, curve A (Fig. 8.1) for ammonia in water at 30°C lies above the corresponding curve for 10°C. At the boiling point of the solvent, provided its vapor pressure is less than that of the gas or vapor solute, the gas solubility will be zero. On the other hand, the solubility of many of the low-molecular-weight gases such as hydrogen, oxygen, nitrogen, methane, and others in water increases with increased temperature above about 100°C and therefore at pressures above atmospheric [29, 35]. This phenomenon can be usefully exploited, e.g., in operations such as certain ore-leaching operations where oxygen-bearing solutions are required. Quantitative treatment of such equilibria is beyond the scope of this book, but excellent treatments are available [27, 34].

Multicomponent Systems

If a mixture of gases is brought into contact with a liquid, under certain conditions the equilibrium solubilities of each gas will be independent of the others, provided, however, that the equilibrium is described in terms of the *partial pressures* in the gas mixture. If all but one of the components of the gas are substantially insoluble, their concentrations in the liquid will be so small that they cannot influence the solubility of the relatively soluble component, and the generalization applies. For example, curve A (Fig. 8.1) will also describe the solubility of ammonia in water when the ammonia is diluted with air, since air is so insoluble in water, provided that the ordinate of the plot is considered as the partial pressure of ammonia in the gas mixture. This is most fortunate, since the amount of experimental work in gathering useful solubility data is thereby considerably reduced. If several components of the mixture are appreciably soluble, the generalization will be applicable only if the solute gases are indifferent to the nature of the liquid, which will be the case only for ideal solutions. For example, a mixture of propane and butane gases will dissolve in a nonvolatile paraffin oil independently since the solutions that result are substantially ideal. On the other hand, the solubility of ammonia in water can be expected to be influenced by the presence of methylamine, since the resulting solutions of these gases are not ideal. The solubility of a gas will also be influenced by the presence of a nonvolatile solute in the liquid, such as a salt in water solution, when such solutions are nonideal.

Ideal Liquid Solutions

When the liquid phase can be considered ideal, we can compute the equilibrium partial pressure of a gas from the solution without resort to experimental determination.

There are four significant characteristics of ideal solutions, all interrelated:

1. The average intermolecular forces of attraction and repulsion in the solution are unchanged on mixing the constituents.
2. The volume of the solution varies linearly with composition.
3. There is neither absorption nor evolution of heat in mixing the constituents. For gases dissolving in liquids, however, this criterion should not include the heat of condensation of the gas to the liquid state.
4. The total vapor pressure of the solution varies linearly with composition expressed as mole fractions.

In reality there are no ideal solutions, and actual mixtures only approach ideality as a limit. Ideality would require that the molecules of the constituents be similar in size, structure, and chemical nature, and the nearest approach to such a condition is perhaps exemplified by solutions of optical isomers of organic compounds. Practically, however, many solutions are so nearly ideal that for engineering purposes they can be so considered. Adjacent or nearly adjacent members of a homologous series of organic compounds particularly fall in this category. So, for example, solutions of benzene in toluene, of ethyl and propyl alcohols, or of the paraffin hydrocarbon gases in paraffin oils can ordinarily be considered as ideal solutions.

When the gas mixture in equilibrium with an ideal liquid solution also follows the ideal-gas law, the partial pressure $\bar{p}^*$ of a solute gas A equals the product of its vapor pressure p at the same temperature and its mole fraction in the solution x. This is *Raoult's law*.

$$\bar{p}^* = px \qquad (8.1)$$

(The asterisk is used to indicate equilibrium.) The nature of the solvent liquid does not enter into consideration except insofar as it establishes the ideality of the solution, and it follows that the solubility of a particular gas in ideal solution in any solvent is always the same.

Illustration 8.1 After long contact with a hydrocarbon oil and establishment of equilibrium, a gas mixture has the following composition at 2×10^5 N/m^2 total pressure, 24°C: methane 60%, ethane 20%, propane 8%, n-butane 6%, n-pentane 6%. Calculate the composition of the equilibrium solution.

SOLUTION The equilibrium partial pressure $\bar{p}^*$ of each constituent in the gas is its volume fraction multiplied by the total pressure. These and the vapor pressures p of the constituents at 24°C are tabulated below. The prevailing pressure is above the critical value for methane, and at this low total pressure its solubility may be considered negligible. For each constituent its mole fraction in the liquid is calculated by Eq. (8.1), $\bar{p}^*/p$, and the last column of the table lists these as the answers to the problem. The remaining liquid, $1 - 0.261 = 0.739$ mole fraction, is the solvent oil.

Component	Equilibrium partial pressure $\bar{p}^*$, $N/m^2 \times 10^{-5}$	Vapor pressure p at 24°C, $N/m^2 \times 10^{-5}$	Mole fraction in the liquid $x = \bar{p}^*/p$
Methane	$1.20 = 0.6(2)$		
Ethane	0.4	42.05	0.0095
Propane	0.16	8.96	0.018
n-Butane	0.12	2.36	0.051
n-Pentane	0.12	0.66	0.182
Total			0.261

For total pressures in excess of those for which the ideal-gas law applies, Raoult's law can frequently be used with fugacities substituted for the pressure terms [27, 34].

Nonideal Liquid Solutions

For liquid solutions which are not ideal, Eq. (8.1) will give highly incorrect results. Line D (Fig. 8.1), for example, is the calculated partial pressure of ammonia in equilibrium with water solutions at 10°C, assuming Raoult's law to be applicable, and it clearly does not represent the data. On the other hand, the straight line E is seen to represent the 10°C ammonia-water data very well up to mole fractions of 0.06 in the liquid. The equation of such a line is

$$y^* = \frac{\bar{p}^*}{p_t} = mx \qquad (8.2)$$

where m is a constant. This is *Henry's law*, † and it is seen to be applicable with different values of m for each of the gases in the figure over at least a modest liquid-concentration range. Failure to follow Henry's law over wide concentration ranges may be the result of chemical interaction with the liquid or electrolytic dissociation, as is the case with ammonia-water, or nonideality in the gas phase. Most gases can be expected to follow the law to equilibrium pressures to about 5×10^5 N/m^2 (≈ 5 std atm), although if solubility is low, as with hydrogen-water, it may be obeyed to pressures as high as 34 N/m^2 [35]. Gases of the vapor type (which are below their critical temperature) will generally

---

† Equation (8.2) is frequently written as $y^* = Kx$, but here we wish to distinguish between the Henry-law constant and mass-transfer coefficients. For conditions under which Eq. (8.2) is inapplicable, it is frequently used in empirical fashion to describe experimental data, but the value of m will then be expected to vary with temperature, pressure, and concentration and should be listed as a function of these variables.

follow the law up to pressures of approximately 50 percent of the saturation value at the prevailing temperature provided no chemical action occurs in the liquid. In any case m must be established experimentally.

The advantages of straight-line plotting for interpolating and extrapolating experimental data are, of course, very great, and an empirical method of wide utility is an extension [32] of the reference-substance vapor-pressure plot described in Chap. 7. As an example of this, Fig. 8.2 shows the data for ammonia-water solutions, covering a wide range of concentrations and temperatures. The coordinates are logarithmic. The abscissa is marked with the vapor pressure of a convenient reference substance, in this case water, and the ordinate is the equilibrium partial pressure of the solute gas. Points are plotted where the corresponding temperatures for the vapor pressure of the reference substance and the partial pressure of solute are identical. For example, at 32.2°C the vapor pressure of water is 36 mmHg, the partial pressure of ammonia for a 10 mole percent solution is 130 mmHg, and these pressures locate point A on the figure. The lines for constant liquid composition are straight with few exceptions. A temperature scale can later be substituted for the reference vapor-pressure scale, using the steam tables where water is the reference substance. A similar plot,

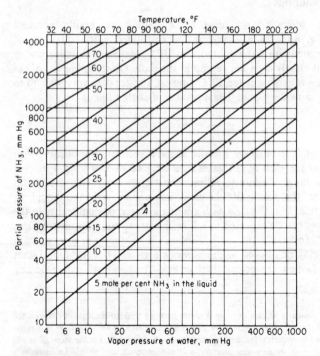

Figure 8.2 Reference-substance plot for gas solubility. The system ammonia-water, water as reference.

with water as reference substance, can also be drawn for the partial pressures of water over these solutions. An important property of such plots is their ability to provide enthalpy data. For substance (solute or solvent) J, let the slope of the line for concentration x_J be m'_{J_r}. Then

$$\overline{H}_J = C_J(t - t_0) + \lambda_{J0} - m'_{J_r}\lambda_{r,t} \tag{8.3}$$

$\overline{H}_J$ = partial enthalpy of component J in solution
C_J = molar heat capacity of gaseous J
λ = appropriate latent heat of vaporization

λ_{J0} is taken as zero if J is a gas at temperature t_0. The enthalpy of the solution relative to pure components at t_0 is then

$$H_L = \sum_J \overline{H}_J x_J \tag{8.4}$$

Choice of Solvent for Absorption

If the principal purpose of the absorption operation is to produce a specific solution, as in the manufacture of hydrochloric acid, for example, the solvent is specified by the nature of the product. If the principal purpose is to remove some constituent from the gas, some choice is frequently possible. Water is, of course, the cheapest and most plentiful solvent, but the following properties are important considerations:

1. *Gas solubility*. The gas solubility should be high, thus increasing the rate of absorption and decreasing the quantity of solvent required. Generally solvents of a chemical nature similar to that of the solute to be absorbed will provide good solubility. Thus hydrocarbon oils, and not water, are used to remove benzene from coke-oven gas. For cases where the solutions formed are ideal, the solubility of the gas is the same in terms of mole fractions for all solvents. But it is greater in terms of weight fractions for solvents of low molecular weight, and smaller weights of such solvents, as measured in pounds, need to be used. A chemical reaction of solvent with the solute will frequently result in very high gas solubility, but if the solvent is to be recovered for reuse, the reaction must be reversible. For example, hydrogen sulfide can be removed from gas mixtures using ethanolamine solutions since the sulfide is readily absorbed at low temperatures and easily stripped at high temperatures. Caustic soda absorbs hydrogen sulfide excellently but will not release it in a stripping operation.
2. *Volatility*. The solvent should have a low vapor pressure since the gas leaving an absorption operation is ordinarily saturated with the solvent and much may thereby be lost. If necessary, a second, less volatile liquid can be used to recover the evaporated portion of the first, as in Fig. 8.3. This is sometimes done, for example, in the case of hydrocarbon absorbers, where a relatively

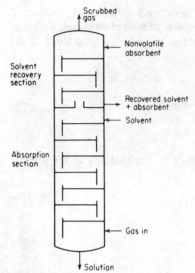

Scrubbed gas

Nonvolatile absorbent

Solvent recovery section

Recovered solvent + absorbent

Solvent

Absorption section

Gas in

Solution

Figure 8.3 Tray absorber with volatile-solvent recovery section.

volatile solvent oil is used in the principal portion of the absorber because of the superior solubility characteristics and the volatilized solvent is recovered from the gas by a nonvolatile oil. Similarly, hydrogen sulfide can be absorbed by a water solution of sodium phenolate, but the desulfurized gas is further washed with water to recover the evaporated phenol.

3. *Corrosiveness.* The materials of construction required for the equipment should not be unusual or expensive.

4. *Cost.* The solvent should be inexpensive, so that losses are not costly, and should be readily available.

5. *Viscosity.* Low viscosity is preferred for reasons of rapid absorption rates, improved flooding characteristics in absorption towers, low pressure drops on pumping, and good heat-transfer characteristics.

6. *Miscellaneous.* The solvent if possible should be nontoxic, nonflammable, and chemically stable and should have a low freezing point.

ONE COMPONENT TRANSFERRED; MATERIAL BALANCES

The basic expressions for material balances and their graphical interpretation were presented for any mass-transfer operation in Chap. 5. Here they are adapted to the problems of gas absorption and stripping.

Countercurrent Flow

Figure 8.4 shows a countercurrent tower which may be either a packed or spray tower, filled with bubble-cap trays or of any internal construction to bring about

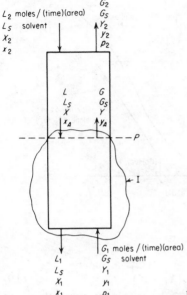

Figure 8.4 Flow quantities for an absorber or stripper.

liquid-gas contact. The gas stream at any point in the tower consists of G total mol/(area of tower cross section) (time), made up of diffusing solute A of mole fraction y, partial pressure $\bar{p}$, or mole ratio Y, and nondiffusing, essentially insoluble gas G_S mol/(area) (time). The relationship between these is

$$Y = \frac{y}{1-y} = \frac{\bar{p}}{p_t - \bar{p}} \tag{8.5}$$

$$G_S = G(1-y) = \frac{G}{1+Y} \tag{8.6}$$

Similarly the liquid stream consists of L total mol/(area) (time), containing x mole fraction soluble gas, or mole ratio X, and essentially nonvolatile solvent L_S mol/(area) (time).

$$X = \frac{x}{1-x} \tag{8.7}$$

$$L_S = L(1-x) = \frac{L}{1+X} \tag{8.8}$$

Since the solvent gas and solvent liquid are essentially unchanged in quantity as they pass through the tower, it is convenient to express the material balance in terms of these. A solute balance about the lower part of the tower

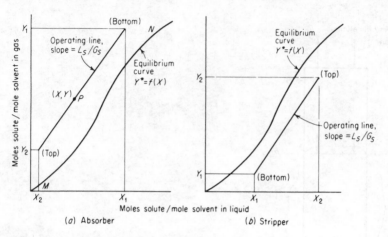

Figure 8.5 Operating lines for absorber and stripper.

(envelope I) is†

$$G_S(Y_1 - Y) = L_S(X_1 - X) \tag{8.9}$$

This is the equation of a straight line (the operating line) on X, Y coordinates, of slope L_S/G_S, which passes through (X_1, Y_1). Substitution of X_2 and Y_2 for X and Y shows the line to pass through (X_2, Y_2), as on Fig. 8.5a for an absorber. This line indicates the relation between the liquid and gas concentration at any level in the tower, as at point P.

The equilibrium-solubility data for the solute gas in the solvent liquid can also be plotted in terms of these concentration units on the same diagram, as curve MN, for example. Each point on this curve represents the gas concentration in equilibrium with the corresponding liquid at its local concentration and temperature. For an absorber (mass transfer from gas to liquid) the operating line always lies above the equilibrium-solubility curve, while for a stripper (mass transfer from liquid to gas) the line is always below, as in Fig. 8.5b.

The operating line is straight only when plotted in terms of the mole-ratio units. In terms of mole fractions or partial pressures the line is curved, as in Fig. 8.6 for an absorber. The equation of the line is then

$$G_S\left(\frac{y_1}{1 - y_1} - \frac{y}{1 - y}\right) = G_S\left(\frac{\bar{p}_1}{p_t - \bar{p}_1} - \frac{\bar{p}}{p_t - \bar{p}}\right) = L_S\left(\frac{x_1}{1 - x_1} - \frac{x}{1 - x}\right)$$

$$\tag{8.10}$$

---

† Equations (8.9) and (8.10) and the corresponding Figs. 8.4 to 8.8 are written as for packed towers, with 1 indicating the streams at the bottom and 2 those at the top. For tray towers, where tray numbers are used as subscripts, as in Fig. 8.12, the same equations apply, but changes in subscripts must be made. Thus Eq. (8.9) becomes, when applied to an entire tray tower,

$$G_S(Y_{N_p+1} - Y_1) = L_S(X_{N_p} - X_0)$$

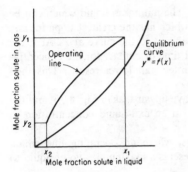

Figure 8.6 Operating line in mole fractions.

The total pressure p_t at any point can ordinarily be considered constant throughout the tower for this purpose.

Minimum Liquid-Gas Ratio for Absorbers

In the design of absorbers, the quantity of gas to be treated G or G_S, the terminal concentrations Y_1 and Y_2, and the composition of the entering liquid X_2 are ordinarily fixed by process requirements, but the quantity of liquid to be used is subject to choice. Refer to Fig. 8.7a. The operating line must pass through point D and must end at the ordinate Y_1. If such a quantity of liquid is used to give operating line DE, the exit liquid will have the composition X_1. If less liquid is used, the exit-liquid composition will clearly be greater, as at point F, but since the driving forces for diffusion are less, the absorption is more difficult. The time of contact between gas and liquid must then be greater, and

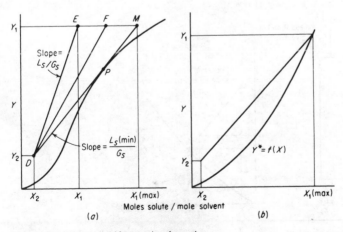

Figure 8.7 Minimum liquid-gas ratio, absorption.

the absorber must be correspondingly taller. The minimum liquid which can be used corresponds to the operating line DM, which has the greatest slope for any line touching the equilibrium curve and is tangent to the curve at P. At P the diffusional driving force is zero, the required time of contact for the concentration change desired is infinite, and an infinitely tall tower results. This then represents the limiting liquid-gas ratio.

The equilibrium curve is frequently concave upward, as in Fig. 8.7b, and the minimum liquid-gas ratio then corresponds to an exit-liquid concentration in equilibrium with the entering gas.

These principles also apply to strippers, where an operating line which anywhere touches the equilibrium curve represents a maximum ratio of liquid to gas and a maximum exit-gas concentration.

Cocurrent Flow

When gas and liquid flow cocurrently, as in Fig. 8.8, the operating line has a negative slope $-L_S/G_S$. There is no limit on this ratio, but an infinitely tall tower would produce an exit liquid and gas in equilibrium, as at (X_e, Y_e). Cocurrent flow may be used when an exceptionally tall tower is built in two sections, as in Fig. 8.9, with the second section operated in cocurrent flow to save on the large-diameter gas pipe connecting the two. It may also be used if the gas to be dissolved in the liquid is a pure substance, where there is no advantage to countercurrent operation, or if a rapid, irreversible chemical reaction with the dissolved solute occurs in the liquid, where only the equivalent of one theoretical stage is required.

Illustration 8.2 A coal gas is to be freed of its light oil by scrubbing with wash oil as an absorbent and the light oil recovered by stripping the resulting solution with steam. The circumstances are as follows.

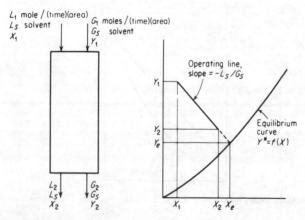

Figure 8.8 Cocurrent absorber.

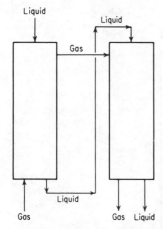

Figure 8.9 Countercurrent-cocurrent arrangement for very tall towers.

Absorber Gas in, 0.250 m³/s (31 800 ft³/h) at 26°C, $p_t = 1.07 \times 10^5$ N/m² (803 mmHg), containing 2.0% by volume of light oil vapors. The light oil will be assumed to be entirely benzene, and a 95% removal is required. The wash oil is to enter at 26°C, containing 0.005 mole fraction benzene, and has an av mol wt 260. An oil circulation rate of 1.5 times the minimum is to be used. Wash oil–benzene solutions are ideal. The temperature will be constant at 26°C.

Stripper The solution from the absorber is to be heated to 120°C and will enter the stripper at 1 std atm pressure. Stripping steam will be at standard atmospheric pressure, superheated to 122°C. The debenzolized oil, 0.005 mole fraction benzene, is to be cooled to 26°C and returned to the absorber. A steam rate of 1.5 times the minimum is to be used. The temperature will be constant at 122°C.

Compute the oil-circulation rate and the steam rate required.

SOLUTION

Absorber Basis: 1 s. Define L, L_S, G, and G_S in terms of kmol/s.

$$G_1 = 0.250 \frac{273}{273 + 26} \frac{1.07 \times 10^5}{1.0133 \times 10^5} \frac{1}{22.41} = 0.01075 \text{ kmol/s}$$

$$y_1 = 0.02 \qquad Y_1 = \frac{0.02}{1 - 0.02} = 0.0204 \text{ kmol benzene/kmol dry gas}$$

$$G_S = 0.01075 (1 - 0.02) = 0.01051 \text{ kmol dry gas/s}$$

For 95% removal of benzene,

$$Y_2 = 0.05 (0.0204) = 0.00102 \text{ kmol benzene/kmol dry gas}$$

$$x_2 = 0.005 \qquad X_2 = \frac{0.005}{1 - 0.005} = 0.00503 \text{ kmol benzene/kmol oil}$$

At 26°C, the vapor pressure of benzene is $p = 100$ mmHg $= 13\,330$ N/m². Equation (8.1) for ideal solutions: $\bar{p}^* = 13\,330x$.

$$y^* = \bar{p}^*/p_t \qquad p_t = 1.07 \times 10^5 \text{ N/m}^2 \qquad Y^* = \frac{y^*}{1 - y^*} \qquad X = \frac{x}{1 - x}$$

Substitution in Eq. (8.1) yields

$$\frac{Y^*}{1 + Y^*} = 0.125 \frac{X^*}{1 + X}$$

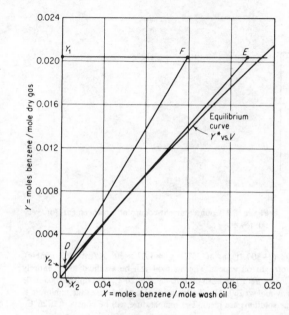

Figure 8.10 Solution to Illustration 8.2, absorption.

which is the equilibrium-curve equation for the absorber, plotted in Fig. 8.10. Operating lines
originate at point D in this figure. For the minimum oil rate, line DE is drawn as the line of
maximum slope which touches the equilibrium curve (tangent to the curve). At $Y_1 = 0.0204$,
$X_1 = 0.176$ kmol benzene/kmol wash oil (point E).

$$\min L_S = \frac{G_S(Y_1 - Y_2)}{X_1 - X_2} = \frac{0.01051\,(0.0204 - 0.00102)}{0.176 - 0.00503} = 1.190 \times 10^{-3}\,\text{kmol oil/s}$$

For 1.5 times the minimum, $L_S = 1.5\,(1.190 \times 10^{-3}) = 1.787 \times 10^{-3}$ kmol/s.

$$X_1 = \frac{G_S(Y_1 - Y_2)}{L_S} + X_2 = \frac{0.01051\,(0.0204 - 0.00102)}{1.787 \times 10^{-3}} + 0.00503$$

$$= 0.1190\,\text{kmol benzene/kmol oil}$$

The operating line is DF.

Stripper At 122°C, the vapor pressure of benzene is 2400 mmHg = 319.9 kN/m². The
equilibrium curve for the stripper is therefore

$$\frac{Y^*}{1 + Y^*} = \frac{319.9}{101.33}\frac{X}{1 + X} = \frac{3.16X}{1 + X}$$

which is drawn in Fig. 8.11. For the stripper, $X_2 = 0.1190$, $X_1 = 0.00503$ kmol benzene/kmol
oil, $Y_1 = 0$ kmol benzene/kmol steam. For the minimum steam rate, line MN is drawn tangent
to the equilibrium curve, and at N the value of $Y_2 = 0.45$ kmol benzene/kmol steam.

$$\min G_S = \frac{L_S(X_2 - X_1)}{Y_2 - Y_1} = \frac{(1.787 \times 10^{-3})(0.1190 - 0.00503)}{0.45 - 0}$$

$$= 4.526 \times 10^{-4}\,\text{kmol steam/s}$$

For 1.5 times the minimum, the steam rate is 1.5 $(4.526 \times 10^{-4}) = 6.79 \times 10^{-4}$ kmol/s,
corresponding to line MP.

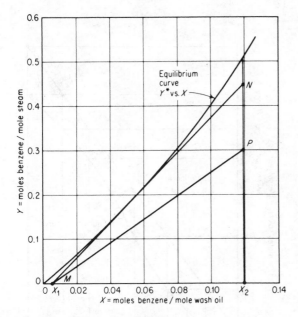

Figure 8.11 Solution to Illustration 8.2, stripping.

COUNTERCURRENT MULTISTAGE OPERATION;
ONE COMPONENT TRANSFERRED

Tray towers and similar devices bring about stepwise contact of the liquid and gas and are therefore countercurrent multistage cascades. On each tray of a sieve-tray tower, for example, the gas and liquid are brought into intimate contact and separated, somewhat in the manner of Fig. 5.14, and the tray thus constitutes a stage. Few of the tray devices described in Chap. 6 actually provide the parallel flow on each tray as shown in Fig. 5.14. Nevertheless it is convenient to use the latter as an arbitrary standard for design and for measurement of performance of actual trays regardless of their method of operation. For this purpose a *theoretical*, or *ideal*, tray is defined as one where the average composition of all the gas leaving the tray is in equilibrium with the average composition of all the liquid leaving the tray.

The number of ideal trays required to bring about a given change in composition of the liquid or the gas, for either absorbers or strippers, can then be determined graphically in the manner of Fig. 5.15. This is illustrated for an absorber in Fig. 8.12, where the liquid and gas compositions corresponding to each tray are marked on the operating diagram. Ideal tray 1, for example, brings about a change in liquid composition from X_0 to X_1 and of gas composition

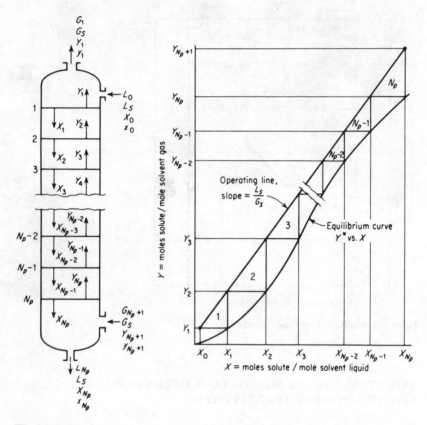

Figure 8.12 Tray absorber.

from Y_2 to Y_1. The step marked 1 on the operating diagram therefore represents this ideal tray. The nearer the operating line to the equilibrium curve, the more steps will be required, and should the two curves touch at any point corresponding to a minimum L_S/G_S ratio, the number of steps will be infinite. The steps can equally be constructed on diagrams plotted in terms of any concentration units, such as mole fractions or partial pressures. The construction for strippers is the same, except, of course, that the operating line lies below the equilibrium curve.

It is usually convenient, for tray towers, to define the flow rates L and G simply as mol/h, rather than to base them on unit tower cross section.

Dilute Gas Mixtures

Where both operating line and equilibrium curve can be considered straight, the number of ideal trays can be determined without recourse to graphical methods.

This will frequently be the case for relatively dilute gas and liquid mixtures. Henry's law [Eq. (8.2)] often applies to dilute solutions, for example. If the quantity of gas absorbed is small, the total flow of liquid entering and leaving the absorber remains substantially constant, $L_0 \approx L_{N_p} \approx L$ total mol/(area) (time), and similarly the total flow of gas is substantially constant at G total mol/(area) (time). An operating line plotted in terms of mole fractions will then be substantially straight. For such cases, the Kremser equations (5.50) to (5.57) and Fig. 5.16 apply.† Small variations in A from one end of the tower to the other due to changing L/G as a result of absorption or stripping or to change in gas solubility with concentration or temperature can be roughly allowed for by using the geometric average of the values af A at top and bottom [17]. For large variations, either more elaborate corrections [13, 20, 33] for A, graphical computations, or tray-to-tray numerical calculations as developed below must be used.

The Absorption Factor A

The absorption factor $A = L/mG$ is the ratio of the slope of the operating line to that of the equilibrium curve. For values of A less than unity, corresponding to convergence of the operating line and equilibrium curve for the lower end of the absorber, Fig. 5.16 indicates clearly that the fractional absorption of solute is definitely limited, even for infinite theoretical trays. On the other hand, for values of A greater than unity, any degree of absorption is possible if sufficient trays are provided. For a fixed degree of absorption from a fixed amount of gas, as A increases beyond unity, the absorbed solute is dissolved in more and more liquid and becomes therefore less valuable. At the same time, the number of trays decreases, so that the equipment cost decreases. From these opposing cost tendencies it follows that in all such cases there will be a value of A, or of L/G, for which the most economical absorption results. This should be obtained generally by computing the total costs for several values of A and observing the minimum. As a rule of thumb for purposes of rapid estimates, it has been

† These become, in terms of mole fractions,

Absorption:
$$\frac{y_{N_p+1} - y_1}{y_{N_p+1} - mx_0} = \frac{A^{N_p+1} - A}{A^{N_p+1} - 1} \tag{5.54a}$$

$$N_p = \frac{\log\left[\frac{y_{N_p+1} - mx_0}{y_1 - mx_0}\left(1 - \frac{1}{A}\right) + \frac{1}{A}\right]}{\log A} \tag{5.55a}$$

Stripping:
$$\frac{x_0 - x_{N_p}}{x_0 - y_{N_p+1}/m} = \frac{S^{N_p+1} - S}{S^{N_p+1} - 1} \tag{5.50a}$$

$$N_p = \frac{\log\left[\frac{x_0 - y_{N_p+1}/m}{x_{N_p} - y_{N_p+1}/m}\left(1 - \frac{1}{S}\right) + \frac{1}{S}\right]}{\log S} \tag{5.51a}$$

where $A = L/mG$, and $S = mG/L$.

frequently found [8] that the most economical A will be in the range from 1.25 to 2.0.

The reciprocal of the absorption factor is called the stripping factor S.

Illustration 8.3 Determine the number of theoretical trays required for the absorber and the stripper of Illustration 8.2.

SOLUTION

Absorber Since the tower will be a tray device, the following notation changes will be made (compare Figs. 8.4 and 8.12):

L_1 is changed to L_{N_p}	G_1 is changed to G_{N_p+1}	x_1 is changed to x_{N_p}
L_2 L_0	G_2 G_1	x_2 x_0
X_1 X_{N_p}	Y_1 Y_{N_p+1}	y_1 y_{N_p+1}
X_2 X_0	Y_2 Y_1	y_2 y_1

The operating diagram is established in Illustration 8.2 and replotted in Fig. 8.13, where the theoretical trays are stepped off. Between 7 and 8 (approximately 7.6) theoretical trays are

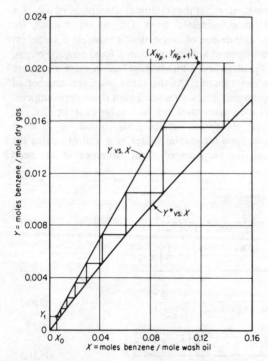

Figure 8.13 Illustration 8.3, the absorber.

required. Alternatively, the number of theoretical trays can be computed analytically.

$$y_{N_p+1} = 0.02 \qquad y_1 = \frac{0.00102}{1 + 0.00102} = 0.00102$$

$$x_0 = 0.005 \qquad m = y^*/x = 0.125$$

$$L_{N_p} = L_S(1 + X_{N_p}) = (1.787 \times 10^{-3})(1 + 0.1190) = 2.00 \times 10^{-3} \text{ kmol/s}$$

$$A_{N_p} = \frac{L_{N_p}}{mG_{N_p}} \approx \frac{L_{N_p}}{mG_{N_p+1}} = \frac{2.00 \times 10^{-3}}{0.125\,(0.01075)} = 1.488$$

$$L_0 = L_S(1 + X_0) = (1.787 \times 10^{-3})\,(1 + 0.00503) = 1.796 \times 10^{-3} \text{ kmol/s}$$

$$G_1 = G_S(1 + Y_1) = 0.01051\,(1 + 0.00102) = 0.01052 \text{ kmol/s}$$

$$A_1 = \frac{L_1}{mG_1} \approx \frac{L_0}{mG_1} = \frac{1.796 \times 10^{-3}}{0.125\,(0.01052)} = 1.366$$

$$A = [1.488\,(1.366)]^{0.5} = 1.424$$

$$\frac{y_1 - mx_0}{y_{N_p+1} - mx_0} = \frac{0.00102 - 0.125\,(0.005)}{0.02 - 0.125\,(0.005)} = 0.0204$$

From Fig. 5.16 or Eq. (5.55) $N_p = 7.7$ equilibrium stages.

Stripper The trays were determined graphically in the same manner as for the absorber and found to be 6.7. Figure 5.16 for this case gave 6.0 trays owing to the relative nonconstancy of the stripping factor, $1/A_{N_p} = S_{N_p} = 1.197$, $1/A_1 = S_1 = 1.561$. The graphical method should be used.

Nonisothermal Operation

Many absorbers and strippers deal with dilute gas mixtures and liquids, and it is frequently satisfactory in these cases to assume that the operation is isothermal. But actually absorption operations are usually exothermic, and when large quantities of solute gas are absorbed to form concentrated solutions, the temperature effects cannot be ignored. If by absorption the temperature of the liquid is raised to a considerable extent, the equilibrium solubility of the solute will be appreciably reduced and the capacity of the absorber decreased (or else much larger flow rates of liquid will be required). If the heat evolved is excessive, cooling coils can be installed in the absorber or the liquid can be removed at intervals, cooled, and returned to the absorber. For stripping, an endothermic action, the temperature tends to fall.

Consider the tray tower of Fig. 8.14. If Q_T is the heat removed per unit time from the entire tower by any means whatsoever, an enthalpy balance for the entire tower is

$$L_0 H_{L0} + G_{N_p+1} H_{G,N_p+1} = L_{N_p} H_{L,N_p} + G_1 H_{G1} + Q_T \qquad (8.11)$$

where H represents in each case the molal enthalpy of the stream at its particular concentration and condition. It is convenient to refer all enthalpies to the condition of pure liquid solvent, pure diluent (or solvent) gas, and pure solute at some base temperature t_0, with each substance assigned zero enthalpy

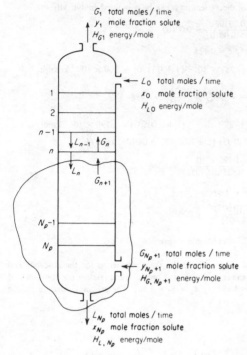

G_1 total moles / time
y_1 mole fraction solute
H_{G1} energy/mole

L_0 total moles / time
x_0 mole fraction solute
H_{L0} energy/mole

1

2

$n-1$

n L_{n-1} G_n

L_n

G_{n+1}

N_p-1

N_p

G_{N_p+1} total moles / time
y_{N_p+1} mole fraction solute
H_{G, N_p+1} energy/mole

L_{N_p} total moles / time
x_{N_p} mole fraction solute
H_{L, N_p} energy/mole

Figure 8.14 Nonisothermal operation.

for its normal state of aggregation at t_0 and 1 atm pressure. Thus, the molal enthalpy of a liquid solution, temperature t_L, composition x mole fraction solute, can be obtained either from Eq. (8.4) or, depending on the nature of the data available, from

$$H_L = C_L(t_L - t_0) + \Delta H_S \qquad (8.12)$$

where the first term on the right represents the sensible heat and the second the molal enthalpy of mixing, or integral heat of solution, at the prevailing concentration and at the base temperature t_0, *per mole of solution*. If heat is evolved on mixing, ΔH_S will be a negative quantity. If the absorbed solute is a gas at t_0, 1 std atm, the gas enthalpy will include only sensible heat. If the absorbed solute is a liquid at the reference conditions, as in the case of many vapors, the enthalpy of the gas stream must also include the latent heat of vaporization of the solute vapor (see Chap. 7). For ideal solutions, ΔH_S for mixing liquids is zero, and the enthalpy of the solution is the sum of the enthalpies of the separate, unmixed constituents. If the ideal liquid solution is formed from a gaseous solute, the heat evolved is the latent heat of condensation of the absorbed solute.

For *adiabatic* operation, Q_T of Eq. (8.11) is zero and the temperature of the streams leaving an absorber will generally be higher than the entering temperatures owing to the heat of solution. The rise in temperature causes a decrease in solute solubility, which in turn results in a larger minimum L/G and a larger number of trays than for isothermal absorption. The design of such absorbers may be done numerically, calculating tray by tray from the bottom to the top. The principle of an ideal tray, that the effluent streams from the tray are in equilibrium both with respect to composition and temperature, is utilized for each tray. Thus, total and solute balances up to tray n, as shown by the envelope, Fig. 8.14, are

$$L_n + G_{N_p+1} = L_{N_p} + G_{n+1} \tag{8.13}$$

$$L_n x_n + G_{N_p+1} y_{N_p+1} = L_{N_p} x_{N_p} + G_{n+1} y_{n+1} \tag{8.14}$$

from which L_n and x_n are computed. An enthalpy balance is

$$L_n H_{L,n} + G_{N_p+1} H_{G,N_p+1} = L_{N_p} H_{L,N_p} + G_{n+1} H_{G,n+1} \tag{8.15}$$

from which the temperature of stream L_n can be obtained. Stream G_n is then at the same temperature as L_n and in composition equilibrium with it. Equations (8.13) to (8.15) are then applied to tray $n - 1$, and so forth. To get started, since usually only the temperatures of the entering streams L_0 and G_{N_p+1} are known, it is usually necessary to estimate the temperature t_1 of the gas G_1 (which is the same as the top tray temperature), and use Eq. (8.11) to compute the temperature of the liquid leaving at the bottom of the tower. The estimate is checked when the calculations reach the top tray, and if necessary the entire computation is repeated. The method is best illustrated by an example.

Illustration 8.4 One kilomole per unit time of a gas consisting of 75% methane, CH_4, and 25% n-pentane vapor, n-C_5H_{12}, 27°C, 1 std atm, is to be scrubbed with 2 kmol/unit time of a nonvolatile paraffin oil, mol wt 200, heat capacity 1.884 kJ/kg · K, entering the absorber free of pentane at 35°C. Compute the number of ideal trays for adiabatic absorption of 98% of the pentane. Neglect solubility of CH_4 in the oil, and assume operation to be at 1 std atm (neglect pressure drop for flow through the trays). The pentane forms ideal solutions with the paraffin oil.

SOLUTION Heat capacities over the range of temperatures to be encountered are

$$CH_4 = 35.59 \text{ kJ/kmol} \cdot K \qquad n\text{-}C_5H_{12} = \begin{cases} 119.75 \text{ kJ/kmol} \cdot K & \text{as vapor} \\ 177.53 \text{ kJ/kmol} \cdot K & \text{as liquid} \end{cases}$$

The latent heat of vaporization of n-C_5H_{12} at 0°C is 27 820 kJ/kmol. Use a base temperature $t_0 = 0$°C. Enthalpies referred to 0°C, liquid pentane, liquid paraffin oil, and gaseous methane, are then

$$H_L = (1 - x)(1.884)(200)(t_L - 0) + x(177.53)(t_L - 0)$$
$$= t_L(376.8 - 199.3x) \text{ kJ/kmol liquid solution}$$
$$H_G = (1 - y)(35.59)(t_G - 0) + y[119.75(t_G - 0) + 27 820]$$
$$= t_G(35.59 + 84.16y) + 27 820y \text{ kJ/kmol gas-vapor mixture}$$

For solutions which follow Raoult's law, Eqs. (8.1) and (8.2) provide

$$y^* = \frac{px}{p_t} = mx$$

where p = vapor pressure of n-pentane. Thus the vapor pressure of n-C_5H_{12} is 400 mmHg (53.32 kN/m^2) at 18.5°C, whence $m = p/p_t = 400/760 = 0.530$. Similarly,

t, °C	20	25	30	35	40	43
m	0.575	0.69	0.81	0.95	1.10	1.25

It is convenient to prepare a graph of these for interpolation.

Basis: unit time. From the given data,

$$G_{N_p+1} = 1 \text{ kmol} \qquad y_{N_p+1} = 0.25$$

$$H_{G, N_p+1} = 27[35.59 + 84.16\,(0.25)] + 27\,820\,(0.25) = 8484 \text{ kJ/kmol}$$

$$L_0 = 2 \text{ kmol} \qquad x_0 = 0 \qquad H_{L0} = 35\,(376.8) = 13\,188 \text{ kJ/kmol}$$

$$n\text{-}C_5H_{12} \text{ absorbed} = 0.98\,(0.25) = 0.245 \text{ kmol}$$

$$n\text{-}C_5H_{12} \text{ in } G_1 = 0.25 - 0.245 = 0.005 \text{ kmol}$$

$$G_1 = 0.75 + 0.005 = 0.755 \text{ kmol} \qquad \text{Required } y_1 = \frac{0.005}{0.755} = 0.00662$$

$$L_{N_p} = 2 + 0.245 = 2.245 \text{ kmol} \qquad x_{N_p} = \frac{0.245}{2.245} = 0.1091$$

Assume $t_1 = 35.6$°C (to be checked later).

$$H_{G1} = 35.6[35.59 + 84.16\,(0.00662)] + 27\,820\,(0.00662) = 1471 \text{ kJ/kmol}$$

Eq. (8.11) with $Q_T = 0$:

$$2(13\,188) + 8484 = L_{N_p}H_{L, N_p} + 0.755\,(1471)$$

$$L_{N_p}H_{L, N_p} = 37\,749 \text{ kJ}$$

$$H_{L, N_p} = \frac{37\,749}{2.245} = t_{N_p}[376.8 - 199.3\,(0.1091)]$$

$$t_{N_p} = 42.3\text{°C}$$

$$m_{N_p} = 1.21 \qquad y_{N_p} = m_{N_p}x_{N_p} = 1.21\,(0.1091) = 0.1330$$

$$G_{N_p} = \frac{0.75}{1 - 0.1330} = 0.865 \text{ kmol}$$

$$H_{G, Np} = 42.3[35.59 + 84.16\,(0.1330)] + 27\,820\,(0.1330) = 5679 \text{ kJ/kmol}$$

Eq. (8.13) with $n = N_p - 1$:

$$L_{N_p-1} + G_{N_p+1} = L_{N_p} + G_{N_p}$$

$$L_{N_p-1} + 1 = 2.245 + 0.865$$

$$L_{N_p-1} = 2.110 \text{ kmol}$$

Eq. (8.14) with $n = N_p - 1$:

$$2.110x_{N_p-1} + 0.25 = 0.245 + 0.865\,(0.1330)$$

$$x_{N_p-1} = 0.0521$$

Eq. (8.15) with $n = N_p - 1$:

$$2.110 H_{L, N_p - 1} + 8484 = 3374 + 0.865 (5679)$$

$$H_{L, N_p - 1} = 14\ 302 \text{ kJ/kmol}$$

$$14\ 302 = t_{N_p - 1} [376.8 - 199.3 (0.0521)]$$

$$t_{N_p - 1} = 39.0°C$$

The computations are continued upward through the tower in this manner until the gas composition falls at least to $y = 0.00662$. The results are:

n = tray no.	t_n, °C	x_n	y_n
$N_p = 4$	42.3	0.1091	0.1320
$N_p - 1 = 3$	39.0	0.0521	0.0568
$N_p - 2 = 2$	36.8	0.0184	0.01875
$N_p - 3 = 1$	35.5	0.00463	0.00450

Figure 8.15 shows the calculated gas composition and tray temperature plotted against tray number. The required $y_1 = 0.00662$ occurs at about 3.75 ideal trays, and the temperature on the top tray is essentially that assumed. Had this not been so, a new assumed value of t_1 and a new

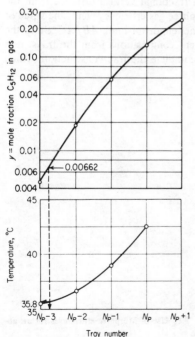

Figure 8.15 Solution to Illustration 8.4.

calculation would have been required. An integral number of trays will require a slightly greater (for $N_p = 3$) or less (for $N_p = 4$) liquid flow, but since a tray efficiency must still be applied to obtain the number of real trays, the nonintegral number is ordinarily accepted.

A graphical solution using an enthalpy-concentration diagram (see Chap. 9) is also possible [43]. When the temperature rise for the liquid is large or for concentrated gases entering and small values of L/G, the rate of convergence of the calculated and assumed t_1 is very slow and the computation is best done on a digital computer. There may be a temperature maximum at some tray other than the bottom one. If the number of trays is fixed, the outlet-gas composition and top-tray temperature must both be found by trial and error. Cases when the carrier gas is absorbed and the solvent evaporates [3, 4] are considered to be multicomponent systems (which see).

Real Trays and Tray Efficiency

Methods for estimating the Murphree tray efficiency corrected for entrainment E_{MGE} for sieve trays are discussed in detail in Chap. 6. For a given absorber or stripper, these permit estimation of the tray efficiency as a function of fluid compositions and temperature as they vary from one end of the tower to the other. Usually it is sufficient to make such computations at only three or four locations and then proceed as in Fig. 8.16. The broken line is drawn between equilibrium curve and operating line at a fractional vertical distance from the operating line equal to the prevailing Murphree gas efficiency. Thus the value of E_{MGE} for the bottom tray is the ratio of the lengths of lines, AB/AC. Since the broken line then represents the real effluent compositions from the trays, it is used instead of the equilibrium curve to complete the tray construction, which now provides the number of real trays.

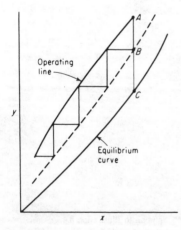

Figure 8.16 Use of Murphree efficiencies for an absorber.

When the Murphree efficiency is constant for all trays, and under conditions such that the operating line and equilibrium curves are straight (Henry's law, isothermal operation, dilute solutions), the overall tray efficiency can be computed and the number of real trays can be determined analytically:

$$E_O = \frac{\text{equilibrium trays}}{\text{real trays}} = \frac{\log\left[1 + E_{MGE}(1/A - 1)\right]}{\log(1/A)} \qquad (8.16)$$

For rough estimates, Fig. 6.24 is useful for both bubble-cap and sieve trays.

Illustration 8.5 A process for making small amounts of hydrogen by cracking ammonia is being considered, and residual, uncracked ammonia is to be removed from the resulting gas. The gas will consist of H_2 and N_2 in the molar ratio 3:1, containing 3% NH_3 by volume, at a pressure of 2 bars and temperature of 30°C.

There is available a sieve-tray tower, 0.75 m diameter, containing 14 cross-flow trays at 0.5 m tray spacing. On each tray, the downflow weir is 0.53 m long and extends 60 mm above the tray floor. The perforations are 4.75 mm in diameter, arranged in triangular pitch on 12.5-mm centers, punched in sheet metal 2.0 mm thick. Assume isothermal operation at 30°C. Estimate the capacity of the tower to remove ammonia from the gas by scrubbing with water.

SOLUTION In this case "capacity" will be taken to mean obtaining a low NH_3 content of the effluent gas at reasonable gas rates.

From the geometry of the tray arrangement, and in the notation of Chap. 6, the following are readily calculated:

T = tower diameter = 0.75 m A_t = tower cross section = 0.4418 m^2

A_d = downspout cross section = 0.04043 m^2 $A_n = A_t - A_d = 0.4014$ m^2

A_o = perforation area = 0.0393 m^2 Z = distance between downspouts = 0.5307 m

W = weir length = 0.53 m h_W = weir height = 0.060 m

t = tray spacing = 0.5 m z = average flow width = $\dfrac{T + W}{2}$ = 64 m

Use a weir crest h_1 = 0.040 m. Eq. (6.34): W_{eff} = 0.4354 m. Eq. (6.32): q = 1.839 (0.4354) $(0.04)^{3/2}$ = 6.406 × 10^{-3} m^3/s liquid flow. *Note*: this is a recommended rate because it produces a liquid depth on the tray of 10 cm.

Liquid density = ρ_L = 996 kg/m^3

Liquid rate = 6.38 kg/s = 0.354 kmol/s

Av mol wt gas in = 0.03 (17.03) + 0.97 (0.25) (28.02) + 0.97 (0.75) (2.02)

= 8.78 kg/kmol

$$\rho_G = \frac{8.78}{22.41}\frac{2.0}{0.986}\frac{273}{273 + 30} = 0.716 \text{ kg/m}^3$$

The flooding gas rate depends upon the liquid rate chosen. With the preceding values, the flooding gas rate is found by simultaneous solution of Eqs. (6.29) and (6.30) by trial and error. For these solutions, $\sigma = 68 \times 10^{-3}$ N/m surface tension. As a final trial, use a gas mass rate = 0.893 kg/s.

$$\frac{L'}{G'}\left(\frac{\rho_G}{\rho_L}\right)^{0.5} = \frac{6.38}{0.893}\left(\frac{0.716}{996}\right)^{0.5} = 0.1916$$

Table 6.2: α = 0.04893, β = 0.0302. Eq. (6.30): C_F = 0.0834. Eq. (6.29): V_F = 3.11 m/s based on A_n. Therefore flooding gas mass rate = 3.11 (0.4014) (0.716) = 0.893 kg/s (check). Use 80%

of the flooding value. $V = 0.8 (3.11) = 2.49$ m/s based on A_n. Gas mass rate $= 0.80 (0.893) = 0.7144$ kg/s $= 0.0814$ kmol/s.

The methods of Chap. 6 provide (Chap. 6 notation): $V_o = 2.49 (0.4014)/0.0393 = 25.43$ m/s, $l = 0.002$ m, $d_o = 0.00475$ m, $C_o = 1.09 (0.00475/0.002)^{0.25} = 1.353$, $l = 0.002$ m, $\mu_G = 1.13$ kg/m · s, $Re_o = d_o V_o \rho_G / \mu_G = 7654$, $f = 0.0082$, $h_D = 0.0413$ m, $V_a = VA_n/A_a = 3.33$ m/s, $h_L = 0.0216$ m, $h_R = 0.0862$ m, $h_G = 0.1491$ m, $h_2 = 3.84 \times 10^{-3}$ m, $h_3 = 0.1529$ m.

Since $h_W + h_1 + h_3 = 0.2529$ m is essentially equal to $t/2$, flooding will not occur. Flow rates are reasonable. $V/V_F = 0.8$, and $(L'/G') (\rho_G/\rho_L)^{0.5} = 0.239$. Fig. 6.17: $E =$ fractional entrainment $= 0.009$.

At the prevailing conditions, the methods of Chap. 2 provide $D_{NH_3(mean)}$ through the N_2-H_2 mixture $= D_G = 2.296 \times 10^{-5}$ m²/s. The gas viscosity is estimated to be 1.122×10^{-5} kg/m · s, whence $Sc_G = \mu_G/\rho_G D_G = 0.683$. The diffusivity of NH_3 in dilute water solutions is $2.421 \times 10^{-9} = D_L$ at 30°C.

For dilute solutions, NH_3-H_2O follows Henry's law, and at 30°C $m = y^*/x = 0.850$. The absorption factor $A = L/mG = 0.354/0.850 (0.0814) = 5.116$. *Note:* "Optimum" values of A such as 1.25 to 2.0 cited earlier apply to new designs but not necessarily to existing towers.

Eq. (6.61): $N_{tG} = 1.987$	Eq. (6.64): $\theta_L = 1.145$ s	
Eq. (6.62): $N_{tL} = 1.691$	Eq. (6.52): $N_{tOG} = 1.616$	
Eq. (6.51): $E_{OG} = 0.801$	Eq. (6.63): $D_E = 0.01344$ m²/s	
Eq. (6.59): Pe $= 18.30$	Eq. (6.58): $\eta = 0.1552$	
Eq. (6.57): $E_{MG} = 0.86$	Eq. (6.60): $E_{MGE} = 0.853$	

Since the quantity of gas absorbed is relatively small with respect to the total flow and the liquid solutions are dilute, the Murphree efficiency will be taken as constant for all trays. For the dilute solutions encountered here, the operating line in terms of mole-fraction concentrations is essentially straight. Consequently, Eq. (8.16):

$$E_O = \frac{\log[1 + 0.853 (1/5.116 - 1)]}{\log(1/5.116)} = 0.710$$

(*Note:* Fig. 6.24, with abscissa $= 1.306 \times 10^{-5}$ for these data, shows $E_O \approx 0.7$ for absorption of NH_3 in water, the point plotted as +.) Eq. (8.16): $N_p = 14 (0.710) = 9.94$ theoretical trays. Fig. 5.16 is off-scale. Therefore, the Kremser equation Eq. (5.54a), gives

$$\frac{y_{N_p+1} - y_1}{y_{N_p+1} - mx_0} = \frac{y_{N_p+1} - y_1}{y_{N_p+1}} = \frac{0.03 - y_1}{0.03} = \frac{(5.116)^{10.94} - 5.116}{(5.116)^{10.94} - 1}$$

$$y_1 = 2.17 \times 10^{-9} \text{ mole fraction } NH_3 \text{ in effluent } \textbf{Ans.}$$

CONTINUOUS-CONTACT EQUIPMENT

Countercurrent packed and spray towers operate in a different manner from plate towers in that the fluids are in contact continuously in their path through the tower, rather than intermittently. Thus, in a packed tower the liquid and gas compositions change continuously with height of packing. Every point on an operating line therefore represents conditions found somewhere in the tower, whereas for tray towers, only the isolated points on the operating line corresponding to trays have real significance.

Height Equivalent to an Equilibrium Stage (Theoretical Plate)

A simple method for designing packed towers, introduced many years ago, ignores the differences between stagewise and continuous contact. In this method the number of theoretical trays or plates required for a given change in concentration is computed by the methods of the previous section. This is then multiplied by a quantity, the height equivalent to a theoretical tray or plate (**HETP**) to give the required height of packing to do the same job. The **HETP** must be an experimentally determined quantity characteristic for each packing. Unfortunately it is found that the **HETP** varies, not only with the type and size of the packing but also very strongly with flow rates of each fluid and for every system with concentration as well, so that an enormous amount of experimental data would have to be accumulated to permit utilization of the method. The difficulty lies in the failure to account for the fundamentally different action of tray and packed towers, and the method has now largely been abandoned.

Absorption of One Component

Consider a packed tower of unit cross section, as in Fig. 8.17. The total effective interfacial surface for mass transfer, as a result of spreading of the liquid in a film over the packing, is S per unit tower cross section. This is conveniently described as the product of a specific interfacial surface, surface per volume of

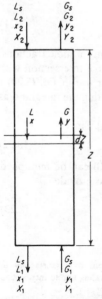

Figure 8.17 Packed tower.

packing, by the packed volume Z volume per unit tower cross section, or height (the quantity a is the a_A of Chap. 6). In the differential volume dZ, the interface surface is

$$dS = a \, dZ \tag{8.17}$$

The quantity of solute A in the gas passing the differential section of the tower under consideration is Gy mol/(area) (time), and the rate of mass transfer is therefore $d(Gy)$ mol A/(differential volume) (time). Since $N_B = 0$ and $N_A/(N_A + N_B) = 1.0$, application of Eq. (5.20) provides

$$N_A = \frac{\text{moles A absorbed}}{(\text{interfacial surface}) (\text{time})} = \frac{d(Gy)}{a \, dZ} = F_G \ln \frac{1 - y_i}{1 - y} \tag{8.18}$$

Both G and y vary from one end of the tower to the other, but G_S, the solvent gas which is essentially insoluble, does not. Therefore,

$$d(Gy) = d\left(\frac{G_S y}{1 - y} \right) = \frac{G_S \, dy}{(1 - y)^2} = \frac{G \, dy}{1 - y} \tag{8.19}$$

Substituting in Eq. (8.18), rearranging, and integrating give

$$Z = \int_0^Z dZ = \int_{y_2}^{y_1} \frac{G \, dy}{F_G a (1 - y) \ln[(1 - y_i)/(1 - y)]} \tag{8.20}$$

The value of y_i can be found by the methods of Chap. 5, using Eq. (5.21) with $N_A/\Sigma N = 1$:

$$\frac{1 - y_i}{1 - y} = \left(\frac{1 - x}{1 - x_i} \right)^{F_L/F_G} = \left(\frac{1 - x}{1 - x_i} \right)^{F_L a/F_G a} \tag{8.21}$$

For any value of (x, y) on the operating curve plotted in terms of mole fractions, a curve of x_i vs. y_i from Eq. (8.21) is plotted to determine the intersection with the equilibrium curve. This provides the local y and y_i for use in Eq. (8.20).† Equation (8.20) can then be integrated graphically after plotting the integrand as ordinate vs. y as abscissa.

However, it is more customary to proceed as follows [6]. Since

$$y - y_i = (1 - y_i) - (1 - y) \tag{8.22}$$

the numerator and denominator of the integral of Eq. (8.20) can be multiplied respectively by the right- and left-hand sides of Eq. (8.22) to provide

$$Z = \int_{y_2}^{y_1} \frac{G(1 - y)_{iM} \, dy}{F_G a (1 - y)(1 - y_i)} \tag{8.23}$$

---

† As demonstrated in Illustration 8.6, for moderately dilute solutions it is satisfactory to determine y_i by a line of slope $-k_x a/k_y a$ drawn from (x, y) on the operating line to intersection with the equilibrium curve, in accordance with Eq. (5.2).

where $(1 - y)_{iM}$ is the logarithmic mean of $1 - y_i$ and $1 - y$. When we define a *height of a gas transfer unit* H_{tG} as

$$H_{tG} = \frac{G}{F_G a} = \frac{G}{k_y a(1 - y)_{iM}} = \frac{G}{k_G a p_t (1 - y)_{iM}} \quad (8.24)$$

Eq. (8.23) becomes

$$Z = \int_{y_2}^{y_1} H_{tG} \frac{(1 - y)_{iM} \, dy}{(1 - y)(y - y_i)} \approx H_{tG} \int_{y_2}^{y_1} \frac{(1 - y)_{iM} \, dy}{(1 - y)(y - y_i)} = H_{tG} N_{tG}$$

$$(8.25)$$

Here advantage is taken of the fact that the ratio $G/F_G a = H_{tG}$ is very much more constant than either G or $F_G a$, and in many cases may be considered constant within the accuracy of the available data. In the integral of Eq. (8.25) containing only the y terms, if we disregard the ratio $(1 - y)_{iM}/(1 - y)$, the remainder is seen to be the number of times the average $y - y_i$ divides into the change of gas concentration $y_1 - y_2$. As in Chap. 7, this is a measure of the difficulty of the absorption, and the integral is called the *number of gas transfer units* N_{tG}. H_{tG} is then the packing height providing one gas transfer unit. Equation (8.25) can be further simplified by substituting the arithmetic average for the logarithmic average $(1 - y)_{iM}$ [9]

$$(1 - y)_{iM} = \frac{(1 - y_i) - (1 - y)}{\ln[(1 - y_i)/(1 - y)]} \approx \frac{(1 - y_i) + (1 - y)}{2} \quad (8.26)$$

which involves little error. N_{tG} then becomes

$$N_{tG} = \int_{y_2}^{y_1} \frac{dy}{y - y_i} + \frac{1}{2}\ln\frac{1 - y_2}{y - y_1} \quad (8.27)$$

which makes for simpler graphical integration. A plot of $1/(y - y_i)$ vs. y for the graphical integration of Eq. (8.27) often covers awkwardly large ranges of the ordinate. This can be avoided [37] by replacing dy by its equal $y \, d \ln y$, so that

$$N_{tG} = 2.303 \int_{\log y_2}^{\log y_1} \frac{y}{y - y_i} d \log y + 1.152 \log\frac{1 - y_2}{1 - y_1} \quad (8.28)$$

For dilute solutions, the second term on the right of Eqs. (8.27) and (8.28) is negligible, $F_G a \approx k_y a$, and y_i can be obtained by plotting a line of slope $-k_x a/k_y a$ from points (x, y) on the operating line to intersection with the equilibrium curve.

The above relationships all have their counterparts in terms of liquid concentrations, derived in exactly the same way

$$Z = \int_{x_2}^{x_1} \frac{L\,dx}{F_L a(1-x)\ln\left[(1-x)/(1-x_i)\right]} = \int_{x_2}^{x_1} \frac{L(1-x)_{iM}\,dx}{F_L a(1-x)(x_i-x)}$$
(8.29)

$$Z = \int_{x_2}^{x_1} H_{tL} \frac{(1-x)_{iM}\,dx}{(1-x)(x_i-x)} \approx H_{tL} \int_{x_2}^{x_1} \frac{(1-x)_{iM}\,dx}{(1-x)(x_i-x)} = H_{tL} N_{tL} \quad (8.30)$$

$$H_{tL} = \frac{L}{F_L a} = \frac{L}{k_x a(1-x)_{iM}} = \frac{L}{k_L ac(1-x)_{iM}}$$
(8.31)

$$N_{tL} = \int_{x_2}^{x_1} \frac{dx}{x_i-x} + \tfrac{1}{2}\ln\frac{1-x_1}{1-x_2}$$
(8.32)

where H_{tL} = height of liquid transfer unit
N_{tL} = number of liquid transfer units
$(1-x)_{iM}$ = logarithmic mean of $1-x$ and $1-x_i$
Either set of equations leads to the same value of Z.

Strippers The same relationships apply as for absorption. The *driving forces* $y - y_i$ and $x_i - x$ which appear in the above equations are then negative, but since for strippers $x_2 > x_1$ and $y_2 > y_1$, the result is a positive Z as before.

Illustration 8.6 The absorber of Illustrations 8.2 and 8.3 is to be a packed tower, 470 mm diameter, filled with 38-mm (1.5-in) Berl saddles. The circumstances are as follows:

Gas Benzene content: in, $y_1 = 0.02$ mole fraction, $Y_1 = 0.0204$ mol/mol dry gas.

Out:
$$y_2 = 0.00102 \qquad Y_2 = 0.00102$$

Nonabsorbed gas:
　　Av mol wt = 11.0
　　Rate in = 0.250 m³/s = 0.01075 kmol/s　　0.01051 kmol/s nonbenzene
　　Temperature = 26°C　　pressure $p_t = 1.07 \times 10^5$ N/m² (803 mmHg)
　　Viscosity = 10^{-5} kg/m · s (0.010 cP)　　$D_{AG} = 1.30 \times 10^{-5}$ m²/s

Liquid Benzene content in, $x_2 = 0.005$ mol fraction, $X_2 = 0.00503$ mol benzene/mol oil.

Out:
$$x_1 = 0.1063 \qquad X_1 = 0.1190$$

Benzene-free oil:
　　Mol wt = 270　　viscosity = 2×10^{-3} kg/m · s (2.0 cp)
　　Density = 840 kg/m³　　rate = 1.787×10^{-3} kmol/s
　　Temperature = 26°C　　$D_{A,L} = 4.77 \times 10^{-10}$ m²/s
　　Surface tension = 0.03 N/m²　　$m = y^*/x = 0.1250$
Compute the depth of packing required.

SOLUTION To plot the operating line, use Eq. (8.9) to calculate X and Y values (or read from Fig. 8.10) and convert to $x = X/(1 + X)$, $y = Y/(1 + Y)$. Thus,

$$0.01051\,(0.0204 - Y) = (1.787 \times 10^{-3})\,(0.1190 - X)$$

X	x	Y	y
0.00503	0.00502	0.00102	0.00102
0.02	0.01961	0.00357	0.00356
0.04	0.0385	0.00697	0.00692
0.06	0.0566	0.01036	0.01025
0.08	0.0741	0.01376	0.01356
0.10	0.0909	0.01714	0.01685
0.1190	0.1063	0.0204	0.0200

Values of y and x (the operating line) are plotted along with the equilibrium line, $y = 0.1250x$, in Fig. 8.18. Although the curvature in this case is not great, the operating line is not straight on mole-fraction coordinates.

The cross-sectional area of the absorber $= \pi(0.47)^2/4 = 0.1746$ m^2. At the bottom,

$$L' = \frac{(1.787 \times 10^{-3})(260) + 0.1190(1.787 \times 10^{-3})(78)}{0.1746} = 2.75 \text{ kg/m}^2 \cdot \text{s}$$

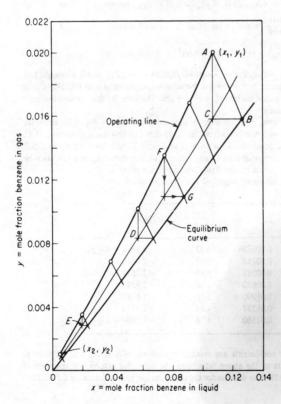

Figure 8.18 Solution to Illustration 8.6.

Similarly at the top, $L' = 2.66 \text{ kg/m}^2 \cdot \text{s}$. Av $L' = 2.71 \text{ kg/m}^2 \cdot \text{s}$. In similar fashion for the gas, at the bottom $G' = 0.761 \text{ kg/m}^2 \cdot \text{s}$; at the top $G' = 0.670 \text{ kg/m}^2 \cdot \text{s}$. Av $G' = 0.716 \text{ kg/m}^2 \cdot \text{s}$.

The flow quantities change so little from one end of the tower to the other that the average values can be used to compute the mass-transfer coefficients, which can then be taken as constant. The circumstances are precisely those of Illustration 6.6, where it was found that

$$F_G a = 0.0719 \text{ kmol/m}^3 \cdot \text{s} \qquad F_L a = 0.01377 \text{ kmol/m}^3 \cdot \text{s}$$

Interface compositions corresponding to points on the operating line of Fig. 8.18 are determined with Eq. (8.21). For example, at ($x = 0.1063$, $y = 0.0200$) on the operating line at point A, Fig. 8.18, Eq. (8.21) becomes

$$\frac{1 - y_i}{1 - 0.0200} = \left(\frac{1 - 0.1063}{1 - x_i} \right)^{0.01377/0.0719}$$

This is plotted (y_i vs. x_i) as curve AB on Fig. 8.18 to intersect the equilibrium curve at B, where $y_i = 0.01580$.

Note: In this case, since the concentrations are low, an essentially correct value of y_i can be obtained by the following less tedious calculation. Table 3.1 shows

$$k_y a = \frac{F_G a}{p_{B,M}/p_t} = \frac{F_G a}{(1 - y)_{iM}} \qquad k_x a = \frac{F_L a}{(1 - x)_{iM}}$$

At point A on the operating line, $1 - y = 1 - 0.02 = 0.98$, $1 - x = 1 - 0.1063 = 0.8937$. As an approximation, these can be taken as $(1 - y)_{iM}$ and $(1 - x)_{iM}$, respectively. Then

$$k_y a = \frac{0.0719}{0.98} = 0.0734 \text{ kmol/m}^3 \cdot \text{s} \cdot \text{(mole fraction)}$$

$$k_x a = \frac{0.01377}{0.9837} = 0.01541 \text{ kmol/m}^3 \cdot \text{s} \cdot \text{(mole fraction)}$$

In accordance with Eq. (5.2), $-k_x a/k_y a = -0.01541/0.0734 = -0.210$, and if a straight line of this slope is drawn from point A, it intersects the equilibrium curve at $y_i = 0.01585$. Curve AB, in other words, is nearly a straight line of slope -0.210. The error by this method becomes even less at lower concentrations.

Where the F's must be used, it is sufficient to proceed as follows. In Fig. 8.18, draw the vertical line AC to intersect with the horizontal line CB, thus locating the intersection at C. Repeat the calculation at two other locations, such as at D and E, and draw the curve EDC. Interfacial concentrations G corresponding to any point F on the operating line can then be obtained by the construction shown on the figure leading to the point G.

In a similar manner the tabulated values of y_i were determined

y	y_i	$\dfrac{y}{y - y_i}$	$\log y$
0.00102	0.000784	4.32	$-2.9999 = \log y_2$
0.00356	0.00285	5.02	-2.4486
0.00692	0.00562	5.39	-2.1599
0.01025	0.00830	5.26	-1.9893
0.01356	0.01090	5.10	-1.8687
0.01685	0.01337	4.84	-1.7734
0.0200	0.01580	4.76	$-1.6990 = \log y_1$

Since the mass-transfer coefficients are essentially constant, Eq. (8.28) will be used to determine N_{tG}. From the data of the above tabulation, the curve for graphical integration of Eq. (8.28) is plotted in Fig. 8.19. The area under the curve (to the zero ordinate) is 6.556. Then

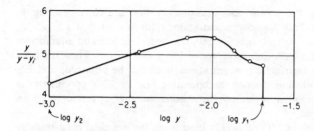

Figure 8.19 Graphical integration for Illustration 8.6.

by Eq. (8.28),

$$N_{tG} = 2.303 \ (6.556) + 1.152 \log \frac{1 - 0.00102}{1 - 0.0200} = 15.11$$

The average gas rate is

$$G = \frac{0.01075 + 0.01051/\ (1 - 0.00102)}{2(0.1746)} = 0.0609 \ \text{kmol/m}^2 \cdot \text{s}$$

Eq. (8.24):

$$H_{tG} = \frac{G}{F_G a} = \frac{0.0609}{0.0719} = 0.847 \ \text{m}$$

$$Z = H_{tG} N_{tG} = 0.847 \ (15.11) = 12.8 \ \text{m packed depth} \quad \textbf{Ans}$$

Note: The circumstances are such that, for this problem, the simpler computations of Illustration 8.7 are entirely adequate. The method used above, however, is suitable for solutions of any concentration and where the equilibrium curve is not straight. If flow rates vary appreciably from top to bottom of the tower, so that mass-transfer coefficients vary, this is easily allowed for in determining y_i. In such cases, Eq. (8.20) would be used to compute Z.

Overall Coefficients and Transfer Units ($m = dy_i/dx_i = $ const)

For cases where the equilibrium distribution curve is straight and the ratio of mass-transfer coefficients is constant, it was shown in Chap. 5 that overall mass-transfer coefficients are convenient. The expressions for the height of packing can then be written

$$Z = N_{tOG} H_{tOG} \tag{8.33}$$

$$N_{tOG} = \int_{y_2}^{y_1} \frac{(1 - y)_{*M} \ dy}{(1 - y)(y - y^*)} \tag{8.34}$$

$$N_{tOG} = \int_{y_2}^{y_1} \frac{dy}{y - y^*} + \frac{1}{2} \ln \frac{1 - y_2}{1 - y_1} \tag{8.35}$$

$$N_{tOG} = \int_{Y_2}^{Y_1} \frac{dY}{Y - Y^*} + \frac{1}{2} \ln \frac{1 + Y_2}{1 + Y_1} \tag{8.36}$$

$$H_{tOG} = \frac{G}{F_{OG} a} = \frac{G}{K_y a(1 - y)_{*M}} = \frac{G}{K_G a p_t (1 - y)_{*M}} \tag{8.37}$$

Here y^* (or Y^*) is the solute concentration in the gas corresponding to equilibrium with the bulk liquid concentration x (or X), so that $y - y^*$ (or $Y - Y^*$) is simply the vertical distance between operating line and equilibrium curve. $(1 - y)_{*M}$ is the logarithmic average of $1 - y$ and $1 - y^*$. These methods are convenient since interfacial concentrations need not be obtained, and Eq. (8.36) is especially convenient since the operating line on X, Y coordinates is straight. N_{tOG} is the *number of overall gas transfer units*, H_{tOG} the *height of an overall gas transfer unit*.

Equations (8.33) to (8.37) are usually used when the principal mass-transfer resistance resides within the gas. For cases where the principal mass-transfer resistance lies within the liquid, it is more convenient to use

$$Z = N_{tOL} H_{tOL} \tag{8.38}$$

$$N_{tOL} = \int_{x_2}^{x_1} \frac{(1 - x)_{*M} \, dx}{(1 - x)(x^* - x)} \tag{8.39}$$

$$N_{tOL} = \int_{x_2}^{x_1} \frac{dx}{x^* - x} + \tfrac{1}{2} \ln \frac{1 - x_1}{1 - x_2} \tag{8.40}$$

$$N_{tOL} = \int_{X_2}^{X_1} \frac{dX}{X^* - X} + \tfrac{1}{2} \ln \frac{1 + X_1}{1 + X_2} \tag{8.41}$$

$$H_{tOL} = \frac{L}{F_{OL} a} = \frac{L}{K_x a (1 - x)_{*M}} = \frac{L}{K_L a c (1 - x)_{*M}} \tag{8.42}$$

Dilute Solutions

The computation of the number of transfer units for dilute mixtures can be greatly simplified. When the gas mixture is dilute, for example, the second term of the definition of N_{tOG} [Eq. (8.35)] becomes entirely negligible and can be discarded

$$N_{tOG} = \int_{y_2}^{y_1} \frac{dy}{y - y^*} \tag{8.43}$$

If the equilibrium curve in terms of mole fractions is linear over the range of compositions x_1 to x_2, then

$$y^* = mx + r \tag{8.44}$$

If the solutions are dilute, the operating line can be considered as a straight line as well

$$y = \frac{L}{G}(x - x_2) + y_2 \tag{8.45}$$

so that the driving force $y - y^*$ is then linear in x

$$y - y^* = qx + s \tag{8.46}$$

where q, r, and s are constants. Therefore Eq. (8.43) becomes

$$N_{tOG} = \frac{L}{G} \int_{x_2}^{x_1} \frac{dx}{qx + s} = \frac{L}{Gq} \ln \frac{(y - y^*)_1}{(y - y^*)_2} = \frac{y_1 - y_2}{\dfrac{(y - y^*)_1 - (y - y^*)_2}{\ln[(y - y^*)_1 / (y - y^*)_2]}}$$

(8.47)

$$N_{tOG} = \frac{y_1 - y_2}{(y - y^*)_M}$$

(8.48)

where $(y - y^*)_M$ is the logarithmic average of the concentration differences at the ends of the tower. This equation is sometimes used in the familiar rate form obtained by substituting the definition of N_{tOG}

$$G(y_1 - y_2) = K_G a Z p_t (y - y^*)_M$$

(8.49)

Dilute Solutions, Henry's Law

If Henry's law applies [r of Eq. (8.44) = 0], by elimination of x between Eqs. (8.44) and (8.45) and substitution of y^* in Eq. (8.43) there results for absorbers [9]

$$N_{tOG} = \frac{\ln\left[\dfrac{y_1 - mx_2}{y_2 - mx_2}\left(1 - \dfrac{1}{A}\right) + \dfrac{1}{A} \right]}{1 - 1/A}$$

(8.50)

where $A = L/mG$, as before. For strippers, the corresponding expression in terms of N_{tOL} is similar

$$N_{tOL} = \frac{\ln\left[\dfrac{x_2 - y_1/m}{x_1 - y_1/m}(1 - A) + A \right]}{1 - A}$$

(8.51)

These are shown in convenient graphical form in Fig. 8.20.

Graphical Construction for Transfer Units [2]

Equation (8.48) demonstrates that one overall gas transfer unit results when the change in gas composition equals the average overall driving force causing the change. Consider now the operating diagram of Fig. 8.21, where line KB has been drawn so as to be everywhere vertically halfway between the operating line and equilibrium curve. The step CFD, which corresponds to one transfer unit, has been constructed by drawing the horizontal line CEF so that line $CE = EF$ and continuing vertically to D. $y_G - y_H$ can be considered as the average driving force for the change in gas composition $y_D - y_F$ corresponding to this step. Since $GE = EH$, and if the operating line can be considered straight, $DF = 2(GE) = GH$, the step CFD corresponds to one transfer unit. In similar fashion the other transfer units were stepped off ($JK = KL$, etc.). For computing N_{tOL},

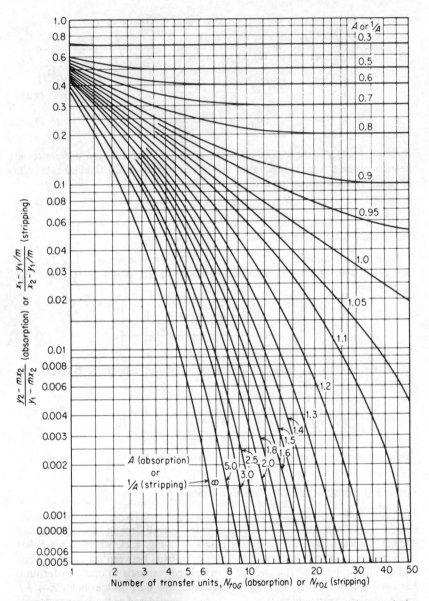

Figure 8.20 Number of transfer units for absorbers or strippers with constant absorption or stripping factor.

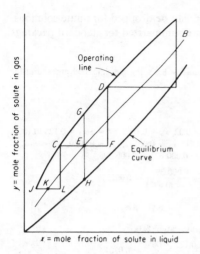

y = mole fraction of solute in gas

x = mole fraction of solute in liquid

Figure 8.21 Graphical determination of transfer units.

the line KB would be drawn horizontally halfway between equilibrium curve and operating line and would bisect the vertical portions of the steps.

Overall Heights of Transfer Units

When overall numbers of transfer units are appropriate, the overall heights of transfer units can be synthesized from those for the individual phases through the relationships developed in Chap. 5. Thus, Eq. (5.25a), with $m' = m'' = m = $ const, can be written

$$\frac{G}{F_{OG}a} = \frac{G(1-y)_{iM}}{F_G a(1-y)_{*M}} + \frac{mG}{L}\frac{L}{F_L a}\frac{(1-x)_{iM}}{(1-y)_{*M}} \tag{8.52}$$

whence, by definition of the heights of transfer units,

$$H_{tOG} = H_{tG}\frac{(1-y)_{iM}}{(1-y)_{*M}} + \frac{mG}{L}H_{tL}\frac{(1-x)_{iM}}{(1-y)_{*M}} \tag{8.53}$$

If the mass-transfer resistance is essentially all in the gas, $y_i \approx y^*$, and

$$H_{tOG} = H_{tG} + \frac{mG}{L}H_{tL}\frac{(1-x)_{iM}}{(1-y)_{*M}} \tag{8.54}$$

and, for dilute solutions, the concentration ratio of the last equation can be dropped. In similar fashion, Eq. (5.26a) yields

$$H_{tOL} = H_{tL}\frac{(1-x)_{iM}}{(1-x)_{*M}} + \frac{L}{mG}H_{tG}\frac{(1-y)_{iM}}{(1-x)_{*M}} \tag{8.55}$$

and if the mass-transfer resistance is essentially all in the liquid,

$$H_{tOL} = H_{tL} + \frac{L}{mG}H_{tG}\frac{(1-y)_{iM}}{(1-x)_{*M}} \tag{8.56}$$

The concentration ratio of the last equation can be dropped for dilute solutions. Data for the individual phase coefficients are summarized for standard packings in Chap. 6.

Illustration 8.7 Repeat the computation of Illustration 8.6 using the simplified procedures for dilute mixtures.

SOLUTION

Number of transfer units (a) Use Eq. (8.48). $y_1 = 0.02$, $x_1 = 0.1063$, $y_1^* = mx_1 = 0.125 (0.1063)$ = 0.01329, $(y - y^*)_1 = 0.02 - 0.01329 = 0.00671$. $y_2 = 0.00102$, $x_2 = 0.0050$, $y_2^* = mx_2 =$ 0.125 (0.005) = 0.000 625, $(y - y^*)_2 = 0.00102 - 0.000 625 = 0.000 395$.

$$(y - y^*)_M = \frac{0.00671 - 0.000\ 395}{\ln (0.00671/0.000\ 395)} = 0.00223$$

$$N_{tOG} = \frac{0.02 - 0.00102}{0.00223} = 8.51 \quad \text{Ans.}$$

(b) Eq. (8.50) or Fig. 8.20: av $A = 1.424$ (Illustration 8.3).

$$\frac{y_2 - mx_2}{y_1 - mx_2} = \frac{0.00102 - 0.125\ (0.005)}{0.02 - 0.125\ (0.005)} = 0.0204$$

From either Eq. (8.50) or Fig. 8.20, $N_{tOG} = 9.16$. **Ans.**

(c) The graphical construction for transfer units is shown in Fig. 8.22. The mole-ratio coordinates are satisfactory for dilute mixtures (even though Fig. 8.21 is drawn for mole fractions), and the operating line and equilibrium curve were redrawn from Fig. 8.10. The line BD was drawn everywhere vertically midway between operating line and equilibrium curve and

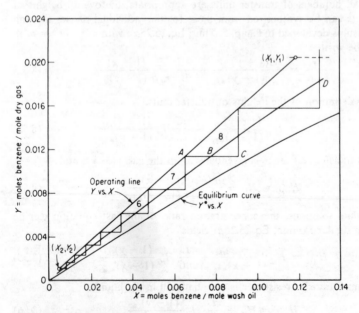

Figure 8.22 Solution to Illustration 8.7.

the transfer-unit steps were constructed by making the horizontal line segments such as AB and BC equal. The number of transfer units required is 8.7. **Ans.**

(d) Eq. (8.36). From Fig. 8.10, for each value of X, values of Y are read from the operating line DF and Y^* from the equilibrium curve (or they can be computed from the equations of these lines).

X	Y	Y^*	$\dfrac{1}{Y - Y^*}$
$X_2 = 0.00502$	0.00102	0.000625	2531
0.02	0.00357	0.00245	893
0.04	0.00697	0.00483	467
0.06	0.01036	0.00712	309
0.08	0.01376	0.00935	227
0.10	0.01714	0.01149	177.0
$X_1 = 0.1190$	0.0204	0.01347	144.3

The integral of Eq. (8.36) can be evaluated graphically from a plot of $1/(Y - Y^*)$ as ordinate vs. Y as abscissa or, following the suggestion of Eq. (8.28), from a plot of $Y/(Y - Y^*)$ as ordinate vs. $\log Y$ as abscissa. The integral term is then 8.63. Then

$$N_{tOG} = 8.63 + \tfrac{1}{2}\ln\frac{1 + 0.00102}{1 + 0.0204} = 8.62 \quad \textbf{Ans.}$$

Height of a transfer unit Since the solutions are dilute, Eq. (8.54) becomes

$$H_{tOG} = H_{tG} + \frac{mG}{L}H_{tL} + H_{tG} + \frac{H_{tL}}{A}$$

From Illustration 6.4, $F_G a = 0.0719$ kmol/m$^3 \cdot$ s, and $F_L a = 0.01377$ kmol/m$^3 \cdot$ s. Av $G = 0.0609$ kmol/m$^2 \cdot$ s (Illustration 8.6). Av $L = (1.787 \times 10^{-3})$ [(1 + 0.1190) + (1 + 0.00503)]/ 2(0.1746) = 0.01087 kmol/m$^2 \cdot$ s. Eq. (8.24):

$$H_{tG} = \frac{G}{F_G a} = \frac{0.0609}{0.0719} = 0.847 \text{ m}$$

Eq. (8.31):

$$H_{tL} = \frac{L}{F_L a} = \frac{0.01087}{0.01377} = 0.789$$

$$H_{tOG} = 0.847 + \frac{0.789}{1.424} = 1.401 \text{ m}$$

$$Z = H_{tOG}N_{tOG} = 1.401 \, (9.16) = 12.83 \text{ m} \quad \textbf{Ans.}$$

Nonisothermal Operation

All the relationships thus far developed for packed towers are correct for either isothermal or nonisothermal operation, but for the nonisothermal case they do not tell the entire story. During absorption, release of energy at the interface due to latent heat and heat of solution raises the interface temperature above that of the bulk liquid. This changes physical properties, mass-transfer coefficients, and equilibrium concentrations. More subtle side effects associated with volume changes and contributions of thermal diffusion can enter the problem [4, 5, 11, 42]. As a practical matter, where the energy release is large, as in absorption of

HCl or nitrogen oxides into water, unless vigorous attempts to remove the heat are made, the temperature rise limits the ultimate concentration of the product solution [23].

For relatively dilute solutions, a conservative approximation can be made by assuming that all the heat evolved on absorption is taken up by the liquid, thus neglecting the temperature rise of the gas. This results in a liquid temperature higher than that which is likely, leading to a tower taller than need be. A somewhat more correct but still approximate approach neglects the thermal- and mass-transfer resistances of the liquid [38]. The method which follows allows for these resistances and also for the simultaneous mass transfer of solvent, which in the warm parts of the tower evaporates and in the cooler parts recondenses [41]. It has been experimentally confirmed [36].

Adiabatic Absorption and Stripping

Because of the possibility of substantial temperature effects, allowance must be made for evaporation and condensation of the liquid solvent. For this problem, therefore, the components are defined as follows:

A = principal transferred solute, present in gas and liquid phases
B = carrier gas, not dissolving, present only in gas
C = principal solvent of liquid, which can evaporate and condense; present in gas and liquid phases

Refer to Fig. 8.23, a schematic representation of a differential section of the packed tower [41]. For convenience, the phases are shown separated by a vertical surface. Enthalpies are referred to the pure unmixed components in their normal state of aggregation at some convenient temperature t_0 and the tower pressure. Liquid enthalpies are given by Eq. (8.12), expressed per mole of solution

$$H_L = C_L(t_L - t_0) + \Delta H_S \tag{8.12}$$

$$dH_L = C_L \, dt_L + d\Delta H_S \tag{8.57}$$

Gas enthalpies are expressed per mole of carrier gas B

$$H_G' = C_B(t_G - t_0) + Y_A[C_A(t_G - t_0) + \lambda_{A0}] + Y_C[C_C(t_G - t_0) + \lambda_{C0}] \tag{8.58}$$

$$dH_G' = C_B \, dt_G + Y_A C_A \, dt_G + [C_A(t_G - t_0) + \lambda_{A0}] \, dY_A + Y_C C_C \, dt_G + [C_C(t_G - t_0) + \lambda_{C0}] \, dY_C \tag{8.59}$$

If solute A is a gas at t_0, λ_{A0} is taken as zero.

Transfer rates The fluxes of mass and heat shown in Fig. 8.23 are taken as positive in the direction gas to liquid, negative in the opposite direction. The relationships developed below are correct for any combination of them. Consider the gas phase: solute A and solvent vapor C may transfer, but $N_B = 0$.

$$N_A a \, dZ = R_A F_{GA}\left(\ln\frac{R_A - y_{A,i}}{R_A - y_A}\right) a \, dZ = -G_B \, dY_A \tag{8.60}$$

and

$$N_C a \, dZ = R_C F_{GC}\left(\ln\frac{R_C - y_{C,i}}{R_C - y_C}\right) a \, dZ = -G_B \, dY_C \tag{8.61}$$

where $R_A = N_A/(N_A + N_C)$, $R_C = N_C/(N_A + N_C)$. Since the fluxes N_A and N_C may have either sign, the R's can be greater or less than 1.0, positive or negative. However $R_A + R_C = 1.0$. The heat-transfer rate for the gas is

$$q_G a \, dZ = h_G' a(t_G - t_i) \, dZ \tag{8.62}$$

where h'_G is the convection coefficient corrected for mass transfer. Similarly for the liquid,

$$N_A a \, dZ = R_A F_{L,A}\left(\ln \frac{R_A - x_A}{R_A - x_{A,i}}\right) a \, dZ \qquad (8.63)$$

$$N_C a \, dZ = R_C F_{L,C}\left(\ln \frac{R_C - x_C}{R_C - x_{C,i}}\right) a \, dZ \qquad (8.64)$$

$$q_L a \, dZ = h_L a (t_i - t_L) \, dZ \qquad (8.65)$$

Mass balance Refer to envelope III, Fig. 8.23.

$$L + dL + G = L + G + dG$$

$$dL = dG$$

Since

$$G = G_B(1 + Y_A + Y_C)$$

then

$$dL = G_B(dY_A + dY_C) \qquad (8.66)$$

Enthalpy balances The balance for the gas employs envelope I, Fig. 8.23, which extends to, but does not include, the interface

$$\text{Rate enthalpy in} = G_B H'_G \qquad (8.67)$$

$$\text{Rate enthalpy out} = G_B(H'_G + dH'_G) - G_B \, dY_A[C_A(t_G - t_0) + \lambda_{A0}]$$

$$- G_B \, dY_C[C_C(t_G - t_0) + \lambda_{C0}] \qquad (8.68)$$

The last two terms are the enthalpies of the transferred mass. At steady state,

$$\text{Rate enthalpy in} - \text{rate enthalpy out} = \text{rate of heat transfer}$$

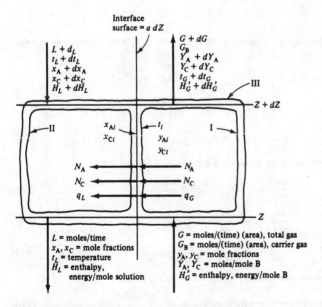

Figure 8.23 Differential section of a packed absorber or stripper.

Substitution of Eqs. (8.62), (8.67), and (8.68), with Eq. (8.59) for H_G, then results in

$$- G_B(C_B + Y_A C_A + Y_C C_C) \, dt_G = h'_G a(t_G - t_i) \, dZ \tag{8.69}$$

Similarly for the liquid. Envelope II extends to, but does not include, the interface

$$\text{Rate enthalpy in} = (L + dL)(H_L + dH_L) + \left(N_A \bar{H}_{A,i} + N_C \bar{H}_{C,i}\right) a \, dZ \tag{8.70}$$

$\bar{H}_{J,i}$ is the partial enthalpy of a component J ($=$ A, C) in solution at temperature t_i, concentration $x_{J,i}$, and the last term is the enthalpy of the substance transferred from the gas

$$\text{Rate enthalpy out} = LH_L \tag{8.71}$$

At steady state,

$$\text{Rate enthalpy out} = \text{rate enthalpy in} + \text{rate of heat transfer}$$

Substituting Eqs. (8.12), (8.57), (8.60), (8.61), (8.65), (8.66), (8.70), and (8.71) and dropping the second-order differential $dH \, dL$ results in

$$L(C_L \, dt_L + d\Delta H_S) = G_B \left\{ - [C_L(t_L - t_0) + \Delta H_S](dY_A + dY_C) + \bar{H}_{A,i} \, dY_A + H_{C,i} \, dY_C \right\}$$
$$- h_L a(t_i - t_L) \, dZ \tag{8.72}$$

Finally, an enthalpy balance using envelope III is

$$\text{Rate enthalpy in} = G_B H'_G + (H_L + dH_L)(L + dL) \tag{8.73}$$

$$\text{Rate enthalpy out} = LH_L + G_B(H'_G + dH'_G) \tag{8.74}$$

At steady state,

$$\text{Rate enthalpy in} = \text{rate enthalpy out}$$

Substituting Eqs. (8.12), (8.57), (8.59), (8.66), (8.73), and (8.74) and ignoring $dH_L \, dL$ produces

$$L(C_L \, dt_L + d\Delta H_S) = G_B\{ C_B \, dt_G + Y_A C_A \, dt_G + [C_A(t_G - t_0) + \lambda_{A0}] \, dY_A$$
$$+ Y_C C_C \, dt_G + [C_C(t_G - t_0) + \lambda_{C0}] \, dY_C$$
$$- (dY_A + dY_C)[C_L(t_L - t_0) + \Delta H_S]\} \tag{8.75}$$

Interface conditions Equating the right sides of Eqs. (8.72) and (8.75) provides the interface temperature

$$t_i = t_L + \frac{G_B}{h_L a} \left\{ \left[\bar{H}_{A,i} - C_A(t_G - t_0) - \lambda_{A0} \right] \frac{dY_A}{dZ} \right.$$

$$+ \left[\bar{H}_{C,i} - C_C(t_G - t_0) - \lambda_{C0} \right] \frac{dY_C}{dZ}$$

$$\left. - (C_B + Y_A C_A + Y_C C_C) \frac{dt_G}{dZ} \right\} \tag{8.76}$$

The gradients for this expression are provided by Eqs. (8.60), (8.61), and (8.69):

$$\frac{dY_A}{dZ} = - \frac{R_A F_{G,A} a}{G_B} \ln \frac{R_A - y_{A,i}}{R_A - y_A} \tag{8.77}$$

$$\frac{dY_C}{dZ} = - \frac{R_C F_{G,C} a}{G_B} \ln \frac{R_C - y_{C,i}}{R_C - y_C} \tag{8.78}$$

$$\frac{dt_G}{dZ} = - \frac{h'_G a(t_G - t_i)}{G_B(C_B + Y_A C_A + Y_C C_C)} \tag{8.79}$$

The interface conditions are obtained from Eqs. (8.60) and (8.63) for A

$$y_{A,i} = R_A - (R_A - y_A) \left(\frac{R_A - x_A}{R_A - x_{A,i}} \right)^{F_{L,A}/F_{G,A}} \tag{8.80}$$

Similarly for C

$$y_{C,i} = R_C - (R_C - y_C)\left(\frac{R_C - x_C}{R_C - x_{C,i}}\right)^{F_{L,C}/F_{G,C}} \tag{8.81}$$

These are solved simultaneously with their respective equilibrium-distribution curves, as described in Chap. 5 (see Illustration 5.1). The calculation is by trial: R_A is assumed, whence $R_C = 1 - R_A$, and the correct R's are those for which $x_{A,i} + x_{C,i} = 1.0$.

The partial enthalpies of Eq. (8.76) can be obtained in a number of ways, e.g., from activity-coefficient data [12] or from integral heats of solution by the method of tangent intercepts [22, 40]. Since, however, the equilibrium partial pressures of components A and C from these solutions are required over a range of temperatures in any event, the simplest procedure is that described earlier in connection with Fig. 8.2 and Eq. (8.3). For ideal solutions, $\overline{H}_{J,i}$ is the enthalpy of pure liquid J at t_i for all $x_{J,i}$.

Mass-transfer coefficients must be those associated with the individual liquid and gas phases: overall coefficients will not serve. The correlations of Chap. 6 will provide these for Berl saddles and Raschig rings, and additional data are summarized in Ref. 14. The k-type coefficients are not suitable because in some cases the transfer is opposite the concentration gradient, but k's can be converted into F's (Table 3.1). Similarly data in the form of H_t's can be converted into Fa [Eqs. (8.24) and (8.31)]. Heat-transfer coefficients can be estimated, if not otherwise available, through the heat- mass-transfer analogy. For the gas, the correction for mass transfer [Eq. (3.70)] provides

$$h_G' a = \frac{-G_B(C_A \, dY_A/dZ + C_C \, dY_C/dZ)}{1 - \exp[G_B(C_A \, dY_A/dZ + C_C \, dY_C/dZ)/h_G A]} \tag{8.82}$$

A corresponding correction for the liquid, which would involve changing ΔH_S with concentration, is not available, but h_L is usually sufficiently large to make a correction unimportant. Mean diffusivities for three-component gas mixtures are estimated through application of Eq. (2.35), which reduces to

$$D_{A,m} = \frac{R_A - y_a}{R_A\left(\dfrac{y_B}{D_{AB}} + \dfrac{y_A + y_C}{D_{AC}}\right) - \dfrac{y_A}{D_{AC}}} \tag{8.83}$$

$$D_{C,m} = \frac{R_C - y_C}{R_C\left(\dfrac{y_B}{D_{CB}} + \dfrac{y_A + y_C}{D_{AC}}\right) - \dfrac{y_C}{D_{AC}}} \tag{8.84}$$

In the design of a tower, the cross-sectional area and hence the mass velocities of gas and liquid are established through pressure-drop considerations (Chap. 6). Assuming that entering flow rates, compositions, and temperatures, pressure of absorption, and percentage absorption (or stripping) of one component are specified, the packed height is then fixed. The problem is then to estimate the packed height and the conditions (temperature, composition) of the outlet streams. Fairly extensive trial and error is required, for which the relations outlined above can best be solved with a digital computer [15, 36]. Computer calculations can also be adapted to the problem of multicomponent mixtures [15]. The procedure for the three-component case is outlined in the following example [41].

Illustration 8.8 A gas consisting of 41.6% ammonia (A), 58.4% air (B) at 20°C, 1 std atm, flowing at the rate 0.0339 kmol/m^2 · s (25 lb mol/ft^2 · h), is to be scrubbed countercurrently with water (C) entering at 20°C at a rate 0.271 kmol/m^2 · s(200 lb mol/ft^2 · h), to remove 99% of the ammonia. The adiabatic absorber is to be packed with 38-mm ceramic Berl saddles. Estimate the packed height.

SOLUTION At 20°C, $C_A = 36\,390$, $C_B = 29\,100$, $C_C = 33\,960$ J/kmol · K. $\lambda_C = 44.24 \times 10^6$ J/kmol. These and other data are obtained from "The Chemical Engineers' Handbook" [31].

Enthalpy base = NH_3 gas, H_2O liquid, air at 1 std atm, $t_0 = 20°C$. $\lambda_{A0} = 0$, $\lambda_{C0} = 44.24 \times 10^6$ J/kmol.

Gas in:

$$G_B = 0.0339(0.584) = 0.01980 \text{ kmol air}/m^2 \cdot s$$

$$y_A = 0.416 \qquad Y_A = \frac{0.416}{1 - 0.416} = 0.7123 \text{ kmol } NH_3/\text{kmol air}$$

$$y_C = Y_C = 0 \qquad H'_G = 0$$

Liquid in:

$$L = 0.271 \text{ kmol}/m^2 \cdot s \qquad x_A = 0 \qquad x_C = 1.0 \qquad H_L = 0$$

Gas out:

$$Y_A = 0.7123(1 - 0.99) = 0.007\,123 \text{ kmol } NH_3/\text{kmol air}$$

Assume outlet gas $t_G = 23.9°C$ (to be checked)

Assume $y_C = 0.0293$ (saturation) (to be checked)

$$y_C = \frac{Y_C}{Y_C + 0.007\,123 + 1} \qquad Y_C = 0.0304 \text{ kmol } H_2O/\text{kmol air}$$

Eq. (8.58):

$$H'_G = 1463 \text{ kJ/kmol air}$$

Liquid out:

$$H_2O \text{ content} = 0.271 - 0.03040(0.01980) = 0.2704 \text{ kmol}/m^2 \cdot s$$

$$NH_3 \text{ content} = 0.01980(0.7123 - 0.007\,123) = 0.01396 \text{ kmol}/m^2 \cdot s$$

$$L = 0.2704 + 0.01396 = 0.2844 \text{ kmol}/m^2 \cdot s$$

$$x_A = \frac{0.01396}{0.2844} = 0.0491 \qquad x_C = 0.9509$$

At $x_A = 0.0491$, $t_0 = 20°C$, $\Delta H_S = -1709.6$ kJ/kmol solution [24]

Conditions at the bottom of the tower Enthalpy balance, entire tower:

Enthalpy gas in + enthalpy liquid in = enthalpy gas out + enthalpy liquid out

$$0 + 0 = 0.01980(1\,463\,000) + 0.2844[C_L(t_L - 20) - 1\,709\,600]$$

The value of C_L, obtained by trial at $(t_L + 20)/2$, equals 75 481 J/kmol · K, whence $t_L = 41.3°C$.

For the gas, $M_{G,\text{av}} = 24.02$, $\rho_G = 0.999$ kg/m^3, $\mu_G = 1.517 \times 10^{-5}$ kg/m · s, k (thermal conductivity) = 0.0261 W/m · K, $C_p = 1336$ J/kg · K; $D_{AB} = 2.297 \times 10^{-5}$ m^2/s, $D_{AC} = 3.084 \times 10^{-5}$ m^2/s, $D_{CB} = 2.488 \times 10^{-5}$ m^2/s, $Pr_G = 0.725$.

For the liquid, $M_{L,\text{av}} = 17.97$, $\rho_L = 953.1$ kg/m^3, $\mu_L = 6.408 \times 10^{-4}$ kg/m · s, $D_{AL} = 3.317 \times 10^{-9}$ m^2/s, k (thermal conductivity) = 0.4777 W/m · K; $Sc_L = 202$, $Pr_L = 5.72$.

The mass velocities at the bottom are $G' = GM_{G,\text{av}} = 0.0339(24.02) = 0.8142$ kg/m^2 · s, and $L' = LM_{L,\text{av}} = 0.2844(17.97) = 5.11$ kg/m^2 · s. From the data of Chap. 6, $d_S = 0.0472$ m, $a = 57.57$ m^2/m^3, $\varphi_{Lt} = 0.054$, $\varepsilon = 0.75$, and $\varepsilon_{L_o} = 0.696$ (Chap. 6 notation).

Eq. (6.72): $k_L = 3.616 \times 10^{-4}$ kmol/m^2 · s · (kmol/m^3). Table 3.1: $F_L = k_L c$, $c = \rho_L/M_L = 953.1/17.97 = 53.04$ kmol/m^3. Therefore $F_L = 0.01918$ kmol/m^2 · s. The heat-mass-transfer analogy used with Eq. (6.72) provides

$$\frac{h_L d_S}{k} = 25.1 \left(\frac{d_S L'}{\mu_L} \right)^{0.45} Pr_L^{0.5}$$

whence, with $k = 0.4777$ W/m · K, $h_L = 8762$ W/m^2 · K. The heat-transfer analog of Eq. (6.69) is

$$\frac{h_G}{C_p G'} Pr_G^{2/3} = 1.195 \left[\frac{d_S G'}{\mu_G(1 - \varepsilon_{L_o})} \right]^{-0.36}$$

from which $h_G = 62.45$ W/m^2 · K.

To obtain the gas mass-transfer coefficients, the mean diffusivities of A and C, and hence R_A and R_C, must be known. Assume $R_A = 1.4$, whence $R_C = 1 - 1.4 = -0.4$. With $y_A = 0.416$, $y_B = 0.584$, $y_C = 0$, Eqs. (8.83) and (8.84) yield $D_{A, m} = 2.387 \times 10^{-5}$ m^2/s and $D_{C, m} = 2.710 \times 10^{-5}$ m^2/s. Therefore $Sc_{G, A} = \mu_G/\rho_G D_{A, m} = 0.636$, $Sc_{G, C} = 0.560$. Eq. (6.69): $F_{G, A} = 2.13 \times 10^{-3}$ kmol/m$^2 \cdot$ s, $F_{G, C} = 2.32 \times 10^{-3}$ kmol/m$^2 \cdot$ s. Eqs. (8.80) and (8.81) then become

$$y_{A, i} = 1.4 - (1.4 - 0.416)\left(\frac{1.4 - 0.0491}{1.4 - x_{A, i}} \right)^{0.01918/2.13 \times 10^{-3}} \tag{8.85}$$

$$y_{C, i} = -0.4 - (-0.4 - 0)\left(\frac{-0.4 - 0.9509}{-0.4 - x_{C, i}} \right)^{0.01918/2.32 \times 10^{-3}} \tag{8.86}$$

These are to be plotted to obtain the intersections with their respective distribution curves at temperature t_i. Equilibrium data for NH_3 and H_2O vapors in contact with solutions are plotted in Figs. 8.2 and 8.24, with water as reference substance.

After one trial, assume $t_i = 42.7°C$, to be checked. The data of Figs. 8.2 and 8.24 are used to draw the equilibrium curves of Fig. 8.25, and Eqs. (8.85) and (8.86) are plotted on their respective portions. Similarly the counterparts of these equations for $R_A = 1.3$, $R_C = -0.7$ and $R_A = 1.5$, $R_C = -0.5$ are plotted. By interpolation, the intersections of the curves with the equilibrium lines satisfy $x_{A, i} + x_{C, i} = 1.0$ for $R_A = 1.38$, $R_C = -0.38$, and $x_{A, i} = 0.0786$, $y_{A, i} = 0.210$, and $x_{C, i} = 0.9214$, $y_{C, i} = 0.0740$.

Eq. (8.77): $dY_A/dZ = -1.706$ (kmol H_2O/kmol air)/m. Eq. (8.78): $dY_C/dZ = 0.4626$ (kmol H_2O/kmol air)/m. Eq. (8.82): $h'_G a = 4154$ W/m$^3 \cdot$ K. Eq. (8.79): $dt_G/dZ = 86.5$ K/m.

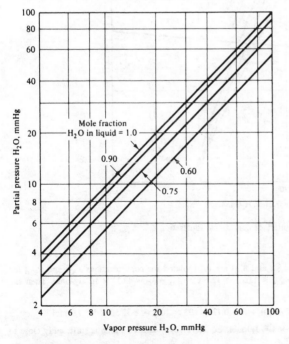

Figure 8.24 Equal-temperature reference-substance graph for partial pressures of water over aqueous ammonia solutions. Reference substance is water.

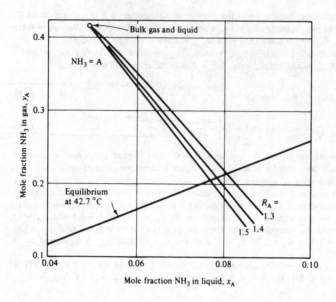

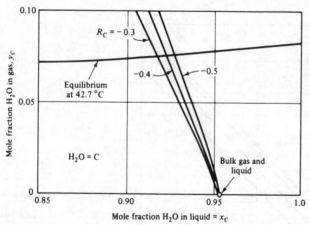

Figure 8.25 Determination of interface conditions, Illustration 8.8.

When the curves of Figs. 8.2 and 8.24 are interpolated for concentrations $x_{A,i}$ and $x_{C,i}$, the slopes are $m'_{A,r} = 0.771$ and $m'_{C,r} = 1.02$. At 42.7°C, $\lambda_{H_2O} = \lambda_C = 43.33 \times 10^6$ J/kmol. Then Eq. (8.3) for A:

$$\overline{H}_{A,i} = 36\,390(42.7 - 20) + 0 - 0.771(43.33 \times 10^6) = -32.58 \times 10^6 \text{ J/kmol}$$

Similarly for C: $H_{C,i} = 0.784 \times 10^6$ J/kmol. Eq. (8.76): $t_i = 42.5$°C, which is sufficiently close to the 42.7°C assumed earlier.

An interval of ΔY_A *up the tower* Take a small increment in ΔY_A. For a desk calculator, $\Delta Y_A =$ -0.05 is suitable (for a digital computer, -0.01 would be more appropriate). Then $\Delta Z =$ $\Delta Y_A/(dY_A/dZ) = -0.05/(-1.706) = 0.0293$ m. At this level,

$Y_{A,\,next} = Y_A + \Delta Y_A = 0.7123 - 0.05 = 0.6623$ kmol/kmol air

$Y_{C,\,next} = Y_C + \Delta Y_C = Y_C + (dY_C/dZ)\,\Delta Z = 0 + 0.4626(0.0293) = 0.01355$ kmol H_2O/kmol air

$t_{G,\,next} = t_G + (dt_G/dZ)\,\Delta Z = 20 + 86.5(0.0293) = 22.53°C$

$L_{next} = L + G_B(\Delta Y_A + \Delta Y_C) = 0.2837$ kmol/m$^2 \cdot$ s

$x_{A,\,next} = \dfrac{G_B\,\Delta Y_A + Lx_A}{L_{next}} = 0.0457$ mole fraction NH_3

$H'_{G,\,next} = 7.33 \times 10^5$ J/kmol air

$H_{L,\,next} = (LH_L)_{top} + \dfrac{G_B(H'_{G,\,next} - H'_{G,\,top})}{L_{next}} = -4.868 \times 10^5$ J/kmol

The previous computations can now be repeated at this level, leading ultimately to a new ΔY_A. The calculations are continued until the specified gas outlet composition is reached, whereupon the assumed outlet gas temperature and water concentration can be checked. The latter are adjusted, as necessary, and the entire computation repeated until a suitable check is obtained. The packed depth required is then the sum of the final ΔZ's. The computations started above lead to $Z = 1.58$ m, $t_G = 23.05°C$. The final results are shown in Fig. 8.26 [26]. Water is stripped in the lower part of the tower and reabsorbed in the upper part.

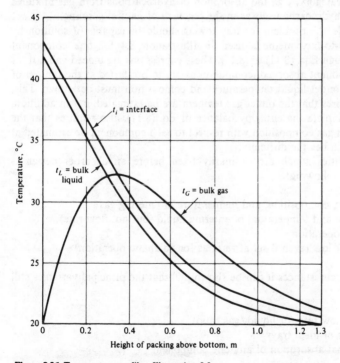

Figure 8.26 Temperature profiles, Illustration 8.8.

In the course of such calculations, it may develop that the direction of mass transfer of the solvent liquid, especially as a vapor in the gas phase, is *against* the concentration gradient. This is entirely possible (see Chap. 2) and is one of the reasons why F-type (rather than k-type) mass-transfer coefficients are essential. The bulk-gas temperature must never be less than the dew point, since this will lead to fog formation for which the computations do not allow. A larger liquid rate, with consequently lower liquid temperatures, will result in less solvent vaporization and less likelihood of fog formation.

MULTICOMPONENT SYSTEMS

Except for the consideration of solvent vaporization in the above discussion of adiabatic packed towers, it has thus far been assumed that only one component of the gas stream has an appreciable solubility. When the gas contains several soluble components or the liquid several volatile ones for stripping, some modifications are needed. The almost complete lack of solubility data for multicomponent systems, except where ideal solutions are formed in the liquid phase and the solubilities of the various components are therefore mutually independent, unfortunately makes estimates of even the ordinary cases very difficult. However, some of the more important industrial applications fall in the ideal-solution category, e.g., the absorption of hydrocarbons from gas mixtures in nonvolatile hydrocarbon oils as in the recovery of natural gasoline.

In principle, the problem for tray towers should be capable of solution by the same tray-to-tray methods used in Illustration 8.4 for one component absorbed, through Eqs. (8.11) to (8.15). These expressions are indeed valid. If, as with one component absorbed, computations are to be started at the bottom of the tower, the outlet liquid temperature and composition must be known. This, as before, requires that the outlet gas temperature be estimated, but in addition, in order to complete the enthalpy balance of Eq. (8.11), also requires that the complete outlet gas composition with respect to each component be estimated at the start. Herein lies the difficulty.

The quantities which are ordinarily fixed before an absorber design is started are the following:

1. Rate of flow, composition, and temperature of entering gas
2. Composition and temperature of entering liquid (but not flow rate)
3. Pressure of operation
4. Heat gain or loss (even if set at zero, as for adiabatic operation)

Under these circumstances it can be shown [30] that the principal variables still remaining are:

1. The liquid flow rate (or liquid/gas ratio)
2. The number of ideal trays
3. The fractional absorption of any one component

Any *two of these last but not all three* can be arbitrarily fixed for a given design. When two have been specified, the third is automatically fixed, as is the extent of absorption of all substances not already specified and the outlet-stream temperatures. For example, if the liquid rate and number of equilibrium trays are specified, the extent of absorption of each substance of the gas is automatically fixed and cannot be arbitrarily chosen. Or if the liquid rate and extent of absorption of one substance are specified, the number of equilibrium trays and the extent of absorption of all components are automatically fixed and cannot be chosen.

As a result, for the tray-to-tray calculations suggested above, not only must the outlet gas temperature be guessed but also the complete outlet gas composition, all to be checked at the end of the calculation. This becomes so hopeless a trial-and-error procedure that it cannot be done practically without some guidance. This is provided through an approximate procedure, either that offered by the Kremser equations, which apply only for constant absorption factor, or through some procedure allowing for variation of the absorption factor with tray number. To establish the latter, we first need an exact expression for the absorber with varying absorption factor. This was first derived by Horton and Franklin [21] as outlined below.

Refer to Fig. 8.27, which shows a multitray absorber or stripper. Since all components can transfer between gas and liquid, there may be no substance which passes through at constant rate in the gas, for example. It is convenient, therefore, to define all gas compositions in terms of the entering gas and similarly all liquid compositions in terms of the entering liquid. Thus, for any component in the liquid leaving any tray n,

$$X'_n = \frac{\text{moles component in } L_n/\text{time}}{L_0} = \frac{x_n L_n}{L_0}$$

and for any component in the gas G_n,

$$Y'_n = \frac{\text{moles component in } G_n/\text{time}}{G_{N_p+1}} = \frac{y_n G_n}{G_{N_p+1}}$$

where x and y are the usual mole fractions.

The equations which follow can all be written separately for each component. Consider the tower of Fig. 8.27 to be an absorber. A material balance for any component about equilibrium tray n is

$$L_0(X'_n - X'_{n-1}) = G_{N_p+1}(Y'_{n+1} - Y'_n) \tag{8.87}$$

The equilibrium relation for the equilibrium tray is†

$$y_n = m_n x_n \tag{8.88}$$

or, in terms of the new concentrations,

$$Y'_n \frac{G_{N_p+1}}{G_n} = m_n X'_n \frac{L_0}{L_n} \tag{8.89}$$

† The ratio y/x at equilibrium is usually written as K, but here we use m to distinguish this from the mass-transfer coefficients.

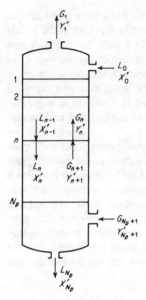

Figure 8.27 Multicomponent absorber or stripper.

Similarly, for tray $n - 1$,

$$Y'_{n-1} \frac{G_{N_p+1}}{G_{n-1}} = m_{n-1} X'_{n-1} \frac{L_0}{L_{n-1}} \qquad (8.90)$$

Solving Eqs. (8.89) and (8.90) for the X's, substituting in Eq. (8.87), and rearranging, we have

$$Y'_n = \frac{Y'_{n+1} + A_{n-1} Y'_{n-1}}{1 + A_n} \qquad (8.91)$$

where $A_n = L_n / m_n G_n$ and $A_{n-1} = L_{n-1} / m_{n-1} G_{n-1}$ are the component absorption factors on the two trays.

If the absorber contained only one tray ($n = 1$), Eq. (8.91) would read

$$Y'_1 = \frac{Y'_2 + A_0 Y'_0}{1 + A_1} \qquad (8.92)$$

From Eq. (8.90)

$$Y'_0 = m_0 X'_0 \frac{L_0}{L_0} \frac{G_0}{G_{N_p+1}} = \frac{m_0 X'_0 G_0}{G_{N_p+1}} \qquad (8.93)$$

and

$$A_0 Y'_0 = \frac{L_0}{m_0 G_0} \frac{m_0 X'_0 G_0}{G_{N_p+1}} = \frac{L_0 X'_0}{G_{N_p+1}} \qquad (8.94)$$

Substituting this into Eq. (8.92) provides

$$Y'_1 = \frac{Y'_2 + L_0 X'_0 / G_{N_p+1}}{1 + A_1} \qquad (8.95)$$

If the absorber contained two trays, Eq. (8.91) with $n = 2$ would become

$$Y'_2 = \frac{Y'_3 + A_1 Y'_1}{1 + A_2} \qquad (8.96)$$

Substituting Y_1' from Eq. (8.95) and rearranging gives

$$Y_2' = \frac{(A_1 + 1)Y_3' + A_1 L_0 X_0' / G_{N_p+1}}{A_1 A_2 + A_2 + 1} \tag{8.97}$$

Similarly, for a three-tray absorber,

$$Y_3' = \frac{(A_1 A_2 + A_2 + 1)Y_4' + A_1 A_2 L_0 X_0' / G_{N_p+1}}{A_1 A_2 A_3 + A_2 A_3 + A_3 + 1} \tag{8.98}$$

and for N_p trays,

$$Y_{N_p}' = \frac{(A_1 A_2 A_3 \cdots A_{N_p-1} + A_2 A_3 \cdots A_{N_p-1} + \cdots + A_{N_p-1} + 1)}{A_1 A_2 A_3 \cdots A_{N_p} + A_2 A_3 \cdots A_{N_p} + \cdots + A_{N_p} + 1}$$
$$\times Y_{N_p+1}' + A_1 A_2 \cdots A_{N_p-1} L_0 X_0' / G_{N_p+1}$$

$$\tag{8.99}$$

In order to eliminate Y_{N_p}', which is inside the absorber, a component material balance about the entire absorber,

$$L_0(X_{N_p}' - X_0') = G_{N_p+1}(Y_{N_p+1}' - Y_1') \tag{8.100}$$

and Eq. (8.89) for $n = N_p$,

$$Y_{N_p}' = m_{N_p} X_{N_p}' \frac{L_0}{L_{N_p}} \frac{G_{N_p}}{G_{N_p+1}} = \frac{L_0 X_{N_p}'}{A_{N_p} G_{N_p+1}} \tag{8.101}$$

are solved simultaneously to eliminate X_{N_p}', the result substituted for Y_{N_p}' in Eq. (8.101), whence rearrangement yields

$$\frac{Y_{N_p+1}' - Y_1'}{Y_{N_p+1}'} = \frac{A_1 A_2 A_3 \cdots A_{N_p} + A_2 A_3 \cdots A_{N_p} + \cdots + A_{N_p}}{A_1 A_2 A_3 \cdots A_{N_p} + A_2 A_3 \cdots A_{N_p} + \cdots + A_{N_p} + 1}$$
$$- \frac{L_0 X_0'}{G_{N_p+1} Y_{N_p+1}'} \frac{A_2 A_3 A_4 \cdots A_{N_p} + A_3 A_4 \cdots A_{N_p} + \cdots + A_{N_p} + 1}{A_1 A_2 A_3 \cdots A_{N_p} + A_2 A_3 \cdots A_{N_p} + \cdots + A_{N_p} + 1}$$

$$\tag{8.102}$$

Equation (8.102) is an expression for the fractional absorption of any component, exact because it is based only upon material balances and the condition of equilibrium which defines an ideal tray.

A similar expression for strippers is

$$\frac{X_0' - X_{N_p}'}{X_0'} = \frac{S_1 S_2 \cdots S_{N_p} + S_1 S_2 \cdots S_{N_p-1} + \cdots + S_1}{S_1 S_2 \cdots S_{N_p} + S_1 S_2 \cdots S_{N_p-1} + \cdots + S_1 + 1}$$
$$- \frac{G_{N_p+1} Y_{N_p+1}'}{L_0 X_0'} \frac{S_1 S_2 \cdots S_{N_p-1} + S_1 S_2 \cdots S_{N_p-2} + \cdots + S_1 + 1}{S_1 S_2 \cdots S_{N_p} + S_1 S_2 \cdots S_{N_p-1} + \cdots + S_1 + 1} \tag{8.103}$$

In order to use Eqs. (8.102) or (8.103), the L/G ratio for each tray and the tray temperatures (which determine the m's) are required to compute the A's or the S's. If the liquids are not ideal, m for any component on any tray additionally depends upon the complete liquid composition on the tray. The same is true for the gas compositions under conditions for which the gas solutions are not ideal. The equations are practically useful, therefore, only for ideal solutions.

As an approximation [21], the gas rate G_n for tray n of an absorber can be estimated on the assumption that the fractional absorption is the same for each tray:

$$\frac{G_n}{G_{n+1}} \approx \left(\frac{G_1}{G_{N_p+1}} \right)^{1/N_p} \tag{8.104}$$

or

$$G_n \approx G_{N_p+1} \left(\frac{G_1}{G_{N_p+1}} \right)^{(N_p+1-n)/N_p} \tag{8.105}$$

The liquid rate L_n can then be obtained by material balance to the end of the tower. Similarly in strippers,

$$L_n \approx L_0 \left(\frac{L_{N_p}}{L_0} \right)^{n/N_p} \tag{8.106}$$

where G_n is determined by material balance. If the molar latent heats and heat capacities are all alike for all components, and if no heat of solution is evolved, the temperature rise on absorption is roughly proportional to the amount of absorption, so that, approximately,

$$\frac{G_{N_p+1} - G_{n+1}}{G_{N_p+1} - G_1} \approx \frac{t_{N_p} - t_n}{t_{N_p} - t_0} \tag{8.107}$$

and similarly for stripping,

$$\frac{L_0 - L_n}{L_0 - L_{N_p}} \approx \frac{t_0 - t_n}{t_0 - t_{N_p}} \tag{8.108}$$

In order further to simplify the computations, Edmister [13] has written the Horton-Franklin equations in terms of average or "effective" absorption and stripping factors instead of the A's and S's for each tray. Thus, for absorption, Eq. (8.102) becomes

$$\frac{Y'_{N_p+1} - Y'_1}{Y'_{N_p+1}} = \left(1 - \frac{L_0 X'_0}{A' G_{N_p+1} Y'_{N_p+1}} \right) \frac{A_E^{N_p+1} - A_E}{A_E^{N_p+1} - 1} \tag{8.109}$$

For a two-tray absorber, it develops that

$$A' = \frac{A_{N_p}(A_1 + 1)}{A_{N_p} + 1} \tag{8.110}$$

and

$$A_E = \left[A_{N_p}(A_1 + 1) + 0.25 \right]^{0.5} - 0.5 \tag{8.111}$$

These are exact, but it is found that Eqs. (8.104) to (8.111) apply reasonably well for any number of trays, provided that unusual temperature profiles (such as a maximum temperature on an intermediate tray) do not develop. If $X'_0 = 0$, Eq. (8.109) is the Kremser equation. It is also convenient to note that $(A_E^{N_p+1} - A_E)/(A_E^{N_p+1} - 1)$ is the Kremser function [Eq. (5.54)], and the ordinate of Fig. 5.16 is 1 minus this function.

Similarly for stripping,

$$\frac{X_0' - X_{N_p}'}{X_0'} = \left(1 - \frac{G_{N_p+1}Y_{N_p+1}'}{S'L_0X_0'}\right)\frac{S_E^{N_p+1} - S_E}{S_E^{N_p+1} - 1} \tag{8.112}$$

with

$$S' = \frac{S_1(S_{N_p} + 1)}{S_1 + 1} \tag{8.113}$$

and

$$S_E = \left[S_1(S_{N_p} + 1) + 0.25\right]^{0.5} - 0.5 \tag{8.114}$$

Equation (8.112) becomes the Kremser equation if $Y_{N_p+1}' = 0$.

Components present only in the entering gas will be absorbed, and those present only in the entering liquid will be stripped. If a component is present in both streams, the terms in the parentheses of the right-hand sides of Eqs. (8.109) and (8.112) are computed and the equation is used which provides the positive quantity.

Equations (8.109) and (8.112) can be used to determine the number of equilibrium trays required to absorb or strip a component to a specified extent and to estimate the extent of absorption or stripping of all other components. This then provides a basis for using the exact equations of Horton and Franklin, Eqs. (8.102) and (8.103). These latter can be used only for an integral number of equilibrium trays. A change of L_0/G_{N_p+1} may be necessary to meet the specifications exactly with such an integral number; alternatively the nearest larger integral number of trays may be accepted.

Illustration 8.9 A gas analyzing 70 mol % CH_4, 15% C_2H_6, 10% n-C_3H_8, and 5% n-C_4H_{10}, at 25°C, 2 std atm, is to be scrubbed in an adiabatic tray absorber with a liquid containing 1 mol % n-C_4H_{10}, 99% nonvolatile hydrocarbon oil, at 25°C, using 3.5 mol liquid/mol entering gas. The pressure is to be 2 std atm, and at least 70% of the C_3H_8 of the entering gas is to be absorbed. The solubility of CH_4 in the liquid will be considered negligible, and the other components form ideal solutions. Estimate the number of ideal trays required and the composition of the effluent streams.

SOLUTION In what follows, CH_4, C_2H_6, C_3H_8, and C_4H_{10} will be identified as C_1, C_2, C_3, and C_4, respectively. Physical properties are:

Component	Average specific heat at 0–40°C, kJ/kmol · K		Latent heat of vaporization at 0°C, kJ/kmol	$m = \dfrac{y^*}{x}$†		
	Gas	Liquid		25°C	27.5°C	30°C
C_1	35.59					
C_2	53.22	105.1	10 032	13.25	13.6	14.1
C_3	76.04	116.4	16 580	4.10	4.33	4.52
C_4	102.4	138.6	22 530	1.19	1.28	1.37
Oil		37.7				

† Values of m from C. L. Depriester, *Chem. Eng. Prog. Symp. Ser.*, **49**(7), 1 (1953); "The Chemical Engineers' Handbook," 5th ed., pp. 13-12 and 13-13.

Basis: 1 kmol entering gas; $t_0 = 0°C$; enthalpies are referred to gaseous C_1 and other components liquid at 0°C. Gas in: $G_{N_p+1} = 1$ kmol, $t_{N_p+1} = 25°C$.

Component	$Y'_{N_p+1} = $ y_{N_p+1}	Enthalpy, kJ/kmol $= H_{G, N_p+1}$ per component	$H_{G, N_p+1} y_{N_p+1}$
C_1	0.70	35.59(25 − 0) = 889.75	622.8
C_2	0.15	53.22(25 − 0) + 10 032 = 16 363	1704.5
C_3	0.10	76.04(25 − 0) + 16 580 = 18 480	1848.0
C_4	0.05	102.4(25 − 0) + 22 530 = 25 090	1254.5
	1.00		5429.8 $= H_{G, N_p+1}$

Liquid in: $L_0 = 3.5$ kmol $t = 25°C$

Component	$X'_0 = x_0$	$L_0 X'_0$	Enthalpy, kJ/kmol $= H_{L0}$ per component	$H_{L0} L_0 X'_0$
C_4	0.01	0.035	138.6(25 − 0) = 3465	121.28
Oil	0.99	3.465	377(25 − 0) = 9425	32 658
	1.00	3.50		32 780 $= H_{L0} L_0$

Preliminary calculations The total absorption is estimated to be 0.15 kmol, and an average temperature 25°C is assumed. A rough value of L/G at the top is then $3.5/(1 − 0.15) = 4.12$, and at the bottom $(3.5 + 0.15)/1.0 = 3.65$, with an average of 3.90. The number of equilibrium trays is fixed by the specified C_3 absorption. For C_3, m at 25°C = 4.10, rough $A = 3.90/4.10 = 0.951$. Equation (8.109) with fractional absorption = 0.7 and $X_0 = 0$ gives

$$0.7 = \frac{0.951^{N_p+1} - 0.951}{0.951^{N_p+1} - 1}$$

This can be solved for N_p directly. Alternatively, since Fig. 5.16 has as its ordinate (1 − Kremser function) = 0.3, then $A = 0.951$ and N_p can be read. Either way, $N_p = 2.53$. Since the Horton-Franklin equation will be used later, an integral number of trays $N_p = 3$ is chosen. Fig. 5.16 with $N_p = 3$, $A = 0.951$, then shows an ordinate of 0.27, or 0.73 fraction C_3 absorbed.

$$\frac{Y'_{N_p+1} - Y'_1}{Y'_{N_p+1}} = \frac{0.10 - Y'_1}{0.1} = 0.73$$

$$Y'_1 = 0.027 \text{ kmol } C_3 \text{ in } G_1$$

For C_2, $m = 13.25$ at 25°C, rough $A = 3.90/13.25 = 0.294$. At low A's, Fig. 5.16 shows the ordinate to be $1 − A$, and the fractional absorption is 0.294.

$$\frac{Y'_{N_p+1} - Y'_1}{Y'_{N_p+1}} = \frac{0.15 - Y'_1}{0.15} = 0.294$$

$$Y'_1 = 0.1059 \text{ kmol } C_2 \text{ in } G_1$$

C_4 is present both in the entering liquid and gas. At 25°C, $m = 1.19$, rough $A = 3.90/1.19 = 3.28$, rough $S = 1/3.28 = 0.305$. Then

$$1 - \frac{L_0 X'_0}{A' G_{N_p+1} Y'_{N_p+1}} = 1 - \frac{3.5(0.01)}{3.28(1.0)(0.05)} = +0.7866$$

$$1 - \frac{G_{N_p+1} Y'_{N_p+1}}{S' L_0 X_0} = 1 - \frac{1(0.05)}{0.305(3.5)(0.01)} = -3.684$$

The C_4 will therefore be absorbed. Eq. (8.109):

$$\frac{0.05 - Y_1'}{0.05} = 0.7866 \frac{3.28^4 - 3.28}{3.28^4 - 1}$$

$$Y_1' = 0.01145 \text{ kmol } C_4 \text{ in } G_1$$

[Alternatively, at $N_p = 3$, $A = 3.28$, Fig. 5.16 provides an ordinate of 0.0198, whence $(0.05 - Y_1')/0.05 = 0.7866(1 - 0.0198)$, and $Y_1' = 0.01145$.]

Edmister method Estimate the top-tray temperature $t_1 = 26°C$ (to be checked). The results of the preliminary calculations then give the following approximate results:

G_1

Component	Y_1'	Enthalpy, kJ/kmol = H_{G1} per component		$H_{G1}Y_1'$
C_1	0.70	$35.59(26 - 0)$	$= 925$	647.5
C_2	0.1059	$53.22(26 - 0) + 10\ 032 = 11\ 420$		1209.4
C_3	0.027	$76.04(26 - 0) + 16\ 580 = 18\ 560$		501.12
C_4	0.01145	$102.4(26 - 0) + 22\ 530 = 25\ 190$		288.4
	$\overline{0.8444} =$			$\overline{2646} = H_{G1}G_1$
	G_1			

L_3

Component	$L_3 x_3$, kmol		Enthalpy, kJ/kmol $= H_{L3}$ per component	$H_{L3}L_3x_3$
C_2	$0.15 - 0.1059 =$	0.0441	$105(t_3 - 0)$	$4.64t_3$
C_3	$0.01 - 0.027 =$	0.073	$116.4(t_3 - 0)$	$8.50t_3$
C_4	$0.035 + 0.05 - 0.01148 =$	0.0736	$138.6(t_3 - 0)$	$10.2t_3$
Oil		3.465	$377(t_3 - 0)$	$1306.3t_3$
	$L_3 =$	$\overline{3.656}$		$\overline{1329t_3}$

Overall enthalpy balance, Eq. (8.11) with $Q_T = 0$, $G_{N_p+1} = 1.0$, $N_p = 3$:

$$32\ 780 + 5429.8 = 1329t_3 + 2646$$

$$t_3 = 26.8°C$$

Equation (8.109) can now be used to obtain a second approximation of the effluent composition. Eq. (8.105) with $n = 2$:

$$G_2 = 1.0 \left(\frac{0.8444}{1.0} \right)^{(3+1-2)/3} = 0.8934$$

Eq. (8.13) with $n = 1$:

$$L_1 + 1.0 = 3.656 + 0.8934$$

$$L_1 = 3.549 \qquad \frac{L_1}{G_1} = \frac{3.549}{0.8444} = 4.203$$

Eq. (8.105) with $n = 3$:

$$G_3 = 1.0 \left(\frac{0.8444}{1.0} \right)^{(3+1-3)/3} = 0.945$$

$$\frac{L_3}{G_3} = \frac{3.656}{0.945} = 3.869$$

For C_4:

$$m \text{ at } 26°C = 1.225 \qquad A_1 = \frac{4.203}{1.225} = 3.43$$

$$m \text{ at } 26.8°C = 1.250 \qquad A_3 = \frac{3.869}{1.250} = 3.095 = A_{N_p}$$

Eq. (8.110):

$$A' = 3.348$$

Eq. (8.111):

$$A_E = 3.236$$

Eq. (8.109):

$$Y_1' = 0.01024 \text{ kmol } C_4 \text{ in } G_1$$

$$C_4 \text{ in } L_3 = 0.035 + 0.05 - 0.01024 = 0.07476 \text{ kmol} = L_3 x_3$$

In similar fashion the Y_1's for the other components, together with the outlet gas and outlet liquid enthalpies, can be computed:

Component	Y_1'	$H_{G1}Y_1'$, 26°C	$L_3 x_3$	$H_3 L_3 x_3$
C_1	0.700	647.5	0	0
C_2	0.1070	1222	0.0430	$4.519 t_3$
C_3	0.0274	508.5	0.0726	$8.45 t_3$
C_4	0.01024	288.4	0.07476	$10.36 t_3$
Oil			3.465	$1306.3 t_3$
	$\overline{0.8446} = G_1$	$\overline{2666.4} = H_{G1}G_1$	$\overline{3.655} = L_3$	$\overline{1329.6 t_3} = H_{L3}L_3$

Eq. (8.11) shows $t_3 = 26.7°C$, sufficiently close to that of the first estimate (26.8°C) for us to proceed with Eq. (8.102).

Horton-Franklin method Equations (8.105), (8.107), and (8.13), used as above with $G_1 = 0.8446$, now show

Tray n	G_n	L_n	L_n/G_n	t_n, °C
1	0.8446	3.549	4.202	26.0
2	0.8935	3.600	4.029	26.1
3	0.9453	3.655	3.866	26.7

The temperatures permit tabulation of m's and the L_n/G_n's permit calculation of A for each component on each tray. Equation (8.102) then gives Y_1', and a material balance gives the moles of each component in L_3:

Component	A_1	A_2	A_3	Y_1' [Eq. (8.102)]	$L_3 x_3$
C_1				0.700	0
C_2	0.314	0.301	0.286	0.1072	0.0428
C_3	1.002	0.959	0.910	0.0273	0.0727
C_4	3.430	3.290	3.093	0.01117	0.0738
Oil					3.465
				$\overline{0.8457} = G_1$	$\overline{3.654} = L_3$

An enthalpy balance again shows $t_3 = 26.7°C$.

For most purposes, the preceding results will be reasonably satisfactory. But they are correct only if the A values for each component on each tray are correct. These can be checked by first verifying the L/G ratios. For the components found in both liquid and gas

$$A_n = \frac{L_n}{m_n G_n} = \frac{L_n}{(y_n/x_n)G_n} \tag{8.115}$$

or

$$G_n y_n = \frac{L_n x_n}{A_n} \tag{8.116}$$

If there are G_S mol/time of nonabsorbed gas, then

$$G_n = G_S + \Sigma \frac{L_n x_n}{A_n} \tag{8.117}$$

and Eq. (8.13) will provide $L_n x_n$, whose sum is L_n. The L_n/G_n ratios can then be compared with those used previously.

Illustration 8.9, continued For tray 3, $G_S = 0.7$ kmol C_1. For C_2, $G_3 y_3 = L_3 x_3/A_3 = 0.0428/0.286 = 0.1497$, and Eq. (8.14) with $n = 2$ provides $L_2 x_2 = 0.0428 + 0.1497 - 0.15 = 0.0425$. Similarly,

Component	$G_3 y_3 = L_3 x_3/A_3$	$L_2 x_2$
C_1	0.700	0
C_2	0.1497	0.0425
C_3	0.0799	0.0526
C_4	0.0239	0.0477
Oil		3.465
	$0.9535 = G_3$	$3.6078 = L_2$

Therefore

$$\frac{L_3}{G_3} = \frac{3.654}{0.9535} = 3.832 \ (3.866 \text{ used previously})$$

Similarly L_2/G_2 and L_1/G_1 can be computed and compared with values used previously.

The procedure then to be followed is to repeat the use of Eq. (8.102) with the new L/G ratios to provide A's, repeating the check of L/G and use of Eq. (8.102) until agreement is reached. This still leaves the temperatures to be checked. For an assumed top-tray temperature, an overall tower enthalpy balance provides t_N. With the compositions and flow rates for each stream as last determined, individual tray enthalpy balances provide the temperature of each tray. These are repeated until the assumed t_1 agrees with that calculated. New m's and A's are then used with Eq. (8.102) and the entire procedure repeated until a completely consistent set of L/G's and temperatures are obtained, at which time the problem is solved. The procedure is very tedious, and methods for reducing the work, including the use of high-speed computers, have been given much study [16, 18, 19, 39].

Use of Reflux—Reboiled Absorbers

Refer again to the preceding illustration. Considering only the three substances absorbed, the wet gas contained these roughly in the proportions $C_2H_6 : C_3H_8 : C_4H_{10} = 50 : 33 : 17$. The rich oil leaving the absorber contained the same substances in the approximate ratio $21 : 36 : 37$. Despite its low solubility, the proportion of absorbed gas which is ethane is high, owing to the relatively high proportion of this substance in the original wet gas. If it is desired to reduce the relative proportions of ethane in the rich oil, the oil just as it leaves the tower must be in contact with a gas relatively leaner in ethane and richer in propane and butane. Such a gas can be obtained by heating a part of the rich oil, which will then evolve a gas of the required low ethane content. This evolved gas can then be returned to an extension of the absorber below the inlet of the wet gas, where in rising past the oil it will strip out the ethane (and also any methane which may have been absorbed as well). The ethane not absorbed will now, of course, leave with the lean gas. The heat exchanger where a portion of the rich oil is heated is called a *reboiler*, and the arrangement of Fig. 8.28 can be used. Calculation methods for multicomponents are available [20, 25].

A stream returned to a cascade of stages, as represented by the trays in the absorber, for the purposes of obtaining an enrichment beyond that obtained by countercurrent contact with the feed to the cascade, is called *reflux*. This principle is used extensively in distillation, liquid extraction, and adsorption, but it is applicable to any countercurrent enrichment operation.

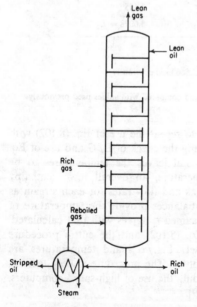

Figure 8.28 Reboiled absorber.

ABSORPTION WITH CHEMICAL REACTION

Many industrial absorption processes are accompanied by chemical reaction. Reaction in the liquid of the absorbed component with a reagent in the absorbing liquid is especially common. Sometimes the reagent and product of reaction are both soluble, as in the absorption of carbon dioxide into an aqueous solution of the ethanolamines or other alkaline solutions. In contrast, furnace gases containing sulfur dioxide can be contacted with slurries of insoluble ground limestone in water, to form insoluble calcium sulfite. There are many other important examples, for which Ref. 28 should be consulted. Reaction of the absorbed solute and a reagent accomplishes two things favorable to the absorption rate: (1) destruction of the absorbed solute as it forms a compound reduces the equilibrium partial pressure of the solute, consequently increasing the concentration difference between the bulk gas and the interface; and the absorption rate is also increased; (2) the liquid-phase mass-transfer coefficient is increased in magnitude, which also contributes to increased absorption rates. These effects have been given extensive theoretical analysis [1, 10, 38] but relatively little experimental verification.

NOTATION FOR CHAPTER 8

Any consistent set of units may be used.

a	specific interface surface, area/packed volume L^2/L^3
A	absorption factor, L/mG, dimensionless
A'	effective absorption factor [Eq. (8.110)], dimensionless
A_E	effective absorption factor [Eq. (8.111)], dimensionless
A, B, C	components A, B, C
c	concentration, moles/volume, mole/L^3
C_J	heat capacity of component J as a gas, FL/mole T
C_L	heat capacity of a liquid, FL/mole T
C_p	heat capacity of a gas at constant pressure, FL/mole T
d_s	equivalent diameter of packing, L
D	diffusivity, L^2/Θ
D_{IJ}	diffusivity of component I in a binary gas IJ mixture, L^2/Θ
D_{Im}	mean diffusivity of I in a multicomponent mixture, L^2/Θ
D_L	liquid diffusivity, L^2/Θ
e	2.7183
E_{MGE}	Murphree gas tray efficiency corrected for entrainment, fractional
E_O	overall tray efficiency, fractional
$F_{G,J}$	gas-phase mass-transfer coefficient for component J, mole/$L^2\Theta$
$F_{L,J}$	liquid-phase mass-transfer coefficient for component J, mole/$L^2\Theta$
F_O	overall mass-transfer coefficient, mole/$L^2\Theta$
G	total gas rate for tray towers, mole/Θ; for packed towers, superficial molar mass velocity, mole/$L^2\Theta$
G_J	gas molar mass velocity of component J, mole/$L^2\Theta$
G'	gas mass velocity M/$L^2\Theta$
h	heat-transfer coefficient, FL/$L^2\Theta T$
h'	heat-transfer coefficient corrected for mass transfer, FL/$L^2\Theta T$

H_G	molar enthalpy of a gas referred to pure substance at temperature t_0, **FL/mole**
H'_G	molar enthalpy of a gas per mole of carrier gas, referred to pure substance at temperature t_0, **FL/mole**
$\bar{H}_{J,i}$	partial enthalpy of component J in solution at concentration x_i, temperature t_i, **FL/mole**
H_L	enthalpy of a liquid solution referred to pure substance at temperature t_0, **FL/mole**
H_t	height of a transfer unit, **L**
H_{tO}	height of an overall transfer unit, **L**
ΔH_S	integral heat of solution per mole of solution, **FL/mole**
k	thermal conductivity, **FL/LTθ**
k_G	gas mass-transfer coefficient, **mole/L$^2\theta$(F/L^2)**
k_L	liquid mass-transfer coefficient, **mole/L$^2\theta$(mole/L^3)**
k_x	liquid mass-transfer coefficient, **mole/L$^2\theta$(mole fraction)**
k_y	gas mass-transfer coefficient, **mole/L$^2\theta$(mole fraction)**
K	overall mass-transfer coefficient (units indicated by subscripts, as for k's)
ln	natural logarithm
log	common logarithm
L	total molar liquid rate, for tray towers, **mole/θ**
L'	total liquid mass rate (tray towers), **M/θ**; total mass rate per unit tower cross section (packed towers), **M/L$^2\theta$**
L_S	solvent mass rate (tray towers), **M/θ**; total mass rate per unit tower cross section (packed towers), **M/L$^2\theta$**
m	slope of the equilibrium curve, dy^*/dx; equilibrium distribution ratio, y^*/x, dimensionless
$m'_{J,r}$	slope of a logarithmic reference-substance plot at concentration x_J, $(\log \bar{p}_J^*)/(\log p_r)$
M	molecular weight, **M/mole**
N_J	mass-transfer flux of component J, **mole/L$^2\theta$**
N_p	number of equilibrium trays
N_t	number of transfer units
N_{tO}	number of overall transfer units
p	vapor pressure of a pure substance, **F/L^2**
$\bar{p}$	partial pressure, **F/L^2**
p_t	total pressure, **F/L^2**
Pr	Prandtl number, $C_p\mu/k$, dimensionless
q	flux of heat transfer, **FL/L$^2\theta$**; a constant
Q_T	rate of total heat removal, **FL/θ**
r	const
R_J	$N_J/\Sigma N$, dimensionless
s	const
S	stripping factor, mG/L, dimensionless
S'	effective stripping factor [Eq. (8.113)], dimensionless
S_E	effective stripping factor [Eq. (8.114)], dimensionless
Sc	Schmidt number, $\mu/\rho D$, dimensionless
t	temperature, **T**
x	concentration in the liquid, mole fraction, **mole/mole**
X	concentration in the liquid, mole/mole solvent, **mole/mole**
X'	concentration in the liquid, mole/mole entering liquid, **mole/mole**
y	concentration in the gas, mole fraction, **mole/mole**
Y	concentration in the gas, mole/mole carrier gas, **mole/mole**
Y'	concentration in the gas, mole/mole entering gas, **mole/mole**
Z	height of packing, **L**
Δ	difference
ε	volume fraction voids, dry packing, **L^3/L^3**
ε_o	volume fraction voids, irrigated packing, **L^3/L^3**
λ_{J0}	molar latent heat of vaporization of J at temperature t_0, **FL/mole**

$\lambda_{r,t}$	molar latent heat of vaporization of reference substance at temperature t, FL/mole
μ	viscosity, $M/L\Theta$
ρ	density, M/L^3
φ_t	fractional total liquid holdup, volume liquid/volume packed space, L^3/L^3

Subscripts

av	average
A, B, C, J	components A, B, C, J
G	gas
i	interface
L	liquid
M	logarithmic mean
n	effluent from tray n
N_p	effluent from tray N_p
O	overall
r	reference substance
0	liquid entering top tray
	reference temperature for enthalpy
1	bottom of a packed tower; effluent from tray 1 (tray tower)
2	top of a packed tower; effluent from tray 2 (tray tower)

Superscript

*	in equilibrium with the other phase

REFERENCES

1. Astarita, G.: "Mass Transfer with Chemical Reaction," Elsevier, Amsterdam, 1966.
2. Baker, T. C.: *Ind. Eng. Chem.*, **27**, 977 (1935).
3. Bourne, J. R., and G. C. Coggan: *Chem. Eng. Sci.*, **24**, 196 (1969).
4. Bourne, J. R., U. von Stockar, and G. C. Coggan: *Ind. Eng. Chem. Process Des. Dev.*, **13**, 115, 124 (1974).
5. Chiang, S. H., and H. L. Toor: *AIChE J.*, **10**, 398 (1964).
6. Chilton, T. H., and A. P. Colburn: *Ind. Eng. Chem.*, **27**, 255 (1935).
7. Coggan, G. C., and J. R. Bourne: *Trans. Inst. Chem. Eng. Lond.*, **47**, T96, T160 (1969).
8. Colburn, A. P.: *Trans. AIChE*, **35**, 211 (1939).
9. Colburn, A. P.: *Ind. Eng. Chem.*, **33**, 111 (1941).
10. Danckwerts, P. V.: "Gas-Liquid Reactions," McGraw-Hill Book Company, New York, 1970.
11. De Lancey, G. B., and S. H. Chiang: *Ind. Eng. Chem. Fundam.*, **9**, 138 (1970).
12. Denbigh, K.: "Principles of Chemical Equilibrium," Cambridge University Press, London, 1961.
13. Edmister, W. C.: *Ind. Eng. Chem.*, **35**, 837 (1943).
14. Fair, J. R., in R. H. Perry and C. H. Chilton (eds.), "The Chemical Engineers' Handbook," 5th ed., pp. 18-32 to 18-39, McGraw-Hill Book Company, New York, 1973.
15. Feintuch, H. M., and R. E. Treybal: *Ind. Eng. Chem. Process Des. Dev.*, in press.
16. Friday, J. R., and B. D. Smith: *AIChE J.*, **10**, 698 (1964).
17. Hachmuth, K. H.: *Chem. Eng. Prog.*, **45**, 716 (1949); **47**, 523, 621 (1951).
18. Holland, C. D.: "Multicomponent Distillation," Prentice-Hall, Englewood Cliffs, N. J., 1963.
19. Holland, C. D.: "Fundamentals and Modeling of Separation Processes," Prentice-Hall, Englewood Cliffs, N.J., 1975.
20. Holland, C. D., G. P. Pendon, and S. E. Gallum: *Hydrocarbon Process.*, **54**(1), 101 (1975).
21. Horton, G., and W. B. Franklin: *Ind. Eng. Chem.*, **32**, 1384 (1940).
22. Hougen, O. A., Watson, K. M., and Ragatz, R. A.: "Chemical Process Principles, Part One," 2d ed., Wiley, New York, 1954.

23. Hulsworth, C. E.: *Chem. Eng. Prog.*, **69**(2), 50 (1973).
24. "International Critical Tables," vol. V, p. 213, McGraw-Hill Book Company, New York, 1929.
25. Ishii, Y., and F. D. Otto: *Can. J. Chem. Eng.*, **51**, 601 (1973).
26. Khurani, M. K.: private communication, Dec. 13, 1969.
27. King, M. B.: "Phase Equilibrium in Mixtures," Pergamon, London, 1969.
28. Kohl, A., and F. Riesenfeld: "Gas Purification," 2d ed., Gulf Publishing, Houston, 1974.
29. Koyashi, R., and D. L. Katz: *Ind. Eng. Chem.*, **45**, 440, 446 (1953).
30. Kwauk, M.: *AIChE J.*, **2**, 240 (1956).
31. Liley, P. E., and W. R. Gambill, in R. H. Perry and C. H. Chilton (eds.), "The Chemical Engineers' Handbook," 5th ed., sec. 3, McGraw-Hill Book Company, New York, 1973.
32. Othmer, D. F., et al.: *Ind. Eng. Chem.*, **36**, 858 (1944); **51**, 89 (1959).
33. Owens, W. R., and R. N. Maddox: *Ind. Eng. Chem.*, **60**(12), 14 (1968).
34. Prausnitz, J. M.: "Molecular Thermodynamics of Fluid Phase Equilibria," Prentice-Hall, Englewood Cliffs, N.J., 1969.
35. Pray, H. A., C. C. Schweickert, and B. H. Minnich: *Ind. Eng. Chem.*, **44**, 1146 (1952).
36. Raal, J. D., and M. K. Khurani: *Can. J. Chem. Eng.*, **51**, 162 (1973); *Process Technol. Int.*, **18**(6-7), 267 (1973).
37. Rackett, H. G.: *Chem. Eng.*, **71**, 108 (Dec. 21, 1964).
38. Sherwood, T. K., R. L. Pigford, and C. R. Wilke: "Mass Transfer," McGraw-Hill Book Company, New York, 1975.
39. Smith, B. D.: "Design of Equilibrium Stage Processes," McGraw-Hill Book Company, New York, 1963.
40. Smith, J. M., and H. C. Van Ness: "Introduction to Chemical Engineering Thermodynamics," 3d ed., McGraw-Hill Book Company, New York, 1975.
41. Treybal, R. E.: *Ind. Eng. Chem.*, **61**(7), 36 (1969).
42. Verma, S. L., and G. B. de Lancey: *AIChE J.*, **21**, 96 (1975).
43. Wankat, C. P.: *Ind. Eng. Chem. Process Des. Dev.*, **11**, 302 (1972).

PROBLEMS

8.1 The equilibrium partial pressures of carbon dioxide over aqueous solutions of monoethanolamine (30 wt%) [Mason and Dodge, *Trans. AIChE*, **32**, 27 (1936)] are:

$\dfrac{\text{mol } CO_2}{\text{mol solution}}$	Partial pressure CO_2, mmHg		
	25°C	50°C	75°C
0.050			65
0.052		7.5	93.5
0.054		13.6	142.6
0.056		25.0	245
0.058	5.6	47.1	600
0.060	12.8	96.0	
0.062	29.0	259	
0.064	56.0		
0.066	98.7		
0.068	155		
0.070	232		

A plant manufacturing dry ice will burn coke in air to produce a flue gas which, when cleaned and cooled, will contain 15% CO_2, 6% O_2, 79.0% N_2. The gas will be blown into a sieve-tray-tower scrubber at 1.2 std atm, 25°C, to be scrubbed with a 30% ethanolamine solution entering at 25°C. The scrubbing liquid, which is recycled from a stripper, will contain 0.058 mol CO_2/mol solution.

The gas leaving the scrubber is to contain 2% CO_2. Assume isothermal operation.

(a) Determine the minimum liquid/gas ratio, mol/mol.

(b) Determine the number of kilograms of solution to enter the absorber per cubic meter of entering gas, for an L/G ratio of 1.2 times the minimum.

Ans.: 18.66.

(c) Determine the number of theoretical trays for the conditions of part (b).

Ans.: 2.5.

(d) The viscosity of the solution is 6.0 cP; sp gr = 1.012. Estimate the average m and the overall tray efficiency to be expected. How many real trays are required?

8.2 (a) Determine the ideal trays for the absorption of Prob. 8.1, assuming adiabatic operation. Use 25.6 kg absorbent solution/m^3 gas in, which will be about 1.2 times the minimum. The heat of solution of CO_2 is 1675 kJ evolved/kg CO_2 absorbed (720 Btu/lb), referred to gaseous CO_2 and liquid solution. The specific heat of the solution is 3.433 kJ/(kg solution) · K (0.82 Btu/lb · °F).

Ans.: 2.6.

(b) Suppose the absorber planned for isothermal operation (L/G and theoretical trays of Prob. 8.1) were operated adiabatically. What concentration of CO_2 in the exit gas could be expected? [Note that this normally requires trial-and-error determination of both the top-tray temperature and effluent gas concentration. However, study of the calculations of part (a) should indicate that in this case the top-tray temperature need not be determined by trial.]

Ans.: 4.59%.

8.3 Derive Eq. (8.16). *Hint*: Start with the definition of E_{MGE} and locate a *pseudo-equilibrium line*, which would be used together with the operating line for graphically constructing steps representing real trays. Then use the Kremser equation (5.55a) with the pseudo-equilibrium line, by moving the origin of the Y, X diagram to the intercept of the pseudo-equilibrium line with the Y axis.

8.4 Carbon disulfide, CS_2, used as a solvent in a chemical plant, is evaporated from the product in a drier into an inert gas (essentially N_2) in order to avoid an explosion hazard. The vapor–N_2 mixture is to be scrubbed with an absorbent hydrocarbon oil, which will be subsequently steam-stripped to recover the CS_2. The CS_2–N_2 mixture has a partial pressure of CS_2 equal to 50 mmHg at 24°C (75°F) and is to be blown into the absorber at essentially standard atmospheric pressure at the expected rate of 0.40 m^3/s (50 000 ft^3/h). The vapor content of the gas is to be reduced to 0.5%. The absorption oil has av mol wt 180, viscosity 2 cP, and sp gr 0.81 at 24°C. The oil enters the absorber essentially stripped of all CS_2, and solutions of oil and CS_2, while not actually ideal, follow Raoult's law [see Ewell et al.: *Ind. Eng. Chem.*, **36**, 871 (1944)]. The vapor pressure of CS_2 at 24°C = 346 mmHg. Assume isothermal operation.

(a) Determine the minimum liquid/gas ratio.

(b) For a liquid/gas ratio of 1.5 times the minimum, determine the kilograms of oil to enter the absorber per hour.

(c) Determine the number of theoretical trays required, both graphically and analytically.

(d) For a conventional sieve-tray tower, estimate the overall tray efficiency to be expected and the number of real trays required.

8.5 Design a suitable sieve tray for the absorber of Prob. 8.4 and compute its hydraulics (which may be considered constant for all the trays) and the number of real trays required, graphically and through Eq. (8.16). Take the surface tension as 30 dyn/cm.

8.6 Determine the number of equilibrium trays for the absorber of Prob. 8.4, assuming adiabatic operation. Use a liquid rate of 2.27 kg/s, which is about 1.5 times the minimum. The specific heats are:

	kJ/kmol · K	Btu/lb mol · °F
Oil	362.2	86.5
CS_2, liquid	76.2	18.2
CS_2, vapor	46.89	11.2

The latent heat of vaporization of CS_2 is 27 910 kJ/kmol at 24°C (12 000 Btu/lb mol).

Ans.: 3.6.

8.7 Starting with Eq. (8.34), replace the y's by the equivalent Y's and derive Eq. (8.36).

8.8 With the help of the Kremser equation and Eq. (8.50), derive the relation between N_p and N_{tOG} for constant absorption factor. Establish the conditions when $N_p = N_{tOG}$.

8.9 Design a tower packed with 50-mm (2-in) ceramic Raschig rings for the carbon disulfide scrubber of Prob. 8.4. Assume isothermal operation and use a liquid/gas ratio of 1.5 times the minimum and a gas-pressure drop not exceeding 327 $(N/m^2)/m$ (0.4 inH_2O/ft) of packing. The liquid surface tension = 30 dyn/cm. A procedure follows.

(a) Determine the tower diameter.

Ans.: 0.725 m.

(b) Using average (top and bottom) flow rates and fluid properties, compute the mass-transfer coefficients $F_G a$ and $F_L a$ and heights of transfer units H_{tG}, H_{tL}, and H_{tOG}.

(c) Compute N_{tG} and with H_{tG}, the packing height.

Ans.: 9.61, 4.73 m.

(d) Compute N_{tOG} through the following methods and the corresponding packing height for each: Eq. (8.36), Eq. (8.48), and Fig. 8.20.

(e) Compare the gas-pressure drop for the full depth of packing with that for all the trays of Prob. 8.5. At a power cost of 15 cents per kilowatt-hour and a blower-motor efficiency of 50%, calculate the annual (350-day) difference in power cost for the two towers.

8.10 For dilute mixtures and cases when Henry's law applies, prove that the number of overall transfer units for *cocurrent* gas absorption in packed towers is given by

$$N_{tOG} = \frac{A}{A+1} \ln \frac{y_1 - mx_1}{y_2 - mx_2}$$

where subscript 1 indicates the top (where gas and liquid enter) and subscript 2 indicates the bottom of the tower.

8.11 Benzene vapor in a coke-oven gas is scrubbed from the gas with wash oil in a countercurrent packed scrubber. The resulting benzene–wash-oil solution is then heated to 125°C and stripped in a tray tower, using steam as the stripping medium. The stripped wash oil is then cooled and recycled to the absorber. Some data relative to the operation follow:

Absorption:

Benzene in entering gas = 1 mol %

Pressure of absorber = 800 mmHg

Oil circulation rate = 2 m^3/1000 m^3 gas at STP (15 U.S. gal/1000 ft^3)

Oil sp gr = 0.88   mol wt = 260

Henry's-law constant $m = \dfrac{y^*}{x}$ for benzene–wash-oil = $\begin{cases} 0.095 & \text{at } 25°C \\ 0.130 & \text{at } 27°C \end{cases}$

N_{tOG} = 5 overall gas transfer units

Stripping:

Pressure = 1 std atm   steam = 1 std atm, 125°C

Henry's-law constant for benzene $m = \dfrac{y^*}{x} = 3.08$ at 125°C

Number of equilibrium stages = 5

(a) In the winter, it is possible to cool the recycled stripped oil to 20°C, at which temperature the absorber then operates. Under these conditions 72.0 kg steam is used in the stripper per 1000 m^3 gas at STP (4.5 lb/1000 ft^3). Calculate the percent benzene recovery from the coke-oven gas in the winter.

Ans.: 92.6%.

(b) In the summer it is impossible to cool the recycled wash oil to lower than 27°C with the available cooling water. It has been suggested that the steam rate to the stripper can be increased so as to obtain the same percentage benzene recovery as in the winter. Assuming that the absorber then operates at 27°C, with the same oil rate, and that N_{tOG} and equilibrium stages remain the same, what steam rate in the summer would be required?

(c) If the oil rate cannot be increased but the steam rate in the summer is increased by 20% over the winter value, what summer recovery of benzene can be expected?
Ans.: 86.2%.

8.12 It is desired to reduce the ammonia content of 0.0472 m^3/s (26.7°C, 1 std atm) (6000 ft^3/h) of an ammonia-air mixture from 5.0 to 0.04% by volume by water scrubbing. There is available a 0.305-m (1-ft)-diameter tower packed with 25-mm Berl saddles to a depth of 3.66 m (12 ft). Is the tower satisfactory, and if so, what water rate should be used? At 26.7°C, ammonia-water solutions follow Henry's law up to 5 mol% ammonia in the liquid, and $m = 1.414$.

8.13 A tower, 0.6 m diameter, packed with 50-mm ceramic Raschig rings to a depth of 1.2 m, is to be used for producing a solution of oxygen in water for certain pollution-control operations. The packed space will be pressurized at 5 std atm abs with pure oxygen gas from a gas cylinder. There is to be no gas outlet, and gas will enter from the cylinder only to maintain pressure as oxygen is dissolved. Water will flow down the packing continuously at 1.50 kg/s. The temperature is to be 25°C. Assuming that the entering water is oxygen-free, what concentration of oxygen can be expected in the water effluent?
Ans.: 1.108×10^{-4} mole fraction.

Data: at 25°C, the diffusivity of O_2 in water = 2.5×10^{-5} cm^2/s, viscosity of water = 0.894 cP, and the solubility of O_2 in H_2O follows Henry's law: $\bar{p} = 4.38 \times 10^4 x$, where $\bar{p}$ = equilibrium partial pressure, atm, and x = mole fraction O_2 in the liquid. Neglect water-vapor content of the gas.

8.14 A system for recovering methanol from a solid product wet with methanol involves evaporation of the methanol into a stream of inert gas, essentially nitrogen. In order to recover the methanol from the nitrogen, an absorption scheme involving washing the gas countercurrently with water in a packed tower is being considered. The resulting methanol-water solution is then to be distilled to recover the methanol. *Note:* An alternative scheme is to wash the gas with refrigerated methanol, whence no distillation is required (see Prob. 7.11). A still different approach is to adsorb the methanol onto activated carbon (see Chap. 11).

The absorption tower will be filled with 38-mm (1.5-in) ceramic Raschig rings. Use a gas-pressure drop not to exceed 400 N/m^2 per meter of packed depth. Tower shells are available in 25-mm-diameter increments.

Partial pressures of methanol and water over aqueous solutions of methanol are available in "The Chemical Engineers' Handbook," 5th ed., p. 3-68. Plot these in the manner of Fig. 8.2, using water as reference substance.

(a) Assume isothermal operation at 26.7°C (80°F). Gas in: 0.70 m^3/s (90 000 ft^3/h) at 1 std atm, 26.7°C; partial pressure methanol = 100 mmHg. Sc_G = 0.783 (MeOH–N$_2$). Gas out: methanol partial pressure = 15 mmHg. Liquid in: water, methanol-free, 26.7°C, 0.15 kg/s (1190 lb/h); neglect evaporation of water. Calculate the depth of packing required.

(b) Assume adiabatic operation. Gas in: 0.70 m^3/s at 1 std atm, 26.7°C, partial pressure methanol = 100 mmHg, partial pressure water = 21.4 mmHg. Gas out: methanol partial pressure = 15 mmHg. Liquid in: water, methanol-free, 26.7°C, 1.14 kg/s (9000 lb/h). Assume that the gas leaves at $t_G = 30$°C, with $\bar{p}_{H_2O} = 26$ mmHg. Do *not* neglect evaporation of water. Calculate the following for the bottom of the tower: t_L, t_i, dt_G/dZ, dY_A/dZ, dY_C/dZ. For $\Delta Y_A = -0.02$, compute the next value of Z, Y_A, Y_S, t_G, and t_L.

Additional data: heat of solution ["International Critical Tables," vol. IV, p. 159]:

x_A, mole fraction MeOH in liquid	0.5	0.10	0.15	0.20	0.25	0.30
ΔH_S, integral heat of solution, kJ/kmol soln at 20°C	−341.7	−597.8	−767.6	−872.3	−914.1	−916.4

Use a base temperature $t_0 = 20°C$. The latent heat of vaporization of methanol at 20°C = 1163 kJ/kg. Heat capacities of liquid solutions: see "Chemical Engineers' Handbook," 5th ed., p. 3-136. Molar heat capacities of gases: N_2, 29.141; MeOH, 52.337; H_2O, 33.579 kJ/kmol · K.

8.15 A gas containing 88% CH_4, 4% C_2H_6, 5% n-C_3H_8, and 3% n-C_4H_{10} is to be scrubbed isothermally at 37.8°C (100°F), 5 std atm pressure, in a tower containing the equivalent of eight equilibrium trays, 80% of the C_3H_8 is to be removed. The lean oil will contain 0.5 mol% of C_4H_{10} but none of the other gaseous constituents. The rest of the oil is nonvolatile. What quantity of lean oil, mol/mol wet gas, should be used, and what will be the composition of rich oil and scrubbed gas? The Henry's-law constants m are

CH_4	C_2H_6	n-C_3H_8	n-C_4H_{10}
32	6.7	2.4	0.74

8.16 An absorber of four theoretical trays, to operate adiabatically at 1.034 kN/m^2 (150 lb$_f$/in^2), is fed with 1 mol per unit time each of liquid and gas, each entering at 32.2°C (90°F), as follows:

Component	Liquid, mole fraction	Gas, mole fraction
CH_4		0.70
C_2H_6		0.12
n-C_3H_8		0.08
n-C_4H_{10}	0.02	0.06
n-C_5H_{12}	0.01	0.04
Nonvolatile oil	0.97	

Estimate the composition and rate (mol/time) of the exit gas.

   Data: Henry's-law constants, enthalpies at 32.2°C relative to saturated liquid at −129°C (−200°F), and molar heat capacities are given in Table 8.1.

Table 8.1 Data for Prob. 8.16

		32.2°C (90°F)				C_p, kJ/kmol · K (Btu/lb mol · °F)	
Component	m	H_G,† kJ/kmol × 10^{-3} (Btu/lb mol)	H_L, kJ/kmol × 10^{-3} (Btu/lb mol)	m 37.8°C (100°F)	43°C (110°F)	Gas	Liquid
CH_4	16.5	12.91 (5510)	9.77 (4200)	17.0	17.8	37.7 (9.0)	50.2 (12.01)
C_2H_6	3.40	22.56 (9700)	15.58 (6700)	3.80	4.03	62.8 (15.0)	83.7 (20.0)
n-C_3H_8	1.16	31.05 (13 350)	16.86 (7250)	1.30	1.44	79.6 (19.0)	129.8 (31.0)
n-C_4H_{10}	0.35	41.05 (17 650)	20.70 (8900)	0.41	0.47	96.3 (23.0)	159.1 (38.0)
n-C_5H_{12}	0.123	50.94 (21 400)	24.66 (10 600)	0.140	0.165	117.2 (28.0)	184.2 (44.0)
Oil							376.8 (90.0)

   † Data from DePriester, *Chem. Eng. Progr. Symp. Ser.*, **49**(7) 1 (1953); "Chemical Engineers' Handbook," 5th ed., pp. 13-12 and 13-13; Maxwell, "Data Book on Hydrocarbons," Van Nostrand, Princeton, N.J., 1950.

8.17 In this chapter it was shown that there is an optimum, or most economical, value of the absorption factor. There are also optimum values for:

(*a*) Outlet gas concentration when there are valuable solutes

(*b*) The solute concentration in the liquid entering the absorber where the absorbent liquid is recirculated through a stripper or fractionator for solute recovery

(*c*) The tray spacing in a tray tower

(*d*) The gas pressure drop per unit height in a packed tower

For each item explain why there is an optimum value.

CHAPTER NINE

DISTILLATION

Distillation is a method of separating the components of a solution which depends upon the distribution of the substances between a gas and a liquid phase, applied to cases where all components are present in both phases. Instead of introducing a new substance into the mixture in order to provide the second phase, as is done in gas absorption or desorption, the new phase is created from the original solution by vaporization or condensation.

In order to make clear the distinction between distillation and the other operations, let us cite a few specific examples. In the separation of a solution of common salt and water, the water can be completely vaporized from the solution without removal of salt since the latter is for all practical purposes quite nonvolatile at the prevailing conditions. This is the operation of evaporation. Distillation, on the other hand, is concerned with the separation of solutions where all the components are appreciably volatile. In this category, consider the separation of the components of a liquid solution of ammonia and water. By contacting the ammonia-water solution with air, which is essentially insoluble in the liquid, the ammonia can be stripped or desorbed by processes which were discussed in Chap. 8, but the ammonia is then mixed with water vapor and air and is not obtained in pure form. On the other hand, by application of heat, we can partially vaporize the solution and thereby create a gas phase consisting of nothing but water and ammonia. And since the gas will be richer in ammonia than the residual liquid, a certain amount of separation will have resulted. By appropriate manipulation of the phases or by repeated vaporizations and condensations it is then ordinarily possible to make as complete a separation as may be desired, recovering both components of the mixture in as pure a state as we wish.

The advantages of such a separation method are clear. In distillation the new phase differs from the original by its heat content, but heat is readily added or removed, although of course the cost of doing this must inevitably be considered. Absorption or desorption operations, on the other hand, which depend upon the introduction of a foreign substance, provide us with a new solution which in turn may have to be separated by one of the diffusional operations unless it happens that the new solution is useful directly.

There are in turn certain limitations to distillation as a separation process. In absorption or similar operations, where it has been agreed to introduce a foreign substance to provide a new phase for distribution purposes, we can ordinarily choose from a great variety of solvents in order to provide the greatest possible separation effect. For example, since water is ineffectual in absorbing hydrocarbon gases from a gas mixture, we choose instead a hydrocarbon oil which provides a high solubility. But in distillation there is no such choice. The gas which can be created from a liquid by application of heat inevitably consists only of the components constituting the liquid. Since the gas is therefore chemically very similar to the liquid, the change in composition resulting from the distribution of the components between the two phases is ordinarily not very great. Indeed, in some cases the change in composition is so small that the process becomes impractical; it may even happen that there is no change in composition whatsoever.

Nevertheless the direct separation, which is ordinarily possible by distillation, into pure products requiring no further processing has made this perhaps the most important of all the mass-transfer operations.

VAPOR-LIQUID EQUILIBRIA

The successful application of distillation methods depends greatly upon an understanding of the equilibria existing between the vapor and liquid phases of the mixtures encountered. A brief review of these is therefore essential. The emphasis here will be on binary mixtures.

Pressure-Temperature-Concentration Phase Diagram

Let us first consider binary mixtures which we shall term *ordinary*; by this is meant that the liquid components dissolve in all proportions to form homogeneous solutions which are not necessarily ideal and that no complications of maximum or minimum boiling points occur. We shall consider component A of the binary mixture A-B as the more volatile, i.e., the vapor pressure of pure A at any temperature is higher than the vapor pressure of pure B. The vapor-liquid equilibrium for each pure substance of the mixture is of course its vapor-pressure-temperature relationship, as indicated in Fig. 7.1. For binary mixtures an additional variable, concentration, must likewise be considered. Mole fractions are the most convenient concentration terms to use, and throughout this discussion x will be the mole fraction of the more volatile substance A in the liquid and y^* the corresponding equilibrium mole fraction of A in the vapor.

Complete graphical representation of the equilibria requires a three-dimensional diagram [29, 47], as in Fig. 9.1. The curve marked p_A is the vapor-pressure curve of A, lying entirely in the nearest composition plane at $x = 1.0$. The curve extends from its critical point C_A to its triple point T_A, but the complications of the solid phase which do not enter into distillation operations will not be considered. Similarly curve p_B is the vapor pressure of pure B, in the far plane at $x = 0$. The liquid and vapor regions at compositions between $x = 0$ and 1.0 are separated by a double surface which extends smoothly from p_A to p_B. The shape of this double surface is most readily studied by considering sections at constant pressure and constant temperature, examples of which are shown in the figure.

Constant-Pressure Equilibria

Consider first a typical section at constant pressure (Fig. 9.2a). The intersection of the double surface of Fig. 9.1 with the constant-pressure plane produces a looped curve without maxima or minima extending from the boiling point of pure B to that of pure A at the pressure in question. The upper curve provides the temperature-vapor composition (t-y^*) relationship, the lower that of the temperature-liquid composition (t-x). Liquid and vapor mixtures at equilibrium are at the same temperature and pressure throughout, so that horizontal *tie lines*

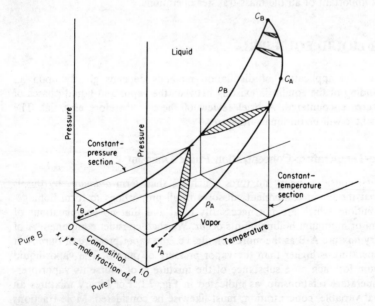

Figure 9.1 Binary vapor-liquid equilibria.

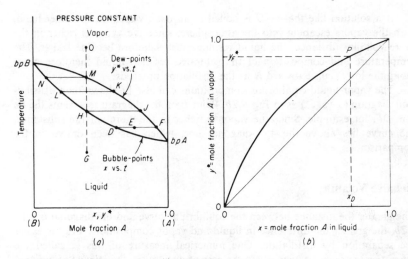

Figure 9.2 Constant-pressure vapor-liquid equilibria.

such as line DF join equilibrium mixtures at D and F. There are an infinite number of such tie lines for this diagram. A mixture on the lower curve, as at point D, is a saturated liquid; a mixture on the upper curve, as at F, is a saturated vapor. A mixture at E is a two-phase mixture, consisting of a liquid phase of composition at D and a vapor phase of composition at F in such proportions that the average composition of the entire mixture is represented by E. The relative amounts of the equilibrium phases are related to the segments of the tie line,

$$\frac{\text{Moles of } D}{\text{Moles of } F} = \frac{\text{line } EF}{\text{line } DE} \tag{9.1}$$

Consider a solution at G in a closed container which can be kept at constant pressure by moving a piston. The solution is entirely liquid. If it is heated, the first bubble of vapor forms at H and has the composition at J, richer in the more volatile substance, and hence the lower curve is called the *bubble-point-temperature curve*. As more of the mixture is vaporized, more of the vapor forms at the expense of the liquid, giving rise, for example, to liquid L and its equilibrium vapor K, although the composition of the entire mass is still the original as at G. The last drop of liquid vaporizes at M and has the composition at N. Superheating the mixture follows the path MO. The mixture has vaporized over a temperature range from H to M, unlike the single vaporization temperature of a pure substance. Thus, the term *boiling point* for a solution ordinarily has no meaning since vaporization occurs over a temperature range, i.e., from the bubble point to the dew point. If the mixture at O is cooled, all the phenomena reappear in reverse order. Condensation, for example, starts at M, whence the upper curve is termed the dew-point curve, and continues to H.

If a solution like that at H is boiled in an open vessel, on the other hand, with the vapors escaping into the atmosphere, since the vapor is richer in the more volatile substance, the liquid residue must therefore become leaner. The temperature and composition of the saturated residual liquid therefore move along the lower curve toward N as the distillation proceeds.

The vapor-liquid equilibrium compositions can also be shown on a distribution diagram (x vs. y^*) as in Fig. 9.2b. Point P on the diagram represents the tie line DF, for example. Since the vapor is richer in the more volatile substance, the curve lies above the 45° diagonal line, which has been drawn in for comparison.

Relative Volatility

The greater the distance between the equilibrium curve and the diagonal of Fig. 9.2b, the greater the difference in liquid and vapor compositions and the easier the separation by distillation. One numerical measure of this is called the *separation factor* or, particularly in the case of distillation, the *relative volatility* α. This is the ratio of the concentration ratio of A and B in one phase to that in the other and is a measure of the separability,

$$\alpha = \frac{y^*/(1 - y^*)}{x/(1 - x)} = \frac{y^*(1 - x)}{x(1 - y^*)} \tag{9.2}$$

The value of α will ordinarily change as x varies from 0 to 1.0. If $y^* = x$ (except at $x = 0$ or 1), $\alpha = 1.0$ and no separation is possible. The larger the value of α above unity, the greater the degree of separability.

Increased Pressures

At higher pressures the sections at constant pressure will of course intersect the double surface of Fig. 9.1 at increased temperatures. The intersections can be projected onto a single plane, as in Fig. 9.3a. It should be noted that not only do the looped curves occur at higher temperatures, but they also usually become narrower. This is readily seen from the corresponding distribution curves of Fig. 9.3b. The relative volatilities and hence the separability therefore usually become less at higher pressures. As the critical pressure of one component is exceeded, there is no longer a distinction between vapor and liquid for that component, and for mixtures the looped curves are therefore shorter, as at pressures above p_{t3}, the critical pressure for A in the figure. Distillation separations can be made only in the region where a looped curve exists.

For particular systems, the critical pressure of the less volatile substance may be reached before that of the more volatile, and it is also possible that the double surface of Fig. 9.1 will extend at intermediate compositions to a small extent beyond the critical pressures of either substance.

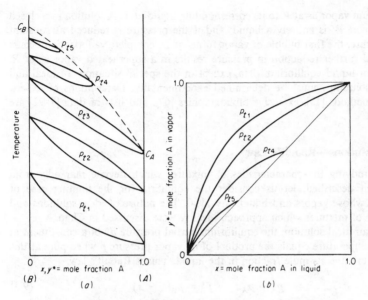

Figure 9.3 Vapor-liquid equilibria at increased pressures.

Constant-Temperature Equilibria

A typical constant-temperature section of the three-dimensional phase diagram is shown in Fig. 9.4. The intersection of the constant-temperature plane with the double surface of Fig. 9.1 provides the two curves which extend without maxima or minima from the vapor pressure of pure B to that of pure A. As before, there are an infinite number of horizontal tie lines, such as TV, which join an

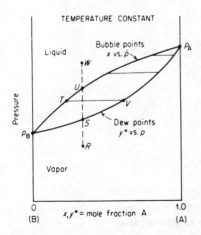

Figure 9.4 Constant-temperature vapor-liquid equilibria.

equilibrium vapor as at V to its corresponding liquid at T. A solution in a closed container at W is entirely a liquid, and if the pressure is reduced at constant temperature, the first bubble of vapor forms at U, complete vaporization occurs at S, and further reduction in pressure results in a superheated vapor as at R.

Vapor-liquid equilibrium data, except in the special situations of ideal and regular solutions, must be determined experimentally. Descriptions of experimental methods [18], extensive bibliographies [67], and lists of data [7, 22] are available.

Ideal Solutions—Raoult's Law

Before studying the characteristics of mixtures which deviate markedly from those just described, let us consider the equilibria for the limiting case of mixtures whose vapors and liquids are ideal. The nature of ideal solutions and the types of mixtures which approach ideality were discussed in Chap. 8.

For an ideal solution, the equilibrium partial pressure $\bar{p}^*$ of a constituent at a fixed temperature equals the product of its vapor pressure p when pure at this temperature and its mole fraction in the liquid. This is Raoult's law.

$$\bar{p}_A^* = p_A x \qquad \bar{p}_B^* = p_B(1 - x) \tag{9.3}$$

If the vapor phase is also ideal,

$$p_t = \bar{p}_A^* + \bar{p}_B^* = p_A x + p_B(1 - x) \tag{9.4}$$

and the total as well as the partial pressures are linear in x at a fixed temperature. These relationships are shown graphically in Fig. 9.5. The equilibrium vapor composition can then be computed at this temperature. For example, the value of y^* at point D on the figure equals the ratio of the

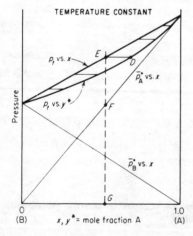

Figure 9.5 Ideal solutions.

distances *FG* to *EG*,

$$y^* = \frac{\bar{p}_A^*}{p_t} = \frac{p_A x}{p_t} \qquad (9.5)$$

$$1 - y^* = \frac{\bar{p}_B^*}{p_t} = \frac{p_B(1 - x)}{p_t} \qquad (9.6)$$

The relative volatility α is, by substitution in Eq. (9.2),

$$\alpha = \frac{p_A}{p_B} \qquad (9.7)$$

For ideal solutions, it is then possible to compute the entire vapor-liquid equilibria from the vapor pressures of the pure substances. For pressures too high to apply the ideal-gas law, fugacities are used instead of pressures [29, 42]. It is also possible to compute the equilibria for the special class of solutions known as *regular* [21]. For all other mixtures, however, it is necessary to obtain the data experimentally.

Illustration 9.1 Compute the vapor-liquid equilibria at constant pressure of 1 std atm for mixtures of *n*-heptane with *n*-octane, which may be expected to form ideal solutions.

SOLUTION The boiling points at 1 std atm of the substances are *n*-heptane (A), 98.4°C and *n*-octane (B), 125.6°C. Computations are therefore made between these temperatures. For example, at 110°C, $p_A = 1050$ mmHg, $p_B = 484$ mmHg, $p_t = 760$ mmHg. Eq. (9.4):

$$x = \frac{p_t - p_B}{p_A - p_B} = \frac{760 - 484}{1050 - 484} = 0.487 \text{ mole fraction heptane in liquid}$$

Eq. (9.5):

$$y^* = \frac{p_A x}{p_t} = \frac{1050(0.487)}{760} = 0.674 \text{ mole fraction heptane in vapor}$$

Eq. (9.7):

$$\alpha = \frac{p_A}{p_B} = \frac{1050}{484} = 2.17$$

In similar fashion, the data of the following table can be computed:

t, °C	p_A, mmHg	p_B, mmHg	x	y^*	α
98.4	760	333	1.0	1.0	2.28
105	940	417	0.655	0.810	2.25
110	1050	484	0.487	0.674	2.17
115	1200	561	0.312	0.492	2.14
120	1350	650	0.1571	0.279	2.08
125.6	1540	760	0	0	2.02

Curves of the type of Fig. 9.2 can now be plotted. Note that although the vapor pressures of the pure substances vary considerably with temperature, α for ideal solutions does not. In this case, an average of the computed α's is 2.16, and substituting this in Eq. (9.2), rearranged,

$$y^* = \frac{\alpha x}{1 + x(\alpha - 1)} = \frac{2.16x}{1 + 1.16x}$$

provides an expression which for many purposes is a satisfactory empirical relation between y^* and x for this system at 1 atm.

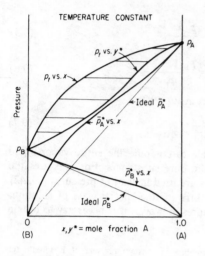

TEMPERATURE CONSTANT

Pressure

p_A

p_t vs. y^*

p_t vs. x

←Ideal $\bar{p}_A^*$

$\bar{p}_A^*$ vs. x

p_B

$\bar{p}_B^*$ vs. x

Ideal $\bar{p}_B^*$

0

(B)

$x, y^* =$ mole fraction A

1.0

(A)

Figure 9.6 Positive deviations from ideality.

Positive Deviations from Ideality

A mixture whose total pressure is greater than that computed for ideality [Eq. (9.4)] is said to show positive deviations from Raoult's law. Most mixtures fall into this category. In these cases the partial pressures of each component are larger than the ideal, as shown in Fig. 9.6.† It should be noted that as the concentration for each component approaches unity mole fraction, the partial pressures for that substance approach ideality tangentially. Raoult's law, in other words, is nearly applicable to the substance present in very large concentrations. This is the case for all substances except where association within the vapor or electrolytic dissociation within the liquid occurs.

The distribution diagram (x vs. y^*) for systems of this type appears much the same as that of Fig. 9.2b.

Minimum-boiling mixtures—azeotropes When the positive deviations from ideality are sufficiently large and the vapor pressures of the two components are not too far apart, the total-pressure curves at constant temperature may rise through a maximum at some concentration, as in Fig. 9.7a. Such a mixture is said to form an *azeotrope,* or *constant-boiling mixture.* The significance of this is more readily seen by study of the constant-pressure section (Fig. 9.7b or c). The liquid- and vapor-composition curves are tangent at point L, the point of

---

† The ratio of the actual equilibrium partial pressure of a component $\bar{p}^*$ to the ideal value px is the activity coefficient referred to the pure substance: $\gamma = p^*/px$. Since γ is greater than unity in these cases and log γ is therefore positive, the deviations are termed positive deviations from ideality. A very extensive science of the treatment of nonideal solutions through activity coefficients has been developed by which, from a very small number of data, all the vapor-liquid equilibria of a system can be predicted [42].

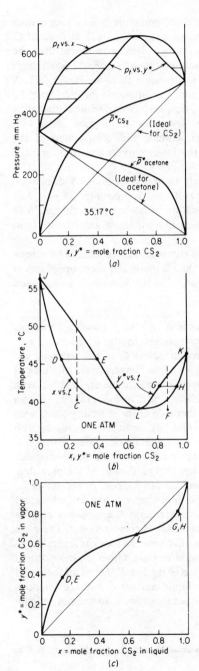

Figure 9.7 Minimum-boiling azeotropism in the system carbon disulfide–acetone: (a) at constant temperature; (b) and (c) at constant pressure.

azeotropism at this pressure, which represents the minimum-boiling temperature for this system. For all mixtures of composition less than L, such as those at C, the equilibrium vapor (E) is richer in the more volatile component than the liquid (D). For all mixtures richer than L, however, such as at F, the equilibrium vapor (G) is less rich in the more volatile substance than the liquid (H). A mixture of composition L gives rise to a vapor of composition identical with the liquid, and it consequently boils at constant temperature and without change in composition. If solutions either at D or H are boiled in an open vessel with continuous escape of the vapors, the temperature and composition of the residual liquids in each case move along the lower curve away from point L (toward K for a liquid at H, and toward J for one at D).

Solutions like these cannot be completely separated by ordinary distillation methods at this pressure, since at the azeotropic composition $y^* = x$ and $\alpha = 1.0$.† The azeotropic composition as well as its boiling point changes with pressure. In some cases, changing the pressure may eliminate azeotropism from the system.

Azeotropic mixtures of this sort are very common, and thousands have been recorded [25]. One of the most important is the ethanol-water azeotrope which at 1 atm occurs at 89.4 mole percent ethanol and 78.2°C. Azeotropism disappears in this system at pressures below 70 mmHg.

Partial liquid miscibility Some substances exhibit such large positive deviations from ideality that they do not dissolve completely in the liquid state, e.g., isobutanol-water (Fig. 9.8). The curves through points C and E represent the solubility limits of the constituents at relatively low temperatures. Mixtures of composition and temperature represented by points within the central area, such as point D, form two liquid phases at equilibrium at C and E, and line CE is a liquid tie line. Mixtures in the regions on either side of the solubility limits such as at F are homogeneous liquids. The solubility ordinarily increases with increased temperature, and the central area consequently decreases in width. If the pressure were high enough for vaporization not to occur, the liquid-solubility curves would continue along the broken extensions as shown. At the prevailing pressure, however, vaporization occurs before this can happen, giving rise to the branched vapor-liquid equilibrium curves. For homogeneous liquids such as that at F, the vapor-liquid equilibrium phenomena are normal, and such a mixture boils initially at H to give the first bubble of vapor of composition J. The same is true of any solution richer than M, except that here the vapor is leaner in the more volatile component. Any two-phase liquid mixture within the composition range from K to M will boil at the temperature of the line KM, and all these give rise to the same vapor of composition L. A liquid mixture of average composition L, which produces a vapor of the same composition, is sometimes called a

---

† For compositions to the right of L (Fig. 9.7) α as usually computed is less than unity, and the reciprocal of α is then ordinarily used.

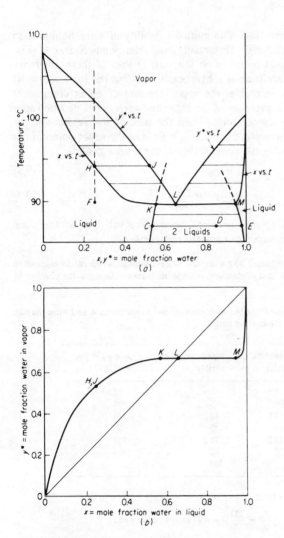

Figure 9.8 Isobutanol-water at 1 std atm.

heteroazeotrope. The corresponding distribution diagram with the tie line *HJ*, solubility limits at the bubble point *K* and *M*, and the azeotropic point *L* is shown in Fig. 9.8*b*.

In relatively few instances the azeotropic composition lies outside the limits of solubility, as in the systems methyl ethyl ketone–water and phenol-water. In others, especially when the components have a very large difference in their boiling points, no azeotrope can form, as for ammonia-toluene and carbon dioxide–water.

Insoluble liquids; steam distillation The mutual solubility of some liquids is so small that they can be considered substantially insoluble: points K and M (Fig. 9.8) are then for all practical purposes on the vertical axes of these diagrams. This is the case for a mixture such as a hydrocarbon and water, for example. If the liquids are completely insoluble, the vapor pressure of either component cannot be influenced by the presence of the other and each exerts its true vapor pressure at the prevailing temperature. When the sum of the separate vapor pressures equals the total pressure, the mixture boils, and the vapor composition is readily computed, assuming the applicability of the simple gas law,

$$p_A + p_B = p_t \tag{9.8}$$

$$y^* = \frac{p_A}{p_t} \tag{9.9}$$

So long as two liquid phases are present, the mixture will boil at the same temperature and produce a vapor of constant composition.

Illustration 9.2 A mixture containing 50 g water and 50 g ethylaniline, which can be assumed to be essentially insoluble, is boiled at standard atmospheric pressure. Describe the phenomena that occur.

SOLUTION Since the liquids are insoluble, each exerts its own vapor pressure, and when the sum of these equals 760 mmHg, the mixture boils.

t, °C	p_A(water), mmHg	p_B(ethylaniline), mmHg	$p_t = p_A + p_B$, mmHg
38.5	51.1	1	52.1
64.4	199.7	5	205
80.6	363.9	10	374
96.0	657.6	20	678
99.15	737.2	22.8	760
113.2	1225	40	1265
204		760	

The mixture boils at 99.15°C.

$$y^* = \frac{p_A}{p_t} = \frac{737.2}{760} = 0.97 \text{ mole fraction water}$$

$$1 - y^* = \frac{p_B}{p_t} = \frac{22.8}{760} = 0.03 \text{ mole fraction ethylaniline}$$

The original mixture contained $50/18.02 = 2.78$ g mol water and $50/121.1 = 0.412$ g mol ethylaniline. The mixture will continue to boil at 99.15°C, with an equilibrium vapor of the indicated composition, until all the water has evaporated together with $2.78(0.03/0.97) = 0.086$ g mol of the ethylaniline. The temperature will then rise to 204°C, and the equilibrium vapor will be pure ethylaniline.

Note that by this method of distillation with steam, so long as liquid water is present, the high-boiling organic liquid can be made to vaporize at a temperature much lower than its normal boiling point without the necessity of a vacuum-pump equipment operating at 22.8 mmHg. If boiled at 204°C, this compound will undergo considerable decomposition. However, the heat requirements of the steam-distillation process are great since such a large amount of water must be evaporated simultaneously. Alternatives would be (1) to operate at a different total pressure in the presence of liquid water where the ratio of the vapor pressures of the substances may be more favorable and (2) to sparge superheated steam (or other insoluble gas) through the mixture in the absence of liquid water and to vaporize the ethylaniline by allowing it to saturate the steam.

Negative Deviations from Ideality

When the total pressure of a system at equilibrium is less than the ideal value, the system is said to deviate negatively from Raoult's law. Such a situation is shown in Fig. 9.9 at constant temperature. Note that in this case, as with positive deviations, where neither vapor association nor liquid dissociation occurs, the partial pressures of the constituents of the solution approach ideality as their concentrations approach 100 percent. The constant-pressure diagram for such a case has the same general appearance as the diagrams shown in Fig. 9.2.

Maximum-boiling mixtures—azeotropes When the difference in vapor pressures of the components is not too great and in addition the negative deviations are large, the curve for total pressure against composition may pass through a minimum, as in Fig. 9.10a. This condition gives rise to a maximum in the boiling temperatures, as at point L (Fig. 9.10b), and a condition of azeotropism. The equilibrium vapor is leaner in the more volatile substance for liquids whose x is less than the azeotropic composition and greater if x is larger. Solutions on either side of the azeotrope, if boiled in an open vessel with escape of the vapor, will ultimately leave a residual liquid of the azeotropic composition in the vessel.

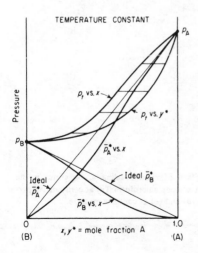

Figure 9.9 Negative deviations from ideality.

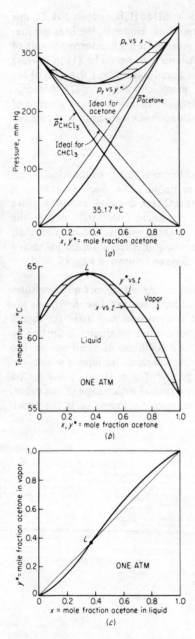

Figure 9.10 Maximum-boiling azeotropism in the system acetone-chloroform: (*a*) at constant temperature; (*b*) and (*c*) at constant pressure.

Maximum-boiling azeotropes are less common than the minimum type. One which is very well known is that of hydrochloric acid–water (11.1 mole percent HCl, 110°C, at 1 std atm), which can be prepared simply by boiling a solution of any strength of the acid in an open vessel. This is one method of standardizing hydrochloric acid.

Enthalpy-Concentration Diagrams

Binary vapor-liquid equilibria can also be plotted on coordinates of enthalpy vs. concentration at constant pressure. Liquid-solution enthalpies include both sensible heat and the heat of mixing the components

$$H_L = C_L(t_L - t_0)M_{av} + \Delta H_S \qquad (9.10)$$

where C_L is the heat capacity of the solution, energy/(mol) (degree), and ΔH_S is the heat of solution at t_0 and the prevailing concentration referred to the pure liquid components, energy/mol *solution*. For saturated liquids, t_L is the bubble point corresponding to the liquid concentration at the prevailing pressure. Heat-of-solution data vary in form, and some adjustment of the units of tabulated data may be necessary. If heat is evolved on mixing, ΔH_S will be negative, and for ideal solutions it is zero. For ideal solutions, the heat capacity is the weighted average of those for the pure components.

For present purposes, saturated-vapor enthalpies can be calculated adequately by assuming that the unmixed liquids are heated separately as liquids to the gas temperature t_G (the dew point), each vaporized at this temperature, and the vapors mixed

$$H_G = y\left[C_{L,A}M_A(t_G - t_0) + \lambda_A M_A\right] + (1 - y)\left[C_{L,B}M_B(t_G - t_0) + \lambda_B M_B\right]$$
$$(9.11)$$

where λ_A and λ_B are latent heats of vaporization of pure substances at t_G in energy per mole and $C_{L,A}$ and $C_{L,B}$ are heat capacities of pure liquids, energy/(mole)(degree).

In the upper part of Fig. 9.11, which represents a typical binary mixture, the enthalpies of saturated vapors at their dew points have been plotted vs. y and those of the saturated liquids at their bubble points vs. x. The vertical distances between the two curves at $x = 0$ and 1 represent, respectively, the molar latent heats of B and A. The heat required for complete vaporization of solution C is $H_D - H_C$ energy/mole solution. Equilibrium liquids and vapors may be joined by tie lines, of which line EF is typical. The relation between this equilibrium phase diagram and the xy plot is shown in the lower part of Fig. 9.11. Here the point G represents the tie line EF, located on the lower plot in the manner shown. Other tie lines, when projected to the xy plot, produce the complete equilibrium-distribution curve.

Characteristics of the Hxy and xy diagrams [46] Let point M on Fig. 9.11 represent M mol of a mixture of enthalpy H_M and concentration z_M, and

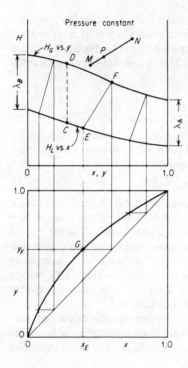

Figure 9.11 Enthalpy-concentration coordinates.

similarly N is N mol of a mixture of properties H_N, z_N. Adiabatic mixing of M and N will produce P mol of a mixture of enthalpy H_P and concentration z_P. A total material balance is

$$M + N = P \tag{9.12}$$

and a balance for component A is

$$Mz_M + Nz_N = Pz_P \tag{9.13}$$

An enthalpy balance is

$$MH_M + NH_N = PH_P \tag{9.14}$$

Elimination of P between Eqs. (9.12) and (9.13) and between (9.12) and (9.14) yields

$$\frac{M}{N} = \frac{z_N - z_P}{z_P - z_M} = \frac{H_N - H_P}{H_P - H_M} \tag{9.15}$$

This is the equation of a straight line on the enthalpy-concentration plot, passing through points (H_M, z_M), (H_N, z_N), and (H_P, z_P). Point P is therefore on the straight line MN, located so that M/N = line NP/line PM. Similarly if mixture N were *removed* adiabatically from mixture P, the mixture M would result.

Consider now mixture $C(H_C, z_C)$ in Fig. 9.12. It will be useful to describe such a mixture in terms of saturated vapors and liquids, since distillation is

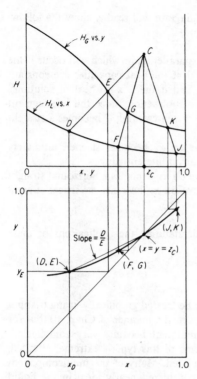

Figure 9.12 Relation between the diagrams.

mostly concerned with such mixtures. C can be considered the result of adiabatically removing saturated liquid D from saturated vapor $E(DE$ is *not* a tie line), and x_D and y_E can be located on the lower diagram as shown. But C can equally well be considered as having been produced by adiabatically subtracting F from G, or J from K, or indeed by such a combination of saturated liquids and vapors given by any line from C which intersects the saturated-enthalpy curves. These, when projected to the lower diagram, form the curve shown there. Thus any point C on the Hxy diagram can be represented by the difference between saturated vapors and liquids and in turn also by a curve on the xy plot. For the combination $E - D = C$, a material balance shows

$$\frac{D}{E} = \frac{z_C - y_E}{z_C - x_D} = \frac{\text{line } CE}{\text{line } CD} \qquad (9.16)$$

This is the equation on the xy diagram of the chord of slope D/E drawn between point (y_E, x_D) and $y = x = z_C$ on the 45° line. Similarly, the ratios F/G and J/K would be shown by the slopes of chords drawn from these points to $y = x = z_C$.

Consideration of the geometry of the diagram will readily show the following:

1. If the $H_G y$ and $H_L x$ curves are straight parallel lines (which will occur if the molar latent heats of A and B are equal, if the heat capacities are constant over the prevailing temperature range, and if there is no heat of solution), then $D/E = F/G = J/K$ for adiabatic subtraction, since the line-segment ratios are then equal, and the curve on xy representing C becomes a straight line.
2. If point C is moved upward, the curve on xy becomes steeper, ultimately coinciding with the 45° line when C is at infinite enthalpy.
3. If point C is on the $H_G y$ curve, the curve on xy becomes a horizontal straight line; if C is on the $H_L x$ curve, the curve on xy becomes a vertical straight line.

These concepts will be useful in understanding the applications of these diagrams.

Multicomponent Systems

Nonideal systems of three components can be treated graphically, using triangular coordinates to express the compositions in the manner of Chap. 10, but for more than three components graphical treatment becomes very complicated. Actually our knowledge of nonideal systems of this type is extremely limited, and relatively few data have been accumulated. Many of the multicomponent systems of industrial importance can be considered nearly ideal in the liquid phase for all practical purposes. This is particularly true for hydrocarbon mixtures of the same homologous series, e.g., those of the paraffin series or the lower-boiling aromatic hydrocarbons.[†] In such cases Raoult's law, or its equivalent in terms of fugacities, can be applied and the equilibria calculated from the properties of the pure components. But it is generally unsafe to predict detailed behavior of a multicomponent system from consideration of the pure components alone, or even from a knowledge of the simple binary systems that may be formed from the components. For example, three-component systems sometimes form ternary azeotropes whose equilibrium vapor and liquid have identical compositions. But the fact that one, two, or three binary azeotropes exist between the components does not make the formation of a ternary azeotrope certain, and a ternary azeotrope need not necessarily coincide with the composition of minimum or maximum bubble-point temperature for the system at constant pressure. The complexities of the systems make the digital computer helpful in dealing with the data [44].

[†] Hydrocarbons of different molecular structure, however, frequently form such nonideal solutions that azeotropism occurs, for example, in the binary systems hexane–methyl cyclopentane, hexane-benzene, and benzene-cyclohexane.

For multicomponent systems particularly, it is customary to describe the equilibrium data by means of the distribution coefficient $m^\dagger$. For component J,

$$m_J = \frac{y_J^*}{x_J} \tag{9.17}$$

where in general m_J depends upon the temperature, pressure, and composition of the mixture. The relative volatility α_{IJ} of component I with respect to J is

$$\alpha_{IJ} = \frac{y_I^*/x_I}{y_J^*/x_J} = \frac{m_I}{m_J} \tag{9.18}$$

For *ideal* solutions at moderate pressure, m_J is independent of composition and depends only upon the temperature (as this affects vapor pressure) and the total pressure:

$$m_J = \frac{p_J}{p_t} \tag{9.19}$$

and

$$\alpha_{IJ} = \frac{p_I}{p_J} \tag{9.20}$$

Bubble point For the bubble-point vapor,

$$\Sigma y_i^* = 1.0 \tag{9.21}$$

or

$$m_A x_A + m_B x_B + m_C x_C + \cdots = 1.0 \tag{9.22}$$

With component J as a reference component,

$$\frac{m_A x_A}{m_J} + \frac{m_B x_B}{m_J} + \frac{m_C x_C}{m_J} + \cdots = \frac{1.0}{m_J} \tag{9.23}$$

or

$$\alpha_{AJ} x_A + \alpha_{BJ} x_B + \alpha_{CJ} x_C + \cdots = \Sigma \alpha_{i,J} x_i = \frac{1}{m_J} \tag{9.24}$$

The equilibrium bubble-point vapor composition is given by

$$y_i = \frac{\alpha_{i,J} x_i}{\Sigma \alpha_{i,J} x_i} \tag{9.25}$$

The liquid composition and total pressure having been fixed, the calculation of the temperature is made by trial to satisfy Eq. (9.24). Convergence is rapid since α's vary only slowly with changing temperature (see Illustration 9.3).

Dew point For the dew-point liquid,

$$\Sigma x_i^* = 1.0 \tag{9.26}$$

$$\frac{y_A}{m_A} + \frac{y_B}{m_B} + \frac{y_C}{m_C} + \cdots = 1.0 \tag{9.27}$$

† K rather than m is generally used to denote the distribution coefficient, but here we shall use K for the overall mass-transfer coefficient.

With component J as a reference component,

$$\frac{m_J y_A}{m_A} + \frac{m_J y_B}{m_B} + \frac{m_J y_C}{m_C} + \cdots = m_J \tag{9.28}$$

$$\frac{y_A}{\alpha_{AJ}} + \frac{y_B}{\alpha_{BJ}} + \frac{y_C}{\alpha_{CJ}} + \cdots = \Sigma \frac{y_i}{\alpha_{i,J}} = m_J \tag{9.29}$$

The dew-point liquid composition is given by

$$x_i = \frac{y_i / \alpha_{i,J}}{\Sigma(y_i / \alpha_{i,J})} \tag{9.30}$$

Illustration 9.3 demonstrates the method of computation.

Illustration 9.3 A solution of hydrocarbons at a total pressure 350 kN/m² (50.8 lb$_f$/in²) has the analysis n-C$_3$H$_8$ = 5.0, n-C$_4$H$_{10}$ = 30.0, n-C$_5$H$_{12}$ = 40.0, n-C$_6$H$_{14}$ = 25.0 mol%. Compute the bubble point and the dew point.

SOLUTION Values of m will be taken from the DePriester nomograph [*Chem. Eng. Prog. Symp. Ser.*, **49**(7), 1 (1953); also "Chemical Engineers' Handbook," 5th ed., fig. 13-6b], which assumes ideal liquid solutions.

Bubble point Column 1 lists the mole-fraction compositions. As a first estimate of the bubble point, 60°C is chosen, and column 2 lists the corresponding m's at 60°C, 350 kN/m². The reference component is chosen to be the pentane, and column 3 lists the relative volatilities (α_{C_3, C_5} = 4.70/0.62 = 7.58, etc.). $\Sigma \alpha_{i, C_5} x_i$ = 1.702 (column 4), whence [Eq. (9.23)] m_{C_5} = 1/1.702 = 0.588. The corresponding temperature, from the nomograph, is 56.8°C (135°F). The calculations are repeated for this temperature in columns 5 to 7. m_{C_5} = 1/1.707 = 0.586, corresponding to t = 56.7°C, which checks the 56.8°C assumed. This is the bubble point, and the corresponding bubble-point vapor composition is given in column 8 (0.391/1.707 = 0.229, etc.).

	x_i (1)	m_i, 60°C (2)	α_{i, C_5} (3)	$\alpha_{i, C_5} x_i$ (4)	m_i, 56.8°C (5)	α_{i, C_5} (6)	$\alpha_{i, C_5} x_i$ (7)	y_i (8)
n-C$_3$	0.05	4.70	7.58	0.379	4.60	7.82	0.391	0.229
n-C$_4$	0.30	1.70	2.74	0.822	1.60	2.72	0.816	0.478
n-C$_5$	0.40	0.62	1.00	0.400	0.588	1.00	0.400	0.234
n-C$_6$	0.25	0.25	0.403	0.1008	0.235	0.40	0.100	0.0586
				1.702			1.707	1.000

	y_i (1)	m_i, 80°C (2)	α_{i, C_5} (3)	$y_i / \alpha_{i, C_5}$ (4)	m_i, 83.7°C (5)	α_{i, C_5} (6)	$y_i / \alpha_{i, C_5}$ (7)	x_i (8)
n-C$_3$	0.05	6.30	6.56	0.00762	6.60	6.11	0.00818	0.0074
n-C$_4$	0.30	2.50	2.60	0.1154	2.70	2.50	0.120	0.1088
n-C$_5$	0.40	0.96	1.00	0.400	1.08	1.00	0.400	0.3626
n-C$_6$	0.25	0.43	0.448	0.558	0.47	0.435	0.575	0.5213
				1.081			1.103	1.000

Dew point The procedure is similar to that for the bubble point. The first estimate is 80°C (176°F), whence $\Sigma y_i / \alpha_{i, C_5} = 1.081$ (column 4 $= m_{C_5}$ [Eq. (9.29)]. The corresponding temperature is 83.7°C (183°F). Repetition leads to $m_{C_5} = 1.103$, corresponding to 84.0°C, which is the dew point. The corresponding dew-point-liquid compositions are listed in column 8 (0.00818/1.103 = 0.0074, etc.).

SINGLE-STAGE OPERATION—FLASH VAPORIZATION

Flash vaporization, or equilibrium distillation as it is sometimes called, is a single-stage operation wherein a liquid mixture is partially vaporized, the vapor allowed to come to equilibrium with the residual liquid, and the resulting vapor and liquid phases are separated and removed from the apparatus. It may be batchwise or continuous.

A typical flowsheet is shown schematically in Fig. 9.13 for continuous operation. Here the liquid feed is heated in a conventional tubular heat exchanger or by passing it through the heated tubes of a fuel-fired furnace. The pressure is then reduced, vapor forms at the expense of the liquid adiabatically, and the mixture is introduced into a vapor-liquid separating vessel. The separator shown is of the cyclone type, where the feed is introduced tangentially into a covered annular space. The liquid portion of the mixture is thrown by centrifugal force to the outer wall and leaves at the bottom, while the vapor rises through the central chimney and leaves at the top. The vapor may then pass to a condenser, not shown in the figure. Particularly for flash vaporization of a volatile substance from a relatively nonvolatile one, operation in the separator can be carried out under reduced pressure, but not so low that ordinary cooling water will not condense the vapor product.

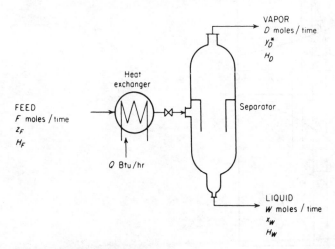

Figure 9.13 Continuous flash vaporization.

The product, D mol/time, richer in the more volatile substance, is in this case entirely a vapor. The material and enthalpy balances are

$$F = D + W \tag{9.31}$$

$$Fz_F = Dy_D + Wx_W \tag{9.32}$$

$$FH_F + Q = DH_D + WH_W \tag{9.33}$$

Solved simultaneously, these yield

$$-\frac{W}{D} = \frac{y_D - z_F}{x_W - z_F} = \frac{H_D - (H_F + Q/F)}{H_W - (H_F + Q/F)} \tag{9.34}$$

On the Hxy diagram, this represents a straight line through points of coordinates (H_D, y_D) representing D, (H_W, x_W) representing W, and $(H_F + Q/F, z_F)$ representing the feed mixture after it leaves the heat exchanger of Fig. 9.13. It is shown on the upper part of Fig. 9.14 as the line DW. The two left-hand members of Eq. (9.34) represent the usual single-stage operating line on distribution coordinates, of negative slope as for all single-stage (cocurrent) operations (see Chap. 5), passing through compositions representing the influent and effluent streams, points F and M on the lower figure. If the effluent streams were in equilibrium, the device would be an equilibrium stage and the products

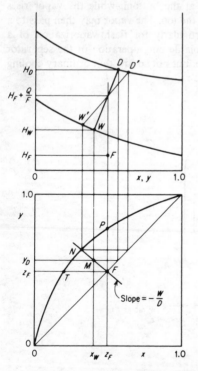

Figure 9.14 Flash vaporization.

D' and W' would be on a tie line in the upper figure and on the equilibrium curve at N on the lower figure. The richest vapor, but infinitesimal in amount, is that corresponding to P at the bubble point of the feed; and the leanest liquid, but also infinitesimal in amount, is that corresponding to T at the dew point of the feed. The compositions of the actual products will be between these limits, depending upon the extent of vaporization of the feed and the stage efficiency.

Partial Condensation

All the equations apply equally well to the case where the feed is a vapor and Q, the heat removed in the heat exchanger to produce incomplete condensation, is taken as negative. On the upper part of Fig. 9.14, point F is then either a saturated or superheated vapor.

Illustration 9.4 A liquid mixture containing 50 mol % n-heptane (A), 50 mol % n-octane (B), at 30°C, is to be continuously flash-vaporized at 1 std atm pressure to vaporize 60 mol % of the feed. What will be the composition of the vapor and liquid and the temperature in the separator for an equilibrium stage?

SOLUTION *Basis*: $F = 100$ mol feed, $z_F = 0.50$. $D = 60$ mol, $W = 40$ mol, $-W/D = -40/60 = -0.667$.

The equilibrium data were determined in Illustration 9.1 and are plotted in Fig. 9.15. The point representing the feed composition is plotted at P, and the operating line is drawn with a slope -0.667 to intersect the equilibrium curve at T, where $y_D^* = 0.575$ mole fraction heptane and $x_W = 0.387$ mole fraction heptane. The temperature at T is 113°C.

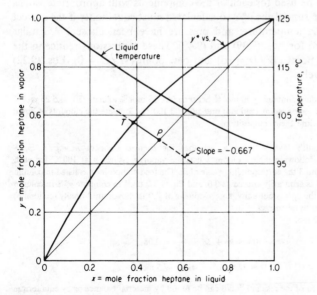

Figure 9.15 Solution to Illustration 9.4.

Multicomponent Systems—Ideal solutions

For mixtures leaving an equilibrium stage containing components A, B, C, etc., the equilibrium relation for any component J can be written

$$y_{J,D}^* = m_J x_{J,W} \tag{9.35}$$

Equation (9.34) also applies for each of the components, and when combined with Eq. (9.35) for any component J, for an equilibrium stage gives

$$\frac{W}{D} = \frac{m_J x_{J,W} - z_{J,F}}{z_{J,F} - x_{J,W}} = \frac{y_{J,D}^* - z_{J,F}}{z_{J,F} - y_{J,D}^*/m_J} \tag{9.36}$$

This provides the following expression, useful for equilibrium vaporization,

$$y_{J,D}^* = \frac{z_{J,F}(W/D + 1)}{1 + W/Dm_J} \tag{9.37}$$

$$\Sigma y_D^* = 1.0 \tag{9.38}$$

and for condensation,

$$x_{J,W} = \frac{z_{J,F}(W/D + 1)}{m_J + W/D} \tag{9.39}$$

$$\Sigma x_W = 1.0 \tag{9.40}$$

Thus Eq. (9.37) can be used for each of the components with appropriate values of m and z_F, and the sum of the y_D^*'s so calculated must equal unity if the correct conditions of W/D, temperature, and pressure have been chosen. A similar interpretation is used for Eqs. (9.39) and (9.40). These expressions reduce in the limit to the bubble point $(W/D = \infty)$ and dew point $(W/D = 0)$, Eqs. (9.22) and (9.30), respectively.

Illustration 9.5 A liquid containing 50 mol % benzene (A), 25 mol % toluene (B), and 25 mol % o-xylene (C) is flash-vaporized at 1 std atm pressure and 100°C. Compute the amounts of liquid and vapor products and their composition.

SOLUTION For Raoult's law, $y^* = px/p_t$, so that for each component $m = p/p_t$. $p_t = 760$ mmHg. In the following table, column 2 lists the vapor pressures p at 100°C for each substance and column 3 the corresponding value of m. The feed composition is listed in column 4. A value of W/D is arbitrarily chosen as 3.0, and Eq. (9.37) used to compute y_D^*'s in column 5. Since the sum of the y_D^*'s is not unity, a new value of W/D is chosen until finally (column 6) $W/D = 2.08$ is seen to be correct.

Basis: $F = 100$ mol.

$$100 = W + D \qquad \frac{W}{D} = 2.08$$

Therefore $\qquad\qquad D = 32.5$ mol $\qquad W = 67.5$ mol

The composition of the residual liquid can be found by material balance or by equilibrium relation, as in column 7.

Substance	$p =$ vapor pressure, mmHg	$m = \dfrac{p}{760}$	z_F	$\dfrac{W}{D} = 3.0$ $\dfrac{z_F(W/D + 1)}{1 + W/D_m} = y_D^*$	$\dfrac{W}{D} = 2.08$ y_D^*	$x_W = \dfrac{Fz_F - Dy_D^*}{W}$ $= \dfrac{y_D^*}{m}$
(1)	(2)	(3)	(4)	(5)	(6)	(7)
A	1370	1.803	0.50	$\dfrac{0.5(3 + 1)}{1 + 3/1.803} = 0.750$	0.715	0.397
B	550	0.724	0.25	0.1940	0.1983	0.274
C	200	0.263	0.25	0.0805	0.0865	0.329
				$\Sigma = 1.0245$	0.9998	1.000

Successive flash vaporizations can be made on the residual liquids in a series of single-stage operations, whereupon the separation will be better than that obtained if the same amount of vapor were formed in a single operation. As the amount of vapor formed in each stage becomes smaller and the total number of vaporizations larger, the operation approaches differential distillation in the limit.

DIFFERENTIAL, OR SIMPLE, DISTILLATION

If during an infinite number of successive flash vaporizations of a liquid only an infinitesimal portion of the liquid were flashed each time, the net result would be equivalent to a differential, or simple, distillation.

In practice this can only be approximated. A batch of liquid is charged to a kettle or still fitted with some sort of heating device such as a steam jacket, as in Fig. 9.16. The charge is boiled slowly, and the vapors are withdrawn as rapidly as they form to a condenser, where they are liquefied, and the condensate (distillate) is collected in the receiver. The apparatus is essentially a large-scale replica of the ordinary laboratory distillation flask and condenser. The first portion of the distillate will be the richest in the more volatile substance, and as distillation proceeds, the vaporized product becomes leaner. The distillate can therefore be collected in several separate batches, called *cuts*, to give a series of distilled products of various purities. Thus, for example, if a ternary mixture contained a small amount of a very volatile substance A, a majority of substance B of intermediate volatility, and a small amount of C of low volatility, the first cut, which would be small, would contain the majority of A. A large second cut would contain the majority of B reasonably pure but nevertheless contaminated with A and C, and the residue left in the kettle would be largely C. While all three cuts would contain all three substances, nevertheless some separation would have been obtained.

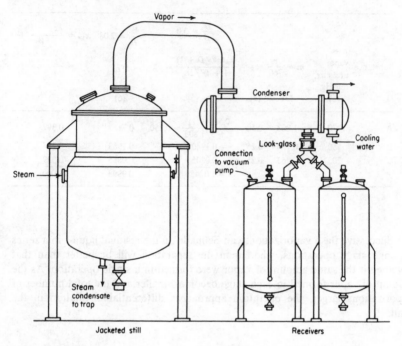

Figure 9.16 Batch still.

For such an operation to approach even approximately the theoretical characteristics of a differential distillation, it would have to proceed infinitely slowly so that the vapor issuing from the liquid would at all times be in equilibrium with the liquid. All entrainment would have to be eliminated, and there could be no cooling and condensation of the vapor before it entered the condenser. Despite the fact that these conditions are substantially impossible to attain, it is nevertheless useful to study the limiting results which a differential distillation could produce as a standard for comparison.

Binary Mixtures

The vapor issuing from a true differential distillation is at any time in equilibrium with the liquid from which it rises but changes continuously in composition. The mathematical approach must therefore be differential. Assume that at any time during the course of the distillation there are L mol of liquid in the still of composition x mole fraction A and that an amount dD mol of distillate is vaporized, of mole fraction y^* in equilibrium with the liquid. Then we have the following material balances:

	Total material	Component A
Moles in	0	0
Moles out	dD	$y^* dD$
Moles accumulated	dL	$d(Lx) = Ldx + xdL$
In—out = accumulation	$0 - dD = dL$	$0 - y^* dD = Ldx + xdL$

The last two equations become

$$y^* dL = Ldx + xdL \tag{9.41}$$

$$\int_W^F \frac{dL}{L} = \ln \frac{F}{W} = \int_{x_W}^{x_F} \frac{dx}{y^* - x} \tag{9.42}$$

where F is the moles of charge of composition x_F and W the moles of residual liquid of composition x_W. This is known as the *Rayleigh equation*, after Lord Rayleigh, who first derived it. It can be used to determine F, W, x_F, or x_W when three of these are known. Integration of the right-hand side of Eq. (9.42), unless an algebraic equilibrium relationship between y^* and x is available, is done graphically by plotting $1/(y^* - x)$ as ordinate against x as abscissa and determining the area under the curve between the indicated limits. The data for this are taken from the vapor-liquid equilibrium relationship. The *composited* distillate composition $y_{D,\, av}$ can be determined by a simple material balance,

$$Fx_F = Dy_{D,\, av} + Wx_W \tag{9.43}$$

Differential Condensation

This is a similar operation where a vapor feed is slowly condensed under equilibrium conditions and the condensate withdrawn as rapidly as it forms. As in distillation, the results can only be approximated in practice. A derivation similar to that above leads to

$$\ln \frac{F}{D} = \int_{y_F}^{y_D} \frac{dy}{y - x^*} \tag{9.44}$$

where F is the moles of feed vapor of composition y_F and D the vaporous residue of composition y_D.

Constant Relative Volatility

If Eq. (9.2) can describe the equilibrium relation at constant pressure by use of some average relative volatility α over the concentration range involved, this can be substituted in Eq. (9.42) to yield

$$\ln \frac{F}{W} = \frac{1}{\alpha - 1} \ln \frac{x_F(1 - x_W)}{x_W(1 - x_F)} + \ln \frac{1 - x_W}{1 - x_F} \tag{9.45}$$

and graphical integration can be avoided. This can be rearranged to another useful form

$$\log \frac{Fx_F}{Wx_W} = \alpha \log \frac{F(1 - x_F)}{W(1 - x_W)} \qquad (9.46)$$

which relates the number of moles of A remaining in the residue, Wx_W, to that of B remaining, $W(1 - x_W)$. These expressions are most likely to be valid for ideal mixtures, for which α is most nearly constant.

Illustration 9.6 Suppose the liquid of Illustration 9.4 [50 mol % n-heptane (A), 50 mol % n-octane (B)] were subjected to a differential distillation at atmospheric pressure, with 60 mol % of the liquid distilled. Compute the composition of the composited distillate and the residue.

SOLUTION *Basis*: $F = 100$ mol. $x_F = 0.50$, $D = 60$ mol, $W = 40$ mol. Eq. (9.42):

$$\ln \frac{100}{40} = 0.916 = \int_{xW}^{0.50} \frac{dx}{y^* - x}$$

The equilibrium data are given in Illustrations 9.1 and 9.4. From these, the following are calculated:

x	0.50	0.46	0.42	0.38	0.34	0.32
y^*	0.689	0.648	0.608	0.567	0.523	0.497
$1/(y^* - x)$	5.29	5.32	5.32	5.35	5.50	5.65

x as abscissa is plotted against $1/(y^* - x)$ as ordinate, and the area under the curve obtained beginning at $x_F = 0.50$. When the area equals 0.916, integration is stopped; this occurs at $x_W = 0.33$ mole fraction heptane in the residue. The composited distillate composition is obtained through Eq. (9.43),

$$100(0.50) = 60y_{D,\text{av}} + 40(0.33)$$

$$y_{D,\text{av}} = 0.614 \text{ mole fraction heptane}$$

Note that, for the same percentage vaporization, the separation in this case is better than that obtained by flash vaporization; i.e., each product is purer in its majority component.

Alternatively, since for this system the average $\alpha = 2.16$ at 1 atm (Illustration 9.1), Eq. (9.46):

$$\log \frac{100(0.5)}{40x_W} = 2.16 \log \frac{100(1 - 0.5)}{40(1 - x_W)}$$

from which by trial and error $x_W = 0.33$.

Multicomponent Systems—Ideal Solutions

For multicomponent systems forming ideal liquid solutions, Eq. (9.46) can be written for any two components. Ordinarily one component is chosen on which to base the relative volatilities, whereupon Eq. (9.46) is written once for each of the others. For example, for substance J, with relative volatility based on substance B,

$$\log \frac{Fx_{J,F}}{Wx_{J,W}} = \alpha_{JB} \log \frac{Fx_{B,F}}{Wx_{B,W}} \qquad (9.47)$$

and $$\Sigma x_W = 1.0 \qquad\qquad (9.48)$$

where $x_{J, F}$ is the mole fraction of J in the feed and $x_{J, W}$ that in the residue.

Illustration 9.7 A liquid containing 50 mol % benzene (A), 25 mol % toluene (B), and 25 mol % o-xylene (C) is differentially distilled at 1 atm, with vaporization of 32.5 mol % of the charge. Raoult's law applies. Compute the distillate and residue compositions. Note that this is the same degree of vaporization as in Illustration 9.5.

SOLUTION The average temperature will be somewhat higher than the bubble point of the feed (95°C, see Illustration 9.3) but is unknown. It will be taken as 100°C. Corrections can later be made by computing the bubble point of the residue and repeating the work at the average temperature, but α's vary little with moderate changes in temperature. The vapor pressures at 100°C are tabulated and α's calculated relative to toluene, as follows:

Substance	p = vapor pressure, 100°C, mmHg	α	x_F
A	1370	1370/550 = 2.49	0.50
B	550	1.0	0.25
C	200	0.364	0.25

Basis: $F = 100$ mol, $D = 32.5$ mol, $W = 67.5$ mol. Eq. (9.47):

For A: $$\log \frac{100(0.50)}{67.5 x_{A, W}} = 2.49 \log \frac{100(0.25)}{67.5 x_{B, W}}$$

For C: $$\log \frac{100(0.25)}{67.5 x_{C, W}} = 0.364 \log \frac{100(0.25)}{67.5 x_{B, W}}$$

Eq. (9.48):
$$x_{A, W} + x_{B, W} + x_{C, W} = 1.0$$

Solving simultaneously by assuming values of $x_{B, W}$, computing $x_{A, W}$ and $x_{C, W}$, and checking their sum until it equals unity gives $x_{A, W} = 0.385$, $x_{B, W} = 0.285$, $x_{C, W} = 0.335$. The sum is 1.005, which is taken as satisfactory.

The composited distillate composition is computed by material balances. For A,

$$100(0.50) = 32.5 y_{A, D, av} + 67.5(0.385) \qquad y_{A, D, av} = 0.742$$

Similarly, $\qquad y_{B, D, av} = 0.178$ and $y_{C, D, av} = 0.075$

Note the improved separation over that obtained by flash vaporization (Illustration 9.5).

CONTINUOUS RECTIFICATION—BINARY SYSTEMS

Continuous rectification, or fractionation, is a multistage countercurrent distillation operation. For a binary solution, with certain exceptions it is ordinarily possible by this method to separate the solution into its components, recovering each in any state of purity desired.

Rectification is probably the most frequently used separation method we have, although it is relatively new. While simple distillation was known in the first century, and perhaps earlier, it was not until about 1830 that Aeneas Coffey of Dublin invented the multistage, countercurrent rectifier for distilling ethanol from fermented grain mash [56]. His still was fitted with trays and down-spouts, and produced a distillate containing up to 95 percent ethanol, the azeotrope composition. We cannot do better today except by special techniques.

The Fractionation Operation

In order to understand how such an operation is carried out, recall the discussion of reboiled absorbers in Chap. 8 and Fig. 8.28. There, because the liquid leaving the bottom of an absorber is at best in equilibrium with the feed and may therefore contain substantial concentrations of volatile component, trays installed below the feed point were provided with vapor generated by a reboiler to strip out the volatile component from the liquid. This component then entered the vapor and left the tower at the top. The upper section of the tower served to wash the gas free of less volatile component, which entered the liquid to leave at the bottom.

So, too, with distillation. Refer to Fig. 9.17. Here the feed is introduced more or less centrally into a vertical cascade of stages. Vapor rising in the section above the feed (called the *absorption, enriching,* or *rectifying* section) is washed with liquid to remove or absorb the less volatile component. Since no extraneous material is added, as in the case of absorption, the washing liquid in this case is provided by condensing the vapor issuing from the top, which is rich in more volatile component. The liquid returned to the top of the tower is called *reflux,* and the material permanently removed is the *distillate,* which may be a vapor or a liquid, rich in more volatile component. In the section below the feed (*stripping* or *exhausting* section), the liquid is stripped of volatile component by vapor produced at the bottom by partial vaporization of the bottom liquid in the reboiler. The liquid removed, rich in less volatile component, is the *residue,* or *bottoms.* Inside the tower, the liquids and vapors are always at their bubble points and dew points, respectively, so that the highest temperatures are at the bottom, the lowest at the top. The entire device is called a *fractionator.*

The purities obtained for the two withdrawn products will depend upon the liquid/gas ratios used and the number of ideal stages provided in the two sections of the tower, and the interrelation of these must now be established. The cross-sectional area of the tower, however, is governed entirely by the quantities of materials handled, in accordance with the principles of Chap. 6.

Overall Enthalpy Balances

In Fig. 9.17, the theoretical trays are numbered from the top down, and subscripts generally indicate the tray from which a stream originates: for example, L_n is mol liquid/time falling from the nth tray. A bar over the quantity indicates that it applies to the section of the column below the point of introduction of the feed. The distillate product may be liquid, vapor, or a mixture. The reflux, however, must be liquid. The molar ratio of reflux to

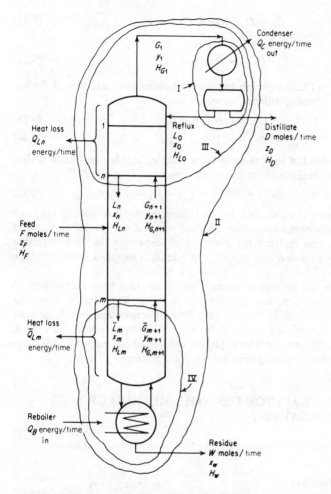

Figure 9.17 Material and enthalpy balances of a fractionator.

withdrawn distillate is the *reflux ratio*, sometimes called the *external reflux ratio*,

$$R = \frac{L_0}{D} \tag{9.49}$$

which is specified in accordance with principles to be established later.†

Consider the condenser, envelope I (Fig. 9.17). A total material balance is

$$G_1 = D + L_0 \tag{9.50}$$

† The ratio L/G is sometimes called the *internal reflux ratio*; L/F is also used for certain reflux correlations.

or
$$G_1 = D + RD = D(R + 1) \tag{9.51}$$

For substance A
$$G_1 y_1 = Dz_D + L_0 x_0 \tag{9.52}$$

Equations (9.50) to (9.52) establish the concentrations and quantities at the top of the tower. An enthalpy balance, envelope I,
$$G_1 H_{G1} = Q_C + L_0 H_{L0} + DH_D \tag{9.53}$$
$$Q_C = D[(R + 1)H_{G1} - RH_{L0} - H_D] \tag{9.54}$$

provides the heat load of the condenser. The reboiler heat is then obtained by a complete enthalpy balance about the entire apparatus, envelope II,
$$Q_B = DH_D + WH_W + Q_C + Q_L - FH_F \tag{9.55}$$

where Q_L is the sum of all the heat losses. Heat economy is frequently obtained by heat exchange between the residue product, which issues from the column at its bubble point, and the feed for purposes of preheating the feed. Equation (9.55) still applies provided that any such exchanger is included inside envelope II.

Two methods will be used to develop the relationship between numbers of trays, liquid/vapor ratios, and product compositions. The first of these, the method of Ponchon and Savarit [41, 46, 50] is rigorous and can handle all situations, but it requires detailed enthalpy data for its application. The second, the method of McCabe and Thiele [36], a simplification requiring only concentration equilibria, is less rigorous yet adequate for many purposes.†

MULTISTAGE (TRAY) TOWERS—THE METHOD OF PONCHON AND SAVARIT

The method will first be developed for the case of negligible heat losses.

The Enriching Section

Consider the enriching section through tray n, envelope III, Fig. 9.17. Tray n is any tray in this section. Material balances for the section are, for total material,
$$G_{n+1} = L_n + D \tag{9.56}$$

and for component A,
$$G_{n+1} y_{n+1} = L_n x_n + Dz_D \tag{9.57}$$
$$G_{n+1} y_{n+1} - L_n x_n = Dz_D \tag{9.58}$$

The left-hand side of Eq. (9.58) represents the difference in rate of flow of

---

† The treatment of each method is complete in itself, independent of the other. For instructional purposes, they may be considered in either order, or one may be omitted entirely.

component A, up − down, or the net flow upward. Since for a given distillation the right-hand side is constant, it follows that the difference, or net rate of flow of A upward, is constant, independent of tray number in this section of the tower, and equal to that permanently withdrawn at the top.

An enthalpy balance, envelope III, with *heat loss negligible*, is

$$G_{n+1}H_{G_{n+1}} = L_nH_{Ln} + Q_C + DH_D \tag{9.59}$$

Let Q' be the heat removed in the condenser and the permanently removed distillate, per mole of distillate. Then

$$Q' = \frac{Q_C + DH_D}{D} = \frac{Q_C}{D} + H_D \tag{9.60}$$

and

$$G_{n+1}H_{G_{n+1}} - L_nH_{L_n} = DQ' \tag{9.61}$$

The left-hand side of Eq. (9.61) represents the difference in rate of flow of heat, up − down, or the net flow upward. Since for a given set of circumstances the right-hand side is constant, the difference, or net rate of flow upward, is constant, independent of tray number in this section of the tower, and equal to that permanently taken out at the top with the distillate and at the condenser.

Elimination of D between Eqs. (9.56) and (9.57) and between Eqs. (9.56) and (9.61) yields

$$\frac{L_n}{G_{n+1}} = \frac{z_D - y_{n+1}}{z_D - x_n} = \frac{Q' - H_{G_{n+1}}}{Q' - H_{L_n}} \tag{9.62}$$

L_n/G_{n+1} is called the *internal reflux ratio*.

On the Hxy diagram, Eq. (9.62) is the equation of a straight line through $(H_{G_{n+1}}, y_{n+1})$ at G_{n+1}, (H_{Ln}, x_n) at L_n, and (Q', z_D) at Δ_D. The last is called a *difference point*, since its coordinates represent differences in rates of flow:

$$\Delta_D \begin{cases} Q' = \dfrac{\text{difference in heat flow, up − down}}{\text{net moles total substance out}} = \dfrac{\text{net heat out}}{\text{net moles out}} \\[3mm] z_D = \dfrac{\text{difference in flow of component A, up − down}}{\text{net moles total substance out}} = \dfrac{\text{net moles A out}}{\text{net moles out}} \end{cases}$$

Δ_D then represents a fictitious stream, in amount equal to the net flow outward (in this case D) and of properties (Q', z_D) such that

$$G_{n+1} - L_n = \Delta_D \tag{9.63}$$

On the xy diagram, Eq. (9.62) is the equation of a straight line of slope L_n/G_{n+1}, through (y_{n+1}, x_n) and $y = x = z_D$. These are plotted on Fig. 9.18, where both diagrams are shown.

Figure 9.18 is drawn for a total condenser. The distillate D and reflux L_0 then have identical coordinates and are plotted at point D. The location shown indicates that they are below the bubble point. If they were at the bubble point, D would be on the saturated-liquid curve. The saturated vapor G_1 from the top tray, when totally condensed, has the same composition as D and L_0. Liquid L_1

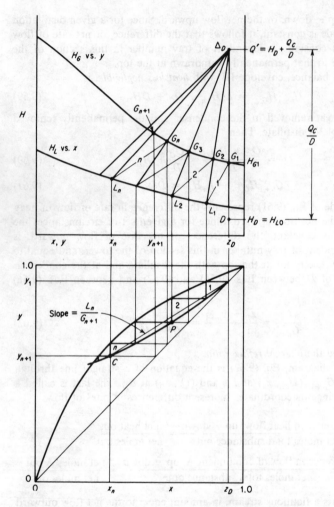

Figure 9.18 Enriching section, total condenser, reflux below the bubble point.

leaving ideal tray 1 is in equilibrium with G_1 and is located at the end of tie line 1. Since Eq. (9.62) applies to all trays in this section, G_2 can be located on the saturated-vapor curve by a line drawn from L_1 to Δ_D; tie line 2 through G_2 locates L_2, etc. Thus, alternate tie lines (each representing the effluents from an ideal tray) and construction lines through Δ_D provide the stepwise changes in concentration occurring in the enriching section. Intersections of the lines radiating from Δ_D with the saturated-enthalpy curves, such as points G_3 and L_2, when projected to the lower diagram, produce points such as P. These in turn produce the operating curve CP, which passes through $y = x = z_D$. The tie lines, when projected downward, produce the equilibrium-distribution curve, and the stepwise nature of the concentration changes with tray number then becomes obvious. The difference point Δ_D is used in this manner for all trays in the enriching section, working downward until the feed tray is reached.

Enriching trays can thus be located on the Hxy diagram alone by alternating construction lines to Δ_D and tie lines, each tie line representing an ideal tray. As an alternative, random lines radiating from Δ_D can be drawn, their intersections with the H_Gy and H_Lx curves plotted on the xy diagram to produce the operating curve, and the trays determined by the step construction typical of such diagrams.

At any tray n (compare Fig. 9.12) the L_n/G_{n+1} ratio is given by the ratio of line lengths $\Delta_D G_{n+1}/\Delta_D L_n$ on the upper diagram of Fig. 9.18 or by the slope of the chord as shown on the lower diagram. Elimination of G_{n+1} between Eqs. (9.56) and (9.62) provides

$$\frac{L_n}{D} = \frac{Q' - H_{G_{n+1}}}{H_{G_{n+1}} - H_{L_n}} = \frac{z_D - y_{n+1}}{y_{n+1} - x_n} \tag{9.64}$$

Applying this to the top tray provides the external reflux ratio, which is usually the one specified:

$$R = \frac{L_0}{D} = \frac{Q' - H_{G1}}{H_{G1} - H_{L0}} = \frac{\text{line } \Delta_D G_1}{\text{line } G_1 L_0} = \frac{\text{line } \Delta_D G_1}{\text{line } G_1 D} \tag{9.65}$$

For a given reflux ratio, the line-length ratio of Eq. (9.65) can be used to locate Δ_D vertically on Fig. 9.18, and the ordinate Q' can then be used to compute the condenser heat load.

In some cases a *partial condenser* is used, as in Fig. 9.19. Here a saturated vapor distillate D is withdrawn, and the condensate provides the reflux. This is frequently done when the pressure required for complete condensation of the vapor G_1, at reasonable condenser temperatures, would be too large. The Δ_D is plotted at an abscissa y_D corresponding to the composition of the withdrawn distillate. Assuming that an equilibrium condensation is realized, reflux L_0 is at the end of the tie line C. G_1 is located by the construction line $L_0 \Delta_D$, etc. In the lower diagram, the line MN solves the equilibrium-condensation problem (compare Fig. 9.14). The reflux ratio $R = L_0/D = $ line $\Delta_D G_1/$line $G_1 L_0$, by application of Eq. (9.65). It is seen that the equilibrium partial condenser provides one equilibrium tray's worth of rectification. However, it is safest not to rely on such complete enrichment by the condenser but instead to provide trays in the tower equivalent to all the stages required.

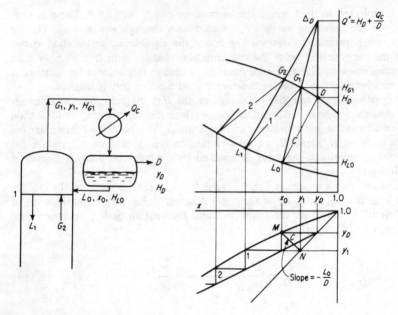

Figure 9.19 Partial condenser.

The Stripping Section

Consider the envelope IV, Fig. 9.17, where tray m is any tray in the stripping section. A balance for total material is

$$\bar{L}_m = \bar{G}_{m+1} + W \tag{9.66}$$

and, for component A,

$$\bar{L}_m x_m = \bar{G}_{m+1} y_{m+1} + W x_W \tag{9.67}$$

$$\bar{L}_m x_m - \bar{G}_{m+1} y_{m+1} = W x_W \tag{9.68}$$

The left-hand side of Eq. (9.68) represents the difference in rate of flow of component A, down − up, or the net flow downward. Since the right-hand side is a constant for a given distillation, the difference is independent of tray number in this section of the tower and equal to the rate of permanent removal of A out the bottom. An enthalpy balance is

$$\bar{L}_m H_{L_m} + Q_B = \bar{G}_{m+1} H_{G_{m+1}} + W H_W \tag{9.69}$$

Define Q'' as the net flow of heat outward at the bottom, per mole of residue

$$Q'' = \frac{W H_W - Q_B}{W} = H_W - \frac{Q_B}{W} \tag{9.70}$$

whence $$\bar{L}_m H_{L_m} - \bar{G}_{m+1} H_{G_{m+1}} = WQ'' \qquad (9.71)$$

The left-hand side of Eq. (9.71) is the difference in rate of flow of heat, down − up, which then equals the constant net rate of heat flow out the bottom for all trays in this section.

Elimination of W between Eqs. (9.66) and (9.67) and between Eqs. (9.66) and (9.71) provides

$$\frac{\bar{L}_m}{\bar{G}_{m+1}} = \frac{y_{m+1} - x_W}{x_m - x_W} = \frac{H_{G_{m+1}} - Q''}{H_{L_m} - Q''} \qquad (9.72)$$

On the Hxy diagram, Eq. (9.72) is a straight line through $(H_{G_{m+1}}, y_{m+1})$ at $\bar{G}_{m+1}$, (H_{L_m}, x_m) at $\bar{L}_m$, and (Q'', x_W) at Δ_W. Δ_W is a *difference* point, whose coordinates mean

$$\Delta_W \begin{cases} Q'' = \dfrac{\text{difference in heat flow, down} - \text{up}}{\text{net moles of total substance out}} = \dfrac{\text{net heat out}}{\text{net moles out}} \\[3mm] x_W = \dfrac{\text{difference in flow of component A, down} - \text{up}}{\text{net moles of total substance out}} = \dfrac{\text{moles A out}}{\text{net moles out}} \end{cases}$$

Thus, Δ_W is a fictitious stream, in amount equal to the net flow outward (in this case W), of properties (Q'', x_W),

$$\bar{L}_m - \bar{G}_{m+1} = \Delta_W \qquad (9.73)$$

On the xy diagram, Eq. (9.72) is a straight line of slope $\bar{L}_m/\bar{G}_{m+1}$, through (y_{m+1}, x_m) and $y = x = x_W$. These straight lines are plotted in Fig. 9.20 for both diagrams.

Since Eq. (9.72) applies to all trays of the stripping section, the line on the Hxy plot of Fig. 9.20 from $\bar{G}_{N_p+1}$ (vapor leaving the reboiler and entering the bottom tray N_p of the tower) to Δ_W intersects the saturated-liquid-enthalpy curve at $\bar{L}_{N_p}$, the liquid leaving the bottom tray. Vapor $\bar{G}_{N_p}$ leaving the bottom tray is in equilibrium with liquid $\bar{L}_{N_p}$ and is located on the tie line N_p. Tie lines projected to the xy diagram produce points on the equilibrium curve, and lines through Δ_W provide points such as T on the operating curve. Substitution of Eq. (9.66) into Eq. (9.72) provides

$$\frac{\bar{L}_m}{W} = \frac{H_{G_{m+1}} - Q''}{H_{G_{m+1}} - H_{L_m}} = \frac{y_{m+1} - x_W}{y_{m+1} - x_m} \qquad (9.74)$$

The diagrams have been drawn for the type of reboiler shown in Fig. 9.17, where the vapor leaving the reboiler is in equilibrium with the residue, the reboiler thus providing an equilibrium stage of enrichment (tie line B, Fig. 9.20). Other methods of applying heat at the bottom of the still are considered later.

Stripping-section trays can thus be determined entirely on the Hxy diagram by alternating construction lines to Δ_W and tie lines, each tie line accounting for an equilibrium stage. Alternatively, random lines radiating from Δ_W can be drawn, their intersections with curves $H_G y$ and $H_L x$ plotted on the xy diagram

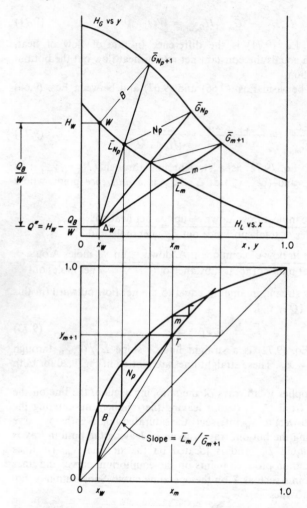

Figure 9.20 Stripping section.

to produce the operating curve, and the stages determined by the usual step construction.

The Complete Fractionator

Envelope II of Fig. 9.17 can be used for material balances over the entire device

$$F = D + W \tag{9.75}$$

$$Fz_F = Dz_D + Wx_W \tag{9.76}$$

Equation (9.55) is a complete enthalpy balance. If, in the absence of heat losses

$(Q_L = 0)$, the definitions of Q' and Q'' are substituted into Eq. (9.55), it becomes

$$FH_F = DQ' + WQ'' \tag{9.77}$$

If F is eliminated from Eqs. (9.75) to (9.77), there results

$$\frac{D}{W} = \frac{z_F - x_W}{z_D - z_F} = \frac{H_F - Q''}{Q' - H_F} \tag{9.78}$$

This is the equation of a straight line on the Hxy diagram, through (Q', z_D) at Δ_D, (H_F, z_F) at F, and (Q'', x_W) at Δ_W, as plotted in Fig. 9.21. In other words,

$$F = \Delta_D + \Delta_W \tag{9.79}$$

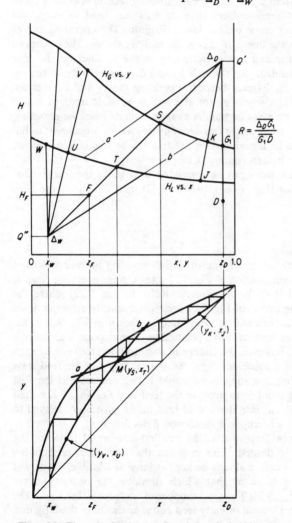

Figure 9.21 The entire fractionator. Feed below the bubble point and a total condenser.

The location of F, representing the feed, on Fig. 9.21 shows the feed in this case to be a liquid below the bubble point. In other situations, F may be on the saturated-liquid or vapor curve, between them, or above the saturated-vapor curve. In any event, the two Δ points and F must lie on a single straight line.

The construction for trays is now clear. After locating F and the concentration abscissas z_D and x_W corresponding to the products on the Hxy diagram, Δ_D is located vertically on line $x = z_D$ by computation of Q' or by the line-length ratio of Eq. (9.65) using the specified reflux ratio R. The line $\Delta_D F$ extended to $x = x_W$ locates Δ_W, whose ordinate can be used to compute Q_B. Random lines such as $\Delta_D J$ are drawn from Δ_D to locate the enriching-section operating curve on the xy diagram, and random lines such as $\Delta_W V$ are used to locate the stripping-section operating curve on the lower diagram. The operating curves intersect at M, related to the line $\Delta_D F \Delta_W$ in the manner shown. They intersect the equilibrium curve at a and b, corresponding to the tie lines on the Hxy diagram which, when extended, pass through Δ_D and Δ_W, respectively, as shown. Steps are drawn on the xy diagram between operating curves and equilibrium curve, beginning usually at $x = y = z_D$ (or at $x = y = x_W$ if desired), each step representing an equilibrium stage or tray. A change is made from the enriching to the stripping operating curve *at the tray on which the feed is introduced*; in the case shown the feed is to be introduced on the tray whose step straddles point M. The step construction is then continued to $x = y = x_W$.

Liquid and vapor flow rates can be computed throughout the fractionator from the line-length ratios [Eqs. (9.62), (9.64), (9.72), and (9.74)] on the Hxy diagram.

Feed-Tray Location

The material and enthalpy balances from which the operating curves are derived dictate that the stepwise construction of Fig. 9.21 must change operating lines at the tray where the feed is to be introduced. Refer to Fig. 9.22, where the equilibrium and operating curves of Fig. 9.21 are reproduced. In stepping down from the top of the fractionator, it is clear that, as shown in Fig. 9.22a, the enriching curve could have been used to a position as close to point a as desired. As point a is approached, however, the change in composition produced by each tray diminishes, and at a a *pinch* develops. As shown, tray f is the feed tray. Alternatively, the stripping operating curve could have been used at the first opportunity after passing point b, to provide the feed tray f of Fig. 9.22b (had the construction begun at x_W, introduction of feed might have been delayed to as near point b as desired, whereupon a pinch would develop at b).

In the design of a new fractionator, the smallest number of trays for the circumstances at hand is desired. This requires that the distance between operating and equilibrium curves always be kept as large as possible, which will occur if the feed tray is taken as that which straddles the operating-curve intersection at M, as in Fig. 9.21. The total number of trays for either Fig. 9.22a or b is of necessity larger. Delayed or early feed entry, as shown in these figures,

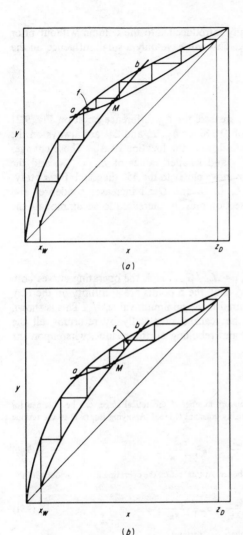

(a)

(b)

Figure 9.22 Delayed and early feed entries.

is used only where a separation is being adapted to an existing tower equipped with a feed-tray entry nozzle on a particular tray, which must then be used.

Consider again the feed tray of Fig. 9.21. It is understood that if the feed is all liquid, it is introduced above the tray in such a manner that it enters the tray along with the liquid from the tray above. Conversely, if the feed is all vapor, it is introduced underneath the feed tray. Should the feed be mixed liquid and vapor, in principle it should be separated outside the column and the liquid portion introduced above, the vapor portion below, the feed tray. This is rarely done, and the mixed feed is usually introduced into the column without prior separation for reasons of economy. This will have only a small influence on the number of trays required [6].

Increased Reflux Ratio

As the reflux ratio $R = L_0/D$ is increased, the Δ_D difference point on Fig. 9.21 must be located at higher values of Q'. Since Δ_D, F, and Δ_W are always on the same line, increasing the reflux ratio lowers the location of Δ_W. These changes result in larger values of L_n/G_{n+1} and smaller values of $\bar{L}_m/\bar{G}_{m+1}$, and the operating curves on the xy diagram move closer to the 45° diagonal. Fewer trays are then required, but Q_C, Q_W, L, $\bar{L}$, G, and $\bar{G}$ all increase; condenser and reboiler surfaces and tower cross section must be increased to accommodate the larger loads.

Total Reflux

Ultimately, when $R = \infty$, $L_n/G_{n+1} = \bar{L}_m/\bar{G}_{m+1} = 1$, the operating curves both coincide with the 45° line on the xy plot, the Δ points are at infinity on the Hxy plot, and the number of trays required is the minimum value, N_m. This is shown in Fig. 9.23. The condition can be realized practically by returning all the distillate to the top tray as reflux and reboiling all the residue, whereupon the feed to the tower must be stopped.

Constant Relative Volatility

A useful analytical expression for the minimum number of theoretical stages can be obtained for cases where the relative volatility is reasonably constant [13, 63]. Applying Eq. (9.2) to the residue product gives

$$\frac{y_W}{1 - y_W} = \alpha_W \frac{x_W}{1 - x_W} \tag{9.80}$$

where α_W is the relative volatility at the reboiler. At total reflux the operating line coincides with the 45° diagonal so that $y_W = x_{N_m}$. Therefore

$$\frac{x_{N_m}}{1 - x_{N_m}} = \alpha_W \frac{x_W}{1 - x_W} \tag{9.81}$$

Similarly for the last tray of the column, where α_{N_m} pertains,

$$\frac{y_{N_m}}{1 - y_{N_m}} = \alpha_{N_m} \frac{x_{N_m}}{1 - x_{N_m}} = \alpha_{N_m} \alpha_W \frac{x_W}{1 - x_W} \tag{9.82}$$

This procedure can be continued up the column until ultimately

$$\frac{y_1}{1 - y_1} = \frac{x_D}{1 - x_D} = \alpha_1 \alpha_2 \cdots \alpha_{N_m} \alpha_W \frac{x_W}{1 - x_W} \tag{9.83}$$

If some average relative volatility α_{av} can be used,

$$\frac{x_D}{1 - x_D} = \alpha_{av}^{N_m + 1} \frac{x_W}{1 - x_W} \tag{9.84}$$

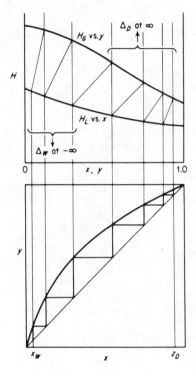

Figure 9.23 Total reflux and minimum stages.

or

$$N_m + 1 = \frac{\log \dfrac{x_D}{1 - x_D} \dfrac{1 - x_W}{x_W}}{\log \alpha_{av}}$$ (9.85)

which is known as *Fenske's equation*. The total minimum number of theoretical stages to produce products x_D and x_W is $N_m + 1$, which then includes the reboiler. For small variations in α, α_{av} can be taken as the geometric average of the values for the overhead and bottom products, $\sqrt{\alpha_1 \alpha_W}$. The expression can be used only with nearly ideal mixtures, for which α is nearly constant.

Minimum Reflux Ratio

The minimum reflux ratio R_m is the maximum ratio which will require an infinite number of trays for the separation desired, and it corresponds to the minimum reboiler heat load and condenser cooling load for the separation.

Refer to Fig. 9.24a, where the lightly drawn lines are tie lines which have been extended to intersect lines $x = z_D$ and $x = x_W$. It is clear that if Δ_D were located at point K, alternate tie lines and construction lines to Δ_D at the tie line k would coincide, and an infinite number of stages would be required to reach tie line k from the top of the tower. The same is true if Δ_W is located at point J.

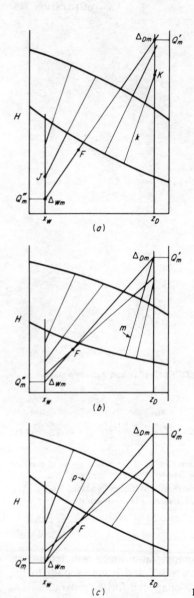

Figure 9.24 Minimum reflux ratio.

Since as Δ_D is moved upward and Δ_W downward the reflux ratio increases, the definition of minimum reflux ratio requires Δ_{D_m} and Δ_{W_m} for the minimum reflux ratio to be located as shown, with Δ_{D_m} at the highest tie-line intersection and Δ_{W_m} at the lowest tie-line intersection. In this case, it is the tie line which, when extended, passes through F, the feed, that determines both, and this is always the case when the xy equilibrium distribution curve is everywhere concave downward.

For some positively deviating mixtures with a tendency to form an azeotrope and for all systems near the critical condition of the more volatile component [66], an enriching-section tie line m in Fig. 9.24b gives the highest intersection with $x = z_D$, not that which passes through F. Similarly, as in Fig. 9.24c for some negatively deviating mixtures, a stripping-section tie line p gives the lowest intersection with $x = x_W$. These then govern the location of Δ_{D_m} as shown. For the minimum reflux ratio, either Δ_{D_m} is located at the highest intersection of an enriching-section tie line with $x = z_D$, or Δ_{W_m} is at the lowest intersection of a stripping-section tie line with $x = x_W$, consistent with the requirements that Δ_{D_m}, Δ_{W_m}, and F all be on the same straight line and Δ_{D_m} be at the highest position resulting in a pinch. Special considerations are necessary for fractionation with multiple feeds and sidestreams [52].

Once Q_m is determined, the minimum reflux ratio can be computed through Eq. (9.65). Some larger reflux ratio must obviously be used for practical cases, whereupon Δ_D is located above Δ_{D_m}.

Optimum Reflux Ratio

Any reflux ratio between the minimum and infinity will provide the desired separation, with the corresponding number of theoretical trays required varying from infinity to the minimum number, as in Fig. 9.25a. Determination of the

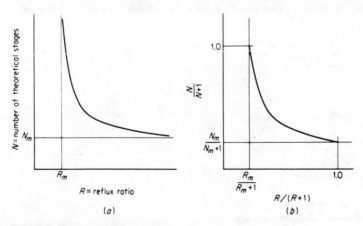

Figure 9.25 Reflux-ratio–stage relation.

number of trays at several values of R, together with the limiting values of N_m and R_m, will usually permit plotting the entire curve with sufficient accuracy for most purposes. The coordinate system of Fig. 9.25*b* [17] will permit locating the ends of the curve readily by avoiding the awkward asymptotes. There have been several attempts at empirically generalizing the curves of Fig. 9.25 [5, 10, 12, 17, 35], but the resulting charts yield only approximate results. An exact relationship for binary distillations, which can also be applied to multicomponent mixtures, is available [59].

The reflux ratio to be used for a new design should be the optimum, or the most economical, reflux ratio, for which the cost will be the least. Refer to Fig. 9.26. At the minimum reflux ratio the column requires an infinite number of trays, and consequently the fixed cost is infinite, but the operating costs (heat for the reboiler, condenser cooling water, power for reflux pump) are least. As R increases, the number of trays rapidly decreases, but the column diameter increases owing to the larger quantities of recycled liquid and vapor per unit quantity of feed. The condenser, reflux pump, and reboiler must also be larger. The fixed costs therefore fall through a minimum and rise to infinity again at total reflux. The heating and cooling requirements increase almost directly with reflux ratio, as shown. The total cost, which is the sum of operating and fixed costs, must therefore pass through a minimum at the optimum reflux ratio. This will frequently but not always occur at a reflux ratio near the minimum value ($1.2R_m$ to $1.5R_m$), on the average probably near the lower limit. A less empirical method for estimating the optimum is available [34].

Illustration 9.8 A methanol (A)–water (B) solution containing 50 wt % methanol at 26.7°C is to be continuously rectified at 1 std atm pressure at a rate of 5000 kg/h to provide a distillate containing 95% methanol and a residue containing 1.0 % methanol (by weight). The feed is to be preheated by heat exchange with the residue, which will leave the system at 37.8°C. The distillate is to be totally condensed to a liquid and the reflux returned at the bubble point. The withdrawn distillate will be separately cooled before storage. A reflux ratio of 1.5 times the minimum will be used. Determine (*a*) quantity of the products, (*b*) enthalpy of feed and of

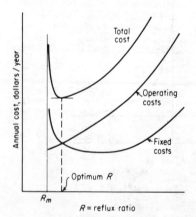

Figure 9.26 Most economical (optimum) reflux ratio.

products, (c) minimum reflux ratio, (d) minimum number of theoretical trays, (e) condenser and reboiler heat loads for specified reflux ratio, (f) number of theoretical trays for specified reflux ratio, and liquid and vapor quantities inside the tower.

SOLUTION (a) Mol wt methanol = 32.04, mol wt water = 18.02. *Basis:* 1 h. Define quantities in terms of kmol/h.

$$F = \frac{5000(0.50)}{32.04} + \frac{5000(0.50)}{18.02} = 78.0 + 138.8 = 216.8 \text{ kmol/h}$$

$$z_F = \frac{78}{216.8} = 0.360 \text{ mole fraction methanol} \qquad M_{av} \text{ for feed} = \frac{5000}{216.8} = 23.1 \text{ kg/kmol}$$

$$x_D = \frac{95/32.04}{95/32.04 + 5/18.02} = \frac{2.94}{3.217} = 0.915 \text{ mole fraction methanol}$$

$$M_{av} \text{ for distillate} = \frac{100}{3.217} = 31.1 \text{ kg/kmol}$$

$$x_W = \frac{1/32.04}{1/32.04 + 99/18.02} = \frac{0.0312}{5.53} = 0.00565 \text{ mole fraction methanol}$$

$$M_{av} \text{ for residue} = \frac{100}{5.53} = 18.08 \text{ kg/kmol}$$

Eq. (9.75):

$$216.8 = D + W$$

Eq. (9.76):

$$216.8(0.360) = D(0.915) + W(0.00565)$$

Solving simultaneously gives

$$D = 84.4 \text{ kmol/h} \qquad 84.4(31.1) = 2620 \text{ kg/h}$$
$$W = 132.4 \text{ kmol/h} \qquad 132.4(18.08) = 2380 \text{ kg/h}$$

(b) The vapor-liquid equilibrium at 1 std atm pressure is given by Cornell and Montana, *Ind. Eng. Chem.*, **25**, 1331 (1933), and by "The Chemical Engineers' Handbook," 4th ed., p. 13-5. Heat capacities of liquid solutions are in the "Handbook," 5th ed., p. 3-136, and latent heats of vaporization of methanol on p. 3-116. Heats of solution are available in "International Critical Tables," vol. V, p. 159, at 19.69°C, which will be used as t_0, the base temperature for computing enthalpies.

To compute enthalpies of saturated liquids, consider the case of $x = 0.3$ mole fraction methanol, $M_{av} = 22.2$. The bubble point = 78.3°C, heat capacity = 3852 J/kg · K, and the heat of solution = 3055 kJ evolved/kmol methanol.

$\Delta H_S = -3055(0.3) = -916.5$ kJ/kmol solution. Therefore, Eq. (9.10):

$$H_L = 3.852(78.3 - 19.69)22.2 - 916.5 = 4095 \text{ kJ/kmol}$$

To compute the enthalpy of saturated vapors, consider the case of $y = 0.665$ mole fraction methanol. The dew point is 78.3°C. At this temperature the latent heat of methanol is 1046.7 kJ/kg, that of water is 2314 kJ/kg. The heat capacity of methanol is 2583, of water 2323 J/kg · K. Eq. (9.11):

$$H_G = 0.665[2.583(32.04)(78.3 - 19.69) + 1046.7(32.04)]$$
$$+ (1 - 0.665)[2.323(18.02)(78.3 - 19.69) + 2314(18.02)]$$
$$= 40\,318 \text{ kJ/kmol}$$

The enthalpy data of Fig. 9.27 were computed in this manner.

From the vapor-liquid equilibria, the bubble point of the residue is 99°C. Heat capacity of the residue is 4179, of the feed 3852 J/kg · K. Enthalpy balance of the feed preheat exchanger:

$$5000(3852)(t_F - 26.7) = 2380(4179)(99 - 37.8)$$

$$t_F = 58.3°C, \text{ temp at which feed enters tower}$$

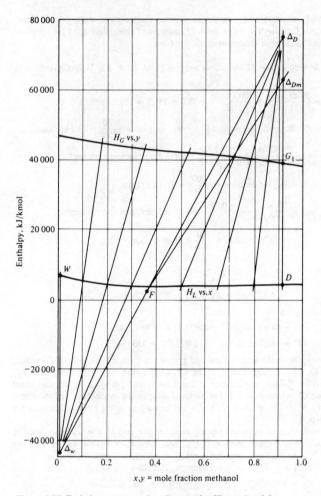

Figure 9.27 Enthalpy-concentration diagram for Illustration 9.8.

(*Note*: The bubble point of the feed is 76.0°C. Had t_F as computed above exceeded the bubble point, the above enthalpy balance would have been discarded and made in accordance with flash-vaporization methods.) For the feed, $\Delta H_S = -902.5$ kJ/kmol. Enthalpy of feed at 58.3°C is

$$H_F = 3.852(58.3 - 19.69)(23.1) - 902.5 = 2533 \text{ kJ/kmol}$$

From Fig. 9.27, $H_D = H_{L0} = 3640$, $H_W = 6000$ kJ/kmol.

(c) Since the xy diagram (Fig. 9.28) is everywhere concave downward, the minimum reflux ratio is established by the tie line in Fig. 9.27 ($x = 0.37$, $y = 0.71$) which, when extended, passes through F, the feed. At Δ_{D_m}, $Q_m = 62\,570$ kJ/kmol. $H_{G1} = 38\,610$ kJ/kmol. Eq. (9.65):

$$R_m = \frac{62\,570 - 38\,610}{38\,610 - 3640} = 0.685$$

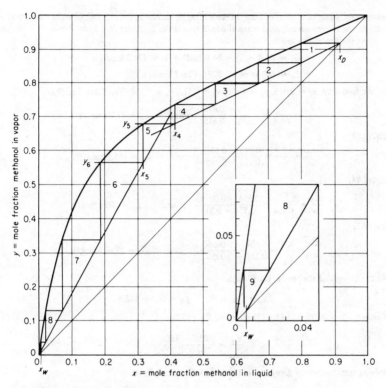

Figure 9.28 xy diagram for Illustration 9.8.

(*d*) The minimum number of trays was determined on the xy diagram in the manner of the lower part of Fig. 9.23, and 4.9 theoretical stages, including the reboiler, were obtained. $N_m = 4.9 - 1 = 3.9$.

(*e*) For $R = 1.5(0.685) = 1.029$, Eq. (9.65) becomes

$$1.029 = \frac{Q' - 38\ 610}{38\ 610 - 3640}$$

$$Q' = 74\ 595 = H_D + \frac{Q_C}{D} = 3640 + \frac{Q_C}{84.4} \qquad Q_C = 5\ 990\ 000\ \text{kJ/h} = 1664\ \text{kW}$$

Eq. (9.77): $216.8(2533) = 84.4(74\ 595) + 132.4Q''$

$$Q'' = -43\ 403 = H_W - \frac{Q_B}{W} = 6000 - \frac{Q_B}{132.4}$$

$$Q_B = 6\ 541\ 000\ \text{kJ/h reboiler heat load} = 1817\ \text{kW}$$

(*f*) In Fig. 9.27, Δ_D at ($x_D = 0.915$, $Q' = 74\ 595$) and Δ_W at ($x_W = 0.00565$, $Q'' = -43\ 403$) are plotted. Random lines from the Δ points, as shown, intersect the saturated-vapor and saturated-liquid curves at values of y and x, respectively, corresponding to points on the operating curve (note that for accurate results a large-scale graph and a sharp pencil are needed). These are plotted on Fig. 9.28 to provide the operating curves, which are nearly, but not exactly, straight. A total of nine theoretical stages including the reboiler, or eight theoretical trays in the tower, are required when the feed tray is the optimum (no. 5) as shown.

At the top of the tower

$$G_1 = D(R + 1) = 84.4(1.029 + 1) = 171.3 \text{ kmol/h}$$

$$L_0 = DR = 84.4(1.029) = 86.7 \text{ kmol/h}$$

At the feed tray, $x_4 = 0.415$, $y_5 = 0.676$, $x_5 = 0.318$, $y_6 = 0.554$ (Fig. 9.28). Eq. (9.64):

$$\frac{L_4}{D} = \frac{L_4}{84.4} = \frac{0.915 - 0.676}{0.676 - 0.415} \quad \text{and} \quad L_4 = 77.2 \text{ kmol/h}$$

Eq. (9.62):

$$\frac{L_4}{G_5} = \frac{77.2}{G_5} = \frac{0.915 - 0.676}{0.915 - 0.415} \quad \text{and} \quad G_5 = 161.5 \text{ kmol/h}$$

Eq. (9.74):

$$\frac{\bar{L}_5}{W} = \frac{\bar{L}_5}{132.4} = \frac{0.554 - 0.00565}{0.554 - 0.318} \quad \text{and} \quad \bar{L}_5 = 308 \text{ kmol/h}$$

Eq. (9.72):

$$\frac{\bar{L}_5}{\bar{G}_6} = \frac{308}{\bar{G}_6} = \frac{0.554 - 0.00565}{0.318 - 0.00565} \quad \text{and} \quad \bar{G}_6 = 175.7 \text{ kmol/h}$$

At the bottom of the tower, Eq. (9.66):

$$\bar{L}_{N_p} = \bar{G}_W + W \qquad \bar{L}_8 = \bar{G}_W + 132.4$$

Further, $y_W = 0.035$, $x_8 = 0.02$ (Fig. 9.28); Eq. (9.72):

$$\frac{\bar{L}_8}{\bar{G}_W} = \frac{0.035 - 0.00565}{0.02 - 0.00565}$$

Solving simultaneously gives $\bar{G}_W = 127.6$, $\bar{L}_8 = 260$ kmol/h.

Reboilers

The heat-exchanger arrangements to provide the necessary heat and vapor return at the bottom of the fractionator may take several forms. Small fractionators used for pilot-plant work may merely require a jacketed kettle, as shown schematically in Fig. 9.29a, but the heat-transfer surface and the corresponding vapor capacity will necessarily be small. The tubular heat exchanger built into the bottom of the tower (Fig. 9.29b) is a variation which provides larger surface, but cleaning requires a shut-down of the distillation operation. This type can also be built with an internal floating head. Both these provide a vapor entering the bottom tray essentially in equilibrium with the residue product, so that the last stage of the previous computations represents the enrichment due to the reboiler.

External reboilers of several varieties are commonly used for large installations, and they can be arranged with spares for cleaning. The kettle reboiler (Fig. 9.29c), with heating medium inside the tubes, provides a vapor to the tower essentially in equilibrium with the residue product and then behaves like a theoretical stage. The vertical thermosiphon reboiler of Fig. 9.29d, with the

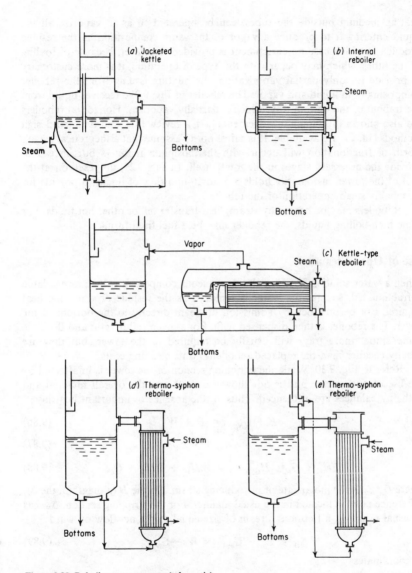

Figure 9.29 Reboiler arrangements (schematic).

heating medium outside the tubes, can be operated so as to vaporize all the liquid entering it to produce a vapor of the same composition as the residue product, in which case no enrichment is provided. However, because of fouling of the tubes, which may occur with this type of operation, it is more customary to provide for only partial vaporization, the mixture issuing from the reboiler comprising both liquid and vapor. The reboiler of Fig. 9.29e receives liquid from the trapout of the bottom tray, which it partially vaporizes. Horizontal reboilers are also known [8]. Piping arrangements [27], a review [37], and detailed design methods [14, 49] are available. It is safest not to assume that a theoretical stage's worth of fractionation will occur with thermosiphon reboilers but instead to provide the necessary stages in the tower itself. In Fig. 9.29, the reservoir at the foot of the tower customarily holds a 5- to 10-min flow of liquid to provide for reasonably steady operation of the reboiler.

Reboilers may be heated by steam, heat-transfer oil, or other hot fluids. For some high-boiling liquids, the reboiler may be a fuel-fired furnace.

Use of Open Steam

When a water solution in which the nonaqueous component is the more volatile is fractionated, so that the water is removed as the residue product, the heat required can be provided by admission of steam directly to the bottom of the tower. The reboiler is then dispensed with. For a given reflux ratio and distillate composition, more trays will usually be required in the tower, but they are usually cheaper than the replaced reboiler and its cleaning costs.

Refer to Fig. 9.30. While the enriching section of the tower is unaffected by the use of open steam and is not shown, nevertheless the overall material and enthalpy balances are influenced. Thus, in the absence of important heat loss,

$$F + \overline{G}_{N_p+1} = D + W \tag{9.86}$$

$$Fz_F = Dz_D + Wx_W \tag{9.87}$$

$$FH_F + \overline{G}_{N_p+1}H_{G, N_p+1} = WH_W + DH_D + Q_C \tag{9.88}$$

where $\overline{G}_{N_p+1}$ is the molar rate of introducing steam. On the Hxy diagram, the Δ_D difference point is located in the usual manner. For the stripping section, Δ_W has its usual meaning, a fictitious stream of size equal to the net flow outward

$$\Delta_W = \overline{L}_m - \overline{G}_{m+1} = W - \overline{G}_{N_p+1} \tag{9.89}$$

of coordinates

$$
\Delta_W \begin{cases}
x_{\Delta W} = \dfrac{\text{net moles A out}}{\text{net moles out}} = \dfrac{Wx_W}{W - \overline{G}_{N_p+1}} \\[3mm]
Q'' = \dfrac{\text{net heat out}}{\text{net moles out}} = \dfrac{WH_W - \overline{G}_{N_p+1}H_{G, N_p+1}}{W - \overline{G}_{N_p+1}}
\end{cases}
$$

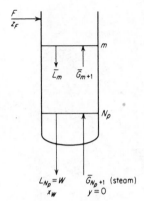

$L_{N_p} = W$ $\overline{G}_{N_p+1}$ (steam)
x_W $y = 0$

Figure 9.30 Use of open steam.

where H_{G, N_p+1} is the enthalpy of the steam. The point is shown on Fig. 9.31. Thus,

$$\overline{L}_m x_m - \overline{G}_{m+1} y_{m+1} = \Delta_W x_W \tag{9.90}$$

$$\overline{L}_m H_{L_m} - \overline{G}_{m+1} H_{G_{m+1}} = \Delta_W Q'' \tag{9.91}$$

and

$$\frac{\overline{L}_m}{\overline{G}_{m+1}} = \frac{y_{m+1} - x_{\Delta W}}{x_m - x_{\Delta W}} = \frac{H_{G_{m+1}} - Q''}{H_{L_m} - Q''} \tag{9.92}$$

The construction is shown in Fig. 9.31. Equation (9.92) is the slope of a chord (not shown) between points P and T. Here, the steam introduced is shown slightly superheated ($H_{G, N_p+1} >$ saturated enthalpy); had saturated steam been used, $\overline{G}_{N_p+1}$ would be located at point M. Note that the operating curve on the x, y diagram passes through the 45° diagonal at T ($x = x_{\Delta W}$) and through the point ($x_W, y = 0$) corresponding to the fluids passing each other at the bottom of the tower.

Illustration 9.9 Open steam, initially saturated at 69 kN/m² (10 lb$_f$/in²) gauge pressure, will be used for the methanol fractionator of Illustration 9.8, with the same distillate rate and composition and the same reflux ratio. Assuming that the feed enters the tower at the same enthalpy as in Illustration 9.8, determine the steam rate, bottoms composition, and the number of theoretical trays.

SOLUTION From Illustration 9.8, $F = 216.8$ kmol/h, $z_F = 0.360$, $H_F = 2533$ kJ/kmol, $D = 84.4$, $z_D = 0.915$, $H_D = 3640$ and $Q_C = 5\,990\,000$ kJ/h. From the steam tables, the enthalpy of saturated steam at 69 kN/m² = 2699 kJ/kg referred to liquid water at 0°C. On expanding adiabatically through a control valve to the tower pressure, it will be superheated at the same enthalpy. The enthalpy of liquid water at 19.7°C (t_0 for Illustration 9.8) = 82.7 kJ/kg referred to 0°C. Therefore $H_{G, N_p+1} = (2699 - 82.7)(18.02) = 47\,146$ kJ/kmol.

Eq. (9.86):

$$216.8 + \overline{G}_{N_p+1} = 84.4 + W$$

Eq. (9.87):

$$216.8(0.360) = 84.4(0.915) + W x_W$$

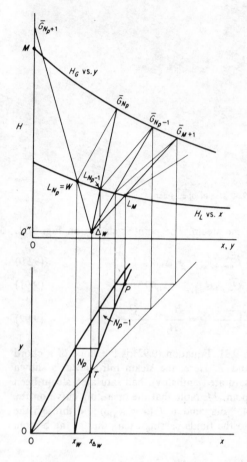

Figure 9.31 Use of open steam.

Eq. (9.88):

$$216.8(2533) + 47\,146\overline{G}_{N_p+1} = WH_W + 84.4(3640) + 5\,990\,000$$

Since the bottoms will be essentially pure water, H_W is tentatively estimated as the enthalpy of saturated water (Fig. 9.27), 6094 kJ/kmol. Solving the equations simultaneously with this value for H_W provides the steam rate as $\overline{G}_{N_p+1} = 159.7$ and $W = 292.1$ kmol/h, with $x_W = 0.00281$. The enthalpy of this solution at its bubble point is 6048, sufficiently close to the 6094 assumed earlier to be acceptable. (Note that had the same interchange of heat between bottoms and feed been used as in Illustration 9.8, with bottoms discharged at 37.8°C, the feed enthalpy would have been changed.) For Δ_W,

$$x_{\Delta W} = \frac{Wx_W}{W - \overline{G}_{N_p+1}} = \frac{292.1(0.00281)}{292.1 - 159.7} = 0.0062$$

$$Q'' = \frac{WH_W - \overline{G}_{N_p+1}H_{G.\,N_p+1}}{W - \overline{G}_{N_p+1}}$$

$$= \frac{292.1(6048) - 159.7(47\,146)}{292.1 - 159.7} = -43\,520 \text{ kJ/kmol}$$

The Hxy and xy diagrams for the enriching section are the same as in Illustration 9.8. For the stripping section, they resemble Fig. 9.31. The number of theoretical stages is $N_p = 9.5$, and they must all be included in the tower.

Condensers and Reflux Accumulators

Reflux may flow by gravity to the tower, in which case the condenser and reflux drum (accumulator) must be elevated above the level of top tray of the tower. Alternatively, especially in order to obviate the need for elevated platforms and supports required for withdrawing the condenser tube bundle for cleaning, the assemblage may be placed at ground level and the reflux liquid pumped up to the top tray. Kern [28] describes the arrangements.

Reflux accumulators are ordinarily horizontal drums, length/diameter = 4 to 5, with a liquid holding time of the order of 5 min. From entrainment considerations, the allowable vapor velocity through the vertical cross section of the space above the liquid can be specified as [55]†

$$V = 0.04\left(\frac{\rho_L - \rho_G}{\rho_G}\right)^{0.5} \tag{9.93}$$

Multiple Feeds

There are occasions when two or more feeds composed of the same substances but of different concentrations require distillation to give the same distillate and residue products. A single fractionator will then suffice for all.

Consider the two-feed fractionator of Fig. 9.32. The construction on the Hxy diagram for the sections of the column above F_1 and below F_2 is the same as for a single-feed column, with the Δ_D and Δ_W points located in the usual manner. For the middle section between the feeds, the difference point Δ_M can be located by consideration of material and enthalpy balances either toward the top, as indicated by the envelope shown on Fig. 9.32, or toward the bottom; the net result will be the same. Consider the envelope shown in the figure, with Δ_M representing a fictitious stream of quantity equal to the net flow upward and out

$$G'_{r+1} - L'_r = D - F_1 = \Delta_M \tag{9.94}$$

whose coordinates are

$$\Delta_M \begin{cases} x_{\Delta M} = \dfrac{\text{net moles A out}}{\text{net moles out}} = \dfrac{Dz_D - F_1 z_{F1}}{D - F_1} \\[2mm] Q_M = \dfrac{\text{net heat out}}{\text{net moles out}} = \dfrac{Q_C + DH_D - F_1 H_{F_1}}{D - F_1} \end{cases}$$

Δ_M may be either a positive or negative quantity.

† V in Eq. (9.93) is expressed as m/s. For V in ft/s, the coefficient is 0.13.

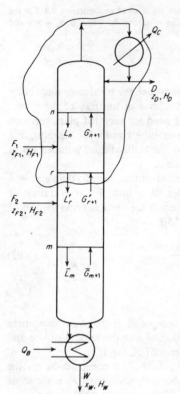

Figure 9.32 Fractionator with two feeds.

Equation (9.94) can be used as a basis for component-A and enthalpy balances

$$G'_{r+1}y_{r+1} - L'_r x_r = \Delta_M x_{\Delta M} \tag{9.95}$$

$$G'_{r+1}H_{G_{r+1}} - L_r H_{L_r} = \Delta_M Q_M \tag{9.96}$$

whence

$$\frac{L'_r}{G'_{r+1}} = \frac{y_{r+1} - x_{\Delta M}}{x_r - x_{\Delta M}} = \frac{H_{G_{r+1}} - Q_M}{H_{L_r} - Q_M} \tag{9.97}$$

Since

$$F_1 + F_2 = D + W = \Delta_D + \Delta_W \tag{9.98}$$

then

$$\Delta_M = F_2 - W \tag{9.99}$$

The construction (both feeds liquid) is shown on Fig. 9.33, where Δ_M lies on the line $\Delta_D F_1$ [Eq. (9.94)] and on the line $\Delta_W F_2$ [Eq. (9.99)]. A solution representing the composited feed, with

$$z_{F,\,\text{av}} = \frac{F_1 z_{F_1} + F_2 z_{F_2}}{F_1 + F_2} \qquad H_{F,\,\text{av}} = \frac{F_1 H_{F_1} + F_2 H_{F_2}}{F_1 + F_2}$$

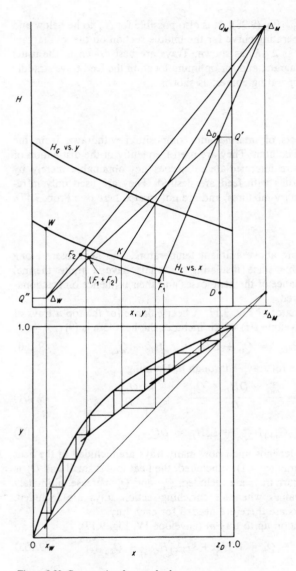

Figure 9.33 Construction for two feeds.

must lie on the line $\Delta_D \Delta_W$ [Eq. (9.98)]. It is also possible for Δ_M to lie below and to the left of Δ_W. The operating curve for the middle section on the xy diagram is located by lines such as $\Delta_M K$, as shown. Trays are best drawn in the usual step fashion on the xy diagram, and for optimum location the feed trays straddle the intersections of the operating curves, as shown.

Side Streams

Side streams are products of intermediate composition withdrawn from the intermediate trays of the column. They are used frequently in the distillation of petroleum products, where intermediate properties not obtainable merely by mixing distillate or bottoms with feed are desired. They are used only infrequently in the case of binary mixtures, and are not treated here (see Prob. 9.17).

Heat Losses

Most fractionators operate above ambient temperature, and heat losses along the column are inevitable since insulating materials have a finite thermal conductivity. The importance of the heat losses and their influence on fractionators will now be considered.

Consider the fractionator of Fig. 9.17. A heat balance for the top n trays of the enriching section (envelope III) which includes the heat loss is [9]

$$G_{n+1} H_{G_{n+1}} = Q_C + D H_D + L_n H_{L_n} + Q_{L_n} \tag{9.100}$$

where Q_{L_n} is the heat loss for trays 1 through n. Defining

$$Q_L' = \frac{Q_C + D H_D + Q_{L_n}}{D} = Q' + \frac{Q_{L_n}}{D} \tag{9.101}$$

we have

$$G_{n+1} H_{G_{n+1}} - L_n H_{L_n} = D Q_L' \tag{9.102}$$

Q_L' is a variable since it depends upon how many trays are included in the heat balance. If only the top tray ($n = 1$) is included, the heat loss is small and Q_L' is nearly equal to Q'. As more trays are included, Q_{L_n} and Q_L' increase, ultimately reaching their largest values when all enriching-section trays are included. Separate difference points are therefore needed for each tray.

For the stripping section up to tray m (envelope IV, Fig. 9.17),

$$\bar{L}_m H_{L_m} + Q_B = W H_W + \bar{G}_{m+1} H_{G_{m+1}} + \bar{Q}_{L_m} \tag{9.103}$$

Letting

$$Q_L'' = \frac{W H_W - Q_B + \bar{Q}_{L_m}}{W} = Q'' + \frac{\bar{Q}_{L_m}}{W} \tag{9.104}$$

results in

$$\bar{L}_m H_{L_m} - \bar{G}_{m+1} H_{G_{m+1}} = W Q_L'' \tag{9.105}$$

If the heat balance includes only the bottom tray, the heat loss is small and Q_L'' nearly equals Q''. As more trays are included, $\bar{Q}_{L_m}$ and therefore Q_L'' increase, reaching their largest values when the balance is made over the entire stripping section. Separate difference points are needed for each tray. An enthalpy balance for the entire tower is

$$FH_F = D\left(H_D + \frac{Q_C}{D} + \frac{Q_L}{D}\right) + W\left(H_W - \frac{Q_B}{W} + \frac{\overline{Q}_L}{W}\right) \qquad (9.106)$$

where Q_L and $\overline{Q}_L$ are the total heat losses for the enriching and stripping sections, respectively. When Eq. (9.106) is solved with the material balances, Eqs. (9.75) and (9.76), the result is

$$\frac{D}{W} = \frac{z_F - x_W}{z_D - z_F} = \frac{H_F - \left(H_W - Q_B/W + \overline{Q}_L/W\right)}{\left(H_D + Q_C/D + Q_L/D\right) - H_F} \qquad (9.107)$$

This is the equation of the line BFT on Fig. 9.34.

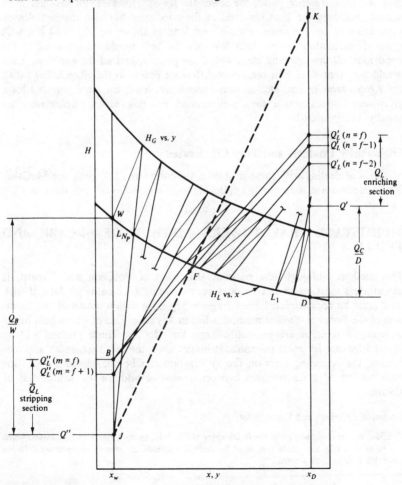

Figure 9.34 Heat losses.

The procedure for design is one of trial and error. For example, as a first trial, heat losses might be neglected and trays calculated with fixed difference points, and after the size of the resulting column has been determined, the first estimate of heat losses for the two column sections can be made by the usual methods of heat-transfer calculations. The heat losses can then be apportioned among the trays and the number of trays redetermined with the appropriate difference points. This leads to a second approximation of the heat loss, and so forth.

As Fig. 9.34 shows, heat losses increase the internal-reflux ratio, and *for a given condenser heat load*, fewer trays for a given separation are required (recall that the higher the enriching-section difference point and the lower the stripping-section difference point, the fewer the trays). However, the reboiler must provide not only the heat removed in the condenser but also the heat losses. Consequently, for the same reboiler heat load as shown on Fig. 9.34 but with complete insulation against heat loss, all the heat would be removed in the condenser, all the stripping trays would use point J, and all the enriching trays would use point K as their respective difference points. It therefore follows that, *for a given reboiler heat load or heat expenditure*, fewer trays are required for a given separation if heat losses are eliminated. For this reason, fractionators are usually well insulated.

High-Purity Products and Tray Efficiencies

Methods of dealing with these problems are considered following the McCabe-Thiele method and are equally applicable to Ponchon-Savarit calculations.

MULTISTAGE TRAY TOWERS—METHOD OF McCABE AND THIELE

This method, although less rigorous than that of Ponchon and Savarit, is nevertheless most useful since it does not require detailed enthalpy data. If such data must be approximated from fragmentary information, much of the exactness of the Ponchon-Savarit method is lost in any case. Except where heat losses or heats of solution are unusually large, the McCabe-Thiele method will be found adequate for most purposes. It hinges upon the fact that, as an approximation, the operating lines on the xy diagram can be considered straight for each section of a fractionator between points of addition or withdrawal of streams.

Equimolal Overflow and Vaporization

Consider the enriching section of the fractionator of Fig. 9.17. In the absence of heat losses, which can be (and usually are) made very small by thermal insulation for reasons of economy if for no other, Eq. (9.61) can be written

$$\frac{L_n}{G_{n+1}} = 1 - \frac{H_{G_{n+1}} - H_{L_n}}{Q' - H_{L_n}} \qquad (9.108)$$

where Q' includes the condenser heat load and the enthalpy of the distillate, per mole of distillate. The liquid enthalpy H_{L_n} is ordinarily small in comparison with Q' since the condenser heat load must include the latent heat of condensation of at least the reflux liquid. If then $H_{G_{n+1}} - H_{L_n}$ is substantially constant, L_n / G_{n+1} will be constant also for a given fractionation [48]. From Eq. (9.11)

$$H_{G_{n+1}} = [y_{n+1}C_{L,A}M_A + (1 - y_{n+1})C_{L,B}M_B](t_{n+1} - t_0) + y_{n+1}\lambda_A M_A + (1 - y_{n+1})\lambda_B M_B$$

$$(9.109)$$

where t_{n+1} is the temperature of the vapor from tray $n + 1$ and the λ's are the latent heats of vaporization at this temperature. If the deviation from ideality of liquid solutions is not great, the first term in brackets of Eq. (9.109) is

$$y_{n+1}C_{L,A}M_A + (1 - y_{n+1})C_{L,B}M_B \approx C_L M_{av} \qquad (9.110)$$

From Eq. (9.10),

$$H_{L_n} = C_L(t_n - t_0)M_{av} + \Delta H_S \qquad (9.111)$$

and

$$H_{G_{n+1}} - H_{L_n} = C_L M_{av}(t_{n+1} - t_n) + y_{n+1}\lambda_A M_A + (1 - y_{n+1})\lambda_B M_B - \Delta H_S \qquad (9.112)$$

For all but unusual cases, the only important terms of Eq. (9.112) are those containing the latent heats. The temperature change between adjacent trays is usually small, so that the sensible-heat term is insignificant. The heat of solution can in most cases be measured in terms of hundreds of kJ/kmol of solution, whereas the latent heats at ordinary pressures are usually in 10^4 kJ/kmol. Therefore, for all practical purposes,

$$H_{G_{n+1}} - H_{L_n} = (\lambda M)_{av} \qquad (9.113)$$

where the last term is the weighted average of the molal latent heats. For many pairs of substances, the molal latent heats are nearly identical, so that averaging is unnecessary. If their inequality is the only barrier to application of these simplifying assumptions, it is possible to assign a fictitious molecular weight to one of the components so that the molal latent heats are then forced to be the same (if this is done, the entire computation must be made with the fictitious molecular weight, including operating lines and equilibrium data). This is, however, rarely necessary.†

If it is sufficiently important, therefore, one can be persuaded that, for all but exceptional cases, the ratio L/G in the enriching section of the fractionator is essentially constant. The same reasoning can be applied to any section of a fractionator between points of addition or withdrawal of streams, although each section will have its own ratio.

Consider next two adjacent trays n and r, between which neither addition nor withdrawal of material from the tower occurs. A material balance provides

$$L_{r-1} + G_{n+1} = L_n + G_r \qquad (9.114)$$

Since $L_{r-1}/G_r = L_n/G_{n+1}$, it follows that $L_n = L_{r-1}$ and $G_{n+1} = G_r$, which is the *principle of equimolal overflow and vaporization*. The rate of liquid flow from each tray in a section of the tower is constant on a molar basis, but since the average molecular weight changes with tray number, the weight rates of flow are different.

It should be noted that, as shown in the discussion of Fig. 9.12, if the $H_G y$ and $H_L x$ lines on the Hxy diagram are straight and parallel, then in the absence of heat loss the L/G ratio for a given tower section will be constant regardless of the relative size of H_{L_n} and Q' in Eq. (9.108).

The general assumptions involved in the foregoing are customarily called the *usual simplifying assumptions*.

---

† If the only enthalpy data available are the latent heats of vaporization of the pure components, an approximate Savarit-Ponchon diagram using these and straight H_L vs x and H_G vs y lines will be better than assuming equal latent heats.

Enriching Section; Total Condenser—Reflux at the Bubble Point

Consider a section of the fractionator entirely above the point of introduction of feed, shown schematically in Fig. 9.35a. The condenser removes all the latent heat from the overhead vapor but does not cool the resulting liquid further. The reflux and distillate product are therefore liquids at the bubble point, and $y_1 = x_D = x_0$. Since the liquid, L mol/h, falling from each tray and the vapor, G mol/h, rising from each tray are each constant if the usual simplifying asssumptions pertain, subscripts are not needed to identify the source of these streams. The compositions, however, change. The trays shown are theoretical trays, so that the composition y_n of the vapor from the nth tray is in equilibrium with the liquid of composition x_n leaving the same tray. The point (x_n, y_n), on x, y coordinates, therefore falls on the equilibrium curve.

A total material balance for the envelope in the figure is

$$G = L + D = D(R + 1) \tag{9.115}$$

For component A,

$$Gy_{n+1} = Lx_n + Dx_D \tag{9.116}$$

from which the enriching-section operating line is

$$y_{n+1} = \frac{L}{G} x_n + \frac{D}{G} x_D \tag{9.117}$$

$$y_{n+1} = \frac{R}{R + 1} x_n + \frac{x_D}{R + 1} \tag{9.118}$$

This is the equation of a straight line on x, y coordinates (Fig. 9.35b) of slope $L/G = R/(R + 1)$, and with a y intercept of $x_D/(R + 1)$. Setting $x_n = x_D$

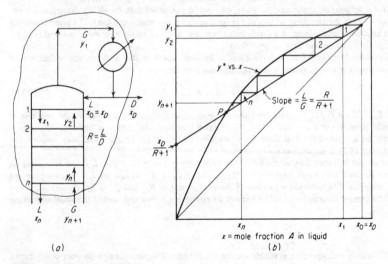

x = mole fraction A in liquid

(a)   (b)

Figure 9.35 Enriching section.

shows $y_{n+1} = x_D$, so that the line passes through the point $y = x = x_D$ on the 45° diagonal. This point and the y intercept permit easy construction of the line. The concentration of liquids and vapors for each tray is shown in accordance with the principles of Chap. 5, and the usual staircase construction between operating line and equilibrium curve is seen to provide the theoretical tray-concentration variation. The construction obviously cannot be carried farther than point P.

In plotting the equilibrium curve of the figure, it is generally assumed that the pressure is constant throughout the tower. If necessary, the variation in pressure from tray to tray can be allowed for after determining the number of real trays, but this will require a trial-and-error procedure. It is ordinarily unnecessary except for operation under very low pressures.

Exhausting Section; Reboiled Vapor in Equilibrium with Residue

Consider next a section of the fractionator below the point of introducing the feed, shown schematically in Fig. 9.36a. The trays are again theoretical trays. The rates of flow $\bar{L}$ and $\bar{G}$ are each constant from tray to tray, but not necessarily equal to the values for the enriching section. A total material balance is

$$\bar{L} = \bar{G} + W \tag{9.119}$$

and, for component A,

$$\bar{L}x_m = \bar{G}y_{m+1} + Wx_W \tag{9.120}$$

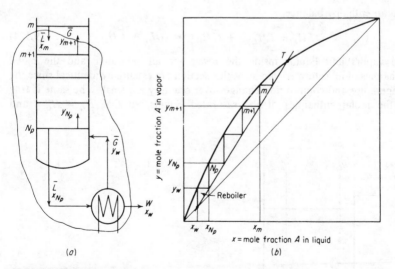

(a)

(b)

Figure 9.36 Exhausting section.

These provide the equation of the exhausting-section operating line,

$$y_{m+1} = \frac{\bar{L}}{\bar{G}} x_m - \frac{W}{\bar{G}} x_W \tag{9.121}$$

$$y_{m+1} = \frac{\bar{L}}{\bar{L} - W} x_m - \frac{W}{\bar{L} - W} x_W \tag{9.122}$$

This is a straight line of slope $\bar{L}/\bar{G} = \bar{L}(\bar{L} - W)$, and since when $x_m = x_W$, $y_{m+1} = x_W$, it passes through $x = y = x_W$ on the 45° diagonal (Fig. 9.36b). If the reboiled vapor y_W is in equilibrium with the residue x_W, the first step of the staircase construction represents the reboiler. The steps can be carried no farther than point T.

Introduction of Feed

It is convenient before proceeding further to establish how the introduction of the feed influences the change in slope of the operating lines as we pass from the enriching to the exhausting sections of the fractionator.

Consider the section of the column at the tray where the feed is introduced (Fig. 9.37). The quantities of the liquid and vapor streams change abruptly at this tray, since the feed may consist of liquid, vapor, or a mixture of both. If, for example, the feed is a saturated liquid, $\bar{L}$ will exceed L by the amount of the added feed liquid. To establish the general relationship, an overall material balance about this section is

$$F + L + \bar{G} = G + \bar{L} \tag{9.123}$$

and an enthalpy balance,

$$F H_F + L H_{L_{f-1}} + \bar{G} H_{G_{f+1}} = G H_{G_f} + \bar{L} H_{L_f} \tag{9.124}$$

The vapors and liquids inside the tower are all saturated, and the molal enthalpies of all saturated vapors at this section are essentially identical since the temperature and composition changes over one tray are small. The same is true of the molal enthalpies of the saturated liquids, so that $H_{G_f} = H_{G_{f+1}}$ and

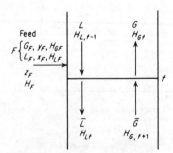

Figure 9.37 Introduction of feed.

$H_{L_{f-1}} = H_{L_f}$. Equation (9.124) then becomes

$$(\bar{L} - L)H_L = (\bar{G} - G)H_G + FH_F \tag{9.125}$$

Combining this with Eq. (9.123) gives

$$\frac{\bar{L} - L}{F} = \frac{H_G - H_F}{H_G - H_L} = q \tag{9.126}$$

The quantity q is thus seen to be the heat required to convert 1 mol of feed from its condition H_F to a saturated vapor, divided by the molal latent heat $H_G - H_L$. The feed may be introduced under any of a variety of thermal conditions ranging from a liquid well below its bubble point to a superheated vapor, for each of which the value of q will be different. Typical circumstances are listed in Table 9.1, with the corresponding range of values of q. Combining Eqs. (9.123) and (9.126), we get

$$\bar{G} - G = F(q - 1) \tag{9.127}$$

which provides a convenient method for determining $\bar{G}$.

The point of intersection of the two operating lines will help locate the exhausting-section operating line. This can be established as follows. Rewriting Eqs. (9.117) and (9.121) without the tray subscripts, we have

$$yG = Lx + Dx_D \tag{9.128}$$

$$y\bar{G} = \bar{L}x - Wx_W \tag{9.129}$$

Subtracting gives

$$(\bar{G} - G)y = (\bar{L} - L)x - (Wx_W + Dx_D) \tag{9.130}$$

Further, by an overall material balance,

$$Fz_F = Dx_D + Wx_W \tag{9.76}$$

Substituting this and Eqs. (9.126) and (9.127) in (9.130) gives

$$y = \frac{q}{q - 1}x - \frac{z_F}{q - 1} \tag{9.131}$$

This, the locus of intersection of operating lines (the q line), is a straight line of slope $q/(q - 1)$, and since $y = z_F$ when $x = z_F$, it passes through the point $x = y = z_F$ on the 45° diagonal. The range of the values of the slope $q/(q - 1)$ is listed in Table 9.1, and the graphical interpretation for typical cases is shown in Fig. 9.38. Here the operating-line intersection is shown for a particular case of feed as a mixture of liquid and vapor. It is clear that, for a given feed condition, fixing the reflux ratio at the top of the column automatically establishes the liquid/vapor ratio in the exhausting section and the reboiler heat load as well.

Table 9.1 Thermal conditions for the feed

Feed condition	G_F, mol/(area)(time)	L_F, mol/(area)(time)	H_{GF}, energy/mol	H_{LF}, energy/mol	H_F, energy/mol	$q = \dfrac{H_G - H_F}{H_G - H_L}$	$\dfrac{q}{q-1}$
Liquid below bubble point	0	F			$H_F < H_L$	>1.0	>1.0
Saturated liquid	0	F		H_F	H_L	1.0	∞
Mixture of liquid and vapor $F = G_F + L_F$	G_F	L_F	H_G	H_L	$H_G > H_F > H_L$	$1.0 > q > 0$	$\dfrac{L_F}{L_F - F}$
Saturated vapor	F	0	H_F		H_G	0	0
Superheated vapor	F	0	H_F		$H_F > H_G$	<0	$1.0 > \dfrac{q}{q-1} > 0$

† In this case the intersection of the q line with the equilibrium curve gives the compositions of the equilibrium liquid and vapor which constitute the feed. The q line is the flash-vaporization operating line for the feed.

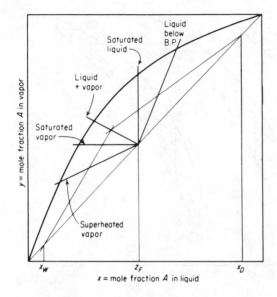

Figure 9.38 Location of q line for typical feed conditions.

Location of the Feed Tray

The q line is useful in simplifying the graphical location of the exhausting line, but the point of intersection of the two operating lines does not necessarily establish the demarcation between the enriching and exhausting sections of the tower. Rather it is the introduction of feed which governs the change from one operating line to the other and establishes the demarcation, and at least in the design of a new column some latitude in the introduction of the feed is available.

Consider the separation shown partially in Fig. 9.39, for example. For a given feed, z_F and the q line are fixed. For particular overhead and residue products, x_D and x_W are fixed. If the reflux ratio is specified, the location of the enriching line DG is fixed, and the exhausting line KC must pass through the q line at E. If the feed is introduced upon the seventh tray from the top (Fig. 9.39a), line DG is used for trays 1 through 6, and, beginning with the seventh tray, the line KC must be used. If, on the other hand, the feed is introduced upon the fourth from the top (Fig. 9.39b), line KC is used for all trays below the fourth. Clearly a transition from one operating line to the other must be made somewhere between points C and D, but anywhere within these limits will serve. The least total number of trays will result if the steps on the diagram are kept as large as possible or if the transition is made at the first opportunity after passing the operating-line intersection, as shown in Fig. 9.39c. In the design of a new column, this is the practice to be followed.

In the adaptation of an existing column to a new separation, the point of introducing the feed is limited to the location of existing nozzles in the column

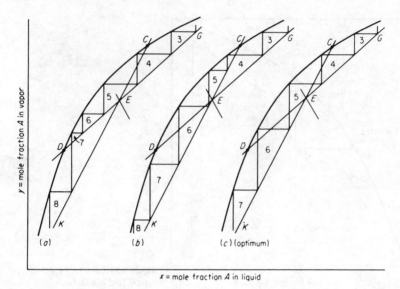

Figure 9.39 Location of feed tray.

wall. The slope of the operating lines (or reflux ratio) and the product compositions to be realized must then be determined by trial and error, in order to obtain numbers of theoretical trays in the two sections of the column consistent with the number of real trays in each section and the expected tray efficiency.

Total Reflux, or Infinite Reflux Ratio

As the reflux ratio $R = L/D$ is increased, the ratio L/G increases, until ultimately, when $R = \infty$, $L/G = 1$ and the operating lines of both sections of the column coincide with the 45° diagonal as in Fig. 9.40. In practice this can be

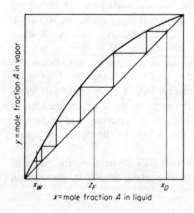

Figure 9.40 Total reflux and minimum trays.

realized by returning all the overhead product back to the column as reflux (total reflux) and reboiling all the residue product, whereupon the forward flow of fresh feed must be reduced to zero. Alternatively such a condition can be interpreted as requiring infinite reboiler heat and condenser cooling capacity for a given rate of feed.

As the operating lines move farther away from the equilibrium curve with increased reflux ratio, the number of theoretical trays required to produce a given separation becomes less, until at total reflux the number of trays is the minimum N_m.

If the relative volatility is constant or nearly so, the analytical expression of Fenske, Eq. (9.85), can be used.

Minimum Reflux Ratio

The minimum reflux ratio R_m is the maximum ratio which will require an infinite number of trays for the separation desired, and it corresponds to the minimum reboiler heat and condenser cooling capacity for the separation. Refer to Fig. 9.41a. As the reflux ratio is decreased, the slope of the enriching

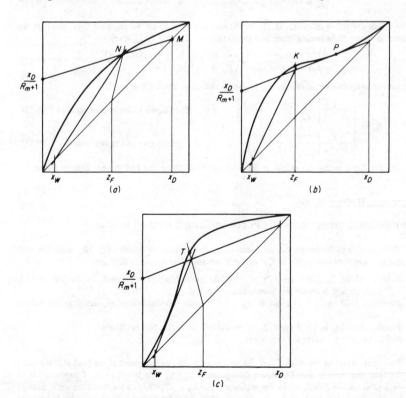

Figure 9.41 Minimum reflux ratio and infinite stages.

operating line becomes less, and the number of trays required increases. Operating line MN, which passes through the point of intersection of the q line and the equilibrium curve, corresponds to the minimum reflux ratio, and an infinite number of trays would be required to reach point N from either end of the tower. In some cases, as in Fig. 9.41b, the minimum-reflux operating line will be tangent to the equilibrium curve in the enriching section, as at point P, while a line through K would clearly represent too small a reflux ratio. It has been pointed out that all systems show concave-upward xy diagrams near the critical condition of the more volatile component [66]. Because of the interdependence of the liquid/vapor ratios in the two sections of the column, a tangent operating line in the exhausting section may also set the minimum reflux ratio, as in Fig. 9.41c.

When the equilibrium curve is always concave downward, the minimum reflux ratio can conveniently be calculated analytically [63]. The required relationship can be developed by solving Eqs. (9.118) and (9.131) simultaneously to obtain the coordinates (x_a, y_a) of the point of intersection of the enriching operating line and the q line. When the tray-number designation in Eq. (9.118) is dropped, these are

$$x_a = \frac{x_D(q - 1) + z_F(R + 1)}{R + q} \qquad y_a = \frac{Rz_F + qx_D}{R + q} \tag{9.132}$$

At the minimum reflux ratio R_m, these coordinates are equilibrium values since they occur on the equilibrium curve. Substituting them into the definition of α, Eq. (9.2), gives

$$\frac{R_m z_F + qx_D}{R_m(1 - z_F) + q(1 - x_D)} = \frac{\alpha[x_D(q - 1) + z_F(R_m + 1)]}{(R_m + 1)(1 - z_F) + (q - 1)(1 - x_D)} \tag{9.133}$$

This conveniently can be solved for R_m for any value of q. Thus, for example:

$$R_m = \begin{cases} \dfrac{1}{\alpha - 1}\left[\dfrac{x_D}{x_F} - \dfrac{\alpha(1 - x_D)}{1 - x_F}\right] & q = 1 \text{ (feed liquid at the bubble point)} \quad (9.134) \\[4mm] \dfrac{1}{\alpha - 1}\left(\dfrac{\alpha x_D}{y_F} - \dfrac{1 - x_D}{1 - y_F}\right) - 1 & q = 0 \text{ (feed vapor at the dew point)} \quad (9.135) \end{cases}$$

In each case, the α is that prevailing at the intersection of the q line and the equilibrium curve.

Optimum Reflux Ratio

The discussion given under the Ponchon-Savarit method applies.

Illustration 9.10 Redesign the methanol-water fractionator of Illustration 9.8, using the simplifying assumptions of the McCabe-Thiele method. The circumstances are:

Feed. 5000 kg/h, 216.8 kmol/h, $z_F = 0.360$ mole fraction methanol, av mol wt = 23.1, temperature entering the fractionator = 58.3°C

Distillate. 2620 kg/h, 84.4 kmol/h, $x_D = 0.915$ mole fraction methanol, liquid at the bubble point

Residue. 2380 kg/h, 132.4 kmol/h, $x_W = 0.00565$ mole fraction methanol

Reflux ratio = 1.5 times the minimum

SOLUTION Refer to Fig. 9.42. From the txy diagram, the bubble point of the feed at 0.360 mole fraction methanol is 76.0°C, and its dew point is 89.7°C. The latent heats of vaporization at 76.0°C are $\lambda_A = 1046.7$ kJ/kg for methanol and $\lambda_B = 2284$ kJ/kg for water. Specific heat of liquid methanol = 2721, of liquid water = 4187, of feed solution = 3852 J/kg K. If heats of

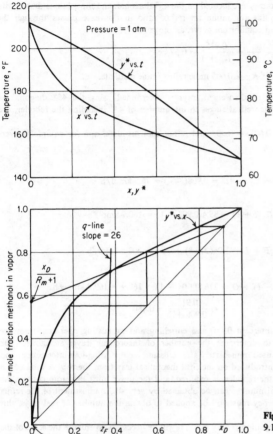

Figure 9.42 Solution to Illustration 9.10. Minimum reflux ratio and minimum trays.

solution are ignored, as is customary with the McCabe-Thiele method, the enthalpy of the feed at 76.0°C (the bubble point) referred to 58.3°C (the feed temperature) is

$$3.852(23.1)(76.0 - 58.3) = 1575 \text{ kJ/kmol}$$

The enthalpy of the saturated vapor at 89.7°C referred to liquids at 58.3°C is

$$0.36[2.721(32.04)(89.7 - 58.3) + 1046.7(32.04)]$$
$$+ (1 - 0.36)[4.187(18.02)(89.7 - 58.3) + 2284(18.02)]$$
$$= 40\,915 \text{ kJ/kmol}$$

$$q = \frac{\text{heat to convert to saturated vapor}}{\text{heat of vaporization}} = \frac{40\,915 - 0}{40\,915 - 1575} = 1.04$$

(Essentially the same value is obtained if the heat of solution is considered.)

$$\frac{q}{q - 1} = \frac{1.04}{1.04 - 1} = 26$$

414 MASS-TRANSFER OPERATIONS

In Fig. 9.42, x_D, x_W, and z_F are located on the 45° diagonal, and the q line is drawn with slope = 26. The operating line for minimum reflux ratio in this case passes through the intersection of the q line and equilibrium curve, as shown.

$$\frac{x_D}{R_m + 1} = \frac{0.915}{R_m + 1} = 0.57$$

$$R_m = 0.605 \text{ mole reflux/mole distillate}$$

The minimum number of theoretical trays is determined using the 45° diagonal as operating line (Fig. 9.42). Theoretical stages to the number of 4.9, including the reboiler, are determined.

For $R = 1.5R_m = 1.5(0.605) = 0.908$ mol reflux/mol distillate, and for equimolal overflow and vaporization:

Eq. (9.49):

$$L = L_0 = RD = 0.908(84.4) = 76.5 \text{ kmol/h}$$

Eq. (9.115):

$$G = D(R + 1) = 85.4(0.908 + 1) = 160.9 \text{ kmol/h}$$

Eq. (9.126):

$$\bar{L} = qF + L = 1.04(216.8) + 76.5 = 302.5 \text{ kmol/h}$$

Eq. (9.127):

$$\bar{G} = F(q - 1) + G = 216.8(1.04 - 1) + 160.9 = 169.7 \text{ kmol/h}$$

$$\frac{x_D}{R + 1} = \frac{0.915}{0.908 + 1} = 0.480$$

Refer to Fig. 9.43. The y intercept 0.480 and enriching and exhausting operating lines are plotted. Steps are drawn to determine the number of theoretical trays, as shown. The exhausting operating line is used immediately after crossing the operating-line intersection, and the feed is therefore to be introduced on the fifth theoretical tray from the top. A total of 8.8 theoretical trays, including the reboiler, is required, and the tower must therefore contain 7.8 theoretical trays. An integral number can be obtained by very slight adjustment of the reflux ratio, but since a tray efficiency must still be applied to obtain the number of real trays, this need not be done.

When the residue composition is very small, it will be necessary to enlarge the scale of the lower left-hand part of the diagram in order to obtain the number of trays. In some cases the graphical determination may still be difficult because of the closeness of the exhausting line and the equilibrium curve. Logarithmic coordinates can then be used to maintain a satisfactory separation of the lines, as in the insert of Fig. 9.43. On such a graph, for very low values of x, the equilibrium curve will be substantially given by $y^* = \alpha x$, which is a straight line of unit slope. The exhausting operating line will, however, be curved and must be plotted from its equation. In this example, the equation is [Eq. (9.121)]

$$y = \frac{302.5}{169.7} x - \frac{132.4}{169.7} 0.00565 = 1.785x - 0.0044$$

The steps representing the theoretical stages are made in the usual manner on these coordinates, continued down from the last made on the arithmetic coordinates (see also page 422).

The diameter of the tower and the tray design are established through the methods of Chap. 6. Note the substantially different liquid loads in the enriching and exhausting sections. A column of constant diameter for all sections is usually desired for simplicity in construction and lower cost. If the discrepancy between liquid or vapor quantities in the two sections is considerable, however, and particularly if expensive alloy or nonferrous metal is used, different diameters for the two sections may be warranted.

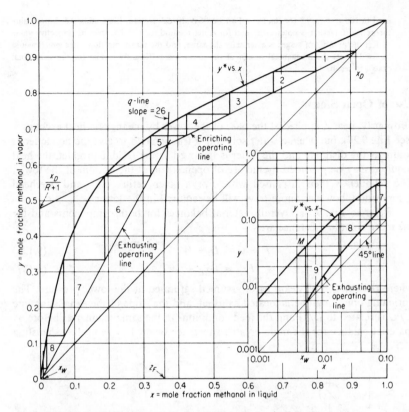

Figure 9.43 Solution to Illustration 9.10. $R = 0.908$ mol reflux/mol distillate product.

Computations for other reflux ratios are easily and quickly made once the diagram and equations have been set up for one value of R. These provide data for determining the most economical R. The following table lists the important quantities for this separation at various values of R:

R	L	G	$\bar{L}$	$\bar{G}$	No. theoretical stages
$R_m = 0.605$	51.0	135.4	277	144.1	∞
0.70	59.0	143.4	285	152.1	11.5
0.80	67.5	151.9	294	160.6	10
0.908	76.5	160.9	303	169.7	8.8
1.029	86.7	171.3	313	180	8.3
2.0	168.8	253	395	262	6.5
4.0	338	422	564	431	5.5
∞	∞	∞	∞	∞	$4.9 = N_m + 1$

Included in the table are the data for the McCabe-Thiele method at $R = 1.029$, the reflux ratio used in Illustration 9.8 for the exact Ponchon-Savarit calculation. Illustration 9.8 showed nine theoretical stages. It is noteworthy that for either method each at 1.5 times its respective value of R_m, the number of stages is essentially the same, and the maximum flow rates which would be used to set the mechanical design of the tower are sufficiently similar for the same final design to result.

Use of Open Steam

Ordinarily, heat is applied at the base of the tower by means of a heat exchanger (see Fig. 9.29), but when a water solution is fractionated to give the nonaqueous solute as the distillate and the water is removed as the residue product, the heat required may be provided by the use of open steam at the bottom of the tower. The reboiler is then dispensed with. For a given reflux ratio and overhead composition, however, more trays will be required in the tower.

Refer to Fig. 9.44. Overall material balances for both components and for the more volatile substance are

$$F + \overline{G} = D + W \tag{9.136}$$

$$Fz_F = Dx_D + Wx_W \tag{9.137}$$

where $\overline{G}$ is mol/h of steam used, assumed saturated at the tower pressure. The enriching operating line is located as usual, and the slope of the exhausting line $\overline{L}/\overline{G}$ is related to L/G and the feed conditions in the same manner as before. A material balance for the more volatile substance below tray m in the exhausting

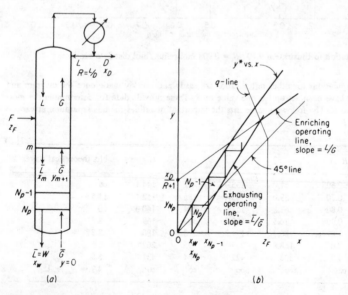

Figure 9.44 Use of open steam.

section is

$$\bar{L}x_m + \bar{G}(0) = \bar{G}y_{m+1} + Wx_W \tag{9.138}$$

and since $\bar{L} = W$ in this case,

$$\frac{\bar{L}}{\bar{G}} = \frac{y_{m+1}}{x_m - x_W} \tag{9.139}$$

The exhausting line therefore passes through the point $(y = 0, x = x_W)$ as shown in Fig. 9.44b. The graphical tray construction must therefore be continued to the x axis of the diagram.

If the steam entering the tower, $\bar{G}_{N_p+1}$, is superheated, it will vaporize liquid on tray N_p to the extent necessary to bring it to saturation, $\bar{G}_{N_p+1}(H_{G,N_p+1} - H_{G,\,\text{sat}})/\lambda M$, where $H_{G,\,\text{sat}}$ is the enthalpy of saturated steam and λM the molar latent heat at the tower pressure.

$$\bar{G} = \bar{G}_{N_p+1}\left(1 + \frac{H_{G,\,N_p+1} - H_{G,\,\text{sat}}}{\lambda M}\right) \tag{9.140}$$

and

$$\bar{L} = \bar{G} - \bar{G}_{N_p+1} + \bar{L}_{N_p} \tag{9.141}$$

from which the internal $\bar{L}/\bar{G}$ ratio can be computed.

Condensers

Condensers are usually conventional tubular heat exchangers, generally arranged horizontally with the coolant inside the tubes, but also vertically with the coolant on either side of the tubes [27]. The condenser may be placed above the tower for gravity flow of the condensed reflux to the top tray, but it is usually more convenient for purposes of construction and cleaning to place the condenser nearer the ground and to return the reflux from an accumulator drum [55] to the top tray by pumping. This procedure also provides more pressure drop for operation of control valves on the reflux line.

The condenser coolant is most frequently water. The pressure of the distillation must then be high enough for the available cooling water to condense the overhead vapor with an adequate temperature difference to provide reasonably rapid heat-transfer rates. The cost of the fractionator, however, will increase with increased pressure of operation, and in the case of very volatile distillates some low-temperature refrigerant may be used as a coolant.

If the condensate is cooled only to the bubble point, the withdrawn distillate is then usually further cooled in a separate heat exchanger to avoid vaporization loss on storage.

Partial Condensers

It is occasionally desired to withdraw a distillate product in the vapor state, especially if its boiling point is low enough to make complete condensation

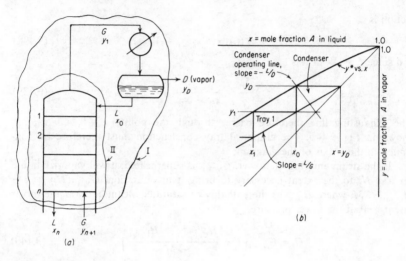

Figure 9.45 Partial equilibrium condenser.

difficult. In this case the overhead vapor is cooled sufficiently to condense the necessary liquid reflux and the residual vapor provides the product, as in Fig. 9.45.

A partial condenser may produce any of several results: (1) If time of contact between vapor product and liquid reflux is sufficient, the two will be in equilibrium with each other and the condenser provides an equilibrium condensation. (2) If the condensate is removed as rapidly as it forms, a differential condensation may occur. (3) If cooling is very rapid, little mass transfer between vapor and condensate results, and the two will have essentially the same composition.

When the first pertains, the condenser acts as one theoretical stage for the separation. The compositions y_D and x_0 can be computed by the methods of equilibrium condensation, as shown in Fig. 9.45b. The enriching operating line is then given as usual by material balances.

Envelope I:
$$G = L + D \tag{9.142}$$

$$Gy_{n+1} = Lx_n + Dy_D \tag{9.143}$$

Envelope II:
$$Gy_{n+1} + Lx_0 = Gy_1 + Lx_n \tag{9.144}$$

In the design of new equipment it is safer to ignore the enrichment which may be obtained by a partial condenser and to include the additional theoretical tray in the column, since it is difficult to ensure that equilibrium condensation will occur.

Cold Reflux

If the overhead vapor is condensed and cooled below its bubble point so that the reflux liquid is cold, vapor G_1 rising from the top tray will be less in quantity than that for the rest of the enriching section since some will be required to condense and heat the reflux to its bubble point. External reflux L_0 will require heat to the extent of $L_0 C_{L0} M_{av}(t_{bp,R} - t_R)$, where $t_{bp,R}$ and t_R are the reflux bubble point and actual temperatures, respectively. An amount of vapor, $L_0 C_{L0} M_{av}(t_{bp,R} - t_R)/(\lambda M)_{av}$ will condense to provide the heat, and the condensed vapor adds to L_0 to provide L, the liquid flow rate below the top tray. Therefore

$$L = L_0 + \frac{L_0 C_{L0} M_{av}(t_{bp,R} - t_R)}{(\lambda M)_{av}} = RD\left[1 + \frac{C_{L0} M_{av}(t_{bp,R} - t_R)}{(\lambda M)_{av}}\right]$$

(9.145)

where R is the usual external reflux ratio, L_0/D. Defining an *apparent* reflux ratio R' by

$$R' = \frac{L}{D} = \frac{L}{G - L}$$

(9.146)

gives

$$R' = R\left[1 + \frac{C_{L0} M_{av}(t_{bp,R} - t_R)}{(\lambda M)_{av}}\right]$$

(9.147)

The enriching operating line becomes

$$y_{n+1} = \frac{R'}{R' + 1} x_n + \frac{x_D}{R' + 1}$$

(9.148)

and it is plotted through $y = x = x_D$, with a y intercept at $x_D/(R' + 1)$ and a slope $R'/(R' + 1)$.

Rectification of Azeotropic Mixtures

Minimum- and maximum-boiling azeotropic mixtures of the type shown in Figs. 9.7 and 9.10 can be treated by the methods already described, except that it will be impossible to obtain two products of compositions which fall on opposite sides of the azeotropic composition. In the rectification of a minimum-boiling azeotrope (Fig. 9.7), for example, the distillate product may be as close to the azeotropic composition as desired. But the residue product will be either rich in A or rich in B depending upon whether the feed is richer or leaner in A than the azeotropic mixture. With maximum-boiling mixtures (Fig. 9.10) the residue product will always approach the azeotropic composition. These mixtures can sometimes be separated completely by addition of a third substance, as described later.

Insoluble mixtures which form two-liquid-phase azeotropes, however, can readily be separated completely provided two fractionators are used. This depends upon the fact that the condensed distillate forms two liquid solutions on opposite sides of the azeotropic composition. Consider the separation of the mixture whose vapor-liquid equilibrium diagram is shown in Fig. 9.46, where the feed has the composition z_F and the solubility limits are $x_{R,\,I}$ and $x_{R,\,II}$ at the boiling point. If the feed is introduced into fractionator I of Fig. 9.47, it is evident that the residue product of composition $x_{W,\,I}$ can be as nearly pure B as desired. The enriching section may contain sufficient trays to produce an overhead vapor approaching the azeotropic composition M, such as vapor $y_{D,\,I}$. This vapor, when totally condensed to mixture K at its boiling point, will form two insoluble liquids of composition $x_{R,\,I}$ and $x_{R,\,II}$, which can be decanted as shown. The layer which is richer in B is returned to the top tray of column I as reflux. The enriching operating line for column I then passes through the point $(y = y_{D,\,I}, x = x_{R,\,I})$, as shown in Fig. 9.46, and its slope will be the liquid/vapor ratio in the enriching section.

The A-rich layer from the decanter (Fig. 9.47) is sent to the top tray of fractionator II, which contains only a stripping or exhausting section. It is clear from Fig. 9.46 that the residue product composition $x_{W,\,II}$ can be as nearly pure A as desired (turn the figure upside down to give it its usual appearance). The overhead vapor from tower II will be of composition $y_{D,\,II}$, which, when totally condensed as mixture N, produces the same two insoluble liquids as the first distillate. Consequently a common condenser can be used for both towers.

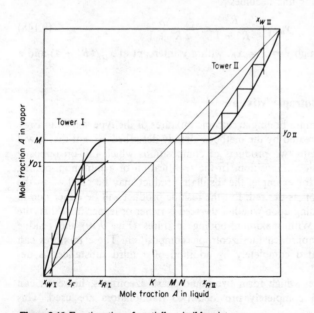

Figure 9.46 Fractionation of partially miscible mixtures.

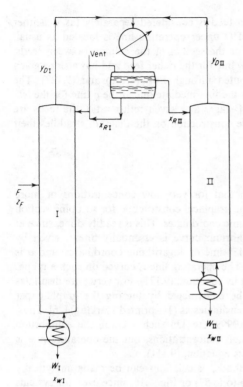

Figure 9.47 Two-tower system for partially miscible mixtures.

In practice it will be desirable to cool the distillate below its bubble point to prevent excessive loss of vapor from the vent of the decanter. This changes the compositions $x_{R,\,I}$ and $x_{R,\,II}$ slightly and provides somewhat larger internal reflux ratios. If the feed itself consists of two insoluble liquids, it can be fed to the decanter, whereupon both fractionators then consist of exhausting sections only. When it is desired to remove the last traces of water from a substance such as a hydrocarbon, it is common practice to use only one tower from which the dry hydrocarbon is removed as the residue product. The hydrocarbon-rich layer from the decanter is returned as reflux, but the water layer, which contains very little hydrocarbon, is normally discarded.

The Ponchon-Savarit method can also be used but the necessary enthalpy data are rarely available for these systems.

Multiple Feeds

If two solutions of different concentrations are to be fractionated to give the same products, both can be handled in the same fractionator. The McCabe-Thiele diagram for a two-feed column will resemble the lower part of Fig. 9.33, with operating lines straight. Each feed is considered separately (as if neither "knew" of the other's presence). The upper operating line is located as usual. The intermediate operating line, for the section of the column between feeds, intersects the enriching line at the q line for the richer feed and has a slope given by the L and G quantities computed through Eqs. (9.126) and (9.127). The lowermost operating line intersects the intermediate one at the q line for the less rich feed. For the least number of stages at a given reflux ratio, the feeds are each introduced at the stage whose construction on the diagram straddles their respective q line.

High-Purity Products

In Illustration 9.10 it was shown that for very low concentrations of more volatile substance in the residue, the graphical construction for stripping-section trays can be completed on logarithmic coordinates. This is readily done, since at very low concentrations the equilibrium curve is essentially linear, given by $y^* = \alpha x$. This plots as a straight 45° line on logarithmic coordinates, and α is taken for very low concentrations. The operating line is curved on such a graph, however, and must be plotted from its equation, (9.121). For very pure distillates a similar logarithmic graph can be constructed by turning the graph paper upside down and re-marking the coordinates as (1—printed marking) [26]. Thus, the printed 0.0001 is marked as 0.9999, etc. On such a graph the equilibrium curve is also straight at 45° for high concentrations, but the operating line is curved and must be plotted from its equation, (9.111).

As an alternative to these methods, calculations can be made analytically using the Kremser equations (5.50) to (5.57) or Fig. 5.16, since even for systems where the McCabe-Thiele assumptions are not generally applicable, the operating lines are straight at the extreme concentrations now under consideration. The *exhausting section* is considered as a stripper for the more volatile component, whence, for a kettle-type reboiler,

$$N_p - m + 1 = \frac{\log\left[\dfrac{x_m - x_W/\alpha}{x_W - x_W/\alpha}(1 - \bar{A}) + \bar{A}\right]}{\log(1/\bar{A})} \quad (9.149)$$

where x_m is the composition of the liquid leaving tray m, the last equilibrium tray obtained by graphical work, and $\bar{A} = \bar{L}/\alpha\bar{G}$. For a thermosiphon reboiler which totally vaporizes the liquid, the left-hand side of Eq. (9.149) is $N_p - m$. For open steam the left-hand side is $N_p - m$, and x_W/α is omitted ($y_{N_p+1} = 0$). The *enriching section* is considered as an absorber for the less volatile component

$$n - 1 = \frac{\log\left[\dfrac{(1 - y_n) - (1 - x_0)/\alpha}{(1 - y_1) - (1 - x_0)/\alpha}\left(1 - \dfrac{1}{A}\right) + \dfrac{1}{A}\right]}{\log A} \quad (9.150)$$

where y_n is the vapor leaving tray n, the last equilibrium tray obtained by graphical construction upward from the feed, x_0 is the reflux composition, and $A = \alpha L / G$.

For many ideal liquids, where α is nearly constant for all concentrations, the analytical calculation of Smoker [58] applies to the entire fractionator.

Tray Efficiencies

Methods for estimating tray efficiencies are discussed in Chap. 6. Murphree vapor efficiencies are most simply used graphically on the xy diagram, whether this is derived from the exact Ponchon-Savarit calculation or with the simplifying assumptions, as shown in Illustration 9.11. Overall efficiencies E_O strictly have meaning only when the Murphree efficiency of all trays is the same and when the equilibrium and operating lines are both straight over the concentrations considered. The latter requirement can be met only over limited concentration ranges in distillation, but nevertheless the empirical correlation of Fig. 6.25 is useful for rough estimates, if not for final designs, with the exception that values of E_O for aqueous solutions are usually higher than the correlation shows.

Illustration 9.11 In the development of a new process, it will be necessary to fractionate 910 kg/h of an ethanol-water solution containing 0.3 mole fraction ethanol, available at the bubble point. It is desired to produce a distillate containing 0.80 mole fraction ethanol, with a substantially negligible loss of ethanol in the residue.

There is available a tower containing 20 identical cross-flow sieve trays, 760 mm diameter, with provision for introducing the feed only on the thirteenth tray from the top. The tower is suitable for use at 1 std atm pressure. The tray spacing is 450 mmHg, the weir on each tray is 530 mm long, extending 60 mm above the tray floor, and the area of the perforations is 0.0462 m²/tray.

Determine a suitable reflux ratio for use with this tower, and estimate the corresponding ethanol loss, assuming that open steam at 69 kN/m² (10 lb$_f$/in²) gauge will be used for heating and that adequate condenser capacity will be supplied.

SOLUTION The McCabe-Thiele method will be used. Mol wt ethanol = 46.05, of water = 18.02. z_F = 0.30.

Basis: 1 h. The feed composition corresponds to 52.27 wt % ethanol. The feed contains 910(0.5227)/46.05 = 10.33 kmol ethanol and 910(1 − 0.5227)/18.02 = 24.10 kmol water, total = 34.43 kmol. If essentially all the ethanol is removed from the residue, the distillate D = 10.33/0.80 = 12.91 kmol/h = 522.2 kg/h. Vapor-liquid equilibrium data are available in "The Chemical Engineers' Handbook," 4th ed., p. 13-5, and in Carey and Lewis, *Ind. Eng. Chem.*, **24**, 882 (1932).

Preliminary calculations indicate that the capacity of the tower is governed by the top vapor rate. At tray 1, av mol wt vapor = 0.8(46.05) + 0.2(18.02) = 40.5 kg/mol, temp = 78.2°C. In Chap. 6 notation,

$$\rho_G = \frac{40.5}{22.41} \frac{273}{273 + 78.2} = 1.405 \text{ kg/m}^3 \qquad \rho_L = 744.9 \text{ kg/m}^3 \qquad \sigma = 0.021 \text{ N/m (est)}$$

Table 6.2: for t = 0.450 m, α = 0.0452, β = 0.0287. The tower cross-sectional area = A_t = $\pi(0.760)^2/4$ = 0.4536 m², and for W = (530/760)T = 0.70T (Table 6.1), the downspout area A_d = 0.0808(0.4536) = 0.0367 m². The active area A_a = 0.4536 − 2(0.0367) = 0.380 m². With A_o = 0.0462 m², and if we tentatively take $(L'/G')(\rho_G/\rho_L)^{0.5}$ = 0.1, Eq. (6.30) provides C_F = 0.0746, and Eq. (6.29) shows V_F = 1.72 m/s as the flooding gas velocity based on $A_n = A_t − A_d$ = 0.4169 m².

Calculations will be made for a reflux ratio corresponding to $G = D(R + 1) = 12.91(3 + 1) = 51.64$ kmol/h vapor at the top, or

$$\frac{51.64(22.41)}{3600} \frac{273 + 78.2}{273} = 0.414 \text{ m}^3/\text{s}$$

The vapor velocity is then $0.414/0.4169 = 0.993$ m/s, or 58 percent of the flooding velocity (amply safe). $L = 3(12.91) = 38.73$ kmol/h or 1567 kg/h. Therefore,

$$\frac{L'}{G'}\left(\frac{\rho_G}{\rho_L}\right)^{0.5} = \frac{1567}{0.414(3600)(1.405)}\left(\frac{1.405}{744.9}\right)^{0.5} = 0.0325$$

Since for values of this quantity less than 0.1, Eq. (6.29) nevertheless uses 0.1, the calculated value of C_F is correct.

Since the feed is at the bubble point, $q = 1$, and Eq. (9.126):

$$\bar{L} = L + qF = 38.73 + 1(34.43) = 73.2 \text{ kmol/h}$$

Eq. (9.127):
$$\bar{G} = G + F(q - 1) = G = 51.64 \text{ kmol/h}$$

The enthalpy of saturated steam, referred to 0°C, at 69 kN/m² is 2699 kN · m/kg, and this will be its enthalpy as it enters the tower if expanded adiabatically to the tower pressure. The enthalpy of saturated steam at 1 std atm (neglecting tower pressure drop) is 2676 kN · m/kg, and the latent heat is 2257 kN · m/kg. Eq. (9.140):

$$51.64 = \bar{G}_{N_p+1}\left[1 + \frac{(26.99 - 26.76)(18.02)}{2257(18.02)}\right]$$

$$\bar{G}_{N_p+1} = 51.1 \text{ kmol steam/h}$$

Eq (9.141):
$$73.2 = 51.64 - 51.1 + L_{N_p}$$

$$L_{N_p} = W = 72.7 \text{ kmol residue/h}$$

Tray efficiencies Consider the situation at $x = 0.5$, $y^* = 0.652$, $t = 79.8$°C, which is in the enriching section. Here (Chap. 6 notation) $\rho_L = 791$, $\rho_G = 1.253$ kg/m³, m (from equilibrium data) = 0.42, and by the methods of Chap. 2, $Sc_G = 0.930$, $D_L = 2.065 \times 10^{-9}$ m²/s. For $L = 38.73$ kmol/h, $q = 4.36 \times 10^{-4}$ m³/s, and for $G = 51.64$ kmol/h, $V_a = 1.046$ m/s. From the tray dimensions, $z = 0.647$ m, $Z = 0.542$ m, and $h_W = 0.060$ m.

Eq. (6.61): $N_{tL} = 0.89$	Eq. (6.38): $h_L = 0.0380$ m
Eq. (6.64): $\theta_L = 30.3$ s	Eq. (6.62): $N_{tL} = 22.8$
Eq. (6.52): $N_{tOG} = 0.871$	Eq. (6.51): $E_{OG} = 0.581$
Eq. (6.63): $D_E = 1.265 \times 10^{-3}$ m²/s	Eq. (6.59): Pe = 7.66
Eq. (6.57): $E_{MG} = 0.72$	

Entrainment is negligible. Similarly,

x	0	0.1	0.3	0.5	0.7
E_{MG}	0.48	0.543	0.74	0.72	0.72

Tray calculations The operating-line intercept = $x_D/(R + 1) = 0.8/(3 + 1) = 0.2$, $q/(q - 1) = \infty$. The operating line for the enriching section and the q line are located as usual (Fig. 9.48). The exhausting-section operating line, on this scale of plot, for all practical purposes passes

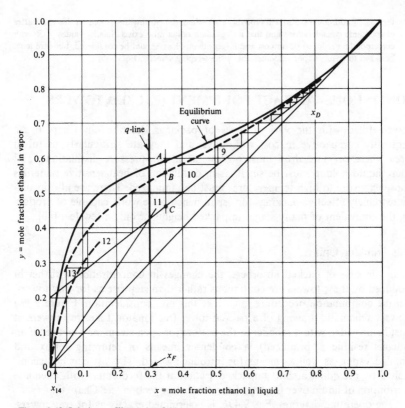

Figure 9.48 Solution to Illustration 9.11.

through the origin. The broken curve is located so that, at each concentration, vertical distances corresponding to lines BC and AC are in the ratio E_{MG}. This curve is used instead of the equilibrium curve to locate real trays, as shown. The feed tray is the thirteenth.

Below tray 14, calculations can be made on logarithmic coordinates as in Illustration 9.10, but that would require trial-and-error location of the operating line since x_W is unknown. It is easier to proceed as follows.

From Fig. 9.48, $x_{14} = 0.0150$, and from this concentration down the equilibrium curve is essentially straight ($\alpha = m = 8.95$), $E_{MG} = 0.48 = $ const, $\bar{A} = \bar{L}/\alpha\bar{G} = 73.2/8.95(51.64) = 0.1584$. Equation (8.16) then provides the overall tray efficiency for the six bottom trays:

$$E_O = \frac{\log[1 + 0.48(1/0.1584 - 1)]}{\log(1/0.1584)} = 0.688$$

The 6 real trays correspond to $6(0.688) = 4.13$ real trays to the bottom of the tower, whence Eq. (9.149) provides, for open steam,

$$4.13 = \frac{\log[(0.015/x_w)(1 - 0.1584) + 0.1584]}{\log(1/0.1584)}$$

$$x_w = 6.25 \times 10^{-6} \text{ mole fraction ethanol in the residue}$$

This corresponds to an ethanol loss of 0.5 kg/day. Larger reflux ratios would reduce this, but the cost of additional steam will probably make them not worthwhile. Note that this is a matter of economic considerations and that an optimum reflux ratio exists. Smaller values of R, with corresponding reduced steam cost and larger ethanol loss, should be considered, but care must be taken to ensure vapor velocities above the weeping velocity, Eq. (6.46).

CONTINUOUS-CONTACT EQUIPMENT (PACKED TOWERS)

Towers filled with the various types of packings described in Chap. 6 are frequently competitive in cost with trays, and they are particularly useful in cases where pressure drop must be low, as in low-pressure distillations, and where liquid holdup must be small, as in distillation of heat-sensitive materials whose exposure to high temperatures must be minimized. There are also available extremely effective packings for use in bench-scale work, capable of producing the equivalent of many stages in packed heights of only a few feet [40].

The Transfer Unit

As in the case of packed absorbers, the changes in concentration with height produced by these towers are continuous rather than stepwise as for tray towers, and the computation procedure must take this into consideration. Figure 9.49a shows a schematic drawing of a packed-tower fractionator. Like tray towers, it must be provided with a reboiler at the bottom (or open steam may be used if an aqueous residue is produced), a condenser, means of returning reflux and reboiled vapor, as well as means for introducing feed. The last can be accomplished by providing a short unpacked section at the feed entry, with adequate distribution of liquid over the top of the exhausting section (see Chap. 6).

The operating diagram, Fig. 9.49b, is determined exactly as for tray towers, using either the Ponchon-Savarit method or, where applicable, the McCabe-Thiele simplifying assumptions. Equations for operating lines and enthalpy balances previously derived for trays are all directly applicable, except that tray-number subscripts can be omitted. The operating lines are then simply the relation between x and y, the bulk liquid and gas compositions, prevailing at each horizontal section of the tower. As before, the change from enriching- to exhausting-section operating lines is made at the point where the feed is actually introduced, and for new designs a shorter column results, for a given reflux ratio, if this is done at the intersection of the operating lines. In what follows, this practice is assumed.

For packed towers, rates of flow are based on unit tower cross-sectional area, mol/(area)(time). As for absorbers, in the differential volume dZ, Fig. 9.49a, the interface surface is $a \, dZ$, where a is the a_A of Chap. 6. The quantity of substance A in the vapor passing through the differential section is Gy mol/(area)(time), and the rate of mass transfer is $d(Gy)$ mol A/(differential volume)(time). Similarly, the rate of mass transfer is $d(Lx)$. Even where the usual simplifying assumptions are not strictly applicable, within a section of the column G and L are both sufficiently constant for equimolar counterdiffusion

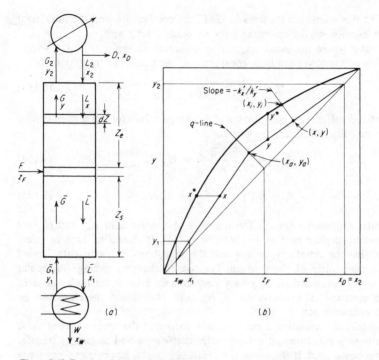

Figure 9.49 Fractionation in a packed tower.

between phases to be practically true: $N_A = -N_B$. Consequently (see Table 3.1), $F_G = k_y'$, $F_L = k_x'$, and the mass-transfer flux is

$$N_A = \frac{d(Gy)}{a\,dZ} = k_y'(y_i - y) = \frac{d(Lx)}{a\,dZ} = k_x'(x - x_i) \qquad (9.151)$$

Therefore $Z_e = \int_0^{Z_e} dZ = \int_{(Gy)_a}^{(Gy)_2} \frac{d(Gy)}{k_y'a(y_i - y)} = \int_{(Lx)_a}^{(Lx)_2} \frac{d(Lx)}{k_x'a(x - x_i)} \qquad (9.152)$

A similar expression, with appropriate integration limits, applies to the stripping section.

For any point (x, y) on the operating line, the corresponding point (x_i, y_i) on the equilibrium curve is obtained at the intersection with a line of slope $-k_x'/k_y' = -k_x'a/k_y'a$ drawn from (x, y), as shown in Fig. 9.49b. For $k_x' > k_y'$, so that the principal mass-transfer resistance lies within the vapor, $y_i - y$ is more accurately read than $x - x_i$. The middle integral of Eq. (9.152) is then best used, evaluated graphically as the area under a curve of $1/k_y'a(y_i - y)$ as ordinate, Gy as abscissa, within the appropriate limits. For $k_x' < k_y'$, it is better to use the last integral. In this manner, variations in G, L, the coefficients, and the interfacial area with location on the operating lines are readily dealt with.

For cases where the usual simplifying assumptions apply, G and L within any section of the tower are both constant, and the heights of transfer units

$$H_{tG} = \frac{G}{k_y'a} \qquad H_{tL} = \frac{L}{k_x'a} \qquad (9.153)$$

are sometimes sufficiently constant (or else average values for the section can be used), so that Eq. (9.152) can be written

$$Z_e = H_{tG}\int_{y_a}^{y_2} \frac{dy}{y_i - y} = H_{tG}N_{tG} \qquad (9.154)$$

$$Z_e = H_{tL}\int_{x_a}^{x_2} \frac{dx}{x - x_i} = H_{tL}N_{tL} \qquad (9.155)$$

with similar expressions for Z_s. The numbers of transfer units N_{tG} and N_{tL} are given by the integrals of Eqs. (9.154) and (9.155). It should be kept in mind, however, that the interfacial area a and the mass-transfer coefficients depend upon the mass rates of flow, which, because of changing average molecular weights with concentration, may vary considerably even if the molar rates of flow are constant. The constancy of H_{tG} and H_{tL} should therefore not be assumed without check.

Ordinarily the equilibrium curve for any section of the tower varies in slope sufficiently to prevent overall mass-transfer coefficients and heights of transfer units from being used. If the curve is essentially straight, however, we can write

$$Z_e = H_{tOG}\int_{y_a}^{y_2} \frac{dy}{y^* - y} = H_{tOG}N_{tOG} \qquad (9.156)$$

$$Z_e = H_{tOL}\int_{x_a}^{x_2} \frac{dx}{x - x^*} = H_{tOL}N_{tOL} \qquad (9.157)$$

where
$$H_{tOG} = \frac{G}{K_y'a} \qquad H_{tOL} = \frac{L}{K_x'a} \qquad (9.158)$$

Here, $y^* - y$ is an overall *driving force* in terms of vapor compositions, and $x - x^*$ is a similar one for the liquid (see Fig. 9.49b). For such cases, Eqs. (8.48), (8.50), and (8.51) can be used to determine the number of overall transfer units without graphical integration. As shown in Chap. 5, with F's equal to the corresponding k''s,

$$\frac{1}{K_y'} = \frac{1}{k_y'} + \frac{m}{k_x'} \qquad (9.159)$$

$$\frac{1}{K_x'} = \frac{1}{mk_y'} + \frac{1}{k_x'} \qquad (9.160)$$

$$H_{tOG} = H_{tG} + \frac{mG}{L}H_{tL} \qquad (9.161)$$

$$H_{tOL} = H_{tL} + \frac{L}{mG}H_{tG} \qquad (9.162)$$

where, in Eqs. (5.27) and (5.28), $m = m' = m'' =$ slope of a straight equilibrium curve.

Although practically all the meaningful da'a on mass-transfer coefficients in packings were obtained at relatively low temperatures, the limited evidence is that they can be used for distillation as well, where the temperatures are normally relatively high. As mentioned in Chap. 6, departure has been noted where surface-tension changes with concentration occur. In the case of *surface-tension-positive systems*, where the more volatile component has the lower surface tension, the surface tension increases down the packing, the film of liquid in the packing becomes more stable, and mass-transfer rates become larger. For *surface-tension-negative systems*, where the opposite surface-tension change occurs, results are as yet unpredictable. Research in this area is badly needed.

Illustration 9.12 Determine suitable dimensions of packed sections of a tower for the methanol-water separation of Illustration 9.8, using 38-mm (1.5-in) Raschig rings.

SOLUTION Vapor and liquid quantities throughout the tower, calculated in the manner of Illustration 9.8, with Eqs. (9.62), (9.64), (9.72), and (9.74), are:

x	t_L, °C	y	t_G, °C	Vapor		Liquid	
				kmol/h	kg/h	kmol/h	kg/h
			Enriching section				
0.915	66	0.915	68.2	171.3	5303	86.7	2723
0.600	71	0.762	74.3	164.0	4684	79.6	2104
0.370	76	0.656	78.7	160.9	4378	76.5	1779
			Stripping section				
0.370	76	0.656	78.7	168.6	4585	301	7000
0.200	82	0.360	89.7	161.6	3721	294	6138
0.100	87	0.178	94.7	160.6	3296	293	5690
0.02	96.3	0.032	99.3	127.6	2360	260	4767

The x and y values are those on the operating line (Fig. 9.28). The temperatures are bubble and dew points for the liquids and vapors, respectively. The operating lines intersect at $x = 0.370$, the dividing point between enriching and stripping sections.

The tower diameter will be set by the conditions at the top of the stripping section because of the large liquid flow at this point. Here (Chap. 8 notation),

$$\rho_G = \frac{4585}{168.6(22.41)} \frac{273}{273 + 78.7} = 0.942 \text{ kg/m}^3 \qquad \rho_L = 905 \text{ kg/m}^3$$

$$\frac{L'}{G'}\left(\frac{\rho_G}{\rho_L}\right)^{0.5} = \frac{7000}{4185}\left(\frac{0.942}{905}\right)^{0.5} = 0.054$$

In Fig. 6.34, choose a gas-pressure drop of 450 $(N/m^2)/m$, whence the ordinate is 0.0825. Table 6.3: $C_f = 95$. $\mu_L = 4.5 \times 10^{-4}$ kg/m · s, $J = 1$, $g_c = 1$.

$$G' = \left[\frac{0.0825(0.942)(905 - 0.942)}{95(4.5 \times 10^{-4})^{0.1}}\right]^{0.5} = 1.264 \text{ kg/m}^2 \cdot \text{s}$$

The tower cross-sectional area = $4185/3600(1.264) = 0.92$ m^2, corresponding to a diameter of 1.1 m^2. $L' = 7000/3600(0.92) = 1.94$ kg/m$^2 \cdot$ s.

Mass-transfer coefficients will be computed for the same location. Table 6.4: $m = 36.5$, $n = -4.69 \times 10^{-3}$, $p = 0.274$, $a_{AW} = 42.36$ m^2/m^3. Table 6.5: $d_s = 0.0530$ m, $\beta = 0.499$, $\varphi_{LsW} = 0.0086$, $\varphi_{LtW} = 0.0279$, $\varphi_{Lo}W = 0.0193$, $\varphi_{Ls} = 0.00362$ (with $\mu_L = 0.00045$ kg/m $\cdot$ s, $\sigma = 0.029$ N/m), $H = 0.91$, $\varphi_{Lo} = 0.0175$, $\varphi_{Lt} = 0.0211$.

Eq. (6.73): $a_A = 38.0$ m^2/m^3 = a. Table 6.3: $\varepsilon = 0.71$; Eq. (6.71): $\varepsilon_0 = 0.71 - 0.0211 = 0.689$. By the methods of Chap. 2, $Sc_G = 1.0$. $G = 0.0466$ kmol/m$^2 \cdot$ s, $\mu_G = 2.96 \times 10^{-5}$ kg/m $\cdot$ s, and by Eq. (6.70), $F_G = k_y' = 1.64 \times 10^{-3}$ kmol/m$^2 \cdot$ s $\cdot$ (mole fraction).

$$D_L = 4.80 \times 10^{-9} \text{ m}^2/\text{s} \qquad Sc_L = 103.5$$

Eq. (6.72): $k_L = 2.66 \times 10^{-4}$ kmol/m$^2 \cdot$ s $\cdot$ (kmol/m^3). Since the molar density of water at 76°C is 53.83 kmol/m^3, $F_L = k_x' = 53.82(2.66 \times 10^{-4}) = 0.0143$ kmol/m$^2 \cdot$ s $\cdot$ (mole fraction).

Similarly values at other concentrations are:

x	G	a	$k_y' \times 10^3$	k_x'
0.915	0.0474	20.18	1.525	0.01055
0.600	0.0454	21.56	1.542	0.00865
0.370	0.0444	21.92	1.545	0.00776
0.370	0.0466	38.00	1.640	0.0143
0.200	0.0447	32.82	1.692	0.0149
0.100	0.0443	31.99	1.766	0.0146
0.02	0.0352	22.25	1.586	0.0150

Figure 9.50 shows the equilibrium curve and operating curves as determined in Illustration 9.8. At ($x = 0.2$, $y = 0.36$) on the operating line, line AB was drawn with slope $= -k_x'/k_y' = 0.0149/(1.692 \times 10^{-3}) = -8.8$. Point B then provides the value of y_i for $y = 0.36$. Similarly the lines were erected at each of the x values of the table. Points such as C and D are joined by the curve CD. Then, at any point M, the corresponding y_i at N is easily located. The enlarged inset of the figure shows y_1 for the vapor entering the packed section to be in equilibrium with the reboiler liquid x_W. Since $k_x' > k_y'$, the middle integral of Eq. (9.152) will be used. The following values of y_i were determined in the manner described, with values of G, k_y', and a obtained from a plot of these against y.

y	y_i	$\dfrac{1}{k_y'a(y_i - y)}$	Gy
$y_2 = 0.915$	0.960	634.0	$0.0433 = (Gy)_2$
0.85	0.906	532.8	0.0394
0.80	0.862	481.1	0.0366
0.70	0.760	499.1	0.0314
$y_a = 0.656$	0.702	786.9	$0.0292 = (Gy)_a$
$y_a = 0.656$	0.707	314.7	$0.0306 = (Gy)_a$
0.50	0.639	124.6	0.0225
0.40	0.580	99.6	0.01787
0.30	0.500	89.0	0.01340
0.20	0.390	92.6	0.00888
0.10	0.230	154.5	0.00416
$y_1 = 0.032$	0.091	481.0	$0.001124 = (Gy)_1$

Graphical integration (not shown) according to Eq. (9.152) provides $Z_e = 7.53$ m for the enriching section, $Z_s = 4.54$ m for the stripping section.

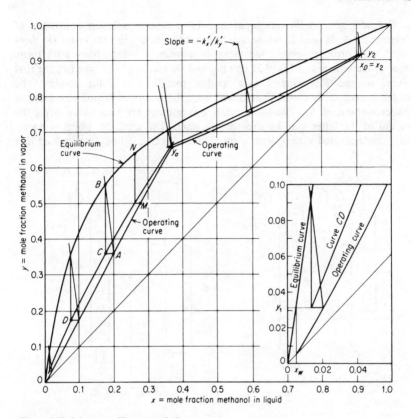

Figure 9.50 Solution to Illustration 9.12.

Berl saddles of the same size, because they provide substantially larger interfacial area, led to packed depths equal to roughly six-tenths of these values, with approximately a 30 percent reduction in gas-pressure drop per meter, at the same mass velocities (although in this instance pressure drop is unimportant). A choice between such packings as these, Pall rings, Intalox saddles, and the like then revolves about their relative cost, not per cubic meter but for the complete tower.

In the above example, H_{tG} for the stripping section varies from $0.0466/(1.640 \times 10^{-3})(38.0) = 0.748$ m at the top to 0.997 m at the bottom. Equation (9.154) with an average $H_{tG} = 0.873$ yields $Z_s = 4.66$ m rather than 4.54, as first calculated. The equilibrium-curve slope varies so greatly that the use of H_{tOG} or overall mass-transfer coefficients is not recommended.

MULTICOMPONENT SYSTEMS

Many of the distillations of industry involve more than two components. While the principles established for binary solutions generally apply to such distillations, new problems of design are introduced which require special consideration.

Consider the continuous separation of a ternary solution consisting of components A, B, and C whose relative volatilities are in that order (A most volatile). In order to obtain the three substances in substantially pure form, either of the schemes of Fig. 9.51 can be used. According to scheme (a), the first column is used to separate C as a residue from the rest of the solution. The residue is necessarily contaminated with a small amount of B and with an even smaller amount of A, although if relative volatilities are reasonably large, the amount of the latter may be exceedingly small. The distillate, which is necessarily contaminated with at least a small amount of C, is then fractionated in a

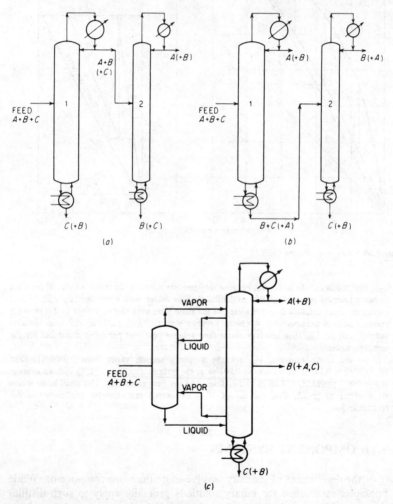

Figure 9.51 Separation of a ternary system.

second column to provide nearly pure A and B. According to scheme (*b*), the first tower provides nearly pure A directly, and the residue is separated in the second. Still a third scheme is shown in (*c*). Which of the schemes is used is a matter of cost, and that may depend upon the relative difficulties of the separations [38]. Generally, scheme (*b*) will be more economical than scheme (*a*) since it requires only one vaporization of substance A. Scheme (*c*) will generally be more economical than either of the others [60].

An important principle to be emphasized is that a single fractionator cannot separate more than one component in reasonably pure form from a multicomponent solution and that a total of $n - 1$ fractionators will be required for complete separation of a system of n components. It might at first be thought, for example, that the component of intermediate volatility B would tend to concentrate in reasonably pure form in the central parts of the first tower, from which it might be withdrawn as a *side stream*, thus allowing pure A and pure C to be withdrawn as distillate and residue, respectively. But this cannot occur. The feed tray of column 1, scheme (*a*), for example, will necessarily contain all three components in proportions not far from those prevailing in the feed itself. Trays immediately above the feed will therefore also contain appreciable quantities of all three substances, with the proportion of C gradually diminishing as we go higher in the enriching section. Similarly, trays immediately below the feed necessarily contain large proportions of all substances, with decreasing amounts of A and B as we penetrate more deeply into the exhausting section. While side streams are indeed sometimes withdrawn from fractionating towers, these streams must be further processed if they are to provide pure products.

The general principles of design of multicomponent fractionators are the same in many respects as those for binary systems, but the dearth of adequate vapor-liquid equilibrium data imposes severe restrictions on their application. These are especially needed for liquids which are not ideal, and the danger of attempting new designs without adequate equilibrium data or pilot-plant study for such solutions cannot be overemphasized. Inadequate methods of dealing with tray efficiencies for multicomponents represent another serious problem still to be solved.

Since the design calculations involve extensive trial and error, high-speed computers are frequently used [1, 24]. Except in extraordinary cases, it is not impossible to carry them out by hand, however, and this may be necessary when only a few designs are to be made. In what follows, hand calculations are assumed and will in any case provide an introduction to the computer methods. The most reliable design procedure is that of Thiele and Geddes [62]. The method assumes that, for a given feed, the number of trays, position of.the feed tray, and the liquid/vapor ratio and temperature for each tray are all known at the start, and it proceeds to compute the resulting distillate and residue products. In most cases, of course, at the beginning of a computation the necessary quantities are ordinarily unknown. The scheme outlined here begins by developing, with a minimum of trials, information necessary to use the Thiele-Geddes method, which then provides the final design. High-speed computer techniques have been developed by grouping equations by equilibrium stages and solving iteratively state to stage, based on the methods of Lewis and Matheson and of Thiele and Geddes. More recently, especially for cases of multiple feeds, a scheme of grouping the equations for each component has been developed.

Specification Limitations

It will be assumed at the beginning that at least the following are established:

1. Temperature, pressure, composition, and rate of flow of the feed
2. Pressure of the distillation (frequently fixed by the temperature of the available cooling water, which must be able to condense the distillate vapor to provide reflux)
3. Feed to be introduced on that tray which will result in the least total number of trays (optimum feed-tray location)
4. Heat losses (even if assumed to be zero)

Under these circumstances it has been shown that there are left to the designer only three additional items to be specified. Having chosen the three, all other characteristics of the fractionator are fixed. The designer can only calculate what they will be, and can arbitrarily assign values to them only temporarily for purposes of trial calculations and later verification. The three can be chosen from the following list, each item counting one:

1. Total number of trays
2. Reflux ratio
3. Reboil ratio, i.e., ratio of vapor produced by the reboiler to residue withdrawn
4. Concentration of one component in one product (a maximum of two may be chosen)
5. Ratio of flow of one component in the distillate to the flow of the same component in the residue, or *split* of the component (a maximum of two may be chosen)
6. Ratio of total distillate to total residue

In what follows, it will be assumed that the reflux ratio and splits of two components are specified. Modifications of the procedures are readily made for other cases. It is clear that, with more than two components in the feed, neither the complete compositions nor the rates of flow of either product are then known.

Key Components

It is convenient first to list the feed components in order of their relative volatility. The more volatile components are called *light*, the less volatile are *heavy*. There will frequently be one component, the *light key component*, which is present in the residue in important amounts while components lighter than the light key are present in only very small amounts. If all components are present in the residue at important concentrations, then the most volatile is the light key. Similarly there will usually be one component, the *heavy key component*, present in the distillate in important amounts while components heavier than the heavy key are present only in small amounts. If all components are found in the distillate at important concentrations, then the least volatile is the heavy key.

The difficulty of the separation, as measured by the number of trays required for a given reflux ratio, is fixed by the key-component concentrations in the products. It is therefore important to establish which are the keys, and they may or may not be those whose splits have been specified.

Relative volatilities will be computed always with respect to the heavy key

$$\alpha_J = \frac{m_J}{m_{hk}} \tag{9.163}$$

where J represents any component and *hk* the heavy key. Thus, $\alpha_{hk} = 1$, α's for components lighter than the heavy key are greater than 1 and for heavier components are less than 1.

Minimum Reflux Ratio

This is the largest reflux ratio requiring an infinite number of trays to separate the key components.

With an infinite number of trays, it is possible to exclude from the distillate all component heavier than the heavy key. Since all components are present at the feed tray, and since it requires several enriching trays to reduce these heavy components to negligible concentrations, the pinch for these components, above which their concentration is zero, lies somewhere above the feed tray. Similarly an infinite number of trays permits exclusion from the residue of all components lighter than the light key, and there is a pinch for these several trays below the feed tray. The situation is therefore different from that for binary distillation, where at minimum reflux ratio there is only one pinch, usually at the feed tray.

Because of the possibility of excluding components from the products, computations at minimum reflux ratio help decide which are the key components. Components between the keys in volatility are found importantly in both products, and they are said to *distribute*. Shiras et al. [54] show that at minimum reflux ratio, approximately

$$\frac{x_{J,D}D}{z_{J,F}F} = \frac{\alpha_J - 1}{\alpha_{lk} - 1}\frac{x_{lkD}D}{z_{lkF}F} + \frac{\alpha_{lk} - \alpha_J}{\alpha_{lk} - 1}\frac{x_{hkD}D}{z_{hkF}F} \tag{9.164}$$

For $x_{J,D}D/z_{J,F}F$ less than -0.01 or greater than 1.01, component J will probably not distribute. For $x_{J,D}D/z_{J,F}F$ between 0.01 and 0.99, component J will undoubtedly distribute. The computations are made with first estimates of what the keys are, to be corrected as necessary.

Many methods of estimating the minimum reflux ratio R_m have been proposed, most of which are tedious to use and not necessarily very accurate. Since the only purpose of obtaining R_m is to estimate the product compositions at R_m and to ensure that the specified R is reasonable, an exact value is not required. Underwood's method [64], which uses constant average α's and assumes constant L/G, is not exact but provides reasonable values without great effort; it is recommended (refer to the original papers for a derivation, which is lengthy). Two equations must be solved:

$$\Sigma \frac{\alpha_J z_{JF} F}{\alpha_J - \phi} = F(1 - q) \tag{9.165}$$

and

$$\Sigma \frac{\alpha_J x_{JD} D}{\alpha_J - \phi} = D(R_m + 1) \tag{9.166}$$

The first of these is written for all J components of the feed and solved for the necessary values of φ. These are the values which lie between the α's of the distributed components. We thus require one more value of φ than there are components between the keys, and these will lie between α_{lk} and $\alpha_{hk} = 1$. Equation (9.166) is then written once for each value of ϕ so obtained, including the heavy key and all lighter components in the summations. These are solved simultaneously for R_m and the unknown $x_{J,D} D$'s. If $x_{J,D} D$ computes to be negative or greater than $z_{J,F} F$, component J will not distribute and the keys have been incorrectly chosen.

The α's are all computed at the average of the distillate dew point and the residue bubble point, which may require a few trial estimates. The method has been extended to columns with multiple feeds [2].

Illustration 9.13 The following feed, at 82°C (180°F), 1035 kN/m² (150 lb$_f$/in²), is to be fractionated at that pressure so that the vapor distillate contains 98% of the C_3H_8 but only 1% of the C_5H_{12}:

Component	CH_4	C_2H_6	n-C_3H_8	n-C_4H_{10}	n-C_5H_{12}	n-C_6H_{14}
z_F, mole fraction	0.03	0.07	0.15	0.33	0.30	0.12

Estimate the minimum reflux ratio and the corresponding products.

SOLUTION The components will be identified as C_1, C_2, etc. Subscript numerals on m's, α's, etc., are Celsius temperatures if larger than 25; otherwise they represent tray numbers. Liquid solutions are assumed to be ideal. Values of m are taken from Depriester, *Chem. Eng. Prog. Symp. Ser.*, 49(7), 1 (1953). Values of H, all as kJ/kmol referred to the saturated liquid at −129°C (−200°F), are taken from Maxwell, "Data Book on Hydrocarbons," D. Van Nostrand Company, Inc., Princeton, N.J., 1950. For H_G for temperatures below the dew point, the enthalpy is that of the saturated vapor. For H_L for temperatures above the critical, the enthalpy is that of the gas in solution. Values of m are conveniently plotted as log m vs. $t\,°C$ for interpolation. Enthalpy can be plotted vs. t on arithmetic coordinates.

	30°C	60°C	90°C	120°C
C_1:				
m	16.1	19.3	21.8	24.0
H_G	12 790	13 910	15 000	16 240
H_L	9 770	11 160	12 790	14 370
C_2:				
m	3.45	4.90	6.25	8.15
H_G	22 440	24 300	26 240	28 140
H_L	16 280	18 140	19 890	21 630
C_3:				
m	1.10	2.00	2.90	4.00
H_G	31 170	33 000	35 800	39 000
H_L	16 510	20 590	25 600	30 900
C_4:				
m	0.35	0.70	1.16	1.78
H_G	41 200	43 850	46 500	50 400
H_L	20 350	25 120	30 000	35 400
C_5:				
m	0.085	0.26	0.50	0.84
H_G	50 500	54 000	57 800	61 200
H_L	24 200	32 450	35 600	41 400
C_6:				
m	0.030	0.130	0.239	0.448
H_G	58 800	63 500	68 150	72 700
H_L	27 700	34 200	40 900	48 150

Basis: 1 kmol feed throughout.

Flash vaporization of the feed Use Eq. (9.37). After several trials, assume $G_F/F = 0.333$, $L_F/G_F = 0.667/0.333 = 2.0$.

Component	z_F	m_{82}	$y = \dfrac{z_F(2 + 1)}{1 + 2/m}$
C_1	0.03	21.0	0.0829
C_2	0.07	5.90	0.1578
C_3	0.15	2.56	0.2530
C_4	0.33	1.00	0.3300
C_5	0.30	0.42	0.1559
C_6	0.12	0.19	0.0312
	1.00		1.0108

Σy is sufficiently close to 1.0. Therefore $q = L_F/F = 0.67$. Tentatively assume C_3 = light key, C_5 = heavy key. Therefore specifications require $x_{lk_D}D = 0.98(0.15) = 0.1470$ kmol, $x_{hk_D}D = 0.01(0.30) = 0.0030$ kmol. Estimate the average temperature to be 80°C (to be checked).

Component	$z_F F$	m_{80}	α_{80}
C_1	0.03	21.0	$53.2 = 21.0/0.395$
C_2	0.07	5.90	14.94
lk C_3	0.15	2.49	6.30
C_4	0.33	0.95	2.405
hk C_5	0.30	0.395	1.0
C_6	0.12	0.180	0.456

Equation (9.164) is to be used for C_1, C_2, and C_6; the others distribute at R_m. Use $y_{J,D}$ instead of $x_{J,D}$ since distillate is a vapor.

C_1:

$$\frac{y_D D}{z_F F} = \frac{53.2 - 1}{6.30 - 1}\frac{0.1470}{0.15} + \frac{6.30 - 53.2}{6.30 - 1}\frac{0.0030}{0.30} = 9.761$$

Similarly, for C_2 and C_6 the values are 2.744 and -0.0892, respectively. None of these components will distribute at R_m, and the chosen keys are correct. The distillate contains 0.03 kmol C_1, 0.07 kmol C_2, and no C_6. The distribution of only C_4 to the products is unknown. Equation (9.165):

$$\frac{53.2(0.03)}{53.2 - \phi} + \frac{14.94(0.07)}{14.94 - \phi} + \frac{6.30(0.15)}{6.30 - \phi} + \frac{2.405(0.33)}{2.405 - \phi} + \frac{1.0(0.30)}{1.0 - \phi} + \frac{0.456(0.12)}{0.456 - \phi} = 1(1 - 0.67)$$

This is to be solved for the two values of ϕ lying between α_{C_3} and α_{C_4}, and α_{C_4} and α_{C_5}. Therefore, $\phi = 4.7760$ and 1.4177. Equation (6.166) is then written for each value of ϕ and for components C_1 through C_5. For $\phi = 4.776$,

$$\frac{53.2(0.03)}{53.2 - 4.776} + \frac{14.94(0.07)}{14.94 - 4.776} + \frac{6.30(0.1470)}{6.30 - 4.776} + \frac{2.405(y_{C_4D}D)}{2.405 - 4.776} + \frac{1.0(0.0030)}{1.0 - 4.776} = D(R_m + 1)$$

This is repeated with $\phi = 1.4177$, and the two solved simultaneously. $y_{C_4D}D = 0.1306$ kmol C_4 in the distillate, $D(R_m + 1) = 0.6099$. For the distillate, try a dew point = 46°C.

Component	$y_D D$	y_D	m_{46}	α_{46}	$\dfrac{y_D}{\alpha_{46}}$
C_1	0.03	0.0789	18.0	100	0.000789
C_2	0.07	0.1840	4.2	23.3	0.00789
lk C_3	0.1470	0.3861	1.50	8.34	0.0463
C_4	0.1306	0.3431	0.500	2.78	0.1234
hk C_5	0.0030	0.0079	0.180	1.0	0.0079
	$D = 0.3806$	1.0000			0.1863

$m_{hk} = m_{C_5} = 0.1863$ at 46.6°C, and the assumed 46°C is satisfactory. For the residue, try a bubble point = 113°C. The amount of residue is obtained from $x_{J, W}W = z_{J, F}F - y_{J, D}D$:

Component	$x_W W$	x_W	m_{113}	α_{113}	$\alpha_{113}x_W$
lk C$_3$	0.0030	0.00484	3.65	5.00	0.0242
C$_4$	0.1994	0.3219	1.60	2.19	0.7050
hk C$_5$	0.2970	0.4795	0.730	1.00	0.4800
C$_6$	0.1200	0.1937	0.380	0.521	0.1010
	W = 0.6194	1.0000			1.3102

$m_{hk} = m_{C_5} = 1/1.3102 = 0.763$ at 114°C, and the assumed 113°C was close enough.

Average temperature = $(114 + 46.6)/2 = 80.3$°C (assumed 80°C is close enough). Therefore,

$$D(R_m + 1) = 0.3806(R_m + 1) = 0.6099$$

$$R_m = 0.58 \text{ mol reflux/mol distillate}$$

Total Reflux

The product compositions change with reflux ratio, and computations at total reflux will help decide the ultimate compositions.

Fenske's equation (9.85) is not limited to binary mixtures and can be applied to the key components to determine the minimum number of trays,

$$N_m + 1 = \frac{\log\left[(x_{lkD}D/x_{hkD}D)(x_{hkW}W/x_{lkW}W)\right]}{\log \alpha_{lk, \text{av}}} \tag{9.167}$$

where $N_m + 1$ is the total number of theoretical stages including the reboiler (and a partial condenser if credit is taken for its fractionating ability). The equation can then be applied to determine the distribution of other components at total reflux,

$$\frac{x_{J, D}D}{x_{J, W}W} = \alpha_{J, \text{av}}^{N_m+1} \frac{x_{hkD}D}{x_{hkW}W} \tag{9.168}$$

The average α is the geometric mean of the values at the distillate dew point and residue bubble point, which may require a few trials to estimate. Winn [68] suggests a method to reduce the number of trials required.

Having obtained N_m and R_m, reference can be made to any of the several empirical correlations mentioned earlier [5, 10, 11, 17, 35] for an estimate of the number of trays at reflux ratio R. These can be unreliable, however, particularly if the majority of the trays are in the exhausting section of the tower. A relationship which is exact for binary mixtures and can be applied to multicomponents yields better results for that case [59]. The result of such an estimate can be a reasonable basis for proceeding directly to the method of Thiele and Geddes.

Illustration 9.13 continued Compute the number of theoretical trays at local reflux and the corresponding products.

SOLUTION Tentatively assume the same distillate and residue temperatures as obtained for the minimum reflux ratio.

Component	$\alpha_{46.6}$	α_{114}	$\alpha_{av} = (\alpha_{46.6}\alpha_{114})^{0.5}$
C_1	100	31.9	56.4
C_2	23.3	10.43	15.6
lk C_3	8.34	5.00	6.45
C_4	2.78	2.19	2.465
hk C_5	1.0	1.0	1.0
C_6	0.415	0.521	0.465

Eq. (9.167):

$$N_m + 1 = \frac{\log[(0.147/0.003)(0.297/0.003)]}{\log 6.45} = 4.55$$

Eq. (9.168) for C_4:

$$\frac{y_{C_4D}D}{x_{C_4W}W} = (2.465)^{4.55}\frac{0.003}{0.297} = 0.611$$

A C_4 balance:

$$y_{C_4D}D + x_{C_4W}W = z_{C_4F}F = 0.33$$

Solving simultaneously gives

$$y_{C_4D}D = 0.1255 \text{ kmol} \qquad x_{C_4W}W = 0.2045 \text{ kmol}$$

Similarly,

C_1: $\qquad\qquad\qquad x_{C_1W}W \approx 0 \qquad y_{C_1D}D = 0.03$

C_2: $\qquad\qquad\qquad x_{C_2W}W \approx 0 \qquad y_{C_2D}D = 0.07$

C_6: $\qquad y_{C_6D}D = 0.00003$ (negligible; assume 0) $\qquad x_{C_4W}W = 0.12$

Therefore, at total reflux,

Component	$y_D D$	$x_W W$
C_1	0.03	Nil
C_2	0.07	Nil
C_3	0.1470	0.0030
C_4	0.1255	0.2045
C_5	0.0030	0.2970
C_6	Nil	0.12
	0.3755 = D	0.6245 = W

The distillate dew point computes to be 46.3°C and the residue bubble point 113°C. The assumed 46 and 114°C are close enough.

Product Compositions

For components between the keys, a reasonable estimate of their distribution at reflux ratio R can be obtained by linear interpolation of $x_{J,D}D$ between R_m and total reflux according to $R/(R + 1)$. For components lighter than the light key and heavier than the heavy key, unless there is clear indication of important distribution at total reflux and unless R is to be very large, it is best at this time to assume they do not distribute. Even very small amounts of light components in the residue or of heavy components in the distillate enormously disturb the computations to follow.

Illustration 9.13 continued A reflux ratio $R = 0.8$ will be used. Estimate the product compositions at this reflux ratio.

SOLUTION Since C_1 and C_2 do not enter the residue (nor C_6 the distillate) appreciably at either total reflux or minimum reflux ratio, it is assumed that they will not at $R = 0.8$. C_3 and C_5 distributions are fixed by the specifications. Only that of C_4 remains to be estimated.

R	$\dfrac{R}{R + 1}$	$y_{C_4D}D$
∞	1.0	0.1255
0.8	0.445	?
0.58	0.367	0.1306

Using a linear interpolation, we get

$$y_{C_4D}D = \frac{1 - 0.445}{1 - 0.367}(0.1306 - 0.1255) + 0.1255 = 0.1300$$

Therefore, for the distillate,

Component	y_DD	y_D	$\dfrac{y_D}{\alpha_{46.6}}$	$x_0 = \dfrac{y_D/\alpha}{\Sigma y_D/\alpha}$
C_1	0.03	0.0789	0.00079	0.00425
C_2	0.07	0.1842	0.0079	0.0425
lk C_3	0.1470	0.3870	0.0464	0.2495
C_4	0.1300	0.3420	0.1230	0.6612
hk C_5	0.0030	0.0079	0.0079	0.0425
	$D = 0.3800$	1.0000	0.1860	1.0000

$m_{C_5} = 0.1860$; $t = 46.6°C$ distillate dew point. The last column of the table shows the liquid reflux in equilibrium with the distillate vapor [Eq. (9.30)]. For the residue,

Component	x_WW	x_W	$\alpha_{114}x_W$	$y_{N_p+1} = \dfrac{\alpha x_W}{\Sigma \alpha x_W}$
lk C_3	0.003	0.00484	0.0242	0.01855
C_4	0.200	0.3226	0.7060	0.5411
hk C_5	0.2970	0.4790	0.4790	0.3671
C_6	0.1200	0.1935	0.0955	0.0732
	$W = 0.6200$	1.0000	1.3047	1.0000

$m_{C_5} = 1/1.3047 = 0.767$; $t = 114°C$, residue bubble point. The last column shows the reboiler vapor in equilibrium with the residue [Eq. (9.25)].

Feed-Tray Location

Just as with binary mixtures, the change from enriching to exhausting section should be made as soon as greater enrichment is thereby produced. This ultimately can be determined only by trial in the Thiele-Geddes calculation, but in the meantime some guidance is required. The following assumes constant L/G, neglects interference of components other than the keys, and assumes that the optimum feed tray occurs at the intersection of operating lines of the key components [48].

Equation (9.116), the operating line for the enriching section, omitting the tray-number designations, can be written for each key component

$$x_{lk} = y_{lk}\frac{G}{L} - \frac{D}{L}x_{lkD} \tag{9.169}$$

$$x_{hk} = y_{hk}\frac{G}{L} - \frac{D}{L}x_{hkD} \tag{9.170}$$

Elimination of L provides

$$y_{lk} = \frac{x_{lk}}{x_{hk}}\left(y_{hk} - \frac{D}{G}x_{hkD}\right) + \frac{D}{G}x_{lkD} \tag{9.171}$$

Similarly, Eq. (9.120) for the exhausting section yields

$$y_{lk} = \frac{x_{lk}}{x_{hk}}\left(y_{hk} + \frac{W}{\overline{G}}x_{hkW}\right) - \frac{W}{\overline{G}}x_{lkW} \tag{9.172}$$

At the operating-line intersection, y_{lk} is the same from Eqs. (9.171) and (9.172) as are y_{hk} and x_{lk}/x_{hk}. Equating the right-hand sides of the two expressions then produces

$$\left(\frac{x_{lk}}{x_{hk}}\right)_{\text{intersection}} = \frac{Wx_{lkW}/\overline{G} + Dx_{lkD}/G}{Wx_{hkW}/\overline{G} + Dx_{hkD}/G} \tag{9.173}$$

Combining Eqs. (9.76) and (9.127) for the light key produces

$$\frac{W}{\overline{G}}x_{lkW} + \frac{D}{G}x_{lkD} = \frac{GFz_{lkF} - DF(1-q)x_{lkD}}{[G - F(1-q)]G} \tag{9.174}$$

and a similar result is obtained for the heavy key. Putting these into Eq. (9.173) then yields, with $G/D = R + 1$,

$$\left(\frac{x_{lk}}{x_{hk}}\right)_{\text{intersection}} = \frac{z_{lkF} - x_{lkD}(1-q)/(R+1)}{z_{hkF} - x_{hkD}(1-q)/(R+1)} \tag{9.175}$$

Recall from the treatment of binaries that the feed tray f is the uppermost step on the exhausting-section operating line and that it rarely happens that the feed step exactly coincides with the operating-line intersection. The feed-tray location is then given by

$$\left(\frac{x_{lk}}{x_{hk}}\right)_{f-1} > \left(\frac{x_{lk}}{x_{hk}}\right)_{\text{intersection}} > \left(\frac{x_{lk}}{x_{hk}}\right)_f \tag{9.176}$$

A method of choosing the feed tray based on minimizing the resulting irreversibility (or increasing the entropy) has been proposed [53]. Ultimately, the best feed-tray location is obtained through the Thiele-Geddes calculation.

Lewis and Matheson Calculation [31]

This establishes the first (and usually the final) estimate of the number of trays required. Although it is possible to allow for variations in the L/G ratio with tray number by tray-to-tray enthalpy balances at the same time, since the products are not yet firmly established, it is best to omit this refinement.

With constant L/G, the McCabe-Thiele operating lines, which are merely material balances, are applicable to each component. Thus, for the enriching section, Eq. (9.116) is

$$y_{J, n+1} = \frac{L}{G} x_{J, n} + \frac{D}{G} z_{J, D} \qquad (9.177)$$

This is used alternately with dew-point (equilibrium) calculations to compute tray by tray for each component from the top down to the feed tray. For components heavier than the heavy key, there will be no $z_{J, D}$ available, and these components cannot be included.

For the exhausting section, it is convenient to solve Eq. (9.120) for x:

$$x_{J, m} = y_{J, m+1} \frac{\overline{G}}{\overline{L}} + \frac{W}{\overline{L}} x_{J, W} \qquad (9.178)$$

This is used alternately with bubble-point (equilibrium) calculations to compute tray by tray for each component from the bottom up to the feed tray. There are no values of $x_{J, W}$ for components lighter than the light key, and they cannot be included.

If, in the first estimate of product compositions, it is found that one product contains all components (lk = most volatile of hk = least volatile component of the feed), the computation can be started with this product and continued past the feed tray to the other end of the column. The operating-line equations appropriate to each section must of course be used. This will avoid the composition corrections discussed later.

Illustration 9.13 continued Determine the number of theoretical trays and the location of the feed tray for $R = 0.8$.

SOLUTION For locating the feed tray, Eq. (9.175):

$$\left(\frac{x_{lk}}{x_{hk}} \right)_{\text{intersection}} = \frac{0.15 - 0.3870(1 - 0.67)/1.8}{0.30 - 0.0079(1 - 0.67)/1.8} = 0.264$$

Enriching section:

$$D = 0.3800 \qquad L = RD = 0.8(0.3800) = 0.3040 \qquad G = L + D = 0.3040 + 0.3800 = 0.6840$$

$$\frac{L}{G} = \frac{0.3040}{0.6840} = 0.445 \qquad \frac{D}{G} = \frac{0.3800}{0.6840} = 0.555$$

Eq. (9.177):

C_1: $\qquad y_{n+1} = 0.445 x_n + 0.555(0.0789) = 0.455 x_n + 0.0438$

Similarly,

C_2: $\qquad\qquad\qquad y_{n+1} = 0.455x_n + 0.1022$

C_3: $\qquad\qquad\qquad y_{n+1} = 0.445x_n + 0.2148$

C_4: $\qquad\qquad\qquad y_{n+1} = 0.445x_n + 0.1898$

C_5: $\qquad\qquad\qquad y_{n+1} = 0.445x_n + 0.0044$

Estimate $t_1 = 57°C$.

Component	x_0	y_1 [Eq. (9.177), $n = 0$]	α_{57}	$\dfrac{y_1}{\alpha}$	$x_1 = \dfrac{y_1/\alpha}{\Sigma y_1/\alpha}$
C_1	0.00425	0.0457	79.1	0.00058	0.00226
C_2	0.0425	0.1211	19.6	0.00618	0.0241
lk C_3	0.2495	0.3259	7.50	0.0435	0.1698
C_4	0.6612	0.4840	2.66	0.1820	0.7104
hk C_5	0.0425	0.0233	1.00	0.0239	0.0933
	1.0000	1.0000		0.2562	1.0000

$m_{C_5} = 0.2562$, $t_1 = 58.4°C$. Liquid x_1's in equilibrium with y_1. $(x_{lk}/x_{hk})_1 = 0.1698/0.0933 = 1.82$. Tray 1 is not the feed tray. Estimate $t_2 = 63°C$.

Component	y_2 [Eq. (9.177), $n = 1$]	α_{63}	$\dfrac{y_2}{\alpha}$	$x_2 = \dfrac{y_2/\alpha}{\Sigma y_2/\alpha}$
C_1	0.0448	68.9	0.00065	0.00221
C_2	0.1129	17.85	0.00632	0.0214
lk C_3	0.2904	6.95	0.0418	0.1419
C_4	0.5060	2.53	0.02000	0.6787
hk C_5	0.0459	1.00	0.0459	0.1557
	1.0000		0.2947	1.0000

$m_{C_5} = 0.2947$, $t_2 = 65°C$. $(x_{lk}/x_{hk})_2 = 0.1419/0.1557 = 0.910$. The tray calculations are continued downward in this manner. The results for trays 5 and 6 are:

Component	x_5	x_6
C_1	0.00210	0.00204
C_2	0.0195	0.0187
lk C_3	0.1125	0.1045
C_4	0.4800	0.4247
hk C_5	0.3859	0.4500
t, °C	75.4	79.2
x_{lk}/x_{hk}	0.292	0.232

Applying Eq. (9.176), it is seen that tray 6 is the feed tray. Exhausting section:

$$\bar{L} = L + qF = 0.3040 + 0.67(1) = 0.9740 \qquad \bar{G} = \bar{L} - W = 0.9740 - 0.6200 = 0.3540$$

$$\frac{\bar{G}}{\bar{L}} = \frac{0.3540}{0.9740} = 0.364 \qquad \frac{W}{\bar{L}} = \frac{0.6200}{0.9740} = 0.636$$

Eq. (9.178):

C_3: $\qquad x_m = 0.364y_{m+1} + 0.636(0.00484) = 0.364y_{m+1} + 0.00308$

Similarly,

C_4: $\qquad x_m = 0.364y_{m+1} + 0.2052$

C_5: $\qquad x_m = 0.364y_{m+1} + 0.3046$

C_6: $\qquad x_m = 0.364y_{m+1} + 0.1231$

Estimate $t_{N_p} = 110°C$.

Component	y_{N_p+1}	x_{N_p} [Eq. (9.178), $m = N_p$]	α_{110}	αx_{N_p}	$y_{N_p} = \dfrac{\alpha x_{N_p}}{\Sigma \alpha x_{N_p}}$
lk C_3	0.01855	0.00983	5.00	0.0492	0.0340
C_4	0.5411	0.4023	2.20	0.885	0.6118
hk C_5	0.3671	0.4382	1.0	0.4382	0.3028
C_6	0.0732	0.1497	0.501	0.0750	0.0514
	1.0000	1.0000		1.4474	1.0000

$m_{C_5} = 1/1.447 = 0.691$, $t_{N_p} = 107.9°C$. y_{N_p} in equilibrium with x_{N_p}. $(x_{lk}/x_{hk})_{N_p} = 0.00983/0.4382 = 0.0224$. N_p is not the feed tray. In like fashion, x_{N_p-1} is obtained from y_{N_p} with Eq. (9.178), and the computations continued up the tower. The results for trays $N_p - 7$ to $N_p - 9$ are:

Component	x_{N_p-7}	x_{N_p-8}	x_{N_p-9}
lk C_3	0.0790	0.0915	0.1032
C_4	0.3994	0.3897	0.3812
hk C_5	0.3850	0.3826	0.3801
C_6	0.1366	0.1362	0.1355
t, °C	95.2	94.1	93.6
x_{lk}/x_{hk}	0.205	0.239	0.272

Application of Eq. (9.176) shows tray $N_p - 8$ as the feed tray. The data for $N_p - 9$ are discarded. Then $N_p - 8 = 6$, and $N_p = 14$ trays.

Composition Corrections

The previous computations provide two feed-tray liquids, computed from opposite directions, which do not agree. In most cases, the number of trays will be satisfactory nevertheless, and for purposes of subsequent calculations (Thiele-Geddes), temperatures and L/G ratios need only be roughly estimated if computer methods are used. If better estimates near the feed tray particularly are desired, the following, due to Underwood [63], is reasonably satisfactory.

Starting with the feed-tray liquid as computed from the enriching section, which shows no heavy components, the mole fractions are all reduced proportionately so that their sum, plus the mole fractions of the missing components (as shown in the feed-tray composition as computed from the exhausting section) is unity. The bubble point is then recalculated. The justification for this somewhat arbitrary procedure is that the *relative* concentrations of the light components in the vapor will remain about the same so long as their relative concentrations in the liquid are unchanged.

The concentrations of the missing heavy components on trays above the feed tray are next estimated. For these, $z_{J,D}$ should be very small, and Eq. (9.177) shows

$$\frac{y_{J,n+1}}{y_{hk,n+1}} = \frac{Lx_{J,n}/G + Dz_{J,D}/G}{Lx_{hk,n}/G + Dz_{hkD}/G} \approx \frac{x_{J,n}}{x_{hkn}} \tag{9.179}$$

Since

$$\frac{y_{J,n+1}}{y_{hk,n+1}} = \frac{m_{J,n+1}x_{J,n+1}}{m_{hk,n+1}x_{hk,n+1}} = \frac{\alpha_{J,n+1}x_{J,n+1}}{x_{hk,n+1}} \tag{9.180}$$

then

$$x_{J,n} = \frac{\alpha_{J,n+1}x_{J,n+1}x_{hkn}}{x_{hk,n+1}} \tag{9.181}$$

The liquid mole fractions on tray n as previously determined are then reduced proportionately to accommodate those of the heavy components, and a new bubble point is calculated. This is continued upward until the concentrations on upper trays as previously computed are no longer changed importantly.

Missing light components are added to trays below the feed in much the same manner. Thus, Eq. (9.178) for these provides

$$\frac{y_{J,m+1}}{y_{lk,m+1}} = \frac{x_{J,m} - Wx_{J,W}/\bar{L}}{x_{lkm} - Wx_{lkW}/\bar{L}} \approx \frac{x_{J,m}}{x_{lkm}} \tag{9.182}$$

where advantage is taken of the fact that $x_{J,W}$ must be small for these components. As before,

$$\frac{y_{J,m+1}}{y_{lk,m+1}} = \frac{m_{J,m+1}x_{J,m+1}}{m_{lk,m+1}x_{lk,m+1}} = \frac{\alpha_{J,m+1}x_{J,m+1}}{\alpha_{lk,m+1}x_{lk,m+1}} \approx \frac{\alpha_{Jm}x_{J,m+1}}{\alpha_{lkm}x_{lk,m+1}} \tag{9.183}$$

The last approximation results from assuming that the ratio of α's stays reasonably constant with small temperature changes. Then

$$x_{J,m+1} = \frac{x_{J,m}\alpha_{lkm}x_{lk,m+1}}{\alpha_{J,m}x_{lkm}} \tag{9.184}$$

The mole fractions of the components on the lower tray are then proportionately reduced to accommodate the light components, and the bubble point is recalculated.

Illustration 9.13 continued Compositions at the feed tray as previously determined are listed in the first two columns. The old x_6's are reduced to accommodate the C_6 from $N_p - 8$, and the new x_6's are determined. The new bubble point is 86.4°C.

Component	x_{N_p-8}	Old x_6	$x_6(1 - 0.1362)$	New x_6	$\alpha_{86.4}$
C_1		0.00204	0.00176	0.00176	46.5
C_2		0.0187	0.0162	0.0162	13.5
lk C_3	0.0915	0.1045	0.0903	0.0903	5.87
C_4	0.3897	0.4247	0.3668	0.3668	2.39
hk C_5	0.3826	0.4500	0.3887	0.3887	1.00
C_6	0.1362			0.1362	0.467
	1.0000	1.0000	0.8638	1.0000	

For tray 5, x_{C_6} is estimated, through Eq. (9.181),

$$x_{C_6, 5} = \frac{0.467(0.1362)(0.3859)}{0.3887} = 0.0631$$

where 0.3859 is the concentration of the heavy key on tray 5 as previously calculated. The old x_5's are all reduced by multiplying them by $1 - 0.0631$, whence their new sum plus 0.0631 for $C_6 = 1$. The new tray 5 has a bubble point of 80°C, and its equilibrium vapor can be obtained from the bubble-point calculation, as usual. In similar fashion, the calculations are continued upward. Results for the top trays and the distillate are:

Component	x_2	y_2	x_1	y_1	x_0	y_D
C_1	0.00221	0.0444	0.00226	0.0451	0.00425	0.0789
C_2	0.0214	0.1111	0.0241	0.1209	0.0425	0.1842
C_3	0.1418	0.2885	0.1697	0.3259	0.2495	0.3870
C_4	0.6786	0.5099	0.7100	0.4840	0.6611	0.3420
C_5	0.1553	0.0458	0.0932	0.0239	0.0425	0.0079
C_6	0.00262	0.00034	0.00079	0.00009	0.00015	0.00001
t, °C	66		58.5		46.6	

To correct tray $N_p - 7 =$ tray 7, use Eq. (9.184) to determine the concentrations of C_1 and C_2 on tray 7. $x_{lk, m+1}$ is taken from the old $N_p - 7$.

$$x_{C_1, 7} = \frac{0.00176(5.87)(0.0790)}{46.5(0.0903)} = 0.000\ 194 \qquad x_{C_2, 7} = \frac{0.0162(5.87)(0.0790)}{13.5(0.0903)} = 0.00615$$

The old $x_{N_p - 7}$ must be reduced by multiplying by $1 - 0.000\ 194 - 0.00615$. The adjusted values together with those above constitute the new x_7's. The new bubble point is 94°C.

The calculations are continued downward in the same fashion. The new tray 6 has $x_{C_1} = 0.000\ 023$, $x_{C_2} = 0.00236$. It is clear that concentrations for these components are reducing so rapidly that there is no need to go further.

Liquid/Vapor Ratios

With the corrected temperatures and compositions, it is now possible to estimate the L/G ratios on the trays reasonably well.

For the enriching section, Eq. (9.59) can be solved simultaneously with Eq. (9.56) to provide

$$G_{n+1} = \frac{Q_C + D(H_D - H_{L_n})}{H_{G_{n+1}} - H_{L_n}} \qquad (9.185)$$

Equation (9.56) then provides L_n; G_n is computed through another application of Eq. (9.185), and hence L_n/G_n is obtained. Similarly for the exhausting section, Eqs. (9.66) and (9.69) provide

$$\overline{G}_{m+1} = \frac{Q_B + W(H_{L_m} - H_W)}{H_{G_{m+1}} - H_{L_m}} \qquad (9.186)$$

and Eq. (9.66) gives the liquid rate. Usually it is necessary to compute only a few such ratios in each section and interpolate the rest from a plot of L/G vs. tray number.

It must be remembered that if the enthalpies are computed from temperatures and concentrations taken from the Lewis-Matheson data, even though corrected in the manner previously shown, the L/G ratios will still be estimates only, since the data upon which they are based assume constant L/G in each section.

Illustration 9.13 continued Compute the condenser and reboiler heat loads and the L/G ratios.

SOLUTION The condenser heat load is given by Eq. (9.54). Values of x_0, y_D, and y_1 are taken from the corrected concentrations previously obtained.

Component	H_D, vapor, 46.6°C	$y_D H_D$	H_{LO}, 46.6°C	$x_0 H_{LO}$	H_{G1}, 58.8°C	$y_1 H_{G1}$
C_1	13 490	1 065	10 470	44.4	13 960	629
C_2	23 380	4 305	17 210	732	24 190	2 926
C_3	32 100	12 420	18 610	4 643	37 260	10 840
C_4	42 330	14 470	22 790	15 100	43 500	21 050
C_5	52 570	415	27 100	1 151	53 900	1 291
C_6	61 480	0.7	31 050	4.7	63 500	9.5
		32 680		21 675		36 745
		$= H_D$		$= H_{LO}$		$= H_{G1}$

Eq. (9.54):

$$Q_C = 0.3800[(0.8 + 1)(36\ 745) - 0.8(21\ 675) - 32\ 680]$$

$$= 6130 \text{ kJ/kmol feed}$$

In similar fashion, $H_W = 39\ 220$, $H_F = 34\ 260$, and Q_B [Eq. (9.55)] = 8606 kJ/kmol feed.

For tray $n = 1$, $G_1 = L_0 + D = D(R + 1) = 0.3800(1.8) = 0.684$ kmol. With x_1 and y_2 from corrected compositions:

Component	H_{G2}, 66°C	$H_{G2} y_2$	H_{L1}, 58.5°C	$H_{L1} x_1$
C_1	14 070	624.5	11 160	25.4
C_2	24 610	2 740	17 910	432
C_3	33 800	9 770	20 470	3 473
C_4	44 100	22 470	24 900	17 680
C_5	54 780	2 510	29 500	2 750
C_6	64 430	21.9	33 840	26.7
		38 136		24 380
		$= H_{G2}$		$= H_{L1}$

Eq. (9.185):

$$G_2 = \frac{6130 + 0.38(32\ 680 - 24\ 380)}{38\ 136 - 24\ 380} = 0.675 \text{ kmol}$$

$$L_2 = G_2 - D = 0.675 - 0.380 = 0.295 \text{ kmol}$$

$$\frac{L_2}{G_2} = \frac{0.295}{0.675} = 0.437$$

Similarly, the calculations can be made for other trays in the enriching section. For tray $N_p = 14$:

Component	$H_{G, 15}$, 113.3°C	$y_{15}H_{G, 15}$	$H_{L, 14}$, 107.9°C	$x_{14}H_{L, 14}$
C_1	Nil			Nil
C_2	Nil			Nil
C_3	38 260	709	29 310	287.7
C_4	49 310	26 700	31 870	12 790
C_5	60 240	22 100	37 680	16 510
C_6	71 640	5 245	43 500	6 470
		54 750		36 058
		$= H_{G, 15}$		$= H_{L, 14}$

Similarly,

$$H_{L, 13} = 36\ 790 \quad \text{and} \quad H_{G, 14} = 52\ 610$$

Eq. (9.186):

$$\bar{G}_{15} = \frac{8606 + 0.62(36\ 058 - 39\ 220)}{54\ 750 - 39\ 220} = 0.358$$

$$\bar{L}_{14} = W + \bar{G}_{15} = 0.620 + 0.358 = 0.978$$

$$\bar{G}_{14} = \frac{8606 + 0.62(36\ 790 - 39\ 220)}{52\ 610 - 36\ 790} = 0.448$$

$$\frac{\bar{L}_{14}}{\bar{G}_{14}} = \frac{0.978}{0.448} = 2.18$$

and similarly for other exhausting-section trays. Thus:

Tray no.	$\dfrac{L}{G}$ or $\dfrac{\bar{L}}{\bar{G}}$	t, °C	Tray no.	$\dfrac{\bar{L}}{\bar{G}}$	t, °C
Condenser	0.80	46.6	8	3.25	96.3
1	0.432	58.4	9	2.88	97.7
2	0.437	66	10	2.58	99
3	0.369	70.4	11	2.48	100
4	0.305	74	12	2.47	102.9
5	0.310	80.3	13	2.42	104.6
6	1.53	86.4	14	2.18	107.9
7	4.05	94.1	Reboiler	1.73	113.5

These values are not final. They scatter erratically because they are based on the temperatures and concentrations computed with the assumption of constant L/G.

Method of Thiele and Geddes

With number of trays, position of the feed tray, temperatures, and L/G ratios, the Thiele-Geddes method allows one to compute the products which will result, thus ultimately permitting a check of all previous calculations. The following is Edmister's [11] variation of the original [62].

All the equations which follow apply separately for each component, and component designations are omitted. For the enriching section, consider first the condenser.

$$G_1 y_1 = L_0 x_0 + D z_D \qquad (9.187)$$

$$\frac{G_1 y_1}{D z_D} = \frac{L_0 x_0}{D z_D} + 1 \qquad (9.188)$$

For a total condenser, $x_0 = z_D = x_D$, and $L_0/D = R$. Therefore

$$\frac{G_1 y_1}{D z_D} = R + 1 = A_0 + 1 \qquad (9.189)$$

For a partial condenser behaving like a theoretical stage, $z_D = y_D$, $z_D/x_0 = y_D/x_0 = m_0$, and

$$\frac{G_1 y_1}{D z_D} = \frac{R}{m_0} + 1 = A_0 + 1 \qquad (9.190)$$

A_0 is therefore either R or R/m_0, depending upon the type of condenser. For tray 1,

$$\frac{L_1 x_1}{G_1 y_1} = \frac{L_1}{G_1 m_1} = A_1 \qquad (9.191)$$

where A_1 is the absorption factor for tray 1. Then

$$L_1 x_1 = A_1 G_1 y_1 \qquad (9.192)$$

For tray 2,

$$G_2 y_2 = L_1 x_1 + D z_D \qquad (9.193)$$

$$\frac{G_2 y_2}{D z_D} = \frac{L_1 x_1}{D z_D} + 1 = \frac{A_1 G_1 y_1}{D z_D} + 1 = A_1(A_0 + 1) + 1 = A_0 A_1 + A_1 + 1 \qquad (9.194)$$

Generally, for any tray,

$$\frac{G_n y_n}{D z_D} = \frac{A_{n-1} G_{n-1} y_{n-1}}{D z_D} + 1 = A_0 A_1 A_2 \cdots A_{n-1} + A_1 A_2 \cdots A_{n-1} + \cdots + A_{n-1} + 1$$

$$(9.195)$$

and for the feed tray,

$$\frac{G_f y_f}{D z_D} = A_0 A_1 A_2 \cdots A_{f-1} + A_1 A_2 \cdots A_{f-1} + \cdots + A_{f-1} + 1 \qquad (9.196)$$

Consider next the exhausting section. For a kettle-type reboiler,

$$\frac{\overline{G}_{N_p+1} y_{N_p+1}}{W x_W} = \frac{\overline{G}_{N_p+1} m_W}{W} = S_W \qquad (9.197)$$

where S_W is the reboiler stripping factor. Since

$$\overline{L}_{N_p} x_{N_p} = \overline{G}_{N_p+1} y_{N_p+1} + W x_W \qquad (9.198)$$

then

$$\frac{\overline{L}_{N_p} x_{N_p}}{W x_W} = \frac{\overline{G}_{N_p+1} y_{N_p+1}}{W x_W} + 1 = S_W + 1 \qquad (9.199)$$

For a thermosiphon reboiler, $y_{N_p+1} = x_W$, and

$$S_W = \frac{\overline{G}_{N_p+1}}{W} \qquad (9.200)$$

S_W is therefore defined by either Eq. (9.197) or (9.200), depending upon the type of reboiler. For the bottom tray,

$$\frac{\overline{G}_{N_p} y_{N_p}}{\overline{L}_{N_p} x_{N_p}} = \frac{\overline{G}_{N_p} m_{N_p}}{\overline{L}_{N_p}} = S_{N_p} \qquad (9.201)$$

$$\overline{G}_{N_p} y_{N_p} = S_{N_p} \overline{L}_{N_p} x_{N_p} \qquad (9.202)$$

Then
$$\bar{L}_{N_p-1}x_{N_p-1} = \bar{G}_{N_p}y_{N_p} + Wx_W \qquad (9.203)$$

and
$$\frac{\bar{L}_{N_p-1}x_{N_p-1}}{Wx_W} = \frac{\bar{G}_{N_p}y_{N_p}}{Wx_W} + 1 = \frac{S_{N_p}\bar{L}_{N_p}x_{N_p}}{Wx_W} + 1 = S_{N_p}(S_W + 1) + 1 = S_{N_p}S_W + S_{N_p} + 1$$
$$(9.204)$$

In general, for any tray,

$$\frac{\bar{L}_m x_m}{Wx_W} = \frac{S_{m+1}\bar{L}_{m+1}x_{m+1}}{Wx_W} + 1 = S_{m+1} \cdots S_{N_p-1}S_{N_p}S_W + S_{m+1} \cdots S_{N_p-1}S_{N_p}$$
$$+ S_{m+1} \cdots S_{N_p-1} + \cdots + S_{m+1} + 1 \qquad (9.205)$$

and at the feed tray,

$$\frac{\bar{L}_f x_f}{Wx_W} = S_{f+1} \cdots S_{N_p}S_W + S_{f+1} \cdots S_{N_p} + \cdots + S_{f+1} + 1 \qquad (9.206)$$

Edmister [11] provides shortcut methods involving "effective" A's and S's for use in Eqs. (9.196) and (9.206), much as for gas absorption.

$$A_f = \frac{\bar{L}_f}{G_f m_f} = \frac{\bar{L}_f x_f}{G_f y_f} \qquad (9.207)$$

and
$$\frac{Wx_W}{Dz_D} \equiv \frac{\bar{L}_f x_f}{G_f y_f} \frac{G_f y_f / Dz_D}{\bar{L}_f x_f / Wx_D} = A_f \frac{G_f y_f / Dz_D}{\bar{L}_f x_f / Wx_W} \qquad (9.208)$$

Since
$$Fz_D = Wx_W + Dz_D \qquad (9.209)$$

$$Dz_D = \frac{Fz_F}{Wx_W / Dz_D + 1} \qquad (9.210)$$

For each component, Eqs. (9.196) and (9.206) are then used to compute the numerator and denominator, respectively, of the term in parentheses of Eq. (9.208), whence Wx_W / Dz_D, or the split of the component, is found. Equation (9.210) then provides Dz_D, and Eq. (9.209) provides Wx_W.

The products thus computed are completely consistent with the number of trays, feed-tray position, and reflux ratio used, provided the A's and S's (or L/G's and temperatures) are correct. For checking these, it is necessary to use the general equations (9.195) and (9.205) for each tray in the appropriate section of the tower. To check tray n in the enriching section,

$$\frac{G_n y_n}{Dz_D} \frac{Dz_D}{G_n} = y_n \qquad (9.211)$$

and this can be obtained from the data already accumulated. If $\Sigma y_{J,n} = 1$, the temperature is satisfactory. If not, a new temperature is obtained by adjusting the y's proportionately until they add to unity and computing the corresponding dew point. To check a tray in the exhausting section,

$$\frac{\bar{L}_m x_m}{Wx_W} \frac{Wx_W}{\bar{L}_m} = x_m \qquad (9.212)$$

which is obtainable from the data. If $\Sigma x_{J,m} = 1$, the temperature is correct. If not, the x's adjusted to add to unity and the bubble point computed. The new temperatures and compositions permit new enthalpy balances to obtain corrected L/G's for each tray, and the Thiele-Geddes calculation can be redone. Problems of convergence of this trial-and-error procedure are considered by Holland [24], Lyster et al. [33], and Smith [57].

The true optimum feed-tray location is obtained by trial, altering the location and observing which location produces the smallest number of trays for the desired products.

Illustration 9.13 concluded Reestimate the products through the Thiele-Geddes method.

SOLUTION The temperature and L/G profiles as previously estimated will be used. Computations will be shown for only one component, C_4; all other components are treated in the same manner.

Using the tray temperatures to obtain m, $A = L/mG$ for the enriching trays and $S = m\bar{G}/\bar{L}$ for the exhausting trays are computed. Since a partial condenser is used, $A_0 = R/m_0$. For the kettle-type reboiler, $S_W = \bar{G}_{15}m_W/W$. Then, for C_4:

Tray	Condenser	1	2	3	4	5	$6 = f$
m	0.50	0.66	0.75	0.81	0.86	0.95	1.07
A	1.600	0.655	0.584	0.455	0.355	0.326	1.431

Tray	$7 = f + 1$	8	9	10	11	12	13	14	Reboiler
m	1.22	1.27	1.29	1.30	1.32	1.40	1.45	1.51	1.65
S	0.301	0.390	0.447	0.503	0.532	0.566	0.599	0.693	0.954

With these data

Eq. (9.196):

$$\frac{G_f y_f}{D z_D} = 1.5778$$

Eq. (9.206):

$$\frac{\bar{L}_f x_f}{W x_W} = 1.5306$$

Eq. (9.208):

$$\frac{W x_W}{D z_D} = 1.431 \frac{1.5778}{1.5306} = 1.475$$

Eq. (9.210):

$$D y_D = \frac{0.33}{1.475 + 1} = 0.1335$$

Eq. (9.209):

$$W x_W = 0.33 - 0.1335 = 0.1965$$

Similarly:

Component	$\dfrac{G_f y_f}{D z_D}$	$\dfrac{L_f x_f}{W x_W}$	$\dfrac{W x_W}{D y_D}$	$D y_D$	$W x_W$
C_1	1.0150	254×10^6	288×10^{-10}	0.03	Nil
C_2	1.0567	8750	298×10^{-5}	0.07	Nil
C_3	1.1440	17.241	0.0376	0.144 7	0.0053
C_4	1.5778	1.5306	1.475	0.133 5	0.1965
C_5	15.580	1.1595	45.7	0.006 43	0.29357
C_6	1080	1.0687	7230	0.000 016 6	0.11998
				0.3846 = D	0.6154 = W

These show that $0.1447(100)/0.15 = 96.3\%$ of the C_3 and 2.14% of the C_5 are in the distillate. These do not quite meet the original specifications. The temperatures and L/G's must still be corrected, however. Thus for tray 2 and component C_4, Eq. (9.195) yields $G_2 y_2/D z_D = 2.705$. From the enthalpy balances, $G_2 = 0.675$. Therefore, by Eq. (9.211),

$$y_2 = \frac{2.705(0.1335)}{0.675} = 0.5349$$

Similarly:

Component	$\dfrac{G_2 y_2}{D z_D}$	y_2	Adjusted y_2
C_1	1.0235	0.0454	0.0419
C_2	1.1062	0.1147	0.1059
C_3	1.351	0.2896	0.2675
C_4	2.705	0.5349	0.4939
C_5	10.18	0.0970	0.0896
C_6	46.9	0.00115	0.00106
		1.0828	1.0000

Since Σy_2 does not equal 1, the original temperature is incorrect. After adjusting the y's to add to unity as shown, the dew point = 71°C instead of the 66°C used. Similarly all tray compositions and temperatures must be corrected, new L/G's obtained by enthalpy balances with the new compositions, temperatures, and L/G's, and the Thiele-Geddes calculations repeated.

Multicomponent fractionation finds its most complex applications in the field of petroleum refining. Petroleum products such as gasoline, naphthas, kerosenes, gas oils, fuel oils, and lubricating oils are each mixtures of hundreds of hydrocarbons, so many that their identity and actual number cannot readily be established. Fortunately, it is not usually specific substances that are desired in these products but *properties*, so that specifications can be made as to boiling range, specific gravity, viscosity, and the like. Fractionators for these products cannot be designed by the detailed methods just described but must be based upon laboratory studies in small-scale equipment. As an example which illustrates the complex nature of some of these separations, consider the schematic diagram of a topping plant for the initial distillation of a crude oil (Fig.

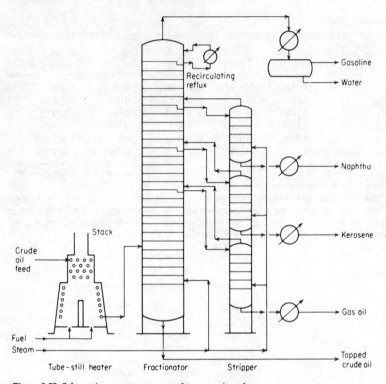

Figure 9.52 Schematic arrangement, petroleum topping plant.

9.52). The crude oil, after preliminary heat exchange with several of the products from the plant, is passed through the tubes of a gas-fired furnace, the tube-still heater. Here a portion of the oil is vaporized, somewhat larger in amount than that ultimately to be taken as vaporized products. The mixture of liquid and vapor then enters the large tray tower. Open steam is introduced at the bottom to strip the last traces of volatile substances from the residue product, and this steam passes up the column, where it lowers the effective pressure and hence the temperature of the distillation. The steam and most volatile substances (crude gasoline) leaving the top tray are condensed, the water separated, and the gasoline sent to storage. Reflux in the scheme shown here is provided by withdrawing a portion of the liquid from the top tray and returning it after it has been cooled. The cold liquid condenses some of the rising vapors to provide internal reflux. Several trays down from the top of the tower, a side stream may be withdrawn which will contain the hydrocarbons characteristic of a desired naphtha product. Since the components of the more volatile gasoline are also present at this point, the liquid is steam-stripped in a short auxiliary tray tower, the steam and vaporized gasoline being sent back to the main fractionator. The stripped naphtha is then withdrawn to storage. In similar fashion kerosene and gas oil cuts may be withdrawn, but each must be separately steam-stripped. The individual steam strippers are frequently built into a single shell, as shown, for economy, so that from the outside the multipurpose nature of the smaller tower is not readily evident. The withdrawn products must ordinarily be further processed before they are considered finished.

The design, method of operation, and number of products from topping units of this sort may vary considerably from refinery to refinery. Indeed, any individual unit will be built for maximum flexibility of operation, with, for example, multiple nozzles for introducing the feed at various trays and multiple side-stream withdrawal nozzles, to allow for variations in the nature of the feed and in the products to be made.

Azeotropic Distillation

This is a special case of multicomponent distillation [23] used for separation of binary mixtures which are either difficult or impossible to separate by ordinary fractionation. If the relative volatility of a binary mixture is very low, the continuous rectification of the mixture to give nearly pure products will require high reflux ratios and correspondingly high heat requirements, as well as towers of large cross section and numbers of trays. In other cases the formation of a binary azeotrope may make it impossible to produce nearly pure products by ordinary fractionation. Under these circumstances a third component, sometimes called an *entrainer*, may be added to the binary mixture to form a new low-boiling azeotrope with one of the original constituents, whose volatility is such that it can easily be separated from the other original constituent.

As an example of such an operation, consider the flowsheet of Fig. 9.53 for the azeotropic separation of acetic acid–water solutions, using butyl acetate as entrainer [39]. Acetic acid can be separated from water by ordinary methods, but only at great expense because of the low relative volatility of the constituents despite their fairly large difference in boiling points at atmospheric pressure (nbp acetic acid = 118.1°C, nbp water = 100°C). Butyl acetate is only slightly soluble in water and consequently forms a heteroazeotrope with it (bp = 90.2°C). Therefore if at least sufficient butyl acetate is added to the top of the distillation column (1) to form the azeotrope with all the water in the binary feed, the azeotrope can readily be distilled from the high-boiling acetic acid, which leaves as a residue product. The heteroazeotrope on condensation forms two insoluble liquid layers, one nearly pure water but saturated with ester, the other nearly pure ester saturated with water. The latter is returned to the top of the column as reflux and is the source of the entrainer in the column. The former can be stripped of its small entrainer content in a second small column (2). The separation of the heteroazeotrope from acetic acid is readily done, so that relatively few trays are required in the principal tower. On the other hand, heat must be supplied, not only to vaporize the water in the overhead distillate, but to vaporize the entrainer as well. The operation can also be done batchwise, in which case sufficient entrainer is charged to the still kettle, together with the feed, to azeotrope the water. The azeotrope is then distilled overhead.

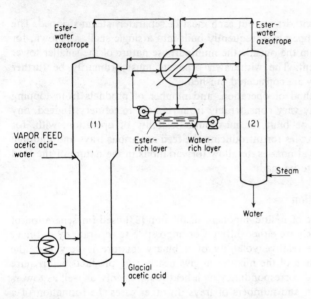

Figure 9.53 Azeotropic distillation of acetic acid–water with butyl acetate.

Sometimes the new azeotrope which is formed contains all three constituents. The dehydration of ethanol-water mixture with benzene as added substance is an example. Dilute ethanol-water solutions can be continuously rectified to give at best mixtures containing 89.4 mole percent ethanol at atmospheric pressure, since this is the composition of the minimum-boiling azeotrope in the binary system. By introducing benzene into the top of a column fed with an ethanol-water mixture, the ternary azeotrope containing benzene (53.9 mol %), water (23.3 mol %), ethanol (22.8 mol %), boiling at 64.9°C, is readily separated from the ethanol (bp = 78.4°C), which leaves as a residue product. In this case also the azeotropic overhead product separates into two liquid layers, one rich in benzene which is returned to the top of the column as reflux, the other rich in water which is withdrawn. Since the latter contains appreciable quantities of both benzene and ethanol, it must be rectified separately. The ternary azeotrope contains nearly equal molar proportions of ethanol and water, and consequently dilute ethanol-water solutions must be given a preliminary rectification to produce substantially the alcohol-rich binary azeotrope which is used as a feed.

In still other cases the new azeotrope which is formed does not separate into two insoluble liquids, and special means for separating it, such as liquid extraction, must be provided, but this is less desirable.

It is clear that the choice of entrainer is a most important consideration. The added substance should preferably form a low-boiling azeotrope with only one of the constituents of the binary mixture it is desired to separate, preferably the constituent present in the minority so as to reduce the heat requirements of the process. The new azeotrope must be of sufficient volatility to make it readily separable from the remaining constituent and so that inappreciable amounts of entrainer will appear in the residue product. It should preferably be lean in entrainer content, to reduce the amount of vaporization necessary in the distillation. It should preferably be of the heterogeneous-liquid type, which then simplifies greatly the recovery of the entrainer. In addition, a satisfactory entrainer must be (1) cheap and readily available, (2) chemically stable and inactive toward the solution to be separated, (3) noncorrosive toward common construction materials, (4) nontoxic, (5) of low latent heat of vaporization, (6) of low freezing point to facilitate storage and outdoor handling, and (7) of low viscosity to provide high tray efficiencies.

The general methods of design for multicomponent distillation apply, the principal difficulty being the paucity of vapor-liquid equilibrium data for these highly nonideal mixtures [23]. Computer programs are available [4].

Extractive Distillation

This is a multicomponent-rectification method similar in purpose to azeotropic distillation. To a binary mixture which is difficult or impossible to separate by ordinary means a third component, termed a *solvent*, is added which alters the relative volatility of the original constituents, thus permitting the separation. The added solvent is, however, of low volatility and is itself not appreciably vaporized in the fractionator.

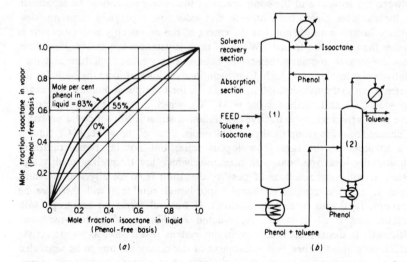

Figure 9.54 Extractive distillation of toluene-isooctane with phenol [*Vapor-liquid equilibria from Drickamer, Brown, and White, Trans. AIChE* **41**, *555 (1945).*]

As an example of such an operation, consider the process of Fig. 9.54. The separation of toluene (bp = 110.8°C) from paraffin hydrocarbons of approximately the same molecular weight is either very difficult or impossible, due to low relative volatility or azeotropism, yet such a separation is necessary in the recovery of toluene from certain petroleum hydrocarbon mixtures. Using isooctane (bp = 99.3°C) as an example of a paraffin hydrocarbon, Fig. 9.54a shows that isooctane in this mixture is the more volatile, but the separation is obviously difficult. In the presence of phenol (bp = 181.4°C), however, the isooctane relative volatility increases, so that, with as much as 83 mole percent phenol in the liquid, the separation from toluene is relatively easy. A flowsheet for accomplishing this is shown in Fig. 9.54b, where the binary mixture is introduced more or less centrally into the extractive distillation tower (1), and phenol as the solvent is introduced near the top so as to be present in high concentration upon most of the trays in the tower. Under these conditions isooctane is readily distilled as an overhead product, while toluene and phenol are removed as a residue. Although phenol is relatively high-boiling, its vapor pressure is nevertheless sufficient for its appearance in the overhead product to be prevented. The solvent-recovery section of the tower, which may be relatively short, serves to separate the phenol from the isooctane. The residue from the tower must be rectified in the auxiliary tower (2) to separate toluene from the phenol, which is recycled, but this is a relatively easy separation. In practice, the paraffin hydrocarbon is a mixture rather than the pure substance isooctane, but the principle of the operation remains the same.

Such a process depends upon the difference in departure from ideality between the solvent and the components of the binary mixture to be separated. In the example given, both toluene and isooctane separately form nonideal liquid solutions with phenol, but the extent of the nonideality with isooctane is greater than that with toluene. When all three substances are present, therefore, the toluene and isooctane themselves behave as a nonideal mixture and their relative volatility becomes high. Considerations of this sort form the basis for the choice of an extractive-distillation solvent. If, for example, a mixture of acetone (bp = 56.4°C) and methanol (bp = 64.7°C), which form a binary azeotrope, were to be separated by extractive distillation, a suitable solvent could probably be chosen from the group of aliphatic alcohols. Butanol (bp = 117.8°C), since it is a member of the same homologous series but not far removed, forms substantially ideal solutions with methanol, which are themselves readily separated. It will form solutions of positive deviation from ideality with acetone, however, and the acetone-methanol vapor-liquid equilibria will therefore be substantially altered in ternary mixtures. If butanol forms no azeotrope with acetone, and if it alters the vapor-liquid equilibrium of acetone-methanol sufficiently to destroy the azeotrope in this system, it will serve as an extractive-distillation solvent. When both substances of the binary mixture to be separated are themselves chemically very similar, a solvent of an entirely different chemical nature will be necessary. Acetone and furfural, for example, are useful as extractive-distillation solvents for separating the hydrocarbons butene-2 and *n*-butane.

Generally the requirements of a satisfactory extractive-distillation solvent are [16, 51, 61]:

1. High selectivity, or ability to alter the vapor-liquid equilibria of the original mixture sufficiently to permit its easy separation, with, however, use of only small quantities of solvent.
2. High capacity, or ability to dissolve the components in the mixture to be separated. It frequently happens that substances which are incompletely miscible with the mixture are very selective; yet if sufficiently high concentrations of solvent cannot be obtained in the liquid phase, the separation ability cannot be fully developed.
3. Low volatility in order to prevent vaporization of the solvent with the overhead product and to maintain high concentration in the liquid phase. Nonvolatile salts can be especially useful [15].
4. Separability; the solvent must be readily separated from the mixture to which it is added, and particularly it must form no azeotropes with the original substances.
5. The same considerations of cost, toxicity, corrosive character, chemical stability, freezing point, and viscosity apply as for entrainers for azeotropic distillation.

The general methods of design for multicomponent distillation apply. Since in most cases all the components of the feed streams are found in the bottoms, the method of Lewis and Matheson can be used, starting at the bottom and computing to the top. There will be an optimum solvent-circulation rate: small solvent rates require many trays, but the column diameters are small; large ratios require fewer trays but larger column diameters and greater solvent-circulation costs.

Comparison of Azeotropic and Extractive Distillation

It is generally true that adding an extraneous substance such as an entrainer or solvent to a process is undesirable. Since it can never be completely removed, it adds an unexpected impurity to the products. There are inevitable losses (ordinarily of the order of 0.1 percent of the solvent-circulation rate), and they may be large since solvent/feed ratios must frequently be greater than 3 or 4 to be effective. An inventory and source of supply must be maintained. Solvent recovery costs can be large, and new problems in choices of materials of construction are introduced. It follows that these processes can be considered only if, despite these drawbacks, the resulting process is less costly than conventional distillation.

Extractive distillation is generally considered to be more desirable than azeotropic distillation since (1) there is a greater choice of added component because the process does not depend upon the accident of azeotrope formation and (2) generally smaller quantities of solvent must be volatilized. However, the latter advantage can disappear if the volatilized impurity is a minor constituent of the feed and the azeotrope composition is favorable, i.e., low in solvent composition. Thus, in the dehydration of ethanol from an 85.6 mole percent ethanol-water solution, azeotroping the water with n-pentane as entrainer is more economical than using ethylene glycol as an extractive distillation solvent [3]. The reverse might be true for a more dilute alcohol feed.

LOW-PRESSURE DISTILLATION

Many organic substances cannot be heated to temperatures which even approach their normal boiling points without chemical decomposition. If such substances are to be separated by distillation, the pressure and the corresponding temperature must be kept low. The time of exposure of the substances to the distillation temperature must also be kept to a minimum, since the extent of thermal decomposition will thereby be reduced. For distillation under absolute pressures of the order of 7 to 35 kN/m^2 (1 to 5 lb_f/in^2), packed towers can be used, bubble-cap and sieve trays can be designed with pressure drops approaching 350 N/m^2 (0.05 lb_f/in^2, or 2.6 mmHg), and other simpler designs, such as the shower tray of Fig. 6.18, for which the pressure drops are of the order of 103 N/m^2 (0.015 lb_f/in^2, or 0.75 mmHg), are possible. Mechanically stirred spray and wetted-wall columns provide even smaller pressure drops [45].

In the distillation of many natural products, such as the separation of vitamins from animal and fish oils, as well as the separation of many synthetic industrial products such as plasticizers, the temperature may not exceed perhaps 200 to 300°C, where the vapor pressures of the substances may be a fraction of a millimeter of mercury. The conventional equipment is, of course, wholly unsuitable for such separations, not only because the pressure drop would result in high temperatures at the bottom of columns but also because of the long exposure time to the prevailing temperatures resulting from high holdup.

Molecular Distillation

This is a form of distillation at very low pressure conducted industrially at absolute pressures of the order of 0.3 to 3 N/m^2 (0.003 to 0.03 mmHg), suitable for the heat-sensitive substances described above.

The rate at which evaporation takes place from a liquid surface is given by the *Langmuir equation*

$$N_A = 1006\bar{p}_A \left(\frac{1}{2\pi M_A R' T} \right)^{0.5} \tag{9.213}$$

where R' is the gas constant. At ordinary pressures the net rate of evaporation is, however, very much less than this, because the evaporated molecules are reflected back to the liquid after collisions occurring in the vapor. By reducing the absolute pressure to values used in molecular distillation, the mean free path of the molecules becomes very large, of the order of 1 cm. If the condensing surface is then placed at a distance from the vaporizing liquid surface not exceeding a few centimeters, very few molecules will return to the liquid, and the net rate of evaporation of each substance of a binary mixture will approach that given by Eq. (9.213). The vapor composition, or the composition of the distillate, will now be different from that given by ordinary equilibrium vaporization, and the ratio of the constituents in the distillate will approach

$$\frac{N_A}{N_B} = \frac{\text{moles of A}}{\text{moles of B}} = \frac{\bar{p}_A/M_A^{0.5}}{\bar{p}_B/M_B^{0.5}} \tag{9.214}$$

If this ratio is to be maintained, however, the surface of the liquid must be rapidly renewed, since otherwise the ratio of constituents in the surface will change as evaporation proceeds. The vigorous agitation or boiling present during ordinary distillations is absent under conditions of molecular distillation, and in most devices the liquid is caused to flow in a thin film over a solid surface, thus continually renewing the surface but at the same time maintaining low holdup of liquid.

Figure 9.55 shows a device which is used industrially for accomplishing a molecular distillation [19, 65]. The degassed liquid to be distilled is introduced continuously at the bottom of the inner surface of the rotor, a rotating conical-shaped surface. The rotor may be as large as 1.5 m in diameter at the top and may revolve at speeds of 400 to 500 r/min. A thin layer of liquid to be distilled, 0.05 to 0.1 mm thick, then spreads over the inner surface and travels rapidly to the upper periphery under the action of centrifugal force. Heat is supplied to the liquid through the rotor by radiant electrical heaters, and the vaporized material is condensed upon the water-cooled louver-shaped condenser. This is maintained at temperatures sufficiently low to prevent reevaporation or reflection of the vaporized molecules. The residue liquid is caught in the collection gutter at the top of the rotor, and the distillate is drained from the collection troughs on the condenser. Each product is pumped from the still body, which is evacuated to the low pressures necessary for molecular distillation, and the time of

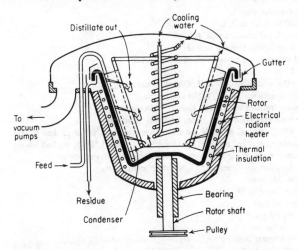

Figure 9.55 Schematic section, centrifugal molecular still of Hickman [19].

residence of the substances in the still may be as low as 1 s or less. Such a device is capable of handling 5×10^{-5} to $25 \times 10^{-5} \text{m}^3/\text{s}$ (50 to 250 gal/h) of liquid to be distilled and gives a separation 80 to 95 percent of that indicated by Eq. (9.214). Multiple distillations are necessary for multistage separation effects [20, 32].

Falling-film stills are constructed of two vertical concentric tubes. The inner tube is heated internally, and the liquid to be distilled flows down the outer wall of the inner tube in a thin film. The outer tube is the condenser, and the annular space is under vacuum. Some thin-film arrangements involve continuous wiping of the film by a rotating blade.

NOTATION FOR CHAPTER 9

Any consistent set of units may be used.

a	specific interfacial area, L^2/L^3
A	more volatile component of a binary mixture absorption factor $= L/mG$, dimensionless
B	less volatile component of a binary mixture
C	molar heat capacity at constant pressure, FL/mole T
d	differential operator
D	distillate rate, mole/Θ for tray towers, $\text{mole}/L^2\Theta$ for packed towers
E_O	overall tray efficiency, fractional
E_{MG}	Murphree vapor tray efficiency, fractional
F	feed rate, mole/Θ for tray towers, $\text{mole}/L^2\Theta$ for packed towers
	quantity of feed, batch distillation, mole
	(with subscript G or L) mass-transfer coefficient, $\text{mole}/L^2\Theta$
g_c	conversion factor, $ML/F\Theta^2$
G	vapor rate, enriching section, mole/Θ for tray towers, $\text{mole}/L^2\Theta$ for packed towers
$\bar{G}$	vapor rate, stripping section, mole/Θ for tray towers, $\text{mole}/L^2\Theta$ for packed towers
H	molar enthalpy, FL/mole
H_{tG}	height of a gas transfer unit, L
H_{tL}	height of a liquid transfer unit, L
H_{tOG}	height of an overall gas transfer unit, L
H_{tOL}	height of an overall liquid transfer unit, L
ΔH_S	integral heat of solution per mole of solution, FL/mole
I, J	components I, J
k'_x	liquid mass-transfer coefficient, $\text{mole}/L^2\Theta(\text{mole fraction})$
k'_y	gas mass-transfer coefficient, $\text{mole}/L^2\Theta(\text{mole fraction})$
K'_x	overall liquid mass-transfer coefficient, $\text{mole}/L^2\Theta(\text{mole fraction})$
K'_y	overall gas mass-transfer coefficient, $\text{mole}/L^2\Theta(\text{mole fraction})$
L	liquid rate, enriching section, mole/Θ for tray towers, $\text{mole}/L^2\Theta$ for packed towers
$\bar{L}$	liquid rate, exhausting section, mole/Θ for tray towers, $\text{mole}/L^2\Theta$ for packed towers
L_0	external reflux rate, mole/Θ for tray towers, $\text{mole}/L^2\Theta$ for packed towers
m	equilibrium distribution coefficient, $y^*/x, y/x^*$, dimensionless
M	molecular weight, M/mole
N	mass-transfer flux, $\text{mole}/L^2\Theta$
N_m	minimum number of trays
N_p	number of theoretical trays
N_{tG}	number of gas transfer units
N_{tL}	number of liquid transfer units
N_{tOG}	number of overall gas transfer units
N_{tOL}	number of overall liquid transfer units

p	vapor pressure, F/L^2
$\bar{p}$	partial pressure, F/L^2
q	quantity defined by Eq. (9.127)
Q	net rate of heating, FL/Θ
Q'	quantity defined by Eq. (9.60), $FL/mole$
Q''	quantity defined by Eq. (9.70), $FL/mole$
Q_B	heat added at the reboiler, FL/Θ for tray towers, $FL/L^2\Theta$ for packed towers
Q_C	heat removed at the condenser, FL/Θ for tray towers, $FL/L^2\Theta$ for packed towers
Q_L	heat loss, FL/Θ for tray towers, $FL/L^2\Theta$ for packed towers
R	external reflux ratio, mole reflux/mole distillate, $mole/mole$
R'	gas constant, $FL/mole$ T
R_m	minimum reflux ratio, moles reflux/mole distillate, $mole/mole$
S	stripping factor, mG/L, dimensionless
t	temperature, T
T	absolute temperature, T
V	allowable vapor velocity, L/Θ
W	residue rate, $mole/\Theta$ for tray towers, $mole/L^2\Theta$ for packed towers
x	concentration (of A in binaries) in the liquid, mole fraction
x^*	x in equilibrium with y, mole fraction
y	concentration (of A in binaries) in the gas, mole fraction
y^*	y in equilibrium with x, mole fraction
z	average concentration in a solution or multiphase mixture, mole fraction
Z	height of packing, L
α	relative volatility, defined by Eq. (9.2), dimensionless
α_{JC}	relative volatility of component J with respect to component C, dimensionless
γ	activity coefficient, dimensionless
Δ	difference point, representing a fictitious stream, $mole/\Theta$ for tray towers, $mole/L^2\Theta$ for packed towers
λ	molar latent heat of vaporization, $FL/mole$
ρ	density, M/L^3
Σ	summation
φ	root of Eq. (9.165)

Subscripts

a	operating-line intersection
av	average
A, B	components A, B
bp	bubble point
D	distillate
e	enriching section
f	feed tray
F	feed
G	gas
hk	heavy key component
i	interface; any component
lk	light key component
L	liquid
m	tray m; minimum
n	tray n
r	tray r
s	stripping section
0	reflux stream; base condition for enthalpy
W	residue
1	bottom of packed tower; tray 1
2	top of tray tower, tray 2

REFERENCES

1. American Institute of Chemical Engineers: Multicomponent Distillation, *Comput. Program Man.* 8; Complex Tower Distillation, *Comput. Program Man.* 4; New York.
2. Barnes, F. J., D. N. Hanson, and C. J. King: *Ind. Eng. Chem. Process Des. Dev.*, **11**, 136 (1972).
3. Black, C., and D. E. Ditsler: in Extractive and Azeotropic Distillation, *Adv. Chem. Ser.*, **115**, 1 (1972).
4. Black, C., R. A. Golding, and D. E. Ditsler: in Extractive and Azeotropic Distillation, *Adv. Chem. Series*, **115**, 64 (1972).
5. Brown, G. G., and H. Z. Martin: *Trans. AIChE*, **35**, 679 (1939).
6. Cavers, S. D.: *Ind. Eng. Chem. Fundam.*, **4**, 229 (1965).
7. Chu, Ju-Chin: "Vapor-Liquid Equilibrium Data," Edwards, Ann Arbor, Mich., 1956.
8. Collins, G. K.: *Chem. Eng.*, **83**, 149 (July 19, 1976).
9. Dodge, B. F.: "Chemical Engineering Thermodynamics," McGraw-Hill Book Company, New York, 1944.
10. Donnell, J. W., and C. M. Cooper: *Chem. Eng.*, **57**, 121 (June, 1950).
11. Edmister, W. C.: *AIChE J.*, **3**, 165 (1957).
12. Erbar, J. H., and R. N. Maddox: *Pet. Refiner*, **40**(5), 183 (1961).
13. Fenske, M. R.: *Ind. Eng. Chem.*, **24**, 482 (1931).
14. Frank, O., and R. D. Prickett: *Chem. Eng.*, **80**, 107 (Sept. 3, 1973).
15. Furter, W. F.: in Extractive and Azeotropic Distillation, *Adv. Chem. Ser.*, **115**, 35 (1972).
16. Gerster, J. A.: *Chem. Eng. Prog.*, **65**(9), 43 (1969).
17. Gilliland, E. R.: *Ind. Eng. Chem.*, **32**, 1220 (1940).
18. Hála, E., J. Pick, V. Fried, and O. Velim: "Vapor-liquid Equilibria," Elsevier, New York, 1973.
19. Hickman, K. C. D.: *Ind. Eng. Chem.*, **39**, 686 (1947).
20. Hickman, K. C. D.: in R. H. Perry and C. H. Chilton (eds.), "The Chemical Engineers' Handbook," 5th ed., p. 13-55, McGraw-Hill Book Company, New York, 1973.
21. Hildebrand, J. H., J. M. Prausnitz, and R. L. Scott: "Regular and Related Solutions," Van Nostrand Reinhold, New York, 1970.
22. Hirata, M., S. Ohe, and K. Nagakama: "Computer-aided Data Book of Vapor-Liquid Equilibria," Elsevier, New York, 1975.
23. Hoffman, E. J.: "Azeotropic and Extractive Distillation," Interscience, New York, 1964.
24. Holland, C. D.: "Multicomponent Distillation," Prentice-Hall, Englewood Cliffs, N.J., 1963; "Fundamentals and Modeling of Separation Processes," Prentice-Hall, Englewood Cliffs, N.J., 1975.
25. Horsley, L. H.: Azeotropic Data III, *Adv. Chem. Ser.*, **116** (1973).
26. Horvath, P. J., and R. F. Schubert: *Chem. Eng.*, **64**, 129 (Feb. 10, 1958).
27. Kern, R.: *Chem. Eng.*, **82**, 107 (Aug. 4, 1975).
28. Kern. R.: *Chem. Eng.*, **82**, 129 (Sept. 15, 1975).
29. King, M. B.: "Phase Equilibrium in Mixtures," Pergamon, New York, 1969.
30. Kwauk, M.: *AIChE J.*, **2**, 240 (1956).
31. Lewis, W. K., and G. L. Matheson: *Ind. Eng. Chem.*, **24**, 494 (1932).
32. Lohwater, R. K.: *Res./Dev.*, **22**(1), 36, 38 (1971).
33. Lyster, W. N., S. L. Sullivan, D. S. Billingsley, and C. D. Holland: *Pet. Refiner*, **38**(6), 221; (7), 151; (10), 139 (1959).
34. Madsen, N.: *Chem. Eng.*, **78**, 73 (Nov. 1, 1971).
35. Mason, W. A.: *Pet. Refiner*, **38**(5), 239 (1959).
36. McCabe, W. L., and E. W. Thiele: *Ind. Eng. Chem.*, **17**, 605 (1925).
37. McKee, H. R.: *Ind. Eng. Chem.*, **62**(12), 76 (1970).
38. Nishimura, H., and Y. Hiraizumi: *Int. Chem. Eng.*, **11**, 188 (1971).
39. Othmer, D. F.: *Chem. Eng. Prog.*, **59**(6), 67 (1963).
40. Perry, E. S., and A. Weissberger (eds.): "Techniques of Organic Chemistry," vol. 4, "Distillation," 1965; vol. 4, "Separation and Purification Methods," 1976, Interscience, New York.
41. Ponchon, M.: *Tech. Mod.*, **13**, 20, 55 (1921).

42. Prausnitz, J. M.: "Molecular Thermodynamics of Fluid-Phase Equilibria," Prentice-Hall, Englewood Cliffs, N.J., 1969.
43. Prausnitz, J. M., and P. L. Chueh: "Computer Calculations for High Pressure Vapor-Liquid Equilibria," Prentice-Hall, Englewood Cliffs, N.J., 1968.
44. Prausnitz, J. M., C. A. Eckert, R. V. Orye, and J. P. O'Connell: "Computer Calculations for Multicomponent Vapor-Liquid Equilibria," Prentice-Hall, Englewood Cliffs, N.J., 1967.
45. Raichle, L., and R. Billet: *Ind. Eng. Chem.*, **57**(4), 52 (1965).
46. Randall, M., and B. Longtin: *Ind. Eng. Chem.*, **30**, 1063, 1188, 1311 (1938); **31**, 908, 1295 (1939); **32**, 125 (1940).
47. Ricci, J. E.: "The Phase Rule and Heterogeneous Equilibria," Van Nostrand, New York, 1951 (also Dover).
48. Robinson, C. S., and E. R. Gilliland: "Elements of Fractional Distillation," 4th ed., McGraw-Hill Book Company, New York, 1950.
49. Sarma, N. V. L. S., P. J. Reddy, and P. S. Murti: *Ind. Eng. Chem. Process Des. Dev.*, **12**, 278 (1973).
50. Savarit, R.: *Arts Metiers*, **1922**, 65, 142, 178, 241, 266, 307.
51. Scheibel, E. G.: *Chem. Eng. Prog.*, **44**, 927 (1948).
52. Scherman, A. D.: *Hydrocarbon Process.*, **48**(9), 187 (1969).
53. Shipman, C. W.: *AIChE J.*, **18**, 1253 (1972).
54. Shiras, R., D. N. Hanson, and C. H. Gibson: *Ind. Eng. Chem.*, **42**, 871 (1950).
55. Sigales, B.: *Chem. Eng.*, **82**, 157 (Mar. 3, 1975); 141 (June 23, 1975); 5, 87 (Sept. 9, 1975).
56. Slater, A. W., in C. Singer, E. J. Holmyard, A. R. Hale, and T. I. Wilhaus (eds.), "A History of Technology," vol. 5, p. 306, Oxford, New York, 1958.
57. Smith, B. D.: "Design of Equilibrium Stage Processes," McGraw-Hill Book Company, New York, 1963.
58. Smoker, E. H.: *Trans. AIChE*, **34**, 165 (1938).
59. Strangio, V. A., and R. E. Treybal: *Ind. Eng. Chem. Process Des. Dev.*, **13**, 279 (1974).
60. Stupin, W. J., and F. J. Lockhart: *Chem. Eng. Prog.*, **68**(10), 71 (1972).
61. Tassios, D. P.: "Extractive and Azeotropic Distillation," *Adv. Chem. Ser.*, **115**, 46 (1972).
62. Thiele, E. W., and R. L. Geddes: *Ind. Eng. Chem.*, **25**, 289 (1933).
63. Underwood, A. J. V.: *Trans. Inst. Chem. Eng. Lond.*, **10**, 112 (1932).
64. Underwood, A. J. V.: *Chem. Eng. Prog.*, **44**, 603 (1948); **45**, 609 (1949).
65. Watt, P. R.: "Molecular Stills," Reinhold, New York, 1963.
66. Wichterle, I., R. Kobayashi, and P. J. Chappelear: *Hydrocarbon Process.*, **50**(11), 233 (1971).
67. Wichterle, I., J. Linek, and E. Hála: "Vapor-Liquid Equilibrium Data Bibliography," Elsevier, New York, 1973, "Supplement I," 1976.
68. Winn, F. W.: *Pet. Refiner*, **37**, 216 (1950).

PROBLEMS

9.1 Solutions of methanol and ethanol are substantially ideal.

(*a*) Compute the vapor-liquid equilibria for this system at 1 and at 5 atm abs pressure, and plot *xy* and *txy* diagrams at each pressure.

(*b*) For each pressure compute relative volatilities, and determine an average value.

(*c*) Using Eq. (9.2) with the average volatilities, compare the values of y^* at each value of x so obtained with those computed directly from the vapor pressures.

9.2 A 1000-kg batch of nitrobenzene is to be steam-distilled from a very small amount of a nonvolatile impurity, insufficient to influence the vapor pressure of the nitrobenzene. The operation is to be carried out in a jacketed kettle fitted with a condenser and distillate receiver. Saturated steam at 35 kN/m² (5.1 lb$_f$/in²) gauge is introduced into the kettle jacket for heating. The nitrobenzene is charged to the kettle at 25°C, and it is substantially insoluble in water.

Liquid water at 25°C is continuously introduced into the nitrobenzene in the kettle, to maintain a liquid water level. The mixture is distilled at atmospheric pressure. (*a*) At what temperature does the distillation proceed? (*b*) How much water is vaporized? (*c*) How much steam must be condensed in the kettle jacket? Neglect the heat required to bring the still up to operating temperature. The heat capacity of nitrobenzene is 1382 J/kg · K, and its latent heat of vaporization can be determined by the methods of Chap. 7.

9.3 A solution has the following composition, expressed as mole fraction: ethane, 0.0025; propane, 0.25; isobutane, 0.185; *n*-butane, 0.560; isopentane, 0.0025. In the following, the pressure is 10 bars (145 lb_f/in^2) abs. Use equilibrium distribution coefficients from the Depriester nomograph (see Illustration 9.3).

(*a*) Calculate the bubble point.

(*b*) Calculate the dew point.

(*c*) The solution is flash-vaporized to vaporize 40 mol % of the feed. Calculate the composition of the products.

(*d*) The solution is differentially distilled to produce a residue containing 0.80 mole fraction *n*-butane. Calculate the complete composition of the residue and the percentage of the feed which is vaporized.

9.4 Vapor-liquid equilibrium data at 1 std atm abs, heats of solution, heat capacities, and latent heats of vaporization for the system acetone-water are as follows:

Mole fraction acetone in liquid, x	Integral heat of solution at 15°C, kJ/kmol soln	Equivalent mole fraction acetone in vapor, y^*	Vapor-liquid temperature, °C	Heat capacity at 17.2°C, kJ/(kg soln) · °C
0.001	0	0.00	100.0	4.187
0.01		0.253	91.7	4.179
0.02	− 188.4	0.425	86.6	4.162
0.05	− 447.3	0.624	75.7	4.124
0.10	− 668.7	0.755	66.6	4.020
0.15	− 770	0.798	63.4	3.894
0.20	− 786	0.815	62.2	3.810
0.30	− 719	0.830	61.0	3.559
0.40	− 509	0.839	60.4	3.350
0.50	− 350.1	0.849	60.0	3.140
0.60	− 252.6	0.859	59.5	2.931
0.70		0.874	58.9	2.763
0.80		0.898	58.2	2.554
0.90		0.935	57.5	2.387
0.95		0.963	57.0	2.303
1.00		1.00	56.5	

t, °C	20	37.8	65.6	93.3	100
Heat capacity acetone, kJ/kg · °C	2.22	2.26	2.34	2.43	
Latent heat of vaporization, kJ/kg	1013	976	917	863	850

Compute the enthalpies of saturated liquids and vapors relative to pure acetone and water at 15°C and plot the enthalpy-concentration diagram, for 1 std atm abs. Retain for use in Probs. 9.5, 9.6, and 9.10.

9.5 A liquid mixture containing 60 mol % acetone, 40 mol % water, at 26.7°C (80°F), is to be continuously flash-vaporized at 1 std atm pressure, to vaporize 30 mol % of the feed.

(a) What will the composition of the products and the temperature in the separator be if equilibrium is established?

(b) How much heat, kJ/mol of feed, is required?

Ans.: 13 590.

(c) If the products are each cooled to 26.7°C, how much heat, kJ/mol of feed, must be removed from each?

9.6 A saturated vapor at 1 std atm pressure, containing 50 mol % acetone and 50 mol % water, is subject to equilibrium condensation to yield 50 mol % of the feed as liquid. Compute the equilibrium vapor and liquid compositions, the equilibrium temperature, and the heat to be removed, J/kmol of feed.

9.7 The liquid solution of Prob. 9.5 is differentially distilled at 1 atm pressure to vaporize 30 mol % of the feed. Compute the composition of the composited distillate and the residue. Compare with the results of Prob. 9.5.

Ans.: $y_D = 0.857$.

9.8 A jacketed kettle is originally charged with 30 mol of a mixture containing 40 mol % benzene, 60 mol % toluene. The vapors from the kettle pass directly to a total condenser, and the condensate is withdrawn. Liquid of the same composition as the charge is continuously added to the kettle at the rate of 15 mol/h. Heat to the kettle is regulated to generate vapor at 15 mol/h, so that the total molar content of the kettle remains constant. The mixture is ideal, and the average relative volatility is 2.51. The distillation is essentially differential.

(a) How long will the still have to be operated before vapor containing 50 mol % benzene is produced, and what is the composition of the composited distillate?

Ans.: 1.466 h, 0.555 mole fraction benzene.

(b) If the rate at which heat is supplied to the kettle is incorrectly regulated so that 18 mol/h of vapor is generated, how long will it take until vapor containing 50 mol % benzene is produced?

Ans.: 1.08 h.

9.9 An open kettle contains 50 kmol of a dilute aqueous solution of methanol, mole fraction methanol = 0.02, at the bubble point, into which steam is continuously sparged. The entering steam agitates the kettle contents so that they are always of uniform composition, and the vapor produced, always in equilibrium with the liquid, is led away. Operation is adiabatic. For the concentrations encountered it may be assumed that the enthalpy of the steam and evolved vapor are the same, the enthalpy of the liquid in the kettle is essentially constant, and the relative volatility is constant at 7.6. Compute the quantity of steam to be introduced in order to reduce the concentration of methanol in the kettle contents to 0.001 mole fraction.

Ans.: 20.5 kmol.

Continuous fractionation: Savarit-Ponchon method

9.10 An acetone-water solution containing 25 wt % acetone is to be fractionated at a rate of 10 000 kg/h at 1 std atm pressure, and it is planned to recover 99.5% of the acetone in the distillate at a concentration of 99 wt %. The feed will be available at 26.7°C (80°F) and will be preheated by heat exchange with the residue product from the fractionator, which will in turn be cooled to 51.7°C (125°F). The distilled vapors will be condensed and cooled to 37.8°C (125°F) by cooling water entering at 26.7°C and leaving at 40.6°C. Reflux will be returned at 37.8°C, at a reflux ratio $L_0/D = 1.8$. Open steam, available at 70 kN/m², will be used at the base of the tower. The tower will be insulated to reduce heat losses to negligible values. Physical property data are given in Prob. 9.4. Determine:

(a) The hourly rate and composition of distillate and reflux.

(b) The hourly condenser heat load and cooling water rate.

(c) The hourly rate of steam and residue and the composition of the residue.

(d) The enthalpy of the feed as it enters the tower and its condition (express quantitatively).

(e) The number of theoretical trays required if the feed is introduced at the optimum location.

Use large-size graph paper and a sharp pencil.

Ans.: 13.1.

(*f*) The rates of flow, kg/h, of liquid and vapor at the top, at $x = 0.6, 0.1, 0.025,$ and at the bottom tray. For a tower of uniform diameter, the conditions on which tray control the diameter, if the criterion is a 75% approach to flooding?

9.11 In a certain binary distillation, the overhead vapor condenser must be run at a temperature requiring mild refrigeration, and the available refrigeration capacity is very limited. The reboiler temperature is very high, and the available high-temperature heat supply is also limited. Consequently the scheme shown in Fig. 9.56 is adopted, where the intermediate condenser E and reboiler S operate at moderate temperatures. Condensers C and E are total condensers, delivering liquids at their bubble points. Reboiler S is a total vaporizer, delivering saturated vapor. Heat losses are negligible, $D = W$, the external reflux ratio $R = L_0/D = 1.0$, and other conditions are shown on the figure. The system is one which follows the McCabe-Thiele assumptions ($H_L x$ and $H_G y$ curves are straight and parallel). Feed is liquid at the bubble point.

(*a*) Per mole of feed, compute the rates of flow of all streams, the relative size of the two condenser heat loads, and the relative size of the two vaporizer heat loads.

(*b*) Establish the coordinates of all difference points. Sketch the Hxy diagram and locate the difference points, showing how they are related to the feed. Show the construction for trays.

(*c*) Sketch the xy diagram, show all operating lines, and locate their intersections with the 45° diagonal. Show the construction for trays.

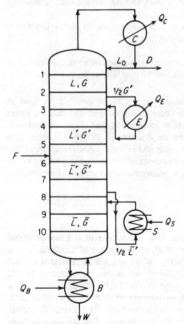

Figure 9.56 Arrangement for Prob. 9.11.

9.12 A distillation column is operated at a temperature below that of the surroundings (as in the distillation of air). Imperfect insulation causes the entire enriching section to be subject to a heat leak inward of Q_E J/s, while the entire stripping section is subject to a heat leak inward of Q_S J/s. Aside from the heat leaks, the system has all the characteristics which satisfy the McCabe-Thiele requirements.

(*a*) Sketch the Hxy and the xy diagrams, labeling the coordinates of all significant points. Are the operating curves on the xy diagram concave upward or downward?

(b) For a given condenser heat load, explain which is better: to design a new column with the inward heat leak minimized, or to take any possible advantage of a heat leak. Consider the reboiler heat load free, available from the surroundings.

(c) Suppose the column with the inward heat leak is operated at full reboiler heat load, limited by the available heat-transfer surface. The column is then insulated thoroughly, essentially eliminating the heat leak, but without provision to alter the heat load of the reboiler or otherwise to alter the operation. Explain whether the separation obtained will be better than, worse than, or the same as, before the insulation was installed.

9.13 A distillation tower for a binary solution was designed with a partial condenser, using $L_0/D = 1.0$, as in Fig. 9.57a. When built, the piping was installed incorrectly, as in Fig. 9.57b. It was then argued that the error was actually a fortunate one, since the richer reflux would provide a withdrawn distillate richer than planned. The mixture to be distilled has all the characteristics of one obeying McCabe-Thiele assumptions. For a reflux ratio = 1.0, reflux at the bubble point, and the

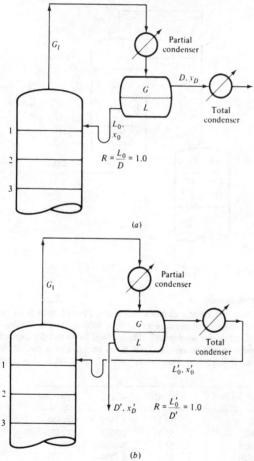

(a)

(b)

Figure 9.57 Arrangements for Prob. 9.13.

arrangement of Fig. 9.57b:

(a) Sketch the Ponchon-Savarit and xy diagrams for the parts of the columns indicated in the figure. Label the indicated vapor and liquid concentrations.

(b) Establish whether or not the distillate will be richer than originally planned.

9.14 Figure 9.58a shows a few of the top trays of a conventionally arranged fractionator. It has been suggested that a richer distillate might be obtained by the scheme shown in Fig. 9.58b, because of the richer reflux to tray 1. All trays are theoretical. The condenser is a total condenser, delivering liquid at the bubble point in both cases. For both arrangements, $L_0/D = L_0/L_R = 1.0$.

(a) For a binary mixture which in every respect follows the McCabe-Thiele simplifying assumptions, sketch fully labeled Hxy and xy diagrams for the scheme of Fig. 9.58b. Mark on the Hxy diagram the location of every stream on Fig. 9.58b and the required difference points. Show construction lines and tie lines.

(b) Prove that for all otherwise similar circumstances the scheme of Fig. 9.58b in fact delivers a *leaner* distillate than that of Fig. 9.58a.

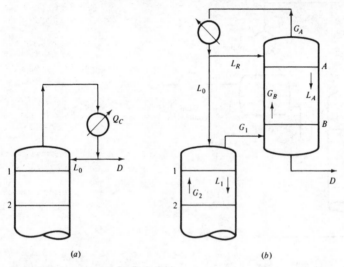

$$(a) \qquad\qquad\qquad (b)$$

Figure 9.58 Arrangements for Prob. 9.14.

Continuous fractionation: McCabe-Thiele method

9.15 A solution of carbon tetrachloride and carbon disulfide containing 50 wt% each is to be continuously fractionated at standard atmospheric pressure at the rate 4000 kg/h. The distillate product is to contain 95 wt % carbon disulfide, the residue 0.5%. The feed will be 30 mol % vaporized before it enters the tower. A total condenser will be used, and reflux will be returned at the bubble point. Equilibrium data ("The Chemical Engineers' Handbook," 4th ed., p. 13-5), x, y = mole fraction CS_2:

t, °C	x	y^*	t, °C	x	y^*
76.7	0	0	59.3	0.3908	0.6340
74.9	0.0296	0.0823	55.3	0.5318	0.7470
73.1	0.0615	0.1555	52.3	0.6630	0.8290
70.3	0.1106	0.2660	50.4	0.7574	0.8780
68.6	0.1435	0.3325	48.5	0.8604	0.9320
63.8	0.2585	0.4950	46.3	1.000	1.000

(a) Determine the product rates, kg/h.

(b) Determine the minimum reflux ratio.

(c) Determine the minimum number of theoretical trays, graphically and by means of Eq. (9.85).

(d) Determine the number of theoretical trays required at a reflux ratio equal to twice the minimum and the position of the feed tray.

Ans.: 12.5 theoretical trays.

(e) Estimate the overall tray efficiency of a sieve-tray tower of conventional design and the number of real trays.

(f) Using the distillate temperature as the base temperature, determine the enthalpy of the feed, the products, and the vapor entering the condenser. Determine the heat loads of the condenser and reboiler. Latent and specific heats are available in "The Chemical Engineers' Handbook," 5th ed., pp. 3-116 and 3-129.

9.16 A solution of carbon tetrachloride and carbon disulfide containing 50 wt % of each is to be continuously fractionated to give a distillate and residue analyzing 99.5 and 0.5 wt % carbon disulfide, respectively. Feed will be available at the bubble point, and reflux will be returned at the bubble point.

There is available a sieve-tray tower of conventional design, suitable for use at 1 std atm pressure abs. The diameter is 0.75 m (30 in), and it contains 26 identical cross-flow trays at a spacing of 0.50 m (20 in). A feed nozzle is available only for the tenth tray from the top. Each tray contains a straight weir, 0.46 m (18 in) long, extending 65 mm (2.5 in) from the tray floor. The perforated area is 12% of the active area of the perforated sheet. Adequate condenser and reboiler will be supplied. Equilibrium data are listed in Prob. 9.15.

The listed products represent the minimum purities acceptable, and higher purities at the expense of additional heat load or reduced capacity are not warranted. Estimate the largest feed rate which the column can reasonably be expected to handle.

9.17 Refer to Fig. 9.46, the McCabe-Thiele diagram for separating partially miscible mixtures. Determine the coordinates of the point of intersection of the enriching-section operating line of tower I and the 45° diagonal of the diagram.

9.18 An aqueous solution of furfural contains 4 mol % furfural. It is to be continuously fractionated at 1 std atm pressure to provide solutions containing 0.1 and 99.5 mol % furfural, respectively. Feed is liquid at the bubble point. Equilibrium data ("The Chemical Engineers' Handbook," 4th ed., p. 13-5), x, y = mole fraction furfural:

t, °C	x	y^*	t, °C	x	y^*
100.	0	0	100.6	0.80	0.11
98.56	0.01	0.055	109.5	0.90	0.19
98.07	0.02	0.080	122.5	0.92	0.32
97.90	0.04	0.092	146.0	0.94	0.64
97.90	0.092	0.092	154.8	0.96	0.81
97.90	0.50	0.092	158.8	0.98	0.90
98.7	0.70	0.095	161.7	1.00	1.00

(a) Arrange a scheme for the separation using kettle-type reboilers, and determine the number of theoretical trays required for a vapor boilup rate of 1.25 times the minimum for an infinite number of trays.

(b) Repeat part (a) but use open steam at essentially atmospheric pressure for the tower delivering the water-rich product, with 1.25 times the minimum steam and vapor boilup rates.

(c) Which scheme incurs the greater furfural loss to the water-rich product?

9.19 An aniline-water solution containing 7 wt % aniline, at the bubble point, is to be steam-stripped at a rate of 1000 kg/h in a tower packed with 38-mm (1.5-in) Berl saddles with open steam to remove 99% of the aniline. The condensed overhead vapor will be decanted at 98.5°C, and the water-rich layer returned to the column. The aniline-rich layer will be withdrawn as distillate product. The steam rate will be 1.3 times the minimum. Design the tower for a pressure drop of 400

$(N/m^2)/m$ at the top. *Note:* For concentrations where the equilibrium curve is nearly straight, it should be possible to use overall numbers and heights of transfer units.
Ans.: 6.5 m of packing.

Data: at 98.5°C, the solubility of aniline in water is 7.02 and 89.90 wt % aniline. The vapor-liquid equilibria at 745 mmHg, at which pressure the tower will be operated, are [Griswold et al., *Ind. Eng. Chem.*, **32**, 878 (1940)]:

x = mole fraction aniline	0.002	0.004	0.006	0.008	0.010	0.012	Two liquid phases, 98.5°C
y^* = mole fraction aniline	0.01025	0.0185	0.0263	0.0338	0.03575	0.03585	0.0360

9.20 A distillation column is to be fed with the following solution:

$$n\text{-}C_3H_8 = 0.06 \text{ mole fraction}$$
$$i\text{-}C_4H_{10} = 0.65$$
$$n\text{-}C_4H_{10} = 0.26$$
$$n\text{-}C_5H_{12} = \frac{0.03}{1.00}$$

The distillate is to contain no more than 0.05 mole fraction n-C_4H_{10} and essentially no C_5H_{12}. The bottoms will contain no more than 0.05 mole fraction i-C_4H_{10} and essentially no C_3H_8. The overhead vapor will be completely condensed at its bubble point, 37.8°C (100°F), with reflux to the top tray at the rate 2 kmol reflux/kmol distillate withdrawn. Compute the column pressure to be used.

9.21 The following feed at the bubble point at 827 kN/m^2 (120 lb$_f$/in^2) abs, is to be fractionated at that pressure to provide no more than 7.45% of the n-C_5H_{12} in the distillate and no more than 7.6% of the n-C_4H_{10} in the bottoms, with a total condenser and reflux returned at the bubble point:

Component	x_F	60°C	90°C	120°C
n-C_3H_8	0.05			
m		2.39	3.59	5.13
H_G		33 730	36 290	39 080
H_L		20 350	25 800	31 050
i-C_4H_{10}	0.15			
m		1.21	1.84	2.87
H_G		40 470	43 960	47 690
H_L		24 050	28 800	34 560
n-C_4H_{10}	0.25			
m		0.80	1.43	2.43
H_G		43 730	46 750	50 360
H_L		24 960	30 010	35 220
i-C_5H_{12}	0.20			
m		0.39	0.72	1.29
H_G		51 870	55 270	59 430
H_L		29 080	34 610	40 470
n-C_5H_{12}	0.35			
m		0.20	0.615	1.13
H_G		54 310	57 680	61 130
H_L		29 560	35 590	41 710

The table provides Henry's-law constants and enthalpies expressed as kJ/kmol, from the same sources and using the same base as the data of Illustration 9.13.

Compute the minimum reflux ratio, the minimum number of theoretical trays, and for a reflux ratio $R = 2.58$ estimate the product analyses, condenser and reboiler heat loads, vapor and liquid rates per mole of feed throughout, and the number of theoretical trays.

Ans.: 8 theoretical trays + reboiler.

9.22 The following feed at 60°C (140°F), 2070 kN/m^2 (300 lb$_f$/in^2) abs, is to be fractionated at that pressure so that the vapor distillate product contains 0.913 percent of the isopentane and the bottoms 0.284 percent of the isobutane:

$$CH_4 = 0.50 \text{ mole fraction}$$
$$C_3H_8 = 0.03$$
$$i\text{-}C_4H_{10} = 0.10$$
$$n\text{-}C_4H_{10} = 0.15$$
$$i\text{-}C_5H_{12} = 0.07$$
$$n\text{-}C_5H_{12} = 0.05$$
$$n\text{-}C_6H_{14} = 0.10$$
$$\overline{1.00}$$

Compute: (a) The minimum reflux ratio and (b) the minimum number of trays. At a reflux ratio $= 2.0$, estimate (c) the product analyses and (d) the number of theoretical trays by the correlations of Gilliland [17], Erbar and Maddox [12], Brown and Martin [5], and Strangio and Treybal [59].

THREE

LIQUID-LIQUID OPERATIONS

Liquid extraction, the only operation in this category, is basically very similar to the operations of gas-liquid contact described in the previous part. The creation of a new insoluble liquid phase by addition of a solvent to a mixture accomplishes in many respects the same result as the creation of a new phase by the addition of heat in distillation operations, for example, or by addition of gas in desorption operations. The separations produced by single stages, the use of countercurrent cascades and reflux to enhance the extent of separation—all have their liquid-extraction counterparts. The similarity between them will be exploited in explaining what liquid extraction can accomplish.

Certain differences in the operations nevertheless make it expedient to provide a separate treatment for liquid extraction. The considerably greater change in mutual solubility of the contacted liquid phases which may occur during passage through a cascade of stages requires somewhat different techniques of computation than are necessary for gas absorption or stripping. The considerably smaller differences in density of the contacted phases and their relatively low interfacial tension, compared with the corresponding gas-liquid systems, require consideration in the design of apparatus. The larger number of variables which apparently influence the rate of mass transfer in countercurrent equipment makes the correlation of design data much more difficult. For these reasons separate treatment is desirable, at least for the present.

Despite the increasingly extensive application of liquid extraction, thanks to its great versatility, and the extensive amount of research that has been done, it is nevertheless a relatively immature unit operation. It is characteristic of such operations that equipment types change rapidly, new designs being proposed frequently and lasting through a few applications, only to be replaced by others. Design principles for such equipment are never fully developed, and a reliance must often be put on pilot-plant testing for new installations.

LIQUID EXTRACTION

Liquid extraction, sometimes called solvent extraction, is the separation of the constituents of a liquid solution by contact with another insoluble liquid. If the substances constituting the original solution distribute themselves differently between the two liquid phases, a certain degree of separation will result, and this can be enhanced by use of multiple contacts or their equivalent in the manner of gas absorption and distillation.

A simple example will indicate the scope of the operation and some of its characteristics. If a solution of acetic acid in water is agitated with a liquid such as ethyl acetate, some of the acid but relatively little water will enter the ester phase. Since at equilibrium the densities of the aqueous and ester layers are different, they will settle when agitation stops and can be decanted from each other. Since now the ratio of acid to water in the ester layer is different from that in the original solution and also different from that in the residual water solution, a certain degree of separation will have occurred. This is an example of stagewise contact, and it can be carried out either in batch or in continuous fashion. The residual water can be repeatedly extracted with more ester to reduce the acid content still further, or we can arrange a countercurrent cascade of stages. Another possibility is to use some sort of countercurrent continuous-contact device, where discrete stages are not involved. The use of reflux, as in distillation, may enhance the ultimate separation still further.

In all such operations, the solution which is to be extracted is called the *feed*, and the liquid with which the feed is contacted is the *solvent*. The solvent-rich product of the operation is called the *extract*, and the residual liquid from which solute has been removed is the *raffinate*.

More complicated processes may use two solvents to separate the components of a feed. For example, a mixture of *p*- and *o*-nitrobenzoic acids can be separated by distributing them between the insoluble liquids chloroform and water. The chloroform preferentially dissolves the para isomer and the water the ortho isomer. This is called *double-solvent*, or *fractional*, extraction.

Fields of Usefulness

Applications of liquid extraction fall into several categories: those where extraction is in direct competition with other separation methods and those where it seems uniquely qualified.

In competition with other mass-transfer operations Here relative costs are important. Distillation and evaporation are direct separation methods, the products of which are composed of essentially pure substances. Liquid extraction, on the other hand, produces new solutions which must in turn be separated, often by distillation or evaporation. Thus, for example, acetic acid can be separated from dilute solution with water, with difficulty by distillation or with relative ease by extraction into a suitable solvent followed by distillation of the extract. For the more dilute solutions particularly, where water must be vaporized in distillation, extraction is more economical, especially since the heat of vaporization of most organic solvents is substantially less than that of water. Extraction may also be attractive as an alternative to distillation under high vacuum at very low temperatures to avoid thermal decomposition. For example, long-chain fatty acids can be separated from vegetable oils by high-vacuum distillation but more economically by extraction with liquid propane. Tantalum and niobium can be separated by very tedious fractional crystallization of the double fluorides with potassium but with relative ease by liquid extraction of the hydrofluoric acid solutions with methyl isobutyl ketone.

As a substitute for chemical methods Chemical methods consume reagents and frequently lead to expensive disposal problems for chemical by-products. Liquid extraction, which incurs no chemical consumption or by-product production, can be less costly. Metal separations such as uranium-vanadium, hafnium-zirconium, and tungsten-molybdenum and the fission products of atomic-energy processes are more economical by liquid extraction. Even lower-cost metals such as copper and inorganic chemicals such as phosphoric acid, boric acid, and the like are economically purified by liquid extraction, despite the fact that the cost of solvent recovery must be included in the final reckoning.

For separations not now possible by other methods In distillation, where the vapor phase is created from the liquid by addition of heat, the vapor and liquid are necessarily composed of the same substances and are therefore chemically very similar. The separations produced then depend upon the vapor pressures of the substances. In liquid extraction, in contrast, the major constituents of the two phases are chemically very different, and this makes separations according to chemical type possible. For example, aromatic and paraffinic hydrocarbons of nearly the same molecular weight are impossible to separate by distillation because their vapor pressures are nearly the same, but they can readily be separated by extraction with any of a number of solvents, e.g., liquid sulfur dioxide, diethylene glycol, or sulfolane. (Extractive distillation is also useful for such operations, but it is merely extraction of the *vapor* phase with a solvent, whereas liquid extraction is extraction of the *liquid* phase. Frequently the same solvents are useful for both, as might be expected.) Many pharmaceutical products, e.g., penicillin, are produced in mixtures so complex that only liquid extraction is a feasible separation device.

LIQUID EQUILIBRIA

Extraction involves the use of systems composed of at least three substances, and although for the most part the insoluble phases are chemically very different, generally all three components appear at least to some extent in both phases. The following notation scheme will be used to describe the concentrations and amounts of these ternary mixtures, for purposes of discussing both equilibria and material balances.

Notation Scheme

1. A and B are pure, substantially insoluble liquids, and C is the distributed solute. Mixtures to be separated by extraction are composed of A and C, and B is the extracting solvent.
2. The same letter will be used to indicate the quantity of a solution or mixture and the location of the mixture on a phase diagram. Quantities are measured by mass for batch operations, mass/time for continuous operation. Thus,

   E = mass/time of solution E, an extract, shown on a phase diagram by point E

   R = mass/time of solution R, a raffinate, shown on a phase diagram by point R

   B = mass/time of solvent B

   Solvent-free (B-free) quantities are indicated by primed letters. Thus,

   E' = mass B-free solution/time, shown on a phase diagram by point E
   $E = E'(1 + N_E)$

3. x = weight fraction C in the solvent-lean (A-rich), or raffinate, liquids
   y = weight fraction C in the solvent-rich (B-rich), or extract, liquids
   $x' = x/(1 - x)$ = mass C/mass non-C in the raffinate liquids
   $y' = y/(1 - y)$ = mass C/mass non-C in the extract liquids
   X = weight fraction C in the raffinate liquids on a B-free basis, mass C/(mass A + mass C)

Y = weight fraction C in the extract liquids on a B-free basis, mass C/(mass A + mass C)

N = weight fraction B on a B-free basis, mass B/(mass A + mass C)

Subscripts identify the solution or mixture to which the concentration terms refer. Stages are identified by number. Thus, x_3 = wt fraction C in the raffinate from stage 3, Y_3 = wt fraction C (B-free basis) in the extract from stage 3, etc. For other solutions identified by a letter on a phase diagram, the same letter is used as an identifying subscript. Thus, x_M = wt fraction C in the mixture M. An asterisk specifically identifies equilibrium concentrations where the condition of equilibrium is emphasized. Thus, y_E^* = wt fraction C in the equilibrium solution E.

4. Throughout the discussion of equilibria, material balances, and stagewise calculations, mole fractions, mole ratios, and kilomoles may be consistently substituted for weight fractions, weight ratios, and kilograms, respectively.

Equilateral-Triangular Coordinates

These are used extensively in the chemical literature to describe graphically the concentrations in ternary systems. It is the property of an equilateral triangle that the sum of the perpendicular distances from any point within the triangle to the three sides equals the altitude of the triangle. We can therefore let the altitude represent 100 percent composition and the distances to the three sides the percentages or fractions of the three components. Refer to Fig. 10.1. Each apex of the triangle represents one of the pure components, as marked. The perpendicular distance from any point such as K to the base AB represents the percentage of C in the mixture at K, the distance to the base AC the percentage of B, and that to the base CB the percentage of A. Thus $x_K = 0.4$. Any point on a side of the triangle represents a binary mixture. Point D, for example, is a binary containing 80 percent A, 20 percent B. All points on the line DC represent mixtures containing the same ratio of A to B and can be considered as

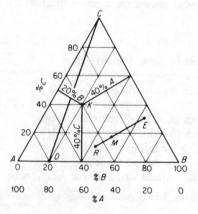

Figure 10.1 Equilateral-triangular coordinates.

mixtures originally at D to which C has been added. If R kg of a mixture at point R is added to E kg of a mixture at E, the new mixture is shown on the straight line RE at point M, such that

$$\frac{R}{E} = \frac{\text{line } ME}{\text{line } RM} = \frac{x_E - x_M}{x_M - x_R} \tag{10.1}$$

Alternatively the composition corresponding to point M can be computed by material balances, as will be shown later. Similarly, if a mixture at M has removed from it a mixture of composition E, the new mixture is on the straight line EM extended in the direction away from E and located at R so that Eq. (10.1) applies.

Equation (10.1) is readily established. Refer to Fig. 10.2, which again shows R kg of mixture at R added to E kg of mixture at E. Let M represent the kilograms of new mixture as well as the composition on the figure. Line RL = weight fraction C in R = x_R, line MO = weight fraction C in M = x_M, and line ET = weight fraction C in E = x_E. A total material balance,

$$R + E = M$$

A balance for component C,

$$R(\text{line } RL) + E(\text{line } ET) = M(\text{line } MO)$$

$$Rx_R + Ex_E = Mx_M$$

Eliminating M, we get

$$\frac{R}{E} = \frac{\text{line } ET - \text{line } MO}{\text{line } MO - \text{line } RL} = \frac{x_E - x_M}{x_M - x_R}$$

But line ET − line MO = line EP, and line MO − line RL = line MK = line PS. Therefore

$$\frac{R}{E} = \frac{\text{line } EP}{\text{line } PS} = \frac{\text{line } ME}{\text{line } RM}$$

The following discussion is limited to those types of systems which most frequently occur in liquid-extraction operations. For a complete consideration of the many types of systems which may be encountered, the student is referred to one of the more comprehensive texts on the phase rule [51].

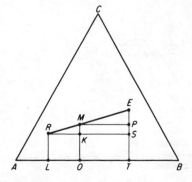

Figure 10.2 The mixture rule.

Systems of Three Liquids—One Pair Partially Soluble

This is the most common type of system in extraction, and typical examples are water (A)–chloroform (B)–acetone (C) and benzene (A)–water (B)–acetic acid (C). The triangular coordinates are used as *isotherms*, or diagrams at constant temperature. Refer to Fig. 10.3*a*. Liquid C dissolves completely in A and B, but A and B dissolve only to a limited extent in each other to give rise to the saturated liquid solutions at L (A-rich) and at K (B-rich). The more insoluble the liquids A and B, the nearer the apexes of the triangle will points L and K be located. A binary mixture J, anywhere between L and K, will separate into two insoluble liquid phases of compositions at L and K, the relative amounts of the phases depending upon the position of J, according to the principle of Eq. (10.1).

Curve $LRPEK$ is the binodal solubility curve, indicating the change in solubility of the A- and B-rich phases upon addition of C. Any mixture outside this curve will be a homogeneous solution of one liquid phase. Any ternary mixture underneath the curve, such as M, will form two insoluble, saturated liquid phases of equilibrium compositions indicated by R (A-rich) and E (B-rich). The line RE joining these equilibrium compositions is a tie line, which must necessarily pass through point M representing the mixture as a whole. There are an infinite number of tie lines in the two-phase region, and only a few are shown. They are rarely parallel and usually change their slope slowly in one direction as shown. In a relatively few systems the direction of the tie-line slope changes, and one tie line will be horizontal. Such systems are said to be *solutropic*. Point P, the *plait point*, the last of the tie lines and the point where the A-rich and B-rich solubility curves merge, is ordinarily not at the maximum value of C on the solubility curve.

The percentage of C in solution E is clearly greater than that in R, and it is said that in this case the distribution of C favors the B-rich phase. This is

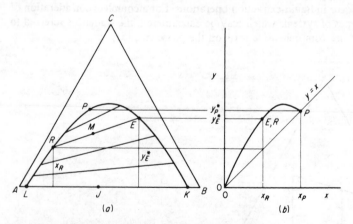

Figure 10.3 System of three liquids, A and B partially miscible.

conveniently shown on the distribution diagram (Fig. 10.3*b*), where the point (E, R) lies above the diagonal $y = x$. The ratio y^*/x, the *distribution coefficient*, is in this case greater than unity. The concentrations of C at the ends of the tie lines, when plotted against each other, give rise to the distribution curve shown. Should the tie lines on Fig. 10.3*a* slope in the opposite direction, with C favoring A at equilibrium, the distribution curve will lie below the diagonal. The distribution curve may be used for interpolating between tie lines when only a few have been experimentally determined. Other methods of interpolation are also available [72].

Effect of temperature To show the effect of temperature in detail requires a three-dimensional figure, as in Fig. 10.4*a*, where temperature is plotted vertically and the isothermal triangles are seen to be sections through the prism. For most systems of this type, the mutual solubility of A and B increases with increasing temperature, and above some temperature t_4, the critical solution temperature, they dissolve completely. The increased solubility at higher temperatures influences the ternary equilibria considerably, and this is best shown by projection of the isotherms onto the base triangle, as in Fig. 10.4*b*. Not only does the area of heterogeneity decrease at higher temperatures, but the slopes of the tie lines may also change. Liquid-extraction operations, which depend upon the formation of insoluble liquid phases, must be carried on at temperatures below t_4. Other, less common, temperature effects are also known [51, 72].

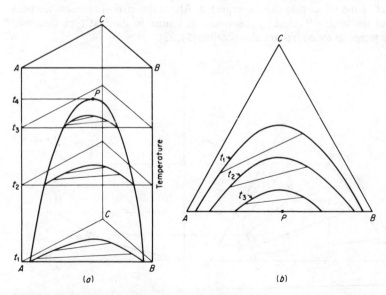

Figure 10.4 Effect of temperature on ternary equilibria.

Effect of pressure Except at very high pressures, the influence of pressure on the liquid equilibrium is so small that it can generally be ignored. All the diagrams shown are therefore to be considered as having been plotted at sufficiently high pressure to maintain a completely condensed system, i.e., above the vapor pressures of the solutions. However, should the pressure be sufficiently reduced to become less than the vapor pressure of the solutions, a vapor phase will appear and the liquid equilibrium will be interrupted. Such an effect on a binary solubility curve of the type APB of Fig. 10.4a is shown in Fig. 9.8.

Systems of Three Liquids—Two Pairs Partially Soluble

This type is exemplified by the system chlorobenzene (A)–water (B)–methyl ethyl ketone (C), where A and C are completely soluble, while the pairs A–B and B–C show only limited solubility. Refer to Fig. 10.5a, a typical isotherm. At the prevailing temperature, points K and J represent the mutual solubilities of A and B and points H and L those of B and C. Curves KRH (A-rich) and JEL (B-rich) are the ternary solubility curves, and mixtures outside the band between these curves form homogeneous single-phase liquid solutions. Mixtures such as M, inside the heterogeneous area, form two liquid phases at equilibrium at E and R, joined on the diagram by a tie line. The corresponding distribution curve is shown in Fig. 10.5b.

Effect of temperature Increased temperature usually increases the mutual solubilities and at the same time influences the slope of the tie lines. Figure 10.6 is typical of the effect that can be expected. Above the critical solution temperature of the binary B–C at t_3, the system is similar to the first type discussed. Other temperature effects are also possible [51, 72].

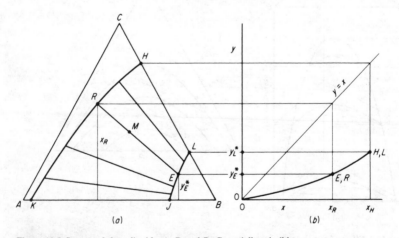

Figure 10.5 System of three liquids, A–B and B–C partially miscible.

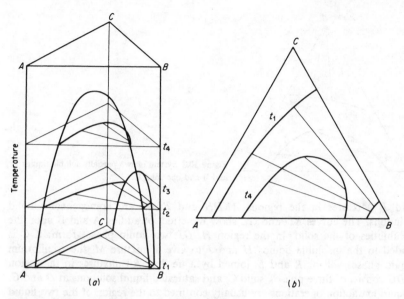

Figure 10.6 Effect of temperature on ternary equilibria.

Systems of Two Partially Soluble Liquids and One Solid

When the solid does not form compounds such as hydrates with the liquids, the system will frequently have the characteristics of the isotherm of Fig. 10.7, an example of which is the system naphthalene (C)–aniline (A)–isooctane (B). Solid C dissolves in liquid A to form a saturated solution at K and in liquid B to give the saturated solution at L. A and B are soluble only to the extent shown at H

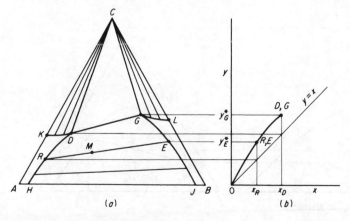

Figure 10.7 System of two partially soluble liquids A and B and one solid C.

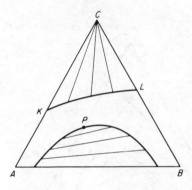

Figure 10.8 System of two partially soluble liquids A and B and one solid C.

and J. Mixtures in the regions $AKDH$ and $BLGJ$ are homogeneous liquid solutions. The curves KD and GL show the effect of adding A and B upon the solubilities of the solid. In the region $HDGJ$ two liquid phases form: if C is added to the insoluble liquids H and J to give a mixture M, the equilibrium liquid phases will be R and E, joined by a tie line. All mixtures in the region CDG consist of three phases, solid C, and saturated liquid solutions at D and G. Liquid-extraction operations are usually confined to the region of the two liquid phases, which is that corresponding to the distribution curve shown.

Increased temperature frequently changes these systems to the configuration shown in Fig. 10.8.

Other Coordinates

Because the equilibrium relationship can rarely be expressed algebraically with any convenience, extraction computations must usually be made graphically on a phase diagram. The coordinate scales of equilateral triangles are necessarily

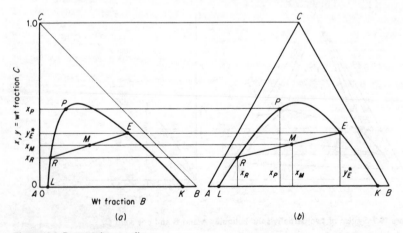

(a)   (b)

Figure 10.9 Rectangular coordinates.

always the same, and in order to be able to expand one concentration scale relative to the other, rectangular coordinates can be used. One of these is formed by plotting concentrations of B as abscissa against concentrations of C (x and y) as ordinate, as in Fig. 10.9a. Unequal scales can be used in order to expand the plot as desired. Equation (10.1) applies for mixtures on Fig. 10.9a, regardless of any inequality of the scales.

Another rectangular coordinate system involves plotting as abscissa the weight fraction C on a B-free basis, X and Y in the A- and B-rich phases, respectively, against N, the B-concentration on a B-free basis, as ordinate, as shown in the upper part of Fig. 10.10. This has been plotted for a system of two partly miscible pairs, such as that of Fig. 10.5.

The similarity of such diagrams to the enthalpy-concentration diagrams of Chap. 9 is clear. In extraction, the two phases are produced by addition of solvent, in distillation by addition of heat, and solvent becomes the analog of heat. This is emphasized by the ordinate of the upper part of Fig. 10.10. Tie lines such as QS can be projected to X, Y coordinates, as shown in the lower figure, to produce a solvent-free distribution graph similar to those of distillation. The mixture rule on these coordinates (see the upper part of Fig. 10.10) is

$$\frac{M'}{N'} = \frac{Y_N - Y_P}{Y_P - Y_M} = \frac{X_N - X_P}{X_P - X_M} = \frac{\text{line } NP}{\text{line } PM} \quad (10.2)$$

where M' and N' are the B-free weights of these mixtures.

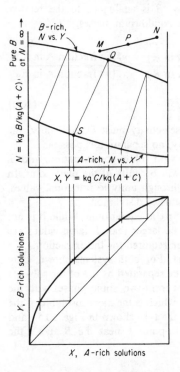

Figure 10.10 Rectangular coordinates, solvent-free basis, for a system of two partly miscible liquid pairs.

Multicomponent Systems

The simplest system of four components occurs when two solutes distribute between two solvents, e.g., the distribution of formic and acetic acids between the partly soluble solvents water and carbon tetrachloride. Complete display of such equilibria requires a three-dimensional graph [72], which is difficult to work with, but frequently we can simplify this to the distribution (xy) curves, one for each solute, such as that for a single solute in Fig. 10.3b. More than four components cannot be conveniently dealt with graphically.

Choice of Solvent

There is usually a wide choice of liquids to be used as solvents for extraction operations. It is unlikely that any particular liquid will exhibit all the properties considered desirable for extraction, and some compromise is usually necessary. The following are the quantities to be given consideration in making a choice:

1. *Selectivity.* The effectiveness of solvent B for separating a solution of A and C into its components is measured by comparing the ratio of C to A in the B-rich phase to that in the A-rich phase at equilibrium. The ratio of the ratios, the separation factor, or selectivity, β is analogous to the relative volatility of distillation. If E and R are the equilibrium phases,

$$\beta = \frac{(\text{wt fraction C in } E)/(\text{wt fraction A in } E)}{(\text{wt fraction C in } R)/(\text{wt fraction A in } R)} = \frac{y_E^*(\text{wt fraction A in } R)}{x_R(\text{wt fraction A in } E)}$$

(10.3)

For all useful extraction operations the selectivity must exceed unity, the more so the better. If the selectivity is unity, no separation is possible.

Selectivity usually varies considerably with solute concentration, and in systems of the type shown in Fig. 10.3 it will be unity at the plait point. In some systems it passes from large values through unity to fractional values, and these are analogous to azeotropic systems of distillation.

2. *Distribution coefficient.* This is the ratio y^*/x at equilibrium. While it is not necessary that the distribution coefficient be larger than 1, large values are very desirable since less solvent will then be required for the extraction.

3. *Insolubility of solvent.* Refer to Fig. 10.11. For both systems shown, only those A–C mixtures between D and A can be separated by use of the solvents B or B', since mixtures richer in C will not form two liquid phases with the solvents. Clearly the solvent in Fig. 10.11a, which is the more insoluble of the two, will be the more useful. In systems of the type shown in Figs. 10.5a and 10.7a, if the solubility of C in B is small (point L near the B apex), the

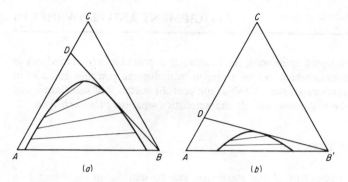

Figure 10.11 Influence of solvent solubility on extraction.

capacity of solvent B to extract C is small and large amounts of solvent are then required.

4. *Recoverability.* It is always necessary to recover the solvent for reuse, and this must ordinarily be done by another of the mass-transfer operations, most frequently distillation. If distillation is to be used, the solvent should form no azeotrope with the extracted solute and mixtures should show high relative volatility for low-cost recovery. That substance in the extract, either solvent or solute, which is present as the lesser quantity should be the more volatile in order to reduce heat costs. If the solvent must be volatilized, its latent heat of vaporization should be small.

5. *Density.* A difference in densities of the saturated liquid phases is necessary, both for stagewise and continuous-contact equipment operation. The larger this difference the better. In systems of the type shown in Fig. 10.3, the density difference for equilibrium phases will become less as C concentrations increase and will be zero at the plait point. It may reverse in sign before reaching the plait point, in which case continuous-contact equipment cannot be specified to operate at the concentrations at which the density difference passes through zero.

6. *Interfacial tension.* The larger the interfacial tension, the more readily coalescence of emulsions will occur but the more difficult the dispersion of one liquid in the other will be. Coalescence is usually of greater importance, and interfacial tension should therefore be high. Interfacial tension between equilibrium phases in systems of the type shown in Fig. 10.3 falls to zero at the plait point.

7. *Chemical reactivity.* The solvent should be stable chemically and inert toward the other components of the system and toward the common materials of construction.

8. *Viscosity, vapor pressure, and freezing point.* These should be low for ease in handling and storage.

9. The solvent should be *nontoxic, nonflammable*, and of *low cost*.

EQUIPMENT AND FLOWSHEETS

As with the gas-liquid operations, liquid extraction processes are carried out in stagewise equipment which can be arranged in multistage cascades and also in differential-contact apparatus. We shall consider the methods of calculation and the nature of the equipment and its characteristics separately for each type.

STAGEWISE CONTACT

Extraction in equipment of the stage type can be carried on according to a variety of flowsheets, depending upon the nature of the system and the extent of separation desired. In the discussion which follows it is to be understood that each state is a *theoretical* or *equilibrium* stage, such that the effluent extract and raffinate solutions are in equilibrium with each other. Each stage must include facilities for contacting the insoluble liquids and separating the product streams. A combination of a mixer and a settler may therefore constitute a stage, and in multistage operation these may be arranged in cascades as desired. In counter-current multistage operation it is also possible to use towers of the multistage type, as described earlier.

Single-Stage Extraction

This may be a batch or a continuous operation. Refer to Fig. 10.12. The flowsheet shows the extraction stage. Feed of mass F (if batch) or F mass/time (if continuous) contains substances A and C at x_F weight fraction C. This is contacted with mass S_1 (or mass/time) of a solvent, principally B, containing y_S weight fraction C, to give the equilibrium extract E_1 and raffinate R_1, each measured in mass or mass/time. Solvent recovery then involves separate removal of solvent B from each product stream (not shown).

The operation can be followed in either of the phase diagrams as shown. If the solvent is pure B ($y_S = 0$), it will be plotted at the B apex, but sometimes it has been recovered from a previous extraction and therefore contains a little A and C as well, as shown by the location of S. Adding S to F produces in the extraction stage a mixture M_1 which, on settling, forms the equilibrium phases E_1 and R_1 joined by the tie line through M_1. A total material balance is

$$F + S_1 = M_1 = E_1 + R_1 \tag{10.4}$$

and point M_1 can be located on line FS by the mixture rule, Eq. (10.1), but it is usually more satisfactory to locate M_1 by computing its C concentration. Thus, a C balance provides

$$Fx_F + S_1 y_S = M_1 x_{M1} \tag{10.5}$$

from which x_{M1} can be computed. Alternatively, the amount of solvent to

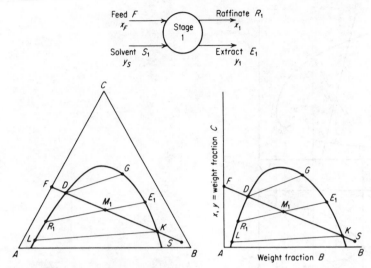

Figure 10.12 Single-stage extraction.

provide a given location for M_1 on the line FS can be computed:

$$\frac{S_1}{F} = \frac{x_F - x_{M1}}{x_{M1} - y_S} \tag{10.6}$$

The quantities of extract and raffinate can be computed by the mixture rule, Eq. (10.1), or by the material balance for C:

$$E_1 y_1 + R_1 x_1 = M_1 x_{M1} \tag{10.7}$$

$$E_1 = \frac{M_1(x_{M1} - x_1)}{y_1 - x_1} \tag{10.8}$$

and R_1 can be determined through Eq. (10.4).

Since two insoluble phases must form for an extraction operation, point M_1 must lie within the heterogeneous liquid area, as shown. The minimum amount of solvent is thus found by locating M_1 at D, which would then provide an infinitesimal amount of extract at G, and the maximum amount of solvent is found by locating M_1 at K, which provides an infinitesimal amount of raffinate at L. Point L also represents the raffinate with the lowest possible C concentration, and if a lower value were required, the recovered solvent S would have to have a lower C concentration.

Computations for systems of two insoluble liquid pairs, or with a distributed solute which is a solid, are made in exactly the same manner, and Eqs. (10.4) to (10.8) all apply.

All the computations can also be made on a solvent-free basis, as in the upper part of Fig. 10.13, if the nature of this diagram makes it convenient. If

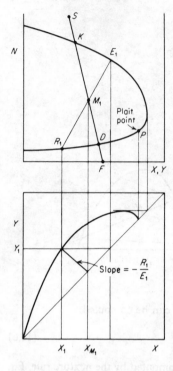

Figure 10.13 Single-stage extraction, solvent-free coordinates.

solvent S is pure B, its N value is infinite and the line FS is then vertical. Products E_1 and R_1 lie on a tie line through M_1 representing the entire mixture in the extractor. Material balances for use with this diagram must be made on a B-free basis. Thus,

$$F' + S' = M_1' = E_1' + R_1' \tag{10.9}$$

where the primes indicate B-free weight (the feed is normally B-free, and $F = F'$). A balance for C is

$$F'X_F + S'Y_S = M_1'X_{M1} = E_1'Y_1 + R_1'X_1 \tag{10.10}$$

and for B

$$F'N_F + X'N_S = M_1'N_{M1} = E_1'N_{E1} + R_1'N_{R1} \tag{10.11}$$

(ordinarily, $N_F = 0$ since the feed contains no B). The coordinates of M_1' can be computed, point M_1 located on line FS, the tie line located, and the B-free weights, E_1' and R_1', computed:

$$E_1' = \frac{M_1'(X_{M1} - X_1)}{Y_1 - X_1} \tag{10.12}$$

and R_1' is obtained through Eq. (10.9). The total weights of the saturated extract and raffinate are then

$$E_1 = E_1'(1 + N_{E1}) \qquad R_1 = R_1'(1 + N_{R1}) \tag{10.13}$$

If the solvent is pure B, whence $N_S = \infty$, these equations still apply, with the simplification that $S' = 0$, $Y_S = 0$, $S'N_S = B$, and $F' = M_1'$. Minimum and maximum amounts of solvent correspond to putting M_1 at D and K on the figure, as before.

Equations (10.9) and (10.10) lead to

$$\frac{R_1'}{E_1'} = \frac{Y_1 - X_{M1}}{X_{M1} - X_1} \tag{10.14}$$

When the equilibrium extract and raffinate are located on the lower diagram of Fig. 10.13, Eq. (10.14) is seen to be that of the operating line shown, of slope $-R_1'/E_1'$. The single stage is seen to be analogous to the flash vaporization of distillation, with solvent replacing heat. If pure B is used as solvent, the operating line on the lower figure passes through the 45° line at X_F, which completes the analogy (see Fig. 9.14).

Multistage Crosscurrent Extraction

This is an extension of single-stage extraction, wherein the raffinate is successively contacted with fresh solvent, and may be done continuously or in batch. Refer to Fig. 10.14, which shows the flow sheet for a three-stage extraction. A

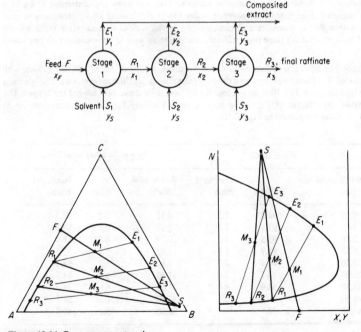

Figure 10.14 Crosscurrent extraction.

single final raffinate results, and the extracts can be combined to provide the composited extract, as shown. As many stages as desired can be used.

Computations are shown on triangular and on solvent-free coordinates. All the material balances for a single stage of course now apply to the first stage. Subsequent stages are dealt with in the same manner, except of course that the feed to any stage is the raffinate from the preceding stage. Thus, for any stage n:

Total balance:
$$R_{n-1} + S_n = M_n = E_n + R_n \tag{10.15}$$

C balance:
$$R_{n-1}x_{n-1} + S_n y_S = M_n x_{Mn} = E_n y_n + R_n x_n \tag{10.16}$$

For the solvent-free coordinates,

A + C balance:
$$R'_{n-1} + S'_n = M'_n = E'_n + R'_n \tag{10.17}$$

C balance:
$$R'_{n-1}X_{n-1} + S'_n Y_S = M'_n X_{M_n} = E'_n Y_n + R'_n X_n \tag{10.18}$$

B balance:
$$R'_{n-1}N_{R_{n-1}} + S'_n N_S = M'_n N_{M_n} = E'_n N_{E_n} + R'_n N_{R_n} \tag{10.19}$$

from which the quantities can be calculated for either type of graph.

Unequal amounts of solvent can be used in the various stages, and even different temperatures, in which case each stage must be computed with the help of a phase diagram at the appropriate temperature. For a given final raffinate concentration, the greater the number of stages the less total solvent will be used.

Illustration 10.1 If 100 kg of a solution of acetic acid (C) and water (A) containing 30% acid is to be extracted three times with isopropyl ether (B) at 20°C, using 40 kg of solvent in each stage, determine the quantities and compositions of the various streams. How much solvent would be required if the same final raffinate concentration were to be obtained with one stage?

SOLUTION The equilibrium data at 20°C are listed below [*Trans. AIChE*, **36**, 628 (1940), with permission]. The horizontal rows give the concentrations in equilibrium solutions. The system is of the type shown in Fig. 10.9, except that the tie lines slope downward toward the B apex. The rectangular coordinates of Fig. 10.9a will be used, but only for acid concentrations up to $x = 0.30$. These are plotted in Fig. 10.15.

	Water layer		Isopropyl ether layer		
Wt % acetic acid, $100x$	Water	Isopropyl ether	Acetic acid, $100y^*$	Water	Isopropyl ether
0.69	98.1	1.2	0.18	0.5	99.3
1.41	97.1	1.5	0.37	0.7	98.9
2.89	95.5	1.6	0.79	0.8	98.4
6.42	91.7	1.9	1.93	1.0	97.1
13.30	84.4	2.3	4.82	1.9	93.3
25.50	71.1	3.4	11.40	3.9	84.7
36.70	58.9	4.4	21.60	6.9	71.5
44.30	45.1	10.6	31.10	10.8	58.1
46.40	37.1	16.5	36.20	15.1	48.7

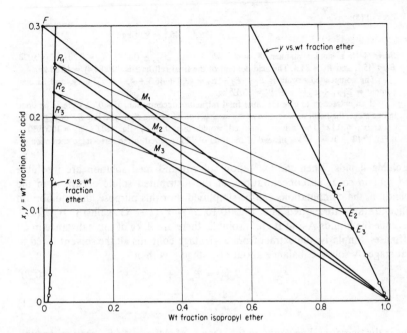

Figure 10.15 Solution to Illustration 10.1.

Stage 1 $F = 100$ kg, $x_F = 0.30$, $y_S = 0$, $S_1 = B_1 = 40$ kg. Eq. (10.4):

$$M_1 = 100 + 40 = 140 \text{ kg}$$

Eq. (10.5):

$$100(0.30) + 40(0) = 140 x_{M1} \qquad x_{M1} = 0.214$$

Point M_1 is located on line FB. With the help of a distribution curve, the tie line passing through M_1 is located as shown, and $x_1 = 0.258$, $y_1 = 0.117$ wt fraction acetic acid. Eq. (10.8):

$$E_1 = \frac{140(0.214 - 0.258)}{0.117 - 0.258} = 43.6 \text{ kg}$$

Eq. (10.4):

$$R_1 = 140 - 43.6 = 96.4 \text{ kg}$$

Stage 2 $S_2 = B_2 = 40$ kg. Eq. (10.15):

$$M_2 = R_1 + B_2 = 96.4 + 40 = 136.4 \text{ kg}$$

Eq. (10.16):

$$96.4(0.258) + 40(0) = 136.4 x_{M2} \qquad x_{M2} = 0.1822$$

Point M_2 is located on line R_1B and the tie line R_2E_2 through M_2. $x_2 = 0.227$, $y_2 = 0.095$. Eq. (10.8) becomes

$$E_2 = \frac{M_2(x_{M2} - x_2)}{y_2 - x_2} = \frac{136.4(0.1822 - 0.227)}{0.095 - 0.227} = 46.3 \text{ kg}$$

Eq. (10.15):

$$R_2 = M_2 - E_2 = 136.4 - 46.3 = 90.1 \text{ kg}$$

Stage 3 In a similar manner, $B_3 = 40$, $M_3 = 130.1$, $x_{M3} = 0.1572$, $x_3 = 0.20$, $y_3 = 0.078$, $E_3 = 45.7$, and $R_3 = 84.4$. The acid content of the final raffinate is $0.20(84.4) = 16.88$ kg.

The composited extract is $E_1 + E_2 + E_3 = 43.6 + 46.3 + 45.7 = 135.6$ kg, and its acid content $= E_1 y_1 + E_2 y_2 + E_3 y_3 = 13.12$ kg.

If an extraction to give the same final raffinate concentration, $x = 0.20$, were to be done in one stage, the point M would be at the intersection of tie line $R_3 E_3$ and line BF of Fig. 10.15, or at $x_M = 0.12$. The solvent required would then be, by Eq. (10.6), $S_1 = 100(0.30 - 0.12)/(0.12 - 0) = 150$ kg, instead of the total of 120 required in the three-stage extraction.

Insoluble liquids When the extraction solvent and feed solution are insoluble and remain so at all concentrations of the distributed solute occurring in the operation, the computations can be simplified. For this purpose, the equilibrium concentrations are plotted as in Fig. 10.16, $x' = x/(1 - x)$ against $y' = y/(1 - y)$. Since the liquids A and B are insoluble, there are A kg of this substance in all raffinates. Similarly, the extract from each stage contains all the solvent B fed to that stage. A solute-C balance about any stage n is then

$$Ax'_{n-1} + B_n y'_S = B_n y'_n + Ax'_n \tag{10.20}$$

$$-\frac{A}{B_n} = \frac{y'_S - y'_n}{x'_{n-1} - x'_n} \tag{10.21}$$

This is the operating-line equation for stage n, of slope $-A/B_n$, passing through points (x'_{n-1}, y'_S) and (x'_n, y'_n). The construction for a three-stage extraction is shown in Fig. 10.16, where for each stage a line is drawn of slope appropriate to that stage. Each operating line intersects the equilibrium curve at the raffinate and extract compositions. No raffinate of concentration smaller than that in equilibrium with the entering solvent is possible.

Solute concentrations for these cases are sometimes also conveniently expressed as mass/volume. Equations (10.20) and (10.21) then apply with x' and y' expressed as mass solute/volume, and A and B as volume/time (or volumes for batch operations).

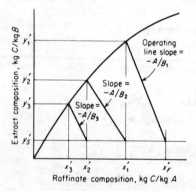

Figure 10.16 Crosscurrent extraction with an insoluble solvent.

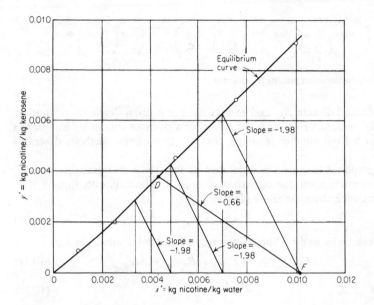

Figure 10.17 Solution to Illustration 10.2.

Illustration 10.2 Nicotine (C) in a water (A) solution containing 1% nicotine is to be extracted with kerosene (B) at 20°C. Water and kerosene are essentially insoluble. (*a*) Determine the percentage extraction of nicotine if 100 kg of feed solution is extracted once with 150 kg solvent. (*b*) Repeat for three theoretical extractions using 50 kg solvent each.

SOLUTION Equilibrium data are provided by Claffey et al., *Ind. Eng. Chem.*, **42**, 166 (1950), and expressed as kg nicotine/kg liquid, they are as follows:

$x' = \dfrac{\text{kg nicotine}}{\text{kg water}}$	0	0.001 011	0.00246	0.00502	0.00751	0.00998	0.0204
$y'^* = \dfrac{\text{kg nicotine}}{\text{kg kerosene}}$	0	0.000 807	0.001961	0.00456	0.00686	0.00913	0.01870

(*a*) $x_F = 0.01$ wt fraction nicotine, $x'_F = 0.01/(1 - 0.01) = 0.0101$ kg nicotine/kg water. $F = 100$ kg. $A = 100(1 - 0.01) = 99$ kg water. $A/B = 99/150 = 0.66$.

Refer to Fig. 10.17, which shows the equilibrium data and the point F representing the composition of the feed. From F, line FD is drawn of slope -0.66, intersecting the equilibrium curve at D, where $x'_1 = 0.00425$ and $y'_1 = 0.00380$ kg nicotine/kg liquid. The nicotine removed from the water is therefore $99(0.0101 - 0.00425) = 0.580$ kg, or 58% of that in the feed.

(*b*) For each stage, $A/B = 99/50 = 1.98$. The construction is started at F, with operating lines of slope -1.98. The final raffinate composition is $x'_3 = 0.0034$, and the nicotine extracted is $99(0.0101 - 0.0034) = 0.663$ kg, or 66.3% of that in the feed.

Continuous Countercurrent Multistage Extraction

The flowsheet for this type of operation is shown in Fig. 10.18. Extract and raffinate streams flow from stage to stage in countercurrent and provide two

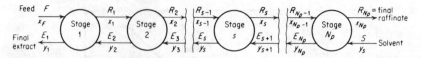

Figure 10.18 Countercurrent multistage extraction.

final products, raffinate R_{N_p} and extract E_1. For a given degree of separation, this type of operation requires fewer stages for a given amount of solvent, or less solvent for a fixed number of stages, than the crosscurrent methods described above.

The graphical treatment is developed in Fig. 10.19 on rectangular coordinates. Construction on the equilateral triangle is identical with this. A total material balance about the entire plant is

$$F + S = E_1 + R_{N_p} = M \tag{10.22}$$

Point M can be located on line FS through a balance for substance C,

$$Fx_F + Sy_S = E_1 y_1 + R_{N_p} x_{N_p} = M x_M \tag{10.23}$$

$$x_M = \frac{Fx_F + Sy_S}{F + S} \tag{10.24}$$

Equation (10.22) indicates that M must lie on line $R_{N_p} E_1$, as shown. Rearrangement of Eq. (10.22) provides

$$R_{N_p} - S = F - E_1 = \Delta_R \tag{10.25}$$

where Δ_R, a difference point, is the net flow outward at the last stage N_p.

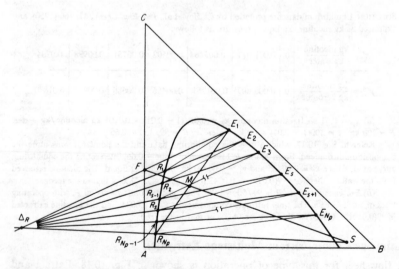

Figure 10.19 Countercurrent multistage extraction.

According to Eq. (10.25), the extended lines E_1F and SR_{N_p} must intersect at Δ_R, as shown in Fig. 10.19. This intersection may in some cases be located at the right of the triangle. A material balance for stages s through N_p is

$$R_{s-1} + S = R_{N_p} + E_s \tag{10.26}$$

or

$$R_{N_p} - S = R_{s-1} - E_s = \Delta_R \tag{10.27}$$

so that the difference in flow rates at a location between any two adjacent stages is constant, Δ_R. Line E_sR_{s-1} extended must therefore pass through Δ_R, as on the figure.

The graphical construction is then as follows. After location of points F, S, M, E_1, R_{N_p}, and Δ_R, a tie line from E_1 provides R_1 since extract and raffinate from the first theoretical stage are in equilibrium. A line from Δ_R through R_1 when extended provides E_2, a tie line from E_2 provides R_2, etc. The lowest possible value of x_{N_p} is that given by the A-rich end of a tie line which, when extended, passes through S.

As the amount of solvent is increased, point M representing the overall plant balance moves toward S on Fig. 10.19 and point Δ_R moves farther to the left. At an amount of solvent such that lines E_1F and SR_{N_p} are parallel, point Δ_R will be at an infinite distance. Greater amounts of solvent will cause these lines to intersect on the right-hand side of the diagram rather than as shown, with point Δ_R nearer B for increasing solvent quantities. The interpretation of the difference point is, however, still the same: a line from Δ_R intersects the two branches of the solubility curve at points representing extract and raffinate from adjacent stages.

If a line from point Δ_R should coincide with a tie line, an infinite number of stages will be required to reach this condition. The maximum amount of solvent for which this occurs corresponds to the minimum solvent/feed ratio which can be used for the specified products. The procedure for determining the minimum amount of solvent is indicated in Fig. 10.20. All tie lines below that marked JK

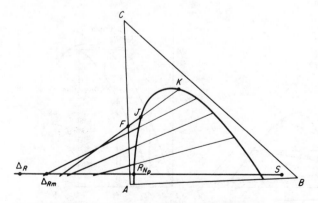

Figure 10.20 Minimum solvent for countercurrent extraction.

are extended to line SR_{N_p}, to give intersections with line SR_{N_p} as shown. The intersection farthest from S (if on the left-hand side of the diagram) or nearest S (if on the right) represents the difference point for minimum solvent, as at point Δ_{R_m} (Fig. 10.20). The actual position of Δ_R must be farther from S (if on the left) or nearer to S (if on the right) for a finite number of stages. The larger the amount of solvent, the fewer the number of stages. Usually, but not in the instance shown, the tie line which when extended passes through F, that is, tie line JK, will locate Δ_{R_m} for minimum solvent.

When the number of stages is very large, the construction indicated in Fig. 10.21 may be more convenient. A few lines are drawn at random from point Δ_R to intersect the two branches of the solubility curve as shown, where the intersections do not now necessarily indicate streams between two actual adjacent stages. The C concentrations x_s and y_{s+1} corresponding to these are plotted on x, y coordinates as shown to provide an operating curve. Tie-line data provide the equilibrium curve y^* vs. x, and the theoretical stages are stepped off in the manner used for gas absorption and distillation.

Figure 10.22 shows the construction for solvent-free coordinates. The B-free material balance for the entire plant is

$$F' + S' = E'_1 + R'_{N_p} = M' \tag{10.28}$$

where ordinarily $F = F'$ since the feed is usually B-free. M' is therefore on the line FS at X_M calculated by a C balance

$$F'X_F + S'Y_S = M'X_M \tag{10.29}$$

If pure B is used as solvent, $S' = 0$, $S'Y_S = 0$, $F' = M'$, $X_M = X_F$, and point M is vertically above F.

Line $E_1R_{N_p}$ must pass through M. Then

$$R'_{N_p} - S' = F' - E'_1 = \Delta'_R \tag{10.30}$$

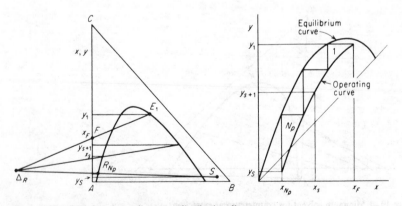

Figure 10.21 Transfer of coordinates to distribution diagram.

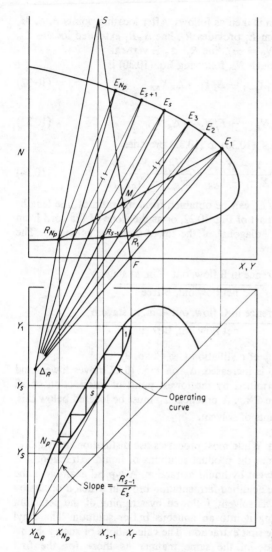

Figure 10.22 Countercurrent extraction on solvent-free coordinates.

The balance for stages s through N_p is

$$R'_{N_p} - S' = R'_{s-1} - E'_s = \Delta'_R \qquad (10.31)$$

where Δ'_R is the difference in solvent-free flow, out minus in, at stage N_p, and the constant difference in solvent-free flows of the streams between any two adjacent stages. Line $E_s R_{s-1}$ extended, where s is any stage, must therefore pass through Δ_R on the graph.

The graphical construction is then as follows. After locating points F, S, M, E_1, R_{N_p}, and Δ_R, a tie line from E_1 provides R_1, line $\Delta_R R_1$ extended locates E_2, etc. If the solvent is pure B ($N_S = \infty$), line $R_{N_p} \Delta_R$ is vertical. A C balance, stages s through N_p, following Eq. (10.30) is

$$R'_{s-1}X_{s-1} - E'_s Y_s = \Delta'_R X_{\Delta R} \tag{10.32}$$

A B balance is

$$R'_{s-1}N_{R_{s-1}} - E'_s N_{E_s} = \Delta'_R N_{\Delta R} \tag{10.33}$$

Elimination of Δ'_R between Eqs. (10.30) to (10.32) provides

$$\frac{R'_{s-1}}{E'_s} = \frac{Y_s - X_{\Delta R}}{X_{s-1} - X_{\Delta R}} = \frac{N_{E_s} - N_{\Delta R}}{N_{R_{s-1}} - N_{\Delta R}} \tag{10.34}$$

Thus, the ratio of flows R'_{s-1}/E'_s can be obtained from the ratio of line lengths $E_s \Delta_R / R_{s-1} \Delta_R$ on the upper part of Fig. 10.22, or as the slope of the cord from (X_{s-1}, Y_s) to $X_{\Delta R}$ on the 45° diagonal of the lower diagram, as shown. The difference-point coordinates are

$$\Delta'_R \begin{cases} N_{\Delta R} = \dfrac{\text{difference in B flow, out } - \text{ in, at stage } N_p}{\text{net flow out, B-free}} \\[2em] X_{\Delta R} = \dfrac{\text{difference in C flow, out } - \text{ in, at stage } N_p}{\text{net flow out, B-free}} \end{cases}$$

and Δ'_R is analogous to the Δ_W of distillation (see Chap. 9).

As the solvent/feed ratio is increased, Δ_R on Fig. 10.22 moves lower, and the minimum solvent is determined by the lowest point of intersection of all extended tie lines with the line SR_{N_p}. A practical Δ_R must be located below this, corresponding to larger amounts of solvent.

Solvent recovery by extraction While most processes use distillation or evaporation to recover the solvent from the product solutions of liquid extraction, it is not uncommon to recover solvent by liquid extraction. A typical example is the recovery of penicillin from the acidified fermentation broth in which it occurs by extraction with amyl acetate as solvent, followed by stripping of the penicillin from the solvent by extracting it into an aqueous buffer solution. The amyl acetate is then returned to the first extraction. The calculations of such solvent-recovery operations are made in the same manner as those for the first extraction.

Illustration 10.3 If 8000 kg/h of an acetic acid (C)–water (A) solution, containing 30% acid, is to be countercurrently extracted with isopropyl ether (B) to reduce the acid concentration to 2% in the solvent-free raffinate product, determine (a) the minimum amount of solvent which can be used and (b) the number of theoretical stages if 20 000 kg/h of solvent is used.

SOLUTION The equilibrium data of Illustration 10.1 are plotted on triangular coordinates in Fig. 10.23. The tie lines have been omitted for clarity.

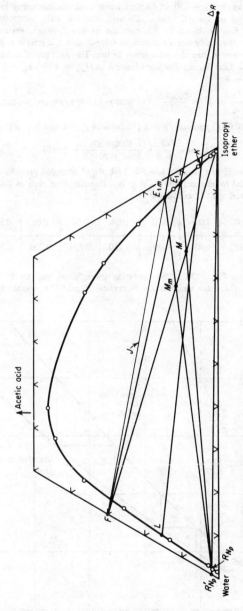

Figure 10.23 Solution to Illustration 10.3.

Isopropyl ether

Acetic acid

Water

(a) $F = 8000$ kg/h; $x_F = 0.30$ wt fraction acetic acid, corresponding to point F on the figure. R'_{N_p} is located on the AC base at 2% acid, and line BR'_{N_p} intersects the water-rich solubility curve at R_{N_p}, as shown. In this case the tie line J, which when extended passes through F, provides the conditions for minimum solvent, and this intersects line $R_{N_p}B$ on the right of the figure nearer B than any other lower tie line. Tie line J provides the minimum E_1 as shown at $y_1 = 0.143$. Line $E_{1m}R_{N_p}$ intersects line FB at M_m, for which $x_M = 0.114$. Eq. (10.24), with $y_S = 0$ and $S = B$:

$$B_m = \frac{Fx_F}{x_m} - F = \frac{8000(0.30)}{0.114} - 8000 = 13\,040 \text{ kg/h, min solvent rate}$$

(b) For $B = 20\,000$ kg solvent/h [Eq. (10.24) with $y_S = 0$ and $S = B$],

$$x_M = \frac{Fx_F}{F + B} = \frac{8000(0.30)}{8000 + 20\,000} = 0.0857$$

and point M is located as shown on line FB. Line $R_{N_p}M$ extended provides E_1 at $y_1 = 0.10$. Line FE_1 is extended to intersect line $R_{N_p}B$ at Δ_R. Random lines such as OKL are drawn to provide y_{s+1} at K and x_s at L, as follows:

y_{s+1}	0	0.01	0.02	0.04	0.06	0.08	$0.10 = y_1$
x_s	0.02	0.055	0.090	0.150	0.205	0.250	$0.30 = x_F$

These are plotted on Fig. 10.24 as the operating curve, along with the tie-line data as the equilibrium curve. There are required 7.6 theoretical stages. The weight of extract can be

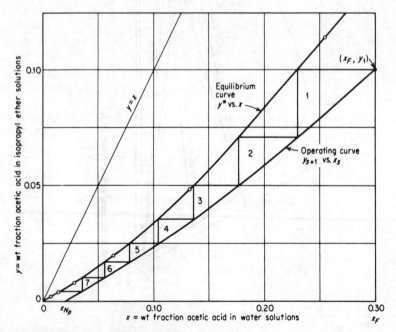

Figure 10.24 Solution to Illustration 10.3.

obtained by an acid balance,

$$E_1 = \frac{M(x_M - x_{N_p})}{y_1 - x_{N_p}} = \frac{28\,000(0.0857 - 0.02)}{0.10 - 0.02} = 23\,000\ \text{kg/h}$$

and $\qquad R_{N_p} = M - E_1 = 28\,000 - 23\,000 = 5000\ \text{kg/h}$

Insoluble liquids When the liquids A and B are insoluble over the range of solute concentrations encountered, the stage computation is made more simply on x', y' coordinates. For this case, the solvent content of all extracts and the A content of all raffinates are constant. An overall plant balance for substance C is

$$By'_S + Ax'_F = Ax'_{N_p} + By'_1 \tag{10.35}$$

or $\qquad \dfrac{A}{B} = \dfrac{y'_1 - y'_S}{x'_F - x'_{N_p}} \tag{10.36}$

which is the equation of a straight line, the operating line, of slope A/B, through points (y'_1, x'_F), (y'_S, x'_{N_p}). For stages 1 through s, similarly,

$$\frac{A}{B} = \frac{y'_1 - y'_{s+1}}{x'_F - x'_s} \tag{10.37}$$

and Fig. 10.25 shows the construction for stages. x' and y' can also be expressed as mass/volume, with A and B as volume/time.

For the special case where the equilibrium curve is of constant slope $m' = y'^*/x'$, Eq. (5.50) applies,

$$\frac{x'_F - x'_{N_p}}{y'_F - y'_S/m'} = \frac{(m'B/A)^{N_p+1} - m'B/A}{(m'B/A)^{N_p+1} - 1} \tag{10.38}$$

where $m'B/A$ is the extraction factor. This can be used in conjunction with Fig. 5.16, with $(x'_{N_p} - y'_S/m')/(x'_F - y'_S/m')$ as ordinate and $m'B/A$ as parameter.

Equation (8.16) for overall efficiency E_O can be used for extraction under the following conditions:

1. Replace E_{MGE} with E_{ME}, and $1/A$ with $m'B/A$
2. Replace E_{MGR} with E_{MR} and A with $A/m'B$

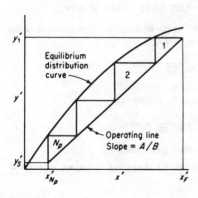

Figure 10.25 Countercurrent extraction, insoluble solvents.

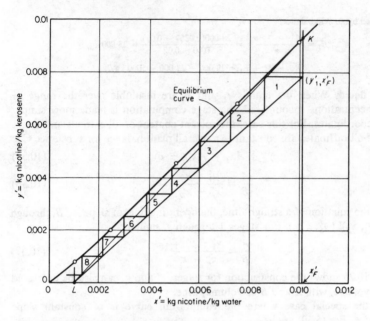

Figure 10.26 Solution to Illustration 10.6.

Illustration 10.4 If 1000 kg/h of a nicotine (C)–water (A) solution containing 1% nicotine is to be countercurrently extracted with kerosene at 20°C to reduce the nicotine content to 0.1%, determine (a) the minimum kerosene rate and (b) the number of theoretical stages required if 1150 kg of kerosene is used per hour.

SOLUTION The equilibrium data of Illustration 10.2 are plotted in Fig. 10.26.

(a) $F = 1000$ kg/h, $x_F = 0.01$, $A = 1000(1 - 0.01) = 990$ kg water/h, $y_S = 0$.

$$x_F' = \frac{0.01}{1 - 0.01} = 0.0101 \text{ kg nicotine/kg water}$$

$$x_{N_p} = 0.001 \qquad x_{N_p}' = \frac{0.001}{1 - 0.001} = 0.001\,001 \text{ kg nicotine/kg water}$$

The operating line starts at point L ($y' = 0$, $x' = 0.001\,001$) and for infinite stages passes through K on the equilibrium curve at x_F'. Since $y_K = 0.0093$,

$$\frac{A}{B_m} = \frac{0.0093 - 0}{0.0101 - 0.001\,001} = 1.021$$

and $B_m = A/1.021 = 990/1.021 = 969$ kg kerosene/h.

(b) $B = 1150$ kg/h, $A/B = 990/1150 = 0.860$. Eq. (10.36):

$$\frac{y_1'}{x_F' - x_{N_p}'} = \frac{y_1'}{0.0101 - 0.001\,001} = 0.860$$

$$y_1' = 0.00782 \text{ kg nicotine/kg kerosene}$$

The operating line is drawn through (y_1', x_F'), and 8.3 theoretical stages are determined graphically.

Alternatively, at the dilute end of the system, $m' = dy'^*/dx' = 0.798$, and

$$\frac{m'B}{A} = \frac{0.798(1150)}{990} = 0.928$$

At the concentrated end, $m' = 0.953$, and $m'B/A = 0.953(1150)/990 = 1.110$. The average is $[0.928(1.110)]^{0.5} = 1.01$. $x_{N_p}'/x_F' = 0.001\ 001/0.0101 = 0.099$, and Fig. 5.16 indicates 8.4 theoretical.stages.

Continuous Countercurrent Extraction with Reflux

Whereas in ordinary countercurrent operation the richest possible extract product leaving the plant is at best only in equilibrium with the feed solution, the use of reflux at the extract end of the plant can provide a product even richer, as in the case of the rectifying section of a distillation column. Reflux is not needed at the raffinate end of the cascade since, unlike distillation, where heat must be carried in from the reboiler by a vapor reflux, in extraction the solvent (the analog of heat) can enter without a carrier stream.

An arrangement for this is shown in Fig. 10.27. The feed to be separated into its components is introduced at an appropriate place into the cascade, through which extract and raffinate liquids are passing countercurrently. The concentration of solute C is increased in the extract-enriching section by countercurrent contact with a raffinate liquid rich in C. This is provided by removing the solvent from extract E_1 to produce the solvent-free stream E', part of which is removed as extract product P_E' and part returned as reflux R_0. The raffinate-stripping section of the cascade is the same as the countercurrent extractor of Fig. 10.18, and C is stripped from the raffinate by countercurrent contact with solvent.

Graphical determination of stages required for such operations is usually inconvenient to carry out on triangular coordinates [72] because of crowding, and only the use of N, X, Y coordinates will be described. For the stages in the raffinate stripping section (Fig. 10.27) the developments for simple countercurrent extraction, Eqs. (10.31) to (10.34) apply, and this section of the plant needs no further consideration.

In the extract-enriching section, an A + C balance about the solvent separator is

$$E_1' = E' = P_E' + R_0' \qquad (10.39)$$

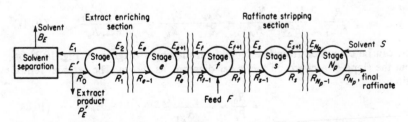

Figure 10.27 Countercurrent extraction with reflux.

Let Δ_E represent the net rate of flow outward from this section. Then, for its A + C content

$$\Delta'_E = P'_E \qquad (10.40)$$

and for its C content

$$X_{\Delta E} = X_{PE} \qquad (10.41)$$

while for its B content

$$B_E = \Delta'_E N_{\Delta E} \qquad (10.42)$$

The point Δ_E is plotted on Fig. 10.28, which is drawn for a system of two partly miscible component pairs. For all stages through e, an A + C balance is

$$E'_{e+1} = P'_E + R'_e = \Delta'_E + R'_e \qquad (10.43)$$

or

$$\Delta'_E = E'_{e+1} - R'_e \qquad (10.44)$$

A C balance is

$$\Delta'_E X_{\Delta E} = E'_{e+1} Y_{e+1} - R'_e X_{R_e} \qquad (10.45)$$

and a B balance is

$$\Delta'_E N_{\Delta E} = E'_{e+1} N_{E_{e+1}} - R'_e N_{R_e} \qquad (10.46)$$

Since e is any stage in this section, lines radiating from point Δ_E cut the solubility curves of Fig. 10.28 at points representing extract and raffinate flowing between any two adjacent stages. Δ_E is therefore a difference point, constant for all stages in this section, whose coordinates mean

$$\Delta'_E \begin{cases} N_{\Delta E} = \dfrac{\text{difference in B flow, out} - \text{in}}{\text{net flow out, B-free}} = \dfrac{\text{B out}}{(\text{A} + \text{C}) \text{ out}} \\[4mm] X_{\Delta E} = \dfrac{\text{difference in C flow, out} - \text{in}}{\text{net flow out, B-free}} = \dfrac{\text{C out}}{(\text{A} + \text{C}) \text{ out}} \end{cases}$$

Alternating tie lines and lines from Δ_E then establish the stages, starting with stage 1 and continuing to the feed stage. Figure 10.28 also shows the raffinate-stripping stages, developed as before for countercurrent extraction.

Solving Eq. (10.43) with (10.45) and (10.46) produces the internal reflux ratio at any stage,

$$\frac{R'_e}{E'_{e+1}} = \frac{N_{\Delta E} - N_{E_{e+1}}}{N_{\Delta E} - N_{R_e}} = \frac{X_{\Delta E} - Y_{e+1}}{X_{\Delta E} - X_{R_e}} = \frac{\text{line } \Delta_E E_{e+1}}{\text{line } \Delta_E R_e} \qquad (10.47)$$

This can be computed from line lengths on the upper diagram of Fig. 10.28 or from the slope of a chord on the lower diagram, as shown. The external reflux ratio is

$$\frac{R'_0}{P'_E} = \frac{R_0}{P_E} = \frac{N_{\Delta E} - N_{E1}}{N_{E1}} \qquad (10.48)$$

which can be used to locate Δ_E for any specified reflux ratio.

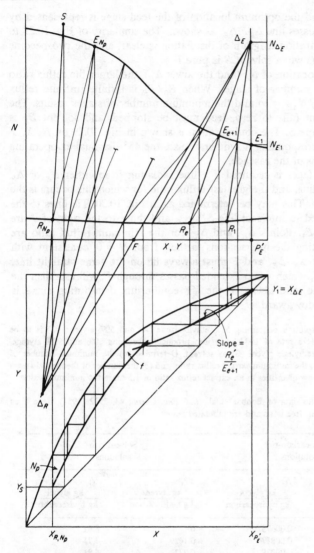

Figure 10.28 Countercurrent extraction with reflux.

Material balances may also be made over the entire plant. For A + C

$$F' + S' = P'_E + R'_{N_p}$$ (10.49)

If Eqs. (10.30) and (10.40) are used, this becomes

$$F' = \Delta'_R + \Delta'_E$$ (10.50)

where normally $F = F'$. Point F must therefore lie on a line joining the two difference points, and the optimum location of the feed stage is represented by the tie line which crosses line $\Delta_R F \Delta_E$, as shown. The similarity of Fig. 10.28 to the enthalpy-concentration diagram of distillation is clear, and the two become completely analogous when solvent S is pure B.

The higher the location of Δ_E (and the lower Δ_R), the larger the reflux ratio and the smaller the number of stages. When R_0/P'_E is infinite (infinite reflux ratio or total reflux), $N_{\Delta E} = \infty$ and the minimum number of stages results. The capacity of the plant falls to zero, feed must be stopped, and solvent B_E is recirculated to become S. The configuration is shown in Fig. 10.29 on N, X, Y coordinates. The corresponding XY diagram uses the 45° diagonal as operating lines for both sections of the cascade.

An infinity of stages is required if a line radiating from either Δ_E or Δ_R coincides with a tie line, and the greatest reflux ratio for which this occurs is the *minimum reflux ratio*. This may be determined as in Fig. 10.30. Tie lines to the left of J are extended to intersect line SR_{N_p+1}, and those to the right of J are extended to line $P'_E E_1$. Points Δ_{E_m} and Δ_{R_m} for the minimum reflux ratio are established by selecting the intersections farthest from $N = 0$, consistent with the requirement that Δ_R, Δ_E, and F must always lie on the same straight line. Frequently tie line J, which when extended passes through F, will establish the minimum reflux ratio, and always if the XY equilibrium distribution curve is everywhere concave downward.

Illustration 10.5 A solution containing 50% ethylbenzene (A) and 50% styrene (C) is to be separated at 25°C at a rate of 1000 kg/h into products containing 10% and 90% styrene, respectively, with diethylene glycol (B) as solvent. Determine (a) the minimum number of theoretical stages, (b) the minimum extract reflux ratio, and (c) the number of theoretical stages and the important flow quantities at an extract reflux ratio of 1.5 times the minimum value.

SOLUTION Equilibrium data of Boobar et al., *Ind. Eng. Chem.*, **43**, 2922 (1951), have been converted to a solvent-free basis and are tabulated below.

Hydrocarbon-rich solutions		Solvent-rich solutions	
X, $\dfrac{\text{kg styrene}}{\text{kg hydrocarbon}}$	N, $\dfrac{\text{kg glycol}}{\text{kg hydrocarbon}}$	Y^*, $\dfrac{\text{kg styrene}}{\text{kg hydrocarbon}}$	N, $\dfrac{\text{kg glycol}}{\text{kg hydrocarbon}}$
0	0.00675	0	8.62
0.0870	0.00817	0.1429	7.71
0.1883	0.00938	0.273	6.81
0.288	0.01010	0.386	6.04
0.384	0.01101	0.480	5.44
0.458	0.01215	0.557	5.02
0.464	0.01215	0.565	4.95
0.561	0.01410	0.655	4.46
0.573	0.01405	0.674	4.37
0.781	0.01833	0.833	3.47
1.00	0.0256	1.00	2.69

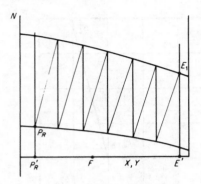

Figure 10.29 Total reflux.

These are plotted on the solvent-free coordinate system of Fig. 10.31. The tie lines corresponding to these points are not drawn in, for clarity. $F = 1000$ kg/h, $X_F = 0.5$ wt fraction styrene, $X_{P'E} = 0.9$, $X_{R, N_p} = 0.1$, all on a solvent-free basis. Point E_1 is located as shown. $N_{E1} = 3.10$.

(a) Minimum theoretical stages are determined by drawing a tie line from E_1, a vertical line from R_1, a tie line from E_2, etc., until the raffinate product is reached, as shown. The minimum number of stages, corresponding to the constructed tie lines, is 9.5.

(b) The tie line which when extended passes through F provides the minimum reflux ratios, since it provides intersections Δ_{E_m} and Δ_{R_m} farthest from the line $N = 0$. From the plot,

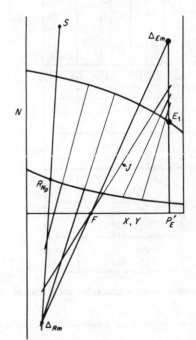

Figure 10.30 Determination of minimum reflux ratio.

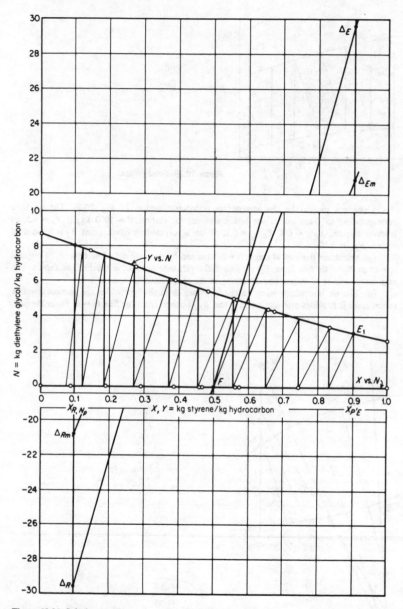

Figure 10.31 Solution to Illustration 10.5. Minimum theoretical stages at total reflux and location of difference points.

$N_{\Delta E_m} = 20.76$. Eq. (10.48):

$$\left(\frac{R_0}{P_E'}\right)_m = \frac{20.76 - 3.1}{3.1} = 5.70 \text{ kg reflux/kg extract product}$$

(c) For $R_0/P_E' = 1.5(5.70) = 8.55$ lb reflux/kg extract product [Eq. (10.48)],

$$8.55 = \frac{N_{\Delta E} - 3.1}{3.1}$$

and $N_{\Delta E} = 29.6$. Point Δ_E is plotted as shown. A straight line from Δ_E through F intersects line $X = 0.10$ at Δ_R. $N_{\Delta R} = -29.6$. Random lines are drawn from Δ_E for concentrations to the right of F, and from Δ_R for those to the left, and intersections of these with the solubility curves provide the coordinates of the operating curve (Fig. 10.32). The tie-line data plotted directly provide the equilibrium curve. The number of theoretical stages is seen to be 15.5, and the feed is to be introduced into the seventh from the extract-product end of the cascade.

From Fig. 10.31, $X_{R, N_p} = 0.10$, $N_{R, N_p} = 0.0082$. On the basis of 1 h, an overall plant balance is

$$F = 1000 = P_E' + R_{N_p}'$$

A C balance is

$$FX_F = 500 = P_E'(0.9) + R_{N_p}'(0.1)$$

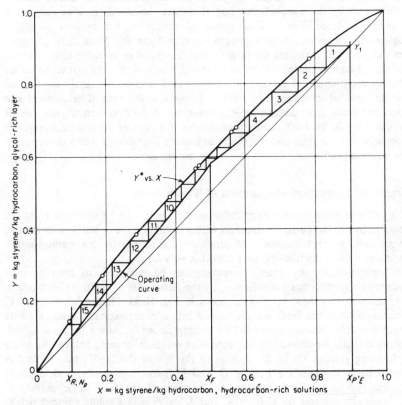

Figure 10.32 Solution to Illustration 10.5.

Solving simultaneously, we get $P'_E = R'_{N_p} = 500$ kg/h

$$R_0 = R'_0 = 8.55P'_E = 8.55(500) = 4275 \text{ kg/h}$$

Eq. (10.39):

$$E'_1 = R'_0 + P'_E = 4275 + 500 = 4775 \text{ kg/h}$$

$$B_E = E'_1 N_{E1} = 4775(3.10) = 14\,800 \text{ kg/h}$$

$$E_1 = B_E + E'_1 = 14\,800 + 4775 = 19\,575 \text{ kg/h}$$

$$R_{N_p} = R'_{N_p}(1 + N_{R, N_p}) = 500(1.0082) = 504 \text{ kg/h}$$

$$S = B_E + R'_{N_p} N_{R, N_p} = 14\,800 + 500(0.0082) = 14\,804 \text{ kg/h}$$

Economic Balances

Several types of economic balance can be made for the various flowsheets just described [71, 72]. For example, the amount of solute extracted for a fixed solvent/feed ratio increases with increased number of stages, and therefore the value of the unextracted solute can be balanced against the cost of the extraction equipment required to recover it. The amount of solvent per unit of feed, or reflux ratio in the case of the last flowsheet described, is also subject to economic balance. For a fixed extent of extraction, the number of stages required decreases as solvent rate or reflux ratio increases. Since the capacity of the equipment for handling the larger liquid flow must at the same time increase, the cost of equipment must then pass through a minimum. The extract solutions become more dilute as solvent rate is increased, and consequently the cost of solvent removal increases. The total cost, which is the sum of investment and operating costs, must pass through a minimum at the optimum solvent rate or reflux ratio. In all such economic balances, the cost of solvent recovery will always be a major item and usually must include consideration of recovery from the saturated raffinate product as well as the extract.

Fractional Extraction—Separation of Two Solutes

If a solution contains two extractable solutes, both can be removed from the solution by countercurrent extraction with a suitable solvent, but it is impossible to produce any great degree of separation of the solutes by this method unless the ratio of their distribution coefficients is very large.

Separation to any extent, however, can be achieved, so long as their distribution coefficients are different, by the techniques of fractional extraction. The simplest flowsheet for this is shown in Fig. 10.33. Here solutes B and C, which constitute the feed, are introduced into a countercurrent cascade where partly miscible solvents A and D flow countercurrently. At the feed stage, both solutes distribute between the solvents, with solute B favoring solvent A, solute C favoring solvent D. In the section to the left of the feed stage, solvent A preferentially extracts the B from D, and D leaves this section with a solute content rich in C. In the section to the right of the feed stage, solvent D preferentially extracts the C from A, and A leaves with a solute content rich in B.

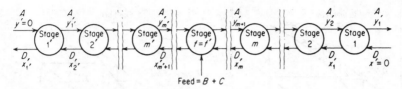

Feed = $B + C$

Figure 10.33 Fractional extraction.

This operation will be considered only in its simplest form, where the solvents A and D are essentially immiscible and enter free of solutes and where solutes B and C distribute independently; i.e., the distribution of each is uninfluenced by the presence of the other. A and D will represent the weight rates of flow of these solvents, mass/time, and concentrations will be expressed as

$$y' = \text{mass solute/mass A } (y'_B \text{ for solute B, } y'_C \text{ for solute C})$$

$$x' = \text{mass solute/mass D } (x'_B \text{ for solute B, } x'_C \text{ for solute C})$$

Consider the stages $1'$ through m'. For *either* solute, a solute material balance is

$$Dx'_{m'+1} = Ay'_{m'} + Dx'_{1'} \qquad (10.51)$$

or

$$\frac{D}{A} = \frac{y'_{m'} - 0}{x'_{m'+1} - x'_{1'}} \qquad (10.52)$$

and we need only add the subscripts B or C to the concentrations to apply these to particular solutes. Similarly, for stages 1 through m, for *either* solute, the balance is

$$Ay'_{m+1} = Dx'_m + Ay'_1 \qquad (10.53)$$

$$\frac{D}{A} = \frac{y'_{m+1} - y'_1}{x'_m - 0} \qquad (10.54)$$

Equations (10.52) and (10.54) represent the operating lines for each solute, one for each section of the cascade, of slope D/A. Figure 10.34 shows the distribution diagrams for the solutes, each of which distributes independently of the other. The operating lines are also drawn on the figure, and the stages stepped off in the usual manner, beginning at stages 1 and $1'$.

To determine the number of stages required in each section of the cascade, a match of concentrations and stage numbers at the point of feed introduction must be made. At the feed stage (which is numbered f and f') for either solute, f must be the same, f' must be the same, and the concentration must be the same when computed from each end of the cascade. The match can be established by plotting the concentrations of both B and C in one of the solvents against stage number, as in Fig. 10.35. The requirements listed above are then met where the rectangle $HJKL$ can be drawn in the manner shown.

It can be seen from Fig. 10.35 that the concentrations of both solutes is greatest at the feed stage, least at stages 1 and $1'$. Sufficient rates of solvent flow must be used to ensure that solute solubilities are not exceeded at the feed stage. Changing the solvent ratio changes the total number of stages and the relative

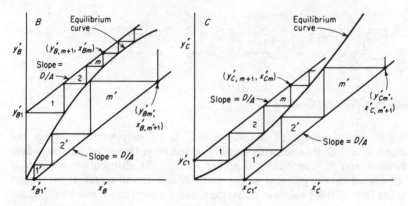

Figure 10.34 Fractional extraction, operating lines and stages.

position of the feed stage, and there is always a solvent ratio for which the total number of stages is the least.

The number of stages can be reduced by use of solute reflux at either or both ends of the cascade. Computation of this effect, treatment of cases where solute distributions are interdependent or solvent miscibility varies, and other special flowsheets are beyond the scope of this book but have been thoroughly worked out [56, 72].

Illustration 10.6 A mixture containing 40% *p*-chloronitrobenzene (B) and 60% *o*-chloronitrobenzene (C) is to be separated at the rate of 100 kg/h into products containing 80 and 15%,

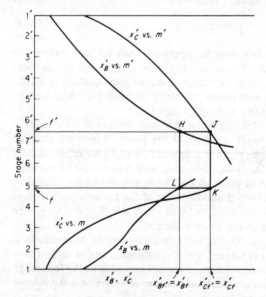

Figure 10.35 Feed-stage matching.

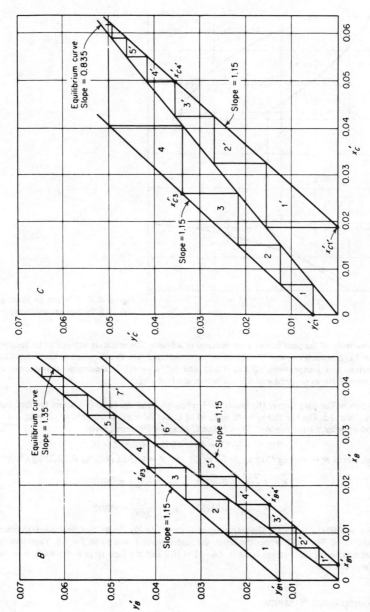

Figure 10.36 Solution to Illustration 10.6.

517

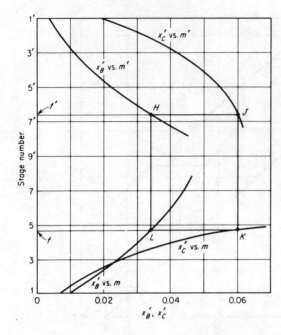

Figure 10.37 Solution to Illustration 10.6, feed-stage matching.

respectively, of the para isomer after removal of solvents. The insoluble solvents to be used are 2400 kg/h heptane (A) and 2760 kg/h aqueous methanol (D). The distribution coefficients are constant and independent, $y_B'^*/x_B' = 1.35$ and $y_C'^*/x_C' = 0.835$. Determine the number of theoretical stages required and the position of the feed stage.

SOLUTION The para isomer (B) favors the heptane (A), and the ortho (C) favors the methanol (D). *Basis:* 1 h. Feed = 100 kg = 40 kg B, 60 kg C. Let W = kg A-rich product, Z = kg D-rich product, after solvent removal. Then B and C balances are, respectively,

$$0.80W + 0.15Z = 40 \qquad 0.20W + 0.85Z = 60$$

from which W = 38.5 kg (30.8 kg B, 7.7 kg C), and Z = 61.5 kg (9.23 kg B, 52.27 kg C).

$$x_{B1}' = \frac{9.23}{2760} = 0.00334 \qquad x_{C1}' = \frac{52.27}{2760} = 0.01895$$

$$y_{B1}' = \frac{30.8}{2400} = 0.01283 \qquad y_{C1}' = \frac{7.7}{2400} = 0.00321$$

D/A = 2760/2400 = 1.15. The stages are constructed on Fig. 10.36, and the feed-stage match is shown on Fig. 10.37. From the latter, the feed stage is f' = 6.6 and f = 4.6. Therefore the total number of ideal stages is 6.6 + 4.6 − 1 = 10.2 and the feed stage is the 4.6th from the solvent-D inlet.

Multicomponent Systems

For most systems containing more than four components, the display of equilibrium data and the computation of stages is very difficult. Some of our most important separations involve hundreds of components whose identity is not even firmly established, as in the extraction of petroleum lubricating oils. In such cases, the stage requirements are best obtained in the laboratory, without detailed study of the equilibria. There are available small-scale mixer-settler extractors of essentially 100 percent stage efficiency which can be used to study extraction processes [72], but they must be run continuously to steady state. Countercurrent flowsheets, which are necessarily for continuous operation, can be simulated with small batches of feed and solvent in separatory funnels, however, and this is frequently most convenient and least costly.

Suppose, for example, it is desired to determine what products will be obtained from a five-stage countercurrent extraction of a complex feed flowing at F kg/h with a solvent flowing at S kg/h. Refer to Fig. 10.38. Here each circle represents a separatory funnel which, when adequately shaken and the liquids allowed to settle, will represent one theoretical stage. Into funnel a are placed suitable amounts of feed and solvent in the ratio F/S, the mixture is shaken,

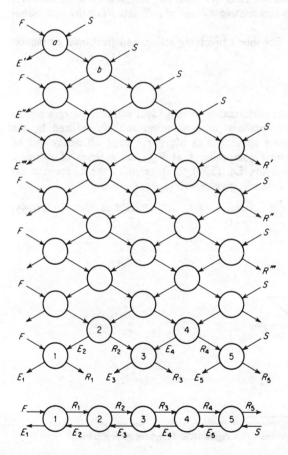

Figure 10.38 Batch simulation of a countercurrent cascade.

settled, and the two phases separated. Extract layer E' is removed and raffinate R_a is contacted in funnel b with precisely the same amount of solvent S as used in funnel a. Raffinates and extracts, feeds and solvents, are moved and contacted in the manner indicated in the figure, care being taken each time to use precisely the same amount of feed and solvent at streams marked F and S as in funnel a.

Clearly extract E' and raffinate R' will not be the same as those produced by a continuous countercurrent extraction. Subsequent raffinates R'', R''', and extracts E'', E''', etc., will approach the desired result, however. One can follow the approach to steady state by observing any conveniently measured property of the raffinates and extracts (density, refractive index, etc.). The properties of R', R'', etc., will approach a constant value, as will those of E', E'', etc. If this has occurred by the time extract E_1 and raffinate R_5 in the upper part of Fig. 10.38 is reached, funnels 1 through 5 represent in every detail the stages of the continuous cascade just below. Thus, for example, raffinate R_3 from funnel 3 will have all the properties and relative volume of raffinate R_3 of the continuous plant.

In a similar manner, flowsheets involving reflux and fractional extraction can also be studied [72].

Stage Efficiency

As in gas-liquid contact, the performance of individual extraction stages can be described in terms of the approach to equilibrium actually realized by the effluent extract and raffinate streams. The Murphree stage efficiency can be expressed in terms of extract compositions as E_{ME} or in terms of raffinate compositions as E_{MR}. Applying Eq. (5.39) to stage m of the countercurrent cascade of Fig. 10.39, for example, gives

$$E_{ME} = \frac{y_m - y_{m+1}}{y_m^* - y_{m+1}} \qquad E_{MR} = \frac{x_{m-1}x_m}{x_{m-1} - x_m^*} \qquad (10.55)$$

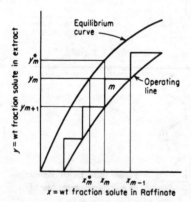

$x =$ wt fraction solute in Raffinate   **Figure 10.39** Murphree stage efficiency.

where x_m and y_m represent the actual average effluent compositions and y_{m+1} and x_{m-1} those of the streams entering the stage. Any other consistent set of concentration units can equally well be used. The overall stage efficiency E_O of a cascade is defined simply as the ratio of the number of theoretical stages to the number of real stages required to bring about a given concentration change.

STAGE-TYPE EXTRACTORS

Stage-type equipment of two major types is used: (1) single-stage mixer-settlers and multistage cascades constructed from them and (2) sieve-tray, multistage towers.

A *mixer-settler* is a single-stage device, ordinarily consisting of two parts, a mixer for contacting the two liquid phases to bring about mass transfer and a settler for their mechanical separation. Operation may be continuous or batchwise.

Mixers are of two types: flow mixers and mixing vessels. *Flow mixers*, or *line mixers*, are devices of very small volume placed in a pipeline, such as a series of orifices or mixing nozzles [72, 76] through which the two liquids to be contacted are pumped cocurrently. The loss in mechanical energy corresponding to the pressure drop is in part used to disperse the liquids in each other. The resulting dispersion then passes to the settler. These devices are useful only for continuous operation and are limited in application: the degree of dispersion produced by a given device depends upon the flow rate, and since the specific interfacial area decays rapidly downstream from the mixer [40, 59] while the holding time is very short, mass transfer can be expected to be poor.

Agitated Vessels

These are of the type shown in the right-hand portion of Fig. 6.4*b* and in Fig. 6.4*c*. If there is a gas-liquid surface, they should be baffled to prevent formation of a vortex. They may be covered, operated full, in which case they may be baffled or not. For continuous operation, the liquids to be contacted may enter at the bottom and leave at the top; in some cascade arrangements the light and heavy liquids enter through the side wall near the top and bottom of the vessel, respectively, and leave through a port in the wall opposite the impeller. For batch operation, the mixing vessel itself may act as the settler after agitation is stopped.

Impellers are usually of the flat-blade turbine type, Fig. 6.3*b* and *c*, either centrally located in the vessel or nearer the bottom entry of the liquids. The impeller–tank-diameter ratio d_i/T is ordinarily in the range 0.25 to 0.33.

Dispersions The mixture of liquids produced consists of droplets of one liquid dispersed in a continuum of the other. Drop diameters are usually in the range 0.1 to 1 mm diameter. If they are too small, difficulty in subsequent settling will result. Ordinarily the liquid flowing at the smaller volume rate will be dispersed in the other, but some control can be exercised. For continuous operation the vessel should first be filled with the liquid to be continuous, and with agitation in progress the two liquids can then be introduced in the desired ratio. For batch operation, the liquid in which the impeller is immersed when at rest is usually continuous. It is ordinarily difficult to maintain ϕ_D, the volume fraction dispersed phase, at values much above 0.6 to 0.7, and attempts to increase it beyond this will frequently result in inversion of the dispersion (continuous phase becomes dispersed) [37, 49, 83].

Power for agitation Power for agitation of two-liquid-phase mixtures by disk flat-blade turbines is given by curves c (baffles) and g (unbaffled, full, no vortex) of Fig. 6.5, provided suitable average density ρ_M and viscosity μ_M are used [35]:

$$\rho_M = \rho_C \phi_C + \rho_D \phi_D \tag{10.56}$$

Baffled vessels:
$$\mu_M = \frac{\mu_C}{\phi_C}\left(1 + \frac{1.5\mu_D\phi_D}{\mu_C + \mu_D}\right) \tag{10.57}$$

Unbaffled vessels, no gas-liquid interface, no vortex:

$$\mu_M = \begin{cases} \dfrac{\mu_w}{\phi_w}\left(1 + \dfrac{6\mu_o\phi_o}{\mu_o + \mu_w}\right) & \phi_w > 0.4 \tag{10.58} \\[3mm] \dfrac{\mu_o}{\phi_o}\left(1 - \dfrac{1.5\mu_w\phi_w}{\mu_o + \mu_w}\right) & \phi_w < 0.4 \tag{10.59} \end{cases}$$

where the subscripts o and w refer to organic and aqueous liquids, respectively [the difference for Eqs. (10.58) and (10.59) may very well be related to which liquid is dispersed]. The viscosity of a mixture can exceed that of either constituent. For other types of turbines, see Ref. 46.

Dispersed-phase holdup At agitator-power levels which are relatively low, the holdup ϕ_D of the dispersed phase in the vessel will be less than its fraction of the feed mixture, $\phi_{DF} = q_{DF}/q_{CF} + q_{DF})$. The value of ϕ_D will approach that of ϕ_{DF} as agitator power increases. For ϕ_D/ϕ_{DF} in the range 0.5 to 0.95, the following provide estimates for the indicated circumstances [81]. Exact prediction cannot be expected because the range of the studies thus far completed is relatively limited. For continuous flow:

Baffled vessels, impeller power/vessel volume > 105 W/m^3 = 2.2 ft lb$_f$/ft$^3 \cdot$ s:

$$\frac{\phi_D}{\phi_{DF}} = 0.764\left(\frac{Pq_D\mu_C^2}{v_L\sigma^3 g_c}\right)^{0.300}\left(\frac{\mu_C^3}{q_D\rho_C^2\sigma g_c}\right)^{0.178}\left(\frac{\rho_C}{\Delta\rho}\right)^{0.0741}\left(\frac{\sigma^3\rho_C g_c^3}{\mu_C^4 g}\right)^{0.276}\left(\frac{\mu_D}{\mu_c}\right)^{0.136} \tag{10.60}$$

Unbaffled vessels, full, no gas-liquid surface, no vortex:

$$\frac{\phi_D}{\phi_{DF}} = 3.39\left(\frac{Pq_D\mu_C^2}{v_L\sigma^3 g_c}\right)^{0.247}\left(\frac{\mu_C^3}{q_D\rho_C^2\sigma g_c}\right)^{0.427}\left(\frac{\rho_C}{\Delta\rho}\right)^{0.430}\left(\frac{\sigma^3\rho_C g_c^3}{\mu_C^4 g}\right)^{0.401}\left(\frac{\mu_D}{\mu_c}\right)^{0.0987} \tag{10.61}$$

The quantities in parentheses are dimensionless. If φ_D/φ_{DF} computes to exceed 1.0, it should be taken as 1.0. Some anomalies can be expected for water dispersed in hydrocarbons.

Specific interfacial area and drop size Interfacial area in extraction vessels has been measured most generally by a light-transmittance technique [10]. It varies considerably with location in the vessel [58, 81]. Intense turbulence at the impeller tip results in small drop sizes there, and a is consequently large. The drops coalesce as they flow to the more quiescent regions removed from the impeller, and the specific area becomes smaller. Average specific area and drop size are related by

$$a = \frac{6\phi_D}{d_p} \tag{6.9}$$

There have been many studies in baffled vessels but relatively few in unbaffled ones. The measurements have been made in the absence of mass transfer, which will probably have an influence through coalescence rates of the drops, and usually with an organic liquid dispersed in an aqueous phase. The following are recommended: for baffled vessels [6]

$$\frac{d_p}{d_p^o} = 1 + 1.18\phi_D \left(\frac{\sigma^2 g_c^2}{d_p^o \mu_C^2 g} \right) \left(\frac{\mu_C^4 g}{\Delta\rho\, \sigma^3 g_c^3} \right)^{0.62} \left(\frac{\Delta\rho}{\rho_C} \right)^{0.05} \tag{10.62}$$

where d_p^o is given by

$$\frac{d_p^{o3} \rho_C^2 g}{\mu_C^2} = 29 \left(\frac{v_L^3 \rho_C^2 \mu_C g^4}{P^3 g_c^3} \right)^{0.32} \left(\frac{\sigma^3 \rho_C g_c^3}{\mu_C^4 g} \right)^{0.14} \tag{10.63}$$

The groups in parentheses are dimensionless. For unbaffled vessels, full, no gas-liquid surface, no vortex [81]†

$$d_p = 10^{-2.066 + 0.732\phi_D} \left(\frac{\mu_C}{\rho_C} \right)^{0.0473} \left(\frac{Pg_c}{v_L \rho_M} \right)^{-0.204} \left(\frac{\sigma g_c}{\rho_C} \right)^{0.274} \tag{10.64}$$

Mass-transfer coefficients There have been relatively few measurements of overall mass-transfer coefficients for liquid-liquid systems in agitated vessels [72, 76] and still fewer studies of k_C for the continuous phase [33, 42, 58], far too few for useful generalizations. On the other hand there have been fairly extensive studies of mass transfer from solid particles. For small solid particles, most closely related in size at least to the droplets in a well-agitated vessel, the recommended relation is [36]

$$Sh_C = \frac{k_{LC} d_p}{D_C} = 2 + 0.47 \left[d_p^{4/3} \left(\frac{Pg_c}{v_L} \right)^{1/3} \frac{\rho_C}{\mu_C}^{2/3} \right]^{0.62} \left(\frac{d_i}{T} \right)^{0.17} \left(\frac{\mu_C}{\rho_C D_C} \right)^{0.36} \tag{10.65}$$

The group in square brackets is a form of particle Reynolds number, as shown in Chap. 11. Values of k_{LC} for liquid particles are known to be larger than those for solids [42, 58], and it is suspected that drop coalescence and breakup may be responsible. In any event, Eq. (10.65) can be expected to provide conservative values.

For small circulating drops, of the sort encountered here, the dispersed-phase coefficient is given by [15]

$$k_{LD} = -\frac{d_p}{6\theta} \ln \left[\frac{3}{8} \sum_{n=1}^{\infty} B_n^2 \exp\left(-\frac{\lambda_n 64 D_D \theta}{d_p^2} \right) \right] \tag{10.66}$$

where θ is the residence time in the vessel. B_n and λ_n are eigenvalues given by Table 10.1. Then, in the absence of any chemical reaction or interfacial resistance (see Chap. 5)

$$\frac{1}{K_{LD}} = \frac{1}{k_{LD}} + \frac{1}{m_{CD} k_{LC}} \tag{10.67}$$

where m_{CD} is the distribution coefficient, dc_C/dc_D.

---

† The coefficient in Eq. (10.64) is for SI units. For units of feet, pounds force, pounds mass, and seconds the coefficient is $10^{-1.812 + 0.732\phi_D}$.

Table 10.1 Eigenvalues for circulating drops †

$k_C d_p / D_C$	λ_1	λ_2	λ_3	B_1	B_2	B_3
3.20	0.262	0.424		1.49	0.107	
5.33	0.386					
8.00	0.534					
10.7	0.680	4.92		1.49	0.300	
16.0	0.860	5.26		1.48	0.382	
21.3	0.982	5.63		1.47	0.428	
26.7	1.082	5.90	15.7	1.49	0.495	0.205
53.3	1.324	7.04	17.5	1.43	0.603	0.298
107	1.484	7.88	19.5	1.39	0.603	0.384
213	1.560	8.50	20.8	1.31	0.588	0.396
320	1.60	8.62	21.3	1.31	0.583	0.391
∞	1.656	9.08	22.2	1.29	0.596	0.386

† From E. L. Elzinga and J. T. Banchero: *Chem. Eng. Prog. Symp. Ser.*, **55**(29), 149 (1959), with permission.

Stage efficiency In view of the uncertainties in the mass transfer data, it is reasonable to use a simplified version of the number of transfer units [compare Eq. (8.43)]. For the dispersed phase,

$$N_{tOD} = \int_{c_{2D}}^{c_{1D}} \frac{dc_D}{c_D - c_D^*} \tag{10.68}$$

where the dispersed-phase concentration changes from c_{1D} to c_{2D} and c_D^* is the dispersed-phase concentration in equilibrium with that in the continuous phase. It has been established that the liquids in a well-agitated vessel are thoroughly back-mixed, so that everywhere the concentrations are constant at the effluent values [58]. Then

$$N_{tOD} = \frac{1}{c_{2D} - c_{2D}^*} \int_{c_{2D}}^{c_{1D}} dc_D = \frac{c_{1D} - c_{2D}}{c_{2D} - c_{2D}^*} \tag{10.69}$$

and
$$N_{tOD} = \frac{Z}{H_{tOD}} = \frac{Z}{V_D / K_{LD} a} \tag{10.70}$$

The Murphree dispersed-phase efficiency E_{MD} is given by Eq. (5.39) adapted to the present situation:

$$
\begin{aligned}
E_{MD} &= \frac{c_{1D} - c_{2D}}{c_{1D} - c_{2D}^*} = \frac{c_{1D} - c_{2D}}{(c_{1D} - c_{2D}) + (c_{2D} - c_{2D}^*)} \\
&= \frac{(c_{1D} - c_{2D})/(c_{2D} - c_{2D}^*)}{(c_{1D} - c_{2D})/(c_{2D} - c_{2D}^*) + 1} = \frac{N_{tOD}}{N_{tOD} + 1}
\end{aligned} \tag{10.71}
$$

In the absence of slow chemical reactions, such as may occur during extraction of metals from ore-leach liquors, and in the absence of interfacial resistance, stage efficiencies of well-agitated vessels can be expected to be high, and relatively short (60 s) holding times are ordinarily ample. The preceding equations, in any event, can be expected to provide rough estimates only, and experimental studies with new systems are essential before designs are completed.

Recycling Especially when the ratio of liquid flows is small, the interfacial area, with the minority liquid dispersed, may be small. Recycling some of the settled liquid back to the mixer may then improve the stage efficiency. Recycling that liquid favored by the solute distribution (frequently the minority liquid) increases the stage efficiency, whereas recycling the other or recycling both reduces it [75].

Illustration 10.7 A covered, baffled vessel, 0.5 m (19.7 in) diameter, 0.5 m deep is to be used for continuous extraction of benzoic acid from water (3×10^{-3} m^3/s = 6.36 ft^3/min) into toluene (3×10^{-4} m^3/s) as solvent. Liquids are to flow cocurrently upward and will fill the vessel; there will be no gas-liquid interface and hence no vortex. A six-blade, flat-blade disk turbine, 0.15 m (6 in) diameter is to be located axially and centrally, turning at 13.3 r/s (800 r/min). The liquid properties are:

	Water solution	Toluene solution
Density ρ	998 kg/m^3	865 kg/m^3
Viscosity μ	0.95×10^{-3} kg/m · s	0.59×10^{-3} kg/m · s
Diffusivity D	2.2×10^{-9} m^2/s	1.5×10^{-9} m^2/s

Interfacial tension, $\sigma = 0.022$ N/m
Distribution coefficient = $c_{\text{toluene}}/c_{\text{water}} = 20.8$

Estimate the stage efficiency to be expected.

SOLUTION In view of the flow ratio, the toluene will be dispersed. Vessel volume = $\pi(0.5)^2(0.5)/4 = 0.09817$ m^3.

$\phi_{DF} = 3 \times 10^{-4}/(3 \times 10^{-3} + 3 \times 10^{-4}) = 0.0909$ vol fraction toluene in the feed mixture. $q_D = 3 \times 10^{-4}$ m^3/s toluene.

$$\frac{\mu_C^3}{q_D \rho_C^2 \sigma g_c} = \frac{(0.95 \times 10^{-3})^3}{(3 \times 10^{-4})(998)^2(0.022)(1)} = 1.304 \times 10^{-10}$$

$$\frac{\rho_C}{\Delta_\rho} = \frac{998}{998 - 865} = 7.504$$

$$\frac{\sigma^3 \rho_C g_c^3}{\mu_C^4 g} = \frac{(0.022)^3(998)(1)}{(0.95 \times 10^{-3})^4(9.807)} = 1.330 \times 10^9$$

$$\frac{\mu_D}{\mu_C} = \frac{0.59 \times 10^{-3}}{0.95 \times 10^{-3}} = 0.621$$

Tentatively take $\phi_D/\phi_{DF} = 0.9$; $\phi_D = 0.9(0.0909) = 0.0818$. Eq. (10.56):

$$\rho_M = 998(1 - 0.0818) + 865(0.0818) = 987 \text{ kg/m}^3$$

Eq. (10.58):

$$\mu_M = \frac{(0.95 \times 10^{-3})}{1 - 0.0818}\left[1 + \frac{6(0.59 \times 10^{-3})(0.0818)}{0.59 \times 10^{-3} + 0.95 \times 10^{-3}}\right] = 1.067 \times 10^{-3} \text{ kg/m} \cdot \text{s}$$

$$\text{Impeller Reynolds number} = \frac{d_i^2 N \rho_M}{\mu_M} = \frac{(0.15)^2(13.3)(987)}{1.067 \times 10^{-3}} = 277\,000$$

Fig. 6.5, curve g:

$$\text{Po} = 0.72$$

$$P = \frac{\text{Po } \rho_M N^3 d_i^5}{g_c} = \frac{0.72(987)(13.3)^3(0.15)^5}{1} = 127 \text{ W impeller power output}$$

$$\frac{P q_D \mu_C^2}{v_L \sigma^3 g_c^2} = \frac{127(3 \times 10^{-4})(0.95 \times 10^{-3})^2}{0.09817(0.022)^3(1)} = 0.0329$$

Eq. (10.61): $\phi_D/\phi_{DF} = 0.907$. This is sufficiently close to the estimated 0.9 to be acceptable.

$$\phi_D = 0.907(0.0909) = 0.0824 \qquad \frac{\mu_C}{\rho_C} = \frac{0.95 \times 10^{-3}}{998} = 9.52 \times 10^{-7} \text{ m/s}^2$$

$$\frac{Pg_c}{v_L\rho_M} = \frac{127(1)}{0.0982(987)} = 1.311 \qquad \frac{\sigma g_c}{\rho_c} = \frac{0.022(1)}{998} = 2.204 \times 10^{-5}$$

Eq. (10.64):

$$d_p = 2.55 \times 10^{-4} \text{ m} \qquad \dot{a} = \frac{6\phi_D}{d_p} = 1940 \text{ m}^2/\text{m}^3$$

$$\text{Sc}_C = \frac{\mu_C}{\rho_C D_C} = \frac{0.95 \times 10^{-3}}{998(2.2 \times 10^{-9})} = 433$$

Eq. (10.65):

$$\text{Sh}_C = \frac{k_{LC}d_p}{D_C} = 65.3$$

$$k_{LC} = \frac{65.3 D_C}{d_p} = 5.45 \times 10^{-4} \text{ kmol/m}^2 \cdot \text{s} \cdot (\text{kmol/m}^3)$$

$$\theta = \frac{v_L}{q_C + q_D} = 29.8 \text{ s}$$

Table (10.1):

$$\lambda_1 = 1.359 \qquad \lambda_2 = 7.23 \qquad \lambda_3 = 17.9$$

$$B_1 = 1.42 \qquad B_2 = 0.603 \qquad B_3 = 0.317$$

n	1	2	3
$B_n^2 \exp(-\lambda_n 64 D_D \theta/d_p^2)$	2.18×10^{-26}	0	0

Eq. (10.66):

$$k_{LD} = 8.57 \times 10^{-5} \text{ kmol/m}^2 \cdot \text{s} \cdot (\text{kmol/m}^3)$$

$$m_{CD} = \frac{1}{20.8} = 0.0481$$

Eq. (10.67):

$$K_{LD} = 2.01 \times 10^{-5} \text{ kmol/m}^2 \cdot \text{s} \cdot (\text{kmol/m}^3)$$

$$Z = 0.5 \text{ m} \qquad V_D = \frac{3 \times 10^{-4}}{\pi(0.5)^2/4} = 1.528 \times 10^{-3} \text{ m/s}$$

Eq. (10.70):

$$N_{tOD} = \frac{0.5}{1.528 \times 10^{-3}/[(2.01 \times 10^{-5})(1940)]} = 12.76$$

Eq. (10.71):

$$\mathbf{E}_{MD} = 0.93 \quad \text{Ans.}$$

Emulsions and Dispersions

The mixture of liquids issuing from any mixing device, an emulsion, consists of small droplets of one liquid dispersed throughout a continuum of the other. The stability, or permanence, of the emulsion is of the utmost importance in liquid extraction, since it is necessary to separate the phases at each extraction stage. Stable emulsions, those which do not settle and coalesce rapidly, must be avoided. For an emulsion to "break," or separate into its phases in bulk, both sedimentation and coalescence of the dispersed phase must occur.

The rate of sedimentation of a quiescent emulsion is the more rapid if the size of the droplets and the density difference of the liquids are large and the viscosity of the continuous phase is small. Stable emulsions, those which settle only over long periods of time, are usually formed when the diameter of the dispersed droplets is of the order of 1 to 1.5 μm, whereas dispersions of particle diameter 1 mm or larger usually sediment rapidly.

Coalescence of the settled droplets is the more rapid the higher the interfacial tension. Interfacial tension is ordinarily low for liquids of high mutual solubility and is lowered by the presence of emulsifying or wetting agents. In addition, high viscosity of the continuous phase hinders coalescence by reducing the rate at which the residual film between drops is removed. Dust particles, which usually accumulate at the interface between liquids, also hinder coalescence.

In an unstable emulsion, after agitation has stopped, the mixture settles and coalesces rapidly into two liquid phases unless the viscosity is high. The appearance of a sharply defined interface between the phases (primary break) is usually very rapid, but one of the phases, ordinarily that in the majority, may remain clouded by a very fine fog or haze, a dispersion of the other phase. The cloud will eventually settle and leave the clouded phase clear (secondary break), but this may take a considerable time. The primary break of an unstable emulsion is usually so rapid that merely stopping agitation for a very short time, a matter of minutes, is sufficient to bring it about. In continuous multistage operation, it is usually impractical to hold the mixture between stages long enough to attain the secondary break.

Settlers

In continuous extraction, the dispersion issuing from the mixer must be passed to a settler, or decanter, where at least the primary break occurs. Typical gravity settler arrangements are shown in Fig. 10.40. The simplest design (Fig. 10.40a) is perhaps the most common: the entering dispersion is prevented from disturbing the vessel contents excessively by the inlet "picket-fence" baffle. The drops then settle in the main part of the vessel, where the velocity must be low enough to prevent turbulent disturbances. At larger flow rates for a given vessel, the dispersion-band thickness increases, and ultimately unsettled dispersion issues from the outlets, which is of course to be avoided. Although a great many fundamental studies have been made [1, 28], design methods are still empirical and arbitrary. There have been three approaches to the design: (1) provision of sufficient residence time based on laboratory observation of settling, (2) estimation of the rate of flow to produce a suitable dispersion-band thickness, and (3) calculation of the time to settle individual drops through a clear liquid above

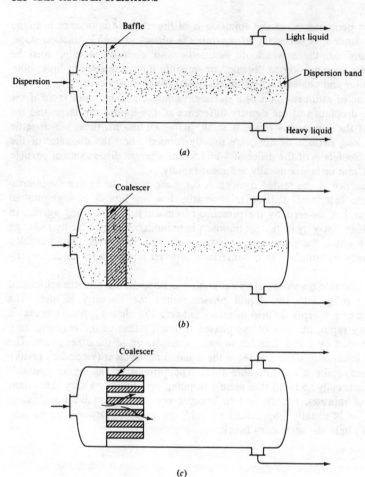

Figure 10.40 Gravity settlers (schematic): (*a*) simple, (*b*) and (*c*) with coalescer.

and below the dispersion band. None of these methods is very reliable [12]. In the absence of any experience with the dispersion at hand, a first estimate of settler size can be made from the empirical expression†

$$T_S = 8.4(q_C + q_D)^{0.5} \qquad (10.72)$$

developed on the basis of typical settling rates for droplets of dispersed phase [18] and a length/diameter ratio for the vessel equal to 4.

† Equation (10.72) is satisfactory only for SI units. For T_S in feet and q in gal/min the coefficient is 0.22.

Settler auxiliaries The dispersion may be passed through a |*coalescer*| in order to increase the size of the droplets and hence their settling rate. Typically, these are relatively thin beds of substances of extended surface and relatively large voids, such as excelsior (especially for coalescing the salt brines accompanying crude petroleum), steel wool, fiber glass, polypropylene cloth, Raschig rings, and the like [20, 66]. One of the best coalescers is a mixture of cotton fibers and glass wool or of cotton and Dynel fibers [53]. A simple arrangement is shown in Fig. 10.40*b*, and a more elaborate but very effective one in *c* [72]. *Separator membranes* are porous membranes, preferentially wet by the continuous phase, which prevent the passage of the nonwetting dispersed phase at low pressure drop but freely pass the continuous liquid. They can be placed in the outlet from the decanter [72].

Mixer-Settler Cascades

A continuous multistage extraction plant contains the required number of stages arranged according to the desired flowsheet. Each stage will consist of at least a mixer and a settler, as in the countercurrent plant of Fig. 10.41. The liquids are generally pumped from one stage to the next, but occasionally gravity flow can be arranged if sufficient headroom is available. Many arrangements have been designed to reduce the amount of interstage piping and the corresponding cost [72]. Figure 10.42 is one such: the mixing vessels are immersed in the large circular settling tanks, heavy liquid flows by gravity, light liquid by air lift, and recycling of settled light liquid to the mixer is accomplished by overflow. Figure 10.43 is typical of several designs of so-called *box extractors*, designed to permit dispensing with intermediate piping. The mixers and settlers are rectangular in cross section, hence "box," and are arranged in alternate positions for adjacent stages, as shown in the plan view of Fig. 10.43. Another arrangement places the stages one above the other in a vertical stack, with mixing impellers on a common shaft, utilizing the impellers as pumps as well as mixing devices [73].

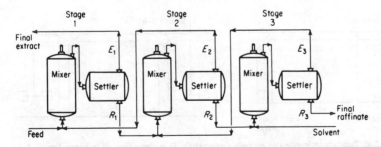

Figure 10.41 Flowsheet of three-stage countercurrent mixer-settler extraction cascade.

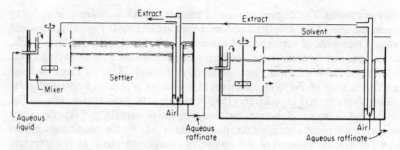

Figure 10.42 Kerr-McGee uranium extractor.

Sieve-Tray (Perforated-Plate) Towers

These multistage, countercurrent towers are very effective, both with respect to liquid-handling capacity and extraction efficiency, particularly for systems of low interfacial tension which do not require mechanical agitation for good dispersion. Their mass-transfer effectiveness results because (1) axial mixing of

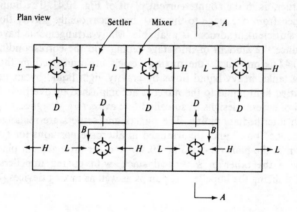

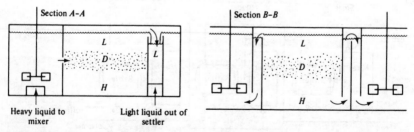

Figure 10.43 Box-type mixer-settler cascade (schematic): D = dispersion, H = heavy liquid, L = light liquid.

the continuous phase is confined to the region between trays and does not spread throughout the tower from stage to stage and (2) the dispersed-phase droplets coalesce and are formed again at each tray, destroying the tendency to establish concentration gradients within the drops which persist for the entire tower height. A tower of simple design is shown in Fig. 10.44, where the general arrangement of plates and downspouts is much the same as for gas-liquid contact except that no weir is required. The figure shows the arrangement for light liquid dispersed. Light liquid passes through the perforations, and the bubbles rise through the heavy continuous phase and coalesce into a layer, which accumulates beneath each plate. The heavy liquid flows across each plate through the rising droplets and passes through the downspouts to the plate below. By turning the tower as shown upside down, the downspouts become "upspouts" and carry the light liquid from plate to plate, while the heavy liquid flows through the perforations and is dispersed into drops. As an alternative, the

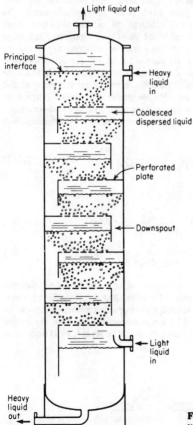

Figure 10.44 Sieve-tray extraction tower, arranged for light liquid dispersed.

heavy liquid can be dispersed in one part of the tower and the light liquid in the other, while the principal interface is maintained in the central portion of the tower. The cross-flow trays of Fig. 10.44 are suitable for relatively small tower diameters (up to roughly 2 m). For large towers, multiple downspouts can be arranged at intervals across the tray.

Sieve-Tray Hydraulics

The flow capacity of the sieve-tray tower depends upon drop-formation characteristics of the system, drop terminal velocities, dispersed-phase holdup, and pressure drop. With these established, the design can be developed and mass-transfer rates estimated.

Drop formation The liquid to be dispersed flows through perforations of diameter 3 to 8 mm, set 12 to 20 mm apart. If the drop liquid preferentially wets the material of the plate, the drop size becomes uncontrollably large and it is best to use small nozzles projecting from the surface in order to avoid this difficulty. The holes can be punched in a flat plate (Fig. 10.45) with the burr left in place facing in the direction of drop formation.

In the absence of mass transfer, the effect of which has not been firmly established [8], when the dispersed phase issues from a perforation in a plate which is not preferentially wet by the dispersed liquid, drop size will be uniform at a given velocity through the opening, while drops form at the orifice. At some moderate velocity, a jet of dispersed liquid issues from the opening and drops are formed by breakup of the jet. The velocity through the orifice when jetting begins can be estimated [54], but as a rule of thumb it will be roughly 0.1 m/s (0.3 ft/s) [19]. For velocities below this, drop sizes can be estimated principally from either of two correlations [19, 54], which, to a large extent but not always, agree reasonably well. For present purposes, estimates are sufficient, so that differences are unimportant, and Fig. 10.46 will serve. At velocities greater than about 0.1 m/s, drop sizes are not uniform. In extraction, maximizing the interface surface of all the drops is important for ensuring rapid mass transfer, and the orifice velocity $V_{o,\,max}$ which does this can be estimated from [11, 64]

$$V_{o,\,max} = 2.69\left(\frac{d_J}{d_o}\right)^2\left[\frac{\sigma}{d_J(0.5137\rho_D + 0.4719\rho_C)}\right]^{0.5} \tag{10.73}$$

The orifice/jet diameter ratio is given by

$$\frac{d_o}{d_J} = \begin{cases} 0.485\left[\dfrac{d_o}{(\sigma gc/\Delta\rho\, g)^{0.5}}\right]^2 + 1 & \text{for} \quad \dfrac{d_o}{(\sigma g_c/\Delta\rho\, g)^{0.5}} < 0.785 & (10.74a) \\[4mm] 1.51\dfrac{d_o}{(\sigma g_c/\Delta\rho\, g)^{0.5}} + 0.12 & \text{for} \quad \dfrac{d_o}{(\sigma g_c/\Delta\rho\, g)^{0.5}} > 0.785 & (10.74b) \end{cases}$$

[Equations (10.73) and (10.74) can be used with any consistent set of units.] The drop diameter corresponding to the above values is $d_p = 2d_J$. The value of σ to be used is intermediate between that for the binary solvent pair and the equilibrium value for the extracting system if that can be estimated (data are scarce). Equations (10.73) and (10.74) are recommended for sieve-tray design, but if the resulting velocity calculates to be less than 0.1 m/s, V_0 should be set at least at from 0.1 to 0.15 m/s and the drop diameter should be estimated from Fig. 10.46.

Figure 10.45 Punched perforations for dispersed phase [39].

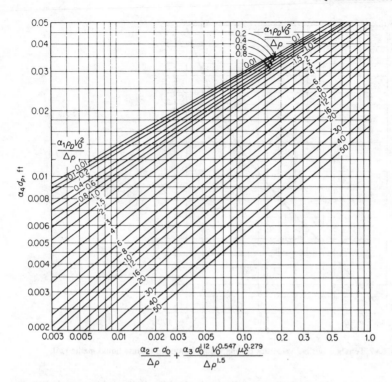

Figure 10.46 Drop diameters for dispersion of insoluble liquids through nozzles and perforations [19].

	α_1	α_2	α_3	α_4
SI	10.76	52 560	1.246×10^6	3.281
lb, lb$_f$, ft, s	1.0	14 600	3040	1.0

Drop terminal velocity It will be necessary to estimate V_t, the terminal velocity of a liquid drop in a liquid medium. This is the free-fall (or rise, depending on the relative density) velocity of a single isolated drop in the gravitational field. The terminal velocities of small drops, which are essentially spherical, are larger than those of solid spheres of the same diameter and density, owing to the mobility and internal circulation within the drop: the surface velocity is not zero, as it is for a solid. With increasing diameter, there occurs a transition drop size, beyond which the drop shape is no longer spherical (although it is nevertheless described with d_p as the diameter of a sphere of the same volume), and the drop oscillates and distorts. The terminal velocity of the transition size is a maximum, and for larger sizes the velocity falls slowly with increased diameter. Dimensional analysis shows Re $= f(C_D,$ We$)$, where Re is the drop Reynolds number at terminal velocity, We is the drop Weber number, and C_D is the usual drag coefficient. For very pure liquids (no surface-active agents, no mass transfer), and continuous phase viscosities less than 0.005 kg/m · s (5 cP), the Hu-Kintner correlation [24] (Fig. 10.47) provides the functional relation. U is defined as

$$U = \frac{4 \, \text{Re}^4}{3 C_D \, \text{We}^3} = \frac{\rho C^2 (\sigma g_c)^3}{g \mu_C^4 \, \Delta\rho} \tag{10.75}$$

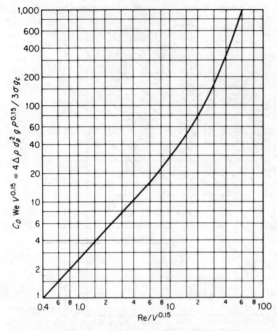

Figure 10.47 Terminal settling velocities of single liquid drops in infinite liquid media [24].

and the transition maximum velocity occurs at an ordinate of Fig. 10.47 equal to approximately 70, or when

$$d_{p,\text{ trans}} = 7.25\left(\frac{\sigma g_c}{g\,\Delta\rho U^{0.15}}\right)^{0.5} \tag{10.76}$$

For larger continuous phase viscosities, but not exceeding 0.030 kg/m · s (30 cP), the ordinate of Fig. 10.47 should be multiplied by $(\mu_W/\mu_C)^{0.14}$, where μ_W is the viscosity of water [30]. Minute amounts of impurities which are frequently present in practice can alter the terminal velocities profoundly (usually to lower values), and the effect of mass transfer is unknown.†

Downspouts The tower will flood if excessive quantities of dispersed-phase droplets are carried through the downspouts (or upspouts) entrained in the continuous phase. The velocity of liquid in the downspouts V_d should therefore be less than the terminal velocity of all except the smallest

---

† The curve of Fig. 10.47 should not be extrapolated to lower values. For values of the ordinate below 1.0, the equation of Klee and Treybal [34] will give more correct results for $\mu_C < 0.002$ kg/m · s (2 cP):

$$\text{Re} = 22.2C_D^{-5.18}\,\text{We}^{-0.169} \tag{10.77a}$$

This is conveniently rearranged to

$$V_t = \frac{0.8364\Delta\rho^{0.5742}d_p^{0.7037}g^{0.5742}}{\rho_C^{0.4446}(\sigma g_c)^{0.01873}\mu_C^{0.11087}} \tag{10.77b}$$

which is applicable for any consistent set of units for values of d_p^- less than $d_{p,\text{ trans}}$.

droplets, e.g., those smaller than 0.6 to 0.8 mm [39], and the downspout cross section A_d is set accordingly. The downspouts are extended well beyond the depth of coalesced dispersed liquid on the tray, to prevent dispersed liquid from flowing through the downspout. Downspouts are installed flush with the tray from which they lead, and no weirs like those needed for gas-liquid towers should be used.

Coalesced liquid on the trays The depth h of coalesced dispersed liquid accumulating on each tray is determined by the pressure drop required for counterflow of the liquids [9]:

$$h = h_C + h_D \tag{10.78}$$

where h_C and h_D are the contributions of each of the liquids. The contribution from the dispersed liquid h_D is that necessary to cause flow through the perforations of the tray h_o plus that necessary to overcome interfacial tension h_σ,

$$h_D = h_o + h_\sigma \tag{10.79}$$

The value of h_o can be computed from the usual orifice equation with a coefficient of 0.67,

$$h_o = \frac{(V_o^2 - V_n^2)\rho_D}{2g(0.67)^2 \, \Delta\rho} \tag{10.80}$$

h_σ, which is important only when the dispersed phase flows slowly, can be computed from

$$h = \frac{6\sigma g_c}{d_{ps} \, \Delta\rho \, g} \tag{10.81}$$

where d_{ps} is the drop diameter produced at perforation velocities $V_o = 0.03$ m/s (0.1 ft/s). At perforation velocities where jets of dispersed liquid issue from the perforations, h_σ can be omitted.

The head required for flow of the continuous phase h_C includes losses owing to (1) friction in the downspout, ordinarily negligible, (2) contraction and expansion upon entering and leaving the downspout, equal to 0.5 and 1.0 velocity heads, respectively, and (3) the two abrupt changes in direction each equivalent to 1.47 velocity heads. The value of h_C is therefore substantially equal to 4.5 velocity heads, or

$$h_C = \frac{4.5 V_D^2 \rho_C}{2g \, \Delta\rho} \tag{10.82}$$

Should h calculate to be small, say 50 mm or less, there is danger that not all perforations will operate unless the tray is installed perfectly level. Under these circumstances, it is best to increase the value of h_C by placing a restriction in the bottom of the downspouts (or upper end of upspouts), the effect of which can be computed as flow through an orifice.

Dispersed-phase holdup The space between a tray and the coalesced layer on the next tray is filled with a dispersion of the dispersed liquid in the continuous liquid. At flow rates below those causing flooding, it has been established [4, 80] that the ratio of slip velocity V_S to terminal velocity V_t of a single particle ($\phi_D = 0$) is a unique function of dispersed-phase holdup ϕ_D for all vertically moving fluid particulate systems, including gas-solid, liquid-solid, and liquid-liquid systems. Slip velocity is the net relative velocity between the two phases, and if they flow countercurrently,

$$V_S = \frac{V_D}{\phi_D} + \frac{V_C}{1 - \phi_D} \tag{10.83}$$

The correlation of Zenz [84] (Fig. 10.48), derived from fluid-solid systems, provides the $(V_S/V_t, \phi_D)$ function, with V_t for solids given by the curve for $\phi_D = 0$. For sieve-tray extractors, V_D becomes V_n, and since the continuous phase flows horizontally, V_C is taken as zero, so that

$$V_S = \frac{V_n}{\phi_D} \tag{10.84}$$

Illustration 10.8 describes the use of the correlation. The specific interfacial area corresponding to the holdup is given by Eq. (6.9).

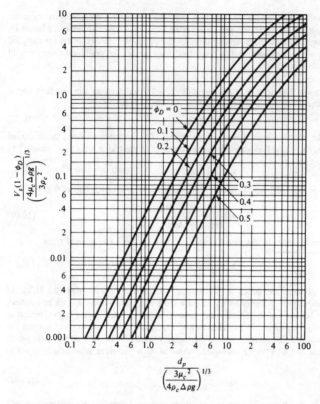

Figure 10.48 Zenz's correlation for fluidized solid particles. [*F. Zenz, Pet. Refiner, 36 (8), 147 (1957), with permission.*]

Sieve-Tray Mass Transfer

Mass transfer occurs during three separate regimes: drop formation and release, drop rise (or fall), and drop coalescence into the layer of coalesced liquid on the tray.

Mass transfer during drop formation There have been a great many studies, for most of which a review is available [21]. There are evidently many influencing phenomena: the rate of drop formation [7]; whether the drops are formed at nozzles or orifices in plates or at the ends of jets; and the presence or absence of interfacial turbulence or surfactants. Although fairly elaborate expressions have been devised to describe some of the data [62, 63], the great divergence of the data at present does not seem to warrant anything more than a simple estimate. The mass-transfer coefficient K_{LDf} can be defined by [76]

$$N_f = K_{LDf}(c_D - c_D^*) \tag{10.85}$$

where N_f is the flux, averaged over the time of drop formation θ_f, and based on the area A_p of the drop at breakaway. If we assume the same mechanism of mass transfer for the liquids on either side of the interface, e.g., surface renewal, penetration theory, or surface stretch, (see Chap. 3), so that

$k_{LCf} = k_{LDf} (D_C/D_D)^{0.5}$, then

$$\frac{1}{K_{LDf}} = \frac{1}{k_{LDf}} + \frac{1}{k_{LCf}m_{CD}} = \frac{1}{k_{LDf}}\left[1 + \frac{1}{m_{CD}}\left(\frac{D_D}{D_C}\right)^{0.5}\right] \qquad (10.86)$$

Theoretical treatment of mass transfer during drop formation generally leads to expressions of the form

$$k_{LDf} = \text{const}\left(\frac{D_D}{\pi\theta_f}\right)^{0.5} \qquad (10.87)$$

with values of the constant in the range 0.857 to 3.43 but mostly in the range 1.3 to 1.8 except in the presence of interfacial turbulence or surfactants.

Mass transfer during drop rise (or fall) Swarms of drops behave differently from single drops [25]. For small drops ($d_p < d_{p,\text{trans}}$) which circulate [52]

$$k_{LCr} = 0.725\left(\frac{d_p V_S \rho_C}{\mu_C}\right)^{-0.43} \text{Sc}^{-0.58}V_S(1 - \phi_D) \qquad (10.88)$$

The dispersed-phase coefficient k_{LDr} is then given by Eq. (10.66) and Table 10.1.

For large drops ($d_p > d_{p,\text{trans}}$) which oscillate as they rise or fall, surface-stretch theory [2] is recommended

$$k_{LDr} = \sqrt{\frac{4D_D\omega}{\pi}}\left(1 + \delta + \tfrac{3}{8}\delta^2\right) \qquad (10.89)$$

where $\omega = \text{oscillation frequency} = \dfrac{1}{2\pi}\sqrt{\dfrac{192\sigma g_c b}{d_p^3(3\rho_D + 2\rho_C)}} \qquad (10.90)$

and† $b = 1.052d_p^{0.225} \qquad (10.91)$

δ is a dimensionless amplitude factor characteristic of the oscillating system, and may be taken as 0.2 in the absence of more specific data. As with drop formation, the same mechanism is assumed to govern the phases on both sides of the interface, so that an equation similar to that of Eq. (10.86) provides K_{LDr}.

Mass transfer during drop coalescence Experimental measurement of this quantity is very difficult, and results are hard to reproduce. The data for only a few systems have been empirically correlated [65]. The mass-transfer coefficient is apparently an order of magnitude less than that for drop formation, and the area on which it is based is difficult to deduce. For the present, it is recommended that $K_{Dc} = 0.1K_{Df}$, based on the area of the drops A_p.

Stage efficiency Refer to Fig. 10.49a, which shows a schematic section through part of a sieve-tray tower. The equilibrium and operating curves corresponding to the entire extractor are shown in Fig. 10.49b. Curve DBE is drawn between the equilibrium and operating curves, everywhere at a fractional distance such as BC/AC equal to the Murphree dispersed-phase stage efficiency E_{MD}. The steps on the diagram then correspond to real stages, as for gas absorption (Fig. 8.16) and distillation (Fig. 9.48)

$$E_{MD} = \frac{c_{D_{n+1}} - c_{D_n}}{c_{D_{n+1}} - c_{D_n}^*} \qquad (10.92)$$

† b is dimensionless, and Eq. (10.91) is written for d_p in meters. For d_p in feet the coefficient is 0.805.

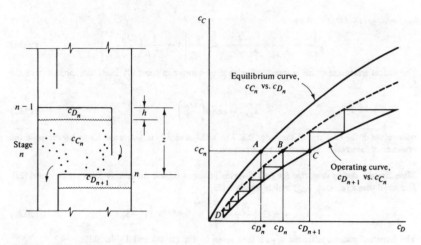

Figure 10.49 Stage efficiency of sieve-tray extractors.

The relatively low velocities prevailing in liquid extractors mean that the continuous phase can be assumed to be thoroughly mixed by the motion of the drops of the dispersed phase and everywhere of solute concentration c_{C_n}. The total rate of extraction occurring in stage n is then

$$q_D(c_{D_{n+1}} - c_{D_n}) = K_{LDf}A_f(c_{D_{n+1}} - c_{D_n}^*) + K_{LDr}A_r(c_D - c_D^*)_M + K_{LDc}(c_{D,n} - c_{D,n}^*) \tag{10.93}$$

Equations (10.92) and (10.93) produce

$$E_{MD} = \frac{K_{LDf}A_p}{q_D} + \frac{K_{LDr}A_r(c_D - c_D^*)_M}{q_D(c_{D_{n+1}} - c_{D_n}^*)} + \frac{K_{LDc}A_p(c_{D_n} - c_{D_n}^*)}{q_D(c_{D_{n+1}} - c_{D_n}^*)} \tag{10.94}$$

A_p is the surface of the N_o drops issuing from the N_o orifices.

$$A_p = \pi d_p^2 N_o \tag{10.95}$$

The interfacial area of the dispersed-phase holdup in the drop-rise region is

$$A_r = a(Z - h)(A_t - A_d) = \frac{6\phi_D}{d_p}(Z - h)A_n \tag{10.96}$$

The mean driving force for the drop-rise region is strictly the logarithmic average, but for present purposes the arithmetic average is adequate:

$$(c_D - c_D^*)_M = \frac{(c_{D_{n+1}} - c_{D_n}^*) + (c_{D_n} - c_{D_n}^*)}{2} \tag{10.97}$$

and, as indicated earlier, K_{LDc} is taken as $0.1 K_{LDf}$. Substitution of this and Eqs. (10.95) to (10.97) into (10.94) and noting that $q_D/A_n = V_n$, $V_o = 4q_D/\pi d_o^2 N_o$, and $1 - E_{MD} = (c_{D_n} - c_{D_n}^*)/(c_{D_{n+1}} - c_{D_n}^*)$ lead to

$$E_{MD} = \frac{\dfrac{4.4 K_{LDf}}{V_o}\left(\dfrac{d_p}{d_o}\right)^2 + \dfrac{6 K_{LDr}\phi_D(Z - h)}{d_p V_n}}{1 + \dfrac{0.4 K_{LDf}}{V_o}\left(\dfrac{d_p}{d_o}\right)^2 + \dfrac{3 K_{LDr}\phi_D(Z - h)}{d_p V_n}} \tag{10.98}$$

Skelland [62] has outlined a similar, more elaborate, procedure, but Eq. (10.98) is believed more than adequate in view of the many uncertainties involved. In any event, it must be understood that we can at best estimate E_{MD} only roughly. The stage efficiency will usually be far less than that obtained for gas-liquid contact, owing to the lower phase velocities resulting from lower density differences and higher viscosities.

Illustration 10.8 Isopropyl ether is to be used to extract acetic acid from water in a sieve-tray tower, ether dispersed. The design conditions agreed upon are those of Illustration 10.3 and Fig. 10.24, stage 2:

	Water solution (continuous)	Isopropyl ether solution (dispersed)
Flow rate	8000 kg/h = 17 600 lb/h	20 000 kg/h = 44 100 lb/h
Acid concentration	$x_{C_n} = 0.175$ mass fraction	$x_{D_{n+1}} = 0.05$ mass fraction
Density	1009 kg/m^3 = 63.0 lb/ft^3	730 kg/m^3 = 45.6 lb/ft^3
Viscosity	3.1×10^{-3} kg/m · s = 3.1 cP	0.9×10^{-3} kg/m · s = 0.9 cP
Acid diffusivity	1.24×10^{-9} m^2/s = 4.8×10^{-5} ft^2/h	1.96×10^{-9} m^2/s = 7.59×10^{-5} ft^2/h

$$\text{Interfacial tension} = 0.013 \text{ N/m} = 13 \text{ dyn/cm}$$
$$\text{Distribution coefficient} = m = \frac{\Delta x_C}{\Delta x_D} = 2.68$$

SOLUTION Mol wt acetic acid = 60.1.

$$c_{C_n} = \frac{0.175(1009)}{60.1} = 2.94 \text{ kmol/m}^3 \qquad c_{D_{n+1}} = \frac{0.05(730)}{60.1} = 0.608 \text{ kmol/m}^3$$

$$m_{CD} = 2.68 \frac{1009}{730} = 3.704 \frac{\text{kmol/m}^3 \text{ in ether}}{\text{kmol/m}^3 \text{ in water}}$$

Perforations Set $d_o = 6$ mm = 0.006 m, arranged in triangular pitch on 15-mm centers. $\rho_D = 730$ kg/m^3; $q_D = 20\,000/730(3600) = 7.61 \times 10^{-3}$ m^3/s; $\sigma = 0.013$ N/m; $g = 9.807$ m/s^2; $g_c = 1.0$ $\rho_C = 1009$ kg/m^3; $\Delta\rho = 1009 - 730 = 279$ kg/m^3.

$$\frac{d_o}{\left(\dfrac{\sigma g_c}{\Delta\rho g}\right)^{0.5}} = 2.75$$

Eq. (10.74b):

$$\frac{d_o}{d_J} = 4.28 \qquad d_J = 1.402 \times 10^{-3} \text{ m}$$

Eq. (10.73):

$$V_{o,\text{max}} = 0.0153 \text{ m/s}$$

Since this is less than 0.1, use $V_o = 0.1$ m/s. Perforation area = $q_D/V_o = 7.61 \times 10^{-2}$ m^2; N_o = perforation area/$(\pi d_o^2/4) = 2690$. Eq. (6.31):

$$\text{Plate area for perforations} = \frac{7.61 \times 10^{-2}}{0.907(0.006/0.015)^2} = 0.5244 \text{ m}^2$$

Downspout Set the continuous-phase velocity V_d equal to the terminal velocity of a dispersed-phase drop, $d_p = 0.7$ mm = 0.0007 m. $\mu_C = 0.0031$ kg/m · s. Eq. (10.75): $U = 8\,852\,000$. Fig.

10.47: ordinate $= 1.515$; abscissa $= 0.62 = d_p V_t \rho_C / \mu_C U^{0.15}$; $V_t = 0.03$ m/s $= V_d$. $q_C = 8000/(1009)(3600) = 2.202 \times 10^{-3}$ m^3/s. $A_d = q_C / V_d = 0.0734$ m^2. Refer to Table 6.2. Allowing for supports and unperforated area,

$$A_t = \frac{0.5244}{0.65} = 0.807 \text{ m}^2 \qquad T = \left(0.807\frac{4}{\pi}\right)^{0.5} = 1.0 \text{ m} \qquad A_n = A_t - A_d = 0.734 \text{ m}^2$$

Drop size Fig. 10.46:

$$\text{Abscissa } (\alpha_2 = 52\,560, \alpha_3 = 1.24 \times 10^6) = 0.0639$$
$$\text{Parameter } (\alpha_1 = 10.76) = 0.282 \qquad \text{ordinate} = 0.024 = 3.281 d_p$$

Therefore

$$d_p = 7.31 \times 10^{-3} \text{ m}$$

Coalesced layer $V_n = q_D / A_n = 0.01037$ m/s. Eq. (10.80): $h_o = 2.94 \times 10^{-3}$ m. h_o is unimportant at the present value of V_o; therefore $h_D = h_o = 2.94 \times 10^{-3}$ m. Eq. (10.82): $h_C = 7.47 \times 10^{-4}$ m; Eq. (10.78): $h = 3.69 \times 10^{-3}$ m. Since this is very shallow, increase it by placing an orifice in the bottom of the downspout. If V_R is the velocity through this restriction and h_R the corresponding depth of coalesced layer that results, the orifice equation provides

$$h_R = \frac{(V_R^2 - V_d^2)\rho_C}{2g_c(0.67)\Delta_\rho}$$

Set $h_R = 0.045$ m, whence $V_R = 0.332$ m/s. This corresponds to a restriction area $= q_C / V_R = 6.633 \times 10^{-3}$ m^2, in turn corresponding to a circular hole 92 mm in diameter. Then $h = 0.045 + 0.00369 = 0.049$ m $= 4.9$ cm, which is satisfactory.

Set the tray spacing at $Z = 0.35$ m, and lead the downspout apron to within 0.1 m of the tray below.

Dispersed-phase holdup Eq. (10.84): $V_S \phi_D = 0.01037$ m/s. For $d_p = 7.31 \times 10^{-3}$ m, $U = 8\,852\,000$; ordinate of Fig. 10.47 $= 165.2$, abscissa $= 30 = d_p V_t \rho_C / \mu_C U^{0.15}$; then $V_t = 0.1389$ m/s.

Fig. 10.48: abscissa $= 52.6$; ordinate $= 44.65 V_S(1 - \phi_D)$ for solid particles.

$\phi_D =$ parameter, Fig. 10.48 (1)	Ordinate, Fig. 10.48 (2)	$V_S(1 - \phi_D)$ for solids (3)	V_S for solids (4)	$\dfrac{V_S}{V_t}$ (5)	V_S for liquids (6)	$V_S \phi_D$ for liquids (7)
0	8.8	0.1971	0.1971 $= V_t$, solids	1.0	0.1389 $= V_t$	0
0.1	5.9	0.1321	0.1468	0.7448	0.1035	0.01035
0.2	4.3	0.0963	0.1204	0.6109	0.0849	0.1698
0.3	3.0	0.0672	0.0960	0.4871	0.0677	0.2031

Example for $\phi_D = 0.1$, ordinate Fig. 10.48 $= 5.9$; column 3 $= 5.9/44.65$; column 4 $= 0.1321/(1 - 0.1)$; column 5 $= 0.1468/0.1971$; column 6 $= 0.7448(0.1389)$; column 7 $= 0.1035(0.1)$.

Interpolation in column 7 for $V_S \phi_D = 0.01037$ provides $\phi_D = 0.10$.

Mass transfer

$$\theta_f = \frac{\pi d_p^3/6}{q_D/N_o} = 0.2498 \text{ s} \qquad D_D = 1.96 \times 10^{-9} \text{ m}^2/\text{s}$$

Eq. (10.87) with const = 1.5: $k_{LDf} = 7.5 \times 10^{-5}$ m/s; Eq. (10.86): $K_{LDf} = 5.6 \times 10^{-5}$ m/s or $\text{kmol/m}^2 \cdot \text{s} \cdot (\text{kmol/m}^3)$.
   The ordinate of Fig. 10.47 for the drops is larger than 70, hence $d_p > d_{p, \text{trans}}$, and the mass-transfer coefficient during drop rise is given by Eq. (10.89).
   Eq. (10.91): $b = 0.348$. Eq. (10.90): $\omega = 3.66$ s^{-1}. Eq. (10.89) with $\delta = 0.2$: $k_{LDr} = 1.053 \times 10^{-4}$ m/s

$$\frac{1}{K_{LDr}} = \frac{1}{k_{LDr}}\left[1 + \frac{1}{m_{CD}}\left(\frac{D_D}{D_C}\right)^{0.5}\right]$$

$$K_{LDr} = 7.86 \times 10^{-5} \text{ m/s or kmol/m}^2 \cdot \text{s} \cdot (\text{kmol/m}^3)$$

Eq. (10.98):

$$E_{MD} = 0.175 \quad \textbf{Ans.}$$

DIFFERENTIAL (CONTINUOUS-CONTACT) EXTRACTORS

When the liquids flow countercurrently through a single piece of equipment, the equivalent of as many theoretical stages may be had as desired. In such devices, the countercurrent flow is produced by virtue of the difference in densities of the liquids, and if the motivating force is the force of gravity, the extractor usually takes the form of a vertical tower, the light liquid entering at the bottom, the heavy liquid at the top. As an alternative, a larger centrifugal force can be generated by rotating the extractor rapidly, in which case the counterflow is radial with respect to the axis of revolution.
   There are several characteristics common to countercurrent extractors which have an important bearing upon their design and performance. It is typical that only one of the liquids can be pumped through the device at any desired rate. The maximum rate for the other will depend, among other things, upon the density difference of the liquids. If an attempt is made to exceed this rate for the second liquid, the extractor will reject one of the liquids, and it is said to be *flooded*. The same is true, of course, for gas-liquid contactors, but since the density difference is much smaller than for a gas and liquid, the flooding velocities of extractors are much lower. For a given volumetric rate of liquids to be handled, the extractor cross section must be large enough to ensure that the flooding velocities are not reached. The more open the cross section, the greater the flow rates before flooding occurs. Internal structures, packing, mechanical agitators, and the like normally reduce the velocities at which flooding occurs.
   The differential extractors are also subject to *axial mixing* (see Chap. 6); it severely reduces extraction rates because of the deterioration of the concentration differences between phases which is the driving force for mass transfer. This is illustrated in Fig. 10.50, where the real (axial-mixing) concentration profiles show a substantially smaller concentration difference than those for plug flow. If the flow ratio of the liquids is not unity, and it very rarely is, dispersing the liquid flowing at the lower rate will lead to small numbers of dispersed-phase drops, small interfacial areas, and small mass-transfer rates. On the other hand, if the majority liquid is dispersed, the axial-mixing problem is exacerbated. The

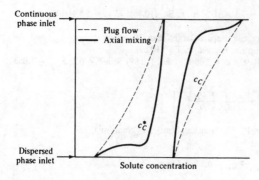

Figure 10.50 Effect of axial mixing on concentration profiles in towers subject to axial mixing.

problem becomes more acute the greater the flow ratio, and this difficulty is common to all differential extractors.

Reviews of calculation methods to account for axial mixing [41] (simplified procedures have been devised [47, 78]) and of axial-mixing data [26] are available.

Spray Towers

These, the simplest of the differential-contact devices, consist merely of an empty shell with provisions at the extremities for introducing and removing the liquids. Because the shell is empty, the extreme freedom of liquid movement makes these towers the worst offenders as far as axial mixing is concerned, so much so that it is difficult to obtain much more than the equivalent of a single stage with them. It is not recommended that they be used. Horizontal baffles (both the segmental and the disk-and-doughnut type) have been used to reduce the axial mixing but with little improvement in results. Spray towers have also been fairly extensively studied for direct-contact heat exchange between two liquids, but the deleterious effects of axial mixing are just as severe in this service as for extraction.

Packed Towers

Towers filled with the same random packings used for gas-liquid contact (Chap. 6) have also been used for liquid extractors. The packing serves to reduce axial mixing somewhat and to jostle and distort the drops of dispersed phase. A typical packed tower is shown schematically in Fig. 10.51, arranged for light liquid dispersed. The void space in the packing is largely filled with the continuous heavy liquid, which flows downward. The remainder of the void space is filled with droplets of light liquid formed at the lower distributor, which rise through the heavy liquid and coalesce at the top into a bulk layer, forming an interface as shown. To maintain the interface in this position, the pressure of liquid in the tower at the bottom must be balanced by a corresponding pressure

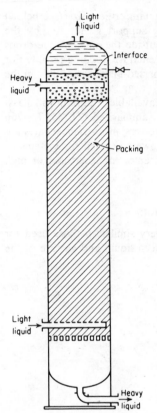

Light
liquid

Interface

Heavy
liquid

Packing

Light
liquid

Heavy
liquid

Figure 10.51 Packed extraction tower, light liquid dispersed.

in the bottom outlet pipe, as set by a control valve. If the pressure drop through that valve is reduced, the weight of the contents of the tower adjusts to a lower value, the light liquid becomes continuous, the heavy liquid dispersed, and the interface falls to below the light-liquid distributor. The interface positions are best regulated by a liquid-level control instrument activating the bottom outlet valve. The valve just above the interface in Fig. 10.51 is for periodic removal of scum and dust particles (*rag*) which accumulate at the interface.

The nature of the liquid flow in such towers requires that the choice of packing and arrangement of dispersed-phase distributor be given careful attention. If the dispersed liquid preferentially wets the packing, it will pass through in rivulets on the packing, not as droplets, and the interfacial area produced will be small. For this reason, the packing material should be preferentially wetted by the continuous phase. Usually, ceramics are preferentially wet by aqueous liquids and carbon and plastics by organic liquids. The packing should be sufficiently small, no greater than one-eighth the tower diameter, for the packing density to be fully developed yet larger than a certain critical size (see below) [48, 72, 76]. Where the material of the packing support is not wet by the dispersed droplets and the distributor is placed outside the packing, the drops will have difficulty in entering the packing and premature flooding results. For this reason it is always desirable to embed the dispersed-phase distributor in the packing, as in Fig. 10.51.

Correlations for estimating flooding rates are available, and data for mass-transfer coefficients and axial mixing have been summarized [23, 43, 72, 76]. Although axial mixing is less severe than in spray towers, mass-transfer rates are poor. It is recommended instead that sieve-tray towers be used for systems of low interfacial tension and mechanically agitated extractors for those of high interfacial tension.

Mechanically Agitated, Countercurrent Extractors

The extraction towers previously described are very similar to those used for gas-liquid contact, where density differences between liquid and gas are of the

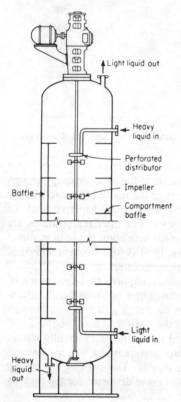

Figure 10.52 Mixco Lightnin CMContractor. *(Mixing Equipment Co.).*

order of 800 kg/m³ (50 lb/ft³) or more, available to provide the energy for good dispersion of one fluid into the other. In liquid extractors, where density differences are likely to be one-tenth as large or less, good dispersion of systems of high interfacial tension is impossible in such towers, and mass-transfer rates are poor. For such systems, dispersion is best brought about by mechanical agitation of the liquids, whereupon good mass-transfer rates are developed. Some examples of such extractors follow. Except for pulse columns, they are proprietary devices for which complete design procedures are not publicly available, and the manufacturers are best consulted.

Mixco Lightnin CMContactor (Oldshue-Rushton Extractor) Refer to Fig. 10.52. This device uses flat-blade disk turbine impellers (Fig. 6.3c) to disperse and mix the liquids and horizontal compartmenting plates to reduce axial mixing. There have been a few mass-transfer and somewhat more extensive axial-mixing studies [5, 13, 16, 17, 27, 44, 45]

Rotating-Disk Contactor (RDC) This (Fig. 10.53) is a somewhat similar device, except that the vertical baffles are omitted and agitation results from rotating disks, which usually turn at much higher speeds than turbine-type impellers [29, 50, 67].

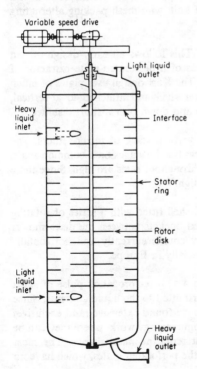

Figure 10.53 Rotating-Disk Contactor (RDC). *(General American Transportation Corp.)*

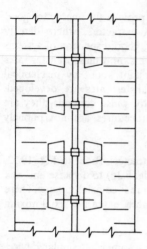

Figure 10.54 Scheibel extractor.

Scheibel Extractor [22, 56, 57] There have been several designs, of which the most recent is shown in Fig. 10.54. The impellers are of the turbine type, and the doughnut-type baffles surrounding them are supported by vertical tie rods, not shown. Earlier designs included sections of knit, wire-mesh packing alternating with sections containing an impeller.

Karr Reciprocating-Plate Extractor [31, 32] This follows an early design of van Dijck [77], who suggested moving the plates of a perforated-plate extractor of the type shown in Fig. 10.55 up and down. The Karr design uses plates of much larger free area, fitting loosely in the tower shell and attached to a vertical, central shaft. They are moved vertically up and down over a short distance.

Treybal Extractor [73] This is in reality a vertical stack of mixer-settlers. The mixers are in a vertical line, and the impellers for the stages on a common shaft. They not only mix but also pump, so that throughput rates are high. Since there is no axial mixing, mass-transfer rates are high.

Graesser Extractor [60] This is a horizontal shell fitted with a series of rotating disks on a central horizontal shaft. C-shaped buckets between the disks shower the liquids, one in the other, as they flow countercurrently and horizontally through the extractor. It has been used especially in Europe.

Pulsed columns [38, 77, 82] A rapid (0.5 to 4 s^{-1}) reciprocating pulse of short amplitude (5 to 25 mm) is hydraulically transmitted to the liquid contents. Since the extractors have no moving parts, they have found extensive (and exclusive) use in processing radioactive solutions in atomic-energy work, where they can be put behind heavy radiation shields without requiring maintenance. The most common arrangement is that of Fig. 10.55; the perforated plates, which have no

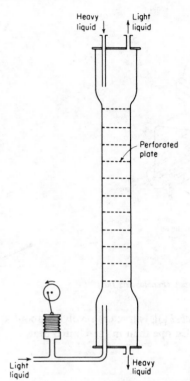

Figure 10.55 Pulsed column.

downspouts, are drilled with very small holes so that ordinarily flow will not occur. The pulsing superimposed upon the liquids alternately forces light and heavy liquids through the perforations. Packed columns, indeed any type of extractor, can also be pulsed. Although the mass-transfer rates are thereby improved at the expense of substantial energy costs, the flow capacities become smaller.

Centrifugal Extractors

The most important of these is the *Podbielniak extractor* [68, 69, 70] (Fig. 10.56). The cylindrical drum containing perforated concentric shells is rapidly rotated on the horizontal shaft (30 to 85 r/s). Liquids enter through the shaft: heavy liquid is led to the center of the drum, light liquid to the periphery. The heavy liquid flows radially outward, displacing the light liquid inwardly, and both are led out through the shaft. These extractors are especially useful for liquids of very small density difference and where very short residence times are essential, as in some pharmaceutical applications, e.g., extraction of penicillin from nutrient broth.

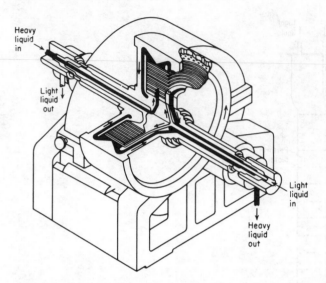

Figure 10.56 Podbielniak centrifugal extractor (schematic). *(Podbielniak, Inc.)*

The *Luwesta extractor* [14] and the *Rotabel* [3], extractors revolving about a vertical shaft, are used more extensively in Europe than in the United States.

Design

Since the change in concentration with height of either liquid as it passes through the extractor is differential, the height of the tower is expressed not in terms of stages or steps but in terms of transfer units.

Consider the continuous-contact tower of Fig. 10.57. Although the raffinate is shown flowing downward as if it were denser, in some instances the solvent-rich, or extract, phase will be denser and will enter at the top. In either case, in what follows subscript 1 will always represent that end of the tower where the raffinate enters and extract leaves, while subscript 2 will indicate where extract enters and raffinate leaves. We are presently unconcerned with which phase is dispersed and which is continuous. If the extractor is fed along the side, Fig. 10.57 and the relationships which follow apply separately to each section above and below the feed inlet.

Throughout this discussion, *unless otherwise specified,* x and y will refer to solute concentrations expressed as mole fractions in the raffinate and extract, respectively, and rates of flow of raffinate R and of extract E will be expressed as mol/(cross-sectional area)(time). Except in special cases, the transfer of solute usually results in changes of mutual solubility of the contacted liquids, so that in general all components of the systems transfer from one phase to the other. The F-type mass-transfer coefficients are capable of handling this problem, but in

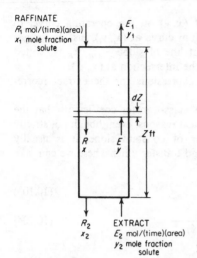

RAFFINATE
R_1 mol/(time)(area)
x_1 mole fraction
solute

E_1
y_1

dZ

R
x

E
y

Z ft

R_2
x_2

EXTRACT
E_2 mol/(time)(area)
y_2 mole fraction
solute

Figure 10.57 Continuous-contact tower.

reality our knowledge of mass-transfer rates in extractors is so poorly developed that we can ordinarily consider only the transfer of a single solute. This discussion is therefore limited to cases where the liquids are insoluble, only solute is transferred, and mutual solubility unchanged. In practice, however, the expressions developed are used for all cases.

It then follows that the equations derived for gas absorption apply. Thus, for the raffinate (which corresponds to the gas in gas absorption), we have the counterparts of Eqs. (8.23) to (8.27)

$$Z = \int_{x_2}^{x_1} \frac{R(1-x)_{iM}dx}{F_R a(1-x)(x-x_i)} = H_{tR} \int_{x_2}^{x_1} \frac{(1-x)_{iM}dx}{(1-x)(x-x_i)} = H_{tR}N_{tR}$$

(10.99)

$$H_{tR} = \frac{R}{F_R a} = \frac{R}{k_R a(1-x)_{iM}}$$

(10.100)

$$N_{tR} = \int_{x_2}^{x_1} \frac{(1-x)_{iM}dx}{(1-x)(x-x_i)} = \int_{x_2}^{x_1} \frac{dx}{x-x_i} + \tfrac{1}{2}\ln\frac{1-x_2}{1-x_1}$$

(10.101)

where x_i = interface concentration of solute
F_R, k_R = transfer coefficients for raffinate phase
H_{tR} = raffinate height of transfer unit
N_{tR} = number of raffinate transfer units
$(1-x)_{iM}$ = logarithmic mean of $1-x$ and $1-x_i$

The interface concentration corresponding to any bulk raffinate concentration x is found through Eq. (5.21) adapted to the present situation

$$\frac{1-x_i}{1-x} = \left(\frac{1-y}{1-y_i}\right)^{F_E/F_R}$$

(10.102)

This equation is plotted, for any value of (x, y) on the operating curve, to determine its intersection with the equilibrium curve at (x_i, y_i), just as for gas absorption. As an approximation, a straight line of slope $-k_R/k_E$ is plotted from (x, y) on the operating line to obtain the intersection at (x_i, y_i).

Similarly, we have the corresponding expressions for the extract (corresponding to the liquid in gas absorption).

In reality, we know so little about the mass-transfer coefficients that the above expressions are of little use. For practical reasons, even though not strictly applicable unless the equilibrium curve is of constant slope, it is usually necessary to deal with overall coefficients and transfer units. These we can take from their gas-absorption counterparts:

$$Z = H_{tOR}N_{tOR} = H_{tOE}N_{tOE} \tag{10.103}$$

$$H_{tOR} = \frac{R}{F_{OR}a} = \frac{R}{K_R a(1 - x)_{*M}} \tag{10.104}$$

$$H_{tOE} = \frac{E}{F_{OE}a} = \frac{E}{K_E a(1 - y)_{*M}} \tag{10.105}$$

$$N_{tOR} = \int_{x_2}^{x_1} \frac{(1 - x)_{*M}dx}{(1 - x)(x - x^*)} = \int_{x_2}^{x_1} \frac{dx}{x - x^*} + \frac{1}{2}\ln\frac{1 - x_2}{1 - x_1} \tag{10.106}$$

$$N_{tOE} = \int_{y_2}^{y_1} \frac{(1 - y)_{*M}dy}{(1 - y)(y^* - y)} = \int_{y_2}^{y_1} \frac{dy}{y^* - y} + \frac{1}{2}\ln\frac{1 - y_1}{1 - y_2} \tag{10.107}$$

$$(1 - x)_{*M} = \frac{(1 - x^*) - (1 - x)}{\ln[(1 - x^*)/(1 - x)]} \tag{10.108}$$

$$(1 - y)_{*M} = \frac{(1 - y) - (1 - y^*)}{\ln[(1 - y)/(1 - y^*)]} \tag{10.109}$$

where x^* is the concentration in equilibrium with y and y^* that in equilibrium with x.

Concentrations in Eqs. (10.103) to (10.109) are in mole fractions. *If x and y are expressed as weight fractions*, for convenience in use with the stage-calculation operating diagrams in terms of weight fractions,

$$N_{tOR} = \int_{x_2}^{x_1} \frac{dx}{x - x^*} + \frac{1}{2}\ln\frac{1 - x_2}{1 - x_1} + \frac{1}{2}\ln\frac{x_2(r - 1) + 1}{x_1(r - 1) + 1} \tag{10.110}$$

$$N_{tOE} = \int_{y_2}^{y_1} \frac{dy}{y^* - y} + \frac{1}{2}\ln\frac{1 - y_1}{1 - y_2} + \frac{1}{2}\ln\frac{y_1(r - 1) + 1}{y_2(r - 1) + 1} \tag{10.111}$$

where r is the ratio of molecular weights of nonsolute to solute. For weight-ratio concentrations,

$$N_{tOR} = \int_{x_2'}^{x_1'} \frac{dx'}{x' - x'^*} + \frac{1}{2}\ln\frac{1 + rx_2'}{1 + rx_1'} \tag{10.112}$$

$$N_{tOE} = \int_{y_2'}^{y_1'} \frac{dy'}{y'^* - y'} + \frac{1}{2}\ln\frac{1 + ry_1'}{1 + ry_2'} \tag{10.113}$$

Dilute Solutions

For dilute solutions, only the integral terms of the above equations for N_{tOE} and N_{tOR} are important. If in addition the equilibrium curve and operating line are straight over the concentration range encountered, it is readily shown in the manner of Chap. 8 that logarithmic averages of the terminal concentration differences are applicable

$$N_{tOR} = \frac{x_1 - x_2}{(x - x^*)_M} \qquad N_{tOE} = \frac{y_1 - y_2}{(y^* - y)_M} \qquad (10.114)$$

The equivalent expressions in terms of mass-transfer coefficients are

$$R(x_1 - x_2) = E(y_1 - y_2) = K_R aZ(x - x^*)_M = K_E aZ(y^* - y)_M \qquad (10.115)$$

If in addition the equivalent of Henry's law applies, so that the equilibrium-distribution curve is a straight line passing through the origin ($m = y^*/x = y/x^* = $ const), a procedure exactly similar to that used previously in the case of gas absorption gives

$$N_{tOR} = \frac{\ln\left[\dfrac{x_1 - y_2/m}{x_2 - y_2/m}\left(1 - \dfrac{R}{mE}\right) + \dfrac{R}{mE}\right]}{1 - R/mE} \qquad (10.116)$$

Figure 8.20 represents a graphical solution, provided $(x_2 - y_2/m)/(x_1 - y_2/m)$ is considered the ordinate and mE/R the parameter. Similarly for the extract-enriching section of a tower used with reflux, the same circumstances provide

$$N_{tOE} = \frac{\ln\left[\dfrac{y_2 - mx_1}{y_1 - mx_1}\left(1 - \dfrac{mE}{R}\right) + \dfrac{mE}{R}\right]}{1 - mE/R} \qquad (10.117)$$

which is also solved graphically in Fig. 8.20 provided $(y_1 - mx_1)/(y_2 - mx_1)$ is the ordinate and R/mE the parameter.

For these dilute solutions Eqs. (10.114), (10.116), and (10.117) can be used with concentrations in terms of weight fractions, in which case m must be defined in these terms as well while E and R are measured as mass/(area)(time). Weight ratios can also be used. Equations (10.114) and (10.115) are frequently used with concentrations expressed as c mol/volume, in which case the extract and raffinate flow rates are measured in terms of V volume/(area)(time) and the appropriate mass-transfer coefficients are K_{LE} and K_{LR} mol/(area)(time) · Δc. The requirements of constant m (in whatever units it is measured) and straight operating lines still apply, of course.

Illustration 10.9 Determine the number of transfer units N_{tOR} for the extraction of Illustration 10.3 if 20 000 kg/h of solvent is used.

SOLUTION Define x and y in terms of weight fractions acetic acid. $x_1 = x_F = 0.30$; $y_2 = 0$; $x_2 = 0.02$; $y_1 = 0.10$. The operating diagram is already plotted in Fig. 10.23. From this plot, values of x and x^* are taken from the operating line and equilibrium curve at various values of

y, as follows:

x	0.30	0.25	0.20	0.15	0.10	0.05	0.02
x^*	0.230	0.192	0.154	0.114	0.075	0.030	0
$\dfrac{1}{x - x^*}$	14.30	17.25	20.75	27.8	40.0	50.0	50.0

The area under a curve of x as abscissa against $1/(x - x^*)$ as ordinate (not shown) between $x = 0.30$ and $x = 0.02$ is determined to be 8.40. In these solutions, the mutual solubility of water and isopropyl ether is very small, so that r can be taken as $18/60 = 0.30$. Eq. (10.110):

$$N_{tOR} = 8.40 + \tfrac{1}{2}\ln\frac{1 - 0.02}{1 - 0.30} + \tfrac{1}{2}\ln\frac{0.02(0.3 - 1) + 1}{0.30(0.3 - 1) + 1} = 8.46$$

The operating and equilibrium curves are nearly parallel in this case, so that N_{tOR} and N_p are nearly the same. The curvature of the lines makes the simplified methods for N_{tOR} inapplicable, however.

Illustration 10.10 Determine the number of transfer units N_{tOR} for the extraction of Illustration 10.4 if 1150 kg/h of kerosene are used.

SOLUTION Use weight-ratio concentrations, as in Illustration 10.4. $x_1' = x_F' = 0.0101$; $y_2' = 0$; $x_2' = 0.001\ 001$; $y_1' = 0.0782$. The calculation can be done through Eq. (10.116) or the equivalent, Fig. 8.20.

$$\frac{x_2' - y_2'/m'}{x_1' - y_2'/m'} = \frac{0.001\ 001}{0.0101} = 0.0909$$

The average $mE/R = m'B/A = 1.01$ (Illustration 10.4). From Fig. 8.20, $N_{tOR} = 8.8$.

Performance of Continuous-Contact Equipment

While over the past 40 years a considerable number of data has been accumulated, taken almost entirely from laboratory-size equipment of a few centimeters diameter, no satisfactory correlation of them has as yet been possible owing to the very large number of variables which influence extraction rates. For the design of new extractors it is essential that pilot-plant experiments be performed under conditions as nearly like those expected in the large-scale equipment as possible [75]. This discussion will therefore be limited to a brief consideration of the important variables, but no data for design will be presented.

The principal difficulty in obtaining an understanding of extractor performance lies with the very large number of variables which influence the performance. The following at least have influence.

1. The liquid system.
   a. Chemical identity and corresponding physical properties. In this category may be included the presence or absence of surface-active agents, finely divided solids, and the like.
   b. Concentration of solute, since this influences physical properties.
   c. Direction of extraction, whether from aqueous to organic, from continuous to dispersed phase.
   d. Total flow rate of the liquids.
   e. Ratio of liquid flows.
   f. What liquid is dispersed.

2. The equipment.
   a. Design, which includes not only the gross and obvious design such as whether the extractor is a packed or mechanically agitated tower, but also such details as size and shape of packing, arrangement of baffles, and the like.
   b. Nature and extent of mechanical agitation, whether rotary or pulsating, fast or slow.
   c. Materials of construction, which influence the relative wetting by the liquids.
   d. Height of the extractor and the end effects.
   e. Diameter of extractor and extent of axial mixing.

It is only recently that a beginning has been made in systematizing axial-mixing data, and reliable correlations of mass-transfer coefficients must wait upon this. Practical applications are advanced far ahead of sound design data.

NOTATION FOR CHAPTER 10

Any consistent set of units may be used, except as noted.

a	specific interfacial surface, L^2/L^3
A	component A, mass/time (continuous), M/Θ; mass (batch), M
A_a	active area of tower devoted to perforations, L^2
A_d	cross-sectional area of downspouts (or upspouts), L^2
A_n	net tower cross-sectional area $= A_t - A_d$, L^2
A_p	surface of drops, L^2
A_r	surface of rising (or falling) drops between trays, L^2
A_t	tower cross-sectional area, L^2
b	constant, dimensionless
B	component B, mass/time (continuous), M/Θ; mass (batch), M
B_n	eigenvalue, dimensionless
c	concentration, especially of solute, $mole/L^3$
c_D^*	dispersed-phase concentration in equilibrium with bulk continuous phase, $mole/L^3$
C	component C, mass/time (continuous), M/Θ; mass (batch), M
C_D	drag coefficient $= 4\Delta\rho\, d_p g/3\rho_C V_t^2$, dimensionless
d_i	impeller diameter, L
d_J	jet diameter, L
d_o	orifice diameter, L
d_p	drop diameter, L
$d_{p,\,trans}$	transition-size drop diameter, L
D	component D, mass/time (continuous), M/Θ; diffusivity, L^2/Θ
E	extract solution, mass/time (staged extractors), M/Θ; mass (batch), M; moles/(area)(time) (continuous-contact extractors), $mole/L^2\Theta$
E'	solvent (B)–free extract, mass/time (continuous), M/Θ; mass (batch), M
E_{MD}	Murphree dispersed-phase stage efficiency, fractional
E_{ME}	Murphree extract stage efficiency, fractional
E_{MR}	Murphree raffinate stage efficiency, fractional
E_O	overall stage efficiency, fractional
F	feed, mass/time (continuous), M/Θ; mass (batch), M
F'	feed, solvent (B)–free basis, mass/time (continuous), M/Θ; mass (batch), M
g	acceleration of gravity, L^2/Θ
g_c	conversion factor, $ML/F\Theta^2$
h	depth of coalesced dispersed liquid accumulating on a tray, L
h_C	depth of dispersed liquid accumulating on a tray owing to flow of continuous liquid, L

h_D	depth of dispersed liquid accumulating on a tray owing to flow of dispersed liquid, L
h_o	head required to cause flow through an orifice, L
h_R	head required to cause flow through a downspout restriction, L
h_σ	head required to overcome effect of interfacial tension, L
H_t	height of a transfer unit, L
k	mass-transfer coefficient, $\text{mole}/L^2\Theta(\text{mole fraction})$
k_L	mass-transfer coefficient, $\text{mole}/L^2\Theta(\text{mole}/L^3)$
K	overall mass-transfer coefficient, $\text{mole}/L^2\Theta(\text{mole fraction})$
K_L	overall mass-transfer coefficient, $\text{mole}/L^2\Theta(\text{mole}/L^3)$
ln	natural logarithm
m	equilibrium distribution coefficient, concentration in extract/concentration in raffinate, dimensionless
m_{CD}	equilibrium distribution coefficient, concentration in continuous phase/concentration in dispersed phase, $(\text{mole}/L^3)/(\text{mole}/L^3)$
M	mixture M, mass/time (continuous), M/Θ; mass (batch), M
M'	solvent (B)–free mixture M, mass/time (continuous), M/Θ; mass (batch), M
N	solvent concentration, solvent (B)–free basis, mass B/mass (A + C), M/M
N_f	time-average flux of mass transfer during drop formation, $\text{mole}/L^2\Theta$
N_o	number of perforations, dimensionless
N_p	number of theoretical stages, dimensionless
N_t	number of transfer units, dimensionless
P	power, FL/Θ
P'_E	solvent (B)–free extract product, mass/time, M/Θ
q	volumetric rate, L^3/Θ
r	ratio of molecular weight, nonsolute to solute, dimensionless
R	raffinate, mass/time (staged extraction), M/Θ; mass (batch), M; moles/(area)(time) (continuous-contact extraction), $\text{mole}/L^2\Theta$
R'	solvent (B)–free raffinate, mass/time (continuous), M/Θ; mass (batch), M
Re	drop Reynolds number, $d_p V_t \rho_C/\mu_C$, dimensionless
S	solvent, mass/time (continuous), M/Θ; mass (batch), M
S'	B–free solvent, mass/time (continuous), M/Θ; mass (batch), M
Sc	Schmidt number, $\mu/\rho D$, dimensionless
Sh	Sherwood number, kd_p/D, dimensionless
T	tower diameter, L
T_S	settler diameter, L
U	dimensionless group defined by Eq. (10.75)
v_L	liquid volume, L^3
V	superficial velocity, L/Θ
V_d	velocity of liquid in downspout, L/Θ
V_n	velocity based on A_n, L/Θ
V_o	velocity through orifice or perforation, L/Θ
$V_{o,\,max}$	velocity through perforation leading to maximum specific interfacial area, L/Θ
V_R	velocity based on a restricted area, L/Θ
V_S	slip velocity [Eqs. (10.83) and (10.84)], L/Θ
V_t	terminal settling velocity, L/Θ
We	drop Weber number, $d_p V_t^2 \rho_C/\sigma g_c$, dimensionless
x	concentration of C in A-rich (raffinate) phase, mass fraction (staged extractors); mole fraction (continuous-contact extractors)
x'	concentration of C in A-rich (raffinate) phase, mass C/mass non-C, M/M concentration of solute in D-rich phase (fractional extraction), mass solute/mass nonsolute, M/M
X	concentration of C in A-rich (raffinate) phase, B-free basis, mass C/mass (A + C), M/M

y	concentration of C in B-rich (extract) phase, mass fraction (staged extractors); mole fraction (continuous extractors)
y'	concentration of C in B-rich (extract) phase, mass C/mass non-C, M/M; concentration of solute in A-rich phase (fractional extraction), mass solute/mass nonsolute, M/M
Y	concentration of C in B-rich (extract) phase, B-free basis, mass C/mass (A + C), M/M
Z	tray spacing, L; depth of liquid in agitated vessel, L
β	selectivity, dimensionless
δ	amplitude factor, dimensionless
Δ'_E	difference in flow, B-free basis [Eq. (10.40)]
Δ_R	difference in flow rate [Eq. (10.25)]
Δ'_R	difference in flow rate, B-free basis [Eq. (10.31)]
θ	time, Θ
θ_f	time for drop formation, Θ
λ_n	eigenvalue, dimensionless
μ	viscosity, M/LΘ
ρ	density, M/L^3
$\Delta\rho$	difference in densities, M/L^3
σ	interfacial tension, F/L
ϕ	volume fraction
ω	oscillation frequency, Θ^{-1}

Subscripts

c	during coalescence
C	continuous phase
D	dispersed phase
e	stage e
E	extract
f	formation
F	feed
m	stage m; minimum
M	mixture M; mean
n	stage n
o	organic; orifice
O	overall
R	raffinate
s	stage s
S	solvent
w	water
1	stage 1; that end of a continuous-contact tower where feed enters
2	stage 2; that end of a continuous-contact tower where solvent enters

Superscripts

$*$	in equilibrium with bulk concentration in the other phase
$'$	solvent-free

REFERENCES

1. Allak, A. M. A., and G. V. Jeffreys: *AIChE J.*, **20**, 564 (1974).
2. Angelo, J. B., E. N. Lightfoot, et al.: *AIChE J.*, **12**, 751 (1966); **14**, 458, 531 (1968); **16**, 771 (1970).

3. Bernard, C., P. Michel, and M. Tarnero: *Proc. Int. Solvent Extn. Conf., The Hague, 1971,* **2,** 1282.
4. Beyaert, B. O., L. Lapidus, and J. C. Elgin: *AIChE J.,* **7,** 46 (1961).
5. Bibaud, R., and R. E. Treybal: *AIChE J.,* **12,** 472 (1966).
6. Bouyatiotis, B. A., and J. D. Thornton: *Ind. Chem.,* **39,** 298 (1963); *Inst. Chem. Eng. Lond. Symp. Ser.,* **26,** 43 (1967).
7. Burkhart, L., P. W. Weathers, and P. C. Sharer: *AIChE J.,* **22,** 1090 (1976).
8. Burkholder, H. C., and J. C. Berg: *AIChE J.,* **20,** 872 (1974).
9. Bussolari, R. S., S. Schiff, and R. E. Treybal: *Ind. Eng. Chem.,* **45,** 2413 (1953).
10. Calderbank, P. H.: in V. W. Uhl and J. B. Gray (eds.), "Mixing," vol. 2, p. 1, Academic, New York.
11. Christiansen, R. M., and A. N. Hixson: *Ind. Eng. Chem.,* **49,** 1017 (1957).
12. Drown, D. C., and W. J. Thomson: *Ind. Eng. Chem. Process Des. Dev.,* **16,** 197 (1977).
13. Dykstra, J., B. H. Thompson, and R. J. Clouse: *Ind. Eng. Chem.,* **50,** 161 (1958).
14. Eisenlohr, H.: *Ind. Chem.,* **27,** 271 (1951).
15. Elzinga, E. R., and J. T. Banchero: *Chem. Eng. Prog. Symp. Ser.,* **55**(29), 149 (1959).
16. Gustison, R. A., R. E. Treybal, and R. C. Capps: *Chem. Eng. Prog. Symp. Ser.,* **58**(39), 8 (1962).
17. Gutoff, E. B.: *AIChE J.,* **11,** 712 (1965).
18. Happel, J., and D. G. Jordan: "Chemical Process Economics," 2d ed., Dekker, New York, 1975.
19. Hayworth, C. B., and R. E. Treybal: *Ind. Eng. Chem.,* **42,** 1174 (1950).
20. Hazlett, R. N.: *Ind. Eng. Chem. Fundam.,* **8,** 625, 633 (1969).
21. Heertjes, P. M., and L. H. de Nie: in C. Hanson (ed.), "Recent Advances in Liquid-Liquid Extraction," chap. 10, Pergamon, New York, 1971.
22. Honnekamp, J. R., and L. E. Burkhart: *Ind. Eng. Chem. Process Des. Dev.,* **1,** 176 (1962).
23. Houlihan, R., and J. Landau: *Can. J. Chem. Eng.,* **52,** 758 (1974).
24. Hu, S., and R. C. Kintner: *AIChE J.,* **1,** 42 (1955).
25. Hughmark, G. A.: *Ind. Eng. Chem. Fundam.,* **6,** 408 (1967).
26. Ingham, J.: in C. Hanson (ed.), "Recent Advances in Liquid-Liquid Extraction," chap. 8, Pergamon, New York, 1971.
27. Ingham, J.: *Trans. Inst. Chem. Eng. Lond.,* **50,** 372 (1972).
28. Jeffreys, G. V., and G. A. Davies: in C. Hanson (ed.), "Recent Advances in Liquid-Liquid Extraction," chap. 14, Pergamon, New York, 1971.
29. Jeffreys, G. V., and C. J. Mumford: *Proc. Int. Solvent Extn. Conf., The Hague, 1971,* **2,** 112.
30. Johnson, A. I., and L. Braida: *Can. J. Chem. Eng.,* **35,** 165 (1957).
31. Karr, A. E.: *AIChE J.,* **5,** 446 (1956).
32. Karr, A. E., and T. C. Lo: *Proc. Int. Solvent Extn. Conf., The Hague, 1971,* **1,** 299; *Ind. Eng. Chem. Process Des. Dev.,* **11,** 495 (1952).
33. Keey, R. B., and J. Glen: *AIChE J.,* **15,** 942 (1969).
34. Klee, A., and R. E. Treybal: *AIChE J.,* **2,** 444 (1956).
35. Laity, D. S., and R. E. Treybal: *AIChE J.,* **3,** 176 (1957).
36. Levins, D. M., and J. J. Glastonbury: *Trans. Inst. Chem. Eng. Lond.,* **50,** 32, 132 (1972).
37. Luhning, R. W., and H. Sawistowski: *Proc. Int. Solvent Extn. Conf., The Hague, 1971,* **2,** 873.
38. Mar, B. W., and A. L. Babb: *Ind. Eng. Chem.,* **51,** 1011 (1959).
39. Mayfield, F. D., and W. L. Church: *Ind. Eng. Chem.,* **44,** 2253 (1952).
40. McDonough, J. A., W. J. Tomme, and C. D. Holland: *AIChE J.,* **6,** 615 (1960).
41. Misek, T., and V. Rod: in C. Hanson (ed.), "Recent Advances in Liquid-Liquid Extraction," chap. 7, Pergamon, New York, 1971.
42. Mok, Y. I., and R. E. Treybal: *AIChE J.,* **17,** 916 (1971).
43. Nemunaitis, R. R., J. S. Eckert, E. H. Foote, and L. Rollison: *Chem. Eng. Prog.,* **67**(11), 60 (1971).
44. Oldshue, J. Y., F. Hodgkinson, and J.-C. Pharamond: *Proc. Int. Solvent Extn. Conf., Lyons, 1974,* **2,** 1651.
45. Oldshue, J. Y., and J. H. Rushton: *Chem. Eng. Prog.,* **48,** 297 (1952).

46. Olney, R. B., and G. J. Carlson: *Chem. Eng. Prog.*, **43**, 473 (1947).
47. Pratt, H. R. C.: *Ind. Eng. Chem. Fundam.*, **10**, 170 (1971); *Ind. Eng. Chem. Process Des. Dev.*, **15**, 544 (1976).
48. Pratt, H. R. C., and coworkers: *Trans. Inst. Chem. Eng. Lond.*, **29**, 89, 110, 126 (1951); **31**, 57, 70, 78 (1953); **35**, 267 (1957).
49. Quinn, J. A., and D. B. Sigloh: *Can. J. Chem. Eng.*, **41**, 15 (1963).
50. Reman, G. H., and R. B. Olney: *Chem. Eng. Prog.*, **51**, 141 (1955).
51. Ricci, J. E.: "The Phase Rule and Heterogeneous Equilibria," Van Nostrand, New York, 1951 (also Dover).
52. Ruby, C. L., and J. C. Elgin: *Chem. Eng. Prog. Symp. Ser.*, **51**(16), 17 (1955).
53. Sareen, S. S., P. M. Rose, R. K. Guden, and R. C. Kintner: *AIChE J.*, **12**, 1045 (1966).
54. Scheele, G. F., and B. J. Meister: *AIChE J.*, **14**, 9 (1968); **15**, 689, 700 (1969).
55. Scheibel, E. G.: *Chem. Eng. Prog.*, **44**, 681, 771 (1948); *AIChE J.*, **2**, 74 (1956).
56. Scheibel, E. G.: *Chem. Eng. Prog.*, **62**(9), 76 (1966).
57. Scheibel, E. G., and A. E. Karr: *Ind. Eng. Chem.*, **42**, 1048 (1950); *Chem. Eng. Prog. Symp. Ser.*, **50**(10), 73 (1954).
58. Schindler, H. D., and R. E. Treybal: *AIChE J.*, **14**, 790 (1968).
59. Scott, L. S., W. B. Hayes, and C. D. Holland: *AIChE J.*, **4**, 346 (1958).
60. Sheikh, M. R., J. C. Ingham, and C. Hanson: *Trans. Inst. Chem. Eng. Lond.*, **50**, 199 (1972).
61. Sigales, B.: *Chem. Eng.*, **82**, 143 (June 23, 1975).
62. Skelland, A. H. P.: "Diffusional Mass Transfer," Wiley, New York, 1974.
63. Skelland, A. H. P., and W. L. Conger: *Ind. Eng. Chem. Process Des. Dev.*, **12**, 448 (1973).
64. Skelland, A. H. P., and K. R. Johnson: *Can. J. Chem. Eng.*, **52**, 732 (1974).
65. Skelland, A. H. P., and S. S. Minhas: *AIChE J.*, **17**, 1316 (1971).
66. Spielman, L. A., and S. L. Goren: *Ind. Eng. Chem. Fundam.*, **11**, 66, 73 (1972); *Ind. Eng. Chem.*, **62**, 10 (1970).
67. Strand, C. P., R. B. Olney, and G. H. Ackerman: *AIChE J.*, **8**, 252 (1962).
68. Todd, D. B.: *Chem. Eng. Prog.*, **62**(8), 119 (1966).
69. Todd, D. B., and G. R. Davies: *Proc. Int. Conf. Solvent Extn., Lyons, 1974*, **3**, 2380.
70. Todd, D. B., and W. J. Podbielniak: *Chem. Eng. Prog.*, **61**(5), 69 (1965).
71. Treybal, R. E.: *AIChE J.*, **5**, 474 (1959).
72. Treybal, R. E.: "Liquid Extraction," 2d ed., McGraw-Hill Book Company, New York, 1963.
73. Treybal, R. E.: *Chem. Eng. Prog.*, **60**(5), 77 (1964).
74. Treybal, R. E.: *Ind. Eng. Chem. Fundam.*, **3**, 185 (1964).
75. Treybal, R. E.: *Chem. Eng. Prog.*, **62**(9), 67 (1966).
76. Treybal, R. E.: in R. H. Perry and C. H. Chilton (eds.), "The Chemical Engineers' Handbook," 5th ed., p. 21-16, McGraw-Hill Book Company, New York, 1973.
77. van Dijck, W. J. D.: U.S. Patent 2,011,186 (1935).
78. Watson, J. S., and H. D. Cochran, Jr.: *Ind. Eng. Chem. Process Des. Dev.*, **10**, 83 (1971).
79. Watson, J. S., L. E. McNeese, J. Day, and P. A. Carroad: *AIChE J.*, **21**, 1080 (1975).
80. Weaver, R. E. C., L. Lapidus, and J. C. Elgin: *AIChE J.*, **5**, 333 (1959).
81. Weinstein, B., and R. E. Treybal: *AIChE J.*, **19**, 304, 851 (1973).
82. Woodfield, F. N., and G. Sege: *Chem. Eng. Prog. Symp. Ser.*, **50**(13), 14 (1954).
83. Yeh, J. C., F. H. Haynes, and R. A. Moses: *AIChE J.*, **10**, 260 (1964).
84. Zenz, F. A.: *Pet. Refiner*, **36**(8), 147 (1957).

PROBLEMS

Problems 10.1 to 10.7 refer to the system water (A)–chlorobenzene (B)–pyridine (C) at 25°C. Equilibrium tie-line data, interpolated from those of Peake and Thompson, *Ind. Eng. Chem.*, **44**,

2439 (1952), in weight percent are:

Pyridine	Chlorobenzene	Water	Pyridine	Chlorobenzene	Water
0	99.95	0.05	0	0.08	99.92
11.05	88.28	0.67	5.02	0.16	94.82
18.95	79.90	1.15	11.05	0.24	88.71
24.10	74.28	1.62	18.90	0.38	80.72
28.60	69.15	2.25	25.50	0.58	73.92
31.55	65.58	2.87	36.10	1.85	62.05
35.05	61.00	3.95	44.95	4.18	50.87
40.60	53.00	6.40	53.20	8.90	37.90
49.0	37.8	13.2	49.0	37.8	13.2

10.1 Plot the equilibrium data on the following coordinate systems: (a) triangular; (b) x and y against weight fraction B; (c) x against y.

10.2 Compute the selectivity of chlorobenzene for pyridine at each tie line, and plot selectivity against concentration of pyridine in water.

10.3 It is desired to reduce the pyridine concentration of 2000 kg of an aqueous solution from 50 to 2% in a single batch extraction with chlorobenzene. What amount of solvent is required? Solve on triangular coordinates.

10.4 A 2000-kg batch of pyridine-water solution, 50% pyridine, is to be extracted with an equal weight of chlorobenzene. The raffinate from the first extraction is to be reextracted with a weight of solvent equal to the raffinate weight, and so on ($B_2 = R_1$, $B_3 = R_2$, etc.). How many theoretical stages and what total solvent will be required to reduce the concentration of pyridine to 2% in the final raffinate? Solve on triangular coordinates.

10.5 A pyridine-water solution, 50% pyridine, is to be continuously and countercurrently extracted at the rate 2.25 kg/s (17 800 lb/h) with chlorobenzene to reduce the pyridine concentration to 2%. Using the coordinate systems plotted in (b) and (c) of Prob. 10.1:

    (a) Determine the minimum solvent rate required.

    (b) If 2.3 kg/s (18 250 lb/h) is used, what are the number of theoretical stages and the saturated weights of extract and raffinate?

    Ans.: 3 theoretical stages.

    (c) Determine the number of transfer units N_{tOR} for the extraction of part (b).

    Ans.: 4.89

10.6 The properties of the solutions have not been completely studied, but from those of the pure constituents, the following properties are estimated:

	Density		Viscosity		Interfacial tension	
	kg/m³	lb/ft³	kg/m · s	cP	N/m	dyn/cm
Feed	994.8	62.1	0.001	1.0	0.008	8
Extract	1041	65	0.0013	1.3		
Raffinate	996.4	62.2	0.00089	0.89	0.035	35
Solvent	1097	68.5	0.00125	1.25		

For the solvent-entrance end of the cascade of Prob. 10.5b, specify the dimensions of a mixing vessel and the size and speed of the impeller and estimate the stage efficiency. Disperse the solvent, and use an average holding time of 30 s.

10.7 Can the separation of Prob. 10.5 be made by distillation at atmospheric pressure?

10.8 Water-dioxane solutions form a minimum-boiling azeotrope at atmospheric pressure and cannot be separated by ordinary distillation methods. Benzene forms no azeotrope with dioxane and can be used as an extraction solvent. At 25°C, the equilibrium distribution of dioxane between water and benzene [*J. Am. Chem. Soc.*, **66**, 282 (1944)] is as follows:

Wt % dioxane in water	5.1	18.9	25.2
Wt % dioxane in benzene	5.2	22.5	32.0

At these concentrations water and benzene are substantially insoluble, and 1000 kg of a 25% dioxane-75% water solution is to be extracted with benzene to remove 95% of the dioxane. The benzene is dioxane-free.

(a) Calculate the solvent requirement for a single batch operation.

(b) If the extraction were done with equal amounts of solvent in five crosscurrent stages, how much solvent would be required?

10.9 A 25% solution of dioxane in water is to be continuously extracted at a rate of 1000 kg/h in countercurrent fashion with benzene to remove 95% of the dioxane. Equilibrium data are given in Prob. 10.8.

(a) What is the minimum solvent requirement, kg/h?

(b) If 900 kg/h of solvent is used, how many theoretical stages are required?

Ans.: 6.1 theoretical stages.

(c) How many transfer units N_{tOR} correspond to the extraction of part (b)?

10.10 An aqueous solution contains 25% acetone by weight together with a small amount of an undesired contaminant. For the purpose of a later process, it is necessary to have the acetone dissolved in water without the impurity. To accomplish this, the solution will be extracted countercurrently with trichloroethane, which extracts the acetone but not the impurity. The extract will then be countercurrently extracted with pure water in a second extractor to give the desired product water solution, and the recovered solvent will be returned to the first extractor. It is required to obtain 98% of the acetone in the final product. Water and trichloroethane are insoluble over the acetone-concentration range involved, and the distribution coefficient (kg acetone/kg trichloroethane)/(kg acetone/kg water) = 1.65 = const.

(a) What is the largest concentration of acetone possible in the recovered solvent?

(b) How many stages would be required in each extractor to obtain the acetone in the final water solution at the same concentration as in the original solution?

(c) If recovered solvent contains 0.005 kg acetone/kg trichloroethane, if 1 kg trichloroethane/kg water is used in the first extractor, and if the same number of stages is used in each extractor, what concentration of acetone in the final product will result?

10.11 A sieve-tray tower is to be designed for the extraction of 90% of the acetic acid from a water solution containing 4.0% acid, using methyl isobutyl ketone, initially free of acid, as solvent, at 25°C. The flow rates are to be 1.6 × 10⁻³ m³/s aqueous (203.4 ft³/h), 3.2 × 10⁻³ m³/s organic. The ketone is to be dispersed.

Physical properties are, for the aqueous solution, viscosity = 0.001 kg/m · s, density = 998 kg/m³, diffusivity of acetic acid = 1.00 × 10⁻⁹ m²/s; for the organic solution, viscosity = 5.70 × 10⁻⁴ kg/m · s, density = 801 kg/m³, diffusivity of acetic acid = 1.30 × 10⁻⁹ m²/s; distribution coefficient, (c in ketone)/(c in water) = 0.545; interfacial tension = 9.1 × 10⁻³ N/m.

Specify the tray dimensions and arrangement, estimate the values of E_{MD} and E_O, and specify the number of real trays required.

Problems 10.12 to 10.14 refer to the system cottonseed oil (A)–liquid propane (B)–oleic acid (C) at 98.5°C, 625 lb/in² abs. Smoothed equilibrium tie-line data of Hixson and Bockelmann, *Trans.*

AIChE, **38**, 891 (1942), in weight percent, are as follows:

Cottonseed oil	Oleic acid	Propane	Cottonseed oil	Oleic acid	Propane
63.5	0	36.5	2.30	0	97.7
57.2	5.5	37.3	1.95	0.76	97.3
52.0	9.0	39.0	1.78	1.21	97.0
46.7	13.8	39.5	1.50	1.90	96.6
39.8	18.7	41.5	1.36	2.73	95.9
31.0	26.3	42.7	1.20	3.8	95.0
26.9	29.4	43.7	1.10	4.4	94.5
21.0	32.4	46.6	1.0	5.1	93.9
14.2	37.4	48.4	0.8	6.1	93.1
8.3	39.5	52.2	0.7	7.2	92.1
4.5	41.1	54.4	0.4	6.1	93.5
0.8	43.7	55.5	0.2	5.5	94.3

10.12 Plot the equilibrium data on the following coordinate systems: (a) N against X and Y; (b) X against Y.

10.13 If 100 kg of a cottonseed oil-oleic acid solution containing 25% acid is to be extracted twice in crosscurrent fashion, each time with 1000 kg of propane, determine the compositions, percent by weight, and the weights of the mixed extracts and the final raffinate. Determine the compositions and weights of the solvent-free products. Make the computations on the coordinates plotted in part (a) of Prob. 10.12.

10.14 If 1000 kg/h of a cottonseed oil-oleic acid solution containing 25% acid is to be continuously separated into products containing 2 and 90% acid (solvent-free compositions) by countercurrent extraction with propane, make the following computations on the coordinate systems of parts (a) and (b) of Prob. 10.12:

    (a) What is the minimum number of theoretical stages required?
    Ans.: 5.
    (b) What is the minimum external extract-reflux ratio required?
    Ans.: 3.08.
    (c) For an external extract-reflux ratio of 4.5, determine the number of theoretical stages, the position of the feed stage, and the quantities, in kg/h, of the following streams: E_1, B_E, E', R_0, R_{Np}, P'_E, and S.
    Ans.: 10.5 theoretical stages.
    (d) What do the equilibrium data indicate as to the maximum purity of oleic acid that could be obtained?

Problems 10.15 and 10.16 refer to the system oxalic and succinic acids distributed between water and *n*-amyl alcohol. The acids distribute practically independently of each other, and for present purposes they will be considered to do so. Equilibrium concentrations, expressed as g acid/l, are:

Either acid in water	0	20	40	60
Oxalic in alcohol	0	6.0	13.0	21.5
Succinic in alcohol	0	12.0	23.5	35.0

The water and amyl alcohol are essentially immiscible. In the calculations, express acid concentrations as g/l and solvent rates as l/time.

10.15 A water solution containing 50 g each of oxalic and succinic acids per liter of water is to be extracted countercurrently with *n*-amyl alcohol (free of acid) to recover 95% of the oxalic acid.

    (a) Compute the minimum solvent required, alcohol/l water.

(b) If an alcohol/water ratio of 3.8 1/1^2 is used, compute the number of theoretical stages required and the solvent-free analysis of the extract product.

Ans.: 9 theoretical stages.

10.16 A mixture of succinic (B) and oxalic (C) acids containing 50% of each is to be separated into products each 90% pure on a solvent-free basis, using n-amyl alcohol (A) and water (D) by a fractional extraction.

(a) Calculate the number of theoretical stages and the location of the feed stage of 9 l of alcohol and 4 l of water are used per 100 g of acid feed.

Ans.: 12.75 stages.

(b) Investigate the effect of solvent ratio on the number of stages, keeping the total liquid flow constant. What is the least number of stages required and the corresponding solvent ratio?

FOUR

SOLID-FLUID OPERATIONS

The mass-transfer operations in this category are adsorption-desorption, drying, and leaching. Adsorption involves contact of solids with either liquids or gases and mass transfer in the direction fluid to solid. Mass transfer is in the opposite direction for its companion operation, desorption, and the combination is analogous to the gas-absorption–stripping pair of Chap. 8. Drying involves gas-solid, and leaching liquid-solid, contact, with mass transfer in each case in the direction solid to fluid, so that these are special cases of desorption.

Theoretically, at least, the same apparatus and equipment useful for gas-solid or liquid-solid contact in adsorption should also be useful in the corresponding operations of drying and leaching. In practice, however, we find special types of apparatus in all three categories. This is probably the result of many years of development of the practical applications of these operations without the realization that basically they are very similar. For example, we find considerable inventive genius applied to the development of equipment for the continuous gas-solid operations of adsorption but little application of the results to the problem of drying. Many clever devices have been developed for continuous leaching of solids with liquids, but there is little application of these to the practical problems of adsorption from liquids. Ideal devices have not yet been invented. But we may reasonably expect a reduction in the number of equipment types and greater interapplication between the three operations as the many problems of solids handling are eventually solved.

It helps considerably in understanding these operations if the strong resemblances to the gas and liquid operations of earlier chapters are kept in mind, although the fixed-bed and fluidized-bed operations of which solids are capable have no gas-liquid counterparts.

PART
FOUR

SOLID-FLUID OPERATIONS

ELEVEN
ADSORPTION AND ION EXCHANGE

The adsorption operations exploit the ability of certain solids preferentially to concentrate specific substances from solution onto their surfaces. In this manner, the components of either gaseous or liquid solutions can be separated from each other. A few examples will indicate the general nature of the separations possible and at the same time demonstrate the great variety of practical applications. In the field of gaseous separations, adsorption is used to dehumidify air and other gases, to remove objectionable odors and impurities from industrial gases such as carbon dioxide, to recover valuable solvent vapors from dilute mixtures with air and other gases, and to fractionate mixtures of hydrocarbon gases containing such substances as methane, ethylene, ethane, propylene, and propane. Typical liquid separations include the removal of moisture dissolved in gasoline, decolorization of petroleum products and aqueous sugar solutions, removal of objectionable taste and odor from water, and the fractionation of mixtures of aromatic and paraffinic hydrocarbons. The scale of operations ranges from the use of a few grams of adsorbent in the laboratory to industrial plants with an adsorbent inventory exceeding 135 000 kg [59].

These operations are all similar in that the mixture to be separated is brought into contact with another insoluble phase, the adsorbent solid, and the unequal distribution of the original constituents between the adsorbed phase on the solid surface and the bulk of the fluid then permits a separation to be made. All the techniques previously found valuable in the contact of insoluble fluids are useful in adsorption. Thus we have batchwise single-stage and continuous multistage separations and separations analogous to countercurrent absorption and stripping in the field of gas-liquid contact and to rectification and extraction with the use of reflux. In addition, the rigidity and immobility of a bed of solid adsorbent particles make possible useful application of semicontinuous methods which are not at all practicable when two fluids are contacted.

Another solid-liquid operation of great importance is ion exchange, the reversible exchange of ions between certain solids and an electrolyte solution, which permits the separation and fractionation of electrolytic solutes. It is, of course, chemical in nature but involves not only the interaction of the ions with the solid but also diffusion of ions within the solid phase. Although the phenomenon may be more complex than adsorption, the general techniques and the results obtained are very similar. The special features of ion exchange are considered separately at the end of this chapter.

Types of Adsorption

We must distinguish at the start between two types of adsorption phenomena, physical and chemical.

Physical adsorption, or *van der Waals adsorption*, a readily reversible phenomenon, is the result of intermolecular forces of attraction between molecules of the solid and the substance adsorbed. When, for example, the intermolecular attractive forces between a solid and a gas are greater than those existing between molecules of the gas itself, the gas will condense upon the surface of the solid even though its pressure may be lower than the vapor pressure corresponding to the prevailing temperature. Such a condensation will be accompanied by an evolution of heat, in amount usually somewhat larger than the latent heat of vaporization and of the order of the heat of sublimation of the gas. The adsorbed substance does not penetrate within the crystal lattice of the solid and does not dissolve in it but remains entirely upon the surface [82]. If, however, the solid is highly porous, containing many fine capillaries, the adsorbed substance will penetrate these interstices if it wets the solid. The equilibrium vapor pressure of a concave liquid surface of very small radius of curvature is lower than that of a large flat surface, and the extent of adsorption is correspondingly increased. In any case, at equilibrium the partial pressure of the adsorbed substance equals that of the contacting gas phase, and by lowering the pressure of the gas phase or by raising the temperature the adsorbed gas is readily removed or desorbed in unchanged form. Industrial adsorption operations of the type we shall consider depend upon this reversibility for recovery of the adsorbent for reuse, for recovery of the adsorbed substance, or for the fractionation of mixtures. Reversible adsorption is not confined to gases but is observed with liquids as well.

Chemisorption, or activated adsorption, is the result of chemical interaction between the solid and the adsorbed substance [13]. The strength of the chemical bond may vary considerably, and identifiable chemical compounds in the usual sense may not actually form, but the adhesive force is generally much greater than that found in physical adsorption. The heat liberated during chemisorption is usually large, of the order of the heat of chemical reaction. The process is frequently irreversible, and on desorption the original substance will often be found to have undergone a chemical change. The same substance which, under conditions of low temperature, will undergo substantially only physical adsorption upon a solid will sometimes exhibit chemisorption at higher temperatures, and both phenomena may occur at the same time. Chemisorption is of particular importance in catalysis but will not be considered here.

Nature of Adsorbents

Adsorbent solids are usually used in granular form, varying in size from roughly 12 mm in diameter to as small as 50 μm. The solids must possess certain engineering properties depending upon the application to which they are put. If they are used in a fixed bed through which a liquid or gas is to flow, for example, they must not offer too great a pressure drop for flow nor must they easily be carried away by the flowing stream. They must have adequate strength and hardness so as not to be reduced in size during handling or crushed in supporting their own weight in beds of the required thickness. If they are to be transported frequently in and out of bins, they should be free-flowing. These are properties which are readily recognized.

The adsorptive ability of solids is quite another matter. Adsorption is a very general phenomenon, and even common solids will adsorb gases and vapors at least to a certain extent. For example, every student of analytical chemistry has observed with annoyance the increase in weight of a dried porcelain crucible on a humid day during an analytical weighing which results from the adsorption of moisture from the air upon the crucible surface. But only certain solids exhibit sufficient specificity and adsorptive capacity to make them useful as industrial adsorbents. Since solids are frequently very specific in their ability to adsorb certain substances in large amounts, the chemical nature of the solid evidently has much to do with its adsorption characteristics. But mere chemical identity is insufficient to characterize its usefulness. In liquid extraction, all samples of pure butyl acetate will extract acetic acid from a water solution with identical ability. The same is not true for the adsorption characteristics of silica gel with respect to water vapor, for example. Much depends on its method of manufacture and on its prior history of adsorption and desorption.

Large surface per unit weight seems essential to all useful adsorbents. Particularly in the case of gas adsorption, the significant surface is not the gross surface of the granular particles which are ordinarily used but the very much larger surface of the internal pores of the particles. The pores are usually very small, sometimes of the order of a few molecular diameters in width, but their large number provides an enormous surface for adsorption. It is estimated, for example, that a typical gasmask charcoal has an effective surface of 1 000 000 m^2/kg [32]. There are many other properties evidently of great importance which are not at all understood, and we must depend largely on empirical observation for recognition of adsorptive ability. The following is a list of the principal adsorbents in general use.

1. *Fuller's earths.* These are natural clays, the American varieties coming largely from Florida and Georgia. They are chiefly magnesium aluminum silicates in the form of the minerals attapulgite

and montmorillonite. The clay is heated and dried, during which operation it develops a porous structure, ground, and screened. Commercially available sizes range from coarse granules to fine powders. The clays are particularly useful in decolorizing, neutralizing, and drying such petroleum products as lubricating oils, transformer oils, kerosenes, and gasolines, as well as vegetable and animal oils. By washing and burning the adsorbed organic matter accumulating upon the clay during use, the adsorbent can be reused many times.

2. *Activated clays.* These are bentonite or other clays which show essentially no adsorptive ability unless activated by treatment with sulfuric or hydrochloric acid. Following such treatment, the clay is washed, dried, and ground to a fine powder. It is particularly useful for decolorizing petroleum products and is ordinarily discarded after a single application.

3. *Bauxite.* This is a certain form of naturally occurring hydrated alumina which must be activated by heating to temperatures varying from 230 to 815°C in order to develop its adsorptive ability. It is used for decolorizing petroleum products and for drying gases and can be reactivated by heating.

4. *Alumina.* This is a hard, hydrated aluminum oxide which is activated by heating to drive off the moisture. The porous product is available as granules or powders, and it is used chiefly as a desiccant for gases and liquids. It can be reactivated for reuse.

5. *Bone char.* This is obtained by the destructive distillation of crushed, dried bones at temperatures in the range 600 to 900°C. It is used chiefly in the refining of sugar and can be reused after washing and burning.

6. *Decolorizing carbons.* These are variously made by (1) mixing vegetable matter with inorganic substances such as calcium chloride, carbonizing, and leaching away the inorganic matter, (2) mixing organic matter such as sawdust, etc., with porous substances such as pumice stone, followed by heating and carbonizing to deposit the carbonaceous matter throughout the porous particles, and (3) carbonizing wood, sawdust, and the like, followed by activation with hot air or steam. Lignite and bituminous coal are also raw materials. They are used for a great variety of purposes, including the decolorizing of solutions of sugar, industrial chemicals, drugs, and dry-cleaning liquids, water purification, refining of vegetable and animal oils, and in recovery of gold and silver from cyanide ore-leach solutions.

7. *Gas-adsorbent carbon.* This is made by carbonization of coconut shells, fruit pits, coal, lignite, and wood. It must be activated, essentially a partial oxidation process, by treatment with hot air or steam. It is available in granular or pelleted form and is used for recovery of solvent vapors from gas mixtures, gas masks, collection of gasoline hydrocarbons from natural gas, and the fractionation of hydrocarbon gases. It is revivified for reuse by evaporation of the adsorbed gas.

8. *Molecular-screening activated carbon.* This is a specially made form with pore openings controlled from 5 to 5.5 Å (those of most activated carbons range from 14 to 60 Å.[†] The pores can admit paraffin hydrocarbons, for example, but reject isoparaffins of large molecular diameters. The product is useful in fractionating acetylene compounds, alcohols, organic acids, ketones, aldehydes, and many others. A good general review of activated carbon is available [37].

9. *Synthetic polymeric adsorbents* [95]. These are porous spherical beads, 0.5 mm diameter, each bead a collection of microspheres, 10^{-4} mm diameter. The material is synthetic, made from polymerizable monomers of two major types. Those made from unsaturated aromatics such as styrene and divinylbenzene are useful for adsorbing nonpolar organics from aqueous solution. Those made from acrylic esters are suitable for more polar solutes. They are used principally for treating water solutions and are regenerated by leaching with low-molecular-weight alcohols or ketones.

10. *Silica gel.* This is a hard, granular, very porous product made from the gel precipitated by acid treatment of sodium silicate solution. Its moisture content before use varies from roughly 4 to 7 percent, and it is used principally for dehydration of air and other gases, in gas masks, and for fractionation of hydrocarbons. It is revivified for reuse by evaporation of the adsorbed matter.

11. *Molecular sieves* [41, 58]. These are porous, synthetic zeolite crystals, metal aluminosilicates.

† The angstrom (Å) is 10^{-10} m.

The "cages" of the crystal cells can entrap adsorbed matter, and the diameter of the passageways, controlled by the crystal composition, regulates the size of the molecules which can enter. The sieves can thus separate according to molecular size, but they also separate by adsorption according to molecular polarity and degree of unsaturation. There are some nine types industrially available, with nominal pore diameters ranging from 3 to 10 Å, in the form of pellets, beads, and powders. They are used for dehydration of gases and liquids, separation of gas and liquid hydrocarbon mixtures, and in a great variety of processes. They are regenerated by heating or elution.

ADSORPTION EQUILIBRIA

The great bulk of the experimental data pertaining to adsorption represents equilibrium measurements. Many of them were gathered in an attempt to provide corroboration for one or another of the many theories which have been advanced in an attempt to explain the adsorption phenomena. No one theory has yet been devised which satisfactorily explains even a majority of the observations, and this discussion is therefore limited simply to a description of the more commonly observed adsorption characteristics. The theories are reviewed elsewhere [13, 18, 44, 82, 102].

SINGLE GASES AND VAPORS

In many respects the equilibrium adsorption characteristics of a gas or vapor upon a solid resemble the equilibrium solubility of a gas in a liquid. Figure 11.1 shows several equilibrium adsorption isotherms for a particular activated carbon as adsorbent, where the concentration of adsorbed gas (the adsorbate) on the solid is plotted against the equilibrium partial pressure $\bar{p}^*$ of the vapor or gas at constant temperature. Curves of this sort are analogous to those of Fig. 8.1. At 100°C, for example, pure acetone vapor at a pressure of 190 mmHg is in equilibrium with an adsorbate concentration of 0.2 kg adsorbed acetone/kg carbon, point A. Increasing the pressure of the acetone will cause more to be adsorbed, as the rising curve indicates, and decreasing the pressure of the system at A will cause acetone to be desorbed from the carbon. While not determined experimentally in this case, it is known that the 100°C isotherm for acetone will continue to rise only to a pressure of 2790 mmHg, the saturation vapor pressure of acetone at this temperature. At higher pressures, no acetone can exist in the vapor state at this temperature but instead will condense entirely to a liquid. It will thus be possible to obtain indefinitely large concentrations of the substance on the solid at pressures higher than the vapor pressure, as at point B (Fig. 11.2). However, concentrations in excess of that corresponding to point B indicate liquefaction but not necessarily adsorption of the vapor. Gases above their critical temperature, of course, do not show this characteristic.

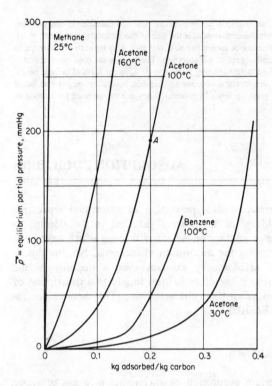

Figure 11.1 Equilibrium adsorption of vapors on activated carbon.

Different gases and vapors are adsorbed to different extents under comparable conditions. Thus benzene (Fig. 11.1) is more readily adsorbed than acetone at the same temperature and gives a higher adsorbate concentration for a given equilibrium pressure. As a general rule, vapors and gases are more readily adsorbed the higher their molecular weight and the lower their critical temperature, although chemical differences such as the extent of unsaturation in the

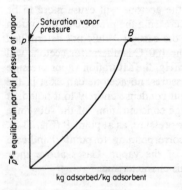

Figure 11.2 Typical complete vapor-adsorption isotherm.

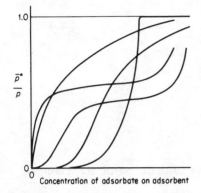

Figure 11.3 Types of adsorption isotherms for vapors.

molecule also influence the extent of adsorption. The so-called permanent gases are usually adsorbed only to a relatively small extent, as the methane isotherm of Fig. 11.1 indicates.

Adsorption isotherms are not always concave to the pressure axis. The shapes shown in Fig. 11.3 have all been observed for various systems. Here the ordinate is plotted as equilibrium partial pressure $\bar{p}^*$ divided by the saturation vapor pressure p of the adsorbed substance (actually, the relative saturation) in order to place all the curves on a comparable basis.

It will be recalled that a change of liquid solvent profoundly alters the equilibrium solubility of a gas except in the case of ideal liquid solutions. In a similar fashion the equilibrium curves for acetone, benzene, and methane on silica gel as adsorbent would be entirely different from those of Fig. 11.1. Indeed, differences in the origin and method of preparation of an adsorbent will result in significant differences in the equilibrium adsorption as well. For this reason, many of the data gathered years ago are no longer of practical value, since methods of preparing adsorbents, and consequently the corresponding adsorbent capacities, have improved greatly over the years. Repeated adsorption and desorption will also frequently alter the characteristics of a particular adsorbent, perhaps as a result of progressive changes in the pore structure within the solid.

There are three commonly used mathematical expressions to describe vapor adsorption equilibria: the Langmuir, the Brunauer-Emmett-Teller (BET), and the Freundlich isotherms [13, 18, 44, 82, 102]. All except the last were derived with a theory in mind, but none is applicable universally, nor can it be predicted which, if any, will apply to a particular case. Since we shall make no application of them, they are not detailed here.

Adsorption Hysteresis

The curves of Fig. 11.1 are true equilibrium curves and therefore represent completely reversible phenomena. The conditions corresponding to point A on

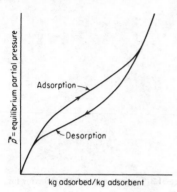

kg adsorbed/kg adsorbent   **Figure 11.4** Adsorption isotherm showing hysteresis.

the figure, for example, can be obtained either by adsorption onto fresh carbon or by desorption of a sample with an initially higher adsorbate concentration. Occasionally, however, different equilibria result, at least over a part of an isotherm, depending upon whether the vapor is adsorbed or desorbed, and this gives rise to the hysteresis phenomenon indicated in Fig. 11.4. This may be the result of the shape of the openings to the capillaries and pores of the solid or of complex phenomena of wetting of the solid by the adsorbate. In any case, when

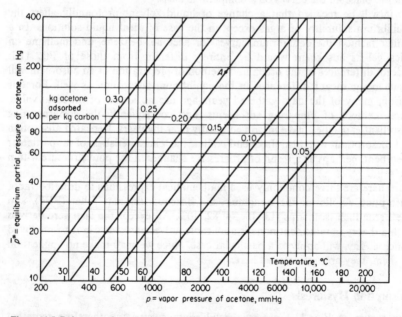

Figure 11.5 Reference-substance plot of equilibrium adsorption of acetone on an activated carbon. [*Data of Josefewitz and Othmer, Ind. Eng. Chem.*, **40**, *739 (1948)*.]

hysteresis is observed, the desorption equilibrium pressure is always lower than that obtained by adsorption.

Effect of Temperature

Since adsorption is an exothermic process, the concentration of adsorbed gas decreases with increased temperature at a given equilibrium pressure, as the several acetone isotherms of Fig. 11.1 indicate.

The reference-substance method of plotting, described in Chap. 8 for gas-liquid solubilities, is conveniently applicable also to adsorption data [81]. As reference substance it is best to use the pure substance being adsorbed, unless the temperatures are above the critical temperature. Figure 11.5, for example, was prepared for the adsorption of acetone vapor on activated carbon with

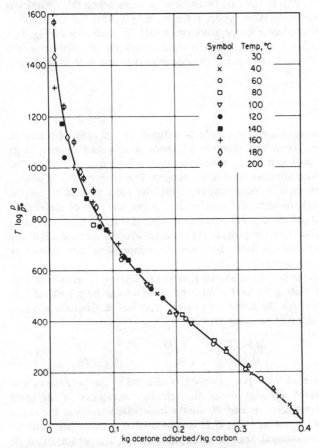

Figure 11.6 Adsorption of acetone on an activated carbon, 30 to 200°C.

acetone as reference substance. The abscissa of the logarithmic coordinates was marked with the vapor pressure of pure acetone, and the corresponding saturation temperatures were also marked. The equilibrium partial pressure of adsorbate was plotted on the ordinate. Points of equal temperature for the vapor pressure of pure acetone and the partial pressure of adsorbate were then plotted. Thus, point A of Fig. 11.1 ($\bar{p}^* = 190$ mmHg, 100°C) was plotted on Fig. 11.5 at A, where $p = 2790$ mmHg, the vapor pressure of acetone at 100°C. Points of constant adsorbate concentration (*isosteres*) form straight lines with few exceptions, and thus only two points are required to establish each.

Figure 11.6 shows all the same data plotted in such a manner as to reduce all the measurements to a single curve, thereby permitting considerable extension of meager data. The free energy of compression of 1 mol of a gas from the equilibrium adsorption pressure $\bar{p}^*$ to the vapor pressure p, the *adsorption potential*, is $RT \ln (p/\bar{p}^*)$. In the case of single substances, when this quantity is plotted against adsorbate concentration, a single curve results for all temperatures, at least over a moderate temperature range [31]. The ordinate of Fig. 11.6 is proportional to this quantity. More complex methods of expressing the adsorbate concentration can be used to provide improved and extended correlations of the data.

Heat of Adsorption

The *differential heat of adsorption* $(-\bar{H})$ is defined as the heat liberated at constant temperature when unit quantity of vapor is adsorbed upon a large quantity of solid already containing adsorbate. Such a large quantity of solid is used that the adsorbate concentration is unchanged. The *integral heat of adsorption* at any concentration X of adsorbate upon the solid is defined as the enthalpy of the adsorbate-adsorbent combination minus the sum of the enthalpies of unit weight of pure solid adsorbent and sufficient pure adsorbed substance (before adsorption) to provide the required concentration X, all at the same temperature. These are both functions of temperature and adsorbate concentration for any system.

Othmer and Sawyer [81] have shown that plots of the type shown in Fig. 11.5 are useful in estimating the heat of adsorption, which can be calculated in a manner similar to that for the latent heat of vaporization of a pure liquid (see Chap. 7). Thus, the slope of an isostere of Fig. 11.5 is

$$\frac{d \ln \bar{p}^*}{d \ln p} = \frac{(-\bar{H})M}{\lambda_r M_r} \qquad (11.1)$$

where $\bar{H}$, energy per mass of vapor adsorbed, is referred to the pure vapor and λ_r is the latent heat of vaporization of the reference substance at the same temperature, energy per mass. M and M_r are the molecular weights of the vapor and reference substance, respectively. If $\bar{H}$ is computed at constant temperature for each isostere, the integral heat of adsorption at this temperature can be computed from the relation

$$\Delta H'_A = \int_0^X \overline{H} \, dX \qquad (11.2)$$

$\Delta H'_A$, energy per mass of adsorbate-free solid, is referred to the pure vapor, and X is the adsorbate concentration, mass adsorbate/mass solid. The integral can be evaluated graphically by determining the area under a curve of $\overline{H}$ vs. X. The integral heat of adsorption referred to solid and the adsorbed substance in the liquid state is $\Delta H_A = \Delta H'_A + \lambda X$, energy per mass solid. The quantities $\overline{H}$, ΔH_A, and $\Delta H'_A$ are negative quantities if heat is evolved during adsorption.

Illustration 11.1 Estimate the integral heat of adsorption of acetone upon activated carbon at 30°C as a function of adsorbate concentration.

SOLUTION Refer to Fig. 11.5. The isosteres for various concentrations are straight on this diagram, and their slopes are measured with the help of a millimeter rule. In the accompanying table, column 1 lists the adsorbate concentrate of each isostere and column 2 the corresponding slope.

			Integral heat of adsorption, kJ/kg carbon	
X, $\dfrac{\text{kg acetone}}{\text{kg carbon}}$	Slope of isostere	Differential heat of adsorption $\overline{H}$, kJ/kg acetone	$\Delta H'_A$, referred to acetone vapor	ΔH_A, referred to acetone liquid
(1)	(2)	(3)	(4)	(5)
0.05	1.170	−640	−29.8	−2.1
0.10	1.245	−686	−63.0	−7.9
0.15	1.300	−716	−97.9	−15.1
0.20	1.310	−721	−134.0	−23.7
0.25	1.340	−740	−170.5	−32.6
0.30	1.327	−730	−207.2	−41.9

Since in this case the adsorbate and reference substance are the same, $M = M_r$ and consequently $\overline{H} = -\lambda_r$ (slope of isostere). At 30°C, $\lambda_r = \lambda$, the latent heat of vaporization of acetone = 551 kJ/kg. Column 3 of the table lists values of $\overline{H}$ calculated in this manner.

Column 3 was plotted as ordinate against column 1 as abscissa (not shown). The area under the curve between $X = 0$ and any value of X is listed in column 4 as the corresponding integral heat of adsorption, referred to acetone vapor. Thus the area under the curve between $X = 0$ and $X = 0.20$ is -134.0. If 0.20 kg acetone vapor at 30°C is adsorbed on 1 kg fresh carbon at 30°C and the product brought to 30°C, 134 kJ will be evolved.

Column 5, the integral heat of adsorption referred to liquid acetone, is computed from the relation $\Delta H_A = \Delta H'_A + \lambda X$. Thus, at $X = 0.20$, $\Delta H_A = -134 + 551(0.20) = -23.7$ kJ/kg carbon.

VAPOR AND GAS MIXTURES

It is necessary to distinguish between mixtures depending upon whether one or several of the components are adsorbed.

One Component Adsorbed

In the case of many mixtures, particularly vapor-gas mixtures, only one component is appreciably adsorbed. This would be the circumstance for a mixture of acetone vapor and methane in contact with activated carbon (Fig. 11.1), for example. In such instances, the adsorption of the vapor will be substantially unaffected by the presence of the poorly adsorbed gas, and the adsorption isotherm for the pure vapor will be applicable provided the equilibrium pressure is taken as the *partial* pressure of the vapor in the vapor-gas mixture. The isotherms for acetone (Fig. 11.1) thus apply for mixtures of acetone with any poorly adsorbed gas such as nitrogen, hydrogen, and the like. This is similar to the corresponding case of gas-liquid solubility.

Binary Gas or Vapor Mixtures, Both Components Appreciably Adsorbed

When both components of a binary gas or vapor mixture are separately adsorbed to roughly the same extent, the amount of either one adsorbed from the mixture will be affected by the presence of the other. Since such systems are composed of three components when the adsorbent is included, the equilibrium data are conveniently shown in the manner used for ternary liquid equilibria in Chap. 10. For this purpose it is convenient to consider the solid adsorbent as being analogous to liquid solvent in extraction operations. However, adsorption is greatly influenced by both temperature and pressure, unlike liquid solubility, which is scarcely affected by pressure under ordinary circumstances. Equilibrium diagrams are consequently best plotted at constant temperature and constant total pressure, and they are therefore simultaneously *isotherms* and *isobars*.

A typical system is shown in Fig. 11.7 on triangular and rectangular coordinates. The properties of these coordinate systems and the relations between them were considered in detail in Chap. 10. Even though mole fraction is generally a more convenient concentration unit in dealing with gases, the figure is plotted in terms of weight-fraction compositions since the molecular weight of the adsorbent is uncertain.† Since the adsorbent is not volatile and does not appear in the gas phase, the equilibrium gas compositions fall upon one axis of either graph, as shown. Points G and H represent the adsorbate concentration for the individual pure gases and the curve GEH that of gas mixtures. Tie lines such as line RE join equilibrium compositions of the gas and adsorbate. The fact that the tie lines do not, when extended, pass through the adsorbent apex indicates that under the prevailing conditions the adsorbent can be used to separate the binary gas mixture into its components. The separation factor, or *relative adsorptivity*, similar to relative volatility in distillation or selectivity in liquid extraction, is obtained by dividing the equilibrium ratio of gas compositions in the adsorbate (as at point E) by the ratio in the gas (as at R). The relative adsorptivity must be larger than unity if the adsorbent is to be useful for

---

† Alternatively, some arbitrary molecular weight could be assigned to the adsorbent.

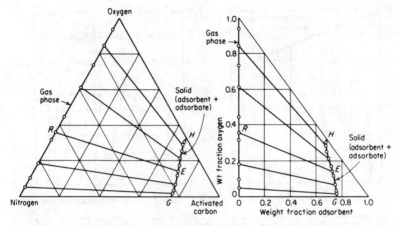

Figure 11.7 System oxygen-nitrogen-activated carbon, $-150°C$, 1 std atm, concentrations expressed as weight fractions. [*Data of Maslan, Altman, and Aberth, J. Phys. Chem., 57, 106 (1953).*]

separating the components of the gas mixture. In the system of Fig. 11.7 the more strongly adsorbed of the two pure gases (oxygen) is also selectively adsorbed from any mixture. This appears to be true for most mixtures, although an inversion of the relative adsorptivity in some systems (analogous to azeotropism in distillation) is certainly a possibility.

Especially when the extent of adsorption is small, it will be more convenient to express compositions on an adsorbent-free basis and to plot them in the manner of Fig. 11.8. Such diagrams are also analogous to those used in liquid extraction (Fig. 10.10, for example), where adsorbent solid and extraction solvent play an analogous role, and to the enthalpy-concentration diagrams of distillation, where heat is analogous to adsorbent. The adsorption characteristics of the binary gas mixture acetylene-ethylene on silica gel for one temperature and pressure are shown in Fig. 11.8a. In the upper portion of this figure, the gas phase appears entirely along the abscissa of the plot owing to the absence of adsorbent in the gas. The adsorbent-free equilibrium compositions corresponding to the tie lines can be plotted in the lower half of the diagram, as at point (R, E), to produce a figure analogous to the xy diagram of distillation. Silica gel selectively adsorbs acetylene from these gas mixtures.

The powerful influence of the adsorbent on the equilibrium is demonstrated with the same gas mixture by Fig. 11.8b, where activated carbon is the adsorbent. Not only is the extent of adsorption greater than for silica gel, so that the curve GH is lower than the corresponding curve for silica gel, but in addition the relative adsorptivity is reversed: ethylene is selectively adsorbed on activated carbon. In both cases, however, that gas which is separately more strongly adsorbed on each adsorbent is selectively adsorbed from mixtures. If a condition corresponding to azeotropism should arise, the curve of the lower half of these figures would cross the 45° diagonal line. In cases like Fig. 11.8b, it will

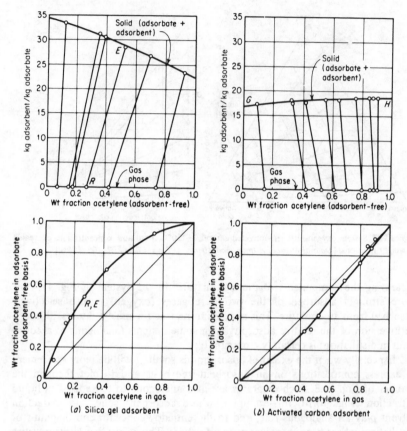

Figure 11.8 Adsorption of acetylene-ethylene on (*a*) silica gel and (*b*) activated carbon, at 25°C, 1 std atm. [*Data of Lewis, et al., J. Am. Chem. Soc., 72, 1157 (1950).*]

generally be preferable to plot compositions in terms of the more strongly adsorbed gas (ethylene), to keep the appearance of the diagram similar to those used previously in liquid extraction.

A beginning has been made on the estimation of adsorbate concentrations from the isotherms of pure vapors, assuming that the adsorbate is ideal and follows an analog of Raoult's law [49, 74, 75]. Some systems follow such a law, such as ethane–propane–carbon black [26] and *n*-butane–ethane–molecular sieves [29]. Others do not. Thus, in the case of benzene–*n*-heptane–silica gel, adsorbates deviate negatively from ideal solutions, whereas in the vapor-liquid equilibria for these hydrocarbons the deviation is positive [90]. An excellent review of the thermodynamics is available [97], including the concept of two-dimensional equations of state for the adsorbate layer, strongly reminiscent of Abbott's "Flatland" [1].

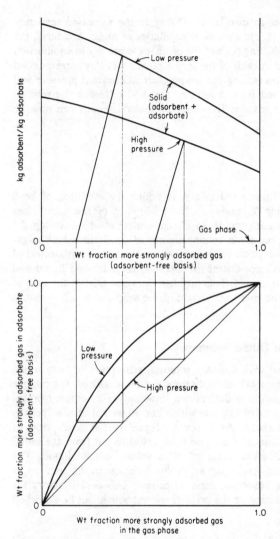

Figure 11.9 Effect of pressure on adsorption isotherms for binary gas mixtures.

Effect of Change of Pressure or Temperature

The available data are meager, and generalizations are very difficult to make. Lowering the pressure will of course reduce the amount of adsorbate upon the adsorbent, as shown in the upper half of Fig. 11.9. In several cases investigated over any appreciable range [24, 62] the relative adsorptivity of paraffin hydrocarbons on carbon decreased at increased pressure, as shown in the lower part of this figure, just as it does in distillation. Owing to the increased tendency toward liquid condensation in the absorbent capillaries at higher pressures, the equilibrium may simply be shifting toward the ordinary vapor-liquid equilibrium with increased pressure, and in each of the investigated cases this corresponded to lower separation factor. Increasing the temperature at constant pressure will decrease the amount adsorbed from a mixture and will influence the relative adsorptivity as well but in a manner for which no generalizations can now be made.

LIQUIDS

When an adsorbent solid is immersed in a pure liquid, the evolution of heat, known as the *heat of wetting*, is evidence that adsorption of the liquid does occur. But immersion does not provide an effective method of measuring the extent of adsorption. No appreciable volume change of the liquid which might be used as a measure of adsorption is ordinarily observed, while withdrawal of the solid and weighing it will not distinguish between the adsorbed liquid and that which is mechanically occluded. This problem does not exist in the case of adsorption of gases, where the change in weight of the solid due to adsorption is readily measured.

Adsorption of Solute from Dilute Solution

When an adsorbent is mixed with a binary solution, adsorption of both solute and solvent occurs. Since the total adsorption cannot be measured, the relative or apparent adsorption of solute is determined instead. The customary procedure is to treat a known volume of solution with a known weight of adsorbent, v volume solution/mass adsorbent. As a result of preferential adsorption of solute, the solute concentration of the liquid is observed to fall from the initial value c_0 to the final equilibrium value c^* mass solute/volume liquid. The apparent adsorption of solute, neglecting any volume change in the solution, is then $v(c_0 - c^*)$ mass solute adsorbed/mass adsorbent. This is satisfactory for dilute solutions when the fraction of the original solvent which can be adsorbed is small.

Correction is sometimes made for the volume of the solute apparently adsorbed. Thus, the initial solvent content of this solution is $v(1 - c_0/\rho)$, and on the assumption that no solvent is adsorbed, the volume of residual solution is $v(1 - c_0/\rho)/(1 - c^*/\rho)$. The apparent solute adsorption is then the difference between initial and final solute content of the liquid, $vc_0 - [v(1 - c_0/\rho)/(1 - c^*/\rho)]c^*$ or $v(c_0 - c^*)/(1 - c^*/\rho)$. This of course, still neglects solvent adsorption.

The apparent adsorption of a given solute depends upon the concentration of solute, the temperature, the solvent, and the type of adsorbent. Typical isotherms are shown in Fig. 11.10. Isotherms of all the indicated forms have been observed, for example, when a given solute is adsorbed on the same

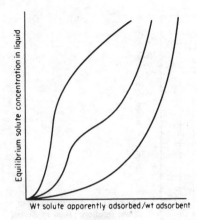

Figure 11.10 Typical adsorption isotherms for dilute solutions.

(y-axis) Equilibrium solute concentration in liquid

(x-axis) Wt solute apparently adsorbed/wt adsorbent

adsorbent, but from different solvents. The extent of adsorption of a given solute practically always decreases at increased temperature and usually is greater the smaller the solubility in the solvent. It is usually reversible, so that the same isotherm results whether solute is desorbed or adsorbed.

The Freundlich Equation

Over a small concentration range, and particularly for dilute solutions, the adsorption isotherms can frequently be described by an empirical expression usually attributed to Freundlich,

$$c^* = k\left[v(c_0 - c^*) \right]^n \tag{11.3}$$

where $v(c_0 - c^*)$ is the apparent adsorption per unit mass of adsorbent and k and n are constants. Other concentration units are frequently used also, and while they will result in different values of k, for the dilute solutions for which the equation is applicable the value of n will be unaffected. The form of the equation indicates that plotting the equilibrium solute concentration as ordinate against adsorbate content of the solid as abscissa on logarithmic coordinates will provide a straight line of slope n and intercept k. Several typical isotherms are plotted in this manner in Fig. 11.11. The effect of the nature of the solvent on adsorption of benzoic acid on silica gel is shown by curves a and b, which follow Eq. (11.3) excellently over the concentration range shown. The adsorption is less strong from benzene solutions, which is to be expected in view of the higher solubility of the acid in this solvent. Curve c of this figure shows the deviation from linearity to be expected over large concentration ranges, although Eq. (11.3) is applicable for the lower concentration ranges. Failure of the data to follow the equation at high solute concentrations may be the result of appreciable adsorption of the solvent which is not taken into account or simply general inapplicability of the expression.

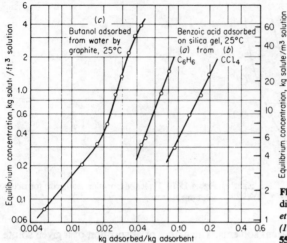

Figure 11.11 Adsorption from a dilute solution. [*Data of Bartell, et al., J. Phys. Chem., 33, 676 (1929); J. Phys. Colloid Chem., 55, 1456 (1951).*]

The Freundlich equation is also frequently useful in cases where the actual identity of the solute is not known, as in the adsorption of colored substances from such materials as sugar solutions and mineral or vegetable oils. In such cases, the concentration of solute can be measured by means of a colorimeter or spectrophotometer and expressed in terms of arbitrary units of color intensity, provided the color scale used varies linearly with the concentration of the responsible solute. Figure 11.12 illustrates the application of this method of plotting the adsorption of colored substances from a petroleum fraction on two adsorbent clays. Here the color concentrations of the solutions are measured on an arbitrary scale, and the concentration of adsorbate on the clay is determined by measuring the change in color expressed in these terms when 100 g of oil is treated with various amounts of clay.

Adsorption from Concentrated Solutions

When the apparent adsorption of solute is determined over the entire range of concentrations from pure solvent to pure solute, curves like those of Fig. 11.13 will result. Curves of the shape marked *a* occur when at all concentrations the solute is adsorbed more strongly relative to the solvent. At increasing solute concentrations, the extent of solute adsorption may actually continue to increase; yet the curve showing apparent solute adsorption necessarily returns to point *E*, since in a liquid consisting of pure solute alone there will be no concentration change on addition of adsorbent. In cases where both solvent and solute are adsorbed to nearly the same extent, the S-shaped curves of type *b* are produced. In the range of concentrations from *C* to *D*, solute is more strongly adsorbed than solvent. At point *D*, both are equally well adsorbed, and the

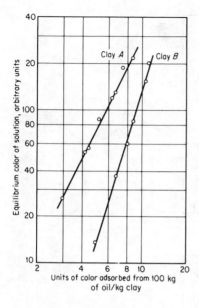

Figure 11.12 Decolorization of cylinder oil with clay. [*Data of Rogers, et al., Ind. Eng. Chem.,* **18,** *164 (1926).*]

apparent adsorption falls to zero. In the range of concentrations from D to E, solvent is more strongly adsorbed. Consequently, on addition of adsorbent to such solutions, the solute concentration of the liquid increases, and the quantity $v(c_0 - c^*)$ indicates an apparent *negative* solute adsorption.

The true adsorption of the substances can be estimated from apparent adsorption data if some mechanism for the process, such as the applicability of the Freundlich isotherm, separately for each

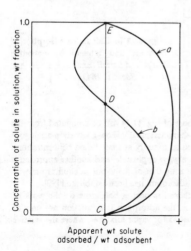

Figure 11.13 Apparent adsorption of solute from solutions.

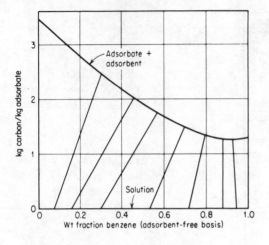

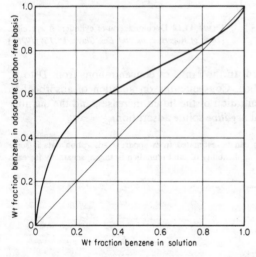

Figure 11.14 Calculated adsorption of benzene and ethanol by activated carbon. [*Bartell and Sloan, J. Am. Chem. Soc.* **51**, *1643 (1929)*.]

component, is assumed [6]. In this manner, curves like those of Fig. 11.14 can be computed from the S-shaped apparent adsorption data (note that such data characteristically show a situation analogous to azeotropism in distillation, with relative adsorptivity equal to unity at some liquid concentration). It has been shown, however, that such isotherms are not always applicable, and another approach is to determine the adsorption of a liquid solution by measuring that of the vapor in equilibrium with the liquid [50, 76]. The thermodynamics of binary liquid adsorption has been established [90].

For adsorption of multiple solutes from solution, there has been a beginning on the work of establishing a theoretical background [72, 83]. In this case, the adsorption is a function not only of the relative, but also of the total, solute concentration, just as for vapor mixtures, where the relative and total pressure are both important.

ADSORPTION OPERATIONS

Adsorption is unique in the very diverse nature of its applications. For example, it is applied to a wide variety of processes such as recovery of vapors from dilute mixture with gases, solute recovery and removal of contaminants from solution, and the fractionation of gas and liquid mixtures. The techniques used include both stagewise and continuous-contacting methods, and these are applied to batch, continuous, and semicontinuous operations. Within each of these categories it is possible to recognize operations which are exactly analogous to those already discussed in previous chapters of this book. Thus, when only one component of a fluid mixture (either a gas or a liquid) is strongly adsorbed, the separation of the mixture is analogous for purposes of calculation to gas absorption, where the added insoluble phase is adsorbent in the present case and liquid solvent in the case of absorption. When both components of the fluid (either gas or liquid) are adsorbed strongly, the separation requires a fractionation procedure. The operation is then conveniently considered as being analogous to liquid extraction, where the added insoluble adsorbent corresponds to the use of solvent in extraction. By this means many simplifications in the treatment become possible.

STAGEWISE OPERATION

Liquid solutions, where the solute to be removed is adsorbed relatively strongly compared with the remainder of the solution, are treated in batch, semicontinuous, or continuous operations in a manner analogous to the mixer-settler operations of liquid extraction (*contact filtration*). Continuous countercurrent cascades can be simulated or actually realized by use of such techniques as fluidized beds.

Gases are treated for solute removal or for fractionation, usually with fluidized-bed techniques.

Contact Filtration of Liquids

Typical process applications include (1) the collection of valuable solutes from dilute solutions e.g., the adsorption onto carbon of iodine from brines after liberation of the element from its salts by oxidation and the collection of insulin from dilute solutions and (2) the removal of undesirable contaminants from a solution.

The extremely favorable equilibrium distribution of solute toward the adsorbent which is frequently possible makes adsorption a powerful tool for the latter purpose, and most industrial applications of stagewise techniques fall into this category. Adsorption of colored substances from aqueous sugar solutions onto carbon, in order to provide a pure product and to assist the crystallization, is a typical example. Similarly, carbon is sometimes used to adsorb odorous substances from potable water, and grease is adsorbed from dry-cleaning liquids. The colors of petroleum and vegetable oils are lightened by treatment with clay.

Equipment and methods As pointed out in Chap. 1, each stage requires the intimate contact of two insoluble phases for a time sufficient for a reasonable approach to equilibrium, followed by physical separation of the phases. The equipment used in applying these principles to adsorption is varied, depending upon the process application. That shown in Fig. 11.15 is very typical of many installations operated in a batchwise fashion. The liquid to be processed and the adsorbent are intimately mixed in the treating tank at the desired temperature for the required period of time, following which the thin slurry is filtered to separate the solid adsorbent and accompanying adsorbate from the liquid. Air agitation, as with sparged vessels (Chap. 6) is also used, particularly in ion exchange. The equipment is readily adaptable to multistage operation by providing additional tanks and filters as necessary. If the operation is to be made continuous, which is sometimes done in the decolorizing of petroleum lubricating oils, for example, centrifuges or a continuous rotary filter can be substituted for the filter press, or the solid can be allowed to settle out by virtue of its higher density when the mixture is passed through a large tank.

The type of adsorbent used depends upon the solution to be treated. Aqueous solutions are frequently treated with activated carbon especially prepared for the purpose at hand, whereas organic liquids such as oils are usually treated with inorganic adsorbents such as clays. Occasionally mixed adsorbents

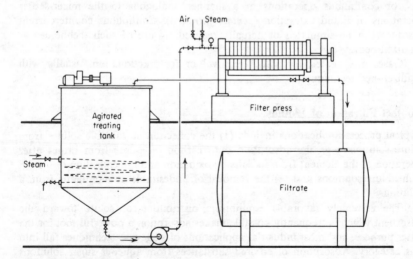

Figure 11.15 Contact filtration. Schematic arrangement for single-stage batch treatment of liquids.

are used. High selectivity for the solute to be removed is desirable in order to reduce the amount of solid to be added. In any case, the adsorbent is applied in the form of a very finely ground powder, usually at least fine enough to pass entirely through a 200-mesh screen and frequently very much finer.

The highest convenient temperature should be used during the mixing, since the resulting decreased liquid viscosity increases both the rate of diffusion of solute and the ease with which the adsorbent particles can move through the liquid. Usually the equilibrium adsorption is decreased to a small extent at higher temperatures, but this is more than compensated for by the increased rate of approach to equilibrium. Operations are sometimes conducted at the boiling point of the liquid if this temperature will cause no injury to the substances involved. In the clay treatment of petroleum-lubricant fractions the adsorbent-oil mixture may be pumped through a tubular furnace to be heated to as much as 120 to 150°C, and for very heavy oils even to 300 to 380°C. If the adsorbed substance is volatile, however, the equilibrium extent of adsorption will be much more strongly affected by temperature and such material is best handled at ordinary temperatures.

Owing to the large quantity of solution usually treated relative to the amount of adsorption occurring, the temperature rise resulting from release of the heat of adsorption can usually be ignored.

The method of dealing with the spent adsorbent depends upon the particular system under consideration. The filter cake is usually washed to displace the solution held within the pores of the cake, but relatively little adsorbate will be removed in this manner. If the adsorbate is the desired product, it can be desorbed by contact of the solid with a solvent other than that which constitutes the original solution, one in which the adsorbate is more soluble. This can be done by washing the cake in the filter or by dispersing the solid into a quantity of the solvent. If the adsorbate is volatile, it can be desorbed by reduction of the partial pressure of the adsorbate over the solid by passage of steam or warm air through the solid. With activated carbon adsorbents, care must be taken to avoid too high temperatures in using air for this purpose, in order to avoid combustion of the carbon. In most decolorizing operations, the adsorbate is of no value and is desorbed with difficulty. The adsorbent can then be revivified by burning off the adsorbate, followed by reactivation. Usually only a limited number of such revivifications is possible before the adsorbent ability is severely reduced, whereupon the solid is discarded.

Single-stage operation The schematic flowsheet for this type of operation in either batch or continuous fashion is shown in the upper part of Fig. 11.16. Here the circle represents all the equipment and procedures constituting one stage. The operation is essentially analogous to a single-stage gas absorption, where the solution to be treated corresponds to a gas and the solid adsorbent to a liquid. Since the amount of adsorbent used is ordinarily very small with respect to the amount of solution treated, and since the solute to be removed is adsorbed much more strongly than the other constituents present, the adsorption of the latter

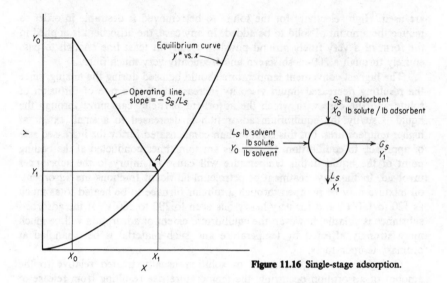

Figure 11.16 Single-stage adsorption.

can be ignored. Furthermore the adsorbent is insoluble in the solution. The solution to be treated contains L_S mass unadsorbed substance or solvent, and the adsorbable solute concentration is reduced from Y_0 to Y_1 mass solute/mass solvent. The adsorbent is added to the extent of S_S mass adsorbate-free solid, and the solute adsorbate content increases from X_0 to X_1 mass solute/mass adsorbent. If fresh adsorbent is used, $X_0 = 0$, and in cases of continuous operation L_S and S_S are measured in terms of mass/time.†

The solute removed from the liquid equals that picked up by the solid,

$$L_S(Y_0 - Y_1) = S_S(X_1 - X_0) \tag{11.4}$$

On X, Y coordinates this represents a straight operating line, through points of coordinates (X_0, Y_0) and (X_1, Y_1) of slope $-S_S/L_S$. If the stage is a theoretical or equilibrium stage, the effluent streams are in equilibrium, so that the point (X_1, Y_1) lies on the equilibrium adsorption isotherm. This is shown on the lower portion of Fig. 11.16. The equilibrium curve should be that obtaining at the final temperature of the operation. If insufficient time of contact is allowed, so that equilibrium is not reached, the final liquid and solid concentrations will correspond to some point such as A (Fig. 11.16), but ordinarily equilibrium is approached very closely.

---

† For the dilute solutions ordinarily used other consistent units can be applied to these terms. Thus Y can be expressed as kg solute/kg solution (or kg solute/m³ solution) and L_S as kilograms (or cubic meters, respectively) of solution. When the adsorbed solute is colored matter whose concentration is measured in arbitrary units, the latter can be considered as Y units of color per kilogram or cubic meter of solution and the adsorbate concentration on the solid X as units of color per kilogram of adsorbent.

ADSORPTION AND ION EXCHANGE **589**

The use of Eq. (11.4) assumes that the amount of liquid mechanically retained with the solid (but not adsorbed) after filtration or settling is negligible. This is quite satisfactory for most adsorption, since the quantity of solid employed is ordinarily very small with respect to that of the liquid treated. If the operation under consideration is *desorption*, and if again the quantity of liquid retained mechanically by the solid is negligible, Eq. (11.4) applies, but the operating line lies below the equilibrium curve on Fig. 11.16. In this case, however, it is much more likely that the quantity of liquid retained mechanically with the solid will be an appreciable portion of the total liquid used, and the methods of calculation described in Chap. 13 for leaching should be used.

Application of the Freundlich equation The Freundlich equation can frequently be applied to adsorption of this type, particularly since small adsorbable solute concentrations are usually involved. This can be written in the following form for the concentration units used here,

$$Y^* = mX^n \tag{11.5}$$

and, at the final equilibrium conditions,

$$X_1 = \left(\frac{Y_1}{m}\right)^{1/n} \tag{11.6}$$

Since the adsorbent used ordinarily contains no initial adsorbate and $X_0 = 0$, substitution in Eq. (11.4) yields

$$\frac{S_S}{L_S} = \frac{Y_0 - Y_1}{(Y_1/m)^{1/n}} \tag{11.7}$$

This permits analytical calculation of the adsorbent/solution ratio for a given change in solution concentration, Y_0 to Y_1.

Refer to Fig. 11.17, where three typical Freundlich isotherms are shown. The isotherm is straight for $n = 1$, concave upward for $n > 1$, and concave

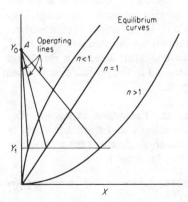

Figure 11.17 Single-stage adsorption, Freundlich equilibrium curves.

downward for $n < 1$. If in each case the solution concentration is to be reduced from Y_0 to Y_1, the three operating lines radiating from point A apply. The slope of the operating line is in each case directly proportional to the adsorbent/solution ratio. It is generally stated [39] that values of n in the range 2 to 10 represent good, 1 to 2 moderately difficult, and less than 1 poor adsorption characteristics (although the value of m is important also). In the last case impractically large adsorbent dosages may be required for appreciable fractional removal of solute.

Multistage crosscurrent operation The removal of a given amount of solute can be accomplished with greater economy of adsorbent if the solution is treated with separate small batches of adsorbent rather than in a single batch, with filtration between each stage. This method of operation, sometimes called *split-feed treatment*, is usually done in batch fashion, although continuous operation is also possible. Economy is particularly important when activated carbon, a fairly expensive adsorbent, is used. The savings are greater the larger the number of batches used but are at the expense of greater filtration and other handling costs. It is therefore seldom economical to use more than two stages. In rare instances, the adsorption may be irreversible, so that separate adsorbent dosages can be applied without intermediate filtration at considerable savings in operating costs [38]. This is by far the exception rather than the rule, and when applied to ordinary reversible adsorption, it will provide the same end result as if all the adsorbent had been used in a single stage.

A schematic flowsheet and operating diagram for a typical operation of two equilibrium stages are shown in Fig. 11.18. The same quantity of solution is treated in each stage by amounts of adsorbent S_{S1} and S_{S2} in the two stages, respectively, to reduce the solute concentration of the solution from Y_0 to Y_2. The material balances are, for stage 1,

$$L_S(Y_0 - Y_1) = S_{S1}(X_1 - X_0) \tag{11.8}$$

and for stage 2

$$L_S(Y_1 - Y_2) = S_{S2}(X_2 - X_0) \tag{11.9}$$

These provide the operating lines shown on the figure, each of a slope appropriate to the adsorbent quantity used in the corresponding stage. The extension to large numbers of stages is obvious. If the amounts of adsorbent used in each stage are equal, the operating lines on the diagram will be parallel. The least total amount of adsorbent will require unequal dosages in each stage except where the equilibrium isotherm is linear, and in the general case this can be established only by a trial-and-error computation.

Application of the Freundlich equation When the Freundlich expression [Eq. (11.5)] describes the adsorption isotherm satisfactorily and fresh adsorbent is used in each stage ($X_0 = 0$), the least total amount of adsorbent for a two-stage system can be computed directly [38, 100]. Thus, for stage 1,

$$\frac{S_{S1}}{L_S} = \frac{Y_0 - Y_1}{(Y_1/m)^{1/n}} \tag{11.10}$$

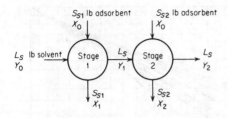

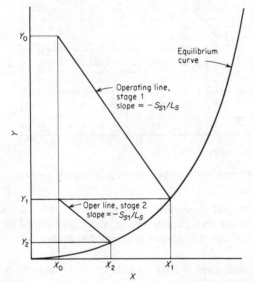

Figure 11.18 Two-stage crosscurrent adsorption.

and for stage 2

$$\frac{S_{S2}}{L_S} = \frac{Y_1 - Y_2}{(Y_2/m)^{1/n}} \tag{11.11}$$

The total amount of adsorbent used is

$$\frac{S_{S1} + S_{S2}}{L_S} = m^{1/n}\left(\frac{Y_0 - Y_1}{Y_1^{1/n}} + \frac{Y_1 - Y_2}{Y_2^{1/n}}\right) \tag{11.12}$$

For minimum total adsorbent, $d[(S_1 + S_2)/L_S]/dY_1$ is set equal to zero, and since for a given case m, n, Y_0, and Y_2 are constants, this reduces to

$$\left(\frac{Y_1}{Y_2}\right)^{1/n} - \frac{1}{n}\frac{Y_0}{Y_1} = 1 - \frac{1}{n} \tag{11.13}$$

Equation (11.13) can be solved for the intermediate concentration Y_1, and the adsorbed quantities calculated by Eqs. (11.10) and (11.11). Figure 11.19 permits solutions of Eq. (11.13) without trial and error.

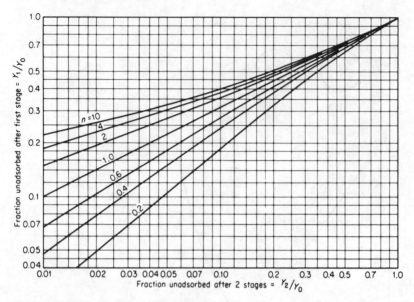

Figure 11.19 Solution to Eq. (11.13). Minimum total adsorbent, two-stage crosscurrent operation.

It has been established [60] that cross-flow operation with a split of the adsorbent, as in Fig. 11.18, is better than the alternative possibility, split of the treated solution, but that countercurrent operation is superior to both.

Multistage countercurrent operation Even greater economy of adsorbent can be obtained by countercurrent operation. When batch methods of treating liquids are used, this can only be simulated, for which the general scheme of batch simulation of countercurrent operations shown in Fig. 10.38 for liquid extraction is actually followed. The flowsheet of Fig. 11.20 then becomes the ultimate, steady-state result, reached only after a number of cycles. However, truly continuous operation has also been used, as in the simultaneous dissolution of gold and silver from finely ground ore by cyanide solution and adsorption of the dissolved metal upon granular activated carbon in a three-stage operation. Coarse screens between agitated vessels separate the large carbon particles from the pulped-ore-liquid mixture [19]. The fluidized-bed tower of Fig. 11.28 for gases and a similar apparatus for liquids, especially for ion exchange, also represent truly continuous stagewise operation.

A solute balance about the N_p stages is

$$L_S(Y_0 - Y_{N_p}) = S_S(X_1 - X_{N_p+1}) \tag{11.14}$$

which provides the operating line on the figure, through the coordinates of the terminal conditions (X_{N_p+1}, Y_{N_p}) and (X_1, Y_0) and of slope S_S/L_S. The number

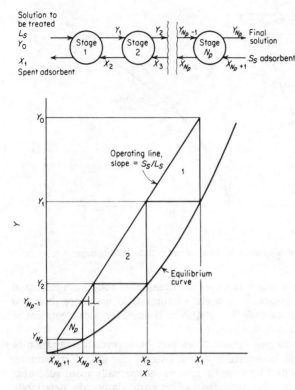

Figure 11.20 Countercurrent multistage adsorption.

of theoretical stages required is found by drawing the usual staircase construction between equilibrium curve and operating line in the manner shown. Alternatively the adsorbent/solution ratio for a predetermined number of stages can be found by trial-and-error location of the operating line. If the operation is a *desorption* (corresponding to stripping in gas-liquid contact), the operating line falls below the equilibrium curve.

The minimum adsorbent/solvent ratio will be the largest which results in an infinite number of stages for the desired change of concentration. This corresponds to the operating line of largest slope which touches the equilibrium curve within the specified range of concentrations. Where the equilibrium isotherm is straight or concave upward, as in Fig. 11.21a, this will cause a *pinch* at the concentrated end of the cascade, as at point A. If the isotherm is concave downward (Fig. 11.21b), the pinch may occur at a point of tangency, as at point B, if Y_0 is sufficiently large. The situations are entirely analogous to those found in gas absorption (Fig. 8.7).

As the number of stages in a cascade is increased, the amount of adsorbent required at first decreases rapidly but approaches the minimum value only

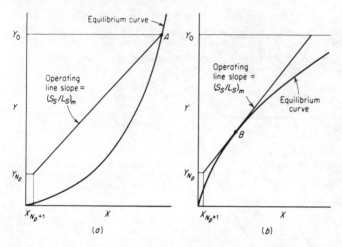

Figure 11.21 Operating line and minimum adsorbent/solvent ratio for infinite stages.

asymptotically. In practice, where intermediate filtration of solid from the liquid must be made between stages, it is rarely economical to use more than two stages in a countercurrent cascade. It will also be realized that, for treatment of gases, the symbol G_S can be substituted for L_S.

In small-scale processing of liquids, there may be appreciable variation in the amounts of solution to be treated from one batch to the next. Furthermore, long periods of time may pass between batches, so that partially spent adsorbent must be stored between stages. Activated carbon particularly may deteriorate during storage through oxidation, polymerization of the adsorbate, or other chemical change, and in such cases the crosscurrent flowsheet may be more practical.

Application of the Freundlich equation Trial-and-error calculation for the adsorbent/solvent ratio can be eliminated if the equilibrium curve can be conveniently described algebraically. If the equilibrium curve is linear, the Kremser equations (5.50) to (5.53) and Fig. 5.16 apply. More frequently the Freundlich expression (11.5) is useful, and fresh adsorbent $(X_{N_p+1} = 0)$ is used in the last stage [86]. For a typical two-stage cascade (Fig. 11.22) a solute material balance for the entire plant is

$$S_S(X_1 - 0) = L_S(Y_0 - Y_2) \tag{11.15}$$

Applying Eq. (11.5) to the effluents from the first equilibrium stage gives

$$X_1 = \left(\frac{Y_1}{m}\right)^{1/n} \tag{11.16}$$

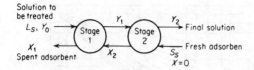

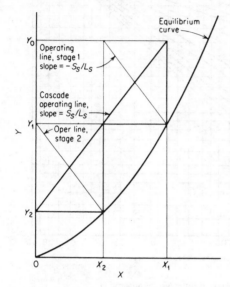

Figure 11.22 Two-stage countercurrent adsorption.

and combining them gives

$$\frac{S_S}{L_S} = \frac{Y_0 - Y_2}{(Y_1/m)^{1/n}}$$ (11.17)

The operating line for the second equilibrium stage is shown on the figure and is given by

$$L_S(Y_1 - Y_2) = S_S X_2 = S_S \left(\frac{Y_2}{m}\right)^{1/n}$$ (11.18)

Eliminating S_S/L_S between Eqs. (11.17) and (11.18) results in

$$\frac{Y_0}{Y_2} - 1 = \left(\frac{Y_1}{Y_2}\right)^{1/n}\left(\frac{Y_1}{Y_2} - 1\right)$$ (11.19)

Equation (11.19) can be solved for the intermediate concentration Y_1 for specified terminal concentrations Y_0 and Y_2, and S_S/L_S is then given by Eq. (11.17). Figure 11.23 will assist in the solution of Eq. (11.19). The greater the value of n the greater the savings in adsorbent by countercurrent operation over single stage.

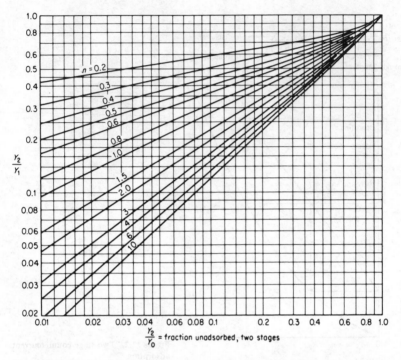

Figure 11.23 Solution to Eq. (11.19). Two-stage countercurrent adsorption.

Illustration 11.2 An aqueous solution containing a valuable solute is colored by small amounts of an impurity. Before crystallization, the impurity is to be removed by adsorption on a decolorizing carbon which adsorbs only insignificant amounts of the principal solute. A series of laboratory tests was made by stirring various amounts of the adsorbent into batches of the original solution until equilibrium was established, yielding the following data at constant temperature:

kg carbon/kg soln	0	0.001	0.004	0.008	0.02	0.04
Equilibrium color	9.6	8.1	6.3	4.3	1.7	0.7

The color intensity was measured on an arbitrary scale, proportional to the concentration of the colored substance. It is desired to reduce the color to 10% of its original value, 9.6. Determine the quantity of fresh carbon required per 1000 kg of solution for a single-stage operation, for a two-stage crosscurrent process using the minimum total amount of carbon, and for a two-stage countercurrent operation.

SOLUTION The experimental data must first be converted to a suitable form for plotting the equilibrium isotherm. For this purpose, define Y as units of color per kilogram of solution and X as units of color adsorbed per kilogram of carbon. The solutions may be considered as dilute in color, so that operating lines will be straight on X, Y coordinates expressed in this manner.

The calculations are made as indicated below.

kg carbon kg soln	Y^* = equilibrium color, units/kg soln	X = adsorbate concentration, units/kg carbon
0	9.6	
0.001	8.6	$(9.6 - 8.6)/0.001 = 1000$
0.004	6.3	$(9.6 - 6.3)/0.004 = 825$
0.008	4.3	663
0.02	1.7	395
0.04	0.7	223

The equilibrium data, when plotted on logarithmic coordinates, provide a straight line, so that the Freundlich equation applies (see Fig. 11.24). The slope of the line is $1.66 = n$, and, at $X = 663$, $Y^* = 4.3$. Therefore [Eq. (11.5)]

$$m = \frac{4.3}{663^{1.66}} = 8.91 \times 10^{-5}$$

The Freundlich equation is therefore

$$Y^* = 8.91 \times 10^{-5} X^{1.66}$$

The equilibrium data can also be plotted on arithmetic coordinates (Fig. 11.25).

Single-stage operation $Y_0 = 9.6$ units of color/kg soln,

$$Y_1 = 0.10(9.6) = 0.96 \text{ unit/kg soln}$$

Let $L_S = 1000$ kg soln. Since fresh carbon is to be used, $X_0 = 0$. In Fig. 11.25, point A representing the initial solution and fresh adsorbent is located, and point B is located on the equilibrium curve at the color concentration of the final solution. At B, $X_1 = 260$. Therefore [Eq. (11.4)]

$$\frac{S_S}{L_S} = \frac{Y_0 - Y_1}{X_1 - X_0} = \frac{9.6 - 0.96}{270 - 0} = 0.032 \text{ kg carbon/kg soln}$$

and

$$S_S = 0.032(1000) = 32.0 \text{ kg carbon/1000 kg soln}$$

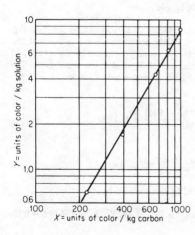

Figure 11.24 Equilibrium data, Illustration 11.2.

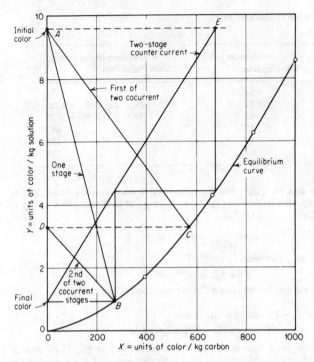

Figure 11.25 Solution to Illustration 11.2.

Alternatively, since the Freundlich equation applies, use Eq. (11.7):

$$\frac{S_S}{L_S} = \frac{Y_0 - Y_1}{(Y_1/m)^{1/n}} = \frac{9.6 - 0.96}{[0.96/(8.91 \times 10^{-5})]^{1/1.66}} = 0.032 \text{ kg carbon/kg soln}$$

$$S_S = 0.032(1000) = 32.0 \text{ kg carbon/1000 kg soln}$$

Two-stage crosscurrent operation The minimum total amount of carbon can be found in Fig. 11.25 by a trial-and-error procedure. Thus, point C on the equilibrium curve is assumed, the operating lines AC and DB drawn, and the values of S_{S1} and S_{S2} computed by Eqs. (11.8) and (11.9). The position of point C is changed until the sum of S_{S1} and S_{S2} is a minimum. The position of C in Fig. 11.25 is the final value, and its coordinates are $(X_1 = 565, Y_1 = 3.30)$. $X_2 = 270$ (at B). Eq. (11.8):

$$S_{S1} = \frac{L_S(Y_0 - Y_1)}{X_1 - X_0} = \frac{1000(9.6 - 3.30)}{565 - 0} = 11.14 \text{ kg}$$

Eq. (11.9):

$$S_{S2} = \frac{L_S(Y_1 - Y_2)}{X_2 - X_0} = \frac{1000(3.30 - 0.96)}{270 - 0} = 8.67 \text{ kg}$$

$$S_{S1} + S_{S2} = 11.14 + 8.67 = 19.81 \text{ kg carbon/1000 kg soln}$$

Alternatively, since the Freundlich equation applies, use Fig. 11.19: $Y_2/Y_0 = 0.96/9.6 = 0.10$,

$n = 1.66$. From the figure, $Y_1/Y_0 = 0.344$. $Y_1 = 0.344(9.6) = 3.30$. Eq. (11.10):

$$\frac{S_{S1}}{L_S} = \frac{Y_0 - Y_1}{(Y_1/m)^{1/n}} = \frac{9.6 - 3.30}{[3.30/(8.91 \times 10^{-5})]^{1/1.66}}$$

$$= 0.01114 \text{ kg carbon/kg soln into 1st stage}$$

Eq. (11.11):

$$\frac{S_{S2}}{L_S} = \frac{Y_1 - Y_2}{(Y_2/m)^{1/n}} = \frac{3.30 - 0.96}{[0.96/(8.91 \times 10^{-5})]^{1/1.66}} = 0.00867 \text{ kg carbon/kg soln into 2d stage}$$

Total carbon required $= (0.1114 + 0.00867)(1000) = 19.81 \text{ kg}/1000 \text{ kg soln}$

Two-stage countercurrent operation $Y_0 = 9.6$, $Y_2 = 0.96$, $X_{N_p+1} = 0$. The operating line is located by trial in Fig. 11.25 until two stages can be drawn between operating line and equilibrium curve, as shown. From the figure at E, $X_1 = 675$. Eq. (11.14):

$$S_S = \frac{L_S(Y_0 - Y_2)}{X_1 - X_{N_p+1}} = \frac{1000(9.6 - 0.96)}{675 - 0} = 12.80 \text{ kg carbon}/1000 \text{ kg soln}$$

Alternatively, $Y_2/Y_0 = 0.96/9.6 = 0.10$; $n = 1.66$. From Fig. 11.24, $Y_2/Y_1 = 0.217$, and $Y_1 = Y_2/0.217 = 0.96/0.217 = 4.42$. Eq. (11.17):

$$\frac{S_S}{L_S} = \frac{Y_0 - Y_2}{(Y_1/m)^{1/n}} = \frac{9.6 - 0.96}{[4.42(8.91 \times 10^{-5})]^{1/1.66}} = 0.01280 \text{ kg carbon/kg soln}$$

and $\quad\quad\quad\quad L = 0.01280(1000) = 12.80 \text{ kg carbon}/1000 \text{ kg soln}$

Agitated Vessels for Liquid-Solid Contact

The vessels are commonly of the proportions described in Chap. 6, with liquid depths in the range 0.75 to 1.5 tank diameters. Four wall baffles, usually $T/12$ wide, are used to eliminate vortices in open vessels and to produce the vertical liquid motion needed to lift the solid particles. They are frequently arranged with a clearance, which may be half the baffle width, between baffle and tank wall to prevent accumulation of solids behind the baffles.

Flat-blade turbines, with or without disk, or pitched-blade turbines, as shown in Fig. 6.3, are most frequently used, arranged on an axially located shaft. Multiple turbines on the same shaft are sometimes used to obtain improved uniformity of the solid-liquid slurry throughout the vessel. The clearance C of the lowest impeller from the bottom of the tank may be $d_i/3$, but it is usually wise in any event to make C larger than the settled depth of the solids in a quiet liquid so that if the impeller stops, it will not become "sanded in." Large impellers operating at relatively low speeds require less power for a given liquid-pumping rate than small, high-speed impellers. Consequently if the solids tend to settle rapidly as a result of a large density or particle size, large impeller diameters are indicated. The recommendations of Lyons [65] for suspensions of solids in water, modified to permit application to liquids of other densities, are shown in Fig. 11.26. The figure shows two ratios T/d_i, one depending on densities and one on particle size; the average T/d_i is chosen. Should the solids consist of mixed sizes and densities, the densest and largest are used. In any event, T/d_i less than 1.6 is not desirable since then the free flow of mixture at

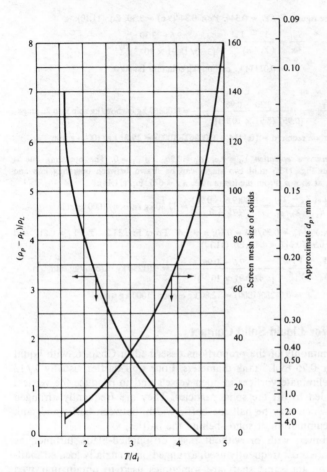

Figure 11.26 Recommended turbine diameters for suspension of solids in liquids of water-like viscosity. [*Modified from E. J. Lyons, in V. W. Uhl and J. B. Gray (eds.), "Mixing," vol. 2, Academic Press, New York, with permission.*]

the vessel wall is interfered with. The choice of impeller diameter can be modified by considerations of torque [79], since the larger impellers require greater torque to turn them and the speed-reducing gear is consequently expensive.

Solid Suspensions

Unless the solids are suspended in the liquid, the mass-transfer rates are seriously impaired. The agitator power P_{min} required just to lift all the particles from the vessel bottom, for particles of density greater than that of the liquid, can be estimated from the expression [101]

$$\frac{P_{\min} g_c}{g v_T V_{tN} \Delta \rho} = 0.092 \left(\frac{\phi_{ST}}{1 - \phi_{ST}} \right)^{1/2} \frac{T}{d_i} e^{5.3 C/T} \tag{11.20}$$

where $\Delta \rho = \rho_p - \rho_L$ and $\rho_p =$ particle density. The latter is calculated from the mass and gross volume of the particle, including the volume of the internal pores. Volume v_T is that for a liquid depth equal to the vessel diameter, $\pi T^3/4$, and ϕ_{ST} is the volume fraction solids based on volume v_T. The limitations of Eq. (11.20) are substantial: particle size and density ranges studied were small; the largest vessel studied was about 0.3 m in diameter. The study was made with flat-blade turbines without disk, but the results agree with the more limited data for disk turbines [84]; it is probable that Eq. (11.20) is reasonably applicable to all types of turbines. Impeller Reynolds numbers Re must be at least 1000. The correlation requires that the terminal velocity V_{tN} of the particles be calculated for single spheres settling with the stagnant-liquid drag coefficient in the Newton's-law range $(d_p V_t \rho_L / \mu_L = 1000$ to 200 000),

$$V_{tN} = 1.74 \left(\frac{g d_p \, \Delta \rho}{\rho_L} \right)^{1/2} \tag{11.21}$$

even though in the agitated vessel it is known that the drag coefficients are substantially larger than in quiet liquids owing to the intense turbulence and the particle acceleration [87]. There are some additional data for very viscous liquids, for which Re will be small [43].

If larger impeller powers are used, the solids are lifted to heights above the bottom depending upon the power applied. The liquid above this height will be clear. It is particularly difficult to suspend large or heavy particles up to the liquid surface because the predominant liquid velocities in the upper regions are horizontal. The power P required to suspend particles of uniform size and density to a height Z' above the midplane of the uppermost impeller, with essentially the same limitations as those of Eq. (11.20), is given by [101]

$$\frac{P g_c}{g n \rho_m v_m V_{tS}} = \phi_{Sm}^{2/3} \left(\frac{T}{d_i} \right)^{1/2} \exp \left(4.35 \frac{Z'}{T} - 0.1 \right) \tag{11.22}$$

where ρ_m, v_m, and ϕ_{Sm} are the properties of the slurry below the height Z' and n is the number of impellers (1, 2, or 3) on the shaft. V_{tS} must be the terminal settling velocity of single spheres computed through Stokes' law,

$$V_{tS} = \frac{g d_p^2 \, \Delta \rho}{18 \mu_L} \tag{11.23}$$

For impeller Reynolds numbers Re less than 25 000, the power P must be multiplied by 4000 Re$^{-0.8}$.

In batch operations, if the solids are of uniform size and density, the slurry produced below height Z of Eq. (11.22) will be of uniform solids concentration [101]. For mixed particles, the slurry can be quite nonuniform with respect to particle size and density [79]. However, for steady-state continuous-flow operation, the effluent slurry for nondissolving solids must be identical in solids size and density with the solids-feed mixture. The residence time of the different particles need not be uniform, however, and their relative concentration in the vessel need not necessarily be identical with that in the feed and effluent streams.

Impeller power For relatively dilute solid-liquid mixtures, except for fibrous solids, the power to agitate at a given speed is essentially the same as for the clear liquid [85]. Concentrated slurries and those containing fibrous solids are likely to be nonnewtonian in character; i.e., the velocity gradient in the moving mass is not directly proportional to shear stress. The viscosity of such fluids must be related to the agitator power, and the work of Metzner and his associates should be consulted [7, 68, 69]. Since such data are not usually available, a rough estimate can be made by using the clear-liquid power-number–Reynolds-number curves of Fig. 6.5, with density and viscosity estimated from [52]

$$\rho_m = \phi_{Sm}\rho_p + (1 - \phi_{Sm})\rho_L \tag{11.24}$$

$$\mu_m = \frac{\mu_L}{(1 - \phi_{Sm}/\phi_{SS})^{1.8}} \tag{11.25}$$

ϕ_{SS} is the volume fraction of solids in the bed of solids after final settling in quiet liquid. In the absence of measured data, this can be estimated roughly as 0.6. Results are poor for impeller Reynolds numbers less than 1000.

Illustration 11.3 A baffled vessel, 1 m diameter, filled to a depth of 1 m with water, is agitated with a 203-mm-diameter (8-in) six-bladed disk turbine, arranged axially 150 mm from the bottom of the vessel and turning at 8.33 r/s (500 r/min), then 50 kg (110 lb) sand, $\rho_S = 2300$ kg/m^3 (143.5 lb/ft^3), $d_p = 0.8$ mm, is added. The temperature is 25°C. What agitator power is required?

SOLUTION From the definition of power number, $g_c P = $ Po $\rho_m d_i^5 N^3$. For one impeller, $n = 1$. If the sand is lifted to a height Z' above the impeller,

$$v_m = \frac{\pi T^2 (Z' + C)}{4}$$

Since the solid volume is S/ρ_p,

$$\phi_{Sm} = \frac{S/\rho_p}{v_m} = \frac{4S}{\pi T^2 (Z' + C)\rho_p}$$

If these are substituted in Eq. (11.22), there results

$$(Z' + C)^{1/3} \exp\frac{4.35Z'}{T - 0.1} = \frac{1.0839 \text{ Po } d^{11/2}N^3\rho_p^{2/3}}{g V_{tS} T^{7/6} S^{2/3}}$$

Tentatively, take Po $= 5$. $d_i = 0.203$ m, $\rho_p = \rho_S = 2300$ kg/m^3, $T = 1$ m, $C = 0.150$ m, $S = 50$ kg, $g = 9.807$ m/s^2, $d_p = 8 \times 10^{-4}$ m, $\mu_L = 8.94 \times 10^{-4}$ kg/m · s, $\rho_L = 998$ kg/m^3, $\Delta\rho = 1302$ kg/m^3, $N = 8.33$ s^{-1}. Eq. (11.23): $V_{tS} = 0.508$ m/s.

Substitution in the above equation yields

$$(Z' + 0.150)^{1/3} \exp(4.35Z' - 0.1) = 1.257$$

By trial, $Z' = 0.104$ m, and the depth to which the solid is lifted $= Z' + C = 0.314$ m. $\phi_{Sm} = 0.0882$, $\rho_m = 1112.8$ kg/m^3. With $\phi_{SS} = 0.6$, Eq. (11.25) will yield $\mu_m = 1.19 \times 10^{-3}$ kg/m · s. The impeller Reynolds number is then

$$\text{Re} = d_i^2 N \rho_m / \mu_m = 321\,000$$

Fig. 6.5: Po $= 5 = Pg_c/\rho_m N^3 d_i^5$, whence $P = 1109$ W (1.5 hp) transmitted to the slurry. **Ans.**

Mass Transfer

Except for dissolving of crystals or crystallization, where the mass transfer is confined to the liquid phase, consideration must generally be given to transfer through the liquid surrounding the particles and to some kind of mass-transfer effect within the solid.

Liquid-phase mass transfer Although many studies have been made of mass-transfer rates in the liquid for solid-liquid slurries, the results are frequently conflicting and difficult to interpret. We do know that once the solids are fully suspended, further expenditure of agitator power does not produce commensurate improvement in the mass-transfer rates [51]. For small particles, the mass-transfer coefficient becomes smaller with increased particle size [36], whereas for large particles there is no effect of particle size [5]. The transition size is about 2 mm [87]. At least for moderate density difference, the coefficients are independent of $\Delta\rho$, but for large differences, e.g., for metal particles in ordinary liquids, there may be an effect. There is no effect of solids concentration, at least up to 10 percent by volume [5], or of the presence or absence of baffles [77] and of agitator design at a given impeller power [9]. The effect of diffusivity is not well known, since large changes in diffusivity usually accompany corresponding changes in viscosity in such a fashion that the influence of the Schmidt number is difficult to establish [5].

It seems reasonable that the mass-transfer coefficients should depend upon some kind of Reynolds number. A great variety of such numbers has been suggested, many using the particle-liquid slip velocity (relative velocity), which is difficult to estimate, or the single-particle terminal settling velocity, which is not appropriate [87]. Refer to Eq. (3.30). The turbulent mass flux is

$$J_{A, \text{ turb}} = E_D \frac{\Delta c}{l} = k_{L, \text{ turb}} \, \Delta c = b_1 u' \, \Delta c$$

so that in an intensely turbulent liquid, as in an agitated vessel, k_L should be closely associated with the fluctuating velocity u'. For a fluctuating velocity over a distance corresponding to the particle size d_p [Eq. (3.26)] the power dissipation would be

$$\frac{P g_c}{v_L} \propto \frac{\rho_L u'^3}{d_p} \tag{11.26}$$

and a turbulent-particle Reynolds number corresponding to this is

$$\text{Re}_p = \frac{d_p u' \rho_L}{\mu_L} \tag{11.27}$$

If we take u' from Eq. (11.26) and ignore proportionality constants in defining dimensionless groups, then

$$\text{Re}_p = d_p^{4/3} \left(\frac{P g_c}{v_L} \right)^{1/3} \frac{\rho_L^{2/3}}{\mu_L} \tag{11.28}$$

where v_L is the volume of liquid in the vessel (not the rate of flow).

Small particles ($d_p < 2$ mm) fully suspended For moderate differences in $\Delta\rho$ [53, 62, 78],

$$\text{Sh}_L = 2 + 0.47 \, \text{Re}_p^{0.62} \left(\frac{d_i}{T} \right)^{0.17} \text{Sc}_L^{0.36} \tag{11.29}$$

Large particles ($d_p > 2$ mm) fully suspended The recommended correlation [11, 12, 71] is

$$\text{Sh}_L = 0.222 \, \text{Re}_p^{3/4} \, \text{Sc}_L^{1/3} \tag{11.30}$$

In this case the exponent on Re_p is chosen so as to make the mass-transfer coefficient independent of particle size, in accordance with observed data. It is important to note that the mass-transfer coefficients contained in the Sherwood numbers of the preceding equations are based on the outside, gross surface of the particles, ignoring the surface of any internal pore structure.

Solid-phase mass transfer The various mass-transfer processes that may occur within the solid can frequently be jointly described in terms of a mass-transfer coefficient k_S (based on external surface) or an effective diffusivity D_S. For a constant or nearly constant equilibrium distribution coefficient and spherical particles, the former can be related to the diffusivity through the approximation [30]

$$k_S a_p = \frac{60 D_S}{d_p^2} \tag{11.31}$$

or, since

$$a_p = \frac{6 \phi_S}{d_p} \tag{11.32}$$

we get

$$k_S = \frac{10 D_S}{d_p \phi_S} \tag{11.33}$$

The liquid- and solid-phase resistances can be added, in the case of constant distribution coefficient,

$$\frac{1}{K_L} = \frac{1}{k_L} + \frac{m}{k_S} \tag{11.34}$$

to produce an overall resistance.

Batch adsorption, negligible solid-phase mass-transfer resistance If the quantity S/mv_L, an adsorption factor, is sufficiently large (large solid doses and equilibrium favorable to adsorption), the solid-phase resistance can be neglected and only k_L need be considered. These are the conditions of what follows.

Let the volume of liquid to be treated be v_L, of initial concentration c_0 mol solute A/volume. The mass of adsorbate-free solid to be used is S_S, of external surface a_{pS} per unit mass of solid, initially of solute concentration X_0 mol solute/mass of solid. The final concentration in the liquid is to be c_1.

At any instant, the adsorbed solute concentration within the solid is X mass solute/mass solute-free solid, and since the solid-phase resistance is negligible, this will be taken as uniform throughout the solid. The distribution coefficient m is then defined as

$$m = \frac{c^* M_A}{X} \tag{11.35}$$

where c^* is the equilibrium concentration in the liquid. Since the solutions involved are usually dilute, the coefficient k_L rather than F_L can be used. Then if c is the concentration in the bulk liquid at any time,

$$N_A = \frac{-dc}{a_{pL} d\theta} = k_L(c - c^*) \tag{11.36}$$

where $a_{pL} = a_{pS} S_S / v_L$. A solute balance is

$$\frac{S}{M_A}(X - X_0) = S_S\left(\frac{c^*}{m} - \frac{X_0}{M_A}\right) = v_L(c_0 - c) \tag{11.37}$$

Substitution of c^* from Eq. (11.37) into (11.36) and rearrangement produces

$$\int_{c_0}^{c_1} \frac{-dc}{c(1 + mv_L/S_S) - (mv_L c_0/S_S + mX_0/M_A)} = k_L a_{pL} \int_0^\theta d\theta \tag{11.38}$$

or

$$\ln \frac{c_0 - mX_0/M_A}{c_1(1 + mv_L/S_S) - (mv_L c_0/S_S + mX_0/M_A)} = \left(1 + \frac{mv_L}{S_S}\right) k_L a_{pL} \theta \tag{11.39}$$

If $X_0 = 0$, this can be rearranged to [28]

$$\frac{c_1}{c_0} = \frac{1}{1 + mv_L/S_S} \exp\left[-\left(1 + \frac{mv_L}{S_S}\right) k_L a_{pL} \theta\right] + \frac{mv_L/S_S}{1 + mv_L/S_S} \tag{11.40}$$

A method of computation for batch processes where both solid and liquid mass-transfer resistances are important is also available [99].

Illustration 11.4 A batch of kerosene containing dissolved water at a concentration 40 ppm $=$ 0.0040 wt percent is to be dried to 5 ppm water by contacting with "dry" silica gel at 25°C. The kerosene density $= 783$ kg/m³, viscosity $= 1.7 \times 10^{-3}$ kg/m · s, and estimated av mol wt $=$ 200. The silica gel, density $\rho_p = 881$ kg/m³, is in the form of 14-mesh spherical beads. The equilibrium distribution ratio for water is $m = c^*/X = 0.522$ (kg water/m³ kerosene)/ (kg water/kg gel). (a) Calculate the minimum solid/liquid ratio which can be used. (b) For a liquid/solid ratio $= 16$ kg gel/m³ kerosene and batches of 1.7 m³ (60 ft³), specify the characteristics of a suitable baffled, agitated vessel and estimate the agitator power and contacting time.

SOLUTION Define c as kg water/m^3 liquid.

(a) Water balance: $v_L(c_0 - c_1) = S_S(X_1 - X_0)$.

$$c_0 = 783(4 \times 10^{-5}) = 0.0313 \text{ kg } H_2O/m^3$$

$$c_1 = 783(5 \times 10^{-6}) = 0.00392 \text{ kg } H_2O/m^3$$

For $S_{S, min}/v_L$, X_1 is in equilibrium with c_1

$$X_1 = \frac{c_1}{m} = \frac{0.00392}{0.522} = 7.51 \times 10^{-7} \text{ kg } H_2O/\text{kg gel}$$

therefore

$$\frac{S_{S, min}}{v_L} = \frac{c_0 - c_1}{X_1 - X_0} = \frac{0.0313 - 0.00392}{7.51 \times 10^{-3} - 0} = 3.65 \text{ kg gel}/m^3 \text{ kerosene} \quad \textbf{Ans.}$$

(b) Basis: 1 batch, 1.7 m^3 kerosene.

$$S_S = 16(1.7) = 27.2 \text{ kg dry gel} \qquad \text{vol solids} = \frac{27.2}{881} = 0.0309 \ m^3$$

Total batch volume = 1.73 m^3

Take $Z = T$. Then $\pi T^2 Z/4 = \pi T^3/4 = 1.73$, and $T = 1.3$ m (4.25 ft). To allow for adequate freeboard, make the vessel height = 1.75 m (5.75 ft).

Use a six-blade disk-turbine impeller. From Fig. 11.26, with d_p corresponding to 14 mesh = 1.4 mm, $T/d_i = 2$. With $(\rho_P - \rho_L)/\rho_L = (881 - 783)/783 = 0.125$, $T/d_i = 4.4$. Average $T/d_i = (4.4 + 2)/2 = 3.2$, $d_i = 1.3/3.2 = 0.406$ m (16 in).

Assuming that the settled volume fraction of solids = 0.6, the settled volume of solids = 0.309/0.6 = 0.0515 m^3. The depth of settled solids = 0.0515/$[\pi(1.3)^2/4] = 0.0388$ m, which is negligible. Locate the turbine 150 mm (6 in) from the bottom of the tank. $C = 0.150$ m.

Power Use sufficient agitator power to lift the solids to 0.6 m above the bottom of the vessel. $Z' = 0.6 - 0.15 = 0.45$ m. The properties of the slurry in the 0.6 m depth to which the solids are suspended are

$$v_m = \frac{0.6\pi(1.3)^2}{4} = 0.796 \ m^3 \qquad \phi_{Sm} = \frac{0.0309}{0.796} = 0.0388 \text{ vol fraction solids}$$

Eq. (11.24): $\rho_m = 786.8$ kg/m^3; Eq. (11.25) with $\phi_{SS} = 0.6$: $\mu_m = 1.917 \times 10^{-3}$ kg/m · s. With $g_c = 1.0$, $g = 9.807$ m/s^2, $\mu_L = 1.7 \times 10^{-3}$ kg/m · s, $d_p = 0.0014$ m, Eq. (11.23): $V_{tS} = 0.0616$ m/s.

With $n = 1$ impeller, Eq. (11.22): $P = 315.8$ W. Try Po = 5. Therefore $N = (g_c P/Po\rho_m d_i^5)^{1/3} = 1.94$ r/s (116 r/min). Use 2 r/s (120 r/min). Re $= d_i^2 N\rho_m/\mu_m = 135\,300$. Therefore Po = 5 (Fig. 6.5), and $P = 315.8(2/1.94)^3 = 346$ W (0.46 hp) applied to the slurry. Power to the motor will be larger, depending upon the efficiency of the motor and speed reducer (see Chap. 6).

Mass transfer Eq. (11.28), with $d_p = 0.0014$ m, $v_L = 1.7 \times 10^{-3} \ m^3$: Re$_p = 44.8$.

D_L will be estimated through Eq. (2.44). $M_B = 200$, $T = 298$ K, $\phi = 1$, $\mu = 1.7 \times 10^{-3}$, $v_A(H_2O) = 0.0756$ (Chap. 2 notation). Therefore $D_L = 1.369 \times 10^{-9}$ m^2/s. Sc$_L = \mu_L/\rho_L D_L = 1586$. Eq. (11.29): Sh$_L = k_L d_p/D_L = 59.8$. Therefore $k_L = 5.85 \times 10^{-5}$ kmol/m^2 · s · (kmol/m^3).

$$\frac{\text{Mass}}{\text{particle}} = \frac{\pi}{6}(0.0014)^3(881) = 1.266 \times 10^{-6} \text{ kg}$$

$$\frac{\text{Surface}}{\text{particle}} = \pi(0.0014)^2 = 6.158 \times 10^{-6} \ m^2$$

Therefore

$$a_{pS} = \frac{6.158 \times 10^{-6}}{1.266 \times 10^{-6}} = 4.864 \ m^2/\text{kg}$$

and $$a_{pL} = 4.864(16) = 77.8 \text{ m}^2/\text{m}^3 \text{ liquid}$$

$$\frac{S_S}{v_L m} = \frac{16}{0.522} = 30.65$$

This is considered sufficiently large for only liquid-phase transfer to require consideration. Eq. (11.40):

$$\theta = \frac{\ln[(c_0/c_1)/(1 + mv_L/S_S - mv_L c_0/S_S c_1)]}{(1 + mv_L/S_S)k_L a_{pL}}$$

= 498 s or 8.3 min contacting time required **Ans.**

Continuous cocurrent adsorption with liquid- and solid-phase mass-transfer resistances For continuous, cocurrent flow of solution and adsorbent through an agitated vessel, the individual particles experience an unsteady-state change of solute concentration during their time of contact. Nevertheless the process as a whole can be steady-state, the concentrations of effluents remaining unchanged with passage of time.

Gröber [33] computed the approach to equilibrium temperature of spherical particles immersed in an agitated fluid of constant temperature. Since the concentration of the solution in a well-stirred vessel in continuous flow is essentially uniform throughout at the effluent value, Gröber's result can be adapted to cocurrent adsorption through the heat- mass-transfer analogy, as in Fig. 11.27. Here the approach of the particles to equilibrium concentration with the effluent liquid is expressed as the Murphree stage efficiency,

$$E_{MS} = \frac{X_1 - X_0}{X_1^* - X_0} = \frac{X_1 - X_0}{M_A c_1/m - X_0} \tag{11.41}$$

X is the average concentration of solute (adsorbate) in the solid. The figure can equally well be used for desorption (drying) and leaching of solute from the solid.

Illustration 11.5 Water containing copper ion Cu^{2+} at a concentration 100 ppm, flowing continuously at the rate 1.1×10^{-4} m³/s (1.75 U.S. gal/min), is to be contacted with 0.0012 kg/s (0.15 lb/min) of an ion-exchange resin to adsorb the Cu^{2+}. The contacting will occur by cocurrent flow through a baffled, agitated vessel 1 m in diameter, 1 m tall, covered. The feed material will enter at the bottom, leave through an opening in the cover, and flow to a filter. The agitator, a six-bladed disk-type turbine, 0.3 m in diameter, will be centered. The temperature will be 25°C. Specify a suitable turbine speed, and estimate the Cu^{2+} concentration in the effluent water.

Data The ion-exchange particles, swollen when wet, are spherical, 0.8 mm diameter, density = 1120 kg/m³. The diffusivity of Cu^{2+} in the solid $D_S = 2 \times 10^{-11}$ m²/s; that in the water = 7.3×10^{-10} m²/s. The equilibrium distribution coefficient is constant: $m = 0.2$ (kg Cu^{2+}/m^3 soln)/(kg Cu^{2+}/kg resin).

SOLUTION The particles will be lifted to the top of the vessel: $Z' = 0.5$ m. $\rho_p = 1120$ kg/m³, $d_p = 8 \times 10^{-4}$ m, $\mu_L =$ viscosity of water, 25°C = 8.94×10^{-4} kg/m · s, $\rho_L =$ density of water = 998 kg/m³, $\Delta\rho = \rho_p - \rho_L = 122$ kg/m³, g = 9.80 m/s². Eq. (11.23): $V_{tS} = 0.0476$ m/s.

$v_m = \pi T^2 Z/4 = \pi(1)^3/4 = 0.785$ m³. $v_L = 1.1 \times 10^{-4}$ m³/s. Volume rate of solid = 0.0012/1120 = 1.07×10^{-6} m³/s.

$\phi_{Sm} = (1.07 \times 10^{-6})/(1.1 \times 10^{-4} + 1.07 \times 10^{-6}) = 9.63 \times 10^{-3}$ vol fraction. Eq. (11.24): $\rho_m = 999$ kg/m³; $n = 1$ impeller, $d_i = 0.3$ m, $g_c = 1.0$. Eq. (11.22): $P = 240.6$ W (0.32 hp).

To estimate the impeller speed, assume the power number Po = 5. Therefore $N = (g_c P/\rho_m d_i^5)^{1/3}$. With $d_i = 0.3$ m, $N = 2.7$ r/s (162 r/min). For this speed, with $\mu_m = \mu_L$ since

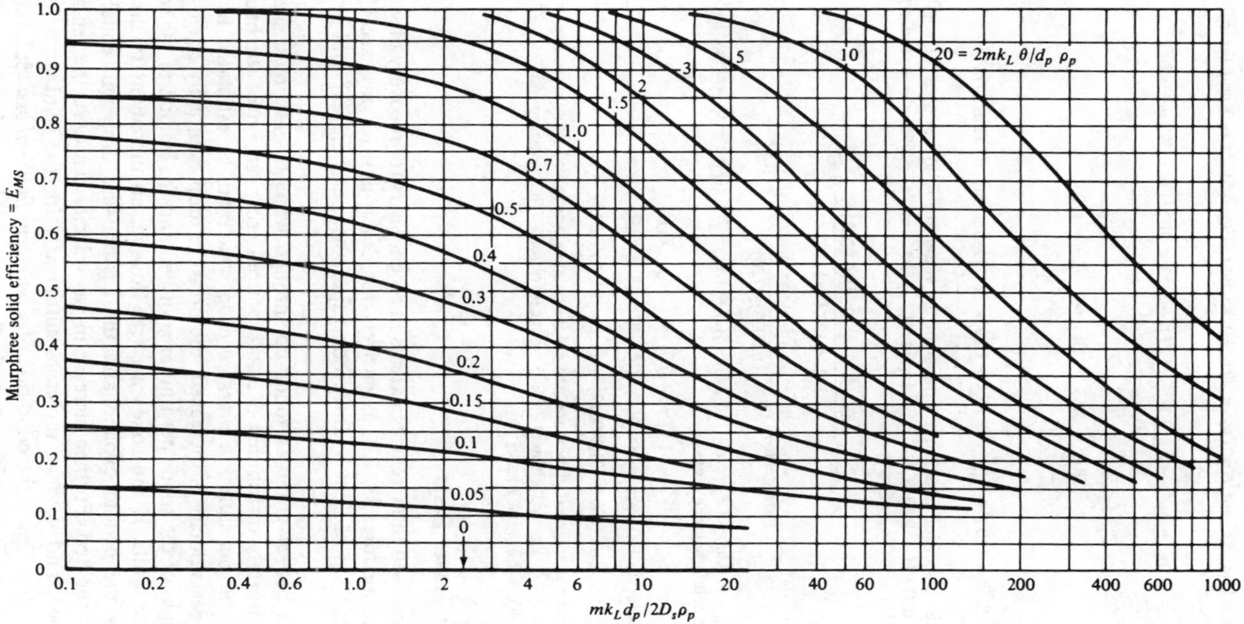
1.0
0.9
0.8
0.7
0.6
0.5
0.4
0.3
0.2
0.1
0
Murphree solid efficiency = E_{MS}
0.1
0.2
0.4
0.6
1.0
2
4
6
10
20
40
60
100
200
400
600
1000
$mk_L d_p / 2D_s \rho_p$
$20 = 2mk_L \theta / d_p \rho_p$
10
5
3
2
1.5
1.0
0.7
0.5
0.4
0.3
0.2
0.15
0.1
0.05
0

Figure 11.27 Stage efficiency for single-stage, continuous, cocurrent adsorption in well-stirred vessels, with spherical particles. [*Adapted from Gröber, Ver. Dtsch. Ing.*, **69**, *705 (1925).*]

the slurry is dilute, $Re = d_t^2 N \rho_m / \mu_L = 271\,500$, so that (Fig. 6.5) $Po = 5$ and the calculated speed is correct.

$$v_L = v_T(1 - \phi_{Sm}) = \frac{\pi}{4}(1)^3(1 - 9.63 \times 10^{-3}) = 0.777 \text{ m}^3$$

Eq. (11.28):

$$Re_p = 56.3 \qquad Sc_L = \frac{\mu_L}{\rho_L D_L} = 1226$$

Eq. (11.29):

$$Sh_L = 130.3 = \frac{k_L d_p}{D_L} \qquad k_L = 1.19 \times 10^{-4} \text{ kmol/m}^2 \cdot \text{s} \cdot (\text{kmol/m}^3)$$

Since the dispersion is uniform throughout the vessel, the residence time for both liquid and solid = 0.777 $(1 - 9.63 \times 10^{-3})$ m^3 liquid/$(1.1 \times 10^{-4}$ m^3/s$) = 6996$ s $= \theta$. Refer to Fig. 11.27.

$$\text{Abscissa} = \frac{0.2(1.19 \times 10^{-4})(0.0008)}{2(2 \times 10^{-11})(1120)} = 0.425$$

$$\text{Parameter} = \frac{2(0.2)(1.19 \times 10^{-4})(6996)}{(0.0008)(1120)} = 0.37$$

Therefore $E_{MS} = 0.63 = (X_1 - X_0)/(c_1/m - X_0) = (X_1 - 0)/(c_1/0.2 - 0)$ where c is defined as kg/m^3. A solute balance gives

$$(X_1 - X_0)S_S = v_L(c_0 - c_1)$$

c_0 corresponding to 100 ppm = $100(998)/10^6 = 0.0998$ kg/m^3.

$$(X_1 - 0)(0.0012) = (1.1 \times 10^{-4})(0.0998 - c_1)$$

The expression for stage efficiency and the solute balance are solved simultaneously, whence $c_1 = 2.82 \times 10^{-3}$ kg/m^3 effluent Cu^{2+} concentration. This corresponds to $(2.82 \times 10^{-3})(10^6)/998 = 2.8$ ppm. **Ans.**

Fluidized and Teeter Beds

These have been used in increasing extents in recent years for recovery of vapors from vapor-gas mixtures [3, 22], extensively for desorption (drying, see Chap. 12), for fractionation of light hydrocarbon vapors with carbon [23], and several other purposes.

Consider a bed of granular solids up through which a gas flows, the gas being fairly uniformly distributed over the bottom cross section of the bed. At low gas rates, the gas suffers a pressure drop which can be estimated by Eq. (6.66). As the gas velocity is increased, the pressure drop eventually equals the sum of the weight of the solids per unit area of bed and the friction of the solids at the container walls. If the solids are free-flowing, an increment in the gas velocity causes the bed to expand and the gas-pressure drop will equal the weight/area of the bed. Further increase in gas velocity causes further enlargement of the bed, and voidage increases sufficiently for the solid particles to move about locally. This is the condition of a *quiescent fluidized bed*. Still further increase in gas velocity further expands the bed, solid particles move about freely (indeed, they are thoroughly circulated throughout the bed), and well-defined bubbles of gas may form and rise through the bed. The bed appears much like a boiling liquid, a distinct interface between the top of the solids and the escaping gas remains evident, and the gas-pressure drop is not much different from that of the quiescent fluidized state. This is the condition used for adsorption. Further increase in gas velocity continues to expand the bed, and eventually the solids are carried away with the gas.

Squires [94] has distinguished between *fluidized* beds and *teeter* beds. Fluidized beds are produced by fine powders, usually smaller than 20-mesh and ranging down to 325-mesh, and will retain a great range of particle sizes within the bed, with little particle attrition. They usually operate with superficial gas velocities of the order of 0.6 m/s (2 ft/s) or less, which may be 10 times that for minimum fluidization. Teeter beds are produced with coarser particles, up to 10-mesh commonly. Gas velocities of from 1.5 to 3 m/s (5 to 10 ft/s) can be used, which may only be twice that for minimum fluidization. There is little gas bubbling, and the size range of particles retained is relatively small. In general, better fluid-solids contacting is obtained in teeter beds, and most adsorption applications are in this category. Solids can also be fluidized by liquids, and these will be ordinarily in the teeter-bed category.

Since the first introduction of these techniques for fluid-solid contact in 1948, with the principal application in the catalysis of gaseous chemical reactions, a very large technology for their design has developed. This is still not in a condition which permits summary for design purposes in the space available here, however, and more extended works should be consulted for such details [16, 54, 61, 103]. The discussion here is limited to the computation of stage requirements and does not include the mechanical design.

Slurry Adsorption of Gases and Vapors

A gas or vapor in a gas can be adsorbed on a solid which is slurried in a liquid. For example, sulfur dioxide can be adsorbed from a mixture with air on activated carbon slurried in water [8]. This procedure was suggested as early as 1910 [25], but interest in it has revived only recently. The reported details thus far are confined to laboratory studies of sparged vessels [73, 88] operated semibatch (gas flow continuous, slurry batch) or cocurrent flow or both [66]. A continuous countercurrent process has been described [4, 27]. The slurry is of course much easier to handle than dry solid, and it has been shown that the capacity of the slurried adsorbent is about the same as for dry solid [73], much larger than that of the liquid solvent alone [96].

Adsorption of a Vapor from a Gas Fluidized Beds

Figure 11.28 shows an arrangement typical of that required for adsorption of a vapor from a nonadsorbed gas, in this case drying of air with silica gel [22]. In the upper part of the tower, the gel is contacted countercurrently with the air to be dried on perforated trays in relatively shallow beds, the gel moving from tray to tray through downspouts. In the lower part of the tower, the gel is regenerated by similar contact with hot gas, which desorbs and carries off the moisture. The

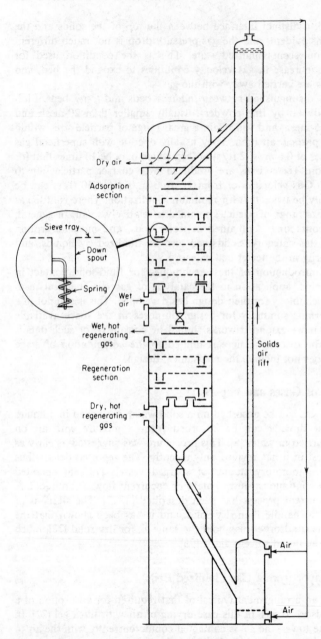

Dry air →

Adsorption
section

Sieve tray

Down
spout

Spring

Wet
air

Wet, hot
regenerating
gas

Solids
air
lift

Regeneration
section

Dry, hot
regenerating →
gas

Air

Air

Figure 11.28 Fluidized bed, multistage countercurrent adsorber with regeneration. *(After Ermenc [22]).*

dried gel is then recirculated by air lift to the top of the tower. If the adsorbed vapor is to be recovered, regeneration might include steam stripping of a carbon adsorbent with distillation or decantation of the condensed water–organic-vapor mixture, followed by air drying of the adsorbent before reuse, and obvious changes in the flowsheet would be necessary.

If adsorption is isothermal, the calculations for theoretical stages can be made in the same manner as for contact filtration of liquids [Fig. 11.20, Eq. (11.14)], and the same is true for desorption-regeneration. Ordinarily, however, the adsorber will operate adiabatically, and there may be a considerable rise in temperature owing to adsorption (or fall, in the case of desorption). Since the equilibrium then changes with stage number, the calculations are done stage by stage, much in the manner of gas absorption. The same procedure will apply to any type of stage device and is not limited to fluidized beds. Thus, for the first n stages of the flowsheet of Fig. 11.20, a balance for adsorbable vapor is

$$G_S(Y_0 - Y_n) = S_S(X_1 - X_{n+1}) \qquad (11.42)$$

and an enthalpy balance for adiabatic operation is

$$G_S(H_{G0} - H_{Gn}) = S_S(H_{S1} - H_{S_{n+1}}) \qquad (11.43)$$

where H_G is the enthalpy of vapor-gas mixture, energy/mass, and H_S the enthalpy of solid plus adsorbate, energy/mass adsorbent. If enthalpies are referred to solid adsorbent, nonadsorbed gas (component C), and the adsorbate (component A) as a liquid, all at a base temperature t_0, then

$$H_G = C_C(t_G - t_0) + Y[C_A(t_G - t_0) + \lambda_{A0}] \qquad (11.44)$$

where C_C = gas heat capacity
  C_A = vapor heat capacity
  λ_{A0} = latent heat of vaporization of A at t_0

$$H_S = C_B(t_S - t_0) + XC_{A,L}(t_S - t_0) + \Delta H_A \qquad (11.45)$$

where C_B = heat capacity of adsorbent (component B)
  $C_{A,L}$ = heat capacity of liquid A
  ΔH_A = integral heat of adsorption at X and t_0, energy/mass adsorbent
Gas and solid leaving a theoretical stage are in thermal and concentration equilibrium. The use of these equations is precisely the same as for gas absorption (Illustration 8.4).

Fractionation of a vapor mixture For the separation of a vapor mixture consisting of, for example, two adsorbable components for which the adsorbent exhibits a relative adsorptivity, calculations are made in the manner for moving beds to locate operating lines and equilibrium curves [Eqs. (11.53) to (11.65) and Fig. 11.33]. Theoretical stages can then be determined by the usual step construction in the lower part of such a figure.

CONTINUOUS CONTACT

In these operations the fluid and adsorbent are in contact throughout the entire apparatus, without periodic separation of the phases. The operation can be carried out in strictly continuous, steady-state fashion, characterized by movement of the solid as well as the fluid. Alternatively, owing to the rigidity of the solid adsorbent particles, it is also possible to operate advantageously in semicontinuous fashion, characterized by a moving fluid but stationary solid. This results in unsteady-state conditions, where compositions in the system change with time.

Steady-State–Moving-Bed Adsorbers

Steady-state conditions require continuous movement of both fluid and adsorbent through the equipment at constant rate, with no change in composition at any point in the system with passage of time. If parallel flow of solid and fluid is used, the net result is at best a condition of equilibrium between the effluent streams, or the equivalent of one theoretical stage. It is the purpose of these applications to develop separations equivalent to many stages, however, and hence only countercurrent operations need be considered. There are applications especially in treating gases and liquids, for purposes of collecting solute and fractionation, through ordinary adsorption and ion exchange.

Equipment It is only in relatively recent years that satisfactory large-scale devices for the continuous contacting of a granular solid and a fluid have been developed. They have had to overcome the difficulties of obtaining uniform flow of solid particles and fluid without channeling or local irregularities, as well as those of introducing and removing the solid continuously into the vessel to be used.

A substantial effort has been expended in developing apparatus for counterflow of gas and solid, with plug flow of solid rather than fluidized beds. The hydraulics and mechanics of such a device were successfully solved, resulting in the *Hypersorber*, used for fractionating hydrocarbon gases with activated gas-adsorbent carbon [8, 48]. The device resembles a countercurrent distilling column with the solid phase, flowing down, analogous to the upward flow of heat in distillation. Several adsorbers on a very large scale were built, but the very brittle carbon was subject to serious attrition losses, and no new continuous-flow, countercurrent device for plug flow of solids and gas is believed to be in operation.

In liquid treating, far less elaborate devices have been used. Solids can be introduced to the top of the tower by a screw feeder or simply from a bin by gravity flow. Withdrawal of solids at the bottom can be accomplished by several types of devices, e.g., a revolving compartmented valve similar to revolving doors for buildings; it is not generally possible to remove solids entirely free of liquid.

The Higgins contactor [35, 42], developed initially for ion exchange but useful for solids-liquid contacting generally, is unique in the intermittent nature of its operation. Figure 11.29 shows it schematically as arranged for simple solute collection. In Fig. 11.29*a*, the temporarily stationary upper bed of solids is contacted with liquid flowing downward, so that fluidization does not occur. In the lower bed, the solid is regenerated by an eluting liquid. After several minutes, the liquid flow is stopped, valves are turned as shown in Fig. 11.29*b*, and the liquid-filled piston pump is moved as shown for a period of several seconds, whereupon solid is moved clockwise hydraulically. In Fig. 11.29*c*, with the valves readjusted to their original position, movement of solid is completed, and liquid flows are started to complete the cycle. Operation, although intermittent and cyclic, is nearly the same as for truly continuous countercurrent operation.

A truly continuous countercurrent scheme for ion-exchange treatment of liquids incorporates the resin in an endless, flexible belt which moves continuously over horizontal rollers and vertically with many passes, through which the solution to be treated must travel [46, 47]. The belt may pass successively from contact with solution to be treated, to regenerating liquid, and back to solution to be treated.

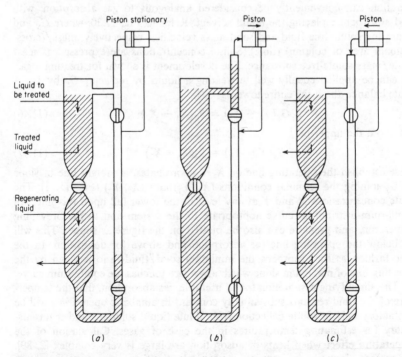

Figure 11.29 Higgins contactor, schematic (solids shown shaded).

MASS-TRANSFER OPERATIONS

ADSORBENT

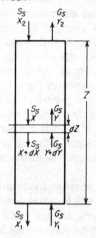

Figure 11.30 Continuous countercurrent adsorption of one component.

One component adsorbed (solute collection) For purposes of computation the operation can conveniently be considered analogous to gas absorption, with solid adsorbent replacing the liquid solvent. Refer to Fig. 11.30, where G_S and S_S are the solute-free fluid and solid mass velocities, respectively, mass/(cross-sectional area of column) (time). Solute concentrations are expressed as mass solute/mass solute-free substance. The development is shown for treating a gas, but can be applied equally well to treating a liquid by replacing G_S by L_S. A solute balance about the entire tower is

$$G_S(Y_1 - Y_2) = S_S(X_1 - X_2) \qquad (11.46)$$

and about the upper part

$$G_S(Y - Y_2) = S_S(X - X_2) \qquad (11.47)$$

These establish the operating line on X, Y coordinates, a straight line of slope S_S/G_S joining the terminal conditions (X_1, Y_1) and (X_2, Y_2) (Fig. 11.31). The solute concentrations X and Y at any level in the tower fall upon this line. An equilibrium-distribution curve appropriate to the system and to the prevailing temperature and pressure can also be plotted on the figure as shown. This will fall below the operating line for adsorption and above for desorption. In the same fashion as for absorbers, the minimum solid/fluid ratio is given by the operating line of maximum slope which anywhere touches the equilibrium curve.

The simplifying assumption found useful in gas absorption, that the temperature of the fluid remains substantially constant in adiabatic operations, will be satisfactory only in solute collection from dilute liquid solutions and is unsatisfactory for estimating temperatures in the case of gases. Calculation of the temperature effect when heats of adsorption are large is very complex [2, 89]. The present discussion is limited to isothermal operation.

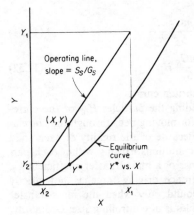

Figure 11.31 Continuous countercurrent adsorption of one component.

The resistance to mass transfer of solute from the fluid to the adsorbed state on the solid will include that residing in the gas surrounding the solid particles, that corresponding to the diffusion of solute through the gas within the pores of the solid, and possibly an additional resistance at the time of adsorption. During physical adsorption, the last of these will probably be negligible. If the remaining resistances can be characterized by an overall gas mass-transfer coefficient $K_Y a_p$ based on a_p, the outside surface of the solid particles, the rate of solute transfer over the differential height of adsorber dZ (Fig. 11.30) can be written in the usual manner as

$$S_S \, dX = G_S \, dY = K_Y a_p (Y - Y^*) \, dZ \qquad (11.48)$$

where Y^* is the equilibrium composition in the gas corresponding to the adsorbate composition X. The driving force $Y - Y^*$ is then represented by the vertical distance between operating line and equilibrium curve (Fig. 11.31). Rearranging Eq. (11.48) and integrating define the number of transfer units N_{tOG},

$$N_{tOG} = \int_{Y_2}^{Y_1} \frac{dY}{Y - Y^*} = \frac{K_Y a_p}{G_S} \int_0^Z dZ = \frac{Z}{H_{tOG}} \qquad (11.49)$$

where
$$H_{tOG} = \frac{G_S}{K_Y a_p} \qquad (11.50)$$

The integral of Eq. (11.49) is ordinarily evaluated graphically and the active height Z determined through knowledge of the height of a transfer unit H_{tOG}, characteristic of the system.

The use of an overall coefficient or overall height of a transfer unit implies that the resistance to mass transfer within the pores of the solid particles can be characterized [21] by an individual mass-transfer coefficient $k_S a_p$ or height of a transfer unit H_{tS}, thus,

$$\frac{G_S}{K_Y a_p} = \frac{G_S}{k_Y a_p} + \frac{mG_S}{S_S} \frac{S_S}{k_S a_p} \qquad (11.51)$$

or

$$H_{tOG} = H_{tG} + \frac{mG_S}{S_S} H_{tS} \qquad (11.52)$$

where $m = dY^*/dX$, the slope of the equilibrium curve.

The resistance within the fluid surrounding the particles H_{tG} or the corresponding coefficient $k_Y a_p$ can be estimated for moving beds through the correlations available for fixed beds (Table 3.3). Because of the rigidity of each solid particle and the unsteady-state diffusional conditions existing within each particle as it travels through the adsorber, the use of a mass-transfer coefficient k_S, with an implied linear concentration-difference driving force, is not strictly correct, and Eqs. (11.49) to (11.52) are sound only when the mass-transfer resistance of the fluid surrounding the particles is of controlling size. Ordinarily, however, the diffusional resistance within the particles is of major importance, and for these conditions, which space does not permit consideration of here, the work of Vermeulen [98, 99] should be consulted.

Illustration 11.6 Eagleton and Bliss [21] have measured the individual resistances to mass transfer residing in the fluid and within the solid during adsorption of water vapor from air by silica gel, using a fixed-bed semicontinuous technique. For low moisture contents of the air they found that

$$k_Y a_p = 31.6 G'^{0.55} \text{ kg H}_2\text{O/m}^3 \cdot \text{s} \cdot \Delta Y$$

and

$$k_S a_p = 0.965 \text{ kg H}_2\text{O/m}^3 \cdot \text{s} \cdot \Delta X$$

where G' is the mass velocity of the gas, kg/m$^2 \cdot$ s. Their silica-gel bed had an apparent bed density 671.2 kg/m^3 and an average particle size 1.727 mm diameter, and the external surface of the particles was 2.167 m^2/kg.

We wish to estimate the height of a continuous countercurrent isothermal adsorber for drying air at 26.7°C, standard atmospheric pressure, from an initial humidity 0.005 to a final humidity 0.0001 kg H$_2$O/kg dry air. The entering gel will be dry. (*Note*: So-called "dry" silica gel must contain a minimum of about 5% water if it is to retain its adsorptive capacity. Moisture measurements as ordinarily reported do not include this.) A gel rate 0.680 kg/m$^2 \cdot$ s (500 lb/ft$^2 \cdot$ h) and an air rate 1.36 kg/m$^2 \cdot$ s (1000 lb/ft$^2 \cdot$ h) will be used. For the moisture contents expected here, the equilibrium adsorption isotherm at 26.7°C, 1 std atm (see Illustration 11.9), can be taken as substantially straight and described by the expression $Y^* = 0.0185X$.

SOLUTION $Y_1 = 0.005$, $Y_2 = 0.0001$ kg H$_2$O/kg dry air. $S_S = 0.680$ kg/m$^2 \cdot$ s, $G_S = 1.36$ kg/m$^2 \cdot$ s, $X_2 = 0$ kg H$_2$O/kg dry gel. Eq. (11.46):

$$X_1 = \frac{G_S(Y_1 - Y_2)}{S_S} + X_2 = \frac{1.36(0.005 - 0.0001)}{0.68} = 0.0098 \text{ kg H}_2\text{O/kg dry gel}$$

$Y_2^* = $ humidity of air in equilibrium with entering gel $= 0$. $Y_1^* = 0.01851X_1 = 0.0001814$ kg H$_2$O/kg dry gel. Since operating line and equilibrium curve will both be straight on X, Y coordinates, the average driving force is the logarithmic average [Eqs. (8.47) and (8.48)]:

$$Y_1 - Y_1^* = 0.005 - 0.0001814 = 0.00482 \qquad Y_2 - Y_2^* = 0.0001 - 0 = 0.0001$$

$$Av \Delta Y = \frac{0.00482 - 0.0001}{\ln(0.00482/0.0001)} = 0.001217$$

$$N_{tOG} = \frac{Y_1 - Y_2}{\Delta Y} = \frac{0.005 - 0.0001}{0.001217} = 4.03$$

If the fixed-bed data are to be used for estimating mass-transfer coefficients for a moving bed of solids, the relative mass velocity of air and solid is appropriate. The linear rate of flow of the solid downward is $0.680/671.2 = 1.013 \times 10^{-3}$ m/s, where 671.2 is the apparent density. The density of the substantially dry air at 26.7°C, 1 std atm, is 1.181 kg/m^3, and its superficial linear velocity upward is $1.36/1.181 = 1.152$ m/s. The relative linear velocity of air and solid is $1.152 + 1.013 \times 10^{-3} = 1.153$ m/s, and the relative mass velocity of the air is $1.153(1.181) = 1.352$ kg/m$^2 \cdot$ s $= G'$.

$$H_{tG} = \frac{G_S}{31.6 G'^{0.55}} = \frac{1.36}{31.6(1.1352)^{0.55}} = 0.0365 \text{ m}$$

$$H_{tS} = \frac{S_S}{k_Y a_p} = \frac{0.680}{0.965} = 0.7047 \text{ m} \qquad \frac{mG_S}{S_S} = \frac{0.0185(1.36)}{0.680} = 0.037$$

Eq. (11.52):

$$H_{tOG} = 0.0365 + 0.037(0.7047) = 0.0625 \text{ m}$$

$$Z = N_{tOG} H_{tOG} = 4.03(0.0625) = 0.25 \text{ m} \quad \textbf{Ans.}$$

Note: The gas-phase mass-transfer coefficient in this case is smaller than that given by item 7, Table 3.3, perhaps because of the very small diameter (16 mm) of the bed in which it was measured. This could lead to an important *wall effect*; however, see Illustration 11.9.

Two components adsorbed; fractionation When several components of a gas mixture are appreciably adsorbed, fractionation is required for their separation. This discussion is confined to binary gas mixtures.

For purposes of computation, it is easiest to recall the similarity between the adsorption operation and continuous countercurrent extraction with reflux. Solid adsorbent as the added insoluble phase is analogous to extraction solvent, the adsorbate is analogous to the solvent-free extract, and the fluid stream is similar to the raffinate. Computations can then be made using the methods and equations of Chap. 10 [Eqs. (10.31) to (10.34) and (10.39) to (10.50) with Fig. 10.28]. Some simplification is possible, however, thanks to the complete insolubility of adsorbent in the mixture to be separated.

Refer to Fig. 11.32, a schematic representation of the adsorber. Feed enters at the rate of F mass/(area)(time), containing components A and C. The adsorbate-free adsorbent enters the top of the tower at the rate of B mass/(area)(time) and flows countercurrent to the gas entering at the bottom at the rate of R_1 mass/(area)(time). Compositions in the gas stream are expressed as x weight fraction C, the more strongly adsorbed substance. E represents the weight of adsorbent-free adsorbate A + C upon the solid, mass/(area)(time), and N the ratio, mass adsorbent/mass adsorbate. At any point in the adsorber, $B \approx NE$. The adsorbate composition is expressed as y weight fraction component C, on an adsorbent-free basis. Solid leaves the adsorber at the bottom and is stripped of its adsorbate in the desorption section, and the desorbed fluid is split into two streams, the reflux R_1 and the C-rich product P_E. At the top the fluid leaving is the A-rich product, R_2 mass/(area)(time). The arrangement is very similar to that shown in Fig. 10.27. Calculations are made on a phase diagram of the type shown in the upper part of Fig. 11.8a, with N plotted as ordinate, x and y as abscissa.

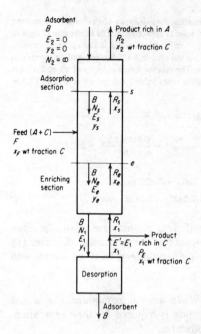

Figure 11.32 Continuous fractionation.

Let Δ_E represent the net adsorbent-free flow downward and out of the adsorber. Then $\Delta_E = P_E$, and the coordinates of point Δ_E on the phase diagram are $(N_{\Delta E} = B/P_E, y_{\Delta E} = x_1)$. At the bottom of the adsorber, $B = N_1 E_1$ and

$$E_1 = E' = P_E + R_1 = \Delta_E + R_1 \tag{11.53}$$

An A-C balance below section e of the enriching section is

$$E_e = P_E + R_e = \Delta_E + R_e \tag{11.54}$$

and, for substance C,

$$E_e y_e = P_E x_1 + R_e x_e = \Delta_E x_1 + R_e x_e \tag{11.55}$$

while for adsorbent it becomes

$$N_e E_e = N_{\Delta E} \, \Delta_E = B \tag{11.56}$$

Δ_E therefore represents the difference $E_e - R_e$ for any level in the enriching section. Equations (11.54) and (11.56) provide a measure of the internal reflux ratio,

$$\frac{R_e}{E_e} = \frac{N_{\Delta E} - N_e}{N_{\Delta E}} = 1 - \frac{N_e}{N_{\Delta E}} \tag{11.57}$$

and at the bottom of the tower the external reflux ratio is

$$\frac{R_1}{P_E} = \frac{R_1}{E_1 - R_1} = \frac{N_{\Delta E} - N_1}{N_1} = \frac{N_{\Delta E}}{N_1} - 1 \tag{11.58}$$

At the top of the adsorber let Δ_R represent the difference between flows of adsorbate and unadsorbed gas, and since $E_2 = 0$, $\Delta_R = -R_2$. The coordinates of Δ_R on the phase diagram are then $(N_{\Delta R} = -B/R_2, x_{\Delta R} = x_2)$. Material balances above section s are then, for A and C,

$$R_s = E_s + R_2 = E_s - \Delta_R \qquad (11.59)$$

for C,

$$R_s x_s = E_s y_s + R_2 x_2 = E_s y_s - \Delta_R x_2 \qquad (11.60)$$

and for adsorbent,

$$B = N_s E_s = \Delta_R N_{\Delta R} \qquad (11.61)$$

Equations (11.59) and (11.61) provide the internal reflux ratio.

$$\frac{R_s}{E_s} = \frac{N_s - N_{\Delta R}}{-N_{\Delta R}} = 1 - \frac{N_s}{N_{\Delta R}} \qquad (11.62)$$

Overall balances about the entire plant are

$$F = R_2 + P_E \qquad (11.63)$$

$$Fx_F = R_2 x_2 + P_E x_1 \qquad (11.64)$$

The definitions of Δ_R and Δ_E, together with Eq. (11.64), provide

$$\Delta_R + F = \Delta_E \qquad (11.65)$$

The graphical interpretation of these relations on the phase diagram is the same as that for the corresponding situations in extraction and is shown in detail in Illustration 11.7 below.

If a stage-type device, e.g., fluidized beds, is used, the number of theoretical stages is readily obtained by the usual step construction on the lower part of a diagram such as Fig. 11.33. For continuous contact,

$$N_{tOG} = \frac{Z}{H_{tOG}} = \int_{\bar{p}_2}^{\bar{p}_1} \frac{d\bar{p}}{\bar{p} - \bar{p}^*} = \int_{x_2}^{x_1} \frac{dx}{x - x^*} - \ln\frac{1 + (r-1)x_1}{1 + (r-1)x_2} \qquad (11.66)$$

where $r = M_A/M_C$ and H_{tOG} is defined by Eq. (11.50). This can be applied separately to the enriching and adsorption sections. It assumes equimolar counterdiffusion of A and C, which is not strictly the case, but refinement at this time is unwarranted in view of the sparsity of the data [45]. It is also subject to the same limitations considered for one component adsorbed (solute collection).

Illustration 11.7 Determine the number of transfer units and the adsorbent circulation rate required to separate a gas containing 60% ethylene C_2H_4 and 40% propane C_3H_8 by volume into products containing 5 and 95% C_2H_4 by volume, isothermally at 25°C and 2.25 std atm, using activated carbon as the adsorbent and a reflux ratio of twice the minimum. *Note:* These adsorbers will customarily operate at reflux ratios closer to the minimum in order to reduce the adsorbent circulation rate. The present value is used in order to make the graphical solution clear.

SOLUTION Equilibrium data for this mixture at 25°C and 2.25 std atm have been estimated from the data of Lewis et al., *Ind. Eng. Chem.*, **42**, 1319, 1326 (1950), and are plotted in Fig.

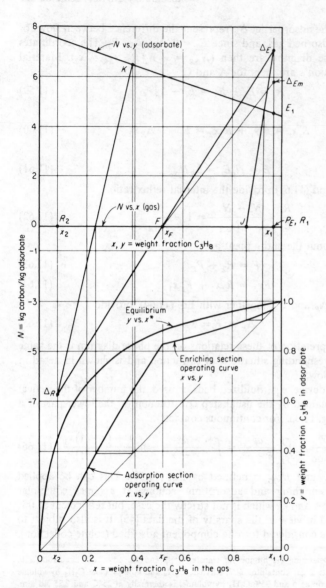

Figure 11.33 Solution to Illustration 11.7.

11.33. C_3H_8 is the more strongly adsorbed component, and compositions in the gas and adsorbate are expressed as weight fraction C_3H_8. Tie lines have been omitted in the upper part of the plot, and the equilibrium compositions are shown instead in the lower part.

Molecular weights are 28.0 for C_2H_4 and 44.1 for C_3H_8. The feed-gas composition is then $x_F = 0.4(44.1)/[0.4(44.1) + 0.6(28.0)] = 0.512$ wt fraction C_3H_8. Similarly $x_1 = 0.967$, and $x_2 = 0.0763$ wt fraction C_3H_8.

Basis: 100 kg feed gas. Eqs. (11.63) and (11.64):

$$100 = R_2 + P_E$$

$$100(0.512) = R_2(0.0763) + P_E(0.967)$$

Solving simultaneously gives $R_2 = 51.1$ kg , $P_E = 48.9$ kg.

Point F at x_F and point E_1 at x_1 are located on the diagram as shown. From the diagram, N_1 (at point E_1) = 4.57 kg carbon/kg adsorbate. The minimum reflux ratio is found as it is for extraction. In this case, the tie line through point F locates $(\Delta_E)_m$ and $(\Delta_E)_m = 5.80$.

$$\left(\frac{R_1}{P_E}\right)_m = \frac{5.80}{4.57} - 1 = 0.269 \text{ kg reflux gas/kg product}$$

$$(R_1)_m = 0.269P_E = 0.269(48.9) = 13.15 \text{ kg}$$

$$(E_1)_m = (R_1)_m + P_E = 13.15 + 48.9 = 62.1 \text{ kg}$$

$$B_m = N_1(E_1)_m = 4.57(62.1) = 284 \text{ kg carbon/100 kg feed}$$

At twice the minimum reflux ratio, $R_1/P_E = 2(0.269) = 0.538$. Eq. (11.58):

$$0.538 = \frac{N_{\Delta E}}{4.57} - 1$$

$$N_{\Delta E} = 7.03 \text{ kg carbon/kg adsorbate}$$

and point Δ_E is located on the diagram.

$$R_1 = 0.538P_E = 0.538(48.9) = 26.3 \text{ kg}$$

$$E_1 = R_1 + P_E = 26.3 + 48.9 = 75.2 \text{ kg}$$

$$B = N_1E_1 = 4.57(75.2) = 344 \text{ kg carbon/100 kg feed}$$

Point Δ_R can be located by extending line $\Delta_E F$ to intersect the line $x = x_2$. Alternatively, $N_{\Delta R} = -B/R_2 = -344/51.1 = -6.74$ kg carbon/kg adsorbate. Random lines such as line $\Delta_R K$ are drawn from Δ_R, and the intersections with the equilibrium curves are projected downward in the manner shown to provide the adsorption-section operating curve. Similarly random lines such as line $\Delta_E J$ are drawn from Δ_E, and intersections are projected downward to provide the enriching-section operating curve. The horizontal distance between operating and equilibrium curves on the lower diagram is the driving force $x - x^*$ of Eq. (11.66). The following were determined from the diagram:

x	x^*	$\dfrac{1}{x - x^*}$	x	x^*	$\dfrac{1}{x - x^*}$
$x_1 = 0.967$	0.825	7.05	$x_F = 0.512$	0.39	8.20
0.90	0.710	5.26	0.40	0.193	4.83
0.80	0.60	5.00	0.30	0.090	4.76
0.70	0.50	5.00	0.20	0.041	6.29
0.60	0.43	5.89	$x_2 = 0.0763$	0.003	13.65

The third column was plotted as ordinate against the first as abscissa, the area under the curve between x_1 and x_F was 2.65, and that between x_F and x_2 was 2.67. Further, $r = 28.0/44.1$

= 0.635. Applying Eq. (11.66) to the enriching section, gives

$$N_{tOG} = 2.65 - \ln\frac{1 + (0.635 - 1)(0.967)}{1 + (0.635 - 1)(0.512)} = 2.52$$

and to the adsorption section,

$$N_{tOG} = 2.67 - \ln\frac{1 + (0.635 - 1)(0.512)}{1 + (0.635 - 1)(0.0763)} = 2.53$$

The total $N_{tOG} = 2.52 + 2.53 = 5.1$.

Simulation of moving beds The arrangement shown in Fig. 11.34 for fractionating liquids avoids the problems of moving beds of solid downward through a tower and transporting the solid up to the top [10]. In the adsorber, divided into a number of sections by sieve plates, the adsorbent is in the form of fixed, i.e., unmoving, beds. The rotary valve, turning counterclockwise periodically, moves the position of introducing the feed successively from positions *a* through *c* and back to *a*. Similarly the desorbent and the withdrawn streams are moved periodically, the topmost stream (A + D) moving from position *a* at the top to *b* and *c* at the bottom, and back to *a*. The recirculating-liquid pump is programmed to change its rate of pumping as needed. The degree of approach to a true countercurrent system increases as the number of compartments increases, at the expense, of course, of complexity of the piping. The adsorber vessel may

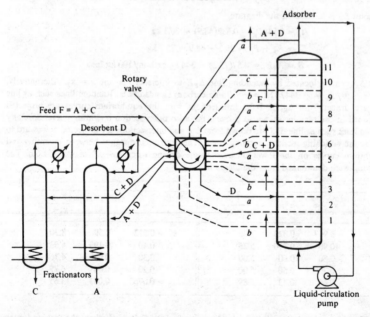

Figure 11.34 Simulation of moving beds. [*Broughton, Chem. Eng. Prog., 64(8), 60 (1968), with permission.*]

be horizontal [80]. Many plants have been built, using molecular sieves for a variety of hydrocarbon separations.

Unsteady State—Fixed-Bed Adsorbers

The inconvenience and relatively high cost of continuously transporting solid particles as required in steady-state operations frequently make it more economical to pass the fluid mixture to be treated through a stationary bed of adsorbent. As increasing amounts of fluid are passed through such a bed, the solid adsorbs increasing amounts of solute, and an unsteady state prevails. This technique is very widely used and finds application in such diverse fields as the recovery of valuable solvent vapors from gases, purifying air as with gas masks, dehydration of gases and liquids, decolorizing mineral and vegetable oils, the concentration of valuable solutes from liquid solutions, and many others.

The adsorption wave Consider a binary solution, either gas or liquid, containing a strongly adsorbed solute at concentration c_0. The fluid is to be passed continuously down through a relatively deep bed of adsorbent initially free of adsorbate. The uppermost layer of solid, in contact with the strong solution entering, at first adsorbs solute rapidly and effectively, and what little solute is left in the solution is substantially all removed by the layers of solid in the lower part of the bed. The effluent from the bottom of the bed is practically solute-free, as at c_a in the lower part of Fig. 11.35. The distribution of adsorbate in the solid bed is indicated in the sketch in the upper part of this figure at a, where the relative density of the horizontal lines in the bed is meant to indicate the relative concentration of adsorbate. The uppermost layer of the bed is practically saturated, and the bulk of the adsorption takes place over a relatively narrow adsorption zone in which the concentration changes rapidly, as shown. As solution continues to flow, the adsorption zone moves downward as a wave, at a rate ordinarily very much slower than the linear velocity of the fluid through the bed. At a later time, as at b in the figure, roughly half the bed is saturated with solute, but the effluent concentration c_b is still substantially zero. At c in the figure the lower portion of the adsorption zone has just reached the bottom of the bed, and the concentration of solute in the effluent has suddenly risen to an appreciable value c_c for the first time. The system is said to have reached the *breakpoint*. The solute concentration in the effluent now rises rapidly as the adsorption zone passes through the bottom of the bed and at d has substantially reached the initial value c_0. The portion of the effluent concentration curve between positions c and d is termed the *breakthrough curve*. If solution continues to flow, little additional adsorption takes place since the bed is for all practical purposes entirely in equilibrium with the feed solution.

If a vapor is being adsorbed adiabatically from a gas mixture in this manner, the evolution of the heat of adsorption causes a temperature wave to flow through the adsorbent bed in a manner somewhat similar to the adsorption wave [56] and the rise in temperature of the bed at the fluid outlet can

624 MASS-TRANSFER OPERATIONS

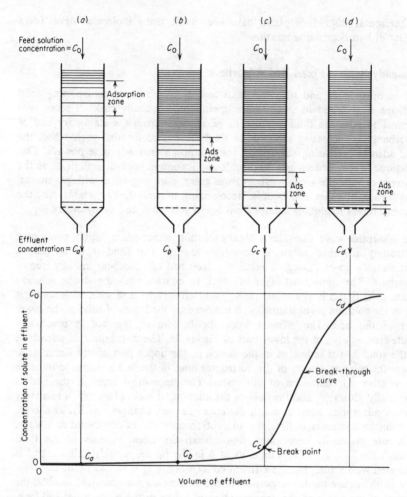

Figure 11.35 The adsorption wave.

sometimes be used as a rough indication of the breakpoint. In adsorption from liquids the temperature rise is usually relatively small.

The shape and time of appearance of the breakthrough curve greatly influence the method of operating a fixed-bed adsorber. The curves generally have an S shape, but they may be steep or relatively flat and in some cases considerably distorted. If the adsorption process were infinitely rapid, the breakthrough curve would be a straight vertical line in the lower part of Fig. 11.35. The actual rate and mechanism of the adsorption process, the nature of the adsorption equilibrium, the fluid velocity, the concentration of solute in the feed, and the length of the adsorber bed (particularly if the solute concentration in the feed is high) all contribute to the shape of the curve produced for any system. The breakpoint is very sharply defined in some cases and in others poorly defined. Generally the breakpoint time decreases with decreased bed height, increased particle size of adsorbent, increased rate of flow of fluid through the bed, and increased initial solute content of the feed. There is a critical minimum bed height below which the solute concentration in the effluent will rise rapidly from the first appearance of effluent. In planning new processes it is best to determine the breakpoint and breakthrough curve for a particular system experimentally under conditions resembling as much as possible those expected in the process.

Adsorption of vapors One of the most important applications of fixed-bed adsorbers is in the recovery of valuable solvent vapors. Solids saturated with solvents such as alcohol, acetone, carbon disulfide, benzene, and others can be dried by evaporation of the solvent into an airstream, and the solvent vapor can be recovered by passing the resulting vapor-gas mixture through a bed of activated carbon. The very favorable adsorption equilibrium provided by a good grade of carbon for vapors of this sort permits substantially complete vapor recovery, 99 to 99.8 percent, from gas mixtures containing as little as 0.5 to 0.05 percent of the vapor by volume [15]. Air-vapor mixtures of concentration well below the explosive limits can thus be handled. In most such adsorption plants it is necessary to operate with a small drop in pressure through the bed of adsorbent in order to keep power costs low. Therefore granular rather than powdered adsorbents are used, and bed depths are relatively shallow (0.3 to 1 m) and large in cross section. The superficial gas velocity may be in the range 0.25 to 0.6 m/s. A typical arrangement of the adsorption vessel is shown in Fig. 11.36.

In a typical operation the air-vapor mixture, if necesssary cooled to 30 to 40°C and filtered free of dust particles which might clog the pores of the adsorbent, is admitted to the adsorbent bed. If the breakthrough curve is steep, the effluent air, substantially free of vapor, may be discharged to the atmosphere until the breakpoint is reached, whereupon the influent stream must be diverted to a second adsorber while the first is regenerated. On the other hand, if the breakthrough curve is flat, so that at the breakpoint a substantial portion of the adsorbent remains unsaturated with adsorbate, the gas may be permitted to flow through a second adsorber in series with the first until the carbon in the first is substantially all saturated. The influent mixture is then passed through the second and a third adsorber in series while the first is regenerated.

After gas flow has been diverted from an adsorber, the carbon is usually desorbed by admission of low-pressure steam. This lowers the partial pressure of the vapor in contact with the solid and provides by condensation the necessary heat of desorption. The steam-vapor effluent from the carbon is condensed and the condensed solvent recovered by decantation from the water if it is water-insoluble or by rectification if an aqueous solution results. When desorption is complete, the carbon is saturated with adsorbed water. This moisture is readily

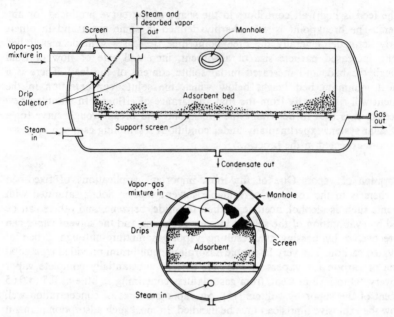

Figure 11.36 Adsorber for solvent vapors at low pressures, schematic. [*Logan, U.S. Patent 2,180,712 (1939).*]

displaced by many vapors and evaporated into the air when the air-vapor mixture is readmitted to the carbon, and indeed much of the heat evolved during adsorption of the vapor can be used in desorbing the water, thus maintaining moderate bed temperatures. If the moisture interferes with vapor adsorption, the bed can first be dried by admitting heated air and then cooled by unheated air, before reuse for vapor recovery. Figure 11.37 shows a typical plant layout arranged for this method. An alternative procedure for regenerating the bed is to heat it electrically or with embedded steam pipes and pump off the adsorbed vapor with a vacuum pump. An arrangement which avoids the intermittent operation usually associated with fixed beds is shown in Fig. 11.38. Operation is reminiscent of continuous filters. As the inner drum rotates, the adsorbent is successively subject to adsorption, regeneration, and drying with cooling. All these, which use heat as a regenerative means, are sometimes referred to as *thermal-swing procedures.*

Moist gases can be dried of their water content by passing them through beds of activated silica gel, alumina, bauxite, or molecular sieves. Especially if the gases are under appreciable pressures, moderately deep beds are used since the pressure drop will still be only a small fraction of the total pressure. Towers containing the adsorbent may be as much as 10 m tall or more, but in such instances the solid is best supported on trays at intervals of 1 to 2 m in order to

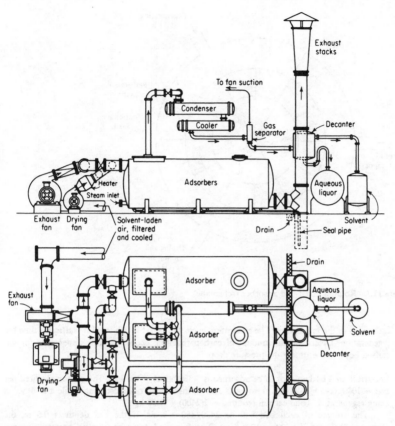

Figure 11.37 Solvent-recovery adsorption plant. [*After Benson and Courouleau, Chem. Eng. Prog., 44, 466 (1948), with permission.*]

minimize the compression of the bed resulting from pressure drop. A vessel for treating gases under pressure can be designed as in Fig. 11.39. After the bed has reached the maximum practical moisture content, the adsorbent can be regenerated, either by application of heat by steam coils embedded in the adsorbent or by admitting heated air or gas. Liquids such as gasoline, kerosene, and transformer oil can also be dehydrated by passage through beds of activated alumina. High-temperature (230°C) steam followed by application of vacuum provided by a steam-jet ejector can be used for regenerating the adsorbent.

Illustration 11.8 A solvent-recovery plant is to recover 0.1 kg/s (800 lb/h) of ethyl acetate vapor from a mixture with air at a concentration of 3 kg vapor/100 m³ (1.9 lb/1000 ft³), 1 std atm pressure. The adsorbent will be activated carbon, 6- to 8-mesh (av particle diam = 2.8 mm), apparent density of individual particles 720 kg/m³ (45 lb/ft³). Apparent density of the packed bed = 480 kg/m³ (30 lb/ft³). The carbon is capable of adsorbing 0.45 kg vapor/kg

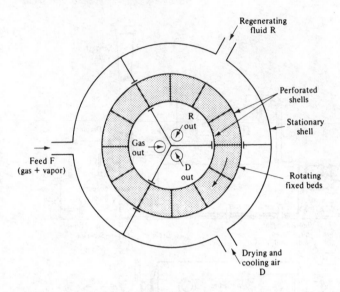

Figure 11.38 Rotating fixed-bed adsorber (schematic).

carbon up to the breakpoint. The adsorption cycle will be set at 3 h to allow sufficient time for regeneration. Determine the amount of carbon required, choose suitable dimensions for the carbon beds, and estimate the pressure drop.

SOLUTION In 3 h, the vapor to be adsorbed is $3(3600)(0.1) = 1080$ kg. The carbon required per bed $= 1080/0.45 = 2400$ kg. Two beds will be necessary, one adsorbing while the other is being regenerated. Total carbon required $= 2(2400) = 4800$ kg.

The volume of each bed will be $2400/480 = 5$ m³. If the bed depth is 0.5 m, the cross-sectional area is $5/0.5 = 10$ m², say 2 by 5 m, and it can be installed horizontally in a suitable vessel, such as that shown in Fig. 11.36.

The pressure drop can be estimated by Eq. (6.66). At 35°C, the viscosity of air $= 0.0182$ cP $= 1.82 \times 10^{-5}$ kg/m · s, and the density $= (29/22.41)(273/308) = 1.148$ kg/m³. Since 1 m³ of the bed, weighing 480 kg, contains $480/720 = 0.667$ m³ of solid particles, the fractional void volume of the bed (space between particles, not including pore volume) $= 1 - 0.667 = 0.333 = \varepsilon$.

The volumetric rate of airflow $= 0.1(100/3) = 3.33$ m³/s, and the superficial mass velocity $= 3.33(1.148)/10 = 0.383$ kg/m² · s (linear velocity $= 0.333$ m/s). In Chap. 6 notation, the particle Reynolds number $\text{Re} = d_p G/\mu = 0.0028(0.383)/(1.82 \times 10^{-5}) = 58.9$, $Z = 0.5$ m, $g_c = 1.0$, and, by Eq. (6.66), $\Delta p = 1421$ N/m² (5.7 in H₂O). **Ans.**

The heatless adsorber (pressure swing) The method of operation shown in Fig. 11.40 for removing a vapor from a gas under pressure uses mechanical work of compression instead of heat to regenerate the adsorbent [92]. Wet gas under pressure passes through four-way valve 3 into adsorber bed 1, and the dry gas leaves at the top. A portion of the product gas is passed through valve 4 and at a lower pressure into bed 2 to evaporate the adsorbate accumulated there. It

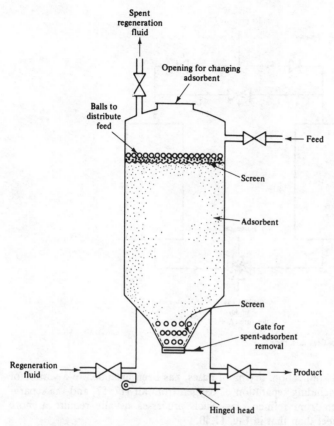

Figure 11.39 Fixed-bed adsorber for vapors at high pressures, and for liquid percolation. [*After Johnston, Chem. Eng., 79, 87 (Nov. 27, 1972).*]

leaves through valve 3. Valves 5 and 6 are check valves allowing flow to the right and left, respectively. After a short interval (30 to 300 s), valve 3 is turned, beds 1 and 2 switch roles, and flow through valve 4 is reversed. The cycle time is short enough to ensure that the adsorption zone always remains in the central part of the beds, the lower portion of the adsorbent being saturated with adsorbate and the upper part being adsorbate-free. The heat of adsorption is confined to a small volume of the bed and is used in purging the bed. The saturated humidity of a gas is given by $Y = [p/(p_t - p)](M_C/M_A)$, where p is the saturation vapor pressure of the more strongly adsorbed solute C. For a given temperature, with p fixed, Y is larger the lower the total pressure p_t. It then follows that to ensure that as much vapor is desorbed as is adsorbed, thus maintaining a steady flow of dry product, the ratio mole of purge gas/moles of feed must exceed the ratio low purge pressure/high feed pressure.

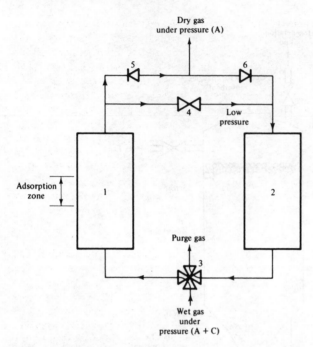

Dry gas
under pressure (A)

Low
pressure

Adsorption
zone

Purge gas

Wet gas
under
pressure (A + C)

Figure 11.40 The heatless adsorber.

The original application, drying of gases, has been extended to a variety of fractionations, including separation of oxygen from air [17, 57] and the separation of hydrogen from refinery gas. Such processes usually require a more elaborate flowsheet than that in Fig. 11.40.

Adsorption of liquids; percolation The dehydration of liquids by stationary beds of adsorbent has already been mentioned. In addition, the colors of petroleum products, such as lubricating oils and transformer oils, and of vegetable oils are commonly reduced by percolation through beds of decolorizing clays; sugar solutions are deashed and decolorized by percolation through bone char; and many other liquid-treating operations use these semibatch methods.

In such service, the bed of adsorbent is commonly called a *filter*. It may be installed in a vertical cylindrical vessel fitted with a dished or conical bottom, of diameter ranging up to as much as 4.5 m and height up to 10 m. In the case of decolorizing clays, the beds may contain as much as 50 000 kg of adsorbent. The granular adsorbent is supported on a screen or blanket, in turn supported by a perforated plate.

The liquid flow is ordinarily downward, either under the force of gravity alone or under pressure from above. At the start of the operation, the adsorbent bed is frequently allowed to soak for a time in the first portions of the feed liquid, in order to displace air from the adsorbent particles before the percolation is begun. In decolorizing operations, the concentration of impurity in the initial effluent liquid is usually much smaller than the specifications of the product require, and the breakthrough curve is frequently rather flat. Consequently it is common practice to allow the effluent liquid to accumulate in a receiving tank below the filter and to blend or composite it until the blended liquid reaches the largest acceptable concentration of impurity. In this way, the largest possible adsorbate concentration can be accumulated upon the solid.

When the solid requires revivification, the flow of liquid is stopped and the liquid in the filter drained. The solid can then be washed in place with an appropriate solvent, e.g., water for sugar-refining filters and naphtha for petroleum products. If necessary the solvent is then removed by admission of steam, following which the solid can be dumped from the filter and reactivated by burning or other suitable procedure, depending upon the nature of the adsorbent.

Elution Desorption of the adsorbed solute by a solvent is called *elution*. The desorption solvent is the *elutant*, and the effluent stream containing the desorbed solute and eluting solvent is the *eluate*. The *elution curve* is a plot of the solute concentration in the eluate against quantity of eluate, as in Fig. 11.41. The initial rise in solute concentration of the eluate, *OA* in the figure, is found when the void spaces between the adsorbent particles are initially filled with fluid remaining from the adsorption. For liquids, if the bed is drained before elution, the elution curve starts at *A*. If elution is stopped after eluate corresponding to *C* has been withdrawn, the area under the curve *OABC* represents the quantity of solute desorbed. For a successful process, this must equal the solute adsorbed during an adsorption cycle; otherwise solute will build up in the bed from one cycle to the next.

Chromatography Imagine a solution containing two solutes, A and B, which are differently adsorbed at equilibrium, A more strongly. A small quantity of this solution, insufficient to saturate all but a small quantity of the adsorbent, is passed through an adsorbent bed, whereupon both solutes are retained in the

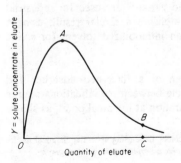

Figure 11.41 Elution of a fixed bed.

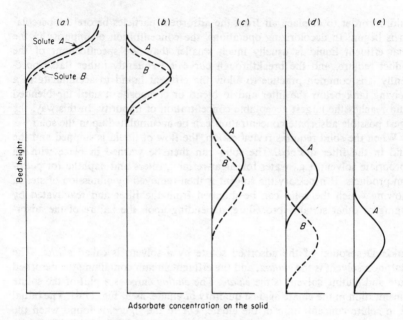

Adsorbate concentration on the solid

Figure 11.42 Chromatographic separation of two solutes.

upper portion of the bed to give rise to adsorbate concentrations as indicated in Fig. 11.42a. A suitable elutant is now passed through the bed, whereupon solute B is more readily desorbed than A. At b, c, and d in the figure, both solutes are seen to be desorbed, only to be readsorbed and redesorbed at lower positions in the bed, but the wave of B concentration moves more rapidly downward than that of A.† At e, solute B has been washed out of the solid, leaving essentially all the A behind. The curves of Fig. 11.42 are idealized, since in reality their shape may change as the concentration waves pass down the column. If the adsorption (or desorption) zone heights are short relative to the height of the bed, and if the selectivity of the adsorbent is sufficiently great, essentially complete separation of the solutes is possible. This technique is the basis of very powerful methods of analysis of mixtures, both gas and liquid, and variants are used for industrial purposes [91]. *Partition chromatography* accomplishes a similar result, using a liquid contained in the pores of a solid as an immobilized solvent for solute separation, rather than adsorption itself.

Rate of adsorption in fixed beds The design of a fixed-bed adsorber and prediction of the length of the adsorption cycle between revivifications require knowledge of the percentage approach to saturation at the breakpoint, as shown

† The adsorption bands in some cases have different colors, depending upon the chemical nature of the solutes, and from this arose the terms *chromatographic adsorption* and *chromatography*.

in Illustration 11.8. This in turn requires the designer to predict the time of the breakpoint and the shape of the breakthrough curve. The unsteady-state circumstances of fixed-bed adsorption and the many factors which influence the adsorption make such computations for the general case very difficult [99]. The following simplified treatment due to Michaels [70] is readily used but is limited to isothermal adsorption from dilute feed mixtures and to cases where the equilibrium adsorption isotherm is concave to the solution-concentration axis, where the adsorption zone is constant in height as it travels through the adsorption column, and where the height of the adsorbent bed is large relative to the height of the adsorption zone. Many industrial applications fall within these restrictions. The development here is in terms of adsorption from a gas, but it is equally applicable to treatment of liquids. Nonisothermal cases have also been treated [56].

Consider the idealized breakthrough curve of Fig. 11.43. This results from the flow of a solvent gas through an adsorbent bed at the rate of G_S mass/(area)(time), entering with an initial solute concentration Y_0 mass solute/mass solvent gas. The total solute-free effluent after any time is w mass/area of bed cross section. The breakthrough curve is steep, and the solute concentration in the effluent rises rapidly from essentially zero to that in the incoming gas. Some low value Y_B is arbitrarily chosen as the breakpoint concentration, and the adsorbent is considered as essentially exhausted when the effluent concentration has risen to some arbitrarily chosen value Y_E, close to Y_0. We are concerned principally with the quantity of effluent w_B at the breakpoint and the shape of the curve between w_B and w_E. The total effluent accumulated during the appearance of the breakthrough curve is $w_a = w_E - w_B$. The adsorption zone, of

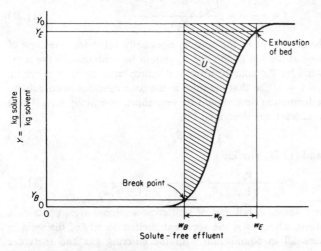

Figure 11.43 Idealized breakthrough curve.

constant height Z_a ft, is that part of the bed in which the concentration change from Y_B to Y_E is occurring at any time.

Let θ_a be the time required for the adsorption zone to move its own height down the column, after the zone has been established. Then

$$\theta_a = \frac{w_a}{G_S} \qquad (11.67)$$

Let θ_E be the time required for the adsorption zone to establish itself and move out of the bed. Then

$$\theta_E = \frac{w_E}{G_S} \qquad (11.68)$$

If the height of the adsorbent bed is Z, and if θ_F is the time required for the formation of the adsorption zone,

$$Z_a = Z \frac{\theta_a}{\theta_E - \theta_F} \qquad (11.69)$$

The quantity of solute removed from the gas in the adsorption zone from the breakpoint to exhaustion is U mass solute/area of bed cross section. This is given by the shaded area of Fig. 11.43, which is

$$U = \int_{w_B}^{w_E} (Y_0 - Y) \, dw \qquad (11.70)$$

If, however, all the adsorbent in the zone were saturated with solute, it would contain $Y_0 w_a$ mass solute/area. Consequently at the breakpoint, when the zone is still within the column, the fractional ability of the adsorbent in the zone still to adsorb solute is

$$f = \frac{U}{Y_0 w_a} = \frac{\int_{w_B}^{w_E} (Y_0 - Y) \, dw}{Y_0 w_a} \qquad (11.71)$$

If $f = 0$, so that the adsorbent in the zone is essentially saturated, the time of formation of the zone at the top of the bed θ_F should be substantially the same as the time θ_a required for the zone to travel a distance equal to its own height. On the other hand, if $f = 1.0$, so that the solid in the zone contains essentially no adsorbate, the zone-formation time should be very short, essentially zero. These limiting conditions, at least, are described by

$$\theta_F = (1 - f)\theta_a \qquad (11.72)$$

Equations (11.69) and (11.72) provide

$$Z_a = Z \frac{\theta_a}{\theta_E - (1 - f)\theta_a} = Z \frac{w_a}{w_E - (1 - f)w_a} \qquad (11.73)$$

The adsorption column, Z tall and of unit cross-sectional area, contains a mass $Z\rho_S$ of adsorbent, where ρ_S is the apparent packed density of the solid in the bed. If this were all in equilibrium with the entering gas and therefore completely saturated at an adsorbate concentration X_T mass adsorbate/mass solid, the adsorbate mass would be $Z\rho_S X_T$. At the breakpoint, the adsorption zone of height Z_a is still in the column at the bottom, but the rest of the column, $Z - Z_a$, is substantially saturated. At the breakpoint, therefore, the adsorbed solute is $(Z - Z_a)\rho_S X_T + Z_a\rho_S(1 - f)X_T$. The fractional approach to saturation of the column at the breakpoint is therefore [34]

$$\text{Degree of saturation} = \frac{(Z - Z_a)\rho_S X_T + Z_a\rho_S(1 - f)X_T}{Z\rho_S X_T} = \frac{Z - fZ_a}{Z}$$

(11.74)

In the fixed bed of adsorbent, the adsorption zone in reality moves downward through the solid, as we have seen. Imagine instead, however, that the solid moves upward through the column countercurrent to the fluid fast enough for the adsorption zone to remain stationary within the column, as in Fig. 11.44a. Here the solid leaving at the top of the column is shown in equilibrium with the entering gas, and all solute is shown as having been removed from the effluent gas. This would, of course, require an infinitely tall column, but our concern will be primarily with the concentrations at the levels corresponding to the extremities of the adsorption zone. The operating line over the entire tower is

$$G_S(Y_0 - 0) = S_S(X_T - 0)$$

(11.75)

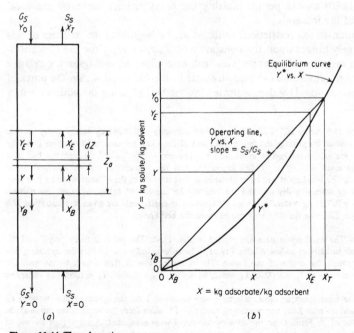

Figure 11.44 The adsorption zone.

636 MASS-TRANSFER OPERATIONS

or
$$\frac{S_S}{G_S} = \frac{Y_0}{X_T} \tag{11.76}$$

Since the operating line passes through the origin of Fig. 11.44b, at any level in the column the concentrations of solute in the gas Y and of adsorbate upon the solid X are then related by
$$G_S Y = S_S X \tag{11.77}$$
Over the differential height dZ, the ratio of adsorption is
$$G_S \, dY = K_Y a_p (Y - Y^*) \, dZ \tag{11.78}$$
For the adsorption zone, therefore,
$$N_{tOG} = \int_{Y_B}^{Y_E} \frac{dY}{Y - Y^*} = \frac{Z_a}{H_{tOG}} = \frac{Z_a}{G_S / K_Y a_p} \tag{11.79}$$

where N_{tOG} is the overall number of gas transfer units in the adsorption zone. For any value of Z less than Z_a, assuming H_{tOG} remains constant with changing concentrations,
$$\frac{Z \text{ at } Y}{Z_a} = \frac{w - w_B}{w_a} = \frac{\int_{Y_B}^{Y} \frac{dY}{Y - Y^*}}{\int_{Y_B}^{Y_E} \frac{dY}{Y - Y^*}} \tag{11.80}$$

Equation (11.80) should permit plotting the breakthrough curve by graphical evaluation of the integrals.

In addition to the restrictions outlined at the beginning, the success of this analysis largely hinges upon the constancy of $K_Y a_p$ or H_{tOG} for the concentrations within the adsorption zone. This will, of course, depend upon the relative constancy of the resistances to mass transfer in the fluid and within the pores of the solid. Illustration 11.9 demonstrates the method of using the equations in a typical case.

Illustration 11.9 Air at 26.7°C, 1 std atm, with a humidity of 0.00267 kg water/kg dry air is to be dehumidified by passage through a fixed bed of the silica gel used in Illustration 11.6. The depth of the adsorbent bed is to be 0.61 m. The air will be passed through the bed at a superficial mass velocity of 0.1295 kg/m² · s, and the adsorption will be assumed to be isothermal. The breakpoint will be considered as that time when the effluent air has a humidity of 0.0001 kg water/kg dry air, and the bed will be considered exhausted when the effluent humidity is 0.0024 kg water/kg dry air. Mass-transfer coefficients are given in Illustration 11.6 for this gel. Estimate the time required to reach the breakpoint.

SOLUTION The equilibrium data are plotted in Fig. 11.45. The gel is initially 'dry," and the effluent air initially of so low a humidity as to be substantially dry, so that the operating line passes through the origin of the figure. The operating line is then drawn to intersect the equilibrium curve at $Y_0 = 0.00267$ kg water/kg dry air. $Y_B = 0.0001$, $Y_E = 0.0024$ kg water/kg dry air.

In the accompanying table, column 1 lists values of Y on the operating line between Y_B and Y_E and column 2 the corresponding values of Y^* taken from the equilibrium curve at the same value of X. From these the data of column 3 were computed. A curve (not shown) of column 1 as abscissa, column 3 as ordinate was prepared and integrated graphically between each value of Y in the table and Y_B, to give in column 4 the numbers of transfer units

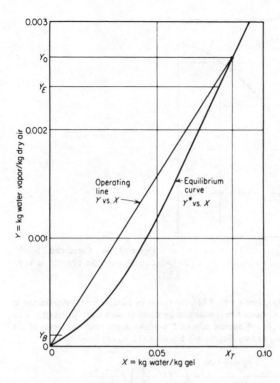

Y = kg water vapor/kg dry air

X = kg water/kg gel

Operating line Y vs. X

Equilibrium curve Y* vs. X

Figure 11.45 Solution to Illustration 11.9.

corresponding to each value of Y (thus, for example, the area under the curve from $Y = 0.0012$ to $Y = 0.0001$ equals 4.438). The total number of transfer units corresponding to the adsorption zone is $N_{tOG} = 9.304$, in accordance with Eq. (11.79).

Y, kg H₂O / kg dry air (1)	Y^*, kg H₂O / kg dry air (2)	$\dfrac{1}{Y - Y^*}$ (3)	$\displaystyle\int_{Y_B}^{Y} \dfrac{dY}{Y - Y^*}$ (4)	$\dfrac{w - w_B}{w_a}$ (5)	$\dfrac{Y}{Y_0}$ (6)
$Y_B = 0.0001$	0.00003	14 300	0	0	0.0374
0.0002	0.00007	7 700	1.100	0.1183	0.0749
0.0004	0.00016	4 160	2.219	0.2365	0.1498
0.0006	0.00027	3 030	2.930	0.314	0.225
0.0008	0.00041	2 560	3.487	0.375	0.300
0.0010	0.00057	2 325	3.976	0.427	0.374
0.0012	0.000765	2 300	4.438	0.477	0.450
0.0014	0.000995	2 470	4.915	0.529	0.525
0.0016	0.00123	2 700	5.432	0.584	0.599
0.0018	0.00148	3 130	6.015	0.646	0.674
0.0020	0.00175	4 000	6.728	0.723	0.750
0.0022	0.00203	5 880	7.716	0.830	0.825
$Y_E = 0.0024$	0.00230	10 000	9.304	1.000	0.899

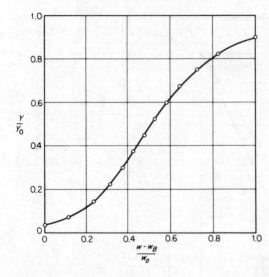

Figure 11.46 Calculated break-through curve for Illustration 11.9.

By dividing each entry in column 4 by 9.304, the values in column 5 were determined in accordance with Eq. (11.80). Column 6 was obtained by dividing each entry in column 1 by $Y_0 = 0.00267$, and column 6 plotted against column 5 provides a dimensionless form of the breakthrough curve between w_B and w_E (Fig. 11.46). Equation (11.74) can be written as

$$f = \frac{\int_{w_B}^{w_E}(Y_0 - Y)\,dw}{Y_0 w_a} = \int_0^{1.0}\left(1 - \frac{Y}{Y_0}\right)d\frac{w - w_B}{w_a}$$

from which it is seen that f equals the entire area above the curve of Fig. 11.46 up to $Y/Y_0 = 1.0$. By graphical integration, $f = 0.530$.

The mass-transfer rates are given in equation form in Illustration 11.6. For a mass velocity of air $G' = 0.1295$ kg/m$^2 \cdot$ s, $k_Y a_p = 31.6(0.1295)^{0.55} = 10.27$ kg H$_2$O/m$^3 \cdot$ s $\cdot \Delta Y$, and $k_S a_p = 0.965$ kg H$_2$O/m$^3 \cdot$ s $\cdot \Delta X$.

From Fig. 11.45, $X_T = 0.0858$ kg H$_2$O/kg gel. Eq. (11.76):

$$S_S = \frac{Y_0 G_S}{X_T} = \frac{0.0026(0.1295)}{0.0858} = 3.92 \times 10^{-3}$$

The average slope of the equilibrium curve is $\Delta Y/\Delta X = 0.0185$, whence $mG_S/S_S = 0.0185(0.1295)/(3.92 \times 10^{-3}) = 0.61$. Eqs. (11.51) and (11.52):

$$H_{tG} = \frac{G_S}{k_Y a_p} = \frac{0.1295}{10.27} = 0.0126 \text{ m} \qquad H_{tS} = \frac{S_S}{k_S a_p} = \frac{3.92 \times 10^{-3}}{0.965} = 4.062 \times 10^{-3} \text{ m}$$

$$H_{tOG} = H_{tG} + \frac{mG_S}{S_S}H_{tS} = 0.0126 + 0.61(4.062 \times 10^{-3}) = 0.0151 \text{ m}$$

Eq. (11.79): $Z_a = N_{tOG}H_{tOG} = 9.304(0.0151) = 0.14$ m. The height of the bed $= Z = 0.61$ m; therefore [Eq. (11.74)]:

$$\text{Degree of saturation at breakpoint} = \frac{0.61 - 0.53(0.14)}{0.61} = 0.878 = 87.8\%$$

The bed contains 0.61 m^3 gel/m^2 of cross section, and since the bed density is 671.2 kg/m^3 (Illustration 11.6), the mass of the gel $= 0.61(671.2) = 409.4$ kg/m^2. At 87.8% of equilibrium with the incoming air, the gel contains $409.4(0.878)(0.0858) = 30.8$ kg H_2O/m^2 cross section. The air introduces $0.1295(0.00267) = 3.46 \times 10^{-4}$ kg $H_2O/m^2 \cdot$ s, and hence the breakpoint occurs at $30.8/(3.46 \times 10^{-4}) = 89\ 140$ s or 24.8 h after air is admitted initially, and $w_B = 89\ 140(0.1295) = 11\ 540$ kg air/m^2 cross section. If the entire bed were in equilibrium with the entering gas, the adsorbed water would be $409.4(0.0858) = 35.1$ kg/m^2, and hence $U = 35.1 - 30.8 = 4.3$ kg/m^2 = ability of the bed still to adsorb. Therefore $w_a = U/f\,Y_0 = 4.3/0.53(0.0026) = 3038$ kg air/m^2.

The circumstances of the calculation correspond to run S2 of Eagleton and Bliss [21], whose observed breakthrough curve (their fig. 5) agrees excellently with Fig. 11.46. However, they observed that 879 kg air/m^2 flowed through while the effluent humidity rose from $Y/Y_0 = 0.1$ to 0.8, whereas the curve of Fig. 11.46 predicts this amount to be $(0.79 - 0.17)(3038) = 1883$ kg/m^2. It is noteworthy that better agreement is obtained if the correlation of Table 3.3 is used for the gas-phase mass-transfer coefficient in the above calculation.

The preceding method is useful if both fluid- and solid-phase mass-transfer coefficients are available. The solid-phase coefficient particularly is not usually available for most industrial developments. In such cases the following extension is a useful approximation, also limited to cases of favorable adsorption equilibrium [14].

Consider Fig. 11.47. If the mass-transfer rate were infinitely rapid, the breakthrough curve would be the vertical line at θ_s, which can be located so that the shaded areas are equal. The adsorption zone of Fig. 11.35 can then be idealized as reduced to a plane, with the length of bed Z_s upstream of the plane at concentration X_T and the length $Z - Z_s$ downstream equal to the *length of unused bed* (LUB). At breakthrough, the length of the bed is taken to be the sum of LUB and a length saturated with solute in equilibrium with the feed stream. If V = velocity of advancement of the "adsorption plane," then at any time, $Z_s = V\theta$, at time θ_s, $Z = V\theta_s$, and at breakthrough, $Z_s = V\theta_B$; therefore

$$\text{LUB} = Z - Z_s = V(\theta_s - \theta_B) = \frac{Z}{\theta_s}(\theta_s - \theta_B) \qquad (11.81)$$

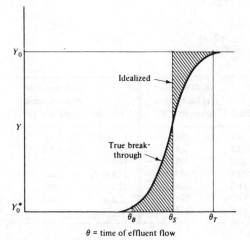

θ = time of effluent flow

Figure 11.47 Idealized breakthrough curve for infinitely rapid mass transfer.

A solute balance is

$$G_S(Y_0 - Y_0^*)\theta = Z_s\rho_s(X_T - X_0) \tag{11.82}$$

where Y_0^* is the fluid concentration in equilibrium with X_0.

In a laboratory experiment at the same G_S to be used on the large scale, the breakthrough curve can be measured, θ_B and θ_s (by equal areas) determined, and LUB calculated from Eq. (11.81). This should be the same on the large scale, provided radial mixing or bed channeling are not important.

If measurement is made on a production adsorber which cannot be operated beyond the breakpoint, the estimate can be made as follows. At θ_s, $Z = Z_s$, and Eq. (11.82) becomes

$$X_T - X_0 = \frac{G_S\theta_s}{Z_s\rho_S}(Y_0 - Y_0^*) \tag{11.83}$$

If we let $Z_s = V\theta$, Eq. (11.82) becomes

$$V = \frac{G_S}{\rho_S}\frac{Y_0 - Y_0^*}{X_T - X_0} \tag{11.84}$$

Elimination of θ_s from Eqs. (11.81) and (11.83) yields

$$\text{LUB} = Z - \frac{G_S\theta_B(Y_0 - Y_0^*)}{\rho_S(X_T - X_0)} \tag{11.85}$$

If the bed depth on the large scale is to be calculated for a given breakpoint time, Z_s can be calculated through Eq. (11.82), letting $\theta = \theta_B$. Then Eq. (11.85) provides LUB.

Illustration 11.10 The following laboratory test was made using type 4A molecular sieves, to remove water vapor from nitrogen (Collins' run B [14]):

Bed depth = 0.268 m (0.88 ft)
Initial water concn in solid = 0.01 kg H_2O/kg solid = X_0
Bulk density of bed = 712.8 kg/m^3 (44.5 lb/ft^3)
Temperature = 28.3°C
Gas pressure = 593 kN/m^2 (86 lb$_f$/in^2)
Nitrogen rate = 4052 kg/m$^2 \cdot$ h (830 lb/ft$^2 \cdot$ h)
Initial H_2O concn in gas = 1440 $\times$ 10^{-6} mole fraction

Time, h	H_2O concn of effluent, mole fraction $\times 10^6$	Time, h	H_2O concn of effluent, mole fraction $\times 10^6$	Time, h	H_2O concn of effluent, mole fraction $\times 10^6$
0	<1	10.2	238	11.75	1235
9 = θ_B	1	10.4	365	12.0	1330
9.2	4	10.6	498	12.5	1410
9.4	9	10.8	650	12.8 = θ_T	1440
9.6	33	11.0	808	13.0	1440
9.8	80	11.25	980	15.0	1440
10.0	142	11.5	1115		

Y_0^* in equilibrium with $X_0 \approx 0$
X_0^* in equilibrium with $Y_0 \approx 0.22$ kg H_2O/kg dry solid

$$Y = \frac{\text{mole fraction } H_2O}{1 - \text{mole fraction } H_2O} \frac{M_{H_2O}}{M_{N_2}} \approx \text{mole fraction } H_2O \frac{M_{H_2O}}{M_{N_2}}$$

The breakthrough data are plotted in the manner of Fig. 11.47, and θ_s is determined (equal areas) to be 10.9 h. Eq. (11.81):

$$LUB = \frac{0.268}{10.9}(10.9 - 9) = 0.0467 \text{ m}$$

It is desired to estimate the depth of bed required for the same molecular sieves operated at the same gas mass velocity, the same Y_0 and X_0, and the same breakpoint concentration, 1×10^{-6} mole fraction water, as in the laboratory test, but the breakpoint time θ_B is to be 15 h.

$$Y_0 = \frac{1440 \times 10^{-6}}{1 - 1440 \times 10^{-6}} \frac{18}{29} = 9.0 \times 10^{-4} \text{ kg } H_2O/\text{kg } N_2 \qquad Y_0^* \approx 0$$

Eq. (11.82):

$$Z_s = \frac{G_S(Y_0 - Y_0^*)\theta_B}{\rho_S(X_T - X_0)} = \frac{4052(9 \times 10^{-4})}{712.8(0.21 - 0.01)} = 0.365 \text{ m}$$

$$Z = LUB + Z_s = 0.0467 + 0.365 = 0.41 \text{ m } (1.35 \text{ ft}) \quad \textbf{Ans.}$$

Note: Collins' run A [14], bed depth = 0.44 m, 26.1°C, $Y_0 = 1490 \times 10^{-6}$, $G_S = 4002$ (conditions not quite the same as above), records a breakpoint at 15 h.

Ion Exchange

Ion-exchange operations are essentially metathetical chemical reactions between an electrolyte in solution and an insoluble electrolyte with which the solution is contacted. The mechanisms of these reactions and the techniques used to bring them about resemble those of adsorption so closely that for most engineering purposes ion exchange can simply be considered as a special case of adsorption.

Principles of ion exchange [20, 40, 55] The ion-exchange solids first used were porous, natural or synthetic minerals containing silica, the zeolites, such as the mineral $Na_2O \cdot Al_2O_3 \cdot 4SiO_2 \cdot 2H_2O$, for example. Positively charged ions (cations) of a solution which are capable of diffusing through the pores will exchange with the Na^+ ions of such a mineral, which is therefore called a *cation exchanger*. For example,

$$Ca^{2+} + Na_2R \rightarrow CaR + Na^+$$

where R represents the residual material of the zeolite. In this manner "hard" water containing Ca^{2+} can be softened by contact with the zeolite, the less objectionable Na^+ replacing the Ca^{2+} in solution and the latter becoming immobilized in the solid. The reaction is reversible, and after saturation with Ca^{2+} the zeolite can be regenerated by contact with a solution of salt,

$$CaR + 2NaCl \rightarrow Na_2R + CaCl_2$$

Later certain carbonaceous cation exchangers were manufactured by treating substances such as coal with reagents such as fuming sulfuric acid, and the like.

The resulting exchangers can be regenerated to a hydrogen form, HR, by treatment with acid rather than salt. Thus, hard water containing $Ca(HCO_3)_2$ would contain H_2CO_3 after removal of the Ca^{2+} by exchange, and since the carbonic acid is readily removed by degasification procedures, the total solids content of the water can be reduced in this manner. Early applications of ion exchangers using these principles were largely limited to water-softening problems.

In 1935, synthetic resinous ion exchangers were introduced. For example, certain synthetic, insoluble polymeric resins containing sulfonic, carboxylic, or phenolic groups can be considered as consisting of an exceedingly large anion and a replaceable or exchangeable cation. These make exchanges of the following type possible,

$$Na^+ + HR \rightleftharpoons NaR + H^+$$

and different cations will exchange with the resin with different relative ease. The Na^+ immobilized in the resin can be exchanged with other cations or with H^+, for example, much as one solute can replace another adsorbed upon a conventional adsorbent. Similarly synthetic, insoluble polymeric resins containing amine groups and anions can be used to exchange anions in solution. The mechanism of this action is evidently not so simple as in the case of the cation exchangers, but for present purposes it can be considered simply as an ion exchange. For example,

$$RNH_3OH + Cl^- \rightleftharpoons RNH_3Cl + OH^-$$
$$H^+ + OH^- \rightarrow H_2O$$

where RNH_3 represents the immobile cationic portion of the resin. Such resins can be regenerated by contact with solutions of sodium carbonate or hydroxide. The synthetic ion-exchange resins are available in a variety of formulations of different exchange abilities, usually in the form of fine, granular solids or beads, 16- to 325-mesh. The individual beads are frequently nearly perfect spheres.

Techniques and applications All the operational techniques ordinarily used for adsorption are used also for ion exchange. Thus we have batch or stagewise treatment of solutions, fluidized- and fixed-bed operations, and continuous countercurrent operations. Fixed-bed percolations are most common. Chromatographic methods have been used for fractionation of multicomponent ionic mixtures. Applications have been made in the treatment of ore slurries ("resin-in-pulp") for collection of metal values.

In addition to the water-softening applications mentioned above, the complete deionization of water can be accomplished by percolation first through a cation exchanger and then through an anion exchanger. By using a bed formed of an intimate mixture of equivalent amounts of a strong cationic and a strong anionic exchange resin, simultaneous removal of all ions at neutrality is possible. For purposes of regeneration, such mixed-bed resins are separated by hydraulic classification through particle-size and density differences for the two resin types, and these are regenerated separately. The ion exchangers have also been used for treatment and concentration of dilute waste solutions. Perhaps the most remarkable application of exchange resins has been to the separation of the rare-earth metals, using chromotographic techniques.

In *ion exclusion*, a resin is presaturated with the same ions as in a solution. It can then reject ions in such a solution but at the same time absorb nonionic organic substances such as glycerin, and the like, which may also be in the solution. The organic matter can later be washed from the resin in an ion-free state.

Equilibria The equilibrium distribution of an ion between an exchange solid and a solution can be described graphically by plotting isotherms in much the same manner used for ordinary adsorption. Various empirical equations for these isotherms, such as the Freundlich equation (11.3), have sometimes been applied to them. It is also possible to apply equations of the mass-action type to the exchange reaction. For example, for the cationic exchange

$$\underset{\text{(soln)}}{Na^+} + \underset{\text{(solid)}}{R - H^+} \rightleftharpoons \underset{\text{(solid)}}{R - Na^+} + \underset{\text{(soln)}}{H^+}$$

the mass-action-law constant is

$$\alpha = \frac{[R - Na^+]_{solid}[H^+]_{soln}}{[R - H^+]_{solid}[Na^+]_{soln}} = \left[\frac{Na^+}{H^+}\right]_{solid}\left[\frac{H^+}{Na^+}\right]_{soln} \tag{11.86}$$

where the square brackets indicate the use of some suitable equilibrium-concentration unit. The quantity α is thus seen to be an expression of relative adsorptivity, in this case of relative adsorptivity of Na^+ to H^+. Since the solution and the solid remain electrically neutral during the exchange process, we can write

$$\alpha = \frac{X}{X_0 - X}\frac{c_0 - c^*}{c^*} = \frac{X/X_0}{1 - X/X_0}\frac{1 - c^*/c_0}{c^*/c_0} \tag{11.87}$$

where c_0 = initial concentration of $Na^+ + H^+$ in solution and consequently their total at any time

c^* = equilibrium Na^+ concentration after exchange

X = equilibrium concentration of Na^+ in solid

X_0 = concentration if all H^+ were replaced by Na^+

All these are expressed as equivalents per unit volume or mass. In the general case for any system, the relative adsorptivity α at a given temperature varies with total cationic concentration c_0 in the solution and also with c. In some cases, α has been found to be essentially constant with varying c at fixed c_0.

Rate of ion exchange The rate of ion exchange depends, as in ordinary adsorption, upon rates of the following individual processes: (1) diffusion of ions from the bulk of the liquid to the external surface of an exchanger particle, (2) inward diffusion of ions through the solid to the site of exchange, (3) exchange of the ions, (4) outward diffusion of the released ions to the surface of the solid, and (5) diffusion of the released ions from the surface of the solid to the bulk of the liquid. In some instances, the kinetics of the exchange reaction (3) may be controlling, but in others the rate of reaction is apparently very rapid in comparison with the rate of diffusion. The diffusion rates can be described by appropriate mass-transfer coefficients for equi-equivalent counterdiffusion through the solid and through the liquid, and in some instances at least it appears that the resistance to diffusion in the liquid phase may be controlling.

When the exchange reactions are rapid in comparison with the rates of mass transfer, the methods of design developed for conventional adsorbers can be applied to ion-exchange operations directly. Some modification of the units of the terms in the various equations may be desirable, owing to the customary use of concentrations expressed as equivalents per unit volume in the cgs system. The following example illustrates this.

Illustration 11.11 A synthetic ion-exchange resin in bead form is to be used for collecting and concentrating the copper in a dilute waste solution. The feed contains $CuSO_4$ at a concentration of 20 milligram equivalents (meq) Cu^{2+}/l and is to be treated at the rate of 37 850 l/h. A continuous system is planned: the solution to be treated and regenerated resin will flow countercurrently through a vertical tower, where 99% of the Cu^{2+} of the feed will be exchanged; the resin will be regenerated in a second tower by countercurrent contact with 2 N sulfuric acid. The necessary data are provided by Selke and Bliss, *Chem. Eng. Prog.*, **47**, 529 (1951).

For collection of Cu^{2+} A superficial liquid velocity of 2.2 l/cm^2 · h will be used, for which the mass-transfer rate is 2.0 meq $Cu^{2+}/(g \text{ resin}) \cdot h \cdot (meq \ Cu^{2+}/l)$. The regenerated resin will contain 0.30 meq Cu^{2+}/g, and 1.2 times the minimum resin/solution ratio will be used.

For regeneration of the resin The superficial liquid velocity will be 0.17 l/cm^2 · h, for which the mass-transfer rate is 0.018 meq $Cu^{2+}/(g \text{ resin}) \cdot h \cdot (meq \ Cu^{2+}/l)$. The acid will be utilized to the extent of 70%.

Compute the necessary rates of flow of resin and the amount of resin holdup in each tower.

SOLUTION Equilibria for the $Cu^{2+}-H^+$ exchange are provided by Selke and Bliss at two concentration levels, 20 and 2000 meq cation/l, shown in Fig. 11.48a and b, respectively.

Collection of Cu^{2+} Feed soln = 37 850 l/h. c_1 = 20 meq Cu^{2+}/l, c_2 = 0.01(20) = 0.20 meq Cu^{2+}/l. Cu^{2+} exchanged = 37 850(20 − 0.20) = 750 000 meq/h.

X_2 = 0.30 meq Cu^{2+}/g. The point (c_2, X_2) is plotted in Fig. 11.48a. For the minimum resin/solution ratio and an infinitely tall tower, the operating line passes through point P at X = 4.9 on this figure, corresponding to equilibrium with c_1. The minimum resin rate is then 750 000/(4.9 − 0.30) = 163 000 g/h. For 1.2 times the minimum the resin rate is 1.2(163 000) = 196 000 g/h. A copper balance is

$$750\ 000 = 196\ 000(X_1 - 0.30)$$

$$X_1 = 4.12 \text{ meq } Cu^{2+}/g \text{ resin}$$

The point (c_1, X_1) is plotted on Fig. 11.48a, and the operating line can be drawn as a straight line at these low concentrations.

The quantity of resin in the tower can be obtained by application of the rate equation written in a form appropriate to the units of the quantities involved. Adapting Eq. (11.48) to this case, we have

$$V \, dc = \frac{K_L' a_p}{\rho_S}(c - c^*) \, d(SZ\rho_S)$$

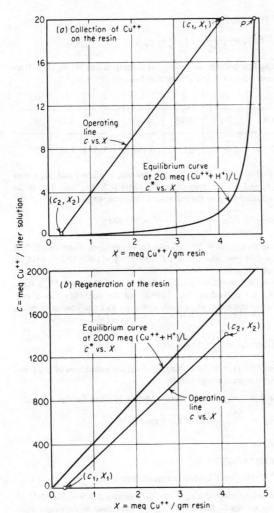

(a) Collection of Cu^{++} on the resin

(c_1, X_1)

p

Operating line c vs. X

Equilibrium curve at 20 meq $(Cu^{++} + H^+)/L$ c^* vs. X

(c_2, X_2)

X = meq Cu^{++}/gm resin

(b) Regeneration of the resin

Equilibrium curve at 2000 meq $(Cu^{++} + H^+)/L$ c^* vs. X

(c_2, X_2)

Operating line c vs. X

(c_1, X_1)

X = meq Cu^{++}/gm resin

c = meq Cu^{++} / liter solution

Figure 11.48 Solution to Illustration 11.11.

where $V = 1$ liquid/h
c = concn Cu^{2+}, meq/l, in soln
c^* = concn Cu^{2+} in soln at equilibrium with resin
$K'_L a_p/\rho_S$ = overall liquid mass-transfer coefficient, meq/(g resin) · h · (meq/l)
K'_L = overall liquid mass-transfer coefficient, meq/cm² · h · (meq/l)
a_p = surface of resin particles, cm²/cm³
ρ_S = packed density of resin, g/cm³
$SZ\rho_S$ = resin in tower, g
S = cross section of tower, cm²
Z = height of tower, cm

Rearranging this equation and integrating gives

$$SZ\rho_S = \frac{V}{K'_L a_p / \rho_S} \int_{c_2}^{c_1} \frac{dc}{c - c^*}$$

For values of c on the operating line between c_1 and c_2, the corresponding values of c^* from the equilibrium curve at the same value of X are obtained as follows:

c	20	16	12	8	4	2	1	0.2
c^*	2.4	1.9	0.5	0.25	0.10	0.05	0.02	0
$\dfrac{1}{c - c^*}$	0.0568	0.0710	0.0870	0.129	0.256	0.513	1.02	5.0

A curve (not shown) of $1/(c - c^*)$ as ordinate, c as abscissa, is plotted and integrated graphically between the limits c_1 and c_2. The area under the curve is 5.72. (*Note:* This is the number of transfer units N_{tOL}.) Substituting in the integrated equation, we get

$$\text{Resin holdup} = SZ\rho_S = \frac{37\,850(5.72)}{2.0} = 108\,300 \text{ g}$$

Regeneration of resin Cu^{2+} to be exchanged $= 750\,000$ meq/h, requiring as many meq H^+/h. For a 70% utilization of acid, the acid feed must contain $750\,000/0.70 = 1\,071\,000$ meq H^+/h, or $1\,071\,000/2000 = 536$ l/h of 2 N acid.
$c_1 = 0$, $c_2 = 750\,000/536 = 1400$ meq Cu^{2+}/l. $X_1 = 0.30$, $X_2 = 4.12$ meq Cu^{2+}/g resin. The points (c_1, X_1) and (c_2, X_2) are plotted on Fig. 11.48b and the operating line drawn. Integration of the rate equation for this case, where both operating and equilibrium lines are straight, provides

$$V(c_2 - c_1) = \frac{K'_L a_p}{\rho_S}(SZ\rho_S)(c^* - c)_m$$

where $(c^* - c)_m$ is the logarithmic average of the driving forces at the extremities of the tower and the other symbols have the same meaning as before.

$$c_1^* - c_1 = 120 - 0 = 120 \qquad c_2^* - c_2 = 1700 - 1400 = 300 \text{ meq } Cu^{2+}/l$$

$$(c^* - c)_m = \frac{300 - 120}{\ln (300/120)} = 196.5 \text{ meq } Cu^{2+}/l$$

Substituting in the rate equation, we get

$$750\,000 = (0.018SZ\rho_S)(196.5)$$

$$SZ\rho_S = 212\,000 \text{ g resin holdup in regeneration tower}$$

The resin should be water-rinsed before reintroducing it into the adsorption tower. The Cu^{2+} in the effluent solution has been concentrated $1400/20 = 70$ times, equivalent to the evaporation of 37 300 l/h of water from the original solution.

NOTATION FOR CHAPTER 11

Any consistent set of units may be used, except as noted.

a_p	surface of solids/volume solid-fluid mixture, L^2/L^3
a_{pL}	surface of solids/volume liquid, L^2/L^3
a_{pS}	surface of solids/mass solids, L^2/M
A	mass velocity of adsorbed solute, $M/L^2\Theta$
B	mass velocity of adsorbate-free adsorbent, $M/L^2\Theta$

c	solute concentration in solution, M/L^3 or $mole/L^3$; for ion exchange, equivalent/L^3
Δc	liquid-phase concentration-difference driving force, $mole/L^3$
C	distance, impeller to bottom of tank, L
	heat capacity, FL/MT
	mass velocity of more strongly adsorbed solute, $M/L^2\Theta$
d_i	impeller diameter, L
d_p	particle diameter, L
D	diffusivity, L^2/Θ
E	mass velocity of adsorbent-free adsorbate, $M/L^2\Theta$
E_D	eddy diffusivity of mass, $L^2\Theta$
E_{MS}	Murphree solid-phase stage efficiency, fractional
f	fractional ability of adsorption zone to adsorb solute, dimensionless
F	mass velocity of feed, $M/L^2\Theta$
F_L	liquid-phase mass-transfer coefficient, $mole/L^2\Theta$
g	acceleration of gravity, L^2/Θ
g_c	conversion factor, $ML/F\Theta^2$
G	total gas mass velocity, $M/L^2\Theta$
G_S	mass velocity of unadsorbed gas based on container cross section, $M/L^2\Theta$
$\overline{H}$	differential heat of adsorption, FL/M
H_G	enthalpy of a gas, solute-free basis, FL/M
H_S	enthalpy of solid and adsorbed solute, per unit mass of adsorbate-free solid, FL/M
H_{tG}	height of a gas transfer unit, L
H_{tOG}	height of an overall gas transfer unit, L
H_{tS}	height of a solid transfer unit, L
ΔH_A	integral heat of adsorption referred to liquid adsorbate, per unit mass of adsorbent, FL/M
$\Delta H'_A$	integral heat of adsorption referred to vapor adsorbate, per unit mass of adsorbent, FL/M
J	mass-transfer flux, $mole/L^2\Theta$
k	a constant
k_L	liquid-phase mass-transfer coefficient, $mole/L^2\Theta(mole/L^3)$
k_S	solid-phase mass-transfer coefficient, $mole/L^2\Theta(mole/L^3)$
k_Y	gas-phase mass-transfer coefficient, $M/L^2\Theta(M/M)$
K_y	overall gas-phase mass-transfer coefficient, $mole/L^2\Theta(mole\ fraction)$
K_Y	overall gas-phase mass-transfer coefficient, $M/L^2\Theta(M/M)$
l	a distance, L
L_S	solvent liquid: mass in a batch process, M; mass velocity in a continuous process, $M/L^2\Theta$
LUB	length of unused portion of fixed bed, L
m	slope of equilibrium adsorption isotherm [Eqs. (11.51) and (11.52)], dY^*/dX, dimensionless; also $(mole/L^3)/(mole/M)$ or $(M/L^3)/(M/M)$ const
M	molecular weight, $M/mole$
n	const
	number of impellers on a shaft, dimensionless
N	rotational speed, Θ^{-1}
	flux of mass transfer based on external solid surface, $mole/L^2\Theta$
	mass adsorbent/mass adsorbate (fractionation), M/M
N_{tOG}	number of overall gas transfer units, dimensionless
p	equilibrium vapor pressure, F/L^2
$\bar{p}$	partial pressure, F/L^2
P	agitator power transmitted to fluid or slurry, FL/Θ

P_E	mass velocity of C-rich product, $M/L^2\Theta$
Po	power number, $Pg_c/d_i^5N^3\rho_m$, dimensionless
r	molecular weight poorly adsorbed gas/molecular weight strongly adsorbed gas, dimensionless
R	mass velocity of reflux to fractionator, $M/L^2\Theta$
	universal gas constant, $FL/mole\ T$
Re	impeller Reynolds number, $d_i^2N\rho_L/\mu_L$, dimensionless
Re_p	particle Reynolds number [Eq. (11.28)], dimensionless
S_S	adsorbate-free solid: mass in a batch process, M; mass velocity in a continuous process, $M/L^2\Theta$
Sc_L	liquid-phase Schmidt number, $\mu_L/\rho_L D_L$, dimensionless
Sh_L	liquid-phase Sherwood number, $k_L d_p/D_L$, dimensionless
T	absolute temperature, T
	tank diameter, L
u'	turbulent fluctuating velocity, L/Θ
U	solute adsorbed in adsorption zone, $M/L^2\Theta$
v	volume of solute/mass adsorbed, L^3/M
v_L	liquid volume in a batch process, L^3; volume rate of liquid in a continuous process, L^3/Θ
v_m	slurry volume, L^3
v_T	volume of liquid in a depth equal to the tank diameter, L^3
V	velocity of advance of adsorption plane, L/Θ
V_{tN}	terminal settling velocity corresponding to Newton's law, L/Θ
V_{tS}	terminal settling velocity corresponding to Stokes' law, L/Θ
w	quantity of effluent from a fixed-bed adsorber, M/L^2
w_a	$w_E - w_B$, M/L^2
w_B	quantity of effluent from a fixed-bed adsorber at breakpoint, M/L^2
w_E	quantity of effluent from a fixed-bed adsorber at bed exhaustion, M/L^2
x	weight fraction of component C in fluid stream, dimensionless
X	adsorbate concentration, mass solute/mass adsorbent, M/M; for ion exchange, equivalent/M or equivalent/L^3
X_T	adsorbate concentration in equilibrium with entering fluid, mass solute/mass adsorbent, M/M; for ion exchange, equivalent/M or equivalent/L^3
y	weight fraction component C in adsorbate, dimensionless
Y	concentration of solute in fluid, mass solute/mass solvent, M/M
Y_B	concentration of solute in effluent at breakpoint, mass solute/mass solvent, M/M
Y_E	concentration of solute in effluent at bed exhaustion, mass solute/mass solid, M/M
Z	height of adsorbent column, L
Z'	height above the midplane of uppermost impeller, L
Z_a	height of adsorption zone, L
Z_s	length of adsorber bed in equilibrium with feed, L
α	mass-action-law constant or relative adsorptivity, dimensionless
Δ	difference
θ	time, Θ
θ_a	time required for the adsorption zone to move a distance Z_a, Θ
θ_E	time required to reach bed exhaustion, Θ
θ_F	time of formation of adsorption zone, Θ
θ_s	time to idealized breakthrough (Fig. 11.47), Θ
θ_T	time to bed saturation, Θ
λ	latent heat of vaporization, FL/M
μ	viscosity, $M/L\theta$
μ_m	viscosity of slurry below Z', $M/L\Theta$
ρ	fluid density, M/L^3
ρ_m	density of slurry below Z', M/L^3

ρ_S	apparent density of adsorbent bed, mass solid/packed volume, M/L^3
$\Delta\rho$	absolute value of density difference, solid and liquid, M/L^3
φ_{Sm}	volume fraction solids in slurry below Z', dimensionless
φ_{SS}	volume fraction solids in fully settled bed, dimensionless
φ_{ST}	volume fraction solids based on v_T, dimensionless

Subscripts

A	weakly adsorbed solute; adsorbate
C	strongly adsorbed solute
e	within the enriching section of a continuous fractionator
L	liquid
min	minimum
n	stage n
p	particle
r	reference substance
s	within the adsorption section of a continuous fractionator
S	solid; adsorbent
0	initial
1	stage 1; bottom of a continuous-contact adsorber
2	stage 2; top of a continuous-contact adsorber

Superscript

*	at equilibrium

REFERENCES

1. Abbott, E. A.: "Flatland," Barnes and Noble, New York, 1963.
2. Amundsen, N. R.: *Ind. Eng. Chem.*, **48**, 26 (1956).
3. Anon.: *Chem. Eng.*, **70**, 92 (Apr. 15, 1963).
4. Astakhov, V. A., V. D. Lukin, P. G. Romankov, and J. A. Tan: *Z. Priklad. Khim.*, **44**, 319 (1971).
5. Barker, J. J., and R. E. Treybal: *AIChE J.*, **6**, 289 (1960).
6. Bartell, F. E., and C. K. Sloan: *J. Am. Chem. Soc.*, **51**, 1643 (1928).
7. Bates, P. L., P. L. Fondy, and J. G. Fenic: in V. Uhl and J. B. Gray (eds.), "Mixing," vol. I, p. 148, Academic, New York, 1966.
8. Berg, C.: *Trans. AIChE*, **42**, 665 (1946); *Chem. Eng. Prog.*, **47**, 585 (1951).
9. Brian, P. L. T., H. B. Hales, and T. K. Sherwood: *AIChE J.*, **15**, 727 (1969).
10. Broughton, D. B.: *Chem. Eng. Prog.* **64**(8), 60 (1968).
11. Calderbank, P. H.: in V. Uhl and J. B. Gray (eds.), "Mixing," vol. II, p. 78, Academic, New York, 1967.
12. Calderbank, P. H., and M. B. Moo-Young: *Chem. Eng. Sci.*, **16**, 39 (1961).
13. Clark, A.: "The Theory of Adsorption and Catalysis," Academic, New York, 1970.
14 Collins, J. J.: *Chem. Eng. Prog. Symp. Ser.*, **63**(74), 31 (1967).
15. Courouleau, P. H. and R. E. Benson: *Chem. Eng.*, **55**(3), 112 (1948).
16. Davidson, J. F., and D. Harrison (eds.): "Fluidization," Academic, New York, 1971.
17. Davis, J. C.: *Chem. Eng.*, **79**, 88 (Oct. 16, 1972).
18. Deitz, V. R.: *Ind. Eng. Chem.*, **57**(5), 49 (1965).
19. Denver Equipment Company: *Deco Trefoil*, January–February 1966, p. 26.
20. Dorfner, K.: "Ion Exchangers, Principles and Applications," Ann Arbor Science Publishers, Ann Arbor, Mich., 1972.
21. Eagleton, L. C., and H. Bliss: *Chem. Eng. Prog.*, **49**, 543 (1953).
22. Ermenc, E. D.: *Chem. Eng.*, **68**, 87 (May 27, 1961).

23. Etherington, L. D., R. J. Fritz, E. W. Nicholson, and H. W. Scheeline: *Chem. Eng. Prog.*, **52**, 274 (1956).
24. Etherington, L. D., R. E. D. Haney, W. A. Herbst, and H. W. Scheeline: *AIChE J.*, **2**, 65 (1956).
25. Findlay, A., and J. M. Greighton: *J. Chem. Soc. Lond.*, **47**, 536 (1910).
26. Friederich, R. O., and J. C. Mullins: *Ind. Eng. Chem. Fundam.*, **11**, 439 (1972).
27. Frost, A. C.: *Chem. Eng. Prog.*, **70**(5), 644 (1974).
28. Furusawa, T., and J. M. Smith: *Ind. Eng. Chem. Fundam.*, **12**, 197 (1973).
29. Glessner, A. J., and A. L. Myers: *AIChE Symp. Ser.*, **65**(96), 73 (1969).
30. Glueckauf, E.: *Trans. Faraday Soc.*, **51**, 1540 (1955).
31. Goldman, F., and M. Polanyi: *Z. Phys. Chem.*, **132**, 321 (1928).
32. Gregg, S. J., and K. S. W. Sing: "Adsorption, Surface Area, and Porosity," Academic, New York, 1967.
33. Gröber, H.: *Z. Ver. Dtsch. Ing.*, **69**, 705 (1925).
34. Halle, E. von: personal communication, 1964.
35. Hancker, C. W., and S. H. Jury: *Chem. Eng. Prog. Symp. Ser.*, **55**(24), 87 (1959).
36. Harriott, P.: *AIChE J.*, **8**, 93 (1962).
37. Hassler, J. W.: "Purification with Activated Carbon," Chemical Publishing, New York, 1974.
38. Helby, W. A.: in J. Alexander (ed.), "Colloid Chemistry," vol. VI, p. 814, Reinhold, New York, 1946.
39. Helby, W. A.: *Chem. Eng.*, **59**(10), 153 (1952).
40. Helferich, F.: "Ion Exchange," McGraw-Hill Book Company, New York, 1962.
41. Hersh, C. K.: "Molecular Sieves," Reinhold, New York, 1961.
42. Higgins, C. W., and J. T. Roberts: *Chem. Eng. Prog. Symp. Ser.*, **50**(14), 87 (1954).
43. Hirsekorn, F. S., and S. A. Miller: *Chem. Eng. Prog.*, **49**, 459 (1953).
44. Hudson, J. B., and S. Ross: *Ind. Eng. Chem.*, **56**(11), 31 (1964).
45. Kapfer, W. H., M. Malow, J. Happel, and C. J. Marsel: *AIChE J.*, **2**, 456 (1956).
46. Karlson, E. L., and S. P. Edkins: *AIChE Symp. Ser.*, **71**(151), 286 (1975).
47. Karlson, E. L., and R. Roman: *Prod. Finish.*, **41**(9), 74 (1977).
48. Kehde, H., R. E. Fairfield, J. C. Frank, and L. W. Zahnstecker: *Chem. Eng. Prog.*, **44**, 575 (1948).
49. Kidnay, A. J., and A. L. Myers: *AIChE J.*, **12**, 981 (1966).
50. Kipling, J. J., and D. A. Tester: *J. Chem. Soc.*, **1952**, 4123.
51. Kneule, F.: *Chem. Ing. Tech.*, **28**, 221 (1956).
52. Kohler, R. H., and J. Estrin: *AIChE J.*, **13**, 179 (1967); personal communication, 1974.
53. Kubai, R., I. Komasawa, T. Otake, and M. Iwasa: *Chem. Eng. Sci.*, **29**, 659 (1974).
54. Kunii, D., and O. Levenspiel: "Fluidization Engineering," Wiley, New York, 1969.
55. Kunin, R.: "Elements of Ion Exchange," Reinhold, New York, 1960.
56. Leavitt, F. W.: *Chem. Eng. Prog.*, **58**(8), 54 (1962).
57. Lee, H., and D. E. Stahl: *AIChE Symp. Ser.*, **69**(134), 1 (1973).
58. Lee, M. N. Y.: in N. N. Li (ed.), "Recent Developments in Separation Science," vol. 1, p. 75, Chemical Publishing, Cleveland, 1972.
59. Lee, M. N. Y., and I. Zwiebel: Adsorption Technology, *AIChE Symp. Ser.*, **67**(117), foreword (1971).
60. Lerch, R. G., and D. A. Rathowsky: *Ind. Eng. Chem. Fundam.*, **6**, 308 (1967).
61. Leva, M.: "Fluidization," McGraw-Hill Book Company, New York, 1959.
62. Levins, D. M., and J. R. Glastonbury: *Trans. Inst. Chem. Eng. Lond.*, **50**, 32, 132 (1972).
63. Levy, C. I., and R. L. Harris: *J. Phys. Chem.*, **58**, 899 (1954).
64. Lewis, W. K., E. R. Gilliland, B. Chertow, and W. P. Cadogen: *Ind. Eng. Chem.*, **42**, 1319, 1326 (1950).
65. Lyons, E. J.: in V. W. Uhl and J. B. Gray (eds.), "Mixing," vol. 2, p. 225, Academic, New York, 1967; personal communication, 1974.
66. Mehta, D. S., and S. Calvert: *Environ. Sci. Technol.*, **1**, 325 (1967).
67. Mehta, D. S., and S. Calvert: *Br. Chem. Eng.*, **15**, 781 (1970).

68. Metzner, A. B., R. H. Feeks, H. L. Ramon, R. E. Otto, and J. D. Tuthill: *AIChE J.*, 7, 3 (1961).
69. Metzner, A. B., and R. E. Otto: *AIChE J.*, 3, 3 (1957).
70. Michaels, A. S.: *Ind. Eng. Chem.*, 44, 1922 (1952).
71. Miller, D. N.: *Ind. Eng. Chem. Process Des. Dev.*, 10, 365 (1971).
72. Minka, C., and A. L. Myers: *AIChE J.*, 19, 453 (1973).
73. Misic, D. M., and J. M. Smith: *Ind. Eng. Chem. Fundam.*, 10, 380 (1971).
74. Myers, A. L., and J. M. Prausnitz: *AIChE J.*, 11, 121 (1965).
75. Myers, A. J.: *Ind. Eng. Chem.*, 60(5), 45 (1968).
76. Myers, A. L., and S. Sircar: *J. Phys. Chem.*, 76, 3415 (1972).
77. Nagata, S., I. Yamaguchi, S. Yabata, and M. Harada: *Mem. Fac. Eng. Kyoto Univ.*, 21(3), 275 (1959); 22(1), 86 (1960).
78. Nienow, A. W.: *Chem. Eng. J.*, 9, 153 (1975).
79. Oldshue, J. Y.: *Ind. Eng. Chem.*, 61(9), 79 (1969).
80. Otavi, S.: *Chem. Eng.*, 80, 106 (Sept. 17, 1973).
81. Othmer, D. F., and F. G. Sawyer: *Ind. Eng. Chem.*, 35, 1269 (1943).
82. Ponic, V., Z. Knor, and S. Černý, "Adsorption on Solids," trans. by D. Smith and N. G. Adams, Butterworth, London, 1974.
83. Radke, C. J., and J. M. Prausnitz: *AIChE J.*, 18, 761 (1972).
84. Rushton, J. H., and D. L. Fridlay: USAEC TID-22073, (1958).
85. Rushton, J. J., and V. W. Smith, Jr.: USAEC TID-22075 (1965).
86. Sanders, M. T.: *Ind. Eng. Chem.*, 20, 791 (1928).
87. Schwartzberg, H. G., and R. E. Treybal: *Ind. Eng. Chem. Fundam.*, 7, 1, 6 (1968).
88. Seaburn, J. T., and A. J. Engel: *AIChE Symp. Ser.*, 69(134), 71 (1973).
89. Siegmund, C. W., W. D. Munro, and N. R. Amundsen: *Ind. Eng. Chem.*, 48, 43 (1956).
90. Sircar, S., and A. L. Myers: *AIChE J.*, 17, 186 (1971).
91. Sircar, S., and A. L. Myers: *AIChE J.*, 19, 159 (1973).
92. Skarstrom, C. W.: *Ann. N.Y. Acad. Sci.*, 72, 751 (1959); in *Recent Dev. Sep. Sci.*, 2, 95 (1972).
93. Snyder, L. R.: "Principles of Adsorption and Chromatography," Dekker, New York, 1968.
94. Squires, A. M.: *Chem. Eng. Prog.*, 58(4), 66 (1962).
95. Stevens, B. W., and J. W. Kerner: *Chem. Eng.*, 82, 84 (Feb. 3, 1975).
96. Uchida, S., K. Korde, and M. Shindo: *Chem. Eng. Sci.*, 30, 644 (1975).
97. Van Ness, H. C.: *Ind. Eng. Chem. Fundam.*, 8, 464 (1969).
98. Vermeulen, T.: *Adv. Chem. Eng.*, 2, (1958).
99. Vermeulen, T., G. Klein, and N. K. Heister: in R. H. Perry and C. H. Chilton (eds.), "The Chemical Engineers' Handbook," 5th ed., sec. 16, McGraw-Hill Book Company, New York, 1973.
100. Walker, W. H., W. K. Lewis, W. H. McAdams, and E. R. Gilliland: "Principles of Chemical Engineering," 3d ed., p. 511, McGraw-Hill Book Company, New York, 1937.
101. Weisman, J., and L. E. Efferding: *AIChE J.*, 6, 414 (1960).
102. Young, D. M., and A. D. Croswell: "Physical Adsorption of Gases," Butterworth, London, 1962.
103. Zenz, F. A., and D. F. Othmer: "Fluidization and Fluid-Particle Systems," Reinhold, New York, 1960.

PROBLEMS

11.1 The equilibrium adsorption of acetone vapor on an activated carbon at 30°C is given by the following data:

g adsorbed/g carbon	0	0.1	0.2	0.3	0.35
Partial pressure acetone, mmHg	0	2	12	42	92

The vapor pressure of acetone at 30°C is 283 mmHg.

A 1-l flask contains air and acetone vapor at 1 std atm and 30°C, with a relative saturation of the vapor of 35%. After 2 g of fresh activated carbon has been introduced into the flask, the flask is sealed. Compute the final vapor concentration at 30°C and the final pressure. Neglect the adsorption of air.

11.2 A solution of washed, raw cane sugar, 48% sucrose by weight, is colored by the presence of small quantities of impurities. It is to be decolorized at 80°C by treatment with an adsorptive carbon in a contact filtration plant. The data for an equilibrium adsorption isotherm were obtained by adding various amounts of the carbon to separate batches of the original solution and observing the equilibrium color reached in each case. The data, with the quantity of carbon expressed on the basis of the sugar content of the solution, are as follows:

kg carbon/kg dry sugar	0	0.005	0.01	0.015	0.02	0.03
Color removed, %	0	47	70	83	90	95

The original solution has a color concentration of 20, measured on an arbitrary scale, and it is desired to reduce the color to 2.5% of its original value.

(a) Convert the equilibrium data to $Y^* =$ color units/kg sugar, $X =$ color units/kg carbon. Do they follow the Freundlich equation? If so, what are the equation constants?

(b) Calculate the necessary dosage of fresh carbon, per 1000 kg of solution, for a single-stage process. Ans.: 20.4 kg.

(c) Calculate the necessary carbon dosages per 1000 kg of solution for a two-stage crosscurrent treatment, using the minimum total amount of fresh carbon. Ans.: 10.54 kg.

(d) Calculate the necessary carbon dosage per 1000 kg of solution for a two-stage countercurrent treatment. Ans.: 6.24 kg.

11.3 The sugar refinery of Prob. 11.2 must treat also a raw cane sugar solution, 48 wt % sucrose, of original color 50, based on the same color scale used in Prob. 11.2. The color scale is such that colors are additive, i.e., equal weights of solution of color 20 and color 50 will give a solution of color $(20 + 50)/2 = 35$. The same adsorption isotherm describes the color removal of the darker solution as that of Prob. 11.2. Equal quantities of the dark solution and that of Prob. 11.2 must both be decolorized to a color 0.5.

(a) In a single-stage process, will it be more economical of carbon first to blend the original solutions and to treat the blend, or to treat each separately to color 0.5 and to blend the finished products?

(b) Repeat for a two-stage crosscurrent treatment, fresh carbon in each stage, arranged for the minimum carbon in each case.

(c) Repeat for a countercurrent two-stage treatment.

(d) The following treating scheme was suggested. The light-colored solution is to be treated in a two-stage countercurrent plant to the final desired color. The spent carbon from this operation is to be used to treat an equal weight of the dark solution, and the carbon is then revivified. The

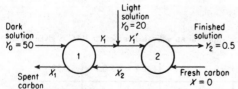

Figure 11.49 Flowsheet for Prob. 11.3e.

residual dark solution is then finished to the desired final color with the necessary amount of fresh carbon. Sketch a flowsheet and an operating diagram (freehand) for the entire process. Determine whether there is any saving of carbon over that for the arrangement of part (c).

(e) Determine whether the scheme of Fig. 11.49 offers any economies of carbon.

11.4 Prove that for crosscurrent two-stage treatment of liquid solutions by contact filtration, when the adsorption isotherm is linear, the least total adsorbent results if the amounts used in each stage are equal.

11.5 For adsorption from dilute liquid solutions in stagewise countercurrent operations, where the Freundlich equation describes the adsorption equilibrium, derive analytical expressions in terms of n, m, Y_0, and Y_{N_p} for the minimum adsorbent/solvent ratio when fresh adsorbent is used.

11.6 A batch of water containing residual chlorine from a treating process, at a concentration 12 ppm, is to be treated with activated carbon at 25°C to reduce the chlorine concentration to 0.5 ppm. The carbon consists of 30-mesh granules, density = 561 kg/m^3 (35 lb/ft^3) = mass of particle/gross volume of particle. Adsorbate diffusional resistance is expected to be small relative to that in the liquid. The equilibrium distribution coefficient = c^*/X = 0.80 (kg Cl$_2$/m^3 liquid)/(kg Cl$_2$/kg C) = 0.05 (lb Cl$_2$/ft^3 liquid)/(lb Cl$_2$/lb C).

(a) Calculate the minimum mass of carbon/unit vol water which can be used. **Ans.**: 18.4 kg/m^3.

(b) A batch of 2 m^3 (528 U.S. gal) is to be contacted with 40 kg (88 lb) of the carbon in an agitated vessel. Specify the dimensions of the vessel, choose a suitable impeller and agitator speed, and estimate the required contact time for the concentration change specified above.

11.7 Refer to Illustration 11.5. (a) For the same concentration of Cu^{2+} in the effluent, what is the minimum solids rate that could be used? **Ans.**: 7.56×10^{-4} kg/s.

(b) If 0.0025 kg/s (0.33 lb/min) resin were used in the vessel with the same agitator speed and the same effluent concentration of Cu^{2+} in the water as in the illustration, what is the maximum volume rate of water that can be treated? **Ans.** 3.2×10^{-4} m^3/s.

11.8 Nitrogen dioxide, NO$_2$, produced by a thermal process for fixation of nitrogen is to be removed from a dilute mixture with air by adsorption on silica gel in a continuous countercurrent adsorber. The gas entering the adsorber at the rate of 0.126 kg/s contains 1.5% NO$_2$ by volume, and 90% of the NO$_2$ is to be removed. Operation is to be isothermal at 25°C, 1 std atm. The entering gel will be free of NO$_2$. The equilibrium adsorption isotherm at this temperature is given by the following data [Foster and Daniels, *Ind. Eng. Chem.*, **43**, 986 (1951)]:

Partial pressure NO$_2$, mmHg	0	2	4	6	8	10	12
kg NO$_2$/100 kg gel	0	0.4	0.9	1.65	2.60	3.65	4.85

(a) Calculate the minimum weight of gel required per hour.

(b) For twice the minimum gel rate, calculate the number of transfer units required.

(c) A superficial air rate of 0.407 kg/m^2 · s is to be used. Assume that the characteristics of the gel are the same as those described in Illustration 11.6. Modify the gas mass-transfer coefficient of Illustration 11.6 so that it will apply to the transfer of NO$_2$ rather than water. Modify the solid-phase mass-transfer coefficient to apply for NO$_2$ on the assumption that the transfer in the pores of the solid is by molecular diffusion through the gas filling the pores. The diffusivity of NO$_2$ in air is estimated to be 1.36×10^{-5} m^2/s at 25°C, 1 std atm.

Estimate the value of H_{tOG}, and calculate the corresponding height of the adsorber. **Ans.**: 3 m.

11.9 Lewis et al., *J. Am. Chem. Soc.*, **72**, 1157 (1950), report the following for the simultaneous adsorption of acetylene and ethylene from mixtures of the two on silica gel at 1 std atm, 25°C (reprinted with permission of the American Chemical Society):

Mole fraction ethylene in adsorbate	Mole fraction ethylene in gas, at equilibrium	Gram moles mixture adsorbed/kg adsorbent
0.0686	0.2422	1.622
0.292	0.562	1.397
0.458	0.714	1.298
0.592	0.814	1.193
0.630	0.838	1.170
0.864	0.932	1.078

A gas containing equimolar amounts of acetylene and ethylene is to be fractionated in a continuous countercurrent adsorber, to yield products containing 98 and 2% acetylene by volume. Assume the temperature to remain constant at 25°C and the pressure to be 1 std atm. Calculate the number of transfer units and the gel circulation rate per 1000 m³ feed gas, using 1.2 times the minimum gel circulation rate. **Ans.:** $N_{tOG} = 15.3$.

11.10 The sulfur content of an oil is to be reduced by percolation through a bed of adsorbent clay. Laboratory tests with the clay and oil in a representative percolation filter show the following instantaneous sulfur contents of the effluent oil as a function of the total oil passing through the filter [adapted from Kaufman, *Chem. Met. Eng.*, **30**, 153 (1924)]:

bbl oil/ton clay	0	10	20	50	100	200	300	400
10^3 m³ oil/kg clay	0	1.752	3.504	8.760	17.52	35.04	52.56	70.08
Sulfur		0.011	0.020	0.041	0.067	0.0935	0.118	0.126	0.129

Assume that the specific gravity of the oil is unchanged during the percolation. The untreated oil has a sulfur content of 0.134%, and a product containing 0.090% sulfur is desired.

(a) If the effluent from the filter is composited, what yield of satisfactory product can be obtained per ton of clay? **Ans.:** 240 bbl/ton = 0.042 m³/kg.

(b) If the effluent from the filter is continually and immediately withdrawn and blended with just sufficient untreated oil to give the desired sulfur content in the blend, what quantity of product can be obtained per ton of clay? **Ans.:** 159.4 bbl/ton = 0.0279 m³/kg.

11.11 A laboratory fixed-bed adsorption column filled with a synthetic sulfonic acid cation-exchange resin in the acid form is to be used to remove Na^+ ions from an aqueous solution of sodium chloride. The bed depth is 33.5 cm, and the solution to be percolated through the bed contains 0.120 meq Na^+/cm³. At saturation, the resin contains 2.02 meq Na^+/cm³ resin. The solution will be passed through the bed at a superficial linear velocity of 0.31 cm/s. For this resin, Michaels [70] reports that the overall liquid mass-transfer rate $K_L' a_p = 0.86 v_L^{0.5}$, where v_L is the superficial liquid velocity, cm/s, and $K_L' a_p$ is expressed as meq Na^+/cm³ · s · (meq/cm³). The relative adsorptivity of Na^+ with respect to H^+ for this resin is $\alpha = 1.20$, and this is constant for the prevailing concentration level. Define the breakpoint concentration as 5% of the initial solution concentration, and assume that practical bed exhaustion occurs when the effluent concentration is 95% of the initial. Estimate the volume of effluent at the breakpoint, per unit bed cross section. **Ans.:** 360 cm³/cm².

Note: For these circumstances, Michaels [70] observed that the adsorption-zone height was 23.8 cm, that the breakpoint occurred after 382 ± 10 cm³ effluent/cm² bed cross section was collected, and that the holdup of liquid in the bed was 14.5 ± 2.5 cm³ solution/cm² bed cross section. Compare the calculated results with these.

The term *drying* refers generally to the removal of moisture from a substance. It is so loosely and inconsistently applied that some restriction in its meaning is necessary in the treatment to be given the subject here. For example, a wet solid such as wood, cloth, or paper can be dried by evaporation of the moisture either into a gas stream or without the benefit of the gas to carry away the vapor, but the mechanical removal of such moisture by expression or centrifuging is not ordinarily considered drying. A solution can be "dried" by spraying it in fine droplets into a hot, dry gas, which results in evaporation of the liquid, but evaporation of the solution by boiling in the absence of a gas to carry away the moisture is not ordinarily considered a drying operation. A liquid such as benzene can be "dried" of any small water content by an operation which is really distillation, but the removal of a small amount of acetone by the same process would not usually be called drying. Gases and liquids containing small amounts of water can be dried by adsorption operations, as discussed in Chap. 11. This discussion will be largely limited to the removal of moisture from solids and liquids by evaporation into a gas stream. In practice, the moisture is so frequently water and the gas so frequently air that this combination will provide the basis for most of the discussion. It is important to emphasize, however, that the equipment, techniques, and relationships are equally applicable to other systems.

EQUILIBRIUM

The moisture contained in a wet solid or liquid solution exerts a vapor pressure to an extent depending upon the nature of the moisture, the nature of the solid, and the temperature. If then a wet solid is exposed to a continuous supply of fresh gas containing a fixed partial pressure of the vapor $\bar{p}$, the solid will either lose moisture by evaporation or gain moisture from the gas until the vapor pressure of the moisture of the solid equals $\bar{p}$. The solid and the gas are then in equilibrium, and the moisture content of the solid is termed its *equilibrium-moisture content* at the prevailing conditions.

Insoluble Solids

A few typical equilibrium-moisture relationships are shown in Fig. 12.1, where the moisture in each case is water. Here the equilibrium partial pressure $\bar{p}$ of the water vapor in the gas stream has been divided by the vapor pressure of pure water p to give the relative saturation, or relative humidity (see Chap. 7), of the gas, since the curves are then applicable over a modest range of temperatures instead of being useful for one temperature only. Consider the curve for wood. If the wood contained initially a very high moisture content, say 0.35 kg water/kg dry solid, and were exposed to a continuous supply of air of 0.6 relative humidity, the wood would lose moisture by evaporation until its equilibrium concentration corresponding to point A on the curve was eventually reached. Further exposure to this air, for even indefinitely long periods, would not bring about additional loss of moisture from the solid. The moisture content could be reduced further, however, by exposure of the solid to air of lower

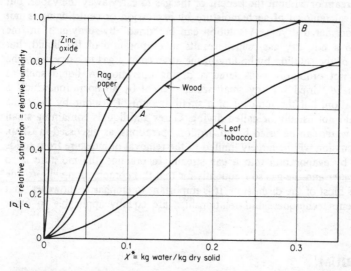

Figure 12.1 Equilibrium water content of some common solids at about 25°C. (*From "International Critical Tables," vol. II, pp. 322–325. with permission.*)

relative humidity, but to remove all the moisture would require exposure to perfectly dry air, corresponding to the origin of the curve. The moisture contained in the wood up to a concentration corresponding to point B in the figure, which exerts a vapor pressure less than that of pure water, may be moisture contained inside the cell walls of the plant structure, moisture in loose chemical combination with the cellulosic material, moisture present as a liquid solution of soluble portions of the solid and as a solid solution, or moisture held in small capillaries and crevasses throughout the solid or otherwise adsorbed upon the surface. Such moisture is called *bound water*. If exposed to saturated air, the wood may have any moisture content greater than 0.3 kg/kg dry solid (point B), and moisture in excess of that at B, *unbound water*, exerts the vapor pressure of pure water at the prevailing temperature.

The equilibrium moisture for a given species of solid may depend upon the particle size or specific surface, if the moisture is largely physically adsorbed. Different solids have different equilibrium-moisture curves, as shown in the figure. Generally, inorganic solids which are insoluble in the liquid and which show no special adsorptive properties, such as the zinc oxide in the figure, show relatively low equilibrium-moisture content, while spongy, cellular materials, especially those of vegetable origin such as the tobacco in the figure, generally show large equilibrium-moisture content. The equilibrium partial pressure for a solid is independent of the nature of the dry gas provided the latter is inert to the solid and is the same in the absence of noncondensable gas also. The same solids, if wet with liquids other than water, will show different equilibrium curves. The effect of large changes in temperature can frequently be shown in the form of a reference-substance plot, as in Fig. 11.5 [57]. It is seen that the equilibrium moisture is similar in many respects to the adsorption equilibria discussed in Chap. 11.

Hysteresis

Many solids exhibit different equilibrium-moisture characteristics depending upon whether the equilibrium is reached by condensation (adsorption) or evaporation (desorption) of the moisture. A typical example is shown in Fig. 12.2, which somewhat resembles the curve of Fig. 11.4. In drying operations, it is the desorption equilibrium which is of particular interest, and this will always show the larger of the two equilibrium-moisture contents for a given partial pressure of vapor. The moisture picked up by a dry solid when exposed to moist air, i.e., the adsorption equilibrium, is sometimes called *regain*, and knowledge of this has practical value in the consideration of drying operations. For example, in the case of Fig. 12.2, it will be of little use to dry the solid to a water content below that corresponding to point A if it is expected to expose the dried material to air of 0.6 relative humidity later. If it is important that the solid be kept at a lower moisture content, it would have to be packaged or stored immediately out of contact with the air in a moisture-impervious container.

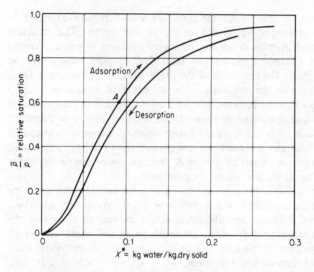

Figure 12.2 Equilibrium water content of a sulfite pulp, showing hysteresis. [*Seborg, Ind. Eng. Chem.*, **29**, *160 (1937).*]

Soluble Solids

Solids which are soluble in the liquid in question ordinarily show insignificant equilibrium-moisture content when exposed to gases whose partial pressure of vapor is less than that of the saturated solution of the solid. Refer to Fig. 12.3, where the characteristics of sodium nitrate–water are shown. A saturated solution of sodium nitrate in water at 25°C exerts a partial pressure of water (B) equal to 17.7 mmHg, and more dilute solutions exert higher partial pressures, as shown by curve *BC*. When exposed to air containing a partial pressure of water less than 17.7 mmHg, a solution will evaporate and the residual solid will retain only a negligible amount of adsorbed moisture, as shown by the curve from the origin to point *A*, and will appear dry. If the solid is exposed to air containing a higher water-vapor content, say 20 mmHg, moisture will be adsorbed to such an extent that the solid will completely dissolve, or *deliquesce*, to produce the corresponding solution at *C*. Solids of very low solubility, when exposed to ordinary atmospheric air, will not deliquesce since the equilibrium partial pressure of their saturated solutions is greater than that ordinarily found in the air.

Hydrated crystals may show more complicated relationships, such as those of Fig. 12.4 for the system copper sulfate–water at 25°C. Three hydrates are formed in this system, as the figure indicates. The anhydrous salt shows a negligible equilibrium-moisture content, which would in any case consist merely of adsorbed water upon the surface of the crystals. If exposed to air containing a partial pressure of water less than 7.8 and more than 5.6 mmHg, the salt will

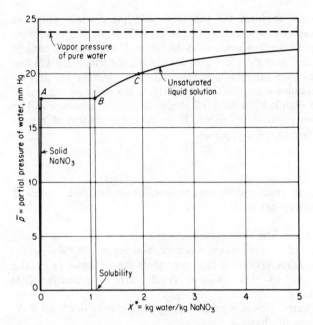

Figure 12.3 Equilibrium moisture content of sodium nitrate at 25°C.

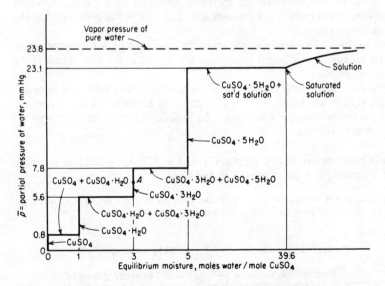

Figure 12.4 Equilibrium moisture of copper sulfate at 25°C (not to scale).

take on sufficient water to form the trihydrate and the crystals will have negligible adsorbed water other than the water of crystallization. The conditions will correspond to a point such as point A on the figure. If the moisture content of the air is then reduced to slightly less than 5.6 mmHg, the trihydrate will lose moisture (*effloresce*) to form the monohydrate, while at 5.6 mmHg any proportion of mono- and trihydrate can coexist. Similarly, if the moisture content of the air is increased to slightly more than 7.8 mmHg, additional moisture will be adsorbed until the pentahydrate is formed. If the moisture content of the air exceeds 23.1 mmHg, the salt will deliquesce.

Definitions

For convenient reference, certain terms used to describe the moisture content of substances are summarized below:

Moisture content, wet basis. The moisture content of a solid or solution is usually described in terms of weight percent moisture, and unless otherwise qualified this is ordinarily understood to be expressed on the wet basis, i.e., as (kg moisture/kg wet solid)100 = [kg moisture/(kg dry solid + kg moisture)]100 = $100X/(1 + X)$.

Moisture content, dry basis. This is expressed as kg moisture/kg dry solid = X. Percentage moisture, dry basis = $100X$.

Equilibrium moisture X^*. This is the moisture content of a substance when at equilibrium with a given partial pressure of the vapor.

Bound moisture. This refers to the moisture contained by a substance which exerts an equilibrium vapor pressure less than that of the pure liquid at the same temperature.

Unbound moisture. This refers to the moisture contained by a substance which exerts an equilibrium vapor pressure equal to that of the pure liquid at the same temperature.

Free moisture. Free moisture is that moisture contained by a substance in excess of the equilibrium moisture: $X - X^*$. Only free moisture can be evaporated, and the free-moisture content of a solid depends upon the vapor concentration in the gas.

These relations are shown graphically in Fig. 12.5 for a solid of moisture content X exposed to a gas of relative humidity A.

Illustration 12.1 A wet solid is to be dried from 80 to 5% moisture, wet basis. Compute the moisture to be evaporated, per 1000 kg of dried product.

SOLUTION

$$\text{Initial moisture content} = \frac{0.80}{1 - 0.80} = 4.00 \text{ kg water/kg dry solid}$$

$$\text{Final moisture content} = \frac{0.05}{1 - 0.05} = 0.0527 \text{ kg water/kg dry solid}$$

$$\text{Dry solid in product} = 1000(0.95) = 950 \text{ kg}$$

$$\text{Moisture to be evaporated} = 950(4 - 0.0527) = 3750 \text{ kg}$$

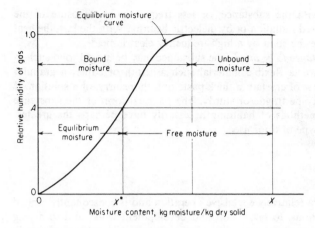

Figure 12.5 Types of moisture.

DRYING OPERATIONS

Drying operations can be broadly classified according to whether they are batch or continuous. These terms are applied specifically from the point of view of the substance being dried. Thus the operation termed *batch drying* is usually in fact a semibatch process wherein a quantity of the substance to be dried is exposed to a continuously flowing stream of air into which the moisture evaporates. In continuous operations, the substance to be dried as well as the gas passes continually through the equipment. No typically stagewise methods are ordinarily used, and all operations involve continuous contact of the gas and the drying substance.

Equipment used for drying can be classified according to equipment type [34] and the nature of the drying process [10]. The following classification is useful for purposes of outlining theories of drying and methods of design. Excellent discussions of equipment and theories of drying are also provided in Refs. 22, 40, and 50.

1. *Method of operation, i.e., batch or continuous.* Batch, or semibatch, equipment is operated intermittently or cyclically under unsteady-state conditions: the drier is charged with the substance, which remains in the equipment until dry, whereupon the drier is emptied and recharged with a fresh batch. Continuous driers are usually operated in steady-state fashion.

2. *Method of supplying the heat necessary for evaporation of the moisture.* In *direct* driers, the heat is supplied entirely by direct contact of the substance with the hot gas into which evaporation takes place. In *indirect* driers, the heat is supplied quite independently of the gas used to carry away the vaporized moisture. For example, heat may be supplied by conduction through a metal

wall in contact with the substance, or less frequently by exposure of the substance to infrared radiation or by dielectric heating. With the last, the heat is generated inside the solid by a high-frequency electric field.

3. *Nature of the substance to be dried.* The substance may be a rigid solid such as wood or fiberboard, a flexible material such as cloth or paper, a granular solid such as a mass of crystals, a thick paste or a thin slurry, or a solution. If it is a solid, it may be fragile or sturdy. The physical form of the substance and the diverse methods of handling necessarily have perhaps the greatest influence on the type of drier used.

BATCH DRYING

Drying in batches is a relatively expensive operation and is consequently limited to small-scale operations, to pilot-plant and development work, and to drying valuable materials whose total cost will be little influenced by added expense in the drying operation.

Direct Driers

The construction of such driers depends greatly upon the nature of the substance being dried. *Tray driers*, also called cabinet, compartment, or shelf driers, are used for drying solids which must be supported on trays, e.g., pasty materials such as wet filter cakes from filter presses, lumpy solids which must be spread upon trays, and similar materials. A typical device, shown schematically in Fig. 12.6, consists of a cabinet containing removable trays on which the solid to be

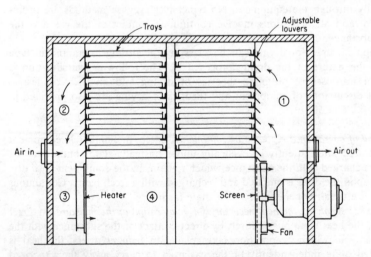

Figure 12.6 Typical tray drier. *(Proctor and Schwartz, Inc.)*

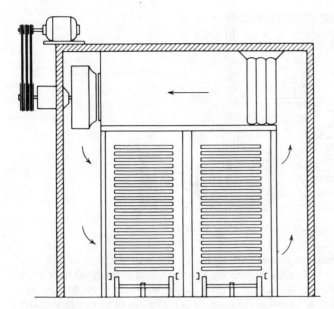

Figure 12.7 Two-truck drier. *(Proctor and Schwartz, Inc.)*

dried is spread. After loading, the cabinet is closed, and steam-heated air is blown across and between the trays to evaporate the moisture (cross-circulation drying). Inert gas, even superheated steam [7, 24] (which has the advantage of a large heat capacity) rather than air can be used if the liquid to be evaporated is combustible or if oxygen is damaging to the solid. When the solid has reached the desired degree of dryness, the cabinet is opened and the trays replaced with a new batch. Figure 12.7 shows a simple modification, a *truck drier*, where the trays are racked upon trucks which can be rolled into and out of the cabinet. Since the trucks can be loaded and unloaded outside the drier, considerable time can be saved between drying cycles. Other obvious modifications of the design are also used, depending upon the nature of the drying substance. Thus, skeins of fibers such as rayon can be hung from poles, and wood or boardlike materials can be stacked in piles, the layers separated from each other by spacer blocks.

With granular materials, the solid can be arranged in thin beds supported on screens so that air or other gas can be passed through the beds. This results in very much more rapid drying. A typical device for this purpose, a batch *through-circulation drier*, is shown schematically in Fig. 12.8. Crystalline solids and materials which are naturally granular such as silica gel can be dried in this manner directly. With others, some sort of preliminary treatment to put them into satisfactory form, *preforming*, is necessary. Pastes, e.g., those resulting from precipitation of pigments or other solids, can be preformed by (1) extrusion into short, spaghettilike rods, (2) granulation, i.e., forcing through screens, or (3) briquetting [31].

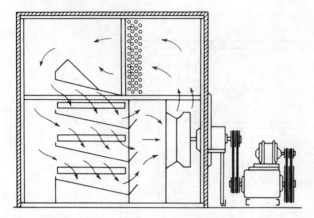

Figure 12.8 Through-circulation drier. *(Proctor and Schwartz, Inc.)*

One of the most important difficulties in the use of driers of the type described is the nonuniformity of moisture content found in the finished product taken from various parts of the drier. This is largely the result of inadequate and nonuniform air movement inside the drier. It is important to eliminate stagnant air pockets and to maintain reasonably uniform air humidity and temperature throughout the drier. In order to do this, large volumes of air must be blown over the trays, if possible at velocities ranging up to 3 or 4 m/s (10 or 20 ft/s) if the solid will not blow from the trays at these air rates. This can be accomplished by blowing large quantities of heated fresh air only once through the drier, but the loss of heat in the discharged air will then usually be prohibitive in cost. Instead, it is the practice to admit only relatively small quantities of fresh air and to recirculate the bulk of it, sometimes as much as 80 to 95 percent [21, 54]. This can be done inside the drier, as shown, for example, in Fig. 12.6, with dampers in the inlet and outlet pipes to regulate the extent of recirculation. The louvers at each tray level can then be adjusted to ensure as nearly uniform air velocity over each tray as possible. Alternatively, the heaters and fans can be installed outside the drier, with ductwork and dampers to permit more careful control of the relative amounts of fresh and recirculated air admitted to the drier itself. It is important also that the trays in such driers be filled level to the brim but not overloaded, so that uniform free space for air movement is available between trays.

The recirculation of large quantities of air necessarily raises the humidity of the air in the drier considerably above that of the fresh air. Low percentage humidity and consequently reasonably rapid drying rates are then obtained by using as high a temperature as practicable. The drier must then be thoroughly insulated, not only to conserve heat but also to maintain the inside walls at temperatures above the dew point of the air to prevent condensation of moisture upon the walls. Specially conditioned, low-humidity air is not used except where low-temperature drying is necessary to avoid damage to the product.

Illustration 12.2 The drier of Fig. 12.6 contains trays arranged in a tier of 10, each on racks 100 mm apart, each tray 38 mm deep and 1 m wide, with 14 m^2 of drying surface. The air entering the trays (position 1 in the figure) is to have a dry-bulb temperature of 95°C and humidity 0.05 kg water/kg dry air. Atmospheric air enters at 25°C, humidity 0.01. The air velocity over the trays at the entrance is to be 3 m/s. When the solid being dried is losing water at a constant rate of 7.5 × 10^{-3} kg/s, determine the percentage recirculation of air and the conditions of the air in the various parts of the drier.

SOLUTION At position 1, Y_1 = 0.05 kg water/ kg dry air, t_{G1} = 95°C, and the humid volume (Table 7.1) is [0.00283 + 0.00456(0.05)](95 + 273) = 1.125 m^3/kg dry air.

Free area for flow between trays = 1[(100 − 38)/1000]11 = 0.682 m^2.

Rate of air flow to trays = 3(0.682) = 2.046 m^3/s, or 2.046/1.125 = 1.819 kg dry air/s (at position 1).

The rate of evaporation is 0.0075 kg water/s, and the humidity at 2 (Fig. 12.6) is therefore 0.05 + 0.0075/1.819 = 0.0541 kg water/kg dry air. Assuming adiabatic drying, the temperature at 2 can be found on the adiabatic-saturation line (Fig. 7.5) drawn through the conditions at 1, and at Y_2 = 0.0541, t_{G2} = 86°C.

The condition of the air at 4, and the discharged air, must be the same as at 1. An overall water balance about the drier therefore is

$$G(0.05 - 0.01) = 0.0075 \text{ kg water evaporated/s}$$

$$G = 0.1875 \text{ kg dry air/s enters and leaves}$$

The rate of airflow at 3 and 4 (Fig. 12.6) is therefore 1.819 + 0.1875 = 2.01 kg dry air/s, and at position 4 the humid volume must be 1.125 m^3/kg dry air. The volumetric rate through the fan is therefore 2.01/1.125 = 1.787 m^3/s. The percentage of air recycled is (1.819/2.01)100 = 90.5%.

The enthalpy of air at 2 (and at 1) is 233 kJ/kg dry air (Fig. 7.5, saturated enthalpy at the adiabatic-saturation temperature, 46.5°C), and that of the fresh air is 50 kJ/kg dry air. Assuming complete mixing, the enthalpy of the air at 3 is, by an enthalpy balance, [233(1.819) + 50(0.1875)]/2.01 = 215.5 kJ/kg dry air. Since its humidity is 0.05, its dry-bulb temperature (Table 7.1) is [215 500 − 2 502 300(0.05)]/[1005 + 1884(0.05)] = 82.2°C. The heater must apply 2.01(233 − 215.5) = 35.2 kW, neglecting heat losses.

The dew-point temperature of the air (Fig. 7.5) at 1, 3, and 4 is 40.4°C, and at 2 it is 41.8°C. The drier should be well enough insulated to ensure that the inside surface will not fall to a temperature of 41.8°C.

The general humidity level in the drier can be altered during the drying cycle. This may be especially important in the case of certain solids which warp, shrink, develop surface cracks, or case harden when dried too rapidly. A cake of soap of high moisture content, for example, if exposed to very dry, hot air, will lose moisture by evaporation from the surface so fast that water will not move rapidly enough from the center of the cake to the surface to keep pace with the evaporation. The surface then becomes hard and impervious to moisture (case hardened), and drying stops even though the average water content of the cake is still very high. With other solids, e.g., wood, shrinkage at the surface may cause cracks or warping. Such substances should be dried slowly at first with air of high humidity, and drier air should be used only after the bulk of the water has been removed.

Driers of the type described are relatively cheap to build and have low maintenance costs. They are expensive to operate, however, owing to low heat economy and high labor costs. Each time the drier is opened for unloading and loading, the temperature of the interior falls and all the metal parts of the drier must be heated again to the operating temperature when operation is resumed. Steam consumption for heating the air will generally not be less than 2.5 kg steam/kg water evaporated and may be as high as 10, especially when the moisture content of the product is reduced to very low levels [25]. The labor requirement for loading, unloading, and supervision of the drying cycle is high.

Indirect Driers

Vacuum shelf driers are tray driers whose cabinets, made of cast-iron or steel plates, are fitted with tightly closing doors so that they can be operated at subatmospheric pressure. No air is blown or recirculated through such driers. The trays containing the solid to be dried rest upon hollow shelves through which warm water or steam is passed to provide the necessary heat for vaporization of moisture. The heat is conducted to the solid through the metal of the shelves and trays. After loading and sealing, the air in the drier is evacuated by a mechanical vacuum pump or steam-jet ejector, and distillation of the moisture proceeds. The vapors usually pass to a condenser, where they are liquefied and collected, and only noncondensable gas is removed by the pump. *Agitated pan driers* [52], which can be used to dry pastes or slurries in small batches, are shallow, circular pans, 1 to 2 m in diameter and 0.3 to 0.6 m deep, with flat bottoms and vertical sides. The pans are jacketed for admission of steam or hot water for heating. The paste, or slurry, in the pan is stirred and scraped by a set of rotating plows, in order to expose new material to the heated surface. Moisture is evaporated into the atmosphere in atmosphere pan driers, or the pan may be covered and operated under vacuum. *Vacuum rotary driers* are steam-jacketed cylindrical shells, arranged horizontally, in which a slurry, or paste, can be dried in vacuum. The slurry is stirred by a set of rotating agitator blades attached to a central horizontal shaft which passes through the ends of the cylindrical shell. Vaporized moisture passes through an opening in the top to a condenser, and noncondensable gas is removed by a vacuum pump. The dried solid is discharged through a door in the bottom of the drier.

Driers of this category are expensive to build and to operate. Consequently they are used only for valuable materials which must be dried at low temperatures or in the absence of air to prevent damage, such as certain pharmaceutical products, or where the moisture to be removed is an expensive or poisonous organic solvent which must be recovered more or less completely.

Freeze drying (sublimation drying) Substances which cannot be heated even to moderate temperatures, such as foodstuffs and certain pharmaceuticals, can be dried by this method [27]. The substance to be dried is customarily frozen by exposure to very cold air and placed in a vacuum chamber, where the moisture sublimes and is pumped off by steam-jet ejectors or mechanical vacuum pumps. An alternative method of freezing is by flash vaporization of part of the moisture under vacuum, although foodstuffs which are not rigid in the unfrozen state may be damaged by this procedure. Some foods, e.g., beef, evidently contain capillary channels, and the water vapor diffuses from the receding ice surface through these channels as drying proceeds [18]. In other cases diffusion through cell walls must occur. In any event, one of the major problems is to supply the heat necessary for sublimation: as the plane of sublimation recedes, heat must be driven through larger thicknesses of dried matter of poor thermal conductivity, requiring increasing temperature differences, which may damage the product. Alternatively, the heat can be introduced through the "back face," i.e., through the surface such as a shelf or tray on which the solid rests, opposite the surface through which the evaporated moisture leaves. This very substantially improves the rate of drying [11]. Radiant heat is sometimes used. Dielectric heat is a possibility although an expensive one: because of the high dielectric constant of water, the heat is liberated directly into the water. Still an additional method, useful for granular products, is through-circulation drying with air instead of pumping off the water by vacuum pump.

The Rate of Batch Drying

In order to set up drying schedules and to determine the size of equipment, it is necessary to know the time required to dry a substance from one moisture content to another under specified conditions. We shall also wish to estimate the influence that different drying conditions will have upon the time for drying. Our knowledge of the mechanism of drying is so incomplete that it is necessary with few exceptions to rely upon at least some experimental measurements for these purposes. Measurements of the rate of batch drying are relatively simple to make and provide much information not only for batch but also for continuous operation.

Drying tests The rate of drying can be determined for a sample of a substance by suspending it in a cabinet or duct, in a stream of air, from a balance. The weight of the drying sample can then be measured as a function of time. Certain precautions must be observed if the data are to be of maximum utility. The sample should not be too small. Further, the following conditions should resemble as closely as possible those expected to prevail in the contemplated large-scale operation: (1) the sample should be similarly supported in a tray or frame; (2) it should have the same ratio of drying to nondrying surface; (3) it should be subjected to similar conditions of radiant-heat transfer; and (4) the air should have the same temperature, humidity, and velocity (both speed and direction with respect to the sample). If possible, several tests should be made on samples of different thicknesses. The dry weight of the sample should also be obtained.

The exposure of the sample to air of constant temperature, humidity, and velocity constitutes drying under *constant drying conditions*.

Rate-of-drying curve From the data obtained during such a test, a curve of moisture content as a function of time (Fig. 12.9) can be plotted. This will be useful directly in determining the time required for drying larger batches under

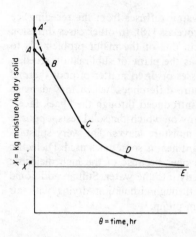

$\theta = \text{time, hr}$

Figure 12.9 Batch drying, constant drying conditions.

the same drying conditions. Much information can be obtained if the data are converted into rates (or fluxes) of drying, expressed as N mass/(area)(time), and plotted against moisture content, as in Fig. 12.10. This can be done by measuring the slopes of tangents drawn to the curve of Fig. 12.9 or by determining from the curve small changes in moisture content ΔX for corresponding small changes in time $\Delta \theta$ and calculating the rate as $N = - S_S \, \Delta X / A \, \Delta \theta$. Here S_S is the mass

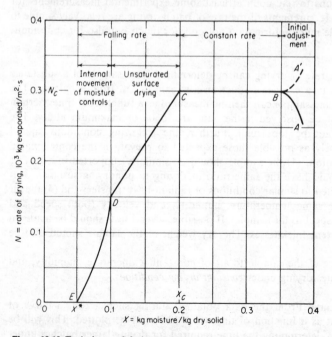

Figure 12.10 Typical rate-of-drying curve, constant drying conditions.

of dry solid, and A is the wet surface over which the gas blows and through which evaporation takes place in the case of cross-air circulation drying. In the case of through-circulation drying, A is the cross section of the bed measured at right angles to the direction of gas flow.

The rate-of-drying curve is sometimes plotted with the ordinate expressed as mass moisture evaporated/(mass dry solid)(time), which in the present notation is $-dX/d\theta$.

There are usually two major parts to the rate curve of Fig. 12.10, a period of constant rate and one of falling rate, as marked on the figure. While different solids and different conditions of drying often give rise to curves of very different shape in the falling-rate period, the curve shown occurs frequently. Some of the differences which may arise will be considered later, but for the present let us briefly review the reasons generally advanced for the various parts of the curve shown [16, 34, 48, 49].

If a solid is initially very wet, the surface will be covered with a thin film of liquid, which we shall assume is entirely unbound moisture. When it is exposed to relatively dry air, evaporation will take place from the surface. The rate at which moisture evaporates can be described in terms of a gas mass-transfer coefficient k_Y and the difference in humidity of the gas at the liquid surface Y_s and in the main stream Y. Thus, for cross-circulation drying

$$N_c = k_Y(Y_s - Y) \tag{12.1}$$

The coefficient k_Y can be expected to remain constant as long as the speed and direction of gas flow past the surface do not change. The humidity Y_s is the saturated humidity at the liquid-surface temperature t_s and will therefore depend upon this temperature. Since evaporation of moisture absorbs latent heat, the liquid surface will come to, and remain at, an equilibrium temperature such that the rate of heat flow from the surroundings to the surface exactly equals the rate of heat absorption. Y_s therefore remains constant. The capillaries and interstices of the solid, filled with liquid, can deliver liquid to the surface as rapidly as it evaporates there. Since in addition Y remains unchanged under constant drying conditions, the rate of evaporation must remain constant at the value N_c, as shown in Figs. 12.9 and 12.10 between points B and C. In the beginning, the solid and the liquid surface are usually colder than the ultimate surface temperature t_s, and the evaporation rate will increase while the surface temperature rises to its ultimate value during the period AB on these curves. Alternatively the equilibrium temperature t_s may be lower than the initial value, which will give rise to a curve $A'B$ while the initial adjustment occurs. The initial period is usually so short that it is ordinarily ignored in subsequent analysis of the drying times.

When the average moisture content of the solid has reached a value X_c, the *critical moisture content* (Fig. 12.10), the surface film of moisture has been so reduced by evaporation that further drying causes dry spots to appear upon the surface; these spots occupy increasingly larger proportions of the exposed surface as drying proceeds. Since, however, the rate N is computed by means of the constant gross surface A, the value of N must fall even though the rate per unit of wet surface remains constant. This gives rise to the first part of the falling-rate period, the period of *unsaturated surface drying*, from points C to D (Figs. 12.9 and 12.10). Ultimately the original surface film of liquid will have entirely evaporated at an average moisture content for the solid corresponding to point D. This part of the curve may be missing entirely, or it may constitute the whole of the falling-rate period. With some textiles, other explanations for the linear falling-rate period have been necessary [39].

On further drying, the rate at which moisture can move through the solid, as a result of concentration gradients existing between the deeper parts and the surface, is the controlling step. As the moisture concentration generally is lowered by the drying, the rate of internal movement of moisture decreases. In some cases, evaporation may take place beneath the surface of the solid in a plane or zone which retreats deeper into the solid as drying proceeds. In any event, the rate of drying falls even more rapidly than before, as from D to E (Fig. 12.10). At point E, the moisture content of the solid has fallen to the equilibrium value X^* for the prevailing air humidity, and drying stops. The moisture distribution within the solid during the falling-rate period has been calculated and displayed graphically [17].

Time of drying If one wishes to determine the time of drying a solid under the same conditions for which a drying curve such as Fig. 12.9 has been completely determined, one need merely read the difference in the times corresponding to the initial and final moisture contents from the curve.

Within limits, it is sometimes possible to estimate the appearance of a rate-of-drying curve such as Fig. 12.10 for conditions different from those used in the experiments. In order to determine the time for drying for such a curve, we proceed as follows. The rate of drying is, by defintion,

$$N = \frac{-S_S \, dX}{A \, d\theta} \tag{12.2}$$

Rearranging and integrating over the time interval while the moisture content changes from its initial value X_1 to its final value X_2 gives

$$\theta = \int_0^\theta d\theta = \frac{S_S}{A} \int_{X_2}^{X_1} \frac{dX}{N} \tag{12.3}$$

1. *The constant-rate period.* If the drying takes place entirely within the constant-rate period, so that X_1 and $X_2 > X_c$ and $N = N_c$, Eq. (12.3) becomes

$$\theta = \frac{S_S(X_1 - X_2)}{AN_c} \tag{12.4}$$

2. *The falling-rate period.* If X_1 and X_2 are both less than X_c, so that drying occurs under conditions of changing N, we proceed as follows:

   a. *General case.* For any shape of falling-rate curve whatsoever, Eq. (12.3) can be integrated graphically by determining the area under a curve of

$1/N$ as ordinate, X as abscissa, the data for which can be obtained from the rate-of-drying curve.

b. *Special case.* N is linear in X, as in the region BC of Fig. 12.10. In this case,

$$N = mX + b \qquad (12.5)$$

where m is the slope of the linear portion of the curve and b is a constant. Substitution in Eq. (12.3) provides

$$\theta = \frac{S_S}{A} \int_{X_2}^{X_1} \frac{dX}{mX + b} = \frac{S_S}{mA} \ln \frac{mX_1 + b}{mX_2 + b} \qquad (12.6)$$

But since $N_1 = mX_1 + b$, $N_2 = mX_2 + b$, and $m = (N_1 - N_2)/(X_1 - X_2)$, Eq. (12.6) becomes

$$\theta = \frac{S_S(X_1 - X_2)}{A(N_1 - N_2)} \ln \frac{N_1}{N_2} = \frac{S_S(X_1 - X_2)}{AN_m} \qquad (12.7)$$

where N_m is the logarithmic average of the rate N_1, at moisture content X_1, and N_2 at X_2.

Frequently the entire falling-rate curve can be taken as a straight line between points C and E (Fig. 12.10). It is often assumed to be so for lack of more detailed data. In this case

$$N = m(X - X^*) = \frac{N_c(X - X^*)}{X_c - X^*} \qquad (12.8)$$

and Eq. (12.7) becomes

$$\theta = \frac{S_S(X_c - X^*)}{N_c A} \ln \frac{X_1 - X^*}{X_2 - X^*} \qquad (12.9)$$

In any particular drying problem, either or both constant- and falling-rate periods may be involved, depending upon the relative values of X_1, X_2, and X_c. The appropriate equations and limits must then be chosen.

Illustration 12.3 A batch of the solid for which Fig. 12.10 is the drying curve is to be dried from 25 to 6% moisture under conditions identical to those for which the figure applies. The initial weight of the wet solid is 160 kg, and the drying surface is 1 $m^2/40$ kg dry weight. Determine the time for drying.

SOLUTION The total weight of the batch is unimportant. $S_S/A = 40$. At 25% moisture, $X_1 = 0.25(1 - 0.25) = 0.333$ kg moisture/kg dry solid. At 6% moisture, $X_2 = 0.06/(1 - 0.06) = 0.064$ kg moisture/kg dry solid. Inspection of Fig. 12.10 shows that both constant- and falling-rate periods are involved. The limits of moisture content in the equations for the different periods will be chosen accordingly.

Constant-rate period This is from $X_1 = 0.333$ to $X_c = 0.200$. $N_c = 0.30 \times 10^{-3}$. Eq. (12.4):

$$\theta = \frac{S_S(X_1 - X_c)}{AN_c} = \frac{40(0.333 - 0.200)}{1(0.30 \times 10^{-3})} = 17\,730 \text{ s}$$

Falling-rate period This is from $X_c = 0.200$ to $X_2 = 0.064$. Use Eq. (12.3). The following table is prepared from data of Fig. 12.10:

x	0.20	0.18	0.16	0.14	0.12	0.10	0.09	0.08	0.07	0.064
$10^3 N$	0.300	0.266	0.239	0.208	0.180	0.150	0.097	0.070	0.043	0.025
$\dfrac{1}{N} \times 10^{-3}$	3.33	3.76	4.18	4.80	5.55	6.67	10.3	14.3	23.3	40.0

A curve, not shown, is prepared of $1/N$ as ordinate, X as abscissa, and the area under the curve between $X = 0.20$ and 0.064 is 1060. Eq. (12.3):

$$\theta = \frac{40}{1}(1060) = 42\,400 \text{ s}$$

The total drying time is therefore $17\,730 + 42\,400 = 60\,130$ s $= 16.7$ h.

Alternatively, since the drying curve is straight from $X = 0.20$ to 0.10, Eq. (12.7) can be used in this range of moisture content,

$$\theta = \frac{S_S(X_c - X_D)}{A(N_c - N_D)} \ln \frac{N_c}{N_D} = \frac{40(0.20 - 0.10)}{1(0.30 - 0.15 \times 10^{-3})} \ln \frac{0.30 \times 10^{-3}}{0.15 \times 10^{-3}} = 18\,500 \text{ s}$$

Graphical integration in the range $X = 0.1$ to 0.064, through Eq. (12.3), provides an additional $23\,940$ s, so that the falling-rate time is $18\,500 + 23\,940 = 42\,440$ s.

As an approximation, the falling rate can be represented by a straight line from C to E (Fig. 12.10). The corresponding falling-rate time is, by Eq. (12.9),

$$\theta = \frac{S_S(X_c - X^*)}{N_c A} \ln \frac{X_c - X^*}{X_2 - X^*} = \frac{40(0.20 - 0.05)}{(0.30 \times 10^{-3})(1)} \ln \frac{0.20 - 0.05}{0.064 - 0.05} = 47\,430 \text{ s}$$

THE MECHANISMS OF BATCH DRYING

We now consider the various portions of the rate-of-drying curve in more detail. Our present knowledge permits us to describe the drying process in the constant-rate period reasonably well, but our understanding of the falling-rate periods is very limited.

Cross-Circulation Drying

The constant-rate period In this period, where surface evaporation of unbound moisture occurs, it has been shown that the rate of drying is established by a balance of the heat requirements for evaporation and the rate at which heat reaches the surface. Consider the section of a material drying in a stream of gas as shown in Fig. 12.11. The solid of thickness z_S is placed on a tray of thickness z_M. The whole is immersed in a stream of drying gas at temperature T_G and humidity Y mass moisture/mass dry gas, flowing at a mass velocity G mass/ (time) (area). The evaporation of moisture takes place from the upper surface, area A, which is at a temperature T_s. The drying surface receives heat from several sources: (1) q_c by convection from the gas stream; (2) q_k by conduction through the solid; (3) q_R by direct radiation from a hot surface at temperature T_R, as shown, all expressed as a flux, energy/(area of solid for heat transfer)

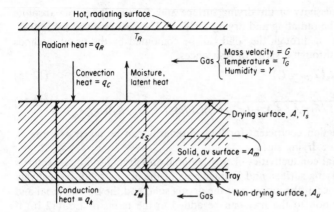

Figure 12.11 Constant-rate drying.

(time). In accordance with the mechanism discussed earlier, the heat arriving at the surface by these methods is removed by the evaporating moisture, so that the surface temperature remains constant at T_s. The entire mechanism resembles the wet-bulb-thermometer process, complicated by the additional source of heat.

The rate of evaporation and the surface temperature can then be obtained by a heat balance [46]. If q represents the total heat arriving at the surface, then

$$q = q_c + q_R + q_k \tag{12.10}$$

If we neglect the heat required to superheat the evaporated moisture to the gas temperature and consider only the latent heat of vaporization λ_s, the flux of evaporation N_c and the flux of heat flow are related,

$$N_c \lambda_s = q \tag{12.11}$$

The heat received at the surface by convection is controlled by the appropriate convection heat-transfer coefficient h_c,

$$q_c = h_c(T_G - T_s) \tag{12.12}$$

The heat received by radiation can be estimated by the usual means [50, 55] and can also be expressed as a heat-transfer coefficient h_R,†

$$q_R = \varepsilon(5.729 \times 10^{-8})(T_R^4 - T_s^4) = h_R(T_R - T_s) \tag{12.13}$$

$$h_R = \frac{\varepsilon(5.729 \times 10^{-8})(T_R^4 - T_s^4)}{T_R - T_s} \tag{12.14}$$

---

† Equations (12.13) and (12.14) are written for SI units: q_R in W/m^2, T in kelvins, h_R in $W/m^2 \cdot K$. For q_R in $Btu/ft^2 \cdot h$, T in degrees Rankine, and h_R in $Btu/ft^2 \cdot h \cdot °F$ the constant is 1.730×10^{-9}. *Note:* The theoretical value of the Stefan-Boltzmann constant is 5.6693×10^{-8} $W/m^2 \cdot K^4$, but the accepted experimental value [50] is 5.729×10^{-8}.

where ε is the emissivity of the drying surface and T_R and T_s are the absolute temperatures of the radiating and drying surfaces. The heat received by convection and conduction through the solid can be computed by the usual methods for heat transfer through a series of resistances,

$$q_k = U_k(T_G - T_s) \qquad (12.15)$$

$$U_k = \frac{1}{(1/h_c)(A/A_u) + (z_M/k_M)(A/A_u) + (z_S/k_S)(A/A_m)} \qquad (12.16)$$

where h_c = convection coefficient for tray; ordinarily can be taken as same as
      that for drying surface
  k_M, k_s = thermal conductivities of tray material and drying solid, respectively
  A_u, A_m = nondrying surface and average area of the drying solid, respectively
A thermal resistance at the junction of the drying solid and the tray material and an effect of radiation to the tray can be added to the terms of Eq. (12.16) if desired.

Combining Eqs. (12.1) and (12.10) to (12.15) permits calculation of the rate of drying,

$$N_c = \frac{q}{\lambda_s} = \frac{(h_c + U_k)(T_G - T_s) + h_R(T_R - T_s)}{\lambda_s} = k_Y(Y_s - Y) \qquad (12.17)$$

The surface temperature must be known in order to use the relationship. It can be obtained by consideration of the left-hand portions of Eq. (12.17), which can be rearranged to read

$$\frac{(Y_s - Y)\lambda_s}{h_c/k_Y} = \left(1 + \frac{U_k}{h_c}\right)(T_G - T_s) + \frac{h_R}{h_c}(T_R - T_s) \qquad (12.18)$$

The ratio h_c/k_Y, applicable to flow of gases past wet-bulb thermometers [Eqs. (7.27) and (7.28)] can be used for present purposes, and for the system air–water vapor this ratio was shown to be substantially the same as the humid heat of the gas C_s. Since Y_s is the saturated humidity of the gas stream corresponding to T_s when unbound moisture is being evaporated, both these quantities can be found by solving Eq. (12.18) simultaneously with the saturated-humidity curve on a psychrometric chart.

If conduction through the solid and radiation effects are absent, Eq. (12.18) reduces to that for the wet-bulb thermometer [Eq. (7.26)] and the surface temperature is the wet-bulb temperature of the gas. The drying surfaces will also be at the wet-bulb temperature if the solid is dried from all surfaces in the absence of radiation. When pans or trays of drying material are placed one above another, as in Figs. 12.6 and 12.7, most of the solid will receive radiation only from the bottom of the pan immediately above it, and unless gas temperatures are very high, this is not likely to be very important. It is essential therefore not to overemphasize the heat received by radiation in conducting drying tests on single pairs of trays.

For flow of gas parallel to a surface and confined between parallel plates, as between the trays of a tray drier, the transfer coefficients h_c and k_Y are described generally by item 3, Table 3.3. Application of the heat-mass-transfer analogy, for $Re_e = 2600$ to $22\,000$, results in

$$j_H = \frac{h_c}{C_p G} Pr^{2/3} = j_D = \frac{k_Y}{G_S} Sc^{2/3} = 0.11\, Re_e^{-0.29} \qquad (12.19)$$

where $Re_e = d_e G/\mu$ and d_e is the equivalent diameter of the airflow space. *With the properties of air at 95° C*, this becomes, in the SI†

$$h_c = 5.90 \frac{G^{0.71}}{d_e^{0.29}} \qquad (12.20)$$

whereas in a detailed study of drying of sand in trays [46], h_c was given (SI) by

$$h_c = 14.3 G^{0.8} \qquad (12.20a)$$

Such discrepancies can be assigned to a number of causes, e.g., different "calming" length ahead of the drying surface [41] or the shape of the leading edge of the drying surface [7]. In the absence of more specific information, Eqs. (12.19) and (12.20) are recommended.

Airflow perpendicular to the surface [37], *for G = 1.08 to 5.04 kg/m² · s (0.9 to 4.5 m/s)*, in the SI,‡

$$h_c = 24.2 G^{0.37} \qquad (12.21)$$

The relationships developed in Eqs. (12.17) to (12.21) permit direct estimates of the rate of drying during the constant-rate period, but they should not be considered as complete substitutes for experimental measurements. Perhaps their greatest value is in conjunction with limited experimental data in order to predict the effect of changing the drying conditions.

Effect of gas velocity. If radiation and conduction through the solid are negligible, N_c is proportional to $G^{0.71}$ for parallel flow of gas and to $G^{0.37}$ for perpendicular flow. If radiation and conduction are present, the effect of gas rate will be less important.

Effect of gas temperature. Increased air temperature increases the quantity $T_G - T_s$ and hence increases N_c. In the absence of radiation effects, if the variation of λ over moderate temperature ranges is neglected, N_c is directly proportional to $T_G - T_s$.

Effect of gas humidity. N_c varies directly as $Y_s - Y$, and consequently increasing the humidity lowers the rate of drying. Usually, changes in Y and T_G involve simultaneous changes in T_s and Y_s, and the effects are best estimated by direct application of Eq. (12.17).

Effect of thickness of drying solid. If heat conduction through the solid occurs, Eqs. (12.15) and (12.16) indicate lowered values of N_c with increased solid

---

† For G in lb/ft² · h, d_e in feet, and h_c in Btu/ft² · h · °F the coefficient of Eq. (12.20) is 0.0135.

‡ For $G = 800$ to 4000 lb/ft² · h (3 to 15 ft/s) and h_c in Btu/ft² · h ·°F the coefficient of Eq. (12.21) becomes 0.37.

thickness. However, conduction of heat through edge surfaces of pans and trays may be an important source of heat which can result in increased rate of drying if the edge surface is large. If nondrying surfaces are heat-insulated, or if drying occurs from all surfaces of the solid, N_c is independent of thickness. The *time* for drying between fixed moisture contents within the constant-rate period will then be directly proportional to thickness.

Illustration 12.4 An insoluble crystalline solid wet with water is placed in a rectangular pan 0.7 m (2.3 ft) by 0.7 m and 25 mm (1 in) deep, made of 0.8-mm-thick ($\frac{1}{32}$-in) galvanized iron. The pan is placed in an airstream at 65°C, humidity 0.01 kg water/kg dry air, flowing parallel to the upper and lower surface at a velocity of 3 m/s (10 ft/s). The top surface of the solid is in direct sight of steam-heated pipes whose surface temperature is 120°C (248°F), at a distance from the top surface of 100 mm (4 in).

(a) Estimate the rate of drying at constant rate.

(b) Reestimate the rate if the pan is thoroughly heat-insulated and there is no radiation from the steam pipes.

SOLUTION (a) $Y = 0.01$ kg water/kg dry air, $t_G = 65°C$. The humid volume of the air (Table 7.1) is $[0.00283 + 0.00456(0.01)] (65 + 273) = 0.972$ m^3/kg dry air.

$$\rho_G = \text{density of gas} = \frac{1.01}{0.972} = 1.04 \text{ kg/m}^3 \qquad G = 3(1.04) = 3.12 \text{ kg/m}^2 \cdot \text{s}$$

$$\text{Estimated } d_e = \frac{4(\text{cross section for flow})}{\text{perimeter}} = \frac{4(0.7)(0.1)}{(0.7 + 0.1)2} = 0.175 \text{ m}$$

Eq. (12.20):

$$h_c = \frac{5.90(3.12)^{0.71}}{(0.175)^{0.29}} = 21.9 \text{ W/m}^2 \cdot \text{K}$$

Take the emissivity of the solid as $\varepsilon = 0.94$. $T_R = 120 + 273 = 393$ K. Tentatively estimate t_s as 38°C, $T_s = 38 + 273 = 311$ K. Eq. (12.14):

$$h_R = \frac{0.94(5.73 \times 10^{-8})(393^4 - 311^4)}{393 - 311} = 9.51 \text{ W/m}^5 \cdot \text{K}$$

Take $A_m = A = (0.7)^2 = 0.49$ m^2. The area of the sides of the pan $= 4(0.7)(0.025) = 0.07$ m^2 and $A_u = (0.7)^2 + 0.07 = 0.56$ m^2 for bottom and sides (this method of including heat transfer through the sides is admittedly an oversimplification but adequate for present purposes). Thermal conductivities are $k_M = 45$ for the metal of the pan and $k_S = 3.5$ for the wet solid [46], both as W/m · K. The latter value must be carefully chosen, and it may bear no simple relation to the conductivity of either the dry solid or its moisture. $z_S = 0.025$ m, $z_M = 0.0008$ m. Eq. (12.16):

$$\frac{1}{U_k} = \frac{0.49}{21.9(0.56)} + \frac{0.0008}{45} \frac{0.49}{0.56} + \frac{0.025}{3.5} \frac{0.49}{0.49}$$

$$U_k = 21.2 \text{ W/m}^2 \cdot \text{K}$$

The humid heat of the air (Table 7.1) is $C_s = 1005 + 1884(0.01) = 1023.8$, and λ_s at the estimated 38°C is 2411.4 kJ/kg. Eq. (12.18):

$$\frac{(Y_s - 0.01)(2411.4 \times 10^3)}{1023.8} = \left(1 + \frac{21.2}{21.9}\right)(65 - t_s) + \frac{9.51}{21.9}(120 - t_s)$$

This reduces to $Y_s = 0.0864 - 10.194 \times 10^{-4}t_s$, which must be solved simultaneously with the saturated-humidity curve of the psychrometric chart for air–water vapor. The line marked a on

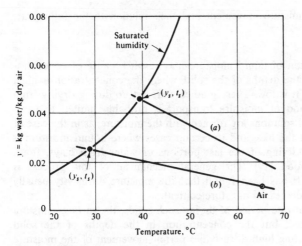

Figure 12.12 Solution to Illustration 12.4.

Fig. 12.12 is the above expression, which intersects the saturated-humidity curve at Y_s = 0.0460, t_s = 39°C, the surface temperature, which is sufficiently close to the 38°C estimated previously to make iteration unnecessary. At 39°C, λ_s = 2409.7 kJ/kg. Eq. (12.17):

$$N_c = \frac{(21.9 + 21.2)(65 - 39) + 9.51(120 - 39)}{2409.7 \times 10^3} = 7.85 \times 10^{-4} \text{ kg water evapd/m}^2 \cdot \text{s}$$

and the evaporation rate is (7.85 × 10^{-4})(0.49) = 3.85 × 10^{-4} kg/s or 3.06 lb/h.

(b) When no radiation or conduction of heat through the solid occurs, the drying surface assumes the wet-bulb temperature of the air. For the system air-water at this humidity, the adiabatic-saturation lines of the psychrometric chart serve as wet-bulb lines, and on Fig. 12.12, line b is the adiabatic-saturation line through the point representing the air (t_G = 65°C, Y = 0.01). The line intersects the saturation-humidity curve at the wet-bulb condition, t_s = 28.5°C, Y_s = 0.025. At this temperature, λ_s = 2435.0 kJ/kg. Eq. (12.17):

$$N_c = \frac{h_c(t_G - t_s)}{\lambda_s} = \frac{21.9(65 - 28.5)}{2435 \times 10^3} = 3.28 \times 10^{-4} \text{ kg/m}^2 \cdot \text{s}$$

and the evaporation rate is (3.28 × 10^{-4})(0.49) = 1.609 × 10^{-4} kg/s or 1.277 lb/h.

When the air suffers a considerable change in temperature and humidity in its passage over the solid, as indicated in Illustration 12.2 for example, the rate of drying at the leading and trailing edge of the solid will differ and this accounts in part for the nonuniform drying frequently obtained in tray driers. This can be counteracted in part by periodic reversal of the airflow.

Movement of moisture within the solid When surface evaporation occurs, there must be a movement of moisture from the depths of the solid to the surface. The nature of the movement influences the drying during the falling-rate periods. In order to appreciate the diverse nature of the falling-rate portions of the drying curve which have been observed, let us review very briefly some of the theories advanced to explain moisture movement and their relation to the falling-rate curves.

Liquid diffusion Diffusion of liquid moisture may result because of concentration gradients between the depths of the solid, where the concentration is high, and the surface, where it is low. These gradients are set up during drying from the surface. This method of moisture transport is probably limited to cases where single-phase solid solutions are formed with the moisture, as in the case of soap, glue, gelatin, and the like, and to certain cases where bound moisture is being dried, as in the drying of the last portions of water from clays, flour, textiles, paper, and wood [19]. The general mechanism of this process is described in Chap. 4. It has been found that the moisture diffusivity usually decreases rapidly with decreased moisture content.

During the constant-rate period of drying such solids, the surface-moisture concentration is reduced, but the concentration in the depths of the solid remains high. The resulting high diffusivities permit movement of the moisture to the surface as fast as it can be evaporated, and the rate remains constant. When dry spots appear because portions of the solid project into the gas film, a period of unsaturated surface evaporation results. The surface eventually dries to the equilibrium-moisture content for the prevailing gas. Further drying occurs at rates which are entirely controlled by the diffusion rates within the solid, since these are slow at low moisture contents. If the initial constant-rate drying is very rapid, the period of unsaturated surface evaporation may not appear, and the diffusion-controlled falling-rate period begins immediately after the constant-rate period is completed [47], as in Fig. 12.13.

For many cases of drying where the diffusion mechanism has satisfactorily explained the rate of drying as a function of average moisture content, the distribution of moisture within the solid at the various stages of drying has not conformed to this mechanism [19, 38]. The superficial applicability of the diffusion mechanism is then apparently accidental.

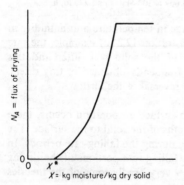

N_A = flux of drying

X = kg moisture/kg dry solid

Figure 12.13 Diffusion-controlled falling rate.

Capillary movement [6, 19, 38, 42, 56] Unbound moisture in granular and porous solids such as clays, sand, paint pigments, and the like, moves through the capillaries and interstices of the solids by a mechanism involving surface tension, the way oil moves through a lamp wick. The capillaries extend from small reservoirs of moisture in the solid to the drying surface. As drying proceeds, at first moisture moves by capillarity to the surface rapidly enough to maintain a uniformly wetted surface and the rate of drying is constant. The water is replaced by air entering the solid through relatively few openings and cracks. The surface moisture is eventually drawn to spaces between the granules of the surface, the wetted area at the surface decreases, and the unsaturated-surface drying period follows. The subsurface reservoirs eventually dry up, the liquid surface recedes into the capillaries, evaporation occurs below the surface in a zone or plane which gradually recedes deeper into the solid, and a second falling-rate period results. During this period, diffusion of vapor within the solid will occur from the place of vaporization to the surface.

With certain pastes dried in pans, the adhesion of the wet cake to the bottom of the pan may not permit ventilation of the subsurface passageways by gas. This can give rise to curves of the sort shown in Fig. 12.14. In this case, the usual constant-rate period prevailed during *a*. When the surface moisture was first depleted, liquid could not be brought to the surface by the tension in the capillaries since no air could enter to replace the liquid, the surface of moisture receded into the capillaries, and the rate fell during *b*. The solid eventually crumpled, admitting air to replace the liquid, whereupon capillary action brought this to the surface and the rate rose again, as at *c*.

Vapor diffusion [42] Especially if heat is supplied to one surface of a solid while drying proceeds from another, the moisture may evaporate beneath the surface and diffuse outward as a vapor. Moisture particles in granular solids which have been isolated from the main portion of the moisture flowing through capillaries may also be evaporated below the surface.

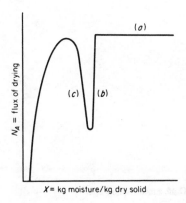

X = kg moisture/kg dry solid

Figure 12.14 Effect of adhesion of drying paste to the pan. [*After Ernst, et al., Ind. Eng. Chem.*, **30** *1119* *(1938).*]

Pressure Shrinkage of the outside layers of a solid on drying may squeeze moisture to the surface.

Usually we can only speculate about which mechanism is appropriate to a particular solid and must rely on more or less empirical treatment of the experimental rates of drying.

Unsaturated-surface drying During such a period the rate of drying N will usually vary linearly with moisture content X. Since the mechanism of evaporation during this period is the same as that in the constant-rate period, the effects of such variables as temperature, humidity, and velocity of the gas and thickness of the solid are the same as for constant-rate drying.

In some cases this period may constitute the whole of the falling-rate drying, giving rise to a curve of the type shown in Fig. 12.15. Equations (12.8) and (12.9) then apply. Combining Eqs. (12.2), (12.8), and (12.17) gives

$$-\frac{dX}{d\theta} = \frac{k_Y A(X - X^*)(Y_s - Y)}{S_S(X_c - X^*)}$$ (12.22)

Noting that $S_S = z_S A\rho_S$, and letting $k_Y = f(G)$, we get

$$-\frac{dX}{d\theta} = \frac{f(G)(X - X^*)(Y_x - Y)}{z_S\rho_S(X_c - X^*)} = \frac{\alpha f(G)(X - X^*)(Y_s - Y)}{z_S}$$ (12.23)

where $\alpha = $ const. This expression is sometimes used as an empirical description of the rates of drying for such cases. Alternatively, when the time θ is defined as that when the moisture content is X, Eq. (12.9) is readily transformed into

$$\ln\frac{X - X^*}{X_1 - X^*} = \frac{-N_c\theta}{\rho_S z_S(X_c - X^*)}$$ (12.24)

which suggests that falling-rate data of this nature will plot as a straight line, line a, on the semilogarithmic coordinates of Fig. 12.16. If drying tests are made

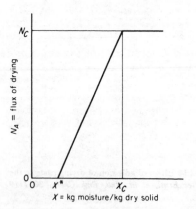

X = kg moisture/kg dry solid

Figure 12.15 Linear falling rate.

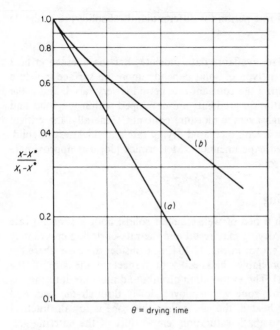

θ = drying time

Figure 12.16 Empirical treatment of falling-rate data: (a) N varies linearly with $\log(X - X^*)$; (b) diffusion-controlled falling rate.

under the same conditions for samples of different thickness, and so that heat is applied to the solid only through the drying surface, the slopes of such lines on this chart should be proportional to $-1/z_S$. The drying time between fixed moisture contents is then directly proportional to the solid thickness.

Internal-diffusion controlling If a period of drying is developed where internal diffusion of moisture controls the rate, it can be expected that variables which influence the gas coefficients of heat or mass transfer will not influence the rate of drying. The drying rates should be independent of gas velocity, and humidity will be of importance only insofar as it controls the equilibrium-moisture concentration. When a semilogarithmic plot (Fig. 12.16) is prepared from drying data of this type, the curve b which results resembles those of Fig. 4.2. For a slab, the curves should be substantially straight at values of the ordinate below 0.6, with slopes proportional to $-1/z_S^2$. The drying time between fixed moisture contents should be proportional to the square of the thickness. Discrepancies may result because the initial moisture distribution is not uniform throughout the solid if a drying period precedes that for which diffusion controls and because the diffusivity varies with moisture content.

Attempts have been made to describe rates of drying in the falling-rate period by means of overall coefficients of heat and mass transfer [35] but these have not been too successful owing to the change in the individual resistances to transfer in the solid during the course of the drying. Various expressions for the time of drying in the falling-rate period for different kinds of solids are available [30].

Critical moisture content The available data indicate that the average critical moisture content for a given type of solid depends upon the surface-moisture concentration. If drying during the constant-rate period is very rapid, and if the solid is thick, steep concentration gradients are developed within the solid and the falling rate begins at high average moisture contents. Generally, the critical moisture content will increase with increased drying rate and thickness of solid. It must usually be measured experimentally. McCormick [34] lists approximate values for many industrial solids.

Through-Circulation Drying

When a gas passes through a bed of wet, granular solids, both a constant-rate and a falling-rate period of drying may result and the rate-of-drying curves may appear very much like that shown in Fig. 12.10 [31]. Consider the case where the bed of solids has an appreciable thickness with respect to the size of the particles, as in Fig. 12.17 [1]. The evaporation of unbound moisture into the gas occurs in a relatively narrow zone which moves slowly through the bed, and unless the bed is internally heated, the gas leaving this zone is for all practical purposes saturated at the adiabatic-saturation temperature of the entering gas. This is also the surface temperature of the wet particles. The rate of drying is constant as long as the zone is entirely within the bed. When the zone first reaches the end of the bed, the rate of drying begins to fall because the gas no longer leaves in a saturated condition. In other words, a *desorption* wave passes through the bed, and the situation is much like that described for elution in fixed beds (Chap. 11). However, the point of view of interest is the moisture content of the solid rather than the concentration changes occurring in the exit gas. In the case of shallow beds composed of large particles, the gas leaves the bed unsaturated from the beginning [31], but as long as each particle surface remains

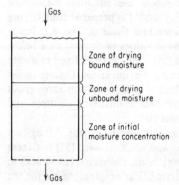

Zone of drying bound moisture

Zone of drying unbound moisture

Zone of initial moisture concentration

Figure 12.17 Through-circulation drying of thick beds of solids.

fully wet, there will still be a constant-rate period. The falling rate then begins when the surface moisture is depleted.

The rate of drying of unbound moisture [1]. Consider a bed of uniform cross section as in Fig. 12.17, fed with a gas of humidity Y_1 at the rate of G_S mass dry gas/(area bed cross section) (time). The maximum rate of drying N_{max} will occur if the gas leaving the bed is saturated at the adiabatic-saturation temperature, with humidity Y_{as},

$$N_{max} = G_S(Y_{as} - Y_1) \qquad (12.25)$$

where N is expressed as mass moisture evaporated/(area bed cross section) (time). In general, the gas will leave the bed at humidity Y_2, and the instantaneous rate of drying is

$$N = G_S(Y_2 - Y_1) \qquad (12.26)$$

For a differential section of the bed where the gas undergoes a change in humidity dY and leaves at a humidity Y, the rate of drying is

$$dN = G_S dY = k_Y dS(Y_{as} - Y) \qquad (12.27)$$

where S is the interfacial surface per unit area of bed cross section. Letting a represent the interfacial surface per unit volume of bed whose thickness is z_S, we get

$$dS = a\, dz_S \qquad (12.28)$$

and Eq. (12.27) becomes

$$\int_{Y_1}^{Y_2} \frac{dY}{Y_{as} - Y} = \int_0^{z_S} \frac{k_Y a\, dz_S}{G_S} \qquad (12.29)$$

$$\ln \frac{Y_{as} - Y_1}{Y_{as} - Y_2} = N_{tG} = \frac{k_Y a z_S}{G_S} \qquad (12.30)$$

where N_{tG} is the number of gas transfer units in the bed. This equation is the same as Eq. (7.57), developed for a somewhat similar situation. The mean driving force for evaporation is then the logarithmic mean of $Y_{as} - Y_1$ and $Y_{as} - Y_2$, in accordance with Eq. (7.61). Combining Eqs. (12.25), (12.26), and (12.30) gives

$$\frac{N}{N_{max}} = \frac{Y_2 - Y_1}{Y_{as} - Y_1} = 1 - \frac{Y_{as} - Y_2}{Y_{as} - Y_1} = 1 - e^{-N_{tG}} = 1 - e^{-k_Y a z_S/G_S}$$

$$(12.31)$$

Equation (12.31) provides the rate of drying N if values of $k_Y a$ or N_{tG} can be determined. These have been established for certain special cases as follows:

1. *Particles small (10- to 200-mesh, or 2.03 to 0.074 mm diameter) with respect to bed depth (greater than 11.4 mm); drying of unbound water from the surface of*

nonporous particles [1]. For this case,† the constant rate is given by N_{max} [Eq. (12.25)]. Equation (12.31) can be used for both constant and falling rates, since at high moisture contents the exponential term becomes negligible. The interfacial surface a varies with moisture content, and it is most convenient to express N_{tG} empirically as‡

$$N_{tG} = \frac{0.273}{d_p^{0.35}} \left(\frac{d_p G}{\mu} \right)^{0.215} (X \rho_S z_S)^{0.64} \qquad (12.32)$$

where d_p is the particle diameter and ρ_S the apparent density of the bed, mass dry solid/volume. Through-drying of such beds in the ordinary equipment may involve a pressure drop for gas flow which is too high for practical purposes, especially if the particles are very small. Drying of such beds is done on continuous rotary filters (crystal-filter driers), however.

2. *Particles large (3.2 to 20 mm diameter) in shallow beds (10 to 64 mm thick); drying of unbound moisture from porous or nonporous particles.* During the constant-rate period the gas leaves the bed unsaturated, and the constant rate of drying is given by Eq. (12.31). For this purpose, k_Y is given by

$$k_Y = \frac{j_D G_S}{Sc^{2/3}} \qquad (12.33)$$

and j_D in turn by the first two entries of item 7, Table 3.3. The interfacial surface may be taken as the surface of the particles. For air drying of water from solids, $Sc = 0.6$. Additional experimental data on a large number of preformed materials are also available [31]. During the falling-rate period, internal resistance to moisture movement may be important, and no general treatment is available. In many cases [31] it is found that semilogarithmic plots of the form of Fig. 12.16 are useful. If the line on this plot is straight, Eqs. (12.8) and (12.9) are applicable.

Drying of bound moisture Presumably some adaptation of the methods used for adsorption in fixed beds (Chap. 11) could be made to describe this. No experimental confirmation is available, however.

Illustration 12.5 A cake of a crystalline precipitate is to be dried by drawing air through the cake. The particles of the cake are nonporous, of average diameter 0.20 mm, and since they are insoluble in water have a negligible equilibrium-moisture content. The cake is 18 mm thick, and the apparent density is 1350 kg dry solid/m^3 (85 lb/ft^3). It is to be dried from 2.5 to 0.1% moisture. The air will enter the cake at 0.24 kg dry air/(m^2 bed cross section) · s (= 177 lb/ft^2 · h), at a dry-bulb temperature 32°C and 50% humidity. Estimate the time for drying.

---

†In beds of finely packed solids containing a large percentage of liquid, the liquid is largely forced from between the solid particles by a mechanical process when gas is forced through the bed. See particularly Brownell and Katz, *Chem. Eng. Prog.*, **43**, 537, 601, 703 (1947). Only the last traces of moisture are removed by the drying process considered here.

‡ Equation (12.32) is empirical, to be used with SI units. For units of feet, pounds mass, and hours the coefficient is 1.14.

SOLUTION $X_1 = 0.025/(1 - 0.025) = 0.0256$ kg water/kg dry solid, $X_2 = 0.001/(1 - 0.001) = 0.001\,001$. From Fig. 7.5, $Y_1 = 0.0153$ kg water/kg dry air, $t_{as} = 24°C$ (the adiabatic-saturation temperature), and $Y_{as} = 0.0190$ kg water/kg dry air. $G_S = 0.24$ kg dry air/m$^2 \cdot$ s.

$$\text{Approx av } G = 0.24 + \frac{0.24(0.0153 + 0.0190)}{2} = 0.244 \text{ kg dry air/m}^2 \cdot \text{s}$$

Eq. (12.26): $N_{max} = 0.24(0.0190 - 0.0153) = 8.88 \times 10^{-4}$ kg evapd/m$^2 \cdot$ s, $z_S = 0.018$ m, $d_p = 2 \times 10^{-4}$ m, $\rho_S = 1350$ kg/m^3. Viscosity of air at $(32 + 24)/2 = 28°C$ is 1.8×10^{-5} kg/m $\cdot$ s.

$$\frac{d_p G}{\mu} = \frac{(2 \times 10^{-4})(0.244)}{1.8 \times 10^{-5}} = 2.71$$

Eq. (12.32):

$$N_{tG} = \frac{0.273}{(2 \times 10^{-4})^{0.35}}(2.71)^{0.215}[X(1350)(0.018)]^{0.64} = 51.37X^{0.64}$$

Eq. (12.31):

$$N = (8.88 \times 10^{-4})[1 - \exp(-51.37X^{0.64})]$$

This provides the following values of N, kg water evapd/m$^2 \cdot$ s, for values of X within the range of interest. They are the coordinates of the rate-of-drying curve.

X	0.0256	0.02	0.015	0.010	0.008	0.006	0.004	0.002	0.001
$N \times 10^4$	8.815	8.747	8.610	8.281	8.022	7.609	6.898	5.488	4.092

Calculation of the time of drying requires a value of the integral $\int dX/N$, in accordance with Eq. (12.3), between X_1 and X_2. This can be evaluated numerically, or graphically from a plot of $1/N$ vs. X. The integral = 31.24. Since $S_S/A = \rho_S z_S = 1350(0.018) = 24.3$ kg dry solid/m^2, the time for drying is, by Eq. (12.3),

$$\theta = 24.3(31.24) = 759 \text{ s} = 12.7 \text{ min}$$

Illustration 12.6 Wet, porous catalyst pellets in the form of small cylinders, 13.5 mm diameter, 13.0 mm long, are to be dried of their water content in a through-circulation drier. The pellets are to be arranged in beds 50 mm deep on screens and dried by air flowing at the rate 1.1 kg dry air/(m^2 bed cross section) $\cdot$ s, entering at 82°C dry-bulb temperature, humidity 0.01 kg water/kg dry air. The apparent density of the bed is 600 kg dry solids/m^3, and the particle surface is 280 m^2/m^3 of bed. Estimate the rate of drying and the humidity and temperature of the air leaving the bed during the constant-rate period.

SOLUTION $Y_1 = 0.01$ kg water/kg dry gas, and from Fig. 7.5, $Y_{as} = 0.031$ at the corresponding adiabatic-saturation temperature, 32°C. $G_S = 1.1$ kg dry air/m$^2 \cdot$ s and the approximate average $G = 1.1 + 1.1(0.031 + 0.01)/2 = 1.123$ kg/m$^2 \cdot$ s. The approximate average air viscosity is 1.9×10^{-5} kg/m $\cdot$ s.

The surface of each particle = $\pi(0.0135)^2(2)/4 + \pi(0.0135)(0.0130) = 8.376 \times 10^{-4}$ m^2. The diameter of a sphere of equal area = $d_p = [(8.376 \times 10^{-4})/\pi]^{0.5} = 0.01633$ mm. $a = 280$ m^2/m^3; $z_S = 0.050$ m.

The Reynolds number for the particles (Re'' of item 7, Table 3.3) = $d_p G/\mu = 0.01633(1.123)/(1.9 \times 10^{-5}) = 965$. ε = fraction voids (see Table 3.3) = $1 - d_p a/6 = 1 - 0.01633(280)/6 = 0.237$. Hence $j_D = (2.06/\varepsilon) \text{Re}''^{-0.575} = 0.1671$. Sc for air-water vapor = 0.6. Eq. (12.33):

$$k_Y = \frac{0.1671(1.1)}{(0.6)^{2/3}} = 0.258 \text{ kg H}_2\text{O/m}^2 \cdot \text{s} \cdot \Delta Y$$

$$N_{tG} = \frac{k_Y a z_S}{G_S} = \frac{0.258(280)(0.050)}{1.1} = 3.28$$

Eq. (12.25):

$$N_{max} = G_S(Y_{as} - Y_1) = 1.1(0.031 - 0.01) = 0.0231 \text{ kg/m}^2 \cdot \text{s}$$

Eq. (12.31):

$$\frac{N}{0.0231} = \frac{Y_2 - 0.01}{0.031 - 0.01} = 1 - \exp(-3.28)$$

Therefore $N = 0.0222$ kg water evaporated/m$^2 \cdot$ s in the constant-rate period, and $Y_2 = 0.0302$ kg water/kg dry air for the outlet air. The corresponding outlet air temperature, from the adiabatic-saturation curve of Fig. 7.5 for the entering air, is 33.0°C. Since

$$\frac{S_S}{A} = \rho_S z_S = 600(0.050) = 30 \text{ kg dry solid/m}^2$$

from Eq. (12.2)

$$\frac{-dX}{d\theta} = \frac{NA}{S_S} = \frac{0.0222}{30} = 7.4 \times 10^{-4} \text{ kg H}_2\text{O/ (kg dry solid)} \cdot \text{s}$$

This must be considered as an estimate only, subject to check by drying-rate tests.

CONTINUOUS DRYING

Continuous drying offers the advantages that usually the equipment necessary is small relative to the quantity of product, the operation is readily integrated with continuous chemical manufacture without intermediate storage, the product has a more uniform moisture content, and the cost of drying per unit of product is relatively small. As in batch drying, the nature of the equipment used is greatly dependent upon the type of material to be dried. Either direct or indirect heating, and sometimes both, may be used.

In many of the direct driers to be described, the solid is moved through a drier while in contact with a moving gas stream. The gas and solid may flow in parallel or in countercurrent, or the gas may flow across the path of the solid. If heat is neither supplied within the drier nor lost to the surroundings, operation is adiabatic and the gas will lose sensible heat and cool down as the evaporated moisture absorbs latent heat of vaporization. By supplying heat within the drier, the gas can be maintained at constant temperature.

In *countercurrent* adiabatic operation, the hottest gas is in contact with the driest solid, and the discharged solid is therefore heated to a temperature which may approach that of the entering gas. This provides the most rapid drying, since especially in the case of bound moisture the last traces are the most difficult to remove, and this is done more rapidly at high temperatures. On the other hand, the dry solid may be damaged by being heated to high temperatures in this manner. In addition, the hot discharged solid will carry away considerable sensible heat, thus lowering the thermal efficiency of the drying operation.

In *parallel* adiabatic operation, the wet solid is contacted with the hottest gas. As long as unbound surface moisture is present, the solid will be heated only to the wet-bulb temperature of the gas, and for this reason even heat-sensitive solids can frequently be dried by fairly hot gas in parallel flow. For example, a typical flue gas resulting from combustion of a fuel, which may have a humidity of 0.03 kg water vapor/kg dry gas at 430°C, has a wet-bulb temperature of only about 65°C. In any event, the wet-bulb temperature can never exceed the boiling point of the liquid at the prevailing pressure. At the outlet of the drier, the gas will have been considerably cooled, and no damage will result to the dry solid. Parallel flow also permits greater control of the moisture content of the discharged solid when the solid must not be completely dried, through control of the quantity of gas passing through the drier and consequently its exit temperature and humidity. For this reason also it is used to avoid case hardening and other problems associated with batch drying.

Tunnel Driers

These direct driers are essentially adaptations of the truck drier to continuous operation. They consist of relatively long tunnels through which trucks, loaded with trays filled with the drying solid, are moved in contact with a current of gas to evaporate the moisture. The trucks may be pulled continuously through the drier by a moving chain, to which they are attached. In a simpler arrangement, the loaded trucks are introduced periodically at one end of the drier, each displacing a truck at the other end. The time of residence in the drier must be long enough to reduce the moisture content of the solid to the desired value. For relatively low-temperature operation the gas is usually steam-heated air, while for higher temperatures and especially for products which need not be kept scrupulously clean, flue gas from the combustion of a fuel can be used. Parallel or countercurrent flow of gas and solid can be used, or in some cases fans placed along the sides of the tunnel blow the gas through the trucks in crossflow. Operation may be essentially adiabatic, or the gas may be heated by steam coils along its path through the drier, and operation may then be substantially at constant temperature. Part of the gas may be recycled, much as in the case of batch driers, for heat economy. Truck-type tunnel driers can be used for any material which can be dried on trays: crystals, filter cakes, pastes, pottery, and the like.

There are many modifications of the tunnel drier which are essentially the same in principle but different in detailed design owing to the nature of the material being dried. For example, skeins of wet yarn may be suspended from poles or racks which move through the tunnel drier. Hides may be stretched on frames which hang from conveyor chains passing through the drier. Material in continuous sheets, such as cloth, may move through the drier under tension, as in a continuous belt over a series of rollers, or be hung in festoons from moving racks if it is to be dried in the absence of tension.

Turbo-Type (Rotating Shelf) Driers

Solids which ordinarily can be dried on trays, such as powdery and granular materials, heavy sludges and pastes, beads and crystalline solids, can be continuously dried in a turbo-type drier, a form of direct drier. The simplest of these is shown in Fig. 12.18. The drier is fitted with a series of annular trays arranged in

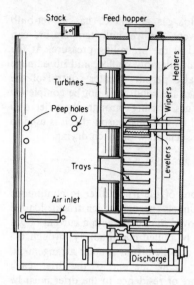

Figure 12.18 Turbo-type drier. *(Wyssmont Co., Inc.)*

a vertical stack. These rotate slowly (from a few to 60 r/s) about a vertical shaft. Each tray is provided with a slot cut into the tray, as well as a leveling rake for spreading the solid. Solid fed in at the top is spread upon the top tray to a uniform thickness, and as the tray revolves, the solid is pushed through the slot by a separate wiper rake, to fall upon the tray beneath. In this way, with overturning and respreading on each tray, the solid progresses to the discharge chute at the bottom of the drier. The drying gas flows upward through the drier, is circulated over the trays by slowly revolving turbine fans, and is reheated by internal heating pipes as shown. The rate of drying will be faster than that experienced in a tray-equipped tunnel drier, thanks to the frequent reloading of the solid on each tray [26]. Alternative arrangements are possible: external heating of the gas, recirculation of the gas, and arrangements for recovery of evaporated solvents may be provided. In some installations the solid is carried upon a moving, endless conveyor, which is wound about the vertical axis of the drier in a close-pitched, screw-type spiral. These driers are regularly built in sizes ranging from 2 to 6 m (6 to 20 ft) in diameter and 2 to 7.5 m (6 to 25 ft) high. A few installations as tall as 18 m (60 ft) have been made.

Similar devices, without the central turbines and built like multiple-hearth furnaces, are known as *plate driers*.

Through-Circulation Driers

Granular solids can be arranged in thin beds for through circulation of the gas, and, if necessary, pastes and filter cakes can be preformed into granules, pellets, or noodles, as described in the case of batch driers. In the continuous through-circulation drier of Fig. 12.19 [20], the solid is spread to a depth of 38 to 50 mm upon a moving endless conveyor which passes through the drier. The conveyor

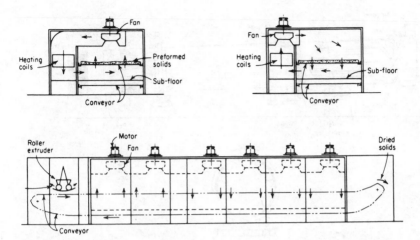

Figure 12.19 Continuous through-circulation (single-conveyor) drier with roller extruder. *(Proctor and Schwartz, Inc.)*

is made of perforated plates or woven wire screens in hinged sections in order to avoid failure from repeated flexing of the screen. Fans blow the heated air through the solid, usually upward through the wet solid and downward after initial drying has occurred. In this way, a more uniform moisture concentration throughout the bed is attained. Much of the gas is usually recycled, and a portion is discarded continuously at each fan position in the drier. For materials which permit the flow of gas through the bed in the manner shown, drying is much more rapid than for tray-type tunnel driers.

Rotary Driers

This is a most important group of driers, suitable for handling free-flowing granular materials which can be tumbled about without concern over breakage. Figure 12.20 shows one form of such a drier, a direct countercurrent hot-air drier. The solid to be dried is continuously introduced into one end of a rotating cylinder, as shown, while heated air flows into the other. The cylinder is installed at a small angle to the horizontal, and the solid consequently moves slowly through the device. Inside the drier, lifting flights extending from the cylinder wall for the full length of the drier lift the solid and shower it down in a moving curtain through the air, thus exposing it thoroughly to the drying action of the gas. This lifting action also assists in the forward motion of the solid. At the feed end of the drier, a few short spiral flights assist in imparting the initial forward motion to the solid before the principal flights are reached. The solid must clearly be one which is neither sticky nor gummy, which might stick to the sides of the drier or tend to ball up. In such cases, recycling of a portion of the dried product may nevertheless permit use of a rotary drier.

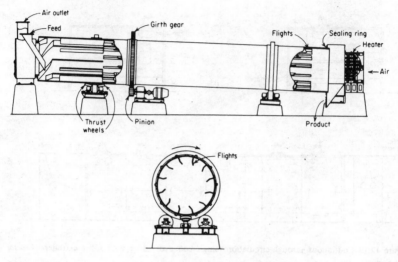

Figure 12.20 Ruggles-Coles XW hot-air drier. *(Hardinge Co., Inc.)*

The drier may be fed with hot flue gas rather than air, and if the gas leaves the drier at a high enough temperature, discharging it through a stack may provide adequate natural draft to provide sufficient gas for drying. Ordinarily, however, an exhaust fan is used to pull the gas through the drier, since this provides more complete control of the gas flow. A dust collector, of the cyclone, filter, or washing type, may be interposed between the fan and the gas exit. A blower may also be provided at the gas entrance, thus maintaining a pressure close to atmospheric in the drier; this prevents leakage of cool air in at the end housings of the drier, and if the pressure is well balanced, outward leakage will also be minimized.

Rotary driers are made for a variety of operations. The following classification includes the major types.

1. *Direct heat, countercurrent flow.* For materials which can be heated to high temperatures, such as minerals, sand, limestone, clays, etc., hot flue gas can be used as the drying gas. For substances which should not be heated excessively, such as certain crystalline chemical products like ammonium sulfate and cane sugar, heated air can be used. The general arrangement is that shown in Fig. 12.20, and if flue gas is used, the heating coils are replaced by a furnace burning gas, oil, or coal.

2. *Direct heat, cocurrent flow.* Solids which can be dried with flue gas without fear of contamination but which must not be heated to high temperatures for fear of damage, such as gypsum, iron pyrites, and organic material such as peat and alfalfa, should be dried in a cocurrent-flow drier. The general construction is much like that of Fig. 12.20, except that the gas and solid both enter at the same end of the drier.

3. *Indirect heat, countercurrent flow.* For solids such as white pigments, and the like, which can be heated to high temperatures but which must remain out of contact with flue gas, the indirect drier indicated schematically in Fig. 12.21*a* can be used. As an alternative construction, the drier may be enclosed in a brick structure and completely surrounded by the hot flue gases. The airflow in such a drier can be kept to a minimum since the heat is supplied by conduction through the shell or central tube, and finely pulverized solids which dust severely can then be handled. For solids which must not be heated to high temperatures and for which indirect heat is desirable such as cattle feed, brewers' grains, feathers, and the like, the steam-tube drier, shown in Fig. 12.21*b*, can be used. This drier may or may not have lifting flights and may be built with one, two, or more concentric rows of steam-heated tubes.

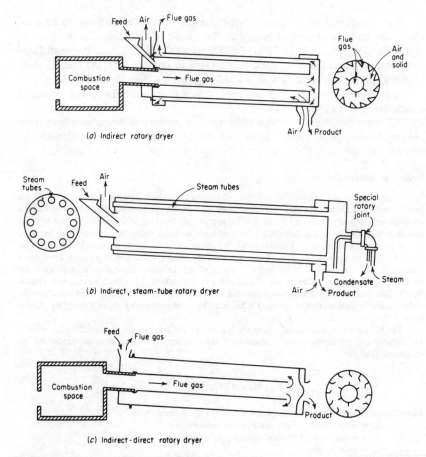

Figure 12.21 Some rotary driers (schematic).

The tubes revolve with the drier, necessitating a special rotary joint where the steam is introduced and the condensate removed. This type of drier is frequently used when recovery of the evaporated liquid is necessary.

4. *Direct-indirect*. These driers, more economical to operate than the direct driers, can be used for solids which can be dried at high temperatures by flue gas, especially when fuel costs are high and when large percentages of moisture must be removed from the solid. A typical schematic arrangement is shown in Fig. 12.21c. In such a drier, the hot gas may enter the center tube at 650 to 980°C (1200 to 1800°F), cool to 200 to 480°C (400 to 900°F) in its first passage through the drier, and on returning through the annular drying space cool further to 60 to 70°C (140 to 170°F) at discharge. Lignite, coal, and coke can be dried in the inert atmosphere of such a drier at relatively high temperatures without danger of burning or dust explosion.

The *Solidaire* drier contains a large number of paddles attached to an axial shaft, the paddles extending nearly to the inside of the shell. These rotate at relatively high speed, 10 to 20 m/s (2000 to 4000 ft/min), and the centrifugal force thus imparted to the solids keeps them in contact with the heated shell. Hot gas flowing cocurrently with the solids provides their forward motion.

High-frequency insonation has been shown to increase the rate of drying in rotary driers substantially [13].

All these driers are available from various manufacturers in standard sizes, ranging from approximately 1 m in diameter by 4 m long to 3 m in diameter by 30 m long.

Holdup in Rotary Driers

The average time of passage, or retention time, of the solid in a drier must equal the required drying time if the solid is to emerge at the desired moisture content. It must be recognized of course that the retention time of individual particles may differ appreciably from the average [36, 45], and this can lead to nonuniformity of product quality. Several agencies bring about the movement of the solid particles through a rotary drier. Flight action is the lifting and dropping of particles by the flights on the drier shell: in the absence of air flow, each time the solid is lifted and dropped, it advances a distance equal to the product of the length of the drop and the slope of the drier. Kiln action is the forward rolling of the particles on top of each other in the bottom of the drier, as in a kiln without flights. The particles also bounce in a forward direction after being dropped from the flight. In addition, the forward motion of the solid is hindered by a counterflowing gas or assisted by parallel flow.

The holdup ϕ_D of solid is defined as the fraction of the drier volume occupied by the solid at any instant, and the average time of retention θ can be computed by dividing the holdup by the volumetric feed rate,

$$\theta = \frac{\phi_D Z \pi T_D^2/4}{(S_S/\rho_S)(\pi T_D^2/4)} = \frac{Z\phi_D\rho_S}{S_S} \tag{12.34}$$

where S_S/ρ_S = volumetric feed rate/cross-sectional area of drier
S_S = mass velocity of dry solids, mass/(area)(time)
ρ_S = apparent solid density, mass dry solid/volume
Z = drier length
T_D = drier diameter

Although the influence of the character of the solids can be considerable [51], Friedman and Marshall [14] found that the holdup of a large number of solids under a variety of typical operating conditions could be expressed simply as

$$\phi_D = \phi_{D0} \pm KG \tag{12.35}$$

where ϕ_{D0} is the holdup with no gas flow and $\pm KG$ is the correction for influence of gas rate, G mass/(area)(time). The plus sign is used for countercurrent flow of gas and solid and the minus sign for cocurrent flow. Holdup for conditions of no gas flow depends to some extent upon flight design and the nature of the solid but under typical conditions and for ϕ_{D0} not exceeding 0.08 their data can be described by†

$$\phi_{D0} = \frac{0.3344 S_s}{\rho_s s N^{0.9} T_D} \tag{12.36}$$

where s = slope of drier, m/m
N = rotational speed, r/s
T_D = drier diameter, m

The constant K is dependent upon the properties of the solid, and for rough estimates it may be taken, for SI units, as‡

$$K = \frac{0.6085}{\rho_s d_p^{1/2}} \tag{12.37}$$

where d_p is the average particle diameter. Holdups in the range 0.05 to 0.15 appear to be best. Higher holdup results in increased kiln action, with consequent poor exposure of the solid to the gas and an increase in power required for operating the drier.

These empirical relationships are applicable only under conditions of reasonable gas rates which do not cause excessive dusting or blowing of the solid particles from the drier. It is ordinarily desirable to keep dust down to 2 to 5 percent of the feed material as a maximum, and the corresponding permissible gas rates depend greatly upon the nature of the solid. From 0.27 to 13.6 kg gas/m² · s (200 to 10 000 lb/ft² · h) is used, depending upon the solid; for most 35-mesh solids (d_p approximately 0.4 mm) 1.36 kg/m² · s (1000 lb/ft² · h) is amply safe [14]. Dusting is less severe for countercurrent than for cocurrent flow, since then the damp feed acts to some extent as a dust collector, and it is also influenced by design of feed chutes and end breechings. In any case, it is best to depend upon actual tests for final design and to use the equations for initial estimates only.

Driers are most readily built with length/diameter ratios $Z/T_D = 4$ to 10. For most purposes, the flights extend from the wall of the drier a distance of 8 to 12 percent of the diameter, and their number range from $6T_D$ to $10T_D$ (T_D in meters). They should be able to lift the entire solids holdup, thus minimizing kiln action, which leads to low retention time. Rates of rotation are such as to provide peripheral speeds of 0.2 to 0.5 m/s (40 to 100 ft/min) and the slopes are usually in the range 0 to 0.08 m/m [14]. Negative slopes are sometimes necessary in cocurrent-flow driers.

Through-Circulation Rotary Driers

Driers of the type indicated in Fig. 12.22 combine the features of the through-circulation and rotary driers. The direct drier shown, the Roto-Louvre [12], consists of a slowly revolving tapered drum fitted with louvers to support the drying solid and to permit entrance of the hot gas beneath the solid. The hot gas is admitted only to those louvers which are underneath the bed of solid. There is

---

† Equation (12.36) is empirical. For units of feet, pounds mass, and hours (N in min⁻¹), the coefficient is 0.0037.
‡ For units of feet, pounds mass, and hours the coefficient is 9.33×10^{-5}.

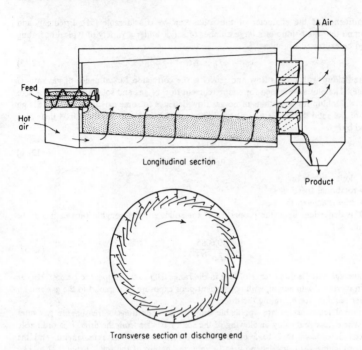

Figure 12.22 Continuous through-circulation rotary drier (Roto-Louvre). *(Link-Belt Co.)*

substantially no showering of the solid through the gas stream, and consequently a minimum of dusting results. The device is satisfactory for both low- and high-temperature drying of the same materials ordinarily treated in a rotary drier.

Drum Driers

Fluid and semifluid materials such as solutions, slurries, pastes, and sludges can be dried on an indirect drier, of which Fig. 12.23, a dip-feed drum drier, gives an example. A slowly revolving internally steam-heated metal drum continuously dips into a trough containing the substance to be dried, and a thin film of the substance is retained on the drum surface. The thickness of the film is regulated by a spreader knife, as shown, and as the drum revolves, moisture is evaporated into the surrounding air by heat transferred through the metal of the drum. The dried material is then continuously scraped from the drum surface by a knife. For such a drier, heat transfer rather than diffusion is the controlling factor. The liquid or solution is first heated to its boiling point; moisture is then evolved by boiling at constant temperature if a solute precipitates from a solution at constant concentration, or at increasing temperatures if the concentration change is gradual; and finally the dried solid is heated to approach the

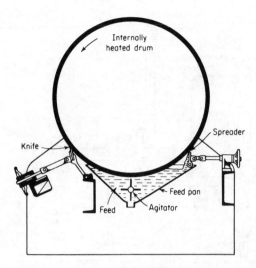

Figure 12.23 Dip-feed single-drum drier. *(Blaw-Knox Co.)*

temperature of the drum surface. With slurries or pastes of insoluble solids, the temperature remains essentially constant at the solvent boiling point as long as the solid is completely wet and increases only during the last stages of drying. The vapors are frequently collected by a ventilated hood built directly over the drier.

The ability of various slurries, solutions, and pastes to adhere to a heated drum varies considerably, and diverse methods of feeding the drum are accordingly resorted to. Slurries of solids dispersed in liquids are frequently fed to the bottom of the drum on an inclined pan, and the excess nonadherent material is recycled to the feed reservoir. Vegetable glues and the like can be pumped against the bottom surface of the drum. The dip-feed arrangement shown in Fig. 12.23 is useful for heavy sludges while for materials which stick to the drum only with difficulty, the feed can be spattered on by a rapidly revolving roll. Double-drum driers, consisting of two drums placed close together and revolving in opposite directions, can be fed from above by admitting the feed into the depression between the drums. The entire drum drier may, on occasion, be placed inside a large evacuated chamber for low-temperature evaporation of the moisture.

Cylinder driers are drum driers used for material in continuous sheet form, such as paper and cloth. The wet solid is fed continuously over the revolving drum, or a series of such drums, each internally heated by steam or other heating fluid.

Spray Driers

Solutions, slurries, and pastes can be dried by spraying them as fine droplets into a stream of hot gas in a spray drier [29, 32]. One such device is shown in

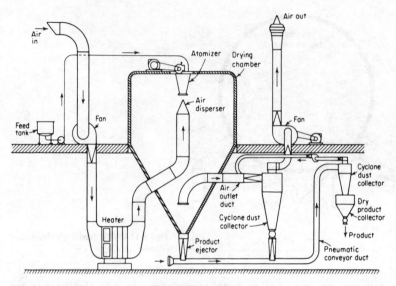

Figure 12.24 Spray drier. *(Nichols Engineering and Research Corp.)*

Fig. 12.24. The liquid to be dried is atomized and introduced into the large drying chamber, where the droplets are dispersed into a stream of heated air. The particles of liquid evaporate rapidly and dry before they can be carried to the sides of the chamber, and the bulk of the dried powder which results falls to the conical bottom of the chamber to be removed by a stream of air to the dust collector. The principal portion of the exit gas is also led to a dust collector, as shown, before being discharged. Many other arrangements are possible, involving both parallel and counterflow of gas and spray [28]. Installations may be very large, as much as 12 m in diameter and 30 m high (40 by 100 ft). Arrangements and detailed designs vary considerably, depending upon the manufacturer. Spray driers are used for a wide variety of products, including such diverse materials as organic and inorganic chemicals, pharmaceuticals, food products such as milk, eggs, and soluble coffee, as well as soap and detergent products.

In order to obtain rapid drying, atomization of the feed must provide small particles of high surface/weight ratio, whose diameter is usually in the range 10 to 60 μm. For this purpose, spray nozzles or rapidly rotating disks can be used. Spray nozzles are of two major types: pressure nozzles, in which the liquid is pumped at high pressure and with a rapid circular motion through a small orifice, and two fluid nozzles, in which a gas such as air or steam at relatively low pressures is used to tear the liquid into droplets. Nozzles are relatively inflexible in their operating characteristics and do not permit even moderate variation in liquid-flow rates without large changes in droplet size. They are also subject to rapid erosion and wear. Rotating disks are therefore favored in the chemical industry. They may be plane, vaned, or cup-shaped, up to about 0.3 m

(12 in) in diameter, and rotate at speeds in the range 50 to 200 r/s. The liquid or slurry is fed onto the disk near the center and is centrifugally accelerated to the periphery, from which it is thrown in an umbrella-shaped spray. Appreciable variation in liquid properties and feed rates can be handled satisfactorily, and even thick slurries or pastes can be atomized without clogging the device provided they can be pumped to the disk.

The drying gas, either flue gas or air, can enter at the highest practical temperature, 80 to 760°C (175 to 1400°F), limited only by the heat sensitivity of the product. Since the contact time for product and gas is so short, relatively high temperatures are feasible. The short time of drying requires effective gas-spray mixing, and attempts to improve upon this account in part for the large number of designs of spray chambers. Cool air is sometimes admitted at the drying-chamber walls in order to prevent sticking of the product to the sides. The effluent gas may convey all the dried product out of the drier or only the fines, but in either case the gas must be passed through some type of dust collector such as cyclones or bag filters, and these are sometimes followed by wet scrubbers for the last traces of dust. Recirculation of hot gas to the drier for purposes of heat economy is not practical, since the dust-recovery operation cannot usually be accomplished without appreciable heat loss.

The drops of liquid reach their terminal velocity in the gas stream quickly, within inches of the atomizing device. Evaporation takes place from the surface of the drops, and with many products solid material may accumulate as an impervious shell at the surface. Since heat is nevertheless rapidly being transmitted to the particles from the hot gas, the entrapped liquid portion of the drop vaporizes and expands the still-plastic wall of the drop to 3 to 10 times the original size, eventually exploding a small blowhole in the wall and escaping, to leave a hollow, dried shell of solid as the product. In other cases, the central liquid core diffuses through the shell to the outside, and the reduced internal pressure causes an implosion. In any event, the dried product is frequently in the form of small hollow beads of low bulk density [8]. Some control over the bulk density is usually possible through control of the particle size during atomization or through the temperature of the drying gas (increased gas temperature causes decreased product bulk density by more extensive expansion of the drop contents). For high-density products, the dried beads can be crushed.

Spray drying offers the advantage of extremely rapid drying for heat-sensitive products, a product particle size and density which are controllable within limits, and relatively low operating costs, especially in large-capacity driers. Extensive data have now been accumulated on drop sizes, drop trajectories, relative velocities of gas and drop, and rates of drying [32]. Logical procedures for design of spray driers with a minimum of experimental work have been outlined in detail in the work of Gauvin and coworkers [3, 15].

Fluidized and Spouted Beds

Granular solids, fluidized (see Chap. 11) by a drying medium such as hot air, can be dried and cooled in a similar fluidized bed [53], shown schematically in

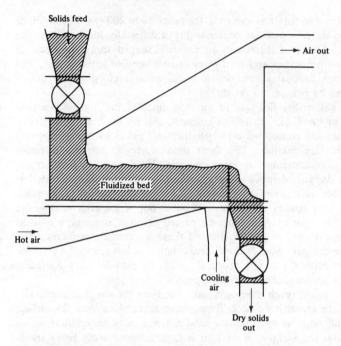

Figure 12.25 Fluidized bed drier (schematic). *(Strong Scott Mfg. Co.)*

Fig. 12.25. The principal characteristics of such beds include cross flow of solid and drying gas, a solids residence time controllable from seconds to hours, and suitability for any gas temperature. It is necessary that the solids be free-flowing, of a size range 0.1 to 36 mm [59]. Since the mass flow rate of gas for thermal requirements is substantially less than that required for fluidization, the bed is most economically operated at the minimum velocity for fluidization. Multistage, cross-flow operation (fresh air for each stage) is a possibility [2], as is a two-stage countercurrent arrangement, as in Fig. 11.28 [58]. A tentative design procedure has been proposed [40].

Coarse solids too large for ready fluidization can be handled in a *spouted bed* [33]. Here the fluid is introduced into the cone-shaped bottom of the container for the solids instead of uniformly over the cross section. It flows upward through the center of the bed in a column, causing a fountainlike spout of solids at the top. The solids circulate downward around the fluid column. Such a bed has found particular use in drying wheat, peas, flax, and the like [43].

Pneumatic (Flash) Driers

If the gas velocity in a fluidized bed is increased to the terminal velocity of the individual solid particles, they are lifted from the bed and are carried along by the fluidizing gas. Such gas–solid-particle mixtures are characteristic of *flash*, or

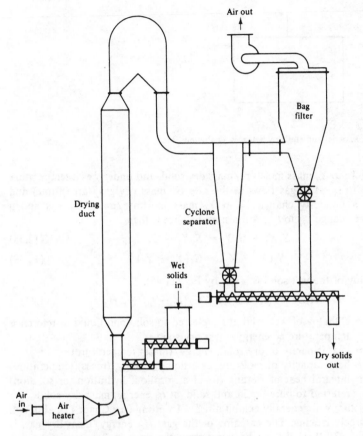

Figure 12.26 Pneumatic drier. *(Strong-Scott Mfg. Co.)*

pneumatic driers. The granular, free-flowing solids are dispersed in a rapidly flowing (~ 25 m/s) hot gas stream, as in Fig. 12.26, with an exposure time of the order of seconds [34]. Such short drying times limit the method to cases of surface moisture only, where internal diffusion of moisture within the solid is unimportant. Although a beginning has been made to systematize the drying-rate parameters [9] and a computer-design procedure has been prepared [5], pilot tests are nevertheless necessary.

Material and Enthalpy Balances

A general flow diagram for a continuous drier, arranged for countercurrent flow, is shown in Fig. 12.27. Solid enters at the rate S_S mass dry solid/(area)(time),† is

---

† For purposes of material and enthalpy balances alone, rates of flow of gas and solid can equally well be expressed as mass/time.

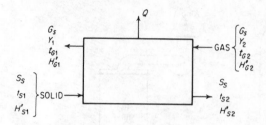

Figure 12.27 Material and enthalpy balances, continuous drier.

dried from X_1 to X_2 mass moisture/mass dry solid, and undergoes a temperature change t_{S1} to t_{S2}. The gas flows at the rate G_S mass dry gas/(area)(time) and undergoes a humidity change Y_2 to Y_1 mass moisture/mass dry gas and a temperature change t_{G2} to t_{G1}. A moisture balance is then

$$S_S X_1 + G_S Y_2 = S_S X_2 + G_S Y_1 \tag{12.38}$$

or

$$S_S(X_1 - X_2) = G_S(Y_1 - Y_2) \tag{12.39}$$

The enthalpy of the wet solid is given by Eq. (11.45),

$$H_S' = C_S(t_S - t_0) + X C_A(t_S - t_0) + \Delta H_A \tag{11.45}$$

where H_S' = enthalpy of wet solid at t_S, referred to solid and liquid at reference temperature t_0, energy/mass dry solid

C_S = heat capacity of dry solid, energy/(mass)(temperature)

C_A = heat capacity of moisture, as a liquid, energy/(mass)(temperature)

ΔH_A = integral heat of wetting (or of adsorption, hydration, or solution) referred to pure liquid and solid, at t_0, energy/mass dry solid

Bound moisture will generally exhibit a heat of wetting (see Chap. 11), although data are largely lacking. The enthalpy of the gas, H_G' energy/mass dry gas, is given by Eq. (7.13). If the net heat lost from the drier is Q energy/time, the enthalpy balance becomes

$$S_S H_{S1}' + G_S H_{G2}' = S_S H_{S2}' + G_S H_{G1}' + Q \tag{12.40}$$

For adiabatic operation, $Q = 0$, and if heat is added within the drier to an extent greater than the heat losses, Q is negative. If the solid is carried on trucks or other support, the sensible heat of the support should also be included in the balance. Obvious changes in the equations can be made for parallel-flow driers.

Illustration 12.7 An uninsulated, hot-air countercurrent rotary drier of the type shown in Fig. 12.20 is to be used to dry ammonium sulfate from 3.5 to 0.2% moisture. The drier is 1.2 m (4 ft) in diameter, 6.7 m (22 ft) long. Atmospheric air at 25°C, 50% humidity, will be heated by passage over steam coils to 90°C before it enters the drier and is expected to be discharged at 32°C. The solid will enter at 25°C and is expected to be discharged at 60°C. Product will be delivered at a rate of 900 kg/h. Estimate the air and heat requirements for the drier.

SOLUTION Define the rates of flow in terms of kg/h. $X_2 = 0.2/(100 - 0.2) = 0.0020$; $X_1 = 3.5/(100 - 3.5) = 0.0363$ kg water/kg dry solid. $S_S = 900(1 - 0.0020) = 898.2$ kg dry solid/h. The rate of drying = $898.2(0.0363 - 0.0020) = 30.81$ kg water evaporated/h.

At 25°C, 50% humidity, the absolute humidity of the available air $= 0.010$ kg water/kg dry air $= Y_2$. Since $t_{G2} = 90°$C, and with $t_0 = 0°$C, the enthalpy of the air entering the drier (Table 7.1) is

$$H'_{G2} = [1005 + 1884(0.01)]90 + 2\ 502\ 300(0.01) = 117\ 200\ \text{J/kg dry air}$$

For the outlet air, $t_{G1} = 32°$C,

$$H'_{G1} = (1005 + 1884Y_2)(32) + 2\ 502\ 300Y_1 = 32\ 160 + 2\ 504\ 200Y_1$$

The heat capacity of dry ammonium sulfate is $C_S = 1507$ and that of water 4187 J/kg · K. ΔH_A will be assumed to be negligible for lack of better information. Taking $t_0 = 0°$C, so that enthalpies of gas and solid are consistent, and since $t_{S1} = 25°$C, $t_{S2} = 60°$C, the solid enthalpies, J/kg dry solid, are [Eq. (11.45)]:

$$H'_{S2} = 1507(60 - 0) + 0.002(4187)(60 - 0) = 90\ 922$$

$$H'_{S1} = 1507(25 - 0) + 0.0363(4187)(25 - 0) = 41\ 475$$

The estimated combined natural convection and radiation heat-transfer coefficient from the drier to the surroundings [23] is 12 W/m² · K. The mean Δt between drier and surroundings is taken as $[(90 - 25) + (32 - 25)]/2 = 36°$C, and the exposed area as $\pi(1.2)(6.7) = 25.3$ m², whence the estimated heat loss is

$$Q = 12(3600)(25.3)(36) = 39\ 350\ \text{kJ/h}$$

Moisture balance [Eq. (12.39)]:

$$898.2(0.0363 - 0.002) = G_S(Y_1 - 0.01)$$

Enthalpy balance [Eq. (12.40)]:

$$898.2(41\ 475) + G_S(117\ 200) = 898.2(90\ 922) + G_S(32\ 160 + 2\ 504\ 200Y_1) + 39\ 350\ 000$$

When solved simultaneously, these yield

$$G_S = 2682\ \text{kg dry air/h} \qquad Y_1 = 0.0215\ \text{kg water/kg dry air}$$

The enthalpy of the fresh air (Fig. 7.5) is 56 kJ/kg dry air, and hence the heat load for the heater is $2682(117\ 200 - 56\ 000) = 164\ 140$ kJ/h. If steam at pressure 70 kN/m² (10.2 lb$_f$/in²) gauge and latent heat 2215 kJ/kg is used, the steam required is $164\ 140/2215 = 74.1$ kg steam/h, or $74.1/30.81 = 2.4$ kg steam/kg water evaporated.

Rate of Drying for Continuous Direct-Heat Driers

Direct-heat driers are best placed in two categories, according to whether high or low temperatures prevail. For operation at temperatures above the boiling point of the moisture to be evaporated, the humidity of the gas has only a minor influence on the rate of drying, and it is easiest to work directly with the rate of heat transfer. At temperatures below the boiling point, mass-transfer driving forces are conveniently established. In any case, it must be emphasized that our imperfect knowledge of the complex drying mechanisms makes experimental testing of the drying necessary. Calculations are useful only for the roughest estimate.

Drying at high temperatures In a typical situation, three separate zones are distinguished in such driers, recognizable by the variation in temperatures of the gas and solid in the various parts of the drier [14]. Refer to Fig. 12.28, where typical temperatures are shown schematically by the solid lines for a countercurrent drier. In zone I, the preheat zone, the solid is heated by the gas until the

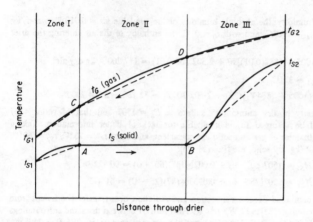

Figure 12.28 Temperature gradients in a continuous countercurrent drier.

rate of heat transfer to the solid is balanced by the heat requirements for evaporation of moisture. Little actual drying will usually occur here. In zone II, the equilibrium temperature of the solid remains substantially constant while surface and unbound moisture are evaporated. At point B, the critical moisture of the solid is reached, and, in zone III, unsaturated surface drying and evaporation of bound moisture occur. Assuming that the heat-transfer coefficients remain essentially constant, the decreased rate of evaporation in zone III results in increased solid temperature, and the discharge temperature of the solid approaches the inlet temperature of the gas.

Zone II represents the major portion for many driers, and it is of interest to consider the temperature-humidity relationship of the gas as it passes through this section. On the psychrometric chart (Fig. 12.29), point D represents the gas conditions at the corresponding point D of Fig. 12.28. If drying is adiabatic, i.e., without addition to, or loss of heat from, the drier, the adiabatic-saturation line DC_1 will represent the variation of humidity and temperature of the gas as it passes through this section of the drier, and the conditions of the gas leaving this zone (point C, Fig. 12.28) are shown at C_1 on Fig. 12.29. The surface temperature of the solid, which can be estimated by the methods described earlier in the case of batch drying, will vary from that at S_1 (corresponding to point B of Fig. 12.28) to S_1' (corresponding to point A). If radiation and conduction through the solid can be neglected, these are the wet-bulb temperatures corresponding to D and C_1, respectively. For the system air-water, whose wet-bulb and adiabatic-saturation temperatures are the same, these will be both given by an extension of the adiabatic-saturation line DC_1 to the saturation-humidity curve. Heat losses may cause the gas to follow some such path as DC_2. On the other hand, if heat is added to the gas in this section, the path will be represented by a line such as DC_3 and, if the gas is kept at constant temperature, by line DC_4. In the case of the last, the surface temperature of the solid will vary from that at S_1 to that at

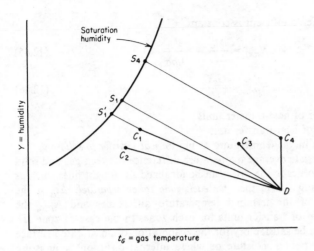

Y = humidity

Saturation
humidity

S_4

S_1

S_1'

C_1

C_2

C_3

C_4

D

t_G = gas temperature

Figure 12.29 Temperature-humidity relations in a continuous drier.

S_4. For any specific drier, the temperatures and humidities can be computed by means of the moisture and enthalpy balances [Eqs. (12.39) and (12.40)] by application of these to each section separately.

Considering only heat transfer from the gas, and neglecting any indirect heat transfer between the solid and the drier itself, we can equate the loss in heat from the gas q_G to that which is transferred to the solid q and the losses Q. For a differential length of the drier, dZ, this becomes

$$dq_G = dq + dQ \qquad (12.41)$$

Rearranging gives

$$dq = dq_G - dQ = U \, dS(t_G - t_S) = Ua(t_G - t_S) \, dZ \qquad (12.42)$$

where U = overall heat-transfer coefficient between gas and solid
$t_G - t_S$ = temperature difference for heat transfer
S = interfacial surface/drier cross section
a = interfacial surface/drier volume

Then $$dq = G_S C_s \, dt_G' = Ua(t_G - t_S) \, dZ \qquad (12.43)$$

where dt_G' is the temperature drop experienced by the gas as a result of transfer of heat to the solid only, exclusive of losses, and C_s is the humid heat.

$$dN_{tOG} = \frac{dt_G'}{t_G - t_S} = \frac{Ua \, dZ}{G_S C_s} \qquad (12.44)$$

and if the heat-transfer coefficient is constant,

$$N_{tOG} = \frac{\Delta t_G'}{\Delta t_m} = \frac{Z}{H_{tOG}} \quad (12.45)$$

$$H_{tOG} = \frac{G_S C_s}{Ua} \quad (12.46)$$

where N_{tOG} = number of heat-transfer units
H_{tOG} = length of heat-transfer unit
$\Delta t_G'$ = change in gas temperature owing to *heat transfer to solid only*
Δt_m = appropriate average temperature difference between gas and solid
If the temperature profiles in the drier can be idealized as straight lines, such as
the broken lines of Fig. 12.28, then *for each zone taken separately* Δt_m is the
logarithmic average of the terminal temperature differences and N_{tOG} the
corresponding number of transfer units for each zone. In the case of zone III,
this simplification will be satisfactory for the evaporation of unsaturated surface
moisture but not for bound moisture or where internal diffusion of moisture
controls the rate of drying.

Tunnel driers In drying solids by cross circulation of air over the surface, as in
the case of materials on trays or solids in sheet form, the surface temperature in
zone II can be estimated through Eq. (12.18). Unless all surfaces are exposed to
heat transfer by radiation, this feature of the heat transfer is better ignored, and
U in Eq. (12.46) can be taken as $h_c + U_k$. This value will also serve in zone I,
and in zone III only for cases of unsaturated surface drying. The quantity a can
be computed from the method of loading the drier.

Rotary driers The surface of the solid exposed to drying gas cannot be conveniently measured, so the group Ua must be considered together. In countercurrent driers, Ua is influenced by changes in holdup due to changes in gas flow
and solid feed rate, but effects of changes in slope or rate of rotation of the drier
are small [14, 44]. The effect of gas mass velocity is somewhat uncertain. In the
absence of experimental data, the value of Ua in Eqs. (12.43) to (12.46) for
commercial driers manufactured in the United States is recommended [34], in SI
units, as†

$$Ua = \frac{237 G^{0.67}}{T_D} \quad (12.47)$$

Here, especially in zone II, the "appropriate" average temperature difference is
the average wet-bulb depression of the gas since the surface of the wet solid is at
the wet-bulb temperature. Use of temperature differences based on the bulk
solid temperature will lead to conservative results. Obviously the character of the
solids must also have an influence, not recognized in Eq. (12.47).

---

† In units of Btu, feet, pounds mass, hours, and degrees Fahrenheit the coefficient of Eq. (12.47)
is 0.5.

Illustration 12.8 A preliminary estimate of the size of a countercurrent, direct-heat, rotary drier for drying an ore-flotation concentrate is to be made. The solid is to be delivered from a continuous filter and introduced into the drier at 8% moisture, 27°C, and is to be discharged from the drier at 150°C, 0.5% moisture. There will be 0.63 kg/s (5000 lb/h) of dried product. The drying gas is a flue gas analyzing 2.5% CO_2, 14.7% O_2, 76.0% N_2, and 6.8% H_2O by volume. It will enter the drier at 480°C. Heat losses will be estimated at 15% of the heat in the entering gas. The ore concentrate is ground to 200-μm average particle diameter and has a bulk density of 1300 kg dry solid/m³ (81.1 lb/ft³) and a heat capacity of 837 J/kg · K (0.2 Btu/lb · °F), dry. The gas rate ought not to exceed 0.70 kg/m² · s to avoid excessive dusting.

SOLUTION $X_1 = 8/(100 - 8) = 0.0870$, $X_2 = 0.5/(100 - 0.5) = 0.00503$ kg water/kg dry solid. Temporarily define S_S and G_S as kg dry substance/s. $S_S = 0.63(1 - 0.00503) = 0.627$ kg dry solid/s. Water to be evaporated = $0.627(0.0870 - 0.00503) = 0.0514$ kg/s.
Basis: 1 kmol gas in. Dry gas = $1 - 0.068 = 0.932$ kmol.

	kmol	kg	Av heat capacity, kJ/kmol · K 480 − 0°C
CO_2	0.025	1.10	45.6
O_2	0.147	4.72	29.9
N_2	0.760	21.3	29.9
Total dry wt		27.1	

Av mol wt dry gas = 27.1/0.932 = 29.1 kg/kmol, nearly the same as that of air.

$$Y_2 = \frac{0.068(18.02)}{0.932(29.1)} = 0.0452 \text{ kg water/kg dry gas} \qquad t_{G2} = 480°C$$

$$\text{Av ht capacity of dry gas} = \frac{0.025(45.6) + (0.147 + 0.760)(29.9)}{0.932(29.1)}$$

$$= 1.042 \text{ kJ/kmol} \cdot K$$

The exit gas temperature will be tentatively taken as 120°C. This is subject to revision after the number of transfer units has been computed. In a manner similar to that above, the average heat capacity of the dry gas, 120 to 0°C, is 1.005 kJ/kmol · K. Eq. (7.13):

$$H'_{G2} = [1.042 + 1.97(0.0452)](480 - 0) + 2502.3(0.0452) = 656.0 \text{ kJ/kg dry gas}$$

$$H'_{G1} = (1.005 + 1.884 Y_1)(120 - 0) + 2502.3 Y_1 = 120.5 + 2728 Y_1$$

Take $\Delta H_A = 0$. Eq. (11.45):

$$H'_{S1} = 0.837(27 - 0) + 0.087(4.187)(27 - 0) = 32.43 \text{ kJ/kg dry solid}$$

$$H'_{S2} = 0.837(150 - 0) + 0.00503(4.187)(150 - 0) = 128.7 \text{ kJ/kg dry solid}$$

$$Q = \text{heat loss} = 0.15(656)G_S = 98.4 G_S \text{ kJ/s}$$

Eq. (12.39):

$$0.0514 = G_S(Y_1 - 0.0452)$$

Eq. (12.40):

$$0.627(32.43) + G_S(656) = 0.627(128.7) + G_S(120.5 + 2728 Y_1) + 98.4 G_S$$

Simultaneous solution:

$$G_S = 0.639 \text{ kg dry gas/s} \qquad Y_1 = 0.1256 \text{ kg water/kg dry gas}$$

$$H'_{G1} = 463 \text{ kJ/kg dry gas} \qquad Q = 62.9 \text{ kJ/s}$$

Assuming that the psychrometric ratio of the gas is the same as that of air, its wet-bulb temperature for zone II [Eq. (7.26)] is estimated to be about 65°C. The surface of the drying solid particles is subject to radiation from the hot walls of the drier, and the surface of the solid in zone II is then estimated to be about 68°C. *Note*: This should be recalculated after the temperature at point D, Fig. 12.28, is known.

For lack of information on the critical moisture content of the solid, which is probably quite low for the conditions found in such a drier, it will be assumed that all moisture is evaporated in zone II at 68°C. Zone I will be taken as a preheat zone for warming the solid to 68°C, without drying. Enthalpy of the solid at 68°C, $X = 0.0870$ (point A, Fig. 12.28) = $0.837(68 - 0) + 0.0870(4.187)(68 - 0) = 81.7$ kJ/kg dry solid. Similarly, enthalpy of the solid at 68°C, $X = 0.00503$ (point B, Fig. 12.28) = 58.3 kJ/kg dry solid.

Assuming that heat losses in the three zones are proportional to the number of transfer units in each zone and to the average temperature difference between the gas and the surrounding air (27°C), the losses are apportioned (by a trial-and-error calculation) as 14% in zone I, 65% in zone II, 21% in zone III.

Calculation for zone III Humid heat of entering gas = $1.042 + 1.97(0.0452) = 1.131$ kJ/(kg dry gas) · K. A heat balance:

$$0.639(1.131)(480 - t_{GD}) = 0.627(128.7 - 58.3) + 0.21(62.9)$$

$$t_{GD} = \text{gas temp at } D \text{ (Fig. 12.28)} = 401°C$$

The change in gas temperature, exclusive of that resulting from heat losses, is

$$\Delta t'_G = \frac{0.627(128.7 - 58.3)}{0.639(1.131)} = 61.1°C$$

Average temp difference between gas and solid = average of $480 - 150$ and $401 - 68 = 332°C$ = Δt_m

$$N_{tOG} = \frac{\Delta t'_G}{\Delta t_m} = \frac{61.1}{332} = 0.18$$

Calculation for zone I Humid heat of outlet gas = $1.005 + 1.884(0.1256) = 1.242$ kJ/(kg dry gas) · K. A heat balance:

$$0.639(1.242)(t_{GC} - 120) = 0.627(81.7 - 32.43) + 0.14(62.9)$$

$$t_{GC} = \text{gas temp at } C \text{ (Fig. 12.28)} = 170°C$$

$$\Delta t'_G = \frac{0.627(81.7 - 32.43)}{0.639(1.242)} = 38.9°C$$

$$\Delta t_m = \text{average of } 170 - 68 \text{ and } 120 - 27 = 98°C$$

$$N_{tOG} = \frac{\Delta t'_G}{\Delta t_m} = \frac{38.9}{98} = 0.39$$

Calculation for zone II

$$\text{Average humid heat of gas} = \frac{1.042 + 1.242}{2} = 1.142 \text{ kJ/kg} \cdot \text{K}$$

$$\text{True change in gas temp} = 401 - 170 = 231°C$$

$$\text{Change in temp resulting from heat loss} = \frac{0.65(62.9)}{0.639(1.142)} = 56°C$$

$\Delta t'_G$ resulting from heat transfer to solid = $231 - 56 = 175°C$

$$\Delta t_m = \frac{(401 - 68) - (170 - 68)}{\ln[(401 - 68)/(170 - 68)]} = 195°C$$

$$N_{tOG} = \frac{\Delta t'_G}{\Delta t_m} = \frac{175}{195} = 0.90$$

$$\text{Total } N_{tOG} = 0.39 + 0.90 + 0.18 = 1.47$$

Size of drier Standard diameters available are 1, 1.2, and 1.4 m, and larger. The 1-m diameter is too small. For $T_D = 1.2$ m, the cross-sectional area $= \pi(1.2)^2/4 = 1.13$ m^2. Expressing the rates of flow as kg dry substance/m$^2 \cdot$ s gives $G_S = 0.639/1.13 = 0.565$, $S_S = 0.627/1.13 = 0.555$ kg/m$^2 \cdot$ s.

$$ \text{Av } G = G_S(1 + Y_{av}) = 0.565\left(1 + \frac{0.1256 + 0.0452}{2}\right) = 0.613 \text{ kg/m}^2 \cdot \text{s} $$

For lack of more specific information, use Eq. (12.47):

$$ Ua = \frac{237(0.613)^{0.67}}{1.2} = 142.3 \text{ W/m}^3 \cdot \text{K} $$

$$ H_{tOG} = \frac{G_S C_s}{Ua} = \frac{0.565(1142)}{142.3} = 4.5 \text{ m} \qquad Z = N_{tOG} H_{tOG} = 1.47(4.5) = 6.6 \text{ m (21.7 ft)} $$

The nearest standard length is 7.5 m (24.6 ft).

Take the peripheral speed as 0.35 m/s (70 ft/min), whence the rate of revolution $= N = 0.35/\pi T_D = 0.093$ s^{-1} (5.6 r/min).

$$ d_p = 200 \times 10^{-6} \text{ m} \qquad \rho_S = 1300 \text{ kg/m}^3 $$

Eq. (12.37):

$$ K = \frac{0.6085}{1300(200 \times 10^{-6})^{0.5}} = 0.0331 $$

Take the holdup $\phi_D = 0.05$. Eq. (12.35): $\phi_{D0} = \phi_D - KG = 0.05 - 0.0331(0.613) = 0.0297$. Eq. (12.36):

$$ s = \frac{0.3344 S_S}{\phi_{D0} \rho_S N^{0.9} T_D} = \frac{0.3344(0.555)}{0.05(1300)(0.093)^{0.9} 1.2} = 0.02 \text{ m/m, drier slope} $$

Drying at low temperatures Continuous driers operating at low temperatures can be divided into zones the same as high-temperature driers. Since the surface moisture will evaporate at a comparatively low temperature in zone II, the preheat zone can generally be ignored and only zones II and III need be considered. Refer to Fig. 12.30, which shows an arrangement for countercurrent flow. In zone II, unbound and surface moisture is evaporated as discussed previously, and the moisture content of the solid falls to the critical value X_c. The rate of drying in this zone would be constant if it were not for the varying conditions of the gas. In zone III, unsaturated-surface drying and evaporation of bound moisture occur, and the gas humidity rises from its initial value Y_2 to Y_c. The latter can be calculated by applying the material-balance relation (12.39) to either zone separately. The retention time can be calculated by integration of Eq. (12.3),

$$ \theta = \theta_{II} + \theta_{III} = \frac{S_S}{A}\left(\int_{X_c}^{X_1} \frac{dX}{N} + \int_{X_2}^{X_c} \frac{dX}{N}\right) \tag{12.48} $$

where A/S_S is the specific exposed drying surface, area/mass dry solid.

Zone II, $X > X_c$ The rate N is given by Eq. (12.17), which, when substituted in the first part of Eq. (12.48), provides

$$ \theta_{II} = \frac{S_S}{A} \frac{1}{k_Y} \int_{X_c}^{X_1} \frac{dX}{Y_s - Y} \tag{12.49} $$

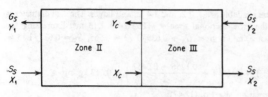

Figure 12.30 Continuous low-temperature countercurrent drier.

Since $G_S \, dY = S_S \, dX$, Eq. (12.49) becomes

$$\theta_{\text{II}} = \frac{G_S}{S_S} \frac{S_S}{A} \frac{1}{k_Y} \int_{Y_c}^{Y_1} \frac{dY}{Y_s - Y} \tag{12.50}$$

Integration of Eq. (12.50) must take into account the variation of Y_s, the humidity of the gas at the solid surface, with Y. If the gas temperature, for example, is held constant in this zone by application of heat, the path of the gas resembles line DC_4 in Fig. 12.29. If, furthermore, radiation and conduction effects can be neglected, Y_s for any value of Y on line DC_4 is the saturated humidity at the corresponding wet-bulb temperature. Equation (12.50) can then be integrated graphically.

For the case where Y_s is constant, as for adiabatic drying of water into air, Eq. (12.50) becomes

$$\theta_{\text{II}} = \frac{G_S}{S_S} \frac{S_S}{A} \frac{1}{k_Y} \ln \frac{Y_s - Y_c}{Y_s - Y_1} \tag{12.51}$$

Zone III, $X < X_c$ Some simplification is necessary for mathematical treatment. For the case where unsaturated-surface drying occurs and the drying rate is dependent strictly upon the conditions prevailing at any instant, independent of the immediate past history of the drying, Eqs. (12.8) and (12.17) apply. These provide

$$N = \frac{N_c(X - X^*)}{X_c - X^*} = \frac{k_Y(Y_s - Y)(X - X^*)}{X_c - X} \tag{12.52}$$

When this is substituted in the second part of Eq. (12.48), the result is

$$\theta_{\text{III}} = \frac{S_S}{A} \frac{X_c - X^*}{k_Y} \int_{X_2}^{X_c} \frac{dX}{(Y_s - Y)(X - X^*)} \tag{12.53}$$

This can be evaluated graphically after determining the relationship between X, X^*, Y_s, and Y. For this purpose, the material balance can be written as

$$Y = Y_2 + (X - X_2) \frac{S_S}{G_S} \tag{12.54}$$

The surface humidity Y_s is found in the manner previously described, and X^* is given by the equilibrium-moisture curve for the appropriate Y.

For the special case where the bound moisture is negligible ($X^* = 0$) and Y_s is constant (adiabatic drying), substitution of Eq. (12.54) and its differential

DRYING **709**

$G_S \, dY = L_S \, dX$ in Eq. (12.53) provides

$$\theta_{III} = \frac{G_S}{S_S} \frac{S_S}{A} \frac{X_c}{k_Y} \int_{Y_2}^{Y_c} \frac{dY}{(Y_s - Y)[(Y - Y_2)G_S/S_S + X_2]} \qquad (12.55)$$

$$\theta_{III} = \frac{G_S}{S_S} \frac{S_S}{A} \frac{X_c}{k_Y} \frac{1}{(Y_s - Y_2)G_S/S_S + X_2} \ln \frac{X_c(Y_s - Y_2)}{X_2(Y_s - Y_c)} \qquad (12.56)$$

These methods must not be applied to solids whose internal resistance to movement of moisture is large, where internal diffusion controls the rate of drying, or where case hardening occurs. In these circumstances the instantaneous rate of drying under variable conditions is not merely a function of the prevailing conditions but depends upon the immediate past drying history as well. For such solids, the time for drying is best determined experimentally in a carefully planned test which simulates the countercurrent action of the continuous drier [4].

In applying Eqs. (12.48) to (12.56), it is clear that, should drying take place only above or only below the critical moisture content, appropriate changes in the limits of moisture content and gas humidities must be made. In parallel-flow driers, where gas enters at humidity Y_1 and leaves at Y_2 while solid enters at moisture content X_1 and leaves at X_2, Eqs. (12.51) and (12.56) become

$$\theta_{II} = \frac{G_S}{S_S} \frac{S_S}{A} \frac{1}{k_Y} \int_{Y_1}^{Y_c} \frac{dY}{Y_s - Y} = \frac{G_S}{S_S} \frac{S_S}{A} \frac{1}{k_Y} \ln \frac{Y_s - Y_1}{Y_s - Y_c} \qquad (12.57)$$

$$\theta_{III} = \frac{S_S}{A} \frac{X_c - X^*}{k_Y} \int_{X_2}^{X_c} \frac{dX}{(Y_s - Y)(X - X^*)}$$

$$= \frac{G_S}{S_S} \frac{S_S}{A} \frac{X_c}{k_Y} \frac{1}{(Y_s - Y_c)G_S/S_S - X_2} \ln \frac{X_c(Y_s - Y_2)}{X_2(Y_s - Y_c)} \qquad (12.58)$$

Illustration 12.9 Wet rayon skeins, after centrifuging, are to be air-dried from 46 to 8.5% water content in a continuous countercurrent tunnel drier. The skeins are hung on poles which travel through the drier. The air is to enter at 82°C, humidity 0.03 kg water/kg dry air, and is to be discharged at a humidity 0.08. The air temperature is to be kept constant at 82°C by heating coils within the drier. The air rate is to be 1.36 kg/m² · s (1000 lb/ft² · h).

The critical moisture content of rayon skeins is 50%, and its percent equilibrium moisture at 82°C can be taken as one-fourth of the percent relative humidity of the air. The rate of drying is then [Simons, Koffolt, and Withrow, *Trans. AIChE*, **39**, 133 (1943)]

$$\frac{-dX}{d\theta} = 0.0137 G^{1.47}(X - X^*)(Y_W - Y)$$

where Y_W is the saturation humidity of the air at the wet-bulb temperature corresponding to Y. *Note*: Here G is expressed as kg/m² · s and θ as seconds; for feet, pounds mass, and hours, the original paper gives the coefficient as 0.003.

Determine the time the rayon should remain in the drier.

SOLUTION $X_1 = 0.46/(1 - 0.46) = 0.852$; $X_2 = 0.085/(1 - 0.085) = 0.093$ kg water/kg dry solid. $Y_1 = 0.08$; $Y_2 = 0.03$ kg water/kg dry air. A water balance [Eq. (12.39)] is

$$\frac{S_S}{G_S} = \frac{0.08 - 0.03}{0.852 - 0.093} = 0.0660 \text{ kg dry solid/kg air}$$

Since the initial moisture content of the rayon is less than the critical, drying takes place entirely within zone III. The form of the rate equation is the same as that of Eq. (12.22), where $k_Y A / S_S (X_c - X^*) = 0.0137 G^{1.47}$. Rearranging the rate equation gives

$$\theta_{III} = \int_0^\theta d\theta = \frac{1}{0.0137 G^{1.47}} \int_{X_2}^{X_1} \frac{dX}{(X - X^*)(Y_W - Y)}$$

which is in the form of Eq. (12.53). On substituting $G = 1.36$, this becomes

$$\theta_{III} = 46.4 \int_{0.093}^{0.852} \frac{dX}{(X - X^*)(Y_W - Y)} \qquad (12.59)$$

Consider that part of the drier where the moisture content of the rayon is $X = 0.4$. Eq. (12.54):

$$Y = 0.03 + (0.4 - 0.093)0.066 = 0.0503 \text{ kg water/kg dry gas}$$

At 82°C, $Y = 0.0503$, the wet-bulb temperature is 45°C, and the corresponding saturation humidity $Y_W = 0.068$ (Fig. 7.5). Eq. (7.8):

$$0.0503 = \frac{\bar{p}}{101\,330 - \bar{p}} \frac{18}{29}$$

$\bar{p}$ = partial pressure of water = 7584 N/m^2

The vapor pressure of water at 82°C = p = 51 780 N/m^2, and the relative humidity of the air = $(7584/51\,780)100 = 14.63\%$. The equilibrium moisture is $14.63/4 = 3.66\%$, and $X^* = 3.66/(100 - 3.66) = 0.038$ kg water/kg dry solid. Therefore

$$\frac{1}{(X - X^*)(Y_W - Y)} = \frac{1}{(0.4 - 0.038)(0.068 - 0.0503)} = 156$$

In similar fashion, other values of this quantity are calculated for other values of X, as follows:

X	Y	Y_W	Relative humidity, %	X^*	$\dfrac{1}{(X - X^*)(Y_W - Y)}$
0.852	0.080	0.0950	22.4	0.0594	84
0.80	0.0767	0.0920	21.5	0.0568	88
0.60	0.0635	0.0790	18.17	0.0488	117
0.40	0.0503	0.0680	14.63	0.0380	156
0.20	0.0371	0.0550	11.05	0.0284	325
0.093	0.030	0.0490	9.04	0.0231	755

The integral of Eq. (12.59) is evaluated either graphically or numerically; the area is 151.6, whence, by Eq. (12.59),

$$\theta_{III} = 46.4(151.6) = 7034 \text{ s} = 1.95 \approx 2 \text{ h} \quad \textbf{Ans.}$$

NOTATION FOR CHAPTER 12

Any consistent set of units can be used, except as noted.

a	specific interfacial surface of the solid, L^2/L^3
A	cross-sectional area perpendicular to the direction of flow for through-circulation drying, L^2
	drying surface for cross-circulation drying, L^2
A_m	average cross-sectional area of drying solid, L^2
b	const

C_A	heat capacity of moisture as a liquid, FL/MT
C_p	heat capacity at constant pressure, FL/MT
C_s	humid heat, heat capacity of wet gas per unit mass dry gas content, FL/MT
C_S	heat capacity of dry solid, FL/MT
d_e	equivalent diameter, (cross-sectional area)/perimeter, L
d_p	particle diameter, L
D	diffusivity, L^2/Θ
G	mass velocity of gas, $M/L^2\Theta$
G_S	mass velocity of dry gas, $M/L^2\Theta$
h_c	heat-transfer coefficient for convection, $FL/L^2T\Theta$
h_R	heat-transfer coefficient for radiation, $FL/L^2T\Theta$
ΔH_A	integral heat of wetting (or of adsorption, hydration, or solution) referred to pure liquid and solid, per unit mass of dry solid, FL/M
H_G'	enthalpy of moist gas per unit mass of dry gas, FL/M
H_S'	enthalpy of wet solid per unit mass of dry solid, FL/M
H_{tOG}	length of an overall gas transfer unit, L
j_D	$k_Y \, Sc^{2/3}/G_S$, dimensionless
j_H	$h_c \, Pr^{2/3}/C_p G$, dimensionless
k_M	thermal conductivity of tray, $FL^2/L^2T\Theta$
k_S	thermal conductivity of drying solid, $FL^2/L^2T\Theta$
k_Y	gas-phase mass-transfer coefficient, mass evaporated/(area)(time) (humidity difference), $M/L^2\Theta(M/M)$
K	const
m	const
N	flux of drying, mass of moisture evaporated/(area)(time), $M/L^2\Theta$
	rate of revolution, Θ^{-1}
N_c	constant flux of drying, $M/L^2\Theta$
N_{tG}	number of gas transfer units, dimensionless
N_{tOG}	number of overall gas transfer units, dimensionless
p	saturation vapor pressure, F/L^2
$\bar{p}$	partial pressure, F/L^2
p_t	total pressure, F/L^2
Pr	Prandtl number $= C_p\mu/k$, dimensionless
q	flux of heat received at the drying surface, batch drying, $FL/L^2\Theta$
	flux of heat received by the solid/area of drier cross section, continuous drying, $FL/L^2\Theta$
q_c	heat flux for convection, $FL/L^2\Theta$
q_G	flux of heat transferred from the gas per unit drier cross section, $FL/L^2\Theta$
q_k	heat flux for conduction, $FL/L^2\Theta$
q_R	heat flux for radiation, $FL/L^2\Theta$
Q	net flux of heat loss, per unit drier cross section, $FL/L^2\Theta$
Re_e	Reynolds number for duct, $d_e G/\mu$, dimensionless
s	slope of drier, L/L
S	interfacial surface of solid/bed cross section, L^2/L^2
S_S	mass of dry solid in a batch, batch drying, M
	mass velocity of dry solid, continuous drying, $M/L^2\Theta$
Sc	Schmidt number, $\mu/\rho D$, dimensionless
t_G	dry-bulb temperature of a gas, T
t_R	temperature of radiating surface, T
t_s	surface temperature, T
t_S	solid temperature, T
t_0	reference temperature, T
T_D	diameter of drier, L
T_G	absolute temperature of a gas, T
T_R	absolute temperature of radiating surface, T

T_s	absolute temperature of surface, T
U	overall heat-transfer coefficient, $FL/L^2T\Theta$
U_k	overall heat-transfer coefficient [Eq. (12.16)], $FL/L^2T\Theta$
X	moisture content of a solid, mass moisture/mass dry solid, M/M
X^*	equilibrium moisture content of a solid, mass moisture/mass dry solid, M/M
z_M	thickness of tray material, L
z_S	thickness of drying solid, L
Z	length of drier, L
Δ	difference
ϵ	emissivity of drying surface, dimensionless
θ	time, Θ
λ_y	latent heat of vaporization at t_s, FL/M
μ	viscosity, $M/L\Theta$
π	3.1416
ρ	density, M/L^3
ρ_S	apparent density of solid, mass dry solid/wet volume, M/L^3
ϕ_D	holdup of solid in a continuous drier, vol solids/vol drier, L^3/L^3
ϕ_{D0}	holdup of solid at no gas flow, L^3/L^3

Subscripts

as	adiabatic saturation
c	at critical moisture content
G	gas
m	average
max	maximum
s	surface
S	dry gas, dry solid
1	at beginning, batch drying; at solids-entering end, continuous drier
2	at end, batch drying; at solids-leaving end, continuous drier
I, II, III	zones I, II, and III in a continuous drier

REFERENCES

1. Allerton, J., L. E. Brownell, and D. L. Katz: *Chem. Eng. Prog.*, **45**, 619 (1949).
2. Beram, Z., and J. Luicha: *Chem. Eng. Lond.*, **1975**, 678.
3. Boltas, L., and W. H. Gauvin: *AIChE J.*, **15**, 764, 772 (1969).
4. Broughton, D. B., and H. S. Mickley: *Chem. Eng. Prog.*, **49**, 319 (1953).
5. Coggan, G. C.: *Chem. Eng. J.*, **2**, 55 (1971).
6. Comings, E. W., and T. K. Sherwood: *Ind. Eng. Chem.*, **26**, 1096 (1934).
7. Chu, J. C., S. Finelt, W. Hoerner, and M. Lin: *Ind. Eng. Chem.*, **51**, 275 (1959).
8. Crosby, E. J., and W. R. Marshall, Jr.: *Chem. Eng. Prog.*, **54**(7), 56 (1958).
9. Debrand, S.: *Ind. Eng. Chem. Process Des. Dev.*, **13**, 396 (1974).
10. Dittman, F. W.: *Chem. Eng.*, **84**, 106 (Jan. 17, 1977).
11. Dyer, D. F., and J. E. Sunderland: *J. Heat Transfer*, **90**, 379 (1968).
12. Erissman, J. L.: *Ind. Eng. Chem.*, **30**, 996 (1938).
13. Fairbanks, H. V., and R. E. Cline: *IEEE Trans. Sonics Ultrason.*, SU-14(4), 175 (1967); *AIME Trans.*, **252**, 70 (1972).
14. Friedman, S. J., and W. R. Marshall, Jr.: *Chem. Eng. Prog.*, **45**, 482, 573 (1949).
15. Gauvin, W. H., and S. Katta: *AIChE J.*, **21**, 143 (1975); **22**, 713 (1976).
16. Gilliland, E. R., and T. K. Sherwood: *Ind. Eng. Chem.*, **25**, 1134 (1933).
17. Harmathy, T. Z.: *Ind. Eng. Chem. Fundam.*, **8**, 92 (1969).
18. Harper, J. C., and A. L. Tappel: *Adv. Food Res.*, **7**, 171 (1957).

19. Hougen, O. A., and H. J. McCauley: Trans. AIChE, 36, 183 (1940).
20. Hurxtal, A. O.: Ind. Eng. Chem., 30, 1004 (1938).
21. Keey, R. B.: Chem. Eng. Sci., 23, 1299 (1968).
22. Keey, R. B.: "Drying, Principles and Practices," Pergamon, Elmsford, N.Y., 1973.
23. Kern, D. Q.: "Process Heat Transfer," McGraw-Hill Book Company, New York, 1950.
24. Lane, A. M., and S. Stern: Mech. Eng., 78, 423 (1956).
25. Lapple, W. C., W. E. Clark, and E. C. Dybdal: Chem. Eng., 62(11), 117 (1955).
26. Lee, D. A.: Am. Ceram. Soc. Bull., 55, 498 (1976).
27. Loesecke, H. W. von: "Drying and Dehydration of Foods," 2d ed., Reinhold, New York, 1955.
28. Lyne, C. W.: Br. Chem. Eng., 16, 370 (1971).
29. Marshall, W. R., Jr.: "Atomization and Spray Drying," Chem. Eng. Prog. Monogr. Ser., 50(2), (1954).
30. Marshall, W. R., Jr.: in H. F. Mark, J. J. McKetta, D. F. Othmer, and A. Standen (eds.), "Encyclopedia of Chemical Technology," 2d ed., vol. 7, p. 326, Interscience, New York, 1965.
31. Marshall, W. R., Jr., and O. A. Hougen: Trans. AIChE, 38, 91 (1942).
32. Masters, K. "Spray Drying," CRC Press, Cleveland, Ohio, 1972.
33. Mathur, K. B., and N. Epstein: "Spouted Beds," Academic, New York, 1974.
34. McCormick, P. Y.: in R. H. Perry and C. H. Chilton (eds.), "Chemical Engineers' Handbook," 5th ed., sec. 20, McGraw-Hill Book Company, New York, 1973.
35. McCready, D. W., and W. L. Mc Cabe: Trans. AIChE, 29, 131 (1933).
36. Miskell, F., and W. R. Marshall, Jr.: Chem. Eng. Prog., 52(1), 35-J (1956).
37. Molstad, M. C., P. Farevaag, and J. A. Farrell: Ind. Eng. Chem., 30, 1131 (1938).
38. Newitt, D. M., et al.: Trans. Inst. Chem. Eng. Lond., 27, 1 (1949); 30, 28 (1952); 33, 52, 64 (1955).
39. Nissan, A. H., W. A. Kaye, and J. R. Bell: AIChE J., 5, 103, 344 (1959).
40. Nonhebel, G., and A. A. H. Moss: "Drying of Solids in the Chemical Industry," Butterworth, London, 1971.
41. Pasquill, F.: Proc. R. Soc., A182, 75 (1944).
42. Pearse, J. G., T. R. Oliver, and D. M. Newitt: Trans. Inst. Chem. Eng. Lond., 27, 1, 9 (1949).
43. Peterson, W. S.: Can. J. Chem. Eng., 40, 226 (1962).
44. Saeman, W. C.: Chem. Eng. Prog., 58(6), 49 (1962).
45. Saeman, W. C., and T. R. Mitchell: Chem. Eng. Prog., 50, 467 (1954).
46. Shepherd, C. B., C. Haddock, and R. C. Brewer: Ind. Eng. Chem., 30, 389 (1938).
47. Sherwood, T. K.: Trans. AIChE, 27, 190 (1931).
48. Sherwood, T. K.: Ind. Eng. Chem., 21, 12, 976 (1929); 22, 132 (1930); 24, 307 (1932).
49. Sherwood, T. K., and E. W. Comings: Ind. Eng. Chem., 25, 311 (1933).
50. Siegel, R., and J. R. Howell: "Thermal Radiation Heat Transfer," McGraw-Hill Book Company, New York, 1972.
51. Spraul, J. R.: Ind. Eng. Chem., 47, 368 (1955).
52. Uhl, V. C., and W. L. Root: Chem. Eng. Prog., 58(6), 37 (1962).
53. Vanáček, V., M. Markvart, and R. Drobohlav: "Fluidized Bed Drying," Hill, London, 1966.
54. Victor, V. P.: Chem. Met. Eng., 52(7), 105 (1945).
55. Welty, J. R.: "Engineering Heat Transfer," Wiley, New York, 1974.
56. Wheat, J. A., and D. A. MacLeod: Can. J. Chem. Eng., 37, 47 (1959).
57. Whitwell, J. C., and R. K. Toner: Text. Res. J., 17, 99 (1947).
58. Williams-Gardner, H.: "Industrial Drying," Hill, London, 1971.
59. Wormwald, D., and E. M. W. Burnell: Br. Chem. Eng., 16, 376 (1971).

PROBLEMS

12.1 A plant wishes to dry a certain type of fiberboard in sheets 1.2 by 2 m by 12 mm (4 by 6 ft by ½ in). To determine the drying characteristics, a 0.3- by 0.3-m (1- by 1-ft) sample of the board, with the edges sealed so that drying took place from the two large faces only, was suspended from a balance in a laboratory cabinet drier and exposed to a current of hot, dry air. The initial moisture content was 75%. The sheet lost weight at the constant rate of 1×10^{-4} kg/s (0.8 lb/h) until the moisture content fell to 60%, whereupon the drying rate fell. Measurements of the rate of drying were discontinued, but after a long period of exposure to this air it was established that the equilibrium moisture content was 10%. The dry mass of the sample was 0.9 kg (2 lb). All moisture contents are on the wet basis.

Determine the time for drying the large sheets from 75 to 20% moisture under the same drying conditions.

12.2 A sample of a porous, manufactured sheet material of mineral origin was dried from both sides by cross circulation of air in a laboratory drier. The sample was 0.3 m square and 6 mm thick, and the edges were sealed. The air velocity over the surface was 3 m/s, its dry-bulb temperature was 52°C, and its wet-bulb temperature 21°C. There were no radiation effects. The solid lost moisture at a constant rate of 7.5×10^{-5} kg/s until the critical moisture content, 15% (wet basis), was reached. In the falling-rate period, the rate of evaporation fell linearly with moisture content until the sample was dry. The equilibrium moisture was negligible. The dry weight of the sheet was 1.8 kg.

Estimate the time for drying sheets of this material 0.6 by 1.2 m by 12 mm thick from both sides, from 25 to 2% moisture (wet basis), using air of dry-bulb temperature 66°C but of the same absolute humidity at a linear velocity over the sheet of 5 m/s. Assume no change in the critical moisture with the changed drying conditions. **Ans.**: 3.24 h.

12.3 Estimate the rate of drying during the constant-rate period for the conditions existing as the air enters the trays of the drier of Illustration 12.2. The solid being dried is a granular material of thermal conductivity when wet = 1.73 W/m · K, and it completely fills the trays. The metal of the trays is stainless steel, 16 BWG (1.65 mm thick). Include in the calculations an estimate of the radiation effect from the undersurface of each tray upon the drying surface.

12.4 A laboratory drying test was made on a 0.1-m² sample of a fibrous boardlike material. The sample was suspended from a balance, its edges were sealed, and drying took place from the two large faces. The air had a dry-bulb temperature of 65°C, wet-bulb temperature 29°C, and its velocity was 1.5 m/s past the sample. The following are the weights recorded at various times during the test:

Time, h	Mass, kg	Time, h	Mass, kg	Time, h	Mass, kg
0	4.820	3.0	4.269	7.0	3.885
0.1	4.807	3.4	4.206	7.5	3.871
0.2	4.785	3.8	4.150	8.0	3.859
0.4	4.749	4.2	4.130	9.0	3.842
0.8	4.674	4.6	4.057	10.0	3.832
1.0	4.638	5.0	4.015	11	3.825
1.4	4.565	5.4	3.979	12	3.821
1.8	4.491	5.8	3.946	14	3.819
2.2	4.416	6.0	3.933	16	3.819
2.6	4.341	6.5	3.905		

The sample was then dried in an oven at 110°C, and the dry mass was 3.765 kg.

(a) Plot the rate-of-drying curve.

(b) Estimate the time required for drying the same sheets from 20 to 2% moisture (wet basis) using air of the same temperature and humidity but with 50% greater air velocity. Assume that the critical moisture remains unchanged. **Ans.**: 8.55 h.

12.5 A pigment material which has been removed wet from a filter press is to be dried by extruding it into small cylinders and subjecting them to through-circulation drying. The extrusions are 6 mm in diameter, 50 mm long, and are to be placed on screens to a depth of 65 mm. The surface of the particles is estimated to be 295 m²/m³ of bed and the apparent density 1040 kg dry solid/m³. Air at a mass velocity 0.95 kg dry air/m² · s will flow through the bed, entering at 120°C, humidity 0.05 kg water/kg dry air.

(a) Estimate the constant rate of drying to be expected. *Note*: For long cylinders it is best to take the equivalent diameter as the actual cylinder diameter. **Ans.**: 0.0282 kg/m$^2 \cdot$ s.

(b) Estimate the constant rate of drying to be expected if the filter cake is dried on trays by cross circulation of the air over the surface at the same mass velocity, temperature, and humidity. Neglect radiation and heat conduction through the solid. **Ans.**: 4.25×10^{-4} kg/m$^2 \cdot$ s.

12.6 A louver-type continuous rotary drier (Fig. 12.22) was used to dry wood chips from 40 to 15% moisture [Horgan, *Trans. Inst. Chem. Eng.*, **6**, 131 (1928)]. The wood was at 33°F (0.56°C), while the dried product was discharged at 100°F (37.8°C) at a rate of 3162 lb/h (0.398 kg/s). The drying medium was the gas resulting from the combustion of fuel, but for the present calculation it may be assumed to have the characteristics of air. It entered the drier at 715°F (380°C), with a humidity 0.038 kg water vapor/kg dry gas, at a rate of 275 lb/min (2.079 kg/s) wet. The gas was discharged at 175°F (77°C). The heat capacity of the dry wood can be taken as 0.42 Btu/lb $\cdot$ °F (1758 J/kg $\cdot$ K), and the heat of wetting can be ignored. Estimate the rate of heat loss.

12.7 A direct-heat, cocurrent-flow rotary drier, 8 ft diameter, 60 ft long (2.44 by 18.3 m), was used to dry chopped alfalfa [see Gutzeit and Spraul, *Chem. Eng. Prog.*, **49**, 380 (1953)]. Over a 5-h test period, the drier delivered an average of 2200 lb/h (0.28 kg/s) of dried product at 11% moisture and 145°F (63°C) when fed with alfalfa containing 79% moisture at 80°F (26.7°C). The drying medium was the combustion products resulting from the burning of 13 074 ft^3/h (80°F, 4 oz/in^2 gauge pressure) [0.1029 m^3/s (26.7°C, 862 N/m^2 gauge pressure)] of natural gas (85% methane, 10% ethane, 5% nitrogen by volume) with air at 80°F (26.7°C), 50% humidity. The gas analyzed 2.9% CO_2, 15.8% O_2, 81.3% N_2 by volume on a dry basis; it entered the drier at 1500°F (816°C) and left at 195°F (91°C). The heat capacity of dry alfalfa is estimated to be 0.37 Btu/lb $\cdot$ °F (1549 J/kg $\cdot$ K), and the heat of wetting can be neglected. Compute the volumetric rate of gas flow through the exhaust fan and the heat losses. **Ans.**: 6.4 m^3/s (13 550 ft^3/min), 1500 kW (5.12 $\times$ 10^6 Btu/h).

12.8 A direct-heat countercurrent rotary hot-air drier is to be chosen for drying an insoluble crystalline organic solid. The solid will enter at 20°C, containing 20% water. It will be dried by air entering at 155°C, 0.01 kg water/kg dry air. The solid is expected to leave at 120°C, with a moisture content 0.3%. Dried product delivered will be 450 kg/h. The heat capacity of the dry solid is 837 J/kg $\cdot$ K, and its average particle size is 0.5 mm. The superficial air velocity should not exceed 1.6 m/s in any part of the drier. The drier will be insulated, and heat losses can be neglected for present purposes. Choose a drier from the following standard sizes and specify the rate of airflow which should be used: 1 by 3 m, 1 by 9 m, 1.2 by 12 m, 1.4 by 9 m, 1.5 by 12 m. **Ans.**: 1.2 by 12 m.

12.9 A manufactured material in the form of sheets 0.6 by 1.2 m by 12 mm is to be continuously dried in an adiabatic countercurrent hot-air tunnel drier at the rate of 100 sheets per hour. The sheets will be supported on a special conveyor carrying the material in tiers 30 sheets high, and they will be dried from both sides. The dry mass of each sheet is 12 kg, and the moisture content will be reduced from 50 to 5% water by air entering at 120°C, humidity 0.01 kg water/kg dry air; 40 kg dry air will be passed through the drier per kilogram dry solid.

In a small-scale experiment, when dried with air at constant drying conditions, dry-bulb temperature 95°C, wet-bulb temperature 50°C, and at the same velocity to be used in the large drier, the constant-drying rate was 3.4×10^{-4} (kg water evaporated)/m$^2 \cdot$ s and the critical moisture content 30%. The equilibrium-moisture content was negligible.

(a) Calculate the value of k_Y from the data of the small-scale experiment. **Ans.**: 0.01545 kg/m$^2 \cdot$ s $\cdot \Delta Y$.

(b) For the large drier, calculate the humidity of the air leaving and at the point where the solid reaches the critical moisture content.

(c) Estimate the time of drying in the large drier. **Ans.**: 9.53 h.

(d) How many sheets of material will be in the drier at all times?

12.10 A continuous countercurrent hot-air tunnel drier is to be designed to dry a filter-press cake of coarse crystals of an inorganic substance, insoluble in water. The filter-press cake will be placed on trays 1.0 m long by 0.9 m wide by 25 mm, 20 trays to a truck, with 50 mm between trays. The tunnel drier will have a cross section 2 m high by 1 m wide. The trays have a reinforced screen bottom, so that drying takes place from both top and bottom of each tray. Production permits introducing one truck load per hour. Each tray contains 30 kg dry solid, which will enter the drier at 25°C, 50% moisture, and will be dried to negligible moisture content. The critical moisture content is 15%, and the equilibrium moisture is negligible. The trucks are steel, each weighing about 135 kg. The air is to enter at 150°C, humidity 0.03 kg water/kg dry air, and the discharged solid is expected to leave at 135°C. The air is to be blown over the trays so that the average velocity at the air entrance is to be 4.5 m/s over the trays. The heat capacity of the dry solid is 1.255 kJ/kg · K. The drier is to be well insulated.

(a) Calculate the length of the drier required. **Ans.**: 8.5 m by Eq. (12.20a), 6.5 m by Eq. (12.20).

(b) The entering air is to be prepared by recycling a portion of the discharged air with atmospheric air (25°C, humidity 0.01 kg water/kg dry air) and heating the mixture to 150°C. Calculate the percentage of discharge air to be recycled and the heat requirement. Calculate the heat also expressed per unit mass of water evaporated. **Ans.**: 3515 kJ/kg.

Leaching is the preferential solution of one or more constituents of a solid mixture by contact with a liquid solvent. This unit operation, one of the oldest in the chemical industries, has been given many names, depending to some extent upon the technique used for carrying it out. *Leaching* originally referred to percolation of the liquid through a fixed bed of the solid, but is now used to describe the operation generally, by whatever means it may be done. *Lixiviation* is used less frequently as a synonym for leaching, although originally it referred specifically to the leaching of alkali from wood ashes. The term *extraction* is also widely used to describe this operation in particular, although it is applied to all the separation operations as well, whether mass-transfer or mechanical methods are involved. *Decoction* refers specifically to the use of the solvent at its boiling temperature. When the soluble material is largely on the surface of an insoluble solid and is merely washed off by the solvent, the operation is sometimes called *elutriation* or *elution*. This chapter will also consider these washing operations, since they are frequently intimately associated with leaching.

The metallurgical industries are perhaps the largest users of the leaching operation. Most useful minerals occur in mixtures with large proportions of undesirable constituents, and leaching of the valuable material is a separation method which is frequently applied. For example, copper minerals are preferentially dissolved from certain of their ores by leaching with sulfuric acid or ammoniacal solutions, and gold is separated from its ores with the aid of sodium cyanide solutions. Leaching similarly plays an important part in the metallurgical processing of aluminum, cobalt, manganese, nickel, and zinc. Many naturally occurring organic products are separated from their original structure by leaching. For example, sugar is leached from sugar beets with hot water, vegetable oils are recovered from seeds such as soybeans and cottonseed by leaching with organic solvents, tannin is dissolved out of various tree barks by leaching with water, and many pharmaceutical products are similarly recovered from plant roots and leaves. Tea and coffee are prepared both domestically and industrially by leaching operations. In addition, chemical precipitates are frequently washed of their adhering mother liquors by techniques and in equipment quite similar to those used in true leaching operations, as in the washing of caustic soda liquor from precipitated calcium carbonate following the reaction between soda ash and lime.

Preparation of the Solid

The success of a leaching and the technique to be used will very frequently depend upon any prior treatment which may be given the solid.

In some instances, small particles of the soluble material are completely surrounded by a matrix of insoluble matter. The solvent must then diffuse into the mass, and the resulting solution must diffuse out, before a separation can result. This is the situation with many metallurgical materials. Crushing and grinding of such solids will greatly accelerate the leaching action, since then the soluble portions are made more accessible to the solvent. A certain copper ore, for example, can be leached effectively by sulfuric acid solutions within 4 to 8 h if ground to pass through a 60-mesh screen, in 5 days if crushed to 6-mm granules, and only in 4 to 6 years if 150-mm lumps are used [43]. Since grinding is expensive, the quality of the ore will have much to do with the choice of size to be leached. For certain gold ores, on the other hand, the tiny metallic particles are scattered throughout a matrix of quartzite which is so impervious to the leaching solvent that it is essential to grind the rock to pass through a 100-mesh screen if leaching is to occur at all. When the soluble substance is more or less uniformly distributed throughout the solid or even in solid solution, the leaching action may provide channels for the passage of fresh solvent and fine grinding may not be necessary. Collapse of the insoluble skeleton which remains after solute removal may present problems, however.

Vegetable and animal bodies are cellular in structure, and the natural products to be leached from these materials are usually found inside the cells. If the cell walls remain intact upon exposure to a suitable solvent, the leaching action involves osmotic passage of the solute through the cell walls. This may be slow, but it is impractical and sometimes undesirable to grind the material small enough to release the contents of individual cells. Thus, sugar beets are cut into thin, wedge-shaped slices called *cossettes* before leaching in order to reduce the time required for the solvent water to reach the individual plant cells. The cells are deliberately left intact, however, so that the sugar will pass through the semipermeable cell walls while the undesirable colloidal and albuminous materials largely remain behind. For many pharmaceutical products recovered from plant roots, stems, and leaves, the plant material is frequently dried before treatment, and this does much toward rupturing the cell walls and releasing the solute for direct action by the solvent. Vegetable seeds and beans, such as soybeans, are usually rolled or flaked to give particles in the size range 0.15 to

0.5 mm. The cells are, of course, smaller than this, but they are largely ruptured by the flaking process, and the oils are then more readily contacted by the solvent.

When the solute is adsorbed upon the surface of solid particles or merely dissolved in adhering solution, no grinding or crushing is necessary and the particles can be washed directly.

Temperature of Leaching

It is usually desirable to leach at as high a temperature as possible. Since higher temperatures result in higher solubility of the solute in the solvent, higher ultimate concentrations in the leach liquor are possible. The viscosity of the liquid is lower and the diffusivities larger at higher temperatures, leading to increased rates of leaching. With some natural products such as sugar beets, however, temperatures which are too high may lead to leaching of excessive amounts of undesirable solutes or chemical deterioration of the solid.

Methods of Operation and Equipment

Leaching operations are carried out under batch and semibatch (unsteady-state) as well as under completely continuous (steady-state) conditions. In each category, both stagewise and continuous-contact types of equipment are to be found. Two major handling techniques are used: spraying or trickling the liquid over the solid, and immersing the solid completely in the liquid. The choice of equipment to be used in any case depends greatly upon the physical form of the solids and the difficulties and cost of handling them. This has led in many instances to the use of very specialized types of equipment in certain industries.

UNSTEADY-STATE OPERATION

The unsteady-state operations include those where the solids and liquids are contacted in purely batchwise fashion and also those where a batch of the solid is contacted with a continually flowing stream of the liquid (semibatch method). Coarse solid particles are usually treated in fixed beds by percolation methods, whereas finely divided solids, which can be kept in suspension more readily, can be dispersed throughout the liquid with the help of some sort of agitator.

In-Place (in Situ) Leaching

This operation, also called *solution mining*, refers to the percolation leaching of minerals in place at the mine, by circulation of the solvent over and through the ore body. It is used regularly in the removal of salt from deposits below the earth's surface by solution of the salt in water which is pumped down to the deposit. It has been applied to the leaching of low-grade copper ores containing as little as 0.2 percent copper [34] and to ores as deep as 335 m (1100 ft) below the surface [3]. In the solution mining of uranium, the ore must be oxidized in place in order to solubilize it in carbonate solutions. The reagent may be injected continuously through one set of pipes drilled down to the ore and the resulting liquor removed through a different set. Alternatively, the reagent may be pumped into the ground intermittently and withdrawn through the same well.

Heap Leaching

Low-grade ores whose mineral values do not warrant the expense of crushing or grinding can be leached in the form of run-of-mine lumps built into huge piles upon impervious ground. The leach liquor is pumped over the ore and collected as it drains from the heap. Copper has been leached from pyritic ores in this manner in heaps containing as much as 2.2×10^7 t of ore, using over 20 000 m^3 (5×10^6 gal) of leach liquor per day. It may require up to 7 or more years to reduce the copper content of such heaps from 2 to 0.3 percent. In a typical case of heap leaching of uranium, after laying a pattern of perforated drain pipe on an impervious clay base, the ore is built into piles on top of the pipes, in heaps some 6 to 8 m tall, trapezoidal in cross section and 120 m wide at the base, as much as 800 m (0.5 mi) long. The leach solution, introduced into ponds in the top of the heap, percolates down to the drain pipes at the base, whence it is led away.

Percolation Tanks

Solids of intermediate size can conveniently be leached by percolation methods in open tanks. The construction of these tanks varies greatly, depending upon the nature of the solid and liquid to be handled and the size of the operation, but they are relatively inexpensive. Small tanks are frequently made of wood if it is not chemically attacked by the leach liquid. The solid particles to be leached rest upon a false bottom, which in the simplest construction consists of a grating of wooden strips arranged parallel to each other and sufficiently close to support the solid. These in turn may rest upon similar strips arranged at right angles, 150 mm or more apart, so that the leach liquor flows to a collection pipe leading from the bottom of the tank. For supporting fairly fine particles, the wood grating may be further covered by a coconut matting and a tightly stretched canvas filter cloth, held in place by caulking a rope into a groove around the periphery of the false bottom. Small tanks may also be made entirely of metal, with perforated false bottoms upon which a filter cloth is placed, as in the leaching of pharmaceutical products from plants. Very large percolation tanks (45 by 34 by 5.5 m deep) for leaching copper ores have been made of reinforced concrete and lined with lead or bituminous mastic. Small tanks may be provided with side doors near the bottom for sluicing away the leached solid, while very large tanks are usually emptied by excavating from the top. Tanks should be filled with solid of as uniform a particle size as practical, since then the percentage of voids will be largest and the pressure drop required for flow of the leaching liquid least. This also leads to uniformity of the extent of leaching individual solid particles and less difficulty with channeling of the liquid through a limited number of passageways through the solid bed. The operation of such a tank may follow any of several procedures. After the tank is filled with solid, a batch of solvent sufficient to immerse the solid completely may be pumped into the tank and the entire mass allowed to steep or soak for a prescribed period of time. During this period the batch of liquid may or may not be circulated over the solid by pumping. The liquid is then drained from the solid by withdrawing it through the false bottom of the tank. This entire operation represents a single stage. Repetition of this process will eventually dissolve all the solute. The only solute then retained is that dissolved in the solution wetting the drained solid. This can be washed out by filling the tank with fresh solvent and repeating the operation as many times as necessary. An alternative method is continuously to admit liquid into the tank and continuously to withdraw the resulting solution, with or without recirculation of a portion of the total flow. Such an operation may be equivalent to many stages. Since the solution which results is usually denser than the solvent, convective mixing is reduced by percolation in the downward direction. Upward flow is sometimes used, nevertheless, in order to avoid clogging the bed or the filter with fines, but this may result in excessive entrainment of the fines in the overflow liquid. Still a further modification, less frequently used, is to spray the liquid continuously over the top and allow it to trickle downward through the solid without fully immersing the solid at any time. An excellent review of the processes, techniques, and chemistry of leaching of ores is available [6].

Retention of Liquid after Drainage

Imagine a bed of granular solids whose void space is completely filled with liquid. When the liquid is allowed to drain under the influence of gravity, with admission of air to the voids from the top of the bed, the rate of liquid flow is at first very rapid. The rate gradually falls, and after a relatively long period of time no additional drainage occurs. The bed still contains liquid, however. The fraction of the void volume still occupied by liquid is termed the *residual saturation* s. Figure 13.1 shows the variation of s with height of the bed [11]. In the upper part of the bed the value of s is constant at s_0, and this represents the liquid which remains in the crevasses and small angles between the particles as fillets, held in place by surface tension. In the lower part of the bed, the liquid is held up in the voids, filling them completely ($s = 1.0$) by capillary action. The drain height Z_D is defined as the height where the value of s is the average in the range s_0 to unity, as shown in the figure. The average value of s for the entire bed will be the area between the ordinate axis and the curve of the figure, divided by the bed height Z,

$$s_{av} = \frac{(Z - Z_D)s_0}{Z} + \frac{Z_D}{Z} \tag{13.1}$$

A large number of measurements of Z_D under a wide variety of conditions showed that, approximately [11],

$$Z_D = \frac{0.275(g_c/g)}{(K/g)^{0.5}(\rho_L/\sigma)} \tag{13.2}$$

where $K =$ "permeability" of bed
$\rho_L =$ liquid density
$\sigma =$ surface tension of liquid
The value of s_0 was found to depend upon the group $(K\rho_L/g\sigma)(g/g_c)$, called the *capillary number*, as

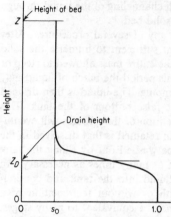

s = fraction of void volume occupied by liquid   Figure 13.1 Drainage of packed beds [11].

follows:

$$S_0 = \begin{cases} 0.075 & \dfrac{K\rho_L}{g\sigma}\dfrac{g}{g_c} < 0.02 \\[4mm] \dfrac{0.0018}{(K\rho_L/g\sigma)(g/g_c)} & \dfrac{K\rho_L}{g\sigma}\dfrac{g}{g_c} > 0.02 \end{cases}$$

(13.3)

(13.4)

In these expressions it is assumed that drainage has occurred under the action of the force of gravity only and that the contact angle between liquid and solid surfaces is 180°.

The permeability K is the proportionality constant in the flow equation for laminar flow through the bed,

$$G = \frac{K\rho_L \, \Delta p}{\mu_L Z}\frac{g}{g_c}$$

(13.5)

where Δp is the drop in pressure across the bed and G is the mass velocity of flow based on the entire cross section of the bed. Equation (6.66) describes the flow through beds of granular solids, and for laminar flow only the first term of the right-hand side of this expression is used. If $\Delta p/Z$ from this equation is substituted in Eq. (13.5), with Re replaced by $d_p G/\mu_L$, simplification leads to

$$K = \frac{d_p^2 \varepsilon^3 g}{150(1 - \varepsilon)^2}$$

(13.6)

where d_p is the diameter of a sphere of the same surface/volume ratio as the particles of the bed and ε is the fraction-void volume. For fibrous material and others whose value of d_p may be difficult to estimate, K can be obtained from Eq. (13.5) after experimental measurement of the pressure drop for laminar flow through the bed.

Illustration 13.1 The sugar remaining in a bed of bone char used for decolorization is leached by flooding the bed with water, following which the bed is drained of the resulting sugar solution. The bed diameter is 1 m, the depth 3.0 m, the temperature is 65°C. The sugar solution which drains has a density 1137 kg/m³ (71 lb/ft³) and a surface tension 0.066 N/m. The bulk density of the char is 960 kg/m³ (60 lb/ft³) and the individual particle density 1762 kg/m³ (110 lb/ft³). The particles have a specific external surface 16.4 m²/kg (80 ft²/lb).

Estimate the mass of solution still retained by the bed after dripping of the solution has stopped. Express this also as mass solution/mass dry bone char.

SOLUTION The fractional void volume $= \varepsilon = 1 -$ (bulk density/particle density) $= 1 - 960/1762 = 0.455$ m^3 voids/m^3 bed. The particle surface $= a_p = (16.4$ m^2/kg)(960 kg/m^3) $= 15\,744$ m^2/m^3 bed. Eq. (6.67):

$$d_p = \frac{6(1 - \varepsilon)}{a_p} = \frac{6(1 - 0.455)}{15\,744} = 2.077 \times 10^{-4}\ \text{m}$$

Eq. (13.6):

$$K = \frac{d_p^2 \varepsilon^3 g}{150(1 - \varepsilon)^2} = \frac{(2.077 \times 10^{-4})^2(0.455)^3(9.807)}{150(1 - 0.455)^2} = 8.94 \times 10^{-10}\ \text{m}^3/\text{s}$$

$$\frac{K\rho_L}{g\sigma}\frac{g}{g_c} = \frac{(8.94 \times 10^{-10})(1137)}{9.81(0.066)}\frac{9.81}{1} = 1.54 \times 10^{-5}$$

Eq. (13.3):

$$s_0 = 0.075$$

Eq. (13.2):

$$Z_D = \frac{0.275(1/9.81)}{[(8.94 \times 10^{-10})/9.81]^{0.5}(1137/0.066)} = 0.1705\ \text{m}$$

$$Z = 3.0\ \text{m}$$

Eq. (13.1):

$$s_{av} = \frac{(3 - 0.1705)(0.075)}{3} + \frac{0.1705}{3} = 0.1276$$

$$\frac{\text{Vol liquid retained}}{\text{Vol bed}} = 0.1276\varepsilon = 0.1276(0.455) = 0.0581\ \text{m}^3/\text{m}^3$$

$$\text{Mass liquid in bed} = 0.0581\frac{\pi(1)^2 3}{4}(1137) = 155.6\ \text{kg}$$

$$\frac{\text{Mass liquid}}{\text{Mass dry solid}} = \frac{0.0581(1137)}{1(960)} = 0.069\ \text{kg/kg}$$

Countercurrent Multiple Contact; the Shanks System

Leaching and washing of the leached solute from the percolation tanks by the crosscurrent methods described above will inevitably result in weak solutions of the solute. The strongest solution will result if a countercurrent scheme is used, wherein the final withdrawn solution is taken from contact with the freshest solid and the fresh solvent is added to solid from which most of the solute has already been leached or washed. In order to avoid moving the solids physically from tank to tank in such a process, the arrangement of Fig. 13.2, shown schematically for a system of six tanks, is used. This Shanks system,† as it is called, is operated in the following manner:

1. Assume at the time of inspecting the system at Fig. 13.2a that it has been in operation for some time. Tank 6 is empty, tanks 1 to 5 are filled with solid, tank 5 most recently and tank 1 for the longest time. Tanks 1 to 5 are also

† Named after James Shanks, who first introduced the system in 1841 into England for the leaching of soda ash from the "black ash" of the Le Blanc process. It was, however, apparently a German development.

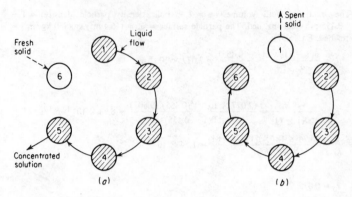

Figure 13.2 Countercurrent multiple contact, Shanks system.

filled with leach liquid, and the most concentrated is in tank 5 since it is in contact with the freshest solid. Fresh solvent has just been added to tank 1.

2. Withdraw the concentrated solution from tank 5, transfer the liquid from tank 4 to tank 5, from 3 to 4, from 2 to 3, and from 1 to 2. Add fresh solid to tank 6.
3. Refer to Fig. 13.2b. Discard the spent solid from tank 1. Transfer the liquid from tank 5 to tank 6, from 4 to 5, from 3 to 4, and from 2 to 3. Add fresh solvent to tank 2. The circumstances are now the same as they were at the start in Fig. 13.2a, except that the tank numbers are each advanced by one.
4. Continue the operation in the same manner as before.

The scheme is identical with the batch simulation of a multistage countercurrent operation shown in Fig. 10.38. After several cycles have been run through in this manner, the concentrations of solution and in the solid in each tank approach very closely the values obtaining in a truly continuous countercurrent multistage leaching. The system can, of course, be operated with any number of tanks, and anywhere from 6 to 16 are common. They need not be arranged in a circle but are better placed in a row, called an *extraction battery*, so that additional tanks can conveniently be added to the system if desired. The tanks may be placed at progressively decreasing levels, so that liquid can flow from one to the other by gravity with a minimum of pumping.

Such leaching tanks and arrangements are used extensively in the metallurgical industries, for recovery of tannins from tree barks and woods, for leaching sodium nitrate from Chilean nitrate-bearing rock (*caliche*), and in many other processes.

Percolation in Closed Vessels

When the pressure drop for flow of liquid is too high for gravity flow, closed vessels must be used and the liquid is pumped through the bed of solid. Such

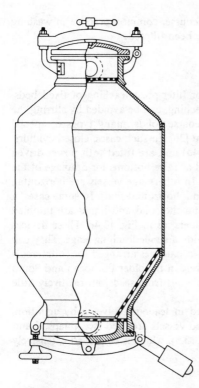

Figure 13.3 Sugar-beet diffuser [12]. *(Courtesy of the Institution of Chemical Engineers.)*

vessels are sometimes called *diffusers*. Closed tanks are also necessary to prevent evaporation losses when the solvent is very volatile or when temperatures above the normal boiling point of the solvent are desired. For example, some tannins are leached with water at 120°C, 345 kN/m^2 (50 lb$_f$/in^2) pressure, in closed percolation tanks.

Designs vary considerably, depending upon the application. In leaching sugar from sugar-beet slices, or cossettes, a diffuser of the type shown in Fig. 13.3 is used. They are arranged in a battery containing up to 16 vessels, and the beets are leached with hot water in the countercurrent fashion of the Shanks system. Heaters are placed between the diffusers to maintain a solution temperature of 70 to 78°C. In this manner 95 to 98 percent of the sugar, in beets containing initially about 18 percent, can be leached to form a solution of 12 percent concentration. Countercurrent, continuous equipment is also used in the sugar-beet industry [4, 20].

Filter-Press Leaching

Finely divided solids, too fine for treatment by percolation in relatively deep percolation tanks, can be filtered and leached in the filter press by pumping the solvent through the press cake. This is, of course, common practice in washing mother liquor from precipitates which have been filtered.

Agitated Vessels

Channeling of the solvent in percolation or filter-press leaching of fixed beds, with its consequent slow and incomplete leaching, can be avoided by stirring the liquid and solid in leaching vessels. For coarse solids, many types of special stirred or agitated vessels have been devised [36]. In such cases, closed cylindrical vessels are arranged vertically (Fig. 13.4a) and are fitted with power-driven paddles or stirrers on vertical shafts, as well as false bottoms for drainage of the leach solution at the end of the operation. In others, the vessels are horizontal, as in Fig. 13.4b, with the stirrer arranged on a horizontal shaft. In some cases, a horizontal drum is the extraction vessel, and the solid and liquid are tumbled about inside by rotation of the drum on rollers, as in Fig. 13.4c. These devices are operated in batchwise fashion and provide a single leaching stage. They can be used singly but frequently are also used in batteries arranged for countercurrent leaching. They have been used extensively in the older European and South American installations for leaching vegetable oils from seeds but relatively little in the United States.

Finely divided solids can be suspended in leaching solvents by agitation, and for batch operation a variety of agitated vessels are used (see Chaps. 6 and 11). The simplest is the Pachuca tank (Fig. 13.5), which is employed extensively

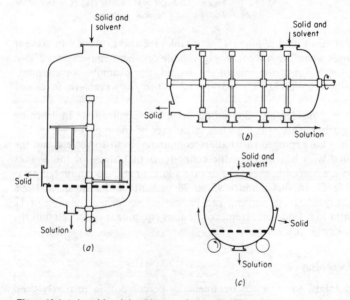

Figure 13.4 Agitated batch leaching vessels (see also Chap. 11).

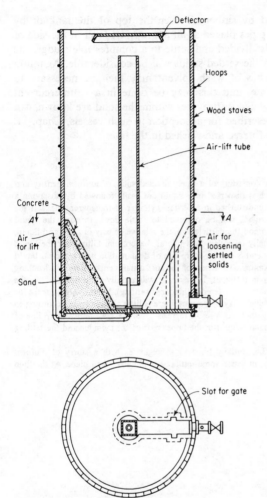

Air
for lift

Air for
loosening
settled
solids

Slot for gate

Section *A-A*

Figure 13.5 Pachuca tank. *(Fran
Liddell, "Handbook of Non-ferrous
Metallurgy," 2d ed., McGraw-Hill
Book Company, New York, 1945. Used
by permission of the publisher.)*

in the metallurgical industries. These tanks are constructed of wood, metal, or
concrete and may be lined with inert metal such as lead, depending upon the
nature of the leaching liquid. Agitation is accomplished by an air lift: the
bubbles of air rising through the central tube cause the upward flow of liquid
and suspended solid in the tube and consequently vertical circulation of the tank
contents [26]. The standard mechanical agitators, with turbine-type impellers, for
example, can also be used to keep the finely divided solids suspended in the
liquid. After the leaching has been accomplished, the agitation is stopped, the
solid is allowed to settle in the same or a separate vessel, and the clear,

supernatant liquid is decanted by siphoning over the top of the tank or by withdrawal through discharge pipes placed at an appropriate level in the side of the tank. If the solids are finely divided and settle to a compressible sludge, the amount of solution retained in the settled solids will be considerable. Agitation and settling with several batches of wash solvent may then be necessary to recover the last traces of solute, and this may be done in a countercurrent fashion. Rates of leaching, provided diffusivities within the solid are known, can be computed in the manner described for adsorption in such vessels, Chap. 11. Alternatively, the solid may be filtered and washed in the filter.

Batch Settling

The settling characteristics of a slurry consisting of a finely divided solid, of uniform density and reasonably uniform particle size, which is dispersed in a liquid are easily followed by observing a sample of the slurry allowed to stand undisturbed in a vertical cylinder of transparent glass. If the slurry is initially very dilute, the particles will be observed to settle down through the liquid individually, each at a rate dependent upon the particle size, the relative density of solid and liquid, and the viscosity of the liquid, eventually to collect in a pile at the bottom. Ultimately the liquid becomes clear, but at no time until the end is there a sharp line of demarcation between clear liquid and the settling slurry. For more concentrated slurries, of the sort usually encountered in leaching and washing operations, the behavior is different, however. It will usually be observed that the particles settle more slowly owing to mutual interference (hindered settling). Furthermore, except for a few particles of relatively large size which may be present, there is little classification according to size, and the particles largely settle together. As a result there is usually a reasonably sharp line of demarcation between the clear, supernatant liquor in the upper part of the cylinder and the settling mass of solids in the lower part.

Consider the cylinder of Fig. 13.6, initially filled to a height Z_0 with a slurry of uniform concentration w_0 weight fraction solids, in which some settling has already taken place. At the time

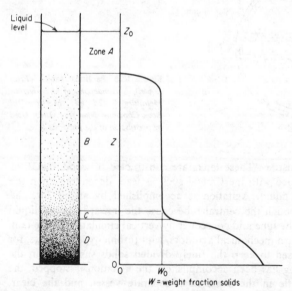

Figure 13.6 Batch settling.

of observation, there is a zone A of clear liquid at the top. Directly beneath this is zone B, throughout which the solids concentration is reasonably uniform at the initial value w_0, as shown by the accompanying graph [10]. In zone D at the bottom, usually called the *compression zone*, the particles accumulating from above have come to rest upon each other, and, owing to their weight, liquid is squeezed out from between the particles. For compressible sludges, this results in increasing solids concentration with depth in this zone, as shown by the curve. Zone C is a transition zone between B and D which may not always be clearly defined. As settling continues beyond the time corresponding to that in the figure, the line of demarcation between zones A and B falls and the height of zone D rises, until eventually zone B disappears and only a compression zone containing all the solids remains. This then slowly subsides to some ultimate height.

The rate of settling is usually followed by plotting the height of the line of demarcation between zones A and B against time, as shown by the solid curve of Fig. 13.7. The broken curve represents the position of the upper level of zone D. The top of zone B settles at constant rate (curve of Z vs. time straight) from the beginning until zone B has nearly disappeared and all the solids are in the compression zone. The rate of settling of the compression zone to its ultimate height Z_∞ is then relatively slow and is not constant. In a few cases, two constant-rate settling periods can be observed, with substantially no compression period. The appearance of the curves depends not only upon the type of slurry (nature and particle size of the solid, and nature of the liquid) but also upon the initial height and concentration of the slurry, as well as the extent of flocculation and whether or not any stirring is done during settling.

Flocculation If the finely divided solid particles are all similarly electrically charged, they repel each other and remain dispersed. If the charge is neutralized by addition, for example, of an electrolyte (flocculating agent) to the mixture, the particles may form aggregates, or flocs. Since the flocs are of larger size, they settle more rapidly. Slurries and suspensions encountered in chemical operations are usually flocculated.

Stirring Very slow stirring, so slow that eddy currents are not formed within the liquid, changes the character of the settling profoundly. The floc structure is altered so that the solids concentration in zone B is no longer uniform at the initial value and zone D may not be clearly defined. The ultimate height of the settled slurry may be only a fraction of that obtained without stirring [21], owing to breakdown of bridged floc structures in the compression zone, and the ultimate concentration of

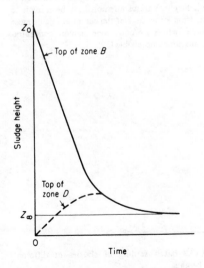

Figure 13.7 Rate of settling.

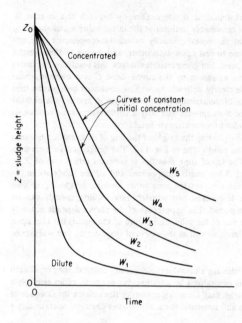

Figure 13.8 Batch settling of slurries. Effect of slurry concentration.

solids in the settled mass is correspondingly greater. Generally, however, the zones of constant- and falling-rate settling are still observed, although the rates will be different from those obtained without stirring [44].

Concentration The rate of settling decreases with increased initial concentration of the solids owing to the increase of the effective density and viscosity of the medium through which the particles settle. Figure 13.8 illustrates the effect usually to be expected when slurries of increasing concentration of the same substance are settled in columns of the same height. Various attempts have been made to predict the effect of concentration on the settling rate, from knowledge of the curves at one or more concentrations. This has been successful only for slurries which are not at one or more concentrations. This has been successful only for slurries which are not compressible [39].

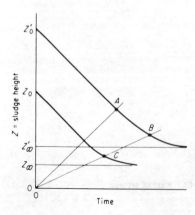

Figure 13.9 Batch settling of slurries at different initial heights.

Height Refer to Fig. 13.9, which shows settling curves for the same slurry begun at different initial heights. The initial constant settling rate is independent of height, and provided some critical minimum value of Z_0 is exceeded, the ultimate value of Z_∞ / Z_0 will apparently also be constant. The constant-settling-rate lines both terminate on a line OA radiating from the origin, and in general any line [44] radiating from the origin such as OB will be cut so that line OC/line $OB = Z_0/Z_0'$. It follows that the time for a slurry to settle to a fixed fractional height Z/Z_0 is proportional to the initial height Z_0. In this way it is possible reasonably well to predict the settling curves for deep tanks from results obtained in small laboratory cylinders. In making such laboratory tests, however, it is important [21] to use cylinders at least 1 m tall and at least 50 mm in diameter and to maintain all other conditions in the laboratory test identical with those expected to prevail on the large scale.

Percolation vs. Agitation

If a solid in the form of large lumps is to be leached, a decision must frequently be made whether to crush it to coarse lumps and leach by percolation or whether to grind it fine and leach by agitation and settling. No general answer can be given to this problem because of the diverse leaching characteristics of the various solids and the values of the solute, but among the considerations are the following. Fine grinding is more costly but provides more rapid and possibly more thorough leaching. It suffers the disadvantages that the weight of liquid associated with the settled solid may be as great as the weight of the solid, or more, so that considerable solvent is used in washing the leached solute free of solute and the resulting solution is dilute. Coarsely ground particles, on the other hand, leach more slowly and possibly less thoroughly but on draining may retain relatively little solution, require less washing, and thus provide a more concentrated final solution.

For more fibrous solids, such as sugar cane, which is leached with water to remove the sugar, it has been shown [35] that leaching is generally more efficient in a thoroughly agitated vessel than by percolation, probably because the large amount of static liquid holdup (see Chap. 6) makes important amounts of solute unavailable.

STEADY-STATE (CONTINUOUS) OPERATION

Equipment for continuous steady-state operations can be broadly classified into two major categories, according to whether it operates in stagewise or in continuous-contact fashion. Stagewise equipment is sometimes assembled in multiple units to produce multistage effects, whereas continuous-contact equipment may provide the equivalent of many stages in a single device.

Leaching during Grinding

As pointed out earlier, many solids require grinding in order to make the soluble portions accessible to the leaching solvents, and if continuous wet grinding is practiced, some of the leaching may be accomplished at this time. As much as 50 to 75 percent of the soluble gold may be dissolved by grinding the ore in the presence of cyanide solution, for example. Similarly, castor seeds are ground in an attrition mill with solvent for the castor oil. The liquid and solid flow through a grinding mill in parallel and consequently tend to come to a concentration equilibrium. Such operations are therefore single-stage leachings and are usually supplemented by additional agitation or washing operations, as described later.

Agitated Vessels

Finely ground solids which can be readily suspended in liquids by agitation may be continuously leached in any of the many types of agitated tanks or vessels. These must be arranged for continuous flow of liquid and solid into and out of the tank and must be carefully designed so that no accumulation of solid occurs. Thanks to the thorough mixing ordinarily obtained, these devices are single-stage in their action, the liquid and solid tending to come to equilibrium within the vessel.

Mechanically agitated vessels may be used, for which the turbine-type agitator is probably most generally suitable (see Chaps. 6 and 11). Pachuca tanks are frequently used in the metallurgical industries. The Dorr agitator (Fig. 13.10) utilizes both the air-lift principle and mechanical raking of the solids and is extensively used in both the metallurgical and the chemical industry for continuous leaching and washing of finely divided solids. The central hollow shaft of the agitator acts as an air lift and at the same time revolves slowly. The arms attached to the bottom of the shaft rake the settled solids toward the center of the tank bottom, where they are lifted by the air lift through the shaft to the revolving launders attached to the top. The launders then distribute the elevated mixture of liquid and solid over the entire cross section of the tank. The rake arms can be lifted to free them of solids which may settle during a shutdown, and they are also provided with auxiliary air lines to assist in freeing them from settled solid. For unevenly sized solids, operation of the agitator can be adjusted so that coarse particles, which may require longer leaching time, remain in the tank longer than the finer. These agitators are regularly built in sizes ranging from 1.5 to 12 m diameter.

The average holding time in an agitated vessel can be calculated by dividing the vessel contents by the rate of flow into the vessel. This can be done separately for solid and liquid, and the holding time for each will be different if the ratio of the amounts of one to the other in the vessel is different from that in the feed. The average holding time of the solid must be sufficient to provide the leaching action required. Individual solid particles, of course, may short-circuit the tank (axial mixing), i.e., pass through in times much shorter than the calculated average, and this will lead to low stage efficiency. Short circuiting can be eliminated by passing the solid-liquid mixture through a series of smaller agitated vessels, one after the other, the sum of whose average holding time is the necessary leach time. This can readily be accomplished with gravity flow of the slurry by placing the individual tanks in the series at progressively lower levels. Three vessels in series are usually sufficient to reduce short circuiting to a negligible amount. It should be noted that since liquid and solid pass through these vessels in parallel flow, the entire series is still equivalent to only a single stage.

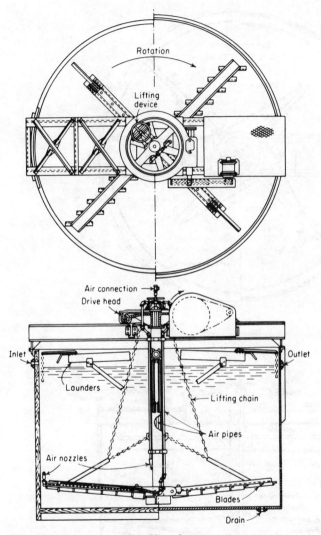

Figure 13.10 Dorr agitator. *(Dorr-Oliver, Inc.)*

The effluent from continuous agitators may be sent to a filter for separating liquid from solid, upon which the solid may be washed free of dissolved solids, or to a series of thickeners for countercurrent washing.

Thickeners

Thickeners are mechanical devices designed especially for continuously increasing the ratio of solid to liquid in a dilute suspension of finely sized particles by

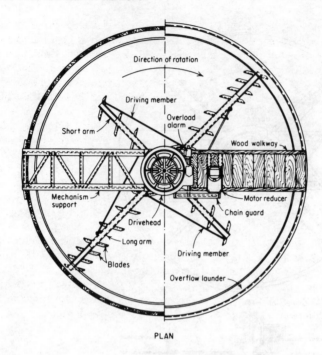

PLAN

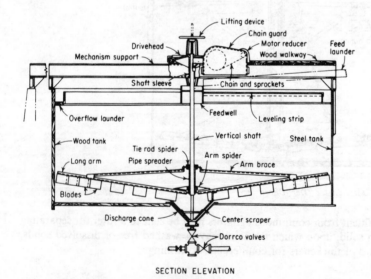

SECTION ELEVATION

Figure 13.11 Dorr thickener. *(Dorr-Oliver, Inc.)*

settling and decanting, producing a clear liquid and a thickened sludge as two separate products. Thickeners may be used before any ordinary filter in order to reduce filtering costs, but since both effluents are pumpable and consequently readily transported, thickeners are frequently used to wash leached solids and chemical precipitates free of adhering solution in a continuous multistage countercurrent arrangement, and it is in this application that they are of interest here.

A typical single-compartment thickener of the Dorr-Oliver Company's design is shown in Fig. 13.11. The thin slurry of liquid and suspended solids enters a large settling tank through a feed well at the top center, in such a manner as to avoid mixing the slurry with the clear liquid at the top of the tank. The solids settle from the liquid which fills the tank, and the settled sludge is gently directed toward the discharge cone at the bottom by four sets of plow blades or rakes, revolving slowly to avoid disturbing the settled solid unduly. The sludge is pumped from the discharge cone by means of a diaphragm pump. The clear, supernatant liquid overflows into a launder built about the upper periphery of the tank. Thickeners are built in sizes ranging from 2 to 180 m (6 to 600 ft) in diameter, for handling granular as well as flocculent solids, and of varying detail design depending upon the size and service. In order to reduce the ground-area requirements, several thickeners operating in parallel and superimposed as in Fig. 13.12 may be used. Such a device delivers a single sludge product.

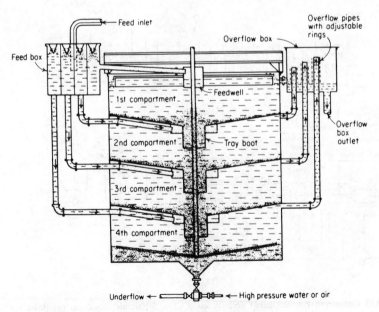

Figure 13.12 Dorr balanced-tray thickener. *(Dorr-Oliver, Inc.)*

The liquid content of the sludge is greatly dependent upon the nature of the solids and liquid and upon the time allowed for settling but in typical cases might be in the range 15 to 75 percent liquid. The less liquid retained, the more efficient the leaching or washing process being carried on.

Continuous Countercurrent Decantation (CCD)

Leaching equipment such as agitators or grinding mills may discharge their effluent into a cascade of thickeners for continuous countercurrent washing of the finely divided solids free of adhering solute. The same type of cascade can also be used to wash the solids formed during chemical reactions, as in the manufacture of phosphoric acid, by treatment of phosphate rock with sulfuric acid, or of blanc fixe, by reaction of sulfuric acid and barium sulfide, or of lithopone.

A simple arrangement is shown in Fig. 13.13a. The solids to be leached (or the reagents for a reaction), together with solution from the second thickener, are introduced into the leaching agitators at the left, and the strong solution thus produced is decanted from the solids by the first thickener. The agitators together with the first thickener then constitute a single stage. The sludge is passed through the cascade to be washed by the solvent in true countercurrent fashion, and the washed solids are discharged at the right. There may, of course,

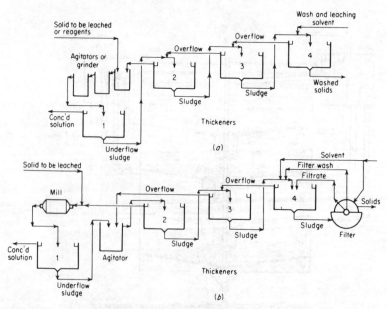

Figure 13.13 Continuous countercurrent decantation (CCD): (a) simple flowsheet; (b) flowsheet with intermediate agitation and filtration of washed solids.

be more or fewer than the four stages shown, and the agitators may be replaced by any continuous-leaching device, such as a grinding mill. Many variations in the flowsheet are regularly made. For example, the sludge from each stage can be *repulped*, or vigorously beaten with the solvent, between stages in order to improve the washing efficiency. Figure 13.13b shows an arrangement whereby the underflow from the first thickener is agitated with overflow from the third, for the purpose of bringing about the additional leaching possible with dilute solution. The sludge from the final stage can be filtered, as shown, when the solid is valuable and is to be delivered reasonably dry or when the solute is valuable and solution adhering to the washed solids must be reduced to a minimum. For successful operation of these plants, very carefully controlled rates of flow of both sludges and solution are necessary to avoid disturbing the steady-state conditions prevailing.

For small decantation plants, where ground area may be limited, it is possible to obtain a countercurrent cascade of thickeners built in superimposed fashion into a single shell.

Continuous Settling

The concentrations existing at the various levels of a continuous thickener under steady-state operation differ considerably from those found in batch settling. The solid curve of Fig. 13.14 shows typical concentrations during normal operation [10a], and four clearly defined zones are found in the thickener corresponding to the various sections of the curve. The feed slurry is diluted as it issues from the feed well of the thickener, and the bulk of the liquid passes upward to overflow into the launder about the thickener periphery. The solid concentration in the top zone is negligible if the overflow is clear. The solids and the remainder of the feed liquid move downward through the lower three zones and leave in the thickened underflow. The solids concentration in the settling zone is much lower than that in the feed, because of the dilution, but rises rapidly in the compression zone immediately below. In the bottom zone, the action of the rake disturbs arched structures which the settling solids may form, the weight of the solids presses out the liquid, and the concentration rises to the value in the underflow. If the feed rate to the thickener is increased, the concentration of solids in the settling zone rises and reaches a constant maximum value not related to the feed concentration when the settling capacity of this zone is exceeded. The excess solids, which cannot settle,

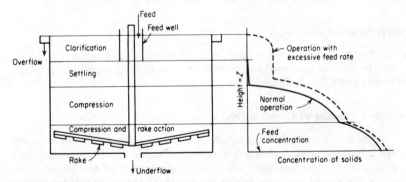

Figure 13.14 Continuous thickener characteristics [10]. *(Courtesy of Industrial and Engineering Chemistry.)*

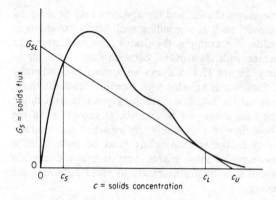

Figure 13.15 Determination of thickener area from flux curve.

c = solids concentration

overflow with the liquid, as indicated by the broken curve of concentrations in Fig. 13.14 for this condition.

The concentration of solids in the underflow sludge for a given rate and concentration of feed can be increased by reducing the rate of withdrawal of sludge. This increases the depth of the compression zone and increases the detention time of the solids within the thickener, although it is important not to raise the level of the compression zone to such an extent that solids appear in the overflow liquid.

The capacity of continuous thickeners, or their cross-sectional area required for a given solids throughput, can be roughly estimated from batch-settling tests [13, 39]. *Initial* settling velocities V for slurries of the solid at various initial uniform concentrations c are determined from the slopes of curves like those of Fig. 13.8, covering the entire range of solids concentration to be dealt with (it is best to determine these curves for slurries made by suspending a given weight of solids in varying amounts of liquid by adding and subtracting liquid). The flux of solids during settling, $G_s = cV$, is then plotted against c to produce a curve like that of Fig. 13.15. The tangent of smallest negative slope is then drawn to the curve from point $(G_s = 0, c = c_U)$, where c_U is the desired under-flow concentration, to intersect the ordinate at G_{SL}, the limiting solids flux. The minimum required thickener cross section for handling W mass/time of solids is then

$$A = \frac{W}{G_{SL}} \tag{13.7}$$

Concentration c_L is that at the point of compression and c_S that in the settling zone.

Hydrocyclones (Hydroclones)

Hydrocyclones, similar to those used for size classification of solids (Fig. 13.16) can also be used as liquid-solid separators in place of thickeners in countercurrent washing of solids in a slurry.

Continuous Leaching of Coarse Solids

Many ingenious devices have been used for moving the solids continuously through a leaching device so that countercurrent action can be obtained. With the exception of the classifiers, which are used mainly in the metallurgical industries, these machines were principally developed for the special solids-handling problems arising in the leaching of sugar beets and of vegetable seeds such as cottonseed, soybeans, and the like. Donald [12] has described many of the

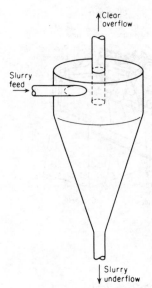

Figure 13.16 Hydrocyclone.

early devices used for sugar beets. Only the more important of the currently used machines can be described here.

Classifiers

Coarse solids can be leached, or more usually washed free of adhering solution or solute, in some types of machinery ordinarily used in the metallurgical industries for classification according to particle size. One such device is shown in Fig. 13.17. The solids are introduced into a tank, made with a sloping bottom and partly filled with the solvent. The rakes, which are given a reciprocating and circular lifting motion by the driving mechanism, rake the solids upward along the bottom of the tank and out of the liquid. In the upper part of the tank the solids are drained and discharged. The liquid overflows at the deep end of the tank. The solute concentration in the liquid is reasonably uniform throughout the tank as a result of the agitation by the rakes, so that the apparatus produces a single-stage action. Several classifiers can be placed in a cascade for continuous multistage countercurrent action, however, in which case they are operated by a single drive mechanism.

Fitch has presented an excellent review of the techniques of separating liquids and solids [14].

Leaching of Vegetable Seeds

Cottonseeds, soybeans, linseeds (flaxseeds), peanuts, rice bran, castor beans, and many other similar products are regularly leached, or *extracted*, with organic

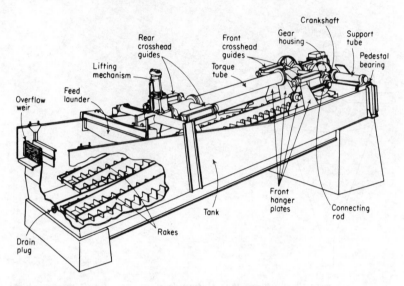

Figure 13.17 Single Dorr classifier for washing coarse solids. *(Dorr-Oliver, Inc.)*

solvents for removing the vegetable oils they contain. The seeds must usually be specially prepared for most advantageous leaching, and this may involve dehulling, precooking, adjustment of the moisture (water) content, and rolling or flaking. Sometimes a portion of the oil is first removed mechanically by expelling or expression. Leaching solvents are usually petroleum naphthas, for most oils a fraction corresponding closely to hexane; chlorinated hydrocarbons leave too toxic a residue for the leached meal to be used as an animal feed. The oil-solvent solution, which usually contains a small amount of finely divided, suspended solids, is called *miscella* and the leached solids *marc*. The various leaching devices are usually called *extractors* in this industry.

The *Rotocel* [2, 27] is essentially a modification of the Shanks system wherein the leaching tanks are continuously moved, permitting continuous introduction and discharge of the solids. Figure 13.18 is a schematic representation of the device, simplified to show the working principle. A circular rotor, containing 18 cells, each fitted with a hinged screen bottom for supporting the solids, slowly revolves above a stationary compartmented tank. As the rotor revolves, each cell passes in turn under a special device for feeding the prepared seeds and then under a series of sprays by which each is periodically drenched with solvent for leaching. After nearly one revolution, the leached contents of each cell are automatically dumped into one of the lower stationary compartments, from which they are continuously conveyed away. The solvent from each spray percolates downward through the solid and the supporting screen into the appropriate compartment of the lower tank, from which it is continuously pumped to the next spray. The leaching is countercurrent, and the strongest

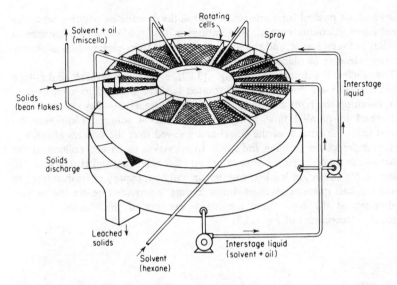

Figure 13.18 Schematic arrangement of the Rotocel.

solution is taken from the freshest seeds. A number of ingenious mechanical devices are necessary for maintaining smooth operation, and the entire machine is enclosed in a vaportight housing to prevent escape of solvent vapors.

The *French stationary-basket extractor* is a variant of this. The flakes are contained in stationary, compartmented beds filled by a rotating spout to feed the solids and are leached with solvent and miscella countercurrently [15, 33].

The *Kennedy* extractor [38], a modern arrangement of which is indicated schematically in Fig. 13.19, is another stagewise device which has been in use since 1927, originally for leaching tannins from tanbark. It is now used for oilseed and other chemical leaching operations. The solids are leached in a series

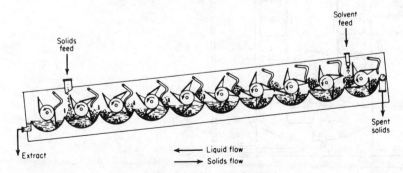

Figure 13.19 Kennedy extractor. *(The Vulcan Copper and Supply Co.)*

of tubs and are pushed from one to the next in the cascade by paddles, while the solvent flows in countercurrent. Perforations in the paddles permit drainage of the solids between stages, and the solids are scraped from each paddle as shown. As many tubs may be placed in a cascade as are required.

The *Bollman* extractor [38] (Fig. 13.20) is one of several basket-type machines. Solids are conveyed in perforated baskets attached to a chain conveyor, down on the right and up on the left in the figure. As they descend, they are leached in parallel flow by a dilute solvent-oil solution (*half miscella*) pumped from the bottom of the vessel and sprayed over the baskets at the top. The liquid percolates through the solids from basket to basket, collects at the bottom as the final strong solution of the oil (*full miscella*), and is removed. On the ascent, the solids are leached countercurrently by a spray of fresh solvent to provide the half miscella. A short drainage time is provided before the baskets are dumped at the top. There are many variants of this device, e.g., the horizontal arrangement of Fig. 13.21.

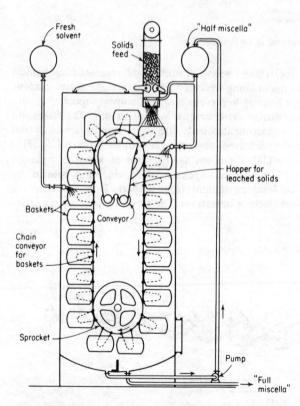

Figure 13.20 Bollman extractor.

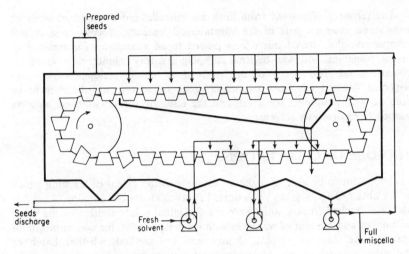

Figure 13.21 Continuous horizontal extractor (schematic).

Continuous tilting-pan filters and horizontal filters [16] are also commonly used. Figure 13.22 shows a typical flowsheet arrangement for a horizontal filter. The filter, in the form of a circular wheel, is divided into a number of sectors and revolves in the horizontal plane. Prepared seeds are slurried with solvent which has already been used for leaching, and the slurry is sent to the filter. The first filtrate is passed again through the filter cake to remove finely divided solids (polishing) before being discharged as miscella. The principle is much the same as that of the Rotocel. Horizontal moving screen-type belts [40] are also used for conveying the solids during leaching.

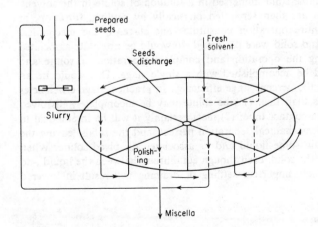

Figure 13.22 Flowsheet for horizontal-filter leaching.

The recovery of solvent from both the miscella and the leached seeds or beans is an essential part of the vegetable-oil leaching process. In a typical arrangement, the filtered miscella is passed to an evaporator for removal of solvent, sometimes followed by final stripping in a tray column, to produce the solvent-free oil. The wet seeds are steamed to remove residual solvent and air-cooled. Vented gas from condensers may be sent to an absorber to be scrubbed with petroleum white oil, and the resulting solvent-white-oil solution stripped to recover any solvent.

METHODS OF CALCULATION

It is important to be able to make an estimate of the extent of leaching which can be obtained for a given procedure, i.e., to calculate the amount of soluble substance leached from a solid, knowing the initial solute content of the solid, the number and amount of washings with leaching solvent, the concentration of solute in the leaching solvent, if any, and the method, whether batch or continuous countercurrent. Alternatively, it may be necessary to compute the number of washings, or number of stages, required to reduce the solute content of the solid to some specified value, knowing the amount and solute concentration of the leaching solvent. The methods of calculation are very similar to those used for liquid extraction.

Stage Efficiency

Consider a simple batch leaching operation, where the solid is leached with more than enough solvent to dissolve all the soluble solute and there is no preferential adsorption of either solvent or solute by the solid. If adequate time of contact of solid and solvent is permitted, all the solute will be dissolved, and the mixture is then a slurry of insoluble solid immersed in a solution of solute in the solvent. The insoluble phases are then separated physically by settling, filtration, or drainage, and the entire operation constitutes one stage. If the mechanical separation of liquid and solid were perfect, there would be no solute associated with the solid leaving the operation and complete separation of solute and insoluble solid would be accomplished with a single stage. This would be an equilibrium stage, of 100 percent stage efficiency. In practice, stage efficiencies are usually much less than this: (1) the solute may be incompletely dissolved because of inadequate contact time; (2) most certainly it will be impractical to make the liquid-solid mechanical separation perfect, and the solids leaving the stage will always retain some liquid and its associated dissolved solute. When solute is adsorbed by the solid, even though equilibrium between the liquid and solid phases is obtained, imperfect settling or draining will result in lowered stage efficiency.

Practical Equilibrium

In the general case, it will be easiest to make calculations graphically, as in other mass-transfer operations, and this will require graphical representation of equilibrium conditions. It is simplest to use practical equilibrium conditions which take stage efficiencies into account directly, either entirely or in part, much as was done in the case of gas absorption and distillation. In the simplest cases, we must deal with three-component systems containing pure solvent (A), insoluble carrier solid (B), and soluble solute (C). Computations and graphical representation can be made on triangular coordinates for any ternary system of this sort, and the details of this have been worked out [17, 37]. Because of frequent crowding of the construction into one corner of such a diagram, it is preferable to use a rectangular coordinate system patterned after that used for fractional adsorption.

The concentration of insoluble solid B in any mixture or slurry will be expressed as N mass B/mass (A + C), whether the solid is wet with liquid solution or not. Solute C compositions will be expressed as weight fractions on a B-free basis: x = wt fraction C in the effluent solution from a stage (B-free basis), and y = wt fraction C in the solid or slurry (B-free basis). The value of y must include all solute C associated with the mixture, including that dissolved in adhering solution as well as undissolved or adsorbed solute. If the solid is dry, as it may be before leaching operations begin, N is the ratio of weights of insoluble to soluble substance, and $y = 1.0$. For pure solvent A, $N = 0$, $x = 0$.

The coordinate system then appears as in Fig. 13.23. Consider first a simple case of a mixture of insoluble solid from which all the solute has been leached, suspended in a solution of the solute in a solvent, as represented by point M_1 on the figure. The concentration of the clear solution is x, and the insoluble solid/solution ratio is N_{M1}. Let the insoluble solid be nonadsorbent. If this mixture is allowed to settle, as in a batch-settling tank, the clear liquid which can be drawn off will be represented by point R_1 and the remaining sludge will consist of the insoluble solid suspended in a small amount of the solution. The composition of the solution in the sludge will be the same as that of the clear liquid withdrawn, so that $y^* = x$. The concentration of solid B in the sludge N_{E1} will depend upon the length of time θ_1 allowed for settling, so that point E_1 then represents the slurry. Line E_1R_1 is a vertical tie line joining the points representing the two effluent streams, clear liquid and slurry. If the circumstances

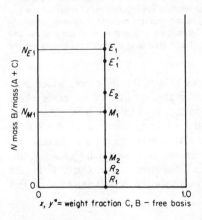

Figure 13.23 Concentrations in leaching and washing.

described are maintained in an actual leaching, points E_1 and R_1 can be taken as the practical conditions of equilibrium for that leaching. Clearly if less time is allowed for settling, say θ_1', the sludge will be less concentrated in insoluble solids and may be represented by point E_1'. There will be some maximum value of N for the sludge, corresponding to its ultimate settled height, in accordance with the description of batch settling given earlier, but usually in practice insufficient time is allowed for this to be attained. Since the concentration of insoluble solid in a sludge settled for a fixed time depends upon the initial concentration in the slurry, a mixture M_2 settled for time θ_1 might result in a sludge corresponding to point E_2. If the solid does not settle to give an absolutely clear solution, if too much solution is withdrawn from the settled sludge so that a small amount of solid is carried with it, or if solid B dissolves to a small extent in the solution, the withdrawn solution will be represented by some point such as R_2, somewhat above the lower axis of the graph. Similar interpretations can be made for compositions obtained when wet solids are filtered or drained of solution rather than settled, or when continuously thickened.

The settling or thickening characteristics of a slurry depend, as shown earlier, upon the viscosity and relative density of the liquid in which the solid is suspended. Since these in turn depend upon the solution composition, it is possible to obtain experimental data showing the variation of compositions of thickened solids with composition of solution and to plot them on the diagram as practical equilibrium conditions. It is evident, however, that in every case they must be obtained under conditions of time, temperature, and concentrations identical with those pertaining in the plant or process for which the calculations are being made. For drained beds of impervious solids, the equilibrium corresponding to the residual saturation after long-time drainage can be estimated by the methods of Illustration 13.1. Data for short-time drainage must be obtained experimentally.

In washing operations where the solute is already dissolved, uniform concentration throughout all the solution is rapidly attained, and reduced stage efficiency is most likely to be entirely the result of incomplete drainage or settling. In leaching an undissolved solute interspersed throughout the solid, on the other hand, lowered stage efficiency may be the result of inadequate time of contact as well as incomplete mechanical separation of liquid and solid. In this case it is possible (but not necessary) to distinguish experimentally between the two effects by making measurements of the amount and composition of liquid retained on the solid after short and after long contact time and to use the latter to establish the equilibrium conditions.

Let us now examine a few of the types of equilibrium curves which may occur. Figure 13.24a represents data which might be obtained for cases where solute C is infinitely soluble in solvent A, so that x and y may have values over the entire range from 0 to 1.0. This would occur in the case of the system soybean oil (C)-soybean meal (B)-hexane (A), where the oil and hexane are infinitely soluble. The curve DFE represents the separated solid under conditions actually to be expected in practice, as discussed above. Curve GHJ, the composition of the withdrawn solution, lies above the $N = 0$ axis, and in this

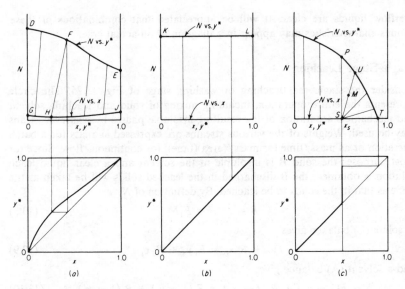

Figure 13.24 Typical equilibrium diagrams.

case, therefore, either solid B is partly soluble in the solvent or an incompletely settled liquid has been withdrawn. The tie lines such as line FH are not vertical, and this will result (1) if insufficient time of contact with leaching solvent to dissolve all solute is permitted, (2) if preferential adsorption of the solute occurs, or (3) if the solute is soluble in the solid B and distributes unequally between liquid and solid phases at equilibrium. The data may be projected upon a plot of x vs. y, as in the manner of adsorption or liquid-extraction equilibria.

Figure 13.24b represents a case where no adsorption of solute occurs, so that withdrawn solution and solution associated with the solid have the same composition and the tie lines are vertical. This results in an xy curve in the lower figure identical with the 45° line, and a distribution coefficient m, defined as y^*/x, equals unity. Line KL is horizontal, indicating that the solids are settled or drained to the same extent at all solute concentrations. It is possible to regulate the operation of continuous thickeners so that this will occur, and the conditions are known as *constant underflow*. The solution in this case contains no substance B, either dissolved or suspended. Figure 13.24c represents a case where solute C has a limited solubility x_S in solvent A. No clear solution stronger than x_S can be obtained, so that the tie lines joining slurry and saturated solution must converge, as shown. In this case any mixture M to the right of line PS will settle to give a clear saturated solution S and a slurry U whose composition depends on the position of M. Point T represents the composition of pure solid solute after drainage or settling of saturated solution. Since the tie lines to the left of PS are shown vertical, no adsorption occurs, and overflow liquids are clear. It will be appreciated that combinations of these various characteristics may appear in a diagram of an actual case.

Single-Stage Leaching

Consider the single real leaching or washing stage of Fig. 13.25. The circle represents the entire operation, including mixing of solid and leaching solvent and mechanical separation of the resulting insoluble phases by whatever means may be used. Weights of the various streams are expressed as mass for a batch operation or as mass/time [or mass/(area)(time)] for continuous flow. Since for most purposes the solid B is insoluble in the solvent and a clear liquid leach solution is obtained, the B discharged in the leached solids will be taken as the same as that in the solids to be leached. By definition of N,

$$B = N_F F = E_1 N_1 \tag{13.8}$$

A solute (C) balance gives

$$F y_F + R_0 x_0 = E_1 y_1 + R_1 x_1 \tag{13.9}$$

and a solvent (A) balance gives

$$F(1 - y_F) + R_0(1 - x_0) = E_1(1 - y_1) + R_1(1 - x_1) \tag{13.10}$$

and a "solution" (solute + solvent) balance gives

$$F + R_0 = E_1 + R_1 = M_1 \tag{13.11}$$

Mixing the solids to be leached and leaching solvent produces a mixture of B-free mass M_1 such that

$$N_{M1} = \frac{B}{F + R_0} = \frac{B}{M_1} \tag{13.12}$$

$$y_{M1} = \frac{y_F F + R_0 x_0}{F + R_0} \tag{13.13}$$

These relations can be shown on the coordinate system of Fig. 13.26. Point F represents the solids to be leached and R_0 the leaching solvent. Point M_1, representing the overall mixture, must fall on the straight line joining R_0 and F, in accordance with the characteristics of these diagrams described in Chap. 9. Points E_1 and R_1, representing the effluent streams, are located at opposite ends of the tie line through M_1, and their compositions can be read from the diagram.

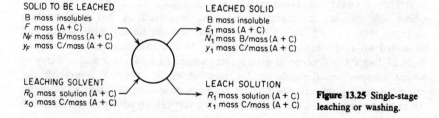

SOLID TO BE LEACHED
B mass insolubles
F mass (A + C)
N_F mass B/mass (A + C)
y_F mass C/mass (A + C)

LEACHED SOLID
B mass insoluble
E_1 mass (A + C)
N_1 mass B/mass (A + C)
y_1 mass C/mass (A + C)

LEACHING SOLVENT
R_0 mass solution (A + C)
x_0 mass C/mass (A + C)

LEACH SOLUTION
R_1 mass solution (A + C)
x_1 mass C/mass (A + C)

Figure 13.25 Single-stage leaching or washing.

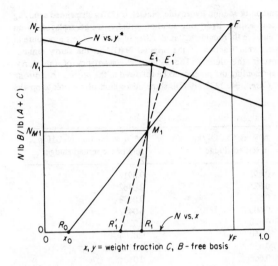

Figure 13.26 Single-stage leaching or washing.

Equation (13.8) permits calculation of the weight of E_1 and Eq. (13.11) that of R_1. Modification to allow for the presence of B in the liquid withdrawn, necessitating an equilibrium diagram of the type shown in Fig. 13.24a, is readily made by analogy with the corresponding problem in liquid extraction.

If the equilibrium data of Fig. 13.26 were obtained experimentally after long contact time of solid and liquid and therefore represent inefficiency of mechanical separation of liquid and solid only, then in a real stage there may be an additional inefficiency owing to short time of contact. The effluent streams can then be represented by points E_1' and R_1' on the figure, and a stage efficiency $(y_F - y_1')/(y_F - y_1)$ can be ascribed to this cause. If the equilibrium curve was obtained under conditions of contact time corresponding to the actual leaching, the tie line $E_1 R_1$ will give the effluent composition directly.

Multistage Crosscurrent Leaching

By contacting the leached solids with a fresh batch of leaching solvent, additional solute can be dissolved or washed away from the insoluble material. The calculations for additional stages are merely repetitions of the procedure for a single stage, with the leached solids from any stage becoming the feed solids to the next. Equations (13.8) to (13.13) apply, with only obvious changes in the subscripts to indicate the additional stages. When the number of stages for reducing the solute content of a solute to some specified value must be determined, it must be recalled that we are dealing with real stages, owing to the use of "practical" equilibrium data, and that the number found must therefore be integral. This may require adjustment by trial of either the amount of solute to be leached or the amount and apportioning of solvent to the stages.

Illustration 13.2 Caustic soda is being made by treatment of slaked lime, $Ca(OH)_2$, with a solution of sodium carbonate. The resulting slurry consists of particles of calcium carbonate,

$CaCO_3$, suspended in a 10% solution of sodium hydroxide, NaOH, 0.125 kg suspended solid/kg solution. This is settled, the clear sodium hydroxide solution withdrawn and replaced by an equal weight of water, and the mixture thoroughly agitated. After repetition of this procedure (a total of two freshwater washes), what fraction of the original NaOH in the slurry remains unrecovered and therefore lost in the sludge? The settling characteristics of the slurry, determined under conditions representing the practice to be followed in the process [Armstrong and Kammermeyer, *Ind. Eng. Chem.*, **34**, 1228 (1942)], show adsorption of the solute on the solid.

x = wt fraction NaOH in clear soln	N = kg $CaCO_3$/kg soln in settled sludge	y^* = wt fraction NaOH in soln of the settled sludge
0.0900	0.495	0.0917
0.0700	0.525	0.0762
0.0473	0.568	0.0608
0.0330	0.600	0.0452
0.0208	0.620	0.0295
0.01187	0.650	0.0204
0.00710	0.659	0.01435
0.00450	0.666	0.01015

SOLUTION The equilibrium data are plotted in Fig. 13.27. *Basis*: 1 kg solution in the original mixture, containing 0.1 kg NaOH (C) and 0.9 kg $H_2O(A)$. B = 0.125 kg $CaCO_3$.

The original mixture corresponds to M_1 with N_{M1} = 0.125 kg $CaCO_3$/kg soln, y_{M1} = 0.10 kg NaOH/kg soln. M_1 is plotted on the figure, and the tie line through this point is drawn. At

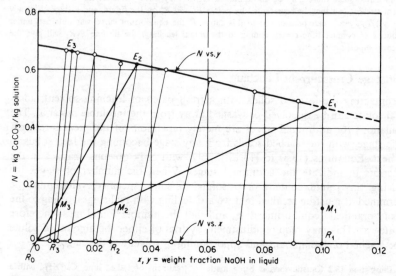

Figure 13.27 Solution to Illustration 13.2.

point E_1 representing the settled sludge, $N_1 = 0.47$, $y_1 = 0.100$. Eq. (13.8):

$$E_1 = \frac{B}{N_1} = \frac{0.125}{0.47} = 0.266 \text{ kg soln in sludge}$$

$$1 - 0.266 = 0.734 \text{ kg clear soln withdrawn}$$

Stage 2 $R_0 = 0.734$ kg water added, $x_0 = 0$ kg NaOH/kg soln. Eq. (13.11) adapted to this stage:

$$M_2 = E_1 + R_0 = E_2 + R_2 = 0.266 + 0.734 = 1.0 \text{ kg liquid}$$

Eq. (13.12):

$$N_{M2} = \frac{B}{E_1 + R_0} = \frac{B}{M_2} = \frac{0.125}{1.0} = 0.125$$

M_2 is located on line $R_0 E_1$ at this value of N, and the tie line through M_2 is drawn. At E_2, $N_2 = 0.62$, $y_2 = 0.035$. Eq. (13.8):

$$E_2 = \frac{B}{N_2} = \frac{0.125}{0.62} = 0.202 \text{ kg}$$

$$1 - 0.202 = 0.789 \text{ kg soln withdrawn}$$

Stage 3 $R_0 = 0.798$ kg water added, $x_0 = 0$. Eq. (13.11):

$$M_3 = E_2 + R_0 = 0.202 + 0.798 = 1.0$$

$$N_{M3} = \frac{B}{M_3} = \frac{0.125}{1} = 0.125$$

Tie line $E_3 R_3$ is located through M_3 as in the case of stage 2, and, at E_3, $N_3 = 0.662$, $y_3 = 0.012$. By Eq. (13.8), $E_3 = B/N_3 = 0.125/0.662 = 0.189$ kg soln in final sludge. $E_3 y_3 = 0.189(0.012) = 0.00227$ kg NaOH in sludge, or $(0.00227/0.1)100 = 2.27\%$ of original.

The process permits an appreciable loss and produces three solutions, two of which (R_2 and R_3) are quite dilute. It should be compared with the countercurrent washing operation of Illustration 13.3.

Multistage Countercurrent Leaching

A general flowsheet for either leaching or washing is shown in Fig. 13.28. Operation must necessarily be continuous for steady-state conditions to prevail, although leaching according to the Shanks system will approach the steady state after a large number of cycles have been worked through. In the flowsheet shown, it is assumed that solid B is insoluble and is not lost in the clear solution,

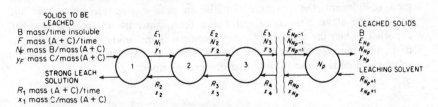

Figure 13.28 Multistage countercurrent leaching or washing.

but the procedure outlined below is readily modified to take care of cases where this may not be true.†

A solvent balance for the entire plant is

$$F + R_{N_p+1} = R_1 + E_{N_p} = M \tag{13.14}$$

and a "solution" (A + C) balance is

$$Fy_F + R_{N_p+1}x_{N_p+1} = R_1x_1 + E_{N_p}y_{N_p} = M y_M \tag{13.15}$$

M represents the hypothetical B-free mixture obtained by mixing solids to be leached and leaching solvent. Refer to Fig. 13.29, the operating diagram for the plant. The coordinates of point M are

$$N_M = \frac{B}{F + R_{N_p+1}} \tag{13.16}$$

$$y_M = \frac{Fy_F + R_{N_p+1}x_{N_p+1}}{F + R_{N_p+1}} \tag{13.17}$$

The points E_{N_p} and R_1, representing the effluents from the cascade, must lie on a line passing through M, and E_{N_p} will be on the "practical" equilibrium curve. Equation (13.14) can be rearranged to read

$$F - R_1 = E_{N_p} - R_{N_p+1} = \Delta_R \tag{13.18}$$

Similarly, a solution balance about any number of stages, such as the first three, can be arranged in this form,

$$F - R_1 = E_3 - R_4 = \Delta_R \tag{13.19}$$

Δ_R represents the constant difference in flow $E - R$ (usually a negative quantity) between each stage. In Fig. 13.29 it can be represented by the intersection of lines FR_1 and $E_{N_p}R_{N_p+1}$ extended, in accordance with the characteristics of these coordinates. Since the effluents from each stage are joined by the practical tie line for the particular conditions which prevail, E_1 is found at the end of the tie line through R_1. A line from E_1 to Δ_R provides R_2, and so forth. Alternatively the stage constructions may be made on the x, y coordinates in the lower part of the figure after first locating the operating line. This can be done by drawing random lines from point Δ_R and projecting their intersections with the equilibrium diagram to the lower curve in the usual manner. The usual staircase construction then establishes the number of stages. The stages are real rather than equilibrium, the practical equilibrium data having already taken into account the stage efficiency, and hence there must be an integral number. Especially when the number of stages required is the unknown quantity, some trial-and-error adjustment of the concentrations of the effluents or amount of solvent will be required to obtain an integral number.

† See Illustration 13.4, for example.

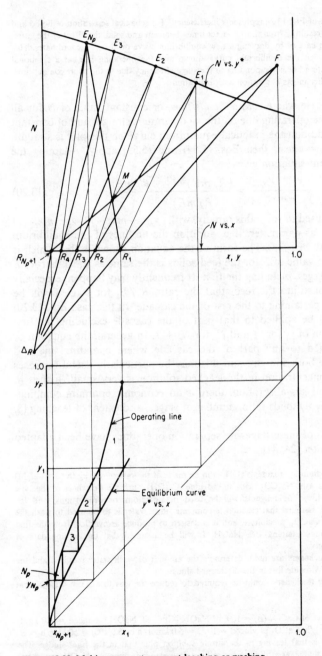

Figure 13.29 Multistage countercurrent leaching or washing.

If the equilibrium curve of Fig. 13.29 represents inefficiency of mechanical separation of liquid and solid only, and not that resulting from short contact time of solvent and solid, the effect of the latter, if known, may be taken care of by drawing a new equilibrium curve on the x, y coordinates. This should be located between the equilibrium curve shown and the operating line, at a fractional distance from the operating line corresponding to the stage efficiency due to the short contact time, in the manner used earlier in gas absorption and distillation.

In the special case where *constant underflow*, or constant value of N for all sludges, pertains, the operating line on the xy diagram is straight and of constant slope R/E. If in addition the practical equilibrium curve on this plot is straight, so that $m = y^*/x = $ const, then Eqs. (5.54) and (5.55) apply. Adapting the former to the present situation gives

$$\frac{y_F - y_{N_p}}{y_F - mx_{N_p+1}} = \frac{(R/mE)^{N_p+1} - R/mE}{(R/mE)^{N_p+1} - 1} \tag{13.20}$$

Figure 5.16 can be used to solve this rapidly, with $(y_{N_p} - mx_{N_p+1})/(y_F - mx_{N_p+1})$ as ordinate, R/mE as parameter. If in addition the tie lines of the equilibrium diagram are vertical, $m = 1.0$. The form of the equation shown is that which is applicable when the value of F for the feed solids is the same as E, so that R/E is constant for all stages, including the first. It frequently may happen, especially when dry solids constitute the feed, that the ratio R_1/E_1 for stage 1 will be different from that pertaining to the rest of the cascade. In this case Eq. (13.20) or Fig. 5.16 should be applied to that part of the cascade excluding the first stage, by substitution of y_1 for y_F and N_p for $N_p + 1$. In general, the equation or chart can be applied to any part of the cascade where operating line and equilibrium line are both straight, and this may be particularly useful for cases where the solute concentration in the leached solution is very small. Just as in liquid extraction and gas absorption, there is an economic optimum combination of treating solvent/solids ratio, number of stages, and extent of leaching [8, 9].

The calculations of these stagewise separation operations have been adapted to the digital computer [24, 41].

Illustration 13.3 Sodium hydroxide, NaOH, is to be made at the rate of 400 kg/h (dry weight) by reaction of soda ash, Na_2CO_3, with slaked lime, $Ca(OH)_2$, using a flowsheet of the type shown in Fig. 13.13a. The reagents will be used in stoichiometric proportions, and for simplicity it will be assumed that reaction is complete. Pure water is to be used to wash the calcium carbonate, $CaCO_3$, precipitate, and it is desired to produce as overflow from the first thickener a solution containing 10% NaOH. It will be assumed that the settling data of Illustration 13.2 apply.

(a) If three thickeners are used, determine the amount of wash water required and the percentage of the hydroxide lost in the discharged sludge.

(b) How many thickeners would be required to reduce the loss to at least 0.1% of that made?

SOLUTION (a) Mol wt of $CaCO_3$ (B) = 100, of NaOH (C) = 40. NaOH produced = 400 kg/h or 400/40 = 10 kmol/h. $CaCO_3$ produced = 10/2 = 5.0 kmol/h or 5.0(100) = 500 kg/h = B. The water required is that leaving in the strong solution plus that in the final sludge. The amount in the final sludge, according to the settling data, depends upon the NaOH concentration in the final sludge, which is not known. After a trial calculation, it is assumed that the solution in the final sludge will contain 0.01 wt fraction NaOH ($y_3 = 0.01$), and the settling data indicate $N_3 = 0.666$ kg $CaCO_3$/kg soln in the final sludge.

$$E = \frac{B}{N_3} = \frac{500}{0.666} = 750 \text{ kg/h soln list} \qquad \text{NaOH lost} = E_3 y_3 = 750(0.01) = 7.50 \text{ kg/h}$$

Water in sludge $= 750 - 7.5 = 742.5$ kg/h   NaOH in overflow $= 400 - 7.5 = 392.5$ kg/h

$x_1 = 0.1$ wt fraction NaOH in overflow $R_1 = \dfrac{392.5}{0.1} = 3925$ kg overflow or strong soln/h

Water in $R_1 = 3925 - 392.5 = 3532.5$ kg/h

Fresh water required $= R_{N_p + 1} = 3532.5 + 742.5 = 4275$ kg/h

For purposes of calculation, it may be imagined that the agitators are not present in the flowsheet and that the first thickener is fed with a dry mixture of the reaction products, $CaCO_3$ and NaOH, together with overflow from the second thickener.

$$F = 400 \text{ kg NaOH/h} \qquad N_F = \frac{B}{F} = \frac{500}{400} = 1.25 \text{ kg } CaCO_3/\text{kg NaOH}$$

$y_F = 1.0$ wt fraction NaOH in dry solid, $CaCO_3$-free basis

Plot points R_1, E_3, $R_{N_p + 1}$, and F on Fig. 13.30, and locate the difference point Δ_R at the intersection of lines FR_1 and $E_3 R_{N_p + 1}$ extended. The coordinates of point Δ_R are $N_{\Delta R} = -0.1419$, $y_{\Delta R} = -0.00213$. (These can be determined analytically, if desired, by simultaneous solution of the equations representing the intersecting lines.) Further computations must be done on an enlarged section of the equilibrium diagram (Fig. 13.31). Point Δ_R is plotted and the stages stepped off in the usual manner. The construction can be projected onto the xy diagram as shown, if desired. Three stages produce a value $y_3 = 0.01$, so that the assumed value of y_3 is correct. The NaOH lost in the sludge $= (7.5/400)100 = 1.87\%$ of that made.

(b)   NaOH lost $= 0.001(400) = 0.4$ kg/h

$$\text{kg } CaCO_3/\text{kg NaOH in final sludge} = \frac{500}{0.4} = 1250 = \frac{N_{N_p}}{y_{N_p}}$$

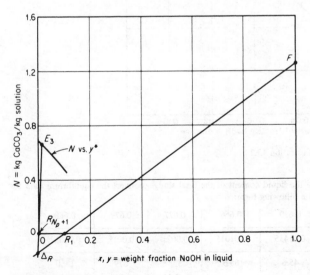

Figure 13.30 Solution to Illustration 13.3.

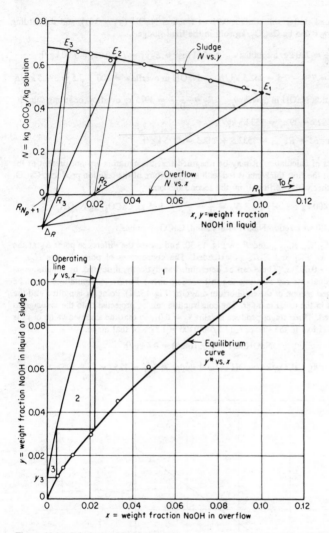

Figure 13.31 Solution to Illustration 13.3.

In order to determine the liquid content of the final sludge, convert the equilibrium data for dilute mixtures into the following form:

N	0.659	0.666	0.677	0.679	0.680
y^*	0.01435	0.01015	0.002†	0.001†	0.0005†
N/y^*	45.6	65.6	338	679	1360

† Estimated values.

By interpolation for $N/y^* = 1250$, $N_{N_p} = 0.680$ kg $CaCO_3$/kg soln, and $y_{N_p} = 0.680/1250 = 0.000544$ wt fraction NaOH in the liquid of the final sludge.

$$E_{N_p} = \frac{B}{N_{N_p}} = \frac{500}{0.680} = 735 \text{ kg/h} \qquad \text{Water in } E_{N_p} = 735 - 0.4 = 734.6 \text{ kg/h}$$

$$\text{NaOH in overflow} = 400 - 0.4 = 399.6 \text{ kg/h} \qquad R_1 = \frac{399.6}{0.1} = 3996 \text{ kg/h}$$

Water in $R_1 = 3996 - 399.6 = 3596$ kg/h   Fresh water $= R_{N_p+1} = 3596 + 734.6 = 4331$ kg/h

On the operating diagram (Fig. 13.32) point Δ_R is located in the same way as before, and the stages are constructed in the usual fashion. It becomes impractical to continue graphical

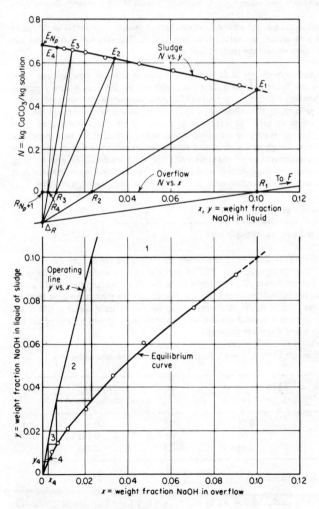

Figure 13.32 Solution to Illustration 13.3.

construction beyond the fourth stage unless considerable magnification of the chart is made, but computations beyond this point can be made with the help of Fig. 5.16. Beyond the fourth stage, the ratio of overflow to liquid in the sludge becomes substantially constant and equal to $R_{N_p+1}/E_{N_p} = 4331/735 = 5.90 = R/E$. This is the initial slope of the operating line on the lower part of Fig. 13.32. The slope of the equilibrium curve at these low concentrations is also substantially constant, $m = y^*/x = 0.01015/0.00450 = 2.26$ and $R/mE = 5.90/2.26 = 2.61$. $x_{N_p+1} = 0$, and $y_4 = 0.007$. Therefore $(y_{N_p} - mx_{N_p+1})/(y_4 - mx_{N_p+1}) = 0.000\ 544/0.007 = 0.0777$. From Fig. 5.16, an additional 2.3 stages beyond the 4 computed graphically are required.

An additional two stages (six thickeners) would make $y_{N_p}/y_4 = 0.099$, or $y_{N_p} = 0.099(0.007) = 0.000\ 693$, corresponding to 0.51 kg NaOH lost/h, while an additional three stages (seven thickeners) would make $y_{N_p} = 0.0365(0.007) = 0.000\ 255$, corresponding to 0.187 kg NaOH lost/h.

It must be emphasized that the cost of these numbers of thickeners probably could not be justified when balanced against the value of the lost NaOH. The very low NaOH loss was specified in order to demonstrate the computation methods.

Illustration 13.4 Flaked soybeans are to be leached with hexane to remove the soybean oil. A 0.3-m-thick layer of the flakes (0.25-mm flake thickness) will be fed onto a slowly moving perforated endless belt which passes under a series of continuously operating sprays [40]. As the solid passes under each spray, it is showered with liquid which percolates through the bed, collects in a trough below the belt, and is recycled by a pump to the spray. The spacing of the sprays is such that the solid is permitted to drain 6 min before it reaches the next spray. The solvent also passes from trough to trough in a direction countercurrent to that of the moving belt, so that a truly continuous countercurrent stagewise operation is maintained, each spraying and draining constituting one stage. Experiments [40] show that the flakes retain solution after 6 min drain time to an extent depending upon the oil content of the solution, as follows:

Wt % oil in soln	0	20	30
kg soln retained/kg insoluble solid	0.58	0.66	0.70

It will be assumed that the retained solution contains the only oil in the drained flakes.

The soybean flakes enter containing 20% oil and are to be leached to 0.5% oil (on a solvent-free basis). The net forward flow of solvent is to be 1.0 kg hexane introduced as fresh solvent per kilogram flakes, and the fresh solvent is free of oil. The solvent draining from the flakes is generally free of solid except in the first stage: the rich miscella contains 10% of the insoluble solid in the feed as a suspended solid, which falls through the perforations of the belt during loading. How many stages are required?

SOLUTION The tie lines are vertical, $x = y^*$. Rearrange the drainage data as follows:

Percent oil in soln $= 100y^*$	$\dfrac{\text{kg soln retained}}{\text{kg insoluble solid}} = \dfrac{1}{N}$	N	$\dfrac{\text{kg oil}}{\text{kg insoluble solid}} = \dfrac{y^*}{N}$
0	0.58	1.725	0
20	0.66	1.515	0.132
30	0.70	1.429	0.210

Basis: 1 kg flakes introduced.

Soybean feed $B = 0.8$ kg insoluble; $F = 0.2$ kg oil; $N_F = 0.8/0.2 = 4.0$ kg insoluble solid/kg oil; $y_F = 1.0$ mass fraction oil, solid-free basis.

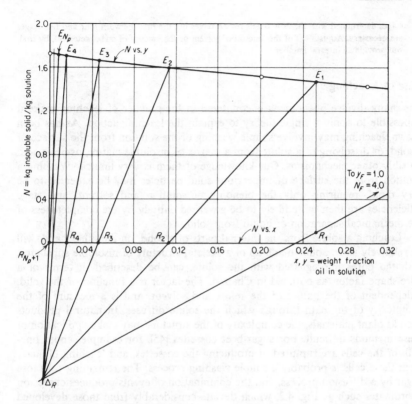

Figure 13.33 Solution to Illustration 13.4.

Solvent R_{N_p+1} = 1.0 kg hexane; x_{N_p+1} = 0 mass fraction oil.

Leached solids kg oil/kg insoluble solid = 0.005/0.995 = 0.00503. By interpolation in the equilibrium data, N_{N_p} = 1.718 kg solid/kg soln.

$$\text{Insoluble solid lost to miscella} = 0.8(0.1) = 0.08 \text{ kg}$$

$$\text{Insoluble solid in leached solids} = 0.8(0.9) = 0.72 \text{ kg}$$

$$E_{N_p} = \frac{0.72}{1.718} = 0.420 \text{ kg soln retained}$$

$$\text{Oil retained} = 0.00503(0.72) = 0.00362 \text{ kg}$$

$$\text{Hexane retained} = 0.420 - 0.00362 = 0.416 \text{ kg}$$

$$y_{N_p} = \frac{0.00362}{0.420} = 0.0086 \text{ mass fraction oil in retained liquid}$$

Miscella Hexane = 1 − 0.416 = 0.584 kg; oil = 0.2 − 0.00362 = 0.196 kg. R_1 = 0.584 + 0.196 = 0.780 kg clear miscella; x_1 = 0.196/0.780 = 0.252 mass fraction oil in liquid. N_{R1} = 0.08/0.780 = 0.1027 kg insoluble solid/kg soln.

The operating diagram is shown in Fig. 13.33. Point R_1 represents the cloudy miscella and is therefore displaced from the axis of the graph at N_{R1}. Point Δ_R is located as usual and the stages determined with the $N = 0$ axis for all stages but the first. Between four and five stages are necessary. Adjustment of the amount of solvent or the amount of unextracted oil, by trial, will provide an integral number.

Rate of Leaching

The many diverse phenomena encountered in the practice of leaching make it impossible to apply a single theory to explain the leaching action. As has been shown, leaching may involve simple washing of the solution from the surface of a solid, or dissolving of a solute from a matrix of insoluble matter, osmosis, and possibly other mechanisms. Our knowledge of them is very limited. Washing a solution from the surface of impervious solid particles may be expected to be very rapid, requiring only the blending of solution and solvent, and stage efficiencies are then quite likely to be governed entirely by the completeness of the mechanical separation of liquid from solid.

Leaching a solute from the internal parts of a solid, on the other hand, will be relatively slow. Solids made up of a skeletal structure of insoluble substance, with the pores impregnated with the solute, can be described in terms of a pore-shape factor, as outlined in Chap. 4. The factor is a function of the solid, independent of the nature of the solute and solvent, and is a measure of the complexity of the path through which the solute diffuses. In natural products such as plant materials, the complexity of the structure may make application of these methods difficult. For sugar-beet cossettes [45], for example, about one-fifth of the cells are ruptured in producing the cossettes, and leaching of sugar from these cells is probably a simple washing process. The remaining cells lose sugar by a diffusion process, and the combination of events produces curves on coordinates such as Fig. 4.2, which deviate considerably from those developed from simple diffusion with constant effective diffusivity or pore-shape factor [25]. Many mechanisms have been considered in an attempt to explain such observations [7]. In another example, wood will show different rates of leaching of an impregnating solute depending upon whether diffusion is in a direction parallel to or across the grain of the wood [30]. If solutes must pass through cell walls by dialysis, it may not be possible to apply the concepts at all. The rates of diffusion of soybean oil from soybean flakes, which do not permit simple interpretation, have been attributed to the presence of several types of structures in the matrix [23] as well as to the presence of a slowly dissolving constituent in the oil [18, 22]. Whole seeds cannot be leached; rolling and flaking evidently crushes cell walls and opens up passageways for penetration of the solvent by capillary action [28, 29]. The fact that the rate of leaching increases with increased surface tension of the solvent-oil solutions and that even for flaked seeds there is a residue of unextractable oil which increases with flake thickness supports this view. That the leached oil is composed of several different substances is evident from the different properties of oil obtained after short and long leaching times. A method of dealing with such differently leached substances has been suggested [19], but these examples serve at least to indicate the complexity of many practical leaching processes. Very little study has been given most of them.

When solids like those described above are immersed in leaching solvents, it is reasonable to suppose that the resistance to mass transfer within the solid itself is likely to be the controlling resistance and that of the liquid surrounding the solid to be quite minor [45]. In such cases increasing the rate of movement of liquid past the solid surface will not appreciably influence the rate of leaching. Applications of the theory of unsteady-state diffusion within solids of various shapes (see Chap. 4) to leaching calculations, including those associated with multistage cascades and those associated with chemical reaction in the solid, have been worked out in considerable detail [1, 31, 32, 42].

NOTATION FOR CHAPTER 13

Any consistent set of units may be used, except as noted.

a	const
q_p	specific surface of particles, area/packed space, L^2/L^3
A	pure leaching solvent
A	thickener cross section, Eq. (13.7), L^2
b	const
B	pure insoluble carrier solid
B	insoluble carrier solid, mass (batch operation), M; mass/time, M/Θ, or mass/(area)(time), $M/L^2\Theta$ (continuous operation)
c	solids concentration in a slurry, M/L^3
c_v	underflow solids concentration, M/L^3
C	soluble solute
d_p	diameter of a sphere of same surface/volume ratio as a particle, L
E	solvent and solute associated with the leached solids, mass (batch operation), M; mass/time, M/Θ, or mass/(area)(time), $M/L^2\Theta$ (continuous operation)
F	solute and solvent in solids to be leached, mass (batch operation), M; mass/time, M/Θ, or mass/(area)(time), $M/L^2\Theta$ (continuous operation)
g	acceleration of gravity, L^2/Θ
g_c	conversion factor, $ML/F\Theta^2$
G	mass velocity, $M/L^2\Theta$
G_S	flux of settling solids, $M/L^2\Theta$
G_{SL}	limiting solids flux, $M/L^2\Theta$
K	permeability, L^3/Θ^2
m	slope of equilibrium curve, dy^*/dx, dimensionless
M	solvent and solute content of a slurry or mixture, mass (batch operation), M; mass/time, M/Θ, or mass/(area)(time), $M/L^2\Theta$ (continuous operation)
N	mass of insoluble solid B/(mass of solute A and solvent C), M/M
N_p	number of stages, dimensionless
Δp	pressure drop, F/L^2
R	solvent and solute in a leaching solution, mass (batch operation), M; mass/time, M/Θ, or mass/(area)(time), $M/L^2\Theta$ (continuous operation)
s	residual saturation of a bed of drained solids, fraction of void volume occupied by liquid, dimensionless
s_0	residual saturation in the upper part of a packed bed, dimensionless
s_{av}	average residual saturation, dimensionless
V	initial settling rate, L/Θ
w	concentration of insoluble solids in a slurry, mass fraction
W	rate of solids flow, M/Θ
x	concentration of solute in solution, mass fraction, B-free basis

y	concentration of solute in a mixture, mass fraction, B-free basis
y^*	y at equilibrium
Z	height of a percolation bed or of settled solid, L
Z_D	drain height, L
Z_∞	ultimate height of settled solids, L
ϵ	fractional void volume of a packed bed, dimensionless
θ	time, Θ
μ_L	liquid viscosity, M/LΘ
ρ_L	liquid density, M/L^3
σ	surface tension, F/L

Subscripts

F	feed; solids to be leached
S	saturated
1, 2, etc.	stage 1, stage 2, etc.

REFERENCES

1. Agarwal, J. C., and I. V. Klumpar: *Chem. Eng.*, **84**, 135 (May 24, 1976).
2. Anderson, E. T., and K. McCubbin: *J. Am. Oil Chem. Soc.*, **31**, 475 (1954).
3. Anon.: *Eng. Min. J.*, **173**(6), 19 (1972).
4. Anon.: *Chem. Eng.*, **82**, 32 (Feb. 3, 1975).
5. Ashley, M. J.: *Chem. Eng. Lond.*, **1974**(286), 368.
6. Bautista, R. G.: *Adv. Chem. Eng.*, **9**, 1 (1974).
7. Brunicke-Olsen, H.: "Solid-Liquid Extraction," NYT Nordisk Forlag Arnold Busch, Copenhagen, 1962.
8. Chen, N. H.: *Chem. Eng.*, **77**, 71 (Aug. 24, 1970).
9. Colman, J. E.: *Chem. Eng.*, **70**, 93 (Mar. 4, 1963).
10. Comings, E. W.: *Ind. Eng. Chem.*, **32**, 663 (1940).
10a. Comings, E. W., C. E. Pruiss, and C. de Bord: *Ind. Eng. Chem.*, **46**, 1164 (1954).
11. Dombrowski, H. S., and L. E. Brownell: *Ind. Eng. Chem.*, **46**, 1267 (1954).
12. Donald, M. B.: *Trans. Inst. Chem. Eng. Lond.*, **15**, 77 (1937).
13. Fitch, B.: *Ind. Eng. Chem.*, **58**(10), 18 (1966); *Chem. Eng.*, **78**, 83 (Aug. 23, 1971).
14. Fitch, B.: *Chem. Eng. Prog.*, **70**(12), 33 (1974).
15. *French Oil Mill Machinery Co. Bull.* 08-10-E, 1977.
16. Gastrock, E. A., et al.: *Ind. Eng. Chem.*, **49**, 921, 930 (1957); *J. Am. Oil Chem. Soc.*, **32**, 160 (1955).
17. George, W. J.: *Chem. Eng.*, **66**, 111 (Feb. 9, 1959).
18. Goss, W. H.: *Oil Soap*, **23**, 348 (1946).
19. Hassett, N. J.: *Br. Chem. Eng.*, **3**, 66, 182 (1958).
20. Havighorst, C. R.: *Chem. Eng.*, **71**, 72 (Mar. 30, 1964).
21. Kammermeyer, K.: *Ind. Eng. Chem.*, **33**, 1484 (1941).
22. Karnofsky, G.: *J. Am. Oil Chem. Soc.*, **26**, 564 (1949).
23. King, C. O., D. L. Katz, and J. C. Brier: *Trans. AIChE*, **40**, 533 (1944).
24. Koenig, D. M.: *Ind. Eng. Chem. Fundam.*, **8**, 537 (1969).
25. Krasuk, J. H., J. L. Lombardi, and D. D. Ostvorsky: *Ind. Eng. Chem. Process Des. Dev.*, **6**, 187 (1967).
26. Lamont, A. G. W.: *Can. J. Chem. Eng.*, **36**, 153 (1958).
27. McCubbins, K., and G. J. Ritz: *Chem. Ind. Lond.*, **66**, 354 (1950).
28. Othmer, D. F., and J. C. Agarwal: *Chem. Eng. Prog.*, **51**, 372 (1955).
29. Othmer, D. F., and W. A. Jaatinen: *Ind. Eng. Chem.*, **51**, 543 (1959).
30. Osburn, J. Q., and D. L. Katz: *Trans. AIChE*, **40**, 511 (1944).

31. Plackko, F. P., and J. H. Krasuk: *Chem. Eng. Sci.*, **27**, 221 (1972); *Ind. Eng. Chem. Process Des. Dev.*, **9**, 419 (1970).
32. Plackko, F. P., and M. E. Lago: *Can. J. Chem. Eng.*, **50**, 611 (1972); *Chem. Eng. Sci.*, **28**, 897 (1973); *Ind. Eng. Chem. Process Des. Dev.*, **15**, 361 (1975).
33. Price, F. C.: *Chem. Eng.*, **67**(7), 84 (1960).
34. Rampacek, C.: *Chem. Eng. Prog.*, **73**(2), 57 (1977).
35. Rein, P. W., and E. T. Woodburn: *Chem. Eng. J.*, 7, 41 (1974).
36. Rushton, J. H., and L. H. Maloney: *J. Met.*, **6**, *AIME Trans.*, **200**, 1199 (1954).
37. Sattler-Dornbacher, E.: *Chem. Ing. Tech.*, **30**, 14 (1958).
38. Scofield, E. P.: *Chem. Eng.*, **58**(1), 127 (1951).
39. Shannon, P. T., E. M. Toy, et al.: *Ind. Eng. Chem.*, **57**(2), 18 (1965); *Ind. Eng. Chem. Fundam.*, **2**, 203 (1963); 3 184, 250 (1964); 4, 195, 367 (1965).
40. Smith, C. T.: *J. Am. Oil Chem. Soc.*, **28**, 274 (1951).
41. Soylemez, S., and W. D. Seider: *AIChE J.*, **19**, 934 (1973).
42. Tettamanti, K., J. Manczinger, J. Hunek, and R. Stomfai: *Acta Chim. Sci. Hung.*, **85**(1), 27 (1975).
43. Van Arsdale, G. D.: "Hydrometallurgy of Base Metals," McGraw-Hill Book Company, New York, 1953.
44. Work, L. T., and A. S. Kohler: *Ind. Eng. Chem.*, **32**, 1329 (1940).
45. Yang, H. H., and J. C. Brier: *AIChE J.*, 4, 453 (1958).

PROBLEMS

13.1 A 1-m-diameter tank fitted with a false bottom and canvas filter is partly filled with 1000 kg (dry weight) of sea sand wet with seawater. The sand is allowed to drain until it stops dripping, whereupon 750 kg fresh water is added and recirculated to reach a uniform salt concentration. The sand is again allowed to drain until dripping stops and is then removed from the tank and dried. Estimate the salt content of the dried sand.

The sand particles have an average size d_p = 0.4 mm, a particle density 2660 kg/m^3, and a bulk density 1490 kg (dry weight)/m^3. Seawater contains 3.5% salt; its density is 1018 kg/m^3 and surface tension 0.0736 N/m. The surface tension of water is 0.728 N/m.

13.2 Derive expressions for the coordinates of point $\Delta_R(y_{\Delta R}, N_{\Delta R})$ (Fig. 13.29), and check the results by determining the numerical values in the case of Illustration 13.3a.

13.3 In order to eliminate the solids in the final miscella of Illustration 13.4, it is decided to pass liquid from stage 3 to stage 1, where the liquid will contact fresh solids. The drained liquid from stage 1, containing the suspended solids, will then be passed to stage 2, where it is filtered by passage of the liquid through the bed of solids in this stage. The final miscella is then withdrawn as a clear solution from stage 2. How many stages will then be required for the same solvent/seeds ratio and the same oil concentration in the discharged solids? *Ans.*: 6.

13.4 A mineral containing 20% elemental sulfur is to be leached with hot gas oil, in which the sulfur is soluble to the extent of 10% by weight. The solvent will be repeatedly pumped over the batch of ground mineral, using 1.5 kg fresh solvent/kg mineral. After no further solution of sulfur is obtained, the liquid will be drained and replaced with a fresh batch of 1.5 kg oil/kg original mineral, and the operation repeated. On drainage, the solid materials retain the solution to the extent of one-tenth the weight of undissolved solid (sulfur and gangue). No preferential adsorption takes place.

(*a*) Calculate the equilibrium data and plot them in the usual manner.

(*b*) Determine the amount of sulfur unextracted and the sulfur concentration of the composited leach liquors.

(*c*) Repeat part (*b*) if a two-stage Shanks system is used, with 3 kg fresh solvent/kg unleached solid. Assume steady state has been reached.

13.5 Aluminum sulfate, $Al_2(SO_4)_3$, is to be produced by action of sulfuric acid, H_2SO_4, on bauxite in a series of agitators, with a cascade of continuous thickeners to wash the insoluble mud free of aluminum sulfate.

$$Al_2O_3 + 3H_2SO_4 \rightarrow Al_2(SO_4)_3 + 3H_2O$$

The flowsheet is similar to that of Fig. 13.13a. The reaction agitators are fed with (1) 25 t bauxite/day, containing 50% Al_2O_3 and the rest insoluble; (2) the theoretical quantity of aqueous acid containing 60% H_2SO_4; and (3) the overflow from the second thickener. Assume the reaction is complete. The strong product solution is to contain 22% $Al_2(SO_4)_3$, and no more than 2% of the $Al_2(SO_4)_3$ produced is to be lost in the washed mud. The last thickener is to be fed with pure wash water. The underflow from each thickener will contain 4 kg liquid/kg insoluble solid, and the concentration of solubles in the liquid of the underflow for each thickener may be assumed to be the same as that in the overflow. Calculate the number of thickeners required and the amount of wash water required per day. **Ans.:** 3 thickeners.

Note: In solving this problem, be certain to account for the water in the acid as well as that produced by the reaction. Adapt Fig. 5.16 to all but the first thickener in the cascade.

13.6 Barium, occurring naturally as the sulfate, $BaSO_4$, is put in water-soluble form by heating with coal, thus reducing the sulfate to the sulfide, BaS. The resulting reaction mixture, barium "black ash" containing 65% soluble BaS, is to be leached with water. Black ash is fed to a tube mill at 100 t/day, together with the overflow from the second of a cascade of thickeners, and the effluent from the mill is fed to the first thickener. All the barium is dissolved in the mill. The strong solution overflowing from the first thickener is to contain 20% BaS by weight. The thickeners will each deliver a sludge containing 1.5 kg liquid/kg insoluble solid. The solution in the overflow and that in the sludge leaving any thickener may be assumed to have the same BaS concentration. It is desired to keep the BaS lost with the final sludge to at most 1 kg/day.

(a) How many thickeners are required? [Adapt Eq. (5.55) to all except the first thickener.] **Ans.:** 6.

(b) It is decided to pass the final leached sludge to a continuous filter, as in Fig. 13.13b, where the liquid content of the filtered solids will be reduced to 15% by weight. The filtrate will be returned to the last thickener, but the filter cake will not be washed. How many thickeners will then be required? **Ans.:** 5.

13.7 In the manufacture of potassium nitrate, KNO_3, potassium chloride, KCl is added to a hot, concentrated aqueous solution of sodium nitrate, $NaNO_3$,

$$KCl + NaNO_3 \rightleftharpoons KNO_3 + NaCl$$

Because of its relatively low solubility, part of the sodium chloride, NaCl, precipitates and is filtered off. A little water is added to the filtrate to prevent further precipitation of NaCl, the mixture is cooled to 20°C, and pure KNO_3 crystallizes. The resulting slurry contains, per 100 kg precipitated KNO_3, 239 kg of a solution analyzing 21.3% KNO_3, 21.6% NaCl, and 57.1% water. The slurry is fed to the first of a cascade of four continuous classifiers, where each 100 kg of crystals is countercurrently washed with 75 kg of a saturated solution of KNO_3, containing 24.0% KNO_3, in order to free them of NaCl. The wet crystals leaving each classifier retain 25% liquid, and the liquid overflows are clear. The washed crystals discharged from the fourth classifier, containing 25% liquid, are sent to a continuous drier. All liquid except that discharged with the washed crystals leaves in the overflow from the first classifier. Equilibrium between solid and liquid is attained in each classifier, and the clear overflows have the same composition as the liquid retained by the crystals. The solubility of KNO_3 in NaCl solutions (KNO_3 is the equilibrium solid phase) at the prevailing temperature is given by the following table:

% NaCl	0	6.9	12.6	17.8	21.6
% KNO₃	24.0	23.3	22.6	22.0	21.3

(a) Plot the equilibrium data [N = kg KNO_3/kg (NaCl + H_2O) for both clear overflow and wet crystals; x and y = kg NaCl/kg (NaCl + H_2O)].

(b) Calculate the percentage NaCl content which can be expected on the dried KNO_3 product.
Ans.: 0.306%.

INDEX

Page numbers in *italic* indicate illustrations or tables.

Absolute humidity:
  mass and molal, defined, 227
  percentage, defined, 229
Absorbers (*see* Gas absorption; Gas-liquid
  operations, equipment for)
Absorption (*see* Gas absorption)
Absorption factor:
  in absorption of multicomponent systems,
    323–326
  defined, 124
  in distillation, 422, 450–453
  estimating most economical, 291–293
  transfer units and, *310*
  in tray absorbers, 291–292, 324–331
Absorption section of fractionators (*see* Enriching
  section of fractionators)
Accumulators, reflux, 397
Acid gases, absorption of, 275, 281–282
Activated adsorption (chemisorption), 566–567
Activated clays (*see* Clays)
Activated diffusion through polymers, 93
Activity (chemical potential):
  in diffusion, 22
  in interphase mass transfers, 105
Adductive crystallization, described, 4
Adhesion in drying, *679*
Adiabatic operations:
  gas absorption\by: interface conditions and,
    316–322, *319–321*
  mass-transfer coefficients for, 314–315
  nonisothermal, 295
  gas-liquid contact in, *236*
  humidification by, 241–263
    dehumidification of air-water vapors, 252
    equipment for, 259, *260–263*
    fundamental relationships in, 242–245,
      243
    general methods of, 255–259, *258*
    recirculating liquid and gas humidification-
      cooling, 252–255, *253*
    of vapor-gas mixtures, *236*, 237

Adiabatic operations:
  humidification by: for water cooling with air,
    245–252, *246*, *249*, *250*
  mass-transfer coefficients for, 261
    absorption, 314–315
    humidification, 261
Adiabatic saturation:
  curves of, *236*, 237
  enthalpy for pure substances in, *225*
Adiabatic-saturation temperature:
  defined, 237
  Lewis relation and, 241
  (*See also* Wet-bulb temperature)
Adsorbents (*see* Adsorption)
Adsorption, 5, 565–641
  by continuous contact, 612–646
    under steady-state conditions, 612–623
      equipment, 612, *613*
      one component absorbed, *614*, *615*, 616–617
      two components absorbed, 617–623, *618*,
        620, *622*
    under unsteady-state conditions, 623–640
      adsorption of liquids by percolation, 630–631
      adsorption of vapors, 625, *626–630*
      adsorption wave, 623, *624*
      by chromatography, 631, *632*
      by elution, *631*
      rate of adsorption, 632–641, *633*, *635*,
        637–639
  defined, 4
  dehumidification by, 263
  equilibrium in, 569–584
    of liquids, 580, *581–584*
    of single gases and vapors, 569–575, *570–573*
    of vapor-gas mixtures, 575–580, *577–579*
  fractional, defined, 4
  by ion exchange, 641–646, *645*
  nature of adsorbents, 567–569
  by stagewise operations, 585–611
    adsorption of vapor from fluidized beds,
      609–611, *610*

Adsorption:
  by stagewise operations: agitated vessels for, 599–608, *600, 607*
  contact filtration of liquids, 585–599, *586, 588, 589, 591–598*
  in fluidized beds, 608–611
  by ion exchange, 642
  of vapors, 609–611, *610*
  slurry adsorption of gases and vapors, 609
  in teeter beds, 608–609
  types of, 566–567
Adsorption rate under unsteady-state conditions, 632–641, *633, 635, 637–639*
Adsorptivity, relative (separation factor), 576–584
Agitated pan driers, 666
Agitated vessels:
  absorption in, 158
  adsorption in, 599–608, *600, 607*
  for gas-liquid operations, 146–158
    for gas-liquid contact, *153–158, 155*
    for single-phase liquids, 146–153, *147, 148, 152*
  leaching in, *726–730*
    under steady-state conditions, 732, *733*
    under unsteady-state conditions, 731
  liquid extraction in, 521–526, *524*
  mass-transfer coefficients in: for adsorption, 602–606
    for gas-liquid contact operations, 156–158
    in stage-type extractors, 523, *524*
Air conditioning, 262, *263*, 664
Air-cooling of water, 245–252, *246, 249, 250*
Air-water vapors:
  dehumidification of, 252
  in humidification operations, 231–236, *232–235*
Alumina as adsorbent, 568, 626, 627
Animal oils:
  adsorption of, 568
  distillation of, 460
  (*See also* Fish oils)
Anion exchange, 642–643
Atmolysis, defined, 6
Atmospheric pressure, diffusivity of gases at standard, *31*
Atomic and molecular volumes of gases, *33*
Axial dispersion, defined, 210
Axial mixing:
  in differential extractors, 541, *542*, 544–546
  in packed towers, 542
    end effects and, 209–210
  in spray towers, 542
  in sieve-tray towers, 530–531
Azeotropic distillation, 455–457, *456*
Azeotropic mixtures (constant-boiling mixtures):
  heteroazeotropic mixtures, 352–353
  maximum-boiling, 355–357, *356*
  minimum-boiling, 350–352, *351*
  rectification of, 419–*421*

Back mixing, 181
  in packing towers, defined, 210
  (*See also* Axial mixing)
Baffled turbulence in open tanks, 149
Ballast tray, 177, *178*
Batch operations:
  adsorption by: ion exchange, 642
    mass-transfer coefficient in, 604–606
  (*See also* Adsorption, by stagewise operations)

Batch operations:
  contact filtration by, *586*
  differential distillation by, 367–371
  stages of interphase mass transfers in, 124
  as unsteady-state operations, 9
  (*See also* Leaching, under unsteady-state conditions)
  (*See also* Drying, batch)
Batch settling in leaching, *728, 729*
Battery, extraction, defined, 724
Bauxite as absorbent, 568, 626
Beds (*see specific types of beds*)
Berl saddles, 189, *190, 198*, 203, *205, 206*, 210, 261
Bernoulli's equation, 145
BET equation (Brunauer-Emmett-Teller equation), 571
Bidisperse catalyst particles, defined, 95
Binary systems:
  adsorption equilibria in binary vapor-gas mixtures, 576–*578*
  continuous rectification of, 371–374
  molecular diffusion in, 24
  (*See also* Distillation, vapor-liquid equilibrium in)
Blowdown, makeup water to replace, 248
Boiling point:
  of liquids, defined, 222
  vapor-liquid equilibrium and, 345–346
Bollman extractors, *742*
  variants of, *743*
Bone char as adsorbent, 568
Bottoms (*see* Residues)
Bound moisture (bound water):
  defined, 657, 660
  drying of, 684–686
Boundary (flux at phase interface):
  defined, 47
  grain, 95
Boundary-layer theory, *64–66*
Box extractors, 529, *530*
Breakpoint, adsorbers, 623, 633
Breakthrough curve, adsorption, 623, *633–641, 638, 639*
Brunauer-Emmett-Teller equation (BET equation), 571
Bubble(s):
  in gas-liquid contact operations, 154
    diameter, gas holdup and interfacial area, 156
  in gas-liquid operations with gas dispersed, 144
    diameter of, 140–141
    rising velocity of single bubbles, 141, *142*
    specific interfacial area of, 144
    swarms of, 142
Bubble-cap trays and towers, 158–160, *159*, 165–*167*
  overall efficiency of, *185*
  (*See also* Tray towers)
Bubble columns (sparge vessels), 140–146, *142, 143*
Bubble point in vapor-liquid equilibrium, 361–363
  computation of, 362–363, 366–367, 436–439
Bubble-point temperature curve, 345
Bubbling promoters, 176, *177*

Cabinet driers (*see* Tray driers)
Capillary movement, drying and, *679*

Capillary number, 721–722
Carbon:
  for adsorption: decolorizing, 568
    equilibrium of, 569–575, *570, 572, 573,*
      584
    gas, 568, 612, 625–628
    liquid adsorption, 586, 587, 592, 594, 596
  diffusion of, 95
  drying of coal, 692
  (*See also* Hydrocarbons)
Cascade(s):
  in interphase mass-transfer operations, *125–127,*
    128–130, *129*
  in mixer-settler operations, *529, 530*
  of stagewise operations, defined, 10
Cascade rings, 189
Case hardening, 665, 709
Catalyst particles, bidisperse, defined, 95
Cation exchange, 641–643, 645, 646
  cation exchanger, defined, 641
CBM (*see* Azeotropic mixtures)
CCD (continuous countercurrent decantation),
  736–738
Centrifugal extractors, 547, *548*
Centrifugal stills, *461,* 462
Centrifugation, 6
Charcoal (*see* Carbon)
Chemical(s):
  purification of, 478
  recovery of: by extraction, 478
    by leaching, 732
  (*See also* Pharmaceuticals)
Chemical methods, liquid extraction as substitute
  for, 478
Chemical potential (*see* Activity)
Chemical reaction, absorption by, 333
Chemical reactivity of solvents, 489
Chemisorption (activated adsorption), 566–567
Chilton-Colburn analogy, 70
Chromatography, 631, *632,* 642, 643
Circular pipes, mass-transfer coefficients and
  turbulent flow in, 70–72, *71*
Circulating drops, eigenvalues for, *524*
Classifiers, leaching in, 739, *740*
Clausius-Clapeyron equation, 222
Clays:
  as adsorbents, 568, 582, *583,* 587
  drying of, 678, 679
Closed tanks, prevention of vortex formation in,
  149
Closed vessels (diffusers), 724, *725*
Coal, drying of, 692
Coalescence:
  in sieve-tray towers, 535
  (*See also* Drops)
Coalescers, *528,* 529
Coarse solids, leaching of, 738–744, *740–743*
Cocurrent flow, gas adsorption in, *286–289*
Cocurrent operations:
  direct-heat, cocurrent-flow drying, 690
  (*See also* Continuous cocurrent operations)
Cold reflux in McCabe-Thiele distillation method,
  419
Collision function, diffusion, *32*
Color removal, 565, 582, *583,* 623
Combination film-surface renewal theory, *62,* 63
Compartment driers (*see* Tray driers)
Compression zone in leaching, defined, 729

Concentrated solutions, adsorption from, 582–*584*
Condensation:
  differential, differential distillation and, 369
  partial, flash vaporization by, *365*
  (*See also* Adsorption; Distillation)
Condensers of fractionators, 377, *378, 381*
  with McCabe-Thiele method, 404–405, 417,
    418
  with Ponchon-Savarit method, 397
Confined fluids, mass transfers to, 47, *48*
Coning, defined, 161
Constant(s), *14*
  force, of gases, as determined from viscosity
    data, *33*
  mass-action-law, 643
Constant-boiling mixtures (CBM; *see* Azeotropic
  mixtures)
Constant drying conditions, defined, 667, *668*
Constant-pressure equilibrium, vapor-liquid,
  344–346
Constant-rate drying, 669, 670
  cross-circulation drying and, *672–677, 673*
  (*See also* Falling-rate period of drying)
Constant-temperature equilibrium:
  solutes and, *93*
  vapor-liquid, *347,* 348
Constant total pressure and diffusion through
  porous solids, 96–99
Constant underflow in leaching, 747, 754
Contact filtration (*see* Mixer-settler operations)
Continuity, equation of, 24–26, *25*
Continuous cocurrent operations:
  adsorption, with liquid- and solid-phase mass-
    transfer resistances, 606–608, *607*
  stages of interphase mass transfers in, 124
Continuous-contact adsorption (*see* Adsorption,
  by continuous contact)
Continuous-contact equipment:
  distillation by fractionation with, 426–*431,*
    427
  (*See also* Gas absorption, in continuous-contact
    equipment; Liquid extraction, in differential
    extractors; Packed towers)
Continuous-contact operations (differential-contact
  operations), 10
  (*See also specific continuous-contact operations*)
Continuous countercurrent operations:
  adsorption of one component, *614, 615*
  continuous countercurrent decantation, *736–738*
  extraction with reflux, *507–514, 509, 511–513*
Continuous operations:
  adsorption by, *618*
  drying (*see* Drying, continuous)
  leaching of coarse solids, 738–744, *740–743*
  (*See also* Distillation, by fractionation)
Convection, natural, Grashof number and, *68*
Cooling and cooling towers (*see* Humidification)
Corrosion in tray and packing towers, 211
Corrosiveness of solvents, 282
Countercurrent cascades:
  in interphase mass-transfers, *126–*130, *127,*
    129
  in mixer-settler operations, *529*
Countercurrent extractors (mechanically-agitated
  extractors), 544–547
Countercurrent operations:
  adiabatic, continuous drying by, 686
  direct-heat, countercurrent-flow drying, 690

Countercurrent operations:
  gas absorption by, 282, *283–285*
    (*See also* Gas absorption, one component
    transferred, in countercurrent multistage
    operations)
  leaching by multiple contact, 723, *724*
  two-stage countercurrent adsorption, *595, 596,
    599*
    (*See also* Continuous countercurrent operations;
    Multistage countercurrent operations)
Counterdiffusion:
  in gases, 29, *30*
  in interphase mass transfers, 113
  in liquids, 34–35
Counterflow trays, 177–178
Critical moisture content, 669, 682
Critical packing size, 544
Critical point (critical state) of vapor-liquid
  equilibrium, 221
Critical pressure of vapor-liquid equilibrium, 221
Critical temperature, vapor-liquid equilibrium
  and, 221
Cross-circulation drying (*see* Drying, batch, by
  cross circulation)
Cross-flow cascades in interphase mass-transfers,
  125
Crosscurrent operations:
  two-stage crosscurrent adsorption, 598–599
    (*See also* Multistage crosscurrent operations)
Crowd-ion mechanism of diffusion through metals,
  95
Crude oil, distillation of, 453–455, *454*
Crystal-filter driers (rotary filters), 684
Crystalline solids, diffusion through, 95
Crystallization, 4
Cut, defined, 367
Cylinder(s):
  analogies between momentum, heat and mass
    transfers in, 70–72, *71*
  diffusion through, 89, *91, 92*
Cylinder driers, 695
Cylinder oil, decolorization of, 582, *583*

Danckwerts' theory of mass transfer, 61–62
Dead-end pores of solids, diffusion through, 95
Decantation:
  continuous-countercurrent, *736–738*
    (*See also* Liquid extraction)
Decoction, defined, 717
Decolorization, 565, 568, 582, *583*, 586, 588, 623
Dehumidification (*see* Adsorption; Humidification)
Dehydration, 623
  (*See also* Drying)
Deliquescence, defined, 660
Density of solvents, 489
Depth of liquids in sieve trays, *169*, 170
Desorption (stripping):
  adiabatic, 314–322, *315, 319–321*
  defined, 3
  interfacial area for, of aqueous liquids, *205*
  through-circulation drying and, 682
    (*See also* Drying)
  (*See also* Adsorption; Strippers; Stripping factor)
Detergents:
  drying of, 696
  separation of, 6

Deviating velocity (fluctuating velocity), defined,
  55
Dew point:
  in humidification of unsaturated vapor-gas
    mixtures, *229*, 230
  in vapor-liquid equilibrium, 361–363
Dialysis, defined, 5–6
Differential condensation, 369
Differential-contact operations (*see* Continuous-
  contact operations)
Differential distillation (simple distillation),
  367–371
Differential energy balance, 26
Differential extractors (*see* Liquid extraction, in
  differential extractors)
Differential heat of adsorption, defined, 574
Differential mass balance, 24–26
Diffusers (closed vessels), 724, *725*
Diffusion:
  in drying: internal-diffusion controlling, 681–682
    of liquid moisture, *678*
    of vapors, 679–680
  eddy, 54–59, *55*
  defined, 22
  gaseous, defined, 5
  in ion exchange, 643–644
  molecular (*see* Molecular diffusion)
  between phases of interphase mass transfers,
    106–117
    average overall coefficients for, 113–117, *114*
    local coefficients for, 111–113
    local overall coefficients, *109*, 110–111, 113
    local two-phase mass transfers, *107–109*
  in solids, 88–103
    through crystalline solids, 95
    Fick's law applied to, 88–93
      unsteady-state diffusion, 89–93, *90, 91*
    through polymers, 93–95
    through porous solids, 95–100
  sweep, 6
  thermal, 6
Diffusion-controlled falling rate of drying, *678*
  internal-diffusion controlling, 681–682
Diffusional sublayer, defined, 59
Diffusivity (diffusion coefficient), 23
  atomic and molecular volumes, *33*
  eddy heat and mass, 58–59
  effective, 29
  of gases, *31–33*, 34
  of liquids, 35–37, *36*
  momentum eddy, 58
  (*See also* Diffusion)
Dilute solutions:
  absorption of, 290–291
    mass-transfer coefficients of, 308–309
  adsorption of solutes from, 580–*582*
  extraction from, 551–552
Dip-feed single-drum driers, *695*
Direct driers, 661–666, *662–664*
  tunnel driers as, 687, 704
Direct-heat drying:
  cocurrent-flow drying, 690
  continuous, rate of, 701–710, *702, 703, 708*
  countercurrent-flow drying, 690
Direct-indirect driers, 692
Direct operations:
  making use of surface phenomena, 7
    (*See also specific direct operations*)

Dispersed-phase holdup, 522, 535, *536*
Dispersion:
  in stage-type extractors, 521–526, *524*
  (*See also* Gas-liquid operations, equipment for,
    with gas dispersed; Gas-liquid operations,
    equipment for, with liquid dispersed)
Distillate, defined, 372
Distillation (fractional distillation) 342–473
  defined, 3
  differential, 367–371
  by flash vaporization, *363–367, 364*
  by fractionation, 371–431
    in continuous-contact equipment, 426–*431,
     427*
    McCabe-Thiele method of, 402–431
     cold reflux in, 419
     condensers in, 404–405, 417, *418*
     enriching section of fractionators, 404–405,
      422–423
     exhausting section of fractionators, *405,*
      406, 422
     feed introduction in, *406–409, 408*
     feed-tray location in, 409, *410*
     high-purity products with, 422–423
     minimum reflux ratio in, *411*, 412, *413*
     multiple feeds in, 421–422
     optimum reflux ratio in, 412–416, *415*
     rectification of azeotropic mixtures in,
      419–*421*
     total reflux ratio in, *410*, 411
     tray efficiency in, 423–426, *425*
     use of open steam with, *416*, 417
    Ponchon-Savarit method of, 374–402
     complete fractionators used, 380–382, *381*
     condensers and reflux accumulators used,
      397
     enriching section of fractionators, 374–378,
      376
     feed-tray location in, 382–384, *383*
     heat loss in, 400–402, *401*
     high-purity products and tray efficiency
      with, 402
     multiple feeds in, 397–400, *398, 399*
     reboilers in, 392–394, *393*
     reflux ratio in (*see* Reflux ratio)
     side streams in, 400
     stripping section in, 378–*380*
     use of open steam in, 394–397, *395, 396*
  low-pressure, 460–462
  of multicomponent systems, 431–460
    azeotropic distillation, 455–457, *456,* 459–460
    composition correctiors in, 445–447
    differential distillation of, 370–371
    extractive distillation of, *457*–460
    feed-tray location in, 442–443
    by flash vaporization, 366–367
    key components in, 434–435
    Lewis and Matheson calculation for, 434,
     443–445, 447, 459
    liquid/vapor ratios in, 447–449
    minimum reflux ratio in, 435–439
    product compositions in, 440–441
    specific limitations of, 434
    of ternary systems, *432*
    Thiele-Geddes method of, 433–434, 439,
     442–443, 445, 449–455, *454*
    total reflux in, 439–440
    vapor-liquid equilibrium in, 360–363

Distillation:
  vapor-liquid equilibrium in, 343–363
    constant-pressure equilibrium, *344–346*
    constant-temperature equilibrium, *347*, 348
    enthalpy in, 357–360, *358, 359*
    at increased pressure, 346, *347*
    in multicomponent systems, 360–363
    pressure-temperature-concentration phase,
     343–344
    Raoult's law of, *348*, 349
     negative deviations from, *355–357*
     positive deviations from, *350–354, 351, 353*
    relative volatility and, 346
Distribution coefficient (*see* Equilibrium distribu-
  tion)
Dobbins theory (combination film-surface-renewal
  theory), *62*, 63
Dorr agitators, 732, *733*
Dorr thickeners, *734*
  balance-tray, *735*
Double-solvent extraction (fractional extraction),
  4, 478, 514, *515–518*
Downspouts (downcomers):
  multiple, 176, *178*
  in sieve-tray towers, 534–535
    backup in, 172, 176
  in tray towers, 163, *164*
Draft towers, 259–261, *260*
Drag coefficient of drops, 150, 533
Drainage from percolation tanks, 721, *722, 723*
Driers (*see* Drying; *and specific types of driers*)
Drift (*see* Liquid entrainment)
Drops:
  in agitated vessels, 521–*524*
  coalescence of, 538–539
  drag coefficient of, 150, 533
  in emulsions and dispersions, 526–527
  in sieve-tray towers: formation of, *532, 533*
    mass transfers during, 536–537
    terminal velocity of, 533, *534*
  (*See also* Liquid extraction)
Drum driers, 694, *695*
Dry-bulb temperature:
  in adiabatic operations, 248
  defined, 228
Drying, 655–716
  batch, 661–686
    by cross circulation, 672–682
     constant-rate period of, *672–677, 673*
     critical moisture content, 682
     internal-diffusion controlling and, 681–682
     movement of moisture within solids in,
      677–680, *678, 679*
     unsaturated-surface drying, 670–*681*
    in direct driers, 661–666, *662–664*
    in indirect driers, 661, 666–667
    rate of, 667–672, *668*
    through-circulation drying, *682–686, 688, 689*
    in rotary driers, 693, *694*
  continuous, 661, 686–710
    in drum driers, 694, *695*
    in fluidized and spouted beds, 697, *698*
    material and enthalpy balances in, 699–701,
     700
    in pneumatic driers, 698, *699*
    rate of drying, 701–710, *702, 703, 708*
    in rotary driers, 689–693, *690, 691*
    rate of drying, 704–707

Drying:
  continuous: in spray driers, 695–697, *696*
    in through-circulation driers, 688, *689*
    in through-circulation rotary driers, 693, *694*
    in tunnel driers, 687
      rate of drying in, 704
    in turbo-type driers, 687, *688*
  defined, 4
  equilibrium in, 655–661
    definitions, 660, *661*
    hysteresis and, 657, *658*
    of insoluble solids, *656*, 657
    of soluble solids, 658–660, *659*
Dumping, defined, 161
"Dusty gas" equation, 98
Dynamic similarity of agitated vessels,
    defined, 150

Economic balances in liquid extraction, 514
Eddy(ies):
  defined, 45
  energy-containing, 55
Eddy diffusion (turbulent diffusion), 54–59, *55*
  defined, 22
Eddy motions, 45, *46*
Edmister method, 329, 449–451
Effective diffusivity, 29
Efficiency (*see* Stage efficiency; Tray efficiency)
Efflorescence, defined, 660
Effusion (*see* Diffusion)
Eigenvalues for circulating drops, *524*
Electrodialysis, defined, 6
Elemental fluid volume, *25*
Eliminators, entrainment, 193
Eluate, defined, 631
Elutant, defined, 631
Elution (elutriation):
  adsorption by, *631*
  defined, 717
Emulsions in stage-type extractors, 526–527
End effects, axial mixing and, in packing towers,
    209–210
Energy balance, differential, 26
Energy-containing eddies, 55
Energy requirements, design principles and, 11–12
Enriching section of fractionators (rectifying
    section of fractionators), 372
  in McCabe-Thiele method of distillation, *404*,
    405, 422–423
  in Ponchon-Savarit method of distillation,
    374–378, *376*
Enthalpy:
  in adiabatic adsorption, *315*, 316
  balance of, in continuous drying, 699–701, *700*
  in distillation, 357–360, *358*, *359*
  fractionation: overall balances of, 372–374, *373*
    in Ponchon-Savarit distillation method, *390*
  in humidification: evaporative cooling, *265*
    for pure substances, 224–226, *225*
    of unsaturated vapor-gas mixtures, 230–231
  (*See also* Heat)
Entrainer, defined, 455
Entrainment (*see* Liquid entrainment)
Equilibrium:
  adsorption, 569–584
    of liquids, 580–*584*
    of single gases and vapors, 569–575, *570–573*
    of vapor-gas mixtures, 575–580, *577–579*

Equilibrium:
  distillation: reboiled vapor in equilibrium with
    residue, 405–406
    (*See also* Distillation, vapor-liquid equilibrium
    in)
  drying (*see* Drying, equilibrium in)
  in interphase mass transfers, 104–106
    departure of bulk-phase concentrations from,
    108
    equilibrium-distribution curve, *105*
  ion exchange, 643
  leaching, 744–748, *745*, *747*
  liquid extraction (*see* Liquid extraction, liquid
    equilibrium in)
  vapor-liquid, for pure substances, 220–*224*, *221*
  (*See also* Henry's law; Multicomponent systems)
Equilibrium distillation (flash vaporization), *363*,
    364, 365–367
Equilibrium distribution (distribution coefficient;
    partition coefficient):
  liquid extraction, 483, 488
  of solutes at constant temperature, *93*
Equilibrium moisture, 656–660
  defined, 656, 660
  hysteresis and, 657, *658*
  in insoluble solids, *656*, 657
  in soluble solids, *658*–660
Equilibrium solubility of gases into liquids (*see*
    Gas absorption, equilibrium solubility of
    gases into liquids and)
Equilibrium stages (ideal stages; theoretical stages):
  in absorption, 301–*307*, *305*
  defined, 10, 123–124
  number of, 11
  in stagewise operations, 10
Equimolal counterdiffusion:
  in gases, 29, *30*
  in interphase mass transfers, 113
  in liquids, 34–35
Equimolal overflow in McCabe-Thiele distillation
    method, 402–403
Equipment (*see specific mass-transfer operations
    and types of equipment*)
Ergun equation, 200
Evaporation:
  distillation compared with, 342
  drying versus, 655
  liquid extraction and, 478
Evaporative cooling, enthalpy in, *245*
Exhausting section of fractionators, 372
  in McCabe-Thiele distillation method, *405*, 406,
    422
  in Ponchon-Savarit distillation method, 378–*380*
Extract:
  defined, 477
  (*See also* Leaching; Liquid extraction)
Extraction:
  defined, 717
  (*See also* Leaching; Liquid extraction)
Extraction battery, defined, 724
Extractive distillation:
  liquid extraction versus, 479
  of multicomponent systems, *457–460*
Extractors (*see* Liquid extraction; *and specific types
    of extractors*)

Falling-film stills, 462
Falling liquid films, gas into, mass-transfer
    coefficients in, *50*–54

Falling-rate period of drying, 669–672, 677–682
    diffusion-controlled, 678
    internal-diffusion controlling, 681–682
    movement of moisture within solids and, 677–679
    in unsaturated-surface drying, *680, 681*
Fanning friction factor, 56, 171
Feed:
    defined, 477
    in McCabe-Thiele distillation method: feed
        introduction, *406–409, 408*
        feed-tray location, 409, *410*
        multiple feeds, 421–422
    in multicomponent systems, feed-tray location,
        442–443
    in Ponchon-Savarit distillation method: feed-tray
        location, 382–384, *383*
        multiple feeds, 397–400, *398, 399*
Fenske equation, 385, 439
Fermentation by gas-liquid operations, 139
Fick's law, 23, 26
    applied to diffusion in solids, 88–93
    unsteady-state diffusion, 89–93, *90, 91*
Fill for towers (*see* Packed towers)
Film(s), gas into falling liquid, mass-transfer
    coefficients of, *50–54*
Film coefficients (*see* Heat-transfer coefficients;
    Mass-transfer coefficients)
Film theory:
    of mass-transfer coefficients in turbulent flow,
        59, 60
    two-film theory, 107, 110
Filter(s):
    percolation, 630
    rotary, 684
    (*See also* Mixer-settler operations)
Filter-press leaching, 725–726
Filtration, contact (*see* Mixer-settler operations)
Fish oils:
    distillation of, 460
    vitamins from, 395
Fixed-bed adsorbers (*see* Adsorption, by continu-
    ous contact, under unsteady-state conditions)
Fixed-bed operations (*see specific mass-transfer
    operations*)
Flash driers (pneumatic driers), 698, *699*
Flash vaporization:
    distillation by, *363–367, 364*
    freezing by, 666
Flat-bed turbines, adsorption in, 599, *600*
Flexirings (Pall rings), 189, *190, 197,* 261
Flocculation, 729
Flooding:
    in packed towers, 194, 195
    in tray towers, 160
Floor loading in tray and packed towers, 211
Flow mixers (line mixers), 190, 521
Fluctuating velocity (deviating velocity), defined,
    55
Fluid(s):
    mass-transfer coefficients on fluid surfaces,
        60–64, 62, 63
    mass transfers to confined, 47, *48*
    mechanics of, and mechanical agitation of
        single-phase liquids, 150–153, *152*
    (*See also* Molecular diffusion; *and specific
        mass-transfer operations*)
Fluid surfaces, mass-transfer coefficients of
    turbulent flow on, *60–64, 62, 63*
Fluid volume, elemental, *25*

Fluidized beds:
    adsorption in, 608–611
        ion exchange, 642
        of vapors, 609–611, *610*
    continuous drying in, 697, 698
Flux:
    at phase interface, defined, 47
    variations in, mass-transfer coefficients and,
        77–78
    (*See also* Reflux; Reflux ratio)
Foam separation, defined, 6
Foaming systems of tray and packing towers, 211
Foodstuffs:
    drying of, 666–667
    (*See also* Fish oils; Sugar and sugar beets;
        Vegetable oils)
Force constants of gases as determined from
    viscosity data, *33*
Fourier's equation, 92
Fractional adsorption, defined, 4
Fractional crystallization, defined, 4
Fractional dialysis, defined, 6
Fractional distillation (*see* Distillation)
Fractional extraction (double-solvent extraction),
    4, 478, 514–518
Fractional sublimation, defined, 3
Fractionation (continuous rectification), 8–9
    adsorption by, 617–623, *618, 620, 622*
        of vapor mixtures, 611
    distillation by (*see* Distillation, by fractionation)
Fractionators, 372
    complete, 380–382, *381*
    condensers of, 377, *378, 381*
        with McCabe-Thiele method, 404–405, 417,
        418
        with Ponchon-Savarit method, 397
    enriching section of, 372
        with McCabe-Thiele method, *404,* 405,
        422–423
        with Ponchon-Savarit method, 374–378, *376*
    exhausting section, 372
        with McCabe-Thiele method, *405,* 406, 422
        with Ponchon-Savarit method, 378–*380*
    (*See also* Distillation, by fractionation)
Francis weir formula, 170
Free moisture, defined, 660
Freeze drying (sublimation drying), 666–667
Freezing point of solvents, 489
French stationary-basket extractors, 741
Freundlich equation, 581–*583,* 597–598
    applied to contact filtration, *589,* 590
    applied to multistage countercurrent adsorption,
        594–599, *595–598*
    applied to multistage crosscurrent adsorption,
        590–*592*
Freundlich isotherms, 571
Froude number, 151
Fuller's earths, 567–568
Fully baffled turbulence in open tanks, 149

Gas(es):
    adsorption equilibrium of single, 569–575,
        570–573
    atomic and molecular volumes of, *33*
    diffusivity of, *31–34*
    effects of humidity and temperature on constant-
        rate drying of, 675
    into falling liquid films, mass-transfer coefficients
        in, *50–54*

Gas(es):
    kinetic theory of, 21, 30, 40
    slurry adsorption of, 609
    steady-state molecular diffusion in, 27, 28–30
    (See also specific mass-transfer operations)
Gas absorption, 275–341
    by chemical reaction, 333
    in continuous-contact equipment, 300–322
        for absorption of one component, 301–304
        height equivalent to theoretical plate, 301–307, 305, 307
        overall coefficients and transfer units in, 307–322
            for dilute solutions, 308–309
            graphical construction for transfer units, 309–311
            for nonisothermal operations, 313–322, 315, 319–321
            overall heights of transfer units, 311–313, 312
    defined, 3
    equilibrium solubility of gases in liquids and, 275–282
        choice of solvents for absorption, 281, 282
        in multicomponent systems, 277–279
        in nonideal liquid solutions, 279–281, 280
        in two-component systems, 276, 277
    interfacial area for, 205
    in multicomponent systems, 322–332, 324,
        equilibrium solubility in, 277–279
    multistage, in agitated vessels, 158
    one component transferred, 282–300
        in cocurrent flow, 286–289
        in countercurrent flow, 282, 283–285
        in countercurrent multistage operations, 289–300
            absorption factor A, 291–293, 292
            dilute gas mixtures, 290–291
            in nonisothermal operations, 293–298, 294, 297
            overall coefficients for, 313–314
            real trays and tray efficiency, 298–300
        minimum liquid-gas ratio, 285, 286
Gas-adsorbent carbon (see Carbon)
Gas bubbles (see Bubbles)
Gas-enthalpy transfer units, number of, 247
Gas-gas operations, 3, 5
Gas holdup, 143, 144, 156
Gas-liquid operations, 3
    equipment for, 139–319
        with gas dispersed, 139–158
            agitated, for single-phase liquids, 146–152, 147, 148
            agitated vessels for gas-liquid contact, 153–158, 155
            sparge vessels, 140–146, 142, 143
        with liquid dispersed, 186–211
            packed towers (see Packed towers, gas-liquid operations in)
            sieve trays (see Sieve trays and sieve-tray towers, gas-liquid operations in)
            spray towers and chambers, 187
            Venturi scrubbers, 186, 187
            wetted-wall towers, 187
    gas-liquid contact adiabatic, 236
    separated by membranes, 5
    (See also Adiabatic operations, humidification by; Desorption; Distillation; Gas absorption; Humidification; Nonadiabatic operations)

Gas-liquid ratio in absorption, 285, 286
Gas-solid operations, 3–4
    (See also Adsorption; Desorption; Drying)
Gaseous diffusion, defined, 5
Gasoline, recovery of, 322
Gelatins, drying of, 678
Geometric similarity of agitated vessels, defined, 150
Glitsch Ballast Tray valve design, 177, 178
Glues, drying of, 678, 695
Goodloe packings, 190
Graesser extractors, 546
Grain boundaries, diffusion through metals along, 95
Grashof number, 68
Grids (hurdles), 190, 259
Grinding, leaching during, 731–732
Grosvenor humidity (mass absolute humidity), 227

Heap leaching, 720
Heat:
    of adsorption, 574–575
        differential, defined, 574
        heat of wetting, 580
    for drying operations, 661–662
        in continuous direct-heat drying, 701–710 702, 703
        in direct driers, 664, 666
        in direct-heat: cocurrent-flow drying, 690
            countercurrent-flow drying, 690
        in freeze drying, 667
    humid: in adiabatic gas-liquid contact, 236–237
        in air-water systems, 235
        defined, 230
    for humidification operations: heat balance, defined, 225
        latent heat of vaporization, defined, 222
        sensible heat, defined, 222
        of unsaturated vapor-gas mixtures, 230
    losses of, in Ponchon-Savarit distillation method, 400, 401, 402
    (See also under Thermal)
Heat capacity, defined, 226
Heat eddy diffusivity, 58–59
Heat transfers and heat-transfer coefficients, 60
    analogies with mass-transfer coefficients, 66–72
        corresponding dimensionless groups of mass and heat transfers and, 68
        turbulent flow in circular pipes and, 70–72, 71
    mass-transfer coefficients for simultaneous mass and, 78–82, 79, 81
Heatless adsorbers (pressure swing), 628–630
Heavy components of multicomponent systems, 434
Heavy key components of multicomponent systems, 435
Height:
    of slurries, in leaching, 731
    of transfer units: in absorption, 311–313, 312
        in adsorption, 615–616
Height equivalent to theoretical plate (HETP), 301–307, 305
Henry's law, 93, 279, 291, 299, 300, 309, 551
Heteroazeotropic mixtures, 352–353

HETP (height equivalent to theoretical plate), *301-307, 305*
Hickman centrifugal molecular stills, *461*
Higbie theory (penetration theory), *60*, 61
Higgins contactors, *613*
High-purity products:
  with McCabe-Thiele distillation method, 422-423
  with Ponchon-Savarit distillation method, 402
Hirschfelder-Bird-Spotz method, Wilke-Lee modification of, 31
Holdup:
  dispersed-phase, 522, 535, *536*
  gas, *143*, 144, 156
  liquid: in packing towers, 202-203, *206*, 207
    in tray towers versus packing towers, 210-211
  solid, 692-693
Horizontal extractors, *743*
Horton-Franklin equation, 323, 326, 327, 330
HTU (*see* Height, of transfer units)
Hu-Kintner correlation, 533
Humid heat:
  in adiabatic gas-liquid contact, 236-237
  in air-water systems, 235
  defined, 230
Humid volume:
  in air-water systems, 231
  defined, 230
Humidification, 220-274
  defined, 3
  enthalpy in: evaporative cooling, *265*
    for pure substances, 224-226, *225*
  of unsaturated vapor-gas mixtures, 230-231
  by gas-liquid contact (*see* Adiabatic operations, humidification by; Nonadiabatic operations)
  of vapor-gas mixtures, 227-241
    with absolute humidity, 227
    by adiabatic operations, *236*, 237
    of air-water vapors, 231, *232-236*
    of saturated vapor-gas mixtures, 227-228
    of unsaturated vapor-gas mixtures, 228-231, *229*
    wet-bulb temperature and, 237-241, *238*
  vapor-liquid equilibrium for pure substances in, 220-224, *221*
Humidity:
  absolute and mass absolute, defined, 227
  relative, defined, 228
  (*See also* Drying; Humidification; Moisture; Wet-bulb temperature)
Hurdles (grids), 190, 259
Hydraulic(s), sieve-tray, *532-536*
Hydraulic head of sieve-trays, 171-172, 176
Hydrocarbons:
  decolorization of, 582, *583*
  as leaching solvents, 740
  separation of: by absorption, 295-298, *297*, 327-331
    by adsorption, 565, 570, 571, 586, 608
    by distillation, 436-439, *457-459*
    by extraction, 478-479, 510-514, *511-513*, 519-520
    by fractionation, 612
Hydrocyclones (hydroclones), leaching in, 738, *739*
Hydrodynamic movement of diffusion through porous solids, 95, 99-100
Hydrogen from waste refinery gases, 94
Hy-Pak rings, 189
Hyperfil packings, 190

Hypersorber, 612
Hysteresis:
  adsorption, 571-573, *572*
  drying, 657, *658*

Ideal-gas law, 145, 222, 279
Ideal solutions (*see* Multicomponent systems; Raoult's law)
Ideal stages (*see* Equilibrium stages)
Ideal trays (theoretical trays), 289-293
Immiscible phases, mass-transfer operations by direct contact of two, 2-5
Impellers:
  for gas-liquid contact, 154, *155-158*
  power of, 601-602
  single-phase liquids agitated by, 146-*148*
In-place leaching (in-situ leaching; solution mining), 719-720
Indirect drying, 661, 666-667
Indirect-heat, countercurrent-flow drying, *691*, 692
Indirect operations making use of surface phenomena, 7
Induced-draft towers, *261*
Inorganic chemicals, purification of, 478
Insoluble liquids:
  continuous multistage countercurrent extraction of, *505-507*
  multistage crosscurrent extraction of, *496*, 497
  steam distillation of, 354-355
Insolubility of solvents, 488, *489*
Insoluble solids, drying equilibrium of, *656*, 657
Intalox saddles, 189, *190*, *199*, 261
Integral heat of adsorption, defined, 574
Interconnected pores of solids, diffusion through, 95
Interfacial absorption, adiabatic, 316-322, *319-321*
Interfacial area:
  in agitated vessels of stage-type extractors, 522-523
  in gas-liquid operations, 144, 156, 204, *205*, 207
Interfacial tension in liquid extraction:
  and drop coalescence, 538-539
  and drop terminal velocity, 533, *534*
  emulsions and, 527
  in flow mixers, 521
  solvents and, 489
Interfacial turbulence (Marangoni effect), 113
Internal-diffusion controlling in batch drying, 681-682
Internal reboilers, *393*
Interphase mass transfers, 104-136
  diffusion between phases of, 106-117
    average overall coefficients for, 113-117, *114*
    local coefficients, 111-113
    local overall coefficients for, *109*-111, 113
    local two-phase mass transfers, *107-109*
  equilibrium in, 104-106
    departure of bulk-phase concentrations from, *108*
    equilibrium-distribution curve, *105*
  material balances in, 117-123
    in steady-state cocurrent processes, *117*-121, *119*, *120*
    in steady-state continuous cocurrent processes, 124
    in steady-state countercurrent processes, *121-124*

Interphase mass transfers:
  stages of, 123–133
  in batch processes, 124
  cascades and, *125–130*
  in continuous cocurrent processes, 124
  mass-transfer rates and, 130–133
  stage defined, 123
Interstitial mechanism of diffusion through metals, 95
Interstitialcy mechanism of diffusion through metals, 95
Ions:
  crowd-ion mechanism, 95
  ion exchange, 641–646, *645*
  ion exclusion, 643
Isolated pores of solids, diffusion through, 95
Isostere, defined, 574

Jacketed kettle reboilers, *393*
j_D, *68*, 70
j_H, *68*, 70

Karr reciprocating-plate extractors, 546
Kennedy extractors, *741*, 742
Kerr-McGee uranium extractors, *530*
Kettle reboilers, 392, *393*
Key components of multicomponent systems, 434–435
Kiln action, described, 692
Kinematic similarity of agitated vessels, defined, 150
Kinetic theory of gases, 21, 31, 40
Kittel trays, 177
Klee-Treybal equation, 534
Knudsen's law, 97–100
Koch Flexitray, 177
Koch-Sulzer packings, 190
Kremser-Brown-Sonders equations (Kremser equations), 128, 291, 323, 326–327, 422, 594

Laminar flow:
  mass-transfer coefficients in, *50–54*
  molecular diffusion in: momentum and heat transfers in, 38–41, *39*
  (*See also* Molecular diffusion, steady-state)
Langmuir equation, 460, 571
Latent heat of vaporization, defined, 222
Leaching, 717–764
  calculation methods for, 744–761
  in multistage countercurrent leaching, *751–760*, *753*, *755–757*, *759*
  in multistage crosscurrent leaching, 749–751, *750*
  rate of leaching, 760–761
  in single-stage leaching, *748*, *749*
  stage efficiency, 744
  defined, 5, 717
  preparation of solids for, 718–719
  under steady-state conditions, 731–744
  in agitated vessels, 732, *733*
  continuous, of coarse solids, 738–744, *740–743*
  by continuous countercurrent decantation, *736–738*
  during grinding, 731–732
  in hydrocyclones, 738, *739*
  in thickeners, 733–*738*, *734*, *735*, *737*

Leaching:
  under unsteady-state conditions, 719–731
  in agitated vessels, *726–730*, 731
  in closed vessels, 724, *725*
  by countercurrent multiple contact operations, 723, *724*, 740
  in filter-press leaching, 725–726
  in heap leaching, 720
  in in-place leaching, 719–720
  in percolation tanks, 720–*722*, 723
Length of unused bed (LUB), 639, 640
Lessing rings, 189, *190*
Leva trays, 177–178
Lewis-Matheson calculation, 434, 443–445, 447, 459
Lewis number, 41, 240, 252
Lewis relation, 240–241, 245
Lewis-Whitman theory (two-film theory; two-resistance theory), *107*, 110
Light components of multicomponent systems, 434
Light key components of multicomponent systems, 434
Lightfoot theory (surface-stretch theory), *63*, 64
Linde trays, 176, *177*
Line mixers (flow mixers), 190, 521
Linear falling-rate period of drying, *680*
Liquid(s):
  boiling point of, defined, 222
  diffusivity of, 35–37, *36*
  saturated, defined, 221
  steady-state molecular diffusion in, 34–35
  (*See also* specific mass-transfer operations)
Liquid entrainment:
  makeup water to replace, 248
  in packing towers, eliminators of, 193
  in sieve-tray towers, 166, *173*, 174, 176
  and tray efficiency, 182
Liquid equilibrium:
  in adsorption, 580, *581–584*
  (*See also* Gas absorption, equilibrium solubility of gases into liquids; Liquid extraction, liquid equilibrium in)
Liquid extraction, 477–561
  defined, 4
  in differential extractors, 541–553
  centrifugal extractors, 547, *548*
  design of, 548–551, *549*
  for dilute solutions, 551–552
  mechanically-agitated extractors, *544–547*
  in packed towers, 542–544, *543*
  performance of, 552–553
  in spray towers, 542
  fields of usefulness of, 478–479
  liquid equilibrium in, 479–489
  choice of solvents and, 488, *489*
  equilateral-triangular coordinates of, *480*, *481*
  in multicomponent systems, 488
  notation scheme for, 479–480
  rectangular coordinates of, *486*, *487*
  in systems of three liquids: with one pair partially soluble, 482–484
  with two pairs partially soluble, *484*, *485*
  in systems of two partially soluble liquids and one solid, *485*, *486*
  stage efficiency in, *520*, 521, 537–539, *538*
  in stage-type extractors, 521–541

Liquid extraction:
in stage-type extractors: agitated vessels, 521–526, *524*
of emulsions and dispersions, 526–527
mixer-settler cascades in, *529, 530*
settlers in, 527–529, *528*
sieve-tray towers, 530–541, *531–534, 536, 538*
by stagewise contact, 490–521
continuous countercurrent with reflux, 507–514, *509, 511–513*
continuous multistage countercurrent extraction, 497–507, *498–501, 503–506*
economic balances in, 514
fractional extraction, 514, *515–518*
in multicomponent systems, 518–520, *519*
multistage crosscurrent extraction, *493–497, 495*
single-stage extraction, 490–493, *491, 492*
stage efficiency in, *520,* 521
Liquid films, gas into falling, mass-transfer coefficients in, 50, *51–54*
Liquid flow in tray towers, 164, *165*
Liquid/gas ratio:
minimum, in absorption, *285,* 286
and use of tray towers, 211
Liquid holdup:
in packing towers, 202–203, *206,* 207
in tray towers versus packing towers, 210–211
Liquid-liquid operations, 4, 476
phases separated by membranes, 5
(*See also* Liquid extraction)
Liquid miscibility, partial, in positive deviations from ideality, 352, *353*
Liquid-solid equilibrium (*see* Adsorption, equilibrium in)
Liquid-solid operations, 4–5
(*See also* Adsorption; Leaching)
Liquid/vapor ratio in distillation of multicomponent systems, 447–449
Lixiviation, defined, 717
Loading in packing towers, 194, 195
Local mass-transfer coefficients, 48
Local two-phase mass transfers, *107–109*
Low-pressure distillation, 460–462
LUB (length of unused bed), 639, 640
Luwesta extractors, 548

McCabe-Thiele method (*see* Distillation, by fractionation, McCabe-Thiele method of)
Macropores, 95
Marangoni effect (interfacial turbulence), 113
Marc, defined, 740
Marine-type propellers, single-phase liquids agitated by, 146, *147*
Mass absolute humidity (Grosvenor humidity), 227
Mass balance in adiabatic absorption, 315
Mass-action-law constant, 643
Mass eddy diffusivity, 58–59
Mass-transfer coefficients, 18–19, 45–87
in adiabatic operations, 261
adiabatic absorption, 314–315
in agitated vessels: adsorption, 602–606
gas-liquid contact, 156–158
in stage-type extractors, 523, *524*

Mass-transfer coefficients:
analogies with heat- and momentum-transfer coefficients, 66–72
corresponding dimensionless groups of mass and heat transfers, *68*
turbulent flow in circular pipes and, 70–72, *71*
eddy motions and, 45, *46*
fractionation, in packed towers, 429–431
in gas absorption (*see* Gas absorption, overall coefficients and transfer units in)
in gas-liquid operations, 202–209, *205, 206*
in interphase mass transfers: average overall coefficients, 113–117, *114*
local coefficients, 111–113
local overall coefficients, *109–111,* 113
stages and, 130–133
in laminar flow, 50–54
in liquid extraction, 536–541, *538*
relations between, *49*
for simple situations, 72–78, *74–75*
for simultaneous mass and heat transfers, 78–82, *79, 81*
in transfers to confined fluids, 47, *48*
in turbulent flow, 54–66
boundary layers and, *64–66*
in circular pipes, 70–72, *71*
eddy diffusion and, 54–59, *55*
film theory of, *59,* 60
flow past solids, 64–66
on fluid surfaces, 60–64, *62, 63*
Mass-transfer operations, 1–18
choice of separation method for, 7–8
(*See also* specific *mass-transfer operations*)
classification of, 2–7
design principles, 11–12
method of conducting, 8–10
unit systems used, 12–18, *13–17*
Mass-transfer resistances in adsorption, 604, 606–608, *607*
Material balances:
in continuous drying, 699–701, *700*
in interphase mass transfers (*see* Interphase mass transfers, material balances in)
Maximum boiling azeotropism, 355–357, *356*
Mechanical agitation (*see* Liquid extraction)
Mechanical-draft towers (induced-draft towers), 259–*261*
Mechanical separation, 8
by drainage, 721–723, *722*
in settlers, 527–529, *528*
Mechanically-agitated extractors (countercurrent extractors), *544–547*
Membrane(s), separator, 529
Membrane operations, 5–6
Metals:
diffusion through, 95
recovery of: by ion exchange, 642–646, *645*
by leaching, 5, 717–720, 724, 727, 731, 732, 739
by liquid extraction, 478
by zone refining, 4
Micropores, 95
Minerals recovery:
choice of method for, 7–8
by leaching, 717
Minimum-boiling azeotropism, 350–352, *351*

Minimum reflux ratio:
  countercurrent extraction and, 510, *511*
  in McCabe-Thiele method, *411-413*
  in multicomponent systems, 435-439
  in Ponchon-Savarit method, 385-387, *386*
Miscella:
  in Bollman extractors, 742, 744
  defined, 740
Miscibility, partial liquid, in positive deviations
    from ideality, 352, *353*
Miscible phases, mass transfers by direct contact
    of, 6
Mixco Lightnin CMContractor (Oldshue-Rushton
    extractor), *544*, 545
Mixer(s):
  static, 190
  (*See also* Liquid extraction, in stage-type
      extractors; *and specific types of mixers*)
Mixer-settler(s), described, 521
Mixer-settler operations (contact filtration):
  adsorption of liquids by, 585-599, *586*, *588*,
      589, *591-598*
  cascades in stage-type extractors, *529*, *530*
Mixing (*see* Axial mixing; Back mixing; *and specific
    mass-transfer operations*)
Mixing length (Prandtl mixing length), 56
Moisture:
  critical, 669, 682
  diffusion of, *678*
  types of, 660, *661*
  unbound, 683-684
  (*See also* Drying; Water)
Moisture content:
  critical, defined, 669, 682
  dry basis, defined, 660
  wet basis, defined, 660
Molal absolute humidity, defined, 227
Molecular diffusion, 21-44
  in binary solutions, *24*
  equation of continuity and, 24-26, *25*
  momentum and heat transfers in laminar flow
      and, 38-41, *39*
  steady-state, 26-38
    applications of, 37-38
    diffusivity of gases and, *31-34*
    diffusivity of liquids and, 35-37, *36*
    in gases, 27-30, *28*
    in liquids, 34-35
Molecular distillation, 460-462, *461*
Molecular-screening activated carbon, 568
Molecular sieves as adsorbents, 568-569
Molecular volumes of gases, *33*
Momentum eddy diffusivity, 58
Momentum-transfer coefficients (*see* Mass-transfer
    coefficients, analogies with heat- and momen-
    tum-transfer coefficients)
Moving-bed adsorbers (*see* Adsorption, by
    continuous contact, under steady-state condi-
    tions)
Multibeam support plates, *192*
Multicomponent absorbers, *324*
Multicomponent strippers, *324*
Multicomponent systems (ideal solutions):
  distillation of (*see* Distillation, of multicompo-
      nent systems)
  gas absorption in, 322-332, *324*
    equilibrium solubility of gases into liquids,
        277-279
  liquid extraction in, 518-520, *519*
    liquid equilibrium, 488

Multicomponent systems (ideal solutions):
  steady-state molecular diffusion in, 29-30
  vapor-liquid equilibrium in, 360-363
  (*See also* Raoult's law)
Multiple downspouts, 176, *178*
Multistage countercurrent operations:
  adsorption by, *592-594*
    in fluidized beds, *610*
    Freundlich equation applied to, 594-599,
        595-598
  continuous countercurrent multistage extraction,
      497-507, *498-501*, *503-506*
  gas absorption by (*see* Gas absorption, one
      component transferred, in countercurrent
      multistage operations)
  leaching by, methods of calculating, *751-760*,
      753, *755-757*, 759
  (*See also* Fractionation)
Multistage crosscurrent operations:
  adsorption by, Freundlich equation applied to,
      590-592
  extraction by, 493-497, *495*
  leaching by, 749-751, *750*
Multistage operations, absorption by, in agitated
    vessels, 158
Multistage tray towers (*see* Distillation, by
    fractionation, McCabe-Thiele method of;
    Distillation, by fractionation, Ponchon-Savarit
    method of)
Murphree stage efficiency, 124, 254, 423
  in adsorption, 606, *607*
  in cross-flow cascades, 125
  in humidification, 254
  in liquid extraction, *520*, 521, 524
  (*See also* Stage efficiency)
Murphree tray efficiency, 181-183
  in absorption, *298-300*

Natural-draft towers (natural-circulation towers),
    259, *260*
Navier-Stokes equation, 51, 56
Neo-Kloss packings, 190
Nonadiabatic operations, 242
  evaporative cooling with, 263-269, *264*, *265*
Nondiffusing liquids:
  mass-transfer coefficients in, 48-50
  steady-state molecular diffusion in, *28*, 29, 34
Nonideal solutions, equilibrium solubility of gases
    in, 279-281, *280*
Nonisothermal operations, gas absorption by,
    293-298, *294*, *297*
  overall coefficients in, 313-322, *315*, *319-321*
Nontoxicity of solvents, 489
Nonvortexing systems in agitated vessels, 151-153,
    152
Normal boiling point of liquids, defined, 222
Nusselt number, 67, 69
Nutter float-valves, 177

Odor removal (*see* Adsorption)
Oils:
  crude, 453-455, *454*
  cylinder, 582, *583*
  (*See also* Animal oils; Fish oils; Vegetable oils)
Oldshue-Rushton extractors (Lixco Lightnin
    CMContractor), *544*, 545
Open steam:
  in McCabe-Thiele distillation method, *416*, 417

Open steam:
  in Ponchon-Savarit distillation method, 394–397,
    395, 396
Open tanks, gas-liquid surfaces and, in prevention
  of vortex formation, 149
Operating line in interphase mass transfers,
  119–121
Optimum reflux ratio:
  in McCabe-Thiele distillation method, 412–416,
    415
  in Ponchon-Savarit distillation method, *387–392,*
    388, 390, 391
Ordinary binary mixtures, vapor-liquid equilibrium
  in, 343–344
Ore(s) (*see* Metals)
Ore slurries, adsorption of, by ion exchange, 642
Osmosis, defined, 6
Overall gas transfer units, number and height of,
  308, 311–313, *312*
Overall mass-transfer coefficients (*see* Mass-trans-
  fer coefficients)
Overall stage efficiency (*see* Stage efficiency)
Overall tray efficiency (*see* Tray efficiency)

Pachuca tanks, 140, 726, *727,* 732
Packed absorbers, *315*
Packed strippers, *315*
Packed towers:
  distillation in, 426–*431, 427*
  gas absorption in, *301,* 302
  gas-liquid operations in, 187–210, *188*
    cocurrent flow of gas and liquid in, 209
    countercurrent flow of liquid and gas in, *194*
    end effects and axial mixing in, 209–210
    flooding and loading in, 194, 195
    liquid distribution in, *192*
    mass-transfer coefficients for, 202–209, *205,*
      206
    packing characteristics, 189
    packing restrainers and entrainment elimina-
      tors, 193
    packing supports, 191, *192*
    pressure drop in: for single-phase flow, 200
      for two-phase flow, 200–201
    random packings in, 189, *190*
      characteristics of, *196–199*
      flooding and pressure drop in, *194, 195*
      size of, and liquid redistribution, 192, *193*
    regular packings in, 190, *191*
    shells of, 191
  for humidification, differential section of, *243*
  leaching in, 721, *722*
  liquid extraction in, 542–544, *543*
  tray towers versus, 210–211
  (*See also* Tray towers)
Paint pigments, drying of, 679
Pall rings (flexirings), 189–191, *190, 197,* 261
Pan driers, agitated, 666
Paper:
  chlorination of, 139
  drying of, 678
Parallel adiabatic operations, continuous drying
  by, 686–687
Parallel-flow trays, 176, *178*
Partial condensation, flash vaporization by, *365*
Partial condensers in McCabe-Thiele distillation
  method, 417, *418*
Partial liquid miscibility in positive deviations from
  ideality, 352, *353*
Partial pressure of gas equilibrium, 277

Partition chromatography, 632
Partition coefficient (*see* Equilibrium distribution)
Partition rings, 189, *190*
Péclet number, *68,* 182
Penetration theory (Higbie theory), *60,* 61
Percentage saturation (percentage absolute humid-
  ity), 229
Percolation and percolation tanks:
  leaching in, 720–723, *722*
  liquid adsorption in, 630–631
Perforated trays (*see* Sieve trays and sieve-tray
  towers)
Permeability:
  diffusion and, 94
  leaching and, 721
Permeation separations, defined, 5
Permissible flow rate, 11
Petroleum (*see* Hydrocarbons)
Petroleum topping plant, *454*
Pharmaceuticals:
  decolorization of, 568
  drying of, 666, 696
  leaching to obtain, 718
  liquid extraction of, 479, 547
Physical adsorption (van der Waals adsorption),
  566
Pitch of impellers, defined, 146
Pitched-blade turbines for adsorption, 599, *600*
Plait point, 482
Plasticizers, defined, *92n.*
Plate(s):
  analogies between mass, heat and momentum
    transfers in, 66–70
  boundary layers in flat, *64*
  momentum transfers in, 38, *39*
  (*See also* Slabs)
Plate driers, 688
Plate towers (*see specific towers*)
Pneumatic driers (flash driers), 698, *699*
Podbielniak extractors, 547, *548*
Point efficiency of trays, 178–181, *179*
Poiseuille's law, 99
Polymers, diffusion through, 93–95
Ponchon-Savarit method (*see* Distillation, by
  fractionation, Ponchon-Savarit method of)
Porous solids, diffusion through, 95–100
Power:
  adsorption: batch adsorption, 605
    in impellers, 601–602
  agitation, 150–*155, 152*
  in stage-type extractors, 522
  supplied to sparge vessels, 145–146
Prandtl mixing length, 56
Prandtl number, 40–41, 56, 70, 71, 73
Preforming of solids, 663
Pressure:
  constant total, and diffusion through porous
    solids, 96–99
  critical, of vapor-liquid equilibrium, 221
  in cross-circulation drying, 680
  diffusivity of gases at standard atmospheric,
    31
  drop in: in packing towers, 200–201
    in random packings, *194, 195*
    in sieve-tray towers, 170–172, 175
    in tray and packing towers, 210
  effects of, on equilibrium of ternary systems
    with one pair partially soluble, 484
  effects of change of, on adsorption equilibrium,
    579, 580

Pressure:
low-pressure distillation, 460–462
partial, of gas equilibrium, 277
vapor-liquid equilibrium and: constant-pressure
vapor-liquid equilibrium, *344–346*
at increased pressure, 346, *347*
Pressure swing (heatless adsorbers), 628–*630*
Pressure-temperature-concentration phase of
vapor-liquid equilibrium, 343–344
Priming in tray towers, 160
Products obtained by distillation:
compositions of, 440–441
with McCabe-Thiele method, 422–423
with Ponchon-Savarit method, 402
Propellers, single-phase liquids agitated by marine-
type propellers, 146, *147*
Proprietary trays, 176–178
Psychometric charts for humidification, *229*, 231,
232–234
Psychometric ratio of wet-bulb temperature, 239
Pulsed columns, 546, *547*

q and *q* line, 407–409, *408*, 412
Quiescent fluidized beds, 608–609

Radial diffusion, 89, 91–92
Raffinate:
defined, 477
(*See also* Liquid extraction)
Random packings, 189, *190*
characteristics of, *196–199*
flooding and pressure drop in, *194*, *195*
size of, and liquid redistribution, 192, *193*
Raoult's law, 578
equilibrium solubility of gases in liquids and,
278–279
of vapor-liquid equilibrium, *348*, 349
(*See also* Multicomponent systems)
Raschig rings, 158, 189–*191*, 203, *205*, *206*, 210
Rate-of-drying curve, 667–670, *668*
Rayleigh equation, 369
Rayleigh number, 157
RDC (rotating-disk contactors), *545*
Reboiled absorbers, *332*
Reboilers, fractionation in, 392–394, *393*
Recirculating liquids and gas-humidification
cooling, 252–255, *253*
Recoverability of solvents, 489
Recovery of solvents by extraction, 502, *503–505*
Rectification, continuous (*see* Distillation, by
fractionation)
Rectifying section of fractionators (*see* Exhausting
section of fractionators)
Recycling:
of gases, in drying, 663–664, 688–689
of liquids, 524
Reflux:
countercurrent extraction with, *507–514*, *509*,
511–513
in McCabe-Thiele distillation method: at bubble
point, 404–405
cold, 419
in multicomponent systems, 332
Reflux accumulators, 397

Reflux ratio, 373
minimum: countercurrent extraction and, 510,
511
in McCabe-Thiele distillation method, *411–413*
in multicomponent systems, 435–439
in Ponchon-Savarit distillation method,
385–387
optimum: in McCabe-Thiele distillation method,
412–416, *415*
in Ponchon-Savarit distillation method,
387–392, *388*, *390*, *391*
total: in countercurrent extraction, *511*, *512*
in McCabe-Thiele distillation method, *410*,
411
in multicomponent systems, 439–440
in Ponchon-Savarit distillation method, 384
385
Regain, defined, 657
Regular packings (stacked packings), 190, *191*
Regular vapor-liquid solutions, equilibrium of,
349
Relative adsorptivity (separation factor), 346,
576–584
Relative saturation (relative humidity), defined,
228
Repulping of sludge, 737
Residual saturation in leaching, 721
Residues (bottoms):
defined, 372
reboiled vapor in equilibrium with, 405–406
Resins for ion exchange, 642–646, *645*
Resting fluids, steady-state molecular diffusion
in (*see* Molecular diffusion, steady-state)
Reversible adsorption, 566
Reynolds number:
in adsorption, 601–603
gas-liquid operations and, 144, 149–153, 156
in humidification, 240
mass-transfer coefficients and, 52, 56, *68–70*
terminal velocity of drops and, 533
Ripple trays, 177
Rising velocity (*see* Terminal velocity)
Rotary-disk contactors (RDC), *545*
Rotary driers, 689–693, *690*, *691*
rate of drying in, 704–707
through-circulation, 693, *694*
Rotary filters (crystal-filter driers), 684
Rotating fixed-bed adsorbers, *628*
Rotating shelves (turbo-type driers), 687, *688*
Rotocel, leaching in, 740, *741*, 743
Ruggles-Coles XW hot-air driers, *690*

Saddle packings (*see* Berl saddles; Intalox saddles)
Sand, drying of, 679
Saturated liquids, defined, 221
Saturated vapor(s), defined, 221
Saturated vapor-gas mixtures, humidification of,
227–228
Saturation:
percentage, defined, 229
relative, defined, 228
residual, in leaching, 721
(*See also* Humidification)
Scheibel correlation, 36
Scheibel extractors, *546*
Schmidt number:
in adsorption, 603

Schmidt number:
  in gas-liquid operations, 151
  in humidification, 240
  mass-transfer coefficients and, 40, 58, *68*, 73
Seal pots, *164*
Selectivity:
  of adsorbents, 567
  of solvents, 488
Semibatch operations, 9
  (*See also* Batch operations; *and specific mass-transfer operations*)
Sensible heat, defined, 222
Separation factor (relative adsorptivity), 346, 576–584
Separator membranes, described, 529
Settlers:
  in stage-type extractors, 527–529, *528*
  (*See also* Mixer-settler operations)
Shanks system of leaching, 723, *724*, 740
Shelf driers (*see* Tray driers)
Shells:
  of packing towers, 191
  of tray towers, 161
Sherwood number, 52, 67, *68*, 70, 72, 151
SI units (*Système International d'Unités*), 12–18, *13–17*
Side streams:
  in distillation of multicomponent systems, 433
  in Ponchon-Savarit distillation method, 400
  removal of, 211
Sieve(s), molecular, 568–569
Sieve trays and sieve-tray towers:
  gas-liquid operations in, 166–176, *168, 169, 171, 173*
    activated area of, 167–*169*
    efficiency of, 183, *184*
    operating characteristics of, *163*
    liquid extraction in, 530–541, *531–534, 536, 538*
Silica gel as adsorbent, 567, 568, 581, 609, 626
Simple distillation (differential distillation), 367–371
Simulation of moving beds, *622, 623*
  leaching and, 724
Single Dorr classifiers, 740
Single gases, adsorption equilibrium for, 569–575, *570–573*
Single-phase liquids, mechanical agitation of, 146–153, *147, 148, 152*
Single-stage operations:
  adsorption by, 587–589, *588*, 597
    Freundlich equation applied to, *589*, 590
  defined, 123
  distillation by (*see* Flash vaporization)
  leaching by, *748, 749*
  liquid extraction by, 490–493, *491, 492*
Single vapors, adsorption equilibrium for, 569–575, *570–573*
Slabs:
  diffusion through, 88, 92–93
    with sealed edges, *90*
  [*See also* Plate(s)]
Slip velocity, *143*
Slotted trays, 176, *177*
Slurries:
  carbonation of lime, 139
  in leaching: effects of concentration of, *730*
    effects of height of, 731

Slurries:
  in leaching: settling of *728*, 729
    settling characteristics, 746
    stirring and, 729
    of vegetable seeds, 743
  slurry adsorption of gases and vapors, 609
Soap, drying of, 678, 696
Solid(s):
  diffusion (*see* Diffusion, in solids)
  mass-transfer coefficients in flow past, 64–66
  preparation of, for leaching, 718–719
    (*See also* Leaching)
  (*See also specific mass-transfer operations*)
Solid-fluid operations (*see* Adsorption; Drying; Leaching)
Solid holdup, 692–693
Solid-solid operations, 5
Solid suspensions, adsorption of, 600–601
Solid-vapor equilibrium (*see* Adsorption, equilibrium in)
Solubility, equilibrium (*see* Gas absorption, equilibrium solubility of gases in liquids and)
Soluble solids, drying of, 658–660, *659*
Solutes, 8–9
  adsorption: from dilute solutions, 580–*582*
    solute collections, *614*–617, *615*
    (*See also* Adsorption)
  separation of two, 515–*518*
    (*See also* Liquid extraction, Membrane operations)
Solution mining (in-place leaching; in-situ leaching), 719–720
Solutropic systems, defined, 482
Solvents:
  choice of: for gas absorption, 281, 282
    (*See also* Gas absorption)
  for liquid extraction, 488, *489*
    (*See also* Liquid extraction)
  cost of, 282, 489
  defined, 457, 477
  recovery of, 502, *503–505*
Spacing of trays, 161, *162*
Sparge vessels (bubble columns), 140–146, *142, 143*
Spheres:
  diffusion through, 89, 91
  of ion-exchange resins, 642
Split-feed treatment, 590–592, *591*
Spouted beds, continuous drying in, 698
Spray chambers, 187
  adiabatic operations in, *262, 263*
Spray driers, 695–697, *696*
Spray ponds for adiabatic operations, 262–263
Spray towers, 187
  liquid extraction in, 542
Square pitch of impellers, defined, 146
Stacked packings (regular packings), 190, *191*
Stage(s):
  defined, 10
  equilibrium, defined, 10, 123–124
    (*See also* Equilibrium stages)
  of interphase mass transfers (*see* Interphase mass transfers, stages of)
  (*See also* Cascade(s); Single-stage operations; Two-stage adsorption: *specific mass-transfer operations and under* Multistage)

Stage efficiency:
  in adsorption, 606, *607*
  defined, 10, 124
  in leaching, 744
  in liquid extraction, *520*, 521, 537–539, *538*
  Murphree (*see* Murphree stage efficiency)
  (*See also* Tray efficiency)
Stage-type extractors (*see* Liquid extraction, in stage-type extractors)
Stagewise equipment (*see* Liquid extraction, in differential extractors; Liquid extraction, in stage-type extractors; Liquid extraction, by stagewise contact)
Stagewise operations, 10
  adsorption by (*see* Adsorption, by stagewise operations)
  ion exchange by, 642
  liquid extraction by (*see* Liquid extraction, by stagewise contact)
Stanton number, *68*
Static mixers, 190
Steady state, 9–10
  interphase mass transfers in: cocurrent, *117*–121, *119*, *120*
    continuous cocurrent, 124
    countercurrent, *121–124*
    (*See also* Adsorption, by continuous contact, under steady-state conditions; Continuous operations; Leaching, under steady-state conditions; Molecular diffusion, steady-state)
Steady-state equimolal counterdiffusion:
  in gases, 29, *30*
  in liquids, 34–35
Steam, open: in McCabe-Thiele distillation method, *416*, 417
  in Ponchon-Savarit distillation method, 394–397, *395*, *396*
Steam distillation, 354–355
Stills:
  falling-film, 462
  Hickman centrifugal molecular, *461*
Stokes' law, 142
Straight operating line, *119*, 120
Streams (*see* Side streams)
Strippers and stripping, *283*, *284*
  in countercurrent multistage absorption operations, 291–293
  HETP applied to, 304–*307*, *305*
  material balances in, 282–*289*, *283*, *284*
  number of transfer units for, *310*
Stripping factor, defined, 124
Stripping section of fractionators (*see* Exhausting section of fractionators)
Sublimation, fractional, defined, 3
Sublimation drying (freeze drying), 666–667
Sugar and sugar beets:
  drying of, 690
  leaching of, 717, 718, 725, 731, 738–739, 760
  production of citric acid from, 139
  refining of, 565, 582, 585, 631
  separation of crystalline substance in solutions of, 5–6
Superheated vapor, defined, 221
Surface(s), mass-transfer coefficients on, 60–64, *62*, *63*
Surface diffusion in porous solids, 95, 99
Surface phenomena, mass-transfer operations making use of, 6–7
Surface-renewal theory, 61–62

Surface stretch theory (lightfoot theory), *63*, 64
Sweep diffusion, described, 6
Synthetic polymeric adsorbents, 568
Système International d' Unites (SI units), 12–18, *13–17*

Tannins, recovery of, by leaching, 724, 725
Teeter beds, 608–609
Tellerettes, *190*
Temperature:
  adsorption equilibrium and, 573–574, *579*, 580
  distillation: bubble-point temperature curve, 345
    (*See also* Constant-temperature equilibrium)
  dry-bulb: in adiabatic operations, 248
    defined, 228
  drying: in constant-rate drying, 675
    rate of drying at high, 701–707, *702*, *703*
    rate of drying at low, 707–710, *708*
  in gas absorption, *321*, 322
  in gas-liquid operations, fluctuations in, 211
  humidification: critical, of vapor-liquid equilibrium, 221
    evaporative cooling, *265*
    (*See also* Adiabatic-saturation temperature)
  leaching, 719
  in liquid extraction: effects on equilibrium of ternary systems: with one pair partially soluble, *483*
    with two pairs partially soluble, 484, *485*
  wet-bulb: in adiabatic operations, 248
    in humidification of vapor-gas mixtures, 237–241, *238*
    Lewis relation and, 241
    (*See also* Heat)
Terminal velocity (rising velocity):
  of drops, 533, *534*
  of single-gas bubbles, 141, *142*
Ternary systems:
  distillation of, *432*
  with one pair partially soluble, liquid equilibrium of, *482–484*
  with two pairs partially soluble, liquid equilibrium of, 484, *485*
Textiles, drying of, 663, 670, 678, 687, 709–710
Theoretical stages (*see* Equilibrium stages)
Theoretical trays (ideal trays), 289–293
Thermal conditions for feed, in McCabe-Thiele distillation method, *408*
Thermal diffusion, described, 6
Thermal-swing procedures, defined, 626
Thermosiphon reboilers, 392–394, *393*
Thickeners, leaching in, 733–738, *734*, *735*, *737*, 747
Thickness of drying solids, effects on constant-rate drying, 675–676
Thiele-Geddes method of distillation, 433–434, 439, 442–443, 445, 449–455, *454*
Through-circulation drying and driers, 663, *664*, *682–686*, 688, *689*
  in rotary driers, 693, *694*
Time:
  drying, 670–672
  as requirement, design and, 11
Total condensers in McCabe-Thiele distillation method, 404–405
Total pressure, constant, diffusion through porous solids and, 96–99

Total reflux ratio (infinite reflux ratio):
  in countercurrent extraction, *511, 512*
  in McCabe-Thiele distillation method, *410*, 411
  in multicomponent systems, 439–440
  in Ponchon-Savarit distillation method, 384, *385*
Towers (*see specific towers and specific mass-transfer operations*)
Transfer units:
  in absorption (*see* Gas absorption, in continuous-contact equipment, overall coefficients and transfer units in)
  in adsorption, 619, 636, 615–616
  in distillation, 426–431, *427*
  in drying, 704–707
  height of (*see* Height, of transfer units)
  in humidification, 247–253, *250*
  in liquid extraction, 548–552
Tray absorbers, *282*, 289, *290*
Tray driers (shelf driers; cabinet driers; compartment driers), *662*, 663
  vacuum shelf driers as, 666
Tray efficiency:
  in distillation, 402, 423–426, *425*
  in gas-liquid operations, 159–160, 178–186, *181, 184, 185*
  Murphree, 181–183
  in absorption, *298–300*
Tray towers:
  adiabatic operations, 262
  for gas-liquid operations: bubble-cap trays, 158, *159*, 165–167, *166*
    proprietary trays, 176–*178*
    (*See also* Sieve trays and sieve-tray towers, gas-liquid operations in)
  general characteristics of, 161, *162–165*
  multistage (*see* Distillation, by fractionation, McCabe-Thiele method of; Distillation, by fractionation, Ponchon-Savarit method of)
  packing towers versus, 210–211
Treybal extractors, 546
Trickle-bed reactors, 209
Truck driers, *663*
Tunnel driers, 687, 704
Turbines:
  flat-bed and pitched-blade, 599, *600*
  single-phase liquids agitated by, *147*
Turbo-Grid, 177
Turbo-type driers (rotating shelves), 687, *688*
Turbulence:
  in circular pipes, 70–72, *71*
  fully-baffled, in open tanks, 149
  interfacial, 113
  mass-transfer coefficients in (*see* Mass-transfer coefficients, in turbulent flow)
Turbulent diffusion (*see* Eddy diffusion)
Turndown ratio, defined, 165
Two-component systems, equilibrium solubility of gases in, *276*, 277
Two-feed fractionators, 397–399, *398*
Two-film theory (Lewis-Whitman theory; two-resistance theory), *107*, 110
Two-stage adsorption:
  countercurrent, *595, 596, 599*
  crosscurrent, 598–599
Two-truck driers, *663*

Unbound moisture, rate of drying of, 683–684
Unbound water, defined, 657

Underflow in leaching, 747, 754
Unit systems, 12–18, *13–17*
Universal range, defined, 55
Unsaturated-surface drying, 670, 680, *681*
Unsaturated vapor-gas mixtures, humidification of, 228–231, *229*
Unsteady state, 9
  adsorption in (*see* Adsorption, by continuous contact, under unsteady-state conditions)
  diffusion in, 89, *90, 91*, 92–93
  leaching in (*see* Leaching, under unsteady-state conditions)
  (*See also* Batch operations)
Uranium recovery, 5, 6
  by extraction, 478, *530*
  by leaching, 720

Vacancy mechanism of diffusion through metals, 95
Vacuum rotary driers, 666
Vacuum shelf driers, 666
Valve trays, 177
Van der Waals adsorption (physical adsorption), 566
Van't Hoff's law, 277
Vapor(s):
  adsorption of: adsorption equilibrium of single vapors, 569–575, *570–573*
    from fluidized beds, 697, *698*
    by fractionation, 611
    under unsteady-state conditions, 625, *626–630*
    (*See also* Adsorption)
  air-water: dehumidification of, 252
    in humidification operations, 231–236, *232–235*
  distillation of, 405–406
  drying of, 679–680
  saturated, defined, 221
  superheated, defined, 221
  (*See also* Gas absorption; Humidification)
Vapor-gas mixtures:
  adsorption equilibrium of, 575–580, *577–579*
  humidification of (*see* Humidification, of vapor-gas mixtures)
Vapor-liquid equilibrium:
  for pure substances, 220–224, *221*
  (*See also* Distillation, vapor-liquid equilibrium in)
Vapor pressure:
  in humidification, *221*, 222
  solvents and, 489
Vaporization:
  flash: distillation by, *363–367, 364*
    freezing by, 666
  latent heat of, defined, 222
  in McCabe-Thiele distillation method, 402–403
  (*See also* Distillation)
Vegetable oils:
  adsorption of, 568
  decolorization of, 586, 623
  recovery of, by leaching, 717–719, 726, 731–732, 738–744, *741–743*, 760
Velocity:
  deviating, defined, 55
  effects of gas, on constant-rate drying, 675
  slip, *143*
  terminal: of drops, 533, *534*
    of single-gas bubbles, 141, *142*

Venturi scrubbers, *186*, 187
Vessels (*see specific types of vessels*)
Viscosity:
  force constants of gases as determined by, *33*
  of solvents, 489
  for gas absorption, 282
Vitamins:
  from fish oils, 395
  low-pressure distillation to obtain, 460
Volatility:
  constant relative, in differential distillation, 369–370
  in Ponchon-Savarit distillation method, 384–385
  relative, vapor-liquid equilibrium and, 346
  of solvents for gas absorption, 281–282
Volumetric overall mass-transfer coefficients, 202
Vortex formation, prevention of, 147–149, *148*

Water:
  air-cooling of, 245–252, *246*, *249*, *250*
  bound and unbound, defined, 657
  deionization of, 642
  deodorization of, 565
  desalinization of, 6, 342
  makeup fresh, in humidification, 248
    (*See also* Humidification)
  softening of, 641, 642
  solubility of gases in, *276*
    (*See also* Drying)

Water-cooling towers, 259, *260*
Weber number, 151, 533
Weeping:
  defined, 161
  in sieve trays, 173
Weir(s):
  of sieve-tray towers, 170, *171*, 175
  of tray towers, 164
Weir-through liquid distributors, *193*
Wet-bulb approach, defined, 248
Wet-bulb depression, defined, 239
Wet-bulb temperature:
  in adiabatic operations, 248
  in humidification of vapor-gas mixtures, 237–241, *238*
  Lewis relation and, 241
Wetted-wall towers, *71*, 187
Wetting, heat of, 580
Wilke-Chang correlation, 35
Wilke-Lee modification of Hirschfelder-Bird-Spotz method, 31
Wood:
  for cooling towers, 259
  drying of, 657, 663, 678
  leaching of, 724
Wood grids (hurdles), 190

Zenz correlation, 535, *536*
Zeolites, 641
Zone refining, described, 4